Secret History

Chapman & Hall/CRC Cryptography and Network Security Series

Series Editors: Douglas R. Stinson and Jonathan Katz

Cryptanalysis of RSA and Its Variants
M. Jason Hinek

Access Control, Security, and Trust: A Logical Approach
Shiu-Kai Chin and Susan Beth Older

Handbook of Financial Cryptography and Security
Burton Rosenberg

Handbook on Soft Computing for Video Surveillance
Sankar K. Pal, Alfredo Petrosino, and Lucia Maddalena

Communication System Security
Lidong Chen and Guang Gong

Introduction to Modern Cryptography, Second Edition
Jonathan Katz and Yehuda Lindell

Group Theoretic Cryptography
Maria Isabel Gonzalez Vasco and Rainer Steinwandt

Guide to Pairing-Based Cryptography
Nadia El Mrabet and Marc Joye

Cryptology: Classical and Modern, Second Edition
Richard Klima and Neil Sigmon

Discrete Encounters
Craig P. Bauer

Data Science for Mathematicians
Nathan Carter

For more information about this series, please visit https://www.routledge.com/
Chapman--HallCRC-Cryptography-and-Network-Security-Series/book-series/CHCRYNETSEC

Secret History

The Story of Cryptology

Second Edition

Craig P. Bauer

York College of Pennsylvania
and
National Security Agency
Center for Cryptologic History
2011–2012 Scholar-in-Residence

CRC Press
Taylor & Francis Group
Boca Raton London New York

CRC Press is an imprint of the
Taylor & Francis Group, an **informa** business

A CHAPMAN & HALL BOOK

Second edition published 2021
by CRC Press
First edition published 2013
by CRC Press

6000 Broken Sound Parkway NW, Suite 300, Boca Raton, FL 33487-2742
and
by CRC Press

2 Park Square, Milton Park, Abingdon, Oxon, OX14 4RN

© 2021 Taylor & Francis Group, LLC

CRC Press is an imprint of Taylor & Francis Group, LLC

ISBN: 978-1-138-06123-1 (hbk)
ISBN: 978-0-367-68574-4 (pbk)
ISBN: 978-1-315-16253-9 (ebk)

Typeset in Adobe Garamond Pro
by KnowledgeWorks Global Ltd.

Visit the Support Materials: https://www.routledge.com/9781138061231

This book is dedicated to all of the authors of
the works cited in the pages that follow.

Contents

New and Improved xv
Note to the Reader xvii
Introduction xix
Acknowledgments xxv

PART I CLASSICAL CRYPTOLOGY

1 Monoalphabetic Substitution Ciphers, *or* MASCs: Disguises for Messages 3
    1.1   Caveman Crypto 3
    1.2   Greek Cryptography 4
            1.2.1   The Skytale Cipher 4
            1.2.2   The Polybius Cipher 5
    1.3   Viking Cryptography 6
    1.4   Early Steganography 7
    1.5   Caesar Cipher 7
    1.6   Other MASC Systems 8
    1.7   Edgar Allan Poe 11
    1.8   Arthur Conan Doyle 14
    1.9   Frequency Analysis 17
    1.10 Biblical Cryptology 19
    1.11 More Frequencies and Pattern Words 20
    1.12 Vowel Recognition Algorithms 23
            1.12.1 Sukhotin's Method 24
    1.13 More MASCs 26
    1.14 Cryptanalysis of a MASC 29
    1.15 Ciphers by a Killer and a Composer 31
    1.16 Affine Ciphers 33
    1.17 Morse Code and Huffman Coding 37
    1.18 MASC Miscellanea 42
    1.19 Nomenclators 44
    1.20 Cryptanalysis of Nomenclators 47
    1.21 Book Codes 48
    References and Further Reading 51

2 Simple Progression to an Unbreakable Cipher 59
    2.1   The Vigenère Cipher 59
    2.2   History of the Vigenère Cipher 61

2.3 Cryptanalysis of the Vigenère Cipher 66
2.4 *Kryptos* 75
2.5 Autokeys 79
2.6 The Running Key Cipher and Its Cryptanalysis 80
2.7 The One-Time Pad or Vernam Cipher 92
2.8 Breaking the Unbreakable 96
2.9 Faking Randomness 98
2.10 An Unsolved Cipher from 1915 101
2.11 OTPs and the SOE 101
2.12 History Rewritten! 102
References and Further Reading 103

3 Transposition Ciphers **107**
3.1 Simple Rearrangements and Columnar Transposition 107
   3.1.1 Rail-Fence Transposition 107
   3.1.2 Rectangular Transposition 108
   3.1.3 More Transposition Paths 110
3.2 Cryptanalysis of Columnar Transposition 111
3.3 Historic Uses 114
3.4 Anagrams 117
3.5 Double Transposition 119
3.6 Word Transposition 120
   3.6.1 Civil War Reenactors 122
3.7 Transposition Devices 122
References and Further Reading ....: 126

4 Shakespeare, Jefferson, and JFK **129**
4.1 Shakespeare vs. Bacon 129
4.2 Thomas Jefferson: President, Cryptographer 134
4.3 Wheel Cipher Cryptanalysis 138
4.4 The Playfair Cipher 147
4.5 Playfair Cryptanalysis 152
   4.5.1 Computer Cryptanalysis 156
4.6 Kerckhoffs's Rules 157
References and Further Reading 158

5 World War I and Herbert O. Yardley **163**
5.1 The Zimmermann Telegram 163
5.2 ADFGX: A New Kind of Cipher 166
5.3 Cryptanalysis of ADFGX 168
5.4 Herbert O. Yardley 182
5.5 Peacetime Victory and a Tell-All Book 185
5.6 The Case of the Seized Manuscript 187
5.7 Cashing In, Again 188
5.8 Herbert O. Yardley—Traitor? 190
5.9 Censorship 192
References and Further Reading 195

6 Matrix Encryption **199**
   6.1   Levine and Hill 199
   6.2   How Matrix Encryption Works 201
   6.3   Levine's Attacks 204
   6.4   Bauer and Millward's Attack 207
   6.5   More Stories Left to Tell 212
   References and Further Reading 213

7 World War II: The Enigma of Germany **217**
   7.1   Rise of the Machines 217
   7.2   How Enigma Works 220
   7.3   Calculating the Keyspace 225
   7.4   Cryptanalysis Part 1: Recovering the Rotor Wirings 226
   7.5   Cryptanalysis Part 2: Recovering the Daily Keys 243
   7.6   After the Break 246
   7.7   Alan Turing and Bletchley Park 247
   7.8   The Lorenz Cipher and Colossus 252
   7.9   What If Enigma Had Never Been Broken? 253
   7.10 Endings and New Beginnings 255
   References and Further Reading 257

8 Cryptologic War against Japan **261**
   8.1   Forewarning of Pearl Harbor? 261
   8.2   Friedman's Team Assembles 263
   8.3   Cryptanalysis of Red, a Japanese Diplomatic Cipher 264
         8.3.1   Orange 267
   8.4   Purple—How It Works 268
   8.5   Purple Cryptanalysis 270
   8.6   Practical Magic 273
   8.7   Code Talkers 276
   8.8   Code Talkers in Hollywood 283
   8.9   Use of Languages as Oral Codes 285
   References and Further Reading 286

9 SIGABA: World War II Defense **291**
   9.1   The Mother of Invention 291
   9.2   Making the Rotors 294
   9.3   Anatomy of a Success 297
   9.4   SIGABA Production 301
   9.5   Keyspace and Modern Cryptanalysis 302
   9.6   Missing or Captured Machines? 304
   9.7   The End of SIGABA 305
   References and Further Reading 307

10 Enciphering Speech **309**
   10.1 Early Voice Encryption 309
   10.2 The Cost of Insecurity 311
   10.3 SIGSALY—A Solution from the Past Applied to Speech 311

10.4 Plan B 319
10.5 SIGSALY in Action 320
10.6 SIGSALY Retires 322
10.7 Voice vs. Text 323
References and Further Reading 324

PART II MODERN CRYPTOLOGY

11 Claude Shannon **327**
11.1 About Claude Shannon 327
11.2 Measuring Information 327
11.3 One More Time… 332
11.4 Unicity Points 335
11.5 Dazed and Confused 335
11.6 Entropy in Religion 336
11.7 Entropy in Literature 337
References and Further Reading 339

12 National Security Agency **345**
12.1 Origins of NSA 346
12.2 TEMPEST 347
12.3 Size and Budget 348
12.4 The *Liberty* and the *Pueblo* 349
12.5 The Church Committee Investigations 352
12.6 Post Cold War Downsizing 355
12.7 The Crypto AG Connection 356
12.8 2000 and Beyond 360
12.9 Interviewing with NSA 362
12.10 Another Betrayal 364
12.11 NSA and the Media 370
12.12 BRUSA, UKUSA, and Echelon 372
References and Further Reading 374

13 The Data Encryption Standard **379**
13.1 How DES Works 379
13.2 Reactions to and Cryptanalysis of DES 390
    13.2.1 Objection 1: Key Size Matters 390
    13.2.2 Objection 2: S-Box Secrecy 393
    13.2.3 S-Boxes Revealed! 394
13.3 EFF vs. DES 395
13.4 A Second Chance 397
13.5 An Interesting Feature 399
    13.5.1 Cryptologic Humor 401
13.6 Modes of Encryption 401
    13.6.1 Levine's Methods 402
    13.6.2 Modern Modes 403
        13.6.2.1 Electronic Code Book Mode 403

13.6.2.2 Cipher Block Chaining Mode 403
13.6.2.3 Cipher Feedback Mode 404
13.6.2.4 Output Feedback Mode 405
13.6.2.5 Counter Mode 406
13.6.2.6 Offset Codebook Mode 406
References and Further Reading 409

14 The Birth of Public Key Cryptography **413**
14.1 A Revolutionary Cryptologist 413
14.2 Diffie–Hellman Key Exchange 414
14.3 RSA: A Solution from MIT 417
    14.3.1 Fermat's Little Theorem (1640) 418
    14.3.2 The Euclidean Algorithm 419
14.4 Government Control of Cryptologic Research 423
14.5 RSA Patented; Alice and Bob Born Free 430
14.6 History Rewritten 432
References and Further Reading 433

15 Attacking RSA **435**
15.1 A Dozen Non-Factoring Attacks 435
    15.1.1 Attack 1. Common Modulus Attack 435
    15.1.2 Attack 2. Man-in-the-Middle 436
    15.1.3 Attack 3. Low Decryption Exponent 437
    15.1.4 Attack 4. Partial Knowledge of p or q 439
    15.1.5 Attack 5. Partial Knowledge of d 439
    15.1.6 Attack 6. Low Encryption Exponent Attack 439
    15.1.7 Attack 7. Common Enciphering Exponent Attack 439
        15.1.7.1 The Chinese Remainder Theorem 440
    15.1.8 Attack 8. Searching the Message Space 442
    15.1.9 Attack 9. Adaptive Chosen Ciphertext Attacks 442
    15.1.10 Attack 10. Timing Attack 443
    15.1.11 Attack 11. Textbook RSA Attack 444
    15.1.12 Attack 12. Ron Was Wrong, Whit Is Right Attack 444
15.2 A Factoring Challenge 446
    15.2.1 An Old Problem 447
15.3 Trial Division and the Sieve of Eratosthenes (c. 284–204 BCE) ... 447
15.4 Fermat's Factorization Method 450
15.5 Euler's Factorization Method 451
15.6 Pollard's $p - 1$ Algorithm 453
15.7 Dixon's Algorithm 454
    15.7.1 The Quadratic Sieve 459
15.8 Pollard's Number Field Sieve 461
    15.8.1 Other Methods 461
    15.8.2 Cryptological Humor 462
References and Further Reading 462

16 Primality Testing and Complexity Theory **465**
   16.1 Some Facts about Primes 465
   16.2 The Fermat Test 468
   16.3 The Miller–Rabin Test 470
      16.3.1 Generating Primes 473
   16.4 Deterministic Tests for Primality 473
      16.4.1 The AKS Primality Test (2002) 473
      16.4.2 GIMPS 476
   16.5 Complexity Classes, **P** vs. **NP**, and Probabilistic vs. Deterministic 477
      16.5.1 Cryptologic Humor 479
   16.6 Ralph Merkle's Public Key Systems 479
   16.7 Knapsack Encryption 483
   16.8 Elgamal Encryption 486
   References and Further Reading 488

17 Authenticity **493**
   17.1 A Problem from World War II 493
   17.2 Digital Signatures (and Some Attacks) 495
      17.2.1 Attack 13. Chosen Ciphertext Attack 495
      17.2.2 Attack 14. Insider's Factoring Attack on the Common Modulus 496
      17.2.3 Attack 15. Insider's Nonfactoring Attack 497
      17.2.4 Elgamal Signatures 497
   17.3 Hash Functions: Speeding Things Up 498
      17.3.1 Rivest's MD5 and NIST's SHA-1, SHA-2, and SHA-3 499
      17.3.2 Hash Functions and Passwords 500
   17.4 The Digital Signature Algorithm 504
   References and Further Reading 506

18 Pretty Good Privacy and Bad Politics **509**
   18.1 The Best of Both Worlds 509
   18.2 The Birth of PGP 510
   18.3 In Zimmermann's Own Words 514
   18.4 The Impact of PGP 518
   18.5 Password Issues 518
   18.6 History Repeats Itself 520
   18.7 A Terrorist and an iPhone 521
   18.8 Another Terrorist and Another iPhone 528
   18.9 Yet Another Attempt at Anti-Crypto Legislation 530
   References and Further Reading 531

19 Stream Ciphers **533**
   19.1 Congruential Generators 533
   19.2 Linear Feedback Shift Registers 535
   19.3 LFSR Attack 537
   19.4 Cell Phone Stream Cipher A5/1 538
   19.5 RC4 539
   References and Further Reading 541

20 Suite B All-Stars **545**
  20.1 Elliptic Curve Cryptography 545
      20.1.1 Elgamal, ECC Style 551
  20.2 Personalities behind ECC 552
  20.3 The Advanced Encryption Standard (AES) 554
      20.3.1 SubBytes 556
      20.3.2 ShiftRows 559
      20.3.3 MixColumns 560
      20.3.4 AddRoundKey 561
      20.3.5 Putting It All Together: How AES-128 Works 563
  20.4 AES Attacks 563
  20.5 Security Guru Bruce Schneier 564
  References and Further Reading 565

21 Toward Tomorrow **569**
  21.1 Quantum Cryptography: How It Works 569
  21.2 Quantum Cryptography: Historical Background 571
  21.3 Quantum Computers and Quantum Distributed Key Networks 576
  21.4 NSA Weighs In 577
  21.5 NIST Responds 578
  21.6 Predictions 579
  21.7 DNA Computing 579
  References and Further Reading 584

Index **589**

New and Improved

The first edition of this book contained chapters devoted to how German and Japanese systems from World War II were cracked. These are extremely important chapters and they were retained in this new edition, but now the other side of this cipher war is told — how the United States was able to come up with systems that were never broken. A new chapter details SIGABA, which enciphered text, and another covers SIGSALY, which was used for voice communications. Despite the addition of these two new chapters, the book you are holding contains only 21 chapters, compared to the first edition's 20. That's because the first two chapters of the original have been combined into the new Chapter 1. So, if you're using this book to teach a course, please make note of this fact! Nothing has been deleted in the process of compressing these first two chapters and the other chapters retain their original order, with the new chapters inserted at the appropriate positions.

New material has come to light and been incorporated throughout the book concerning various eras in cryptology's long history. Some of the "history" is so recent that it should be referred to as current events. For example, much has happened concerning political aspects of cryptology since the first edition was completed. The still unfolding story is updated in this new edition. The final chapter includes the impact of quantum computers, which is both a current event and an extremely important part of the future. We are living in interesting times!

Note to the Reader

This book was intentionally written in an informal entertaining style. It can be used as leisure reading for anyone interested in learning more about the history or mathematics of cryptology. If you find any material confusing, feel free to skip over it. The history alone constitutes a book and can be enjoyed by itself. Others, especially high school teachers and college professors in fields such as mathematics, history, and computer science, will find the book useful as a reference that provides many fascinating topics for use in the classroom. Although the book assumes no previous knowledge of cryptology, its creation required making use of a tremendous number of sources, including other books, research papers, newspaper articles, letters, original interviews, and previously unexamined archival materials. Thus, even the experts will likely find much that is new.

The purpose of the book is to give as complete a picture of cryptology as is possible in a single volume, while remaining accessible. The most important historical and mathematical topics are included. A major goal is that the reader will fall in love with the subject, as I have, and seek out more cryptologic reading. The References and Further Reading sections that close every chapter make this much easier.

I've used this book for two completely different classes I teach. One is a 100-level general elective titled "History of Codes and Ciphers." It is populated by students with a wide range of majors and has no prerequisites. The other is a 300-level math and computer science elective with a prerequisite of Calculus I. This prerequisite is only present to make sure that the students have a bit of experience with mathematics; I don't actually use calculus. In the 100-level class, much of the mathematics is skipped, but for the upper-level class, the minimal prerequisite guarantees that all of the material is within the students' reach.

Anyone else wanting to use this book as a text for a cryptology class will also want to take advantage of supplementary material provided at https://www.routledge.com/9781138061231. This material includes hundreds of exercises that can be taken as challenges by any of the readers. Many of the exercises offer real historic ciphers for the reader to test his or her skill against. These were originally composed by diverse groups such as spies, revolutionists, lovers, and criminals. There's even one created by Wolfgang Amadeus Mozart. Other exercises present ciphers that played important roles in novels and short stories. Some of the exercises are elementary, while others are very difficult. In some cases, it is expected that the reader will write a computer program to solve the problem or use other technology; however, the vast majority of the problems may be solved without knowledge of a programming language. For the problems that are not historic ciphers, the plaintexts were carefully chosen or created to offer some entertainment value to the decipherer, as a reward for the effort required to find them. Not all of the exercises involve breaking ciphers. For those so inclined, there are numerous exercises testing the reader's mastery of the mathematical concepts that are key components of the various systems covered. Sample syllabi and suggested paths through the book for various level classes are also present. The website will also feature a (short, I hope!) list of errata. Should you wish to correspond with the author, he may be reached at cryptoauthor@gmail.com.

Introduction

This brief introduction defines the necessary terminology and provides an overview of the book. The reader should feel free to flip back to this material, or make use of the detailed index, to find the first appearances of definitions that might require refreshing over the course of the book.

Codes are a part of everyday life, from the ubiquitous Universal Price Code (UPC) to postal Zip Codes. They need not be intended for secrecy. They generally use groups of letters (sometimes pronounceable code *words*) or numbers to represent other words or phrases. There is typically no mathematical rule to pair an item with its representation in code. Codes intended for secrecy are usually changed often. In contrast to these, there's no harm in nonsecret codes, like Zip Codes, staying the same for decades. In fact, it is more convenient that way.

The invention of the telegraph and the desire to decrease the length of messages sent in that manner, in order to minimize costs (they did not go for free like e-mails!), led to nonsecret *commercial codes* in which phrases were replaced by small groups of letters (see Figure I.1). These provide an early example of data compression, a topic that will arise again in Section 1.17.

Of course, codes were used in wars too numerous to list here, as well as in countless intrigues and conspiracies throughout history, and they are still with us. For now, we only consider one more example. A code is produced by many copiers and printers on each page they turn out that identifies the machine that was used. Few users are even aware that it is there, as you need a blue light, magnifying glass, or microscope to make out the dots that conceal the information.[1] Although it was not how he was identified, such a code appears to have been part of the evidence used to hunt for serial killer BTK (Dennis Rader), who used a copier at Wichita State University for one of his taunting letters to the police.

The copier and printer code is also an example of *steganography*, in which the very presence of a message is intended to be concealed. Other forms that steganography may take include invisible inks and microdots.

Examples of codes and steganography appear from place to place in this book, but they are not the main focus. The focus is on *ciphers*, which typically work on individual letters, bits, or groups of these through substitution, transposition (reordering), or a combination of both. In contrast to codes, modern ciphers are usually defined in terms of mathematical rules and operations. In the early days, however, this was not the case. In fact, Charles Babbage (1791–1871), the great pioneer in computer science, is often credited as being the first to model ciphers using mathematics. This isn't correct; there were earlier efforts, but they didn't catch on.[2] In any case, it's not until the twentieth century that cryptology really became mathematical. A few more definitions will make what follows easier.

Cryptography is the science of creating cipher systems. The word comes from the Greek κρυπτός, meaning "hidden," and γραφία, meaning "writing." *Cryptanalysis* is the science and art

---

[1] The Electronic Frontier Foundation (EFF) has a nice webpage on this topic that includes many relevant links. See http://www.eff.org/issues/printers.

[2] Buck, Friederich Johann, *Mathematischer Beweiß: daß die Algebra zur Entdeckung einiger verborgener Schriften bequem angewendet werden könne*, Königsberg, 1772, available online at https://web.archive.org/web/20070611153102/http://www-math.uni-paderborn.de/~aggathen/Publications/buc72.pdf. This is the first known work on algebraic cryptology.

49

The Signal Letters on the right denote the Universal Signals of Part I.

	CREDENTIALS—*cont.*		CQWT	—Crop of ———,
	Have you (*or has person indicated*) the necessary credentials or certificate? - - - - - - - NWB			A good crop of ———. Crops look well. - - - - - - - - NSM
				Crops not much injured yet. NSP
	CREDIT-S. CREDIT ON. - - PJD			Crops have suffered severely. NSQ
	Can you get credit? - - - PJF			A short crop of ———. - - NSR
	Have you a credit on ———? PJG		CQWV	CROSS-ES-ING.
	I will give you credit for ———. PJM			Cross jack-yard. - - - - JLS
CQVW	—Letters of Credit.		CRBD	—Cross trees.
	CREDITOR-S. - - - - - PJR			Cross heads. - - - - - KHP
CQWB	CREEK-S-ER.		CRBF	—Cross ways.
CQWD	CREEP. CREEPERS.		CRBG	—The Victoria Cross.
	CREW. (*See* HANDS.) - - - DHN		CRBH	CROW-S. Crow-bar.
	Boat's crew. - - - - - JBP		CRBJ	CROWD-S-ED-ING.
	By the crew. - - - - - DHW			A crowd. - - - - - - WRL
	Full crew. Hands enough. DJN		CRBK	CROWN-S-ED-ING. CORONET.
	Crew (*number to be shown*) have left the ship. - - - - DJP		CRBL	CRUEL-LY-TY-TIES.
	Crew not all on board. - - DJQ		CRBM	CRUISE-ING—OFF ———.
CQWF	—Native crew.			CRUISER-S. - - - - - - CHQ
CQWG	—Foreign crew.		CRBN	—Cruisers are very vigilant.
	Is your crew all on board? DJV		CRBP	—Enemy's cruisers.
	Crew not heard of. - - - DKC		CRBQ	CRUSH-ES-ED-ING.
	Part of the crew (*indicate the number* ———). - - - - - DKF		CRBS	CRUTCH-ES.
	Crew will not pass. - - - DKG		CRBT	CRYSTAL-LINE-IZE-ES-ED.
	Not safe to go on with the crew as at present. - - - - - DKH			CUBIC CONTENTS. - - - - VNK
	Crew will not leave the vessel. DKJ			Cubic foot—feet. - - - - VNL
	Crew sick. - - - - - - DKL		CRBV	CUDDY-IES.
	Crew healthy. - - - - - DKM			Cuddy passenger-s. - - - NQT
	Crew discontented, will not work. DKN		CRBW	CULPABLE-ILITY.
	Crew deserted. - - - - - DKP		CRDB	CULTIVATE-S-D-ING-ION-URE.
CQWH	—Crew have appealed to the authorities.		CRDF	CUNN (*or* CONN)-S-ED-ING.
CQWJ	—Some squabble or fight on shore with crew.		CRDG	CUNNING-LY-NESS.
	Crew imprisoned. - - - - DKQ		CRDH	CUP-S.
				CUPOLA SHIPS. - - - - - SCP
CQWK	CRIME-S-INAL-LY.		CRDJ	CURE-S-D-ING-ABLE.
CQWL	CRIMP-S.		CRDK	CURIOUS-LY. CURIOSITY.
CQWM	CRIMSON.		CRDL	CURB-S-ED-ING.
CQWN	CRINGLE.		CRDM	CURL-S-ED-ING.
CQWP	CRIPPLE-S-D-ING.		CRDN	CURRANT-S.
	CRISIS. - - - - - - - PCH		CRDP	CURRENT-S-LY.
	Has not reached the crisis. PCJ			What current (*rate and direction*) do you expect? - - - MJN
	Crisis is over. - - - - - PCK			
CQWR	CRITICAL-LY.		CRDQ	—Do we feel any current? What is the current?
	CROCKERY. - - - - - - NPC			
CQWS	CROOKED-LY-NESS.		CRDS	—Try the current.
	CROPS. - - - - - - - NSK			Current will run very strong (*indicate miles per hour if necessary*). MKL
	What is the opinion of the ——— crop? - - - - - - NSL			

Figure I.1 A page from an 1875 code book includes short code groups for commonly used phrases such as "Some squabble or fight on shore with crew. Crew Imprisoned." (From Greene, B. F., editor, *The International Code of Signals for All Nations, American Edition,* published under the authority of the Secretary of the Navy by the Bureau of Navigation, U.S. Government Printing Office, Washington, DC, 1875, p. 49. Courtesy of the National Cryptologic Museum.)

of breaking ciphers (deciphering without the key). *Cryptology,* the most general term, embraces both cryptography and cryptanalysis. Most books on ciphers are on cryptology, as one cannot determine the security of a cipher without attempting to break it, and weaknesses in one system must be understood to appreciate the strengths in another. That is, it doesn't make sense to study cryptography without studying cryptanalysis. Nevertheless, the term *cryptography* is used more frequently and taken to mean cryptology.

Encipher and *encrypt* both refer to the process of converting a message into a disguised form, *ciphertext,* using some cryptographic algorithm. *Decipher* and *decrypt* refer to the reverse process that reveals the original message or *plaintext* (sometimes called *cleartext*). The International Organization for Standardization (ISO[3]), a group that offers over 22,800 voluntary standards for technology and business, even has a standard (7498-2) regarding these terms; *encipher* and *decipher* are the ones to use, as the terms *encrypt* and *decrypt* are considered offensive by some cultures because they refer to dead bodies.[4]

Modern expectations of encrypted communications include not only the inability of an eavesdropper to recover the original messages, but much more. It is expected, for example, that any changes made to a message in transit can be detected. This is referred to as *data integrity*. Suppose an encrypted order to buy 500 shares of a given stock at a particular price is sent. Someone could intercept the ciphertext and replace a certain portion of it with alternate characters. This is possible without an ability to decipher the message. It only requires knowledge of which positions in the message correspond to the stock, the number of shares, and the price. Altering any of these will result in a different order going through. If unauthorized alterations to a message can be detected, this sort of mischief will be less of a problem. Another important property is *authentication,* the ability to determine if the message really originated from the person indicated. Billions of dollars can be lost when the expectations of authenticity and data integrity are not met.

Encryption protects both individual privacy and industrial secrets. Financial transactions from your own ATM withdrawals and online credit card purchase on up to fund transfers in international banking and the major deals of transnational corporations can all be intercepted and require protection. Encryption has never protected more data than it does now.

In today's world, a *cryptosystem* is a collection of algorithms that attempts to address the concerns outlined above. One algorithm takes care of the actual encryption, but many others play an important role in the security of the system. In the pages that follow, I'll sometimes refer to incredibly simple ciphers as cryptosystems. At such times, I only mean to distinguish them from other ciphers, not to imply that they have modern features.

Presumably, no one reading this needs to be talked into pursuing the subject, but if you want more reasons to study cryptology:

1. "In Russia, no private encryption is allowed, at least not without a license from Big Brother."[5]
2. "In France, cryptography is considered a weapon and requires a special license."[6] (This book was written over many years — I am leaving this quote in because it is interesting and was

---

[3] "Because 'International Organization for Standardization' would have different acronyms in different languages (IOS in English, OIN in French for *Organisation internationale de normalisation*), our founders decided to give it the short form ISO. ISO is derived from the Greek 'isos', meaning equal. Whatever the country, whatever the language, we are always ISO." — quoted from https://www.iso.org/about-us.html.

[4] Schneier, Bruce, *Applied Cryptography*, second edition, Wiley, New York, 1996, p. 1.

[5] Kippenhahn, Rudolph, *Code Breaking: A History and Exploration*, The Overlook Press, New York, 1999, p. 209.

[6] Kippenhahn, Rudolph, *Code Breaking: A History and Exploration*, The Overlook Press, New York, 1999, p. 209.

once true, but in 1998 and 1999 France repealed her anti-crypto laws.[7] In general, nations that are members of the European Union place fewer restrictions on cryptology than other nations.)

3. The *Kama Sutra* lists secret writing as one of the 64 arts that women should know and practice.[8] (It is #45.)

4. "No one shall be subjected to arbitrary interference with his privacy, family, home or correspondence, nor to attacks upon his honor and reputation. Everyone has the right to the protection of the law against such interference or attacks." (Article 12, *Universal Declaration of Human Rights*, United Nations, G.A. res. 217A (III), U.N. Doc A/810 at 71, 1948).[9]

A Roadmap to the Book

The time period represented by World War II is a major turning point in the history of cryptology, as it marks the introduction of the computer. Computers were, in fact, invented in order to break ciphers. Of course, they soon found other uses, like finding roots of the Riemann Zeta function and playing video games. The systems introduced before World War II comprise classical cryptography. Many of these systems are still used by amateurs, but for the most part they have been replaced by methods that make use of computers (modern cryptology). I consider World War II-era ciphers to be classical, because, although they mechanized the processes of enciphering and deciphering, the algorithms used are direct descendants from previous eras. On the other hand, the ciphers of modern cryptology are true children of the computer age and almost always operate on bits or blocks of bits.

Thus, the material presented in these pages is split into two parts. The first part examines classical cryptology, including World War II cipher systems, and the second part examines modern cryptology. The order of presentation was determined by striking a compromise between a strictly chronological ordering and the logical order suggested by the development of the concepts. For example, the idea of the one-time pad follows naturally from and is presented immediately after the running key cipher, although many other ciphers came into being between these two.

Part I

Chapter 1 begins by detailing some systems used by the ancient Greeks and the Vikings, as well as the impact steganography had in the history of ancient Greece. It continues with a close look at monoalphabetic substitution ciphers (MASCs), including historical uses, as well as appearances in fictional works created by Edgar Allan Poe, Sir Arthur Conan Doyle (creator of Sherlock Holmes), J. R. R. Tolkien, and others. Important ideas such as modular arithmetic are first presented in this chapter, and sophisticated modern attacks on MASCs are included, along with sections on data compression, nomenclators (systems that make use of both a cipher and a code), and book codes. Chapter 2, as already mentioned, shows the logical progression from the Vigenère cipher (which uses multiple substitution alphabets) to the running key cipher, and on to the unbreakable

---

[7] Singh, Simon, *The Code Book*, Doubleday, New York, p. 311.

[8] Vatsyayana, *Kama Sutra: The Hindu Ritual of Love*, Castle Books, New York, 1963, p. 14.

[9] Taken here from Garfinkel, Simson, *Database Nation: the death of privacy in the 21st century*, O'Reilly, Sebastopol, California, p. 257.

one-time pad. Historical uses are provided from the U.S. Civil War (for the Vigenère cipher) and World War II (for the one-time pad). Chapter 3 shifts gears to take a look at transposition ciphers, in which letters or words are rearranged rather than replaced by others. Most modern systems combine substitution and transposition, so this helps to set the stage for later chapters such as 5, 13, and 20. In Chapter 4, we examine a steganographic system alleged to reveal Francis Bacon as the true author of William Shakespeare's plays. Although I hope to convince the reader that such arguments are not valid, the system has been used elsewhere. This chapter also examines Thomas Jefferson's cipher wheel, which saw use as recently as World War II, and looks at how John F. Kennedy's life hung in the balance during that war, dependent on the security of the 19th-century Playfair cipher. Again, modern attacks on these older systems are examined. Stepping back a bit, Chapter 5 examines the great impact cryptology had on World War I and looks closely at the fascinating cryptologic figure Herbert O. Yardley, who may be accurately described as the "Han Solo of cryptology." This chapter also includes a brief description of censorship, with emphasis on censorship of writings dealing with cryptology. Linear algebra shows its importance in Chapter 6, where matrix encryption is examined. Two attacks on this system that, prior to the first edition of this work, had never before appeared in book form are presented. Electromechanical machines come on the scene in Chapter 7, as the Germans attempt to protect their World War II-era secrets with Enigma. This chapter contains a detailed look at how the Poles broke these ciphers, aided by machines of their own. Following the invasion of Poland, the setting shifts to Bletchley Park, England, and the work of Alan Turing, the great pioneer in computer science. A brief look is taken at the Nazi Lorenz ciphers and the computer the British used to break them. Chapter 8 shifts to the Pacific Theater of World War II and takes a close look at Japanese diplomatic ciphers and naval codes and the effect their cryptanalysis had on the war. The role the Navajo code talkers played in securing the Allies' victory closes out this chapter. Having seen how weaknesses in cipher machine design can be exploited by cryptanalysts, Chapter 9 details SIGABA, a World War II-era machine used by the United States that was never broken. Chapter 10 moves away from text and looks at systems used to encipher speech prior to and during World War II.

Part II

Chapter 11 leads off Part II with a look at how Claude Shannon's ideas helped shape the information age, as well as modern cryptology. His method of measuring the information content of a message (using the terms *entropy* and *redundancy*) is explained and simple ways to calculate these values are provided, along with the impact such concepts had in other fields. A history of the National Security Agency is given in Chapter 12. Included with this is a discussion of how electromagnetic emanations can cause an otherwise secure system to be compromised. TEMPEST technology seeks to protect systems from such vulnerabilities. A betrayal of the agency is looked at in some detail and the new Crypto AG revelations are covered. The chapter on NSA is followed by a close examination in Chapter 13 of a cipher (DES) that the Agency had a role in designing. The controversy over DES's key size and its classified design criteria is given full coverage, as is the Electronic Frontier Foundation's attack on the system. The revolutionary concept of public key cryptography, where people who have never met to agree on a key can nevertheless communicate securely despite the presence of eavesdroppers (!), is introduced in Chapter 14 with Diffie-Hellman key exchange and RSA. The mathematical background is given, along with historical context and a look at the colorful personalities of the key players. Attempts by the U.S. government to exert control over cryptologic research and the reaction in the academic world receive a thorough treatment as well. Chapter 15 is devoted to attacks on RSA. A dozen are presented that don't involve factoring, then a series of more and more sophisticated factoring algorithms are examined. In Chapter 16 practical consideration such as how to quickly find primes of the size needed for RSA encryption are examined, as well as the important field of complexity theory. The public key systems of Ralph Merkle and Taher Elgamal are included in this chapter. Although this chapter is more technical than many of the others, some of the key work described was first done, in whole or in part, by undergraduates (Merkle, Kayal, Saxena). Chapter 17 opens with the trouble that the lack of authenticity caused during World War II. It then shows how RSA and Elgamal offer the possibility of attaining authenticity by allowing the sender to sign messages. Unlike traditional letters, the time required to sign a digital message increases with the size of the message, if signing is done in the most straightforward manner! To fight back against this problem, hash functions condense messages to shorter representations that may then be signed quickly. Thus, a discussion of hash functions naturally appears in this chapter. Chapter 18 covers PGP and shows how such hybrid systems can securely combine the speed of a traditional encryption algorithm with the convenience of a (slower) public key system. Many political issues already seen in Chapters 13 and 14 are expanded upon in this chapter, which is mainly historical. The unbreakable one-time pad isn't very practical, so we have more convenient stream ciphers to approximate them. These are detailed in Chapter 19, and the most modern examples are used to encrypt cellphone conversations in real time. Finally, Chapter 20 looks at elliptic curve cryptography and the Advanced Encryption Standard, two of the very best (unclassified) modern systems. These systems were endorsed by NSA. Chapter 21 closes Part II, and the book, with a look at quantum cryptography, quantum computers, and DNA computers. Cryptographers have spent years preparing for the threat posed by quantum computers, but post-quantum cryptography still looks like the wild west.

Enjoy!

Acknowledgments

Thanks to Chris Christensen and Robert Lewand for reading the entire manuscript of the first edition and providing valuable feedback; to Brian J. Winkel for his comments on several chapters and much help and great advice over the years, as well as unbelievably generous cryptologic gifts; to René Stein, who researched numerous queries on my behalf at the National Cryptologic Museum; to Robert Simpson, the new librarian at the National Cryptologic Museum, who is off to a fabulous start; to David Kahn for his inspiration and generosity; to everyone at the National Security Agency's Center for Cryptologic History (the best place to work!) for my wonderful year as the 2011–2012 scholar in residence; to the editorial board of *Cryptologia*, for sharing their expertise; and to Jay Anderson, for getting me hooked on the subject in the first place.

Thanks also to the American Cryptogram Association, Steven M. Bellovin, Paolo Bonavoglia Gilles Brassard, Jay Browne, Stephen Budiansky, Jan Bury, Kieran Crowley, John W. Dawson, Jr., John Dixon, Sarah Fortener, Benjamin Gatti, Josh Gross, Sam Hallas, Robert E. Hartwig, Martin Hellman, Regan Kladstrup, Neal Koblitz, George Lasry, Susan Landau, Harvey S. Leff, Robert Lord, Andrea Meyer, Victor S. Miller, Adam Reifsneider, Karen Rice-Young, Barbara Ringle, Kenneth Rosen, Neelu Sahu, Klaus Schmeh, Mary Shelly at Franklin & Marshall College Library, William Stallings, Bob Stern, Ernie Stitzinger, Dave Tompkins, Sir Dermot Turing, Patrick Weadon, Bob Weiss, Avi Wigderson, Betsy Wollheim (president of DAW Books), John Young, and Philip Zimmermann.

Thank you all!

CLASSICAL
CRYPTOLOGY

I

Chapter 1

Monoalphabetic Substitution Ciphers, *or* MASCs: Disguises for Messages

Secrets are the very root of cool.

— **William Gibson,** *Spook Country*

A monoalphabetic substitution cipher (MASC) is a system in which a given letter is consistently replaced by the same ciphertext letter. While examples of these go back to ancient times, there are other ancient systems that are not of this type and they are also detailed in this chapter, which begins with some of the oldest known techniques.

1.1 Caveman Crypto

Various records from cultures before that of classical Greece are sometimes included in texts as examples of cryptography, but the term must be used loosely, as the methods are extremely primitive. How far back one looks for the origins of cryptology depends on how far one is willing to stretch definitions. Most authors would agree that Henry E. Langen went back a bit too far in his *Cryptanalytics—A Course in Cryptography*:

> Early cavemen may have developed a system of secret oral grunts or mimetic signs to convey messages to one another.

We shall be content to begin in ancient Sumer with an example of "proto-cryptography."

The Sumerians recognized many gods, but only 12 were part of the "Great Circle." Six of these were male and six were female.

Male	Female
60 – Anu	55 – Antu
50 – Enlil	45 – Ninlil
40 – Ea/Enki	35 – Ninki
30 – Nanna/Sin	25 – Ningal
20 – Utu/Shamash	15 – Inanna/Ishtar
10 – Ishkur/Adad	5 – Ninhursag

The number paired with the name of the god was sometimes used instead of the name;[1] thus, we have a substitution cipher. In general, though, as explained in the Introduction, when entire words or names are swapped out for numbers or letters, we refer to it as a *code*, rather than a *cipher*.

It seems that every culture that develops writing (which itself offers some secrecy if nearly everyone is illiterate) develops cryptography soon thereafter. Although many more examples could be included, we now move on to the ancient Greeks' use of cryptography.

1.2 Greek Cryptography

1.2.1 The Skytale Cipher

The *skytale* (pronounced to rhyme with "Italy"), shown in Figure 1.1, was used in 404 BCE by the Spartans. The message sent to Lysander warned of a planned attack by Pharnabazus of Persia.[2] This warning provided Lysander with sufficient time to prepare a successful defense.

Although this historical example has been recounted by many authors, a dissenting view has been offered by Thomas Kelly, a professor of Greek history.[3] Kelly examined the earliest extant appearances of the word "skytale" and concluded that it merely stood for "either a plaintext message or a device for keeping records." He claimed that later usage indicating a device for enciphering was a misinterpretation and that the skytale never had such a purpose.

In any case, the idea is to wrap a strip of paper[4] around a staff and then write the message out in rows aligned with the staff. When the paper is taken off the staff, the order of the letters changes. This is a *transposition cipher* (the letters are not changed—only their order, in contrast to a *substitution cipher*). Anyone with a staff of diameter equal to the original one can recover the message. An alternate spelling of this simple device is "scytale". These variations are natural when one considers that the word is of Greek origin. The original is σκυτάλε.

The American Cryptogram Association (which celebrated its 90th anniversary in 2020) uses a skytale in their logo (Figure 1.2). It has been conjectured that the skytale marks the origin of the commander's baton, which became purely symbolic.

---

[1] Röllig, Werner, "Götterzahlen," in Ebeling, Erich and Bruno Meissner, editors, *Reallexikon der Assyriologie und Vorderasiatischen Archäologie*, Vol. 3, Walter de Gruyter & Co., Berlin, Germany, 1971, pp. 499–500.

[2] Singh, Simon, *The Code Book*, Doubleday, New York, 1999, p. 9.

[3] Kelly, Thomas, "The Myth of the Skytale," *Cryptologia*, Vol. 22, No. 3, July 1998, pp. 244–260.

[4] The Greeks reportedly used a leather strip, sometimes disguised as a belt.

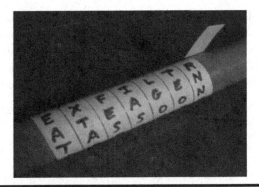

Figure 1.1   A skytale reveals a message.

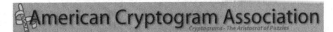

Figure 1.2   The American Cryptogram Association logo (www.cryptogram.org/).

1.2.2 The Polybius Cipher[5]

Another example of Greek cryptography is the Polybius cipher. This cipher has the disadvantage of doubling the length of the message.

	1	2	3	4	5
1	A	B	C	D	E
2	F	G	H	I&J	K
3	L	M	N	O	P
4	Q	R	S	T	U
5	V	W	X	Y	Z

In the Polybius cipher, each letter is replaced by the position in which it appears in the square, using first the row number and then the column number. An example is

```
 T  H  I  S  I  S  E  A  S  Y  T  O  B  R  E  A  K
44 23 24 43 24 43 15 11 43 54 44 34 12 42 15 11 25
```

This system was originally intended for signaling over long distances. To send the first letter, T, one would hold four torches in the right hand and four in the left hand. To send the next letter, H, one would hold two torches in the right hand and three in the left hand. We will see how to break such a cipher later in this chapter. It's a special case of a general class of ciphers known as *monoalphabetic substitution ciphers*, mentioned at the beginning of this chapter, where a given letter is always substituted for with the same symbol wherever it appears.

The five-by-five square forced us to combine I and J as 24. The Greeks, using a smaller alphabet, did not have this inconvenience. For us, though, decipherment is not unique, but context should make clear which choice is correct. Of course, we could use a six-by-six square, which would allow for 26 letters and 10 digits. Alternatively, numbers can be spelled out. A six-by-six

---

[5] Polybius wrote about this system in Chapter 46 of his tenth book of history, *circa* 170 BCE.

square would be used for languages written in the Cyrillic alphabet. The Hawaiian alphabet, on the other hand, contains only 12 letters: 5 vowels (a, e, i, o, u) and 7 consonants (h, k, l, m, n, p, w). Thus, for the Hawaiian alphabet, a four-by-four square would suffice.

The Polybius Square is sometimes called a *Polybius checkerboard*. The letters may be placed in the square in any order; for example, the keyword DERANGEMENT could be used to rearrange the letters like so:

	1	2	3	4	5
1	D	E	R	A	N
2	G	M	T	B	C
3	F	H	I&J	K	L
4	O	P	Q	S	U
5	V	W	X	Y	Z

Observe that repeated letters of the keyword are left out when they reappear. Once the keyword is used up, the remaining letters of the alphabet are filled in the square.

The term derangement has a technical meaning, in that it refers to a reordering where no object occupies its original position. Thus, the scrambling provided above is not a derangement, since the letters U, V, W, X, Y, and Z are left in their original locations.

1.3 Viking Cryptography

Figure 1.3   Example of Viking cryptography. (Redrawn from Franksen, Ole Immanuel, *Mr. Babbage's Secret: The Tale of a Cypher—and APL*, Prentice Hall, Englewood Cliffs, New Jersey, 1984.)

Markings on the Swedish Rotbrunna stone, reproduced in Figure 1.3, may appear meaningless, but they're actually an example of Viking cryptography.

If we note the numbers of long strokes and short strokes in each group, we get the numbers 2, 4, 2, 3, 3, 5, 2, 3, 3, 6, 3, 5. Pairing the numbers up gives 24, 23, 35, 23, 36, 35. Now consider Figure 1.4.

Figure 1.4   Examples of Pre-Viking (left) and Viking (right) cryptography.

The Vikings used diagrams like these to translate the numbers to runes; for example, 24 indicates the second row and the fourth column. In the diagram on the left, this gives the rune for J. Thus, this Viking encryption system was essentially a Polybius cipher. This is just one of the Viking systems; there are others. Secrecy must have been important for these people from the beginning, for the word rune means "secret" in Anglo-Saxon.

1.4 Early Steganography

The Greeks also made use of steganography. In 480 BCE, they received advance warning of an attack by the Persians launched by King Xerxes. Herodotus explained how this was achieved.[6]

> As the danger of discovery was great, there was only one way in which he could contrive to get the message through: this was by scraping the wax off a pair of wooden folding tablets, writing on the wood underneath what Xerxes intended to do, and then covering the message over with wax again. In this way the tablets, being apparently blank, would cause no trouble with the guards along the road. When the message reached its destination, no one was able to guess the secret until, as I understand, Cleomenes' daughter Gorgo, who was the wife of Leonidas, discovered it and told the others that, if they scraped the wax off, they would find something written on the wood underneath. This was done; the message was revealed and read, and afterwards passed on to the other Greeks.

The warning, combined with the 300 Spartans who held off the Persians for three days, allowed time to prepare a successful defense. Leonidas was among those who died trying to gain the time necessary for the Greeks to build a stronger defense. Without the advance warning given in this instance (or for the later attack mentioned previously, where the skytale was used to convey the warning), it's conceivable that the lack of preparation could have led to a victory for Persia, in which case there would have been no "cradle of western civilization."

Another steganographic trick was carried out by Histiaeus. In 499 BCE, he had a slave's head shaved for the purpose of tattooing a message on it, encouraging Aristagoras of Miletus to revolt against the Persian King. When the slave's hair grew back, concealing the message, the slave was dispatched with instructions to tell the intended recipient to shave his head. This is not a good method, though, if time is of the essence or if the message is long and the slave's head small![7]

1.5 Caesar Cipher

MASCs have been used for thousands of years. It would be impossible to include all of the diverse examples in this chapter. Readers wishing to delve deeper will find many examples from history and literature in the online exercises that accompany this book. However, a MASC used by Julius Caesar deserves to be detailed here.

The Caesar Cipher[8] was described by Suetonius in his biography of Caesar and by Caesar himself.[9] Each letter is replaced by the third letter to follow it alphabetically. Upon reaching the end of our ciphertext alphabet, we jot down the A, B, C that weren't yet used.

---

[6] Herodotus, *The Histories*, Book 7: *Polymnia*, Section 239.

[7] Herodotus, *The Histories*, Book 5: *Terpsichore*, Section 35.

[8] Actually, this is just one of Caesar's ciphers. In another, Roman letters were replaced by Greek. This was done during the Gallic Wars. "The letter he sent written in Greek characters, lest by intercepting it the enemy might get to know of our designs." (*The Gallic War*, Book V, by Caesar) Some of Caesar's other cipher systems have apparently been lost to history.

[9] Kahn, David, *The Codebreakers*, second edition, Scribner, New York, 1996, p. 83.

```
A B C D E F G H I J K L M N O P Q R S T U V W X Y Z plaintext
D E F G H I J K L M N O P Q R S T U V W X Y Z A B C ciphertext
```

For example,

ET TU BRUTE?	**message**

becomes

HW WX EUXWH?	**ciphertext**

(*Note:* These were not Caesar's last words. They are the last words of Caesar in Shakespeare's play, which is not completely historically accurate.)

We don't have to shift by three. We could shift by some other integer value K. If we think in terms of the numbers 0 through 25 representing the letters A through Z, the enciphering process may be viewed mathematically as $C = M + K \pmod{26}$, where C is the ciphertext letter, M is the plaintext letter, and K is the key. The "mod 26" part (short for "modulo 26") simply means that if the sum $M + K$ is greater than or equal to 26, we subtract 26 from this number to get our result. The keyspace (defined as the set of possible choices for K) has 25 elements, because the identity, $K = 0$, leaves the message unchanged, as does $K = 26$. Only values strictly between 0 and 26 offer distinct encipherments. For those of you who have had a semester of abstract algebra, we are now working with elements from the group Z_{26}. Just as Brutus helped kill Caesar, a brute force attack (trying all possible keys) quickly destroys his cipher. One requirement for a strong cryptosystem is a big keyspace.

Perhaps in an act of rebellion against more modern ciphers, during the U.S. Civil War, General Albert S. Johnston (fighting for the Confederacy), agreed with his second in command, General Pierre Beauregard, to use a Caesar shift cipher![10]

1.6 Other MASC Systems

The substitutions made by a MASC needn't be obtained by a shift. Examples where the substitutions are random appear in many newspapers alongside the comics and are very easy to break despite having a keyspace of 26! elements. So, although a large keyspace is a necessary condition for security, it is not sufficient.

Note: $n!$ is read as "*n* factorial" and is used to indicate the product of the integers 1 through n; for example, $4! = 1 \cdot 2 \cdot 3 \cdot 4 = 24$. Christian Kramp introduced the ! notation in 1808 because $n!$ grows surprisingly fast for small values of n.[11] It was not, as comedian Steven Wright claimed, an attempt to make mathematics look exciting. An arbitrary MASC can make use of any of 26 possible letters to represent A and any of the remaining 25 letters to represent B, and so on, until only 1 choice is left to represent Z, so the size of the keyspace is $26! = 403{,}291{,}461{,}126{,}605{,}635{,}584{,}000{,}000$. Try double-checking this with your calculator! Alternatively, type 26! into Mathematica or Maple or another computer algebra system. Do you get all of the digits?

Stirling's formula[12] provides a handy approximation of $n!$ for large n:

$$n! \sim \left(\sqrt{2\pi n}\right)\left(n^n\right)\left(e^{-n}\right)$$

---

[10] Kahn, David, *The Codebreakers*, second edition, Scribner, New York, 1996, p. 216.

[11] Ghahramani, Saeed, *Fundamentals of Probability*, Prentice-Hall, Upper Saddle River, New Jersey, 1996, p. 45.

[12] Discovered by James Stirling (1692–1770) in 1730. You will see the ~ notation again in Section 16.1, where the importance of prime numbers in cryptology is discussed. For a closer look at this formula, see Bauer, Craig P., *Discrete Encounters*, CRC/Chapman & Hall, Boca Raton, Florida, 2020, pp. 200-201.

where $\pi \approx 3.1415$ and $e \approx 2.71828$. The ~ is read as "asymptotically approaches" and means

$$\lim_{n \to \infty} \frac{n!}{\left(\sqrt{2\pi n}\right)\left(n^n\right)\left(e^{-n}\right)} = 1$$

So, we have a nice compact formula that relates the concept of factorials to some of the most important constants in mathematics!

In classical cryptography, it has often been desirable to have a key that can be memorized, because if the key is written down, it is susceptible to seizure. It would be time-consuming to memorize A goes to H, B goes to Q, C goes to R, etc. for all 26 letters. This leads us to *keyword ciphers*.[13] For example, using the keyword PRIVACY, we have

```
A B C D E F G H I J K L M N O P Q R S T U V W X Y Z plaintext
P R I V A C Y B D E F G H J K L M N O Q S T U W X Z ciphertext
```

Letters that are not used in the keyword follow it in alphabetical order when writing out the ciphertext alphabet. Thus, we have a cipher superior to the Caesar shift, but still weakened by some order being retained in the cipher alphabet. Such ciphers may be further complicated by having the keyword placed somewhere other than the start of the alphabet. For example, we may use the two-part key (PRIVACY, H) and encipher with

```
A B C D E F G H I J K L M N O P Q R S T U V W X Y Z plaintext
Q S T U W X Z P R I V A C Y B D E F G H J K L M N O ciphertext
```

Also, long key phrases may be used to determine the order of the substitutions. If a letter is repeated, simply ignore it when it reappears,[14] as you write the cipher alphabet out. Key phrases may be chosen that contain all of the letters of the alphabet. For example,

The quick brown fox jumps over a lazy dog. (33 letters)

```
A B C D E F G H I J K L M N O P Q R S T U V W X Y Z plaintext
T H E Q U I C K B R O W N F X J M P S V A L Z Y D G ciphertext
```

Richard Lederer provides several shorter phrases using all 26 letters:[15]

Pack my box with five dozen liquor jugs.	(32 letters)
Jackdaws love my big sphinx of quartz.	(31 letters)
How quickly daft jumping zebras vex.	(30 letters)
Quick wafting zephyrs vex bold Jim.	(29 letters)
Waltz, nymph, for quick jigs vex Bud.	(28 letters)
Bawds jog, flick quartz, vex nymphs.	(27 letters)
Mr. Jock, TV quiz Ph.D., bags few lynx.	(26 letters, a minimum)

---

[13] The Argentis were the first to use keywords in this manner (during the late 1580s). See Kahn, David, *The Codebreakers*, second edition, Scribner, New York, 1996, p. 113.

[14] An enciphered message sent to Edgar Allan Poe was made much more difficult to break because the sender did not do this (more on this later).

[15] Lederer, Richard, *Crazy English*, Pocket Books, New York, 1989, pp. 159–160.

A few devices have been invented to ease translation from plaintext to ciphertext and back again. We now take a look at two of these.

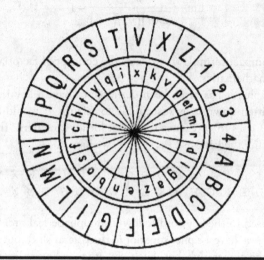

Figure 1.5 Leon Battista Alberti's cipher disk. (Courtesy of the David Kahn Collection, National Cryptologic Museum, Fort Meade, Maryland.)

The manner in which Leon Battista Alberti's cipher disk (Figure 1.5) can be used to encrypt and decrypt should be immediately clear and require no explanation. The inner alphabet is on a separate disk that may be rotated with respect to the larger disk in order to form other substitutions. Alberti (1404–1472) invented this disk in 1466 or 1467.[16]

Another device is the St.-Cyr Slide (Figure 1.6).[17] Here again, one alphabet (the one written twice in the figure) may be moved relative to the other. This device is only good for performing a Caesar type shift, if using a straight alphabet, as pictured below, and keeping the slide in a fixed position.

```
ABCDEFGHIJKLMNOPQRSTUVWXYZ
ABCDEFGHIJKLMNOPQRSTUVWXYZABCDEFGHIJKLMNOPQRSTUVWXYZ
```

Figure 1.6 St.-Cyr slide.

Yet, even with a completely mixed alphabet, it is a trivial matter to break such a cipher. MASCs have often been used in fiction. The hero of the tale typically succeeds in his attempt at cryptanalysis and explains how he did it.[18] One of the best of these tales was penned by Edgar Allan Poe. His connection with cryptology is examined next. The following section examines Sherlock Holmes's encounters with ciphers and we then conclude our treatment of MASCs with a look at solution techniques unknown to Poe or Arthur Conan Doyle's famous detective.

---

[16] Kahn, David, *The Codebreakers*, second edition, Scribner, New York, 1996, pp. 126–128.

[17] This device is named after the French national military academy.

[18] For an overview of codes and ciphers in fiction see *Codes and Ciphers in Fiction: An Overview* by John F. Dooley in the October 2005 issue of *Cryptologia*, Vol. 29, No. 4, pp. 290–328. This paper contains an annotated bibliography with 132 entries. For an updated list, go to https://www.johnfdooley.com/ and follow the link "Crypto Fiction." As of October 7, 2020, this list contains 420 examples.

Figure 1.7 Edgar Allan Poe. (Courtesy of Karen's Whimsy, http://karenswhimsy.com/public-domain-images/.)

1.7 Edgar Allan Poe

The first known reference that Edgar Allan Poe (Figure 1.7) made to cryptography was in the December 18, 1839 issue of *Alexander's Weekly Messenger* in an article titled "Enigmatical and Conundrum-ical." After explaining how an MASC worked, Poe challenged the readers:

> Let any one address us a letter in this way, and we pledge ourselves to read it forthwith—however unusual or arbitrary may be the characters employed.

He did insist that word spacing be preserved. Poe solved the ciphers that were sent in, with a few exceptions. Some wiseguys *apparently* sent in fake ciphers that never contained a message, but rather consisted of random letters. Several readers made Poe's work easier by enciphering well-known writings. For example, Poe only needed to glance at one such ciphertext to know that it was the Lord's Prayer. Poe had articles on cryptography in a total of 15 issues,[19] yet he didn't reveal his method of deciphering, despite the pleas of readers.

Poe quit *Alexander's* in May of 1840 and began serving as editor for *Graham's* magazine a year later. He repeated his cipher challenge, although this time it was buried in a review, "Sketches of Conspicuous Living Characters" (April 1841). Poe's longest article on the subject, titled, oddly enough, "A Few Words on Secret Writing," appeared in the July 1841 issue. The articles were once again very popular, yet Poe still did not reveal his method. The suspense was good for sales.

---

[19] These articles are all available at http://www.eapoe.org/works/misc/index.htm#awmcrypto.

Polyphonic ciphers were among the systems Poe solved in *Graham's*. In these, more than one letter may be enciphered as the same character. This makes deciphering more difficult for the cryptanalyst, as well as for the intended recipient. An example is shown below.

> Ofoiioiiaso ortsiii sov eodisoioe afduiostifoi ft iftvi
> si tri oistoiv oiniafetsorit ifeov rsri inotiiiiv ridiiot,
> irio rivvio eovit atrotfetsoria aioriti iitri tf oitovin
> tri aetifei ioreitit sov usttoi oioittstifo dfti afdooitior
> trso iteov tri dfit otftfeov softriedi ft oistoiv
> oriofiforiti suitteii viireiiitifoi ft tri iarfoisiti, iiti
> trir uet otiiiotiv uitfti rid io tri eoviieeiiiv rfasueostr
> tf rii dftrit tfoeei.

At a glance, we can tell that this ciphertext wasn't obtained by simply replacing each letter with another in a one-to-one manner; for example, the second to last word in the second line, inotiiiiv, has four copies of the same letter grouped together, but there is no such word in English! Nevertheless, Poe broke this cipher. He determined the key phrase (in Latin) to be *Suaviter in modo, fortiter in re*. So, the substitutions were determined as follows.

```
A B C D E F G H I J K L M N O P Q R S T U V W X Y Z  plaintext
S U A V I T E R I N M O D O F O R T I T E R I N R E  ciphertext
```

The letter i in the ciphertext may represent E, I, S, or W. Other letters provide multiple possibilities as well. Intentionally or not, the encipherer made things even more difficult through the presence of misspellings or typos. Also, he didn't exactly use a simple vocabulary. Poe stated the solution as

> Nonsensical phrases and unmeaning combinations of words, as the learned lexicographer would have confessed himself, when hidden under cryptographic ciphers, serve to perpdex [sic] the curious enquirer, and baffle penetration more completely than would the most profound apothems [sic] of learned philosophers. Abstruse disquisitions of the scholiasts, were they but presented before him in the undisguised vocabulary of his mother tongue.

In both periodicals, Poe made a claim that he repeated in his short story "The Gold Bug" (1843).[20] The phrasing differed. It is stated here in the form seen in the story.

> Yet it may be roundly asserted that human ingenuity cannot concoct a cipher which human ingenuity cannot resolve.

This is Poe's most famous quote concerning cryptology. It is in error, as we will see later; there *is* a theoretically unbreakable cipher.

In "The Gold Bug," Poe finally gave his readers what they were clamoring for. He revealed his method. It became his most popular short story. I believe there is a tendency to read fiction more passively than nonfiction. Perhaps this is a result of the willing suspension of disbelief. In most

---

[20] Available online at http://eserver.org/books/poe/goldbug.html and elsewhere.

references to "The Gold Bug," a glaring error in cryptanalysis is not mentioned. The hero of the tale, Legrand, begins his cryptanalysis of the message by observing that word divisions are not present in the ciphertext, yet one of the symbols could have stood for a blank space. This symbol would have the greatest frequency—larger than that of e. Okay, no big deal, but when Legrand says,

> Now, in English, the letter which most frequently occurs is e. Afterward, the succession runs thus: a o i d h n r s t u y c f g l m w b k p q x z. E predominates so remarkably, that an individual sentence of any length is rarely seen in which it is not the prevailing character.

we have a more serious error. Do you see it? If not, look again before reading on.

In a fantastic computer science text,[21] William Bennett pointed out the errors that Poe made.[22] He titled this section "Gold Bug or Humbug?" and referred to the origin of the frequency table above as "the most fundamental mystery of the entire short story." The letter t should be the second most frequent. Why did Poe place it tenth? This was not an error made by the printer, as Legrand's decipherment is consistent with the erroneous frequency table. Identifying the second most frequent cipher symbol (which stood for t) as a, according to the table, would lead to a dead end. Hence, Legrand discards the table and uses the fact that the is the most common word.

Bennett's solution to the mystery is that the story was originally written to have the cipher in Spanish and when changed to English, at the request of the publisher, the frequency table was left unaltered. However, before setting out on a search for an early draft of the story to confirm Bennett's conjecture, a search of the cryptologic literature reveals a better solution.

Although Bennett wasn't aware of it, the mystery had been solved decades before the appearance of his text. Raymond T. Bond, in the role of editor, assembled a collection of short stories involving codes and ciphers and penned an introduction for each.[23] In the introduction to Poe's tale, he described an article on ciphers in Abraham Rees's *Cyclopædia or, Universal Dictionary of Arts, Sciences, and Literature* (1819). This article, which was written by William Blair, discussed letter frequencies, but handled consonants and vowels separately. Blair split the consonants into four groups, according to frequency and didn't attempt to rank the letters within each group, but rather ordered each group alphabetically:

```
d h n r s t      c f g l m w      b k p      q x z
```

In describing the vowels, Blair stated that e was the most frequent, followed by o, a, and i, and then pointed out that some consonants, such as s and t, are more frequent than u and y. The relatively recent letters v and j were not included!

Upon reading this, Poe placed u and y after s and t in the ordering given above and then put the other vowels at the start, accidentally switching the order of o and a to get

```
e a o i d h n r s t u y c f g l m w b k p q x z
```

---

21 Bennett, Jr., William Ralph, *Scientific and Engineering Problem-Solving with the Computer*, Prentice-Hall, Upper Saddle River, New Jersey, 1976, pp. 159–160.

22 Poe also made errors concerning wines in his excellent short story "The Cask of Amontillado." Hey Edgar, Amontillado *is* a Sherry! For more errors in "The Cask of Amontillado" see Fadiman, Clifton, editor, *Dionysus: a Case of Vintage Tales About Wine*, McGraw-Hill, New York, 1962.

23 Bond, Raymond T., *Famous Stories of Code and Cipher*, Rinehart and Company, New York, 1947, pp. 98–99. The contents of the paperback edition differ from the hardcover by one story, but "The Gold Bug" is present in both.

Because he created the genre of detective fiction, I'd like to think Poe would have enjoyed the work that went into solving the "Mystery of the Incorrect Letter Frequencies."

David Kahn pointed out several other errors in *The Gold Bug* that have nothing to do with cryptography.[24] Read the whole story and see how many you can find.

Other authors have made mistakes in their explanations of cryptanalysis. I came across an amusing example when I thought I was reading something far removed from crypto. If you can understand a little German, check out *Das Geheimnis im Elbtunnel* by Heinrich Wolff (National Textbook Company, 1988). A familiarity with vowel recognition algorithms (see Section 1.12) shows that the detective in this tale isn't much better than Poe's protagonist.

In Honoré de Balzac's *The Physiology of Marriage* (published in 1829 as part of his *Human Comedy*), the author includes four pages of nonsense.[25] This consisted of meaningless combinations of letters, as well as upside-down letters and punctuation. In the years since it first appeared, it has never been deciphered and it is widely believed among cryptanalysts that Balzac might have been blowin' a little smoke. In fact, the nonsense pages changed from one edition to the next.

This nitpicking isn't meant to detract from Poe's greatness. Generations of cryptologists trace their interest in this subject back to reading "The Gold Bug" as children, and I will have more to say about Poe's cipher challenge in Chapter 2. For now, note that Poe placed a hidden message in the following sonnet. Can you find it? The solution is given at the end of this chapter.

AN ENIGMA (1848)

by Edgar Allan Poe

"Seldom we find," says Solomon Don Dunce,
"Half an idea in the profoundest sonnet.
Through all the flimsy things we see at once
As easily as through a Naples bonnet-
Trash of all trash!—how can a lady don it?
Yet heavier far than your Petrarchan stuff-
Owl-downy nonsense that the faintest puff
Twirls into trunk-paper the while you con it."
And, veritably, Sol is right enough.
The general tuckermanities are arrant
Bubbles—ephemeral and so transparent-
But this is, now—you may depend upon it-
Stable, opaque, immortal—all by dint
Of the dear names that he concealed within't.

1.8 Arthur Conan Doyle

Among cryptologists, the best known Sherlock Holmes story is "The Adventure of the Dancing Men" (1903).[26] In this story, the cipher messages that Holmes attempts to crack have the letters replaced by dancing men; hence, the title of the tale. The ciphertexts are reproduced in Figure 1.8.

---

[24] Kahn, David, *The Codebreakers*, second edition, Scribner, New York, 1996, pp. 790–791.

[25] Kahn, David, *The Codebreakers*, second edition, Scribner, New York, 1996, p. 781.

[26] Available online at http://sherlock-holm.es/stories/pdf/a4/1-sided/danc.pdf.

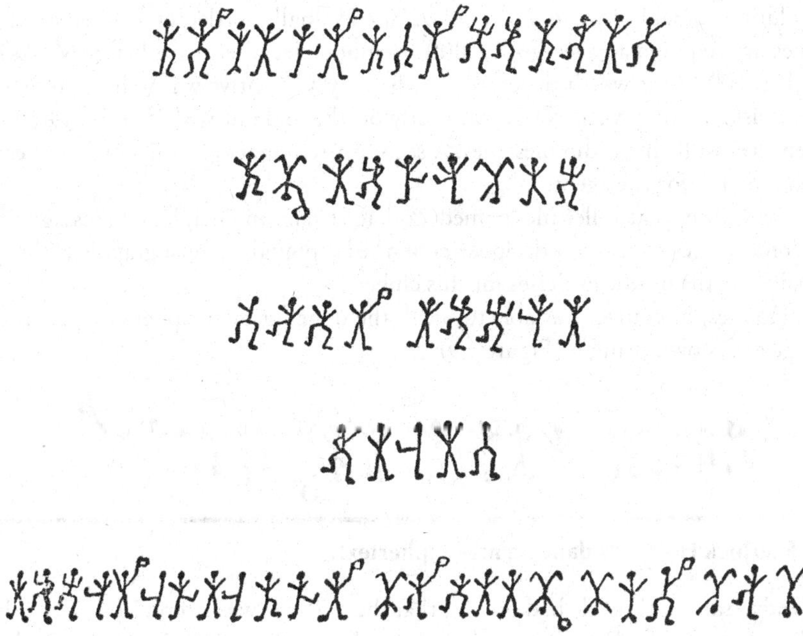

Figure 1.8   Ciphertexts found in "The Adventure of the Dancing Men."

We'll develop techniques for breaking such ciphers later in this chapter, but let's make a few simple observations now. First, observe that either the messages consist of very long words or word spacing has been disguised. Assuming the latter, either we have spacing eliminated completely or a special character represents the space. If a special symbol designates the space, it should be of very high frequency, but the highest frequency symbol in the ciphertexts sometimes appears at the end of a message. This would be pointless if it really represented a space.

Finally, we hit upon the idea that the flags represent the spaces. This makes our task much easier, as we now have word divisions. However, there are still some complicating factors. The messages are short, turn out to include proper names and locations (not just words you'd find in a dictionary) *and* there's at least one typo present! The number of typos varies from edition to edition, but all have at least one!

Holmes never mentioned the typos, so one may assume that they were errors made by Doyle and/or the printers and not intended to make things more difficult. Anyone attempting to decipher a message must be able to deal with typos. These occur frequently in real-world messages. They can be caused by a careless encipherer or by problems in transmission. "Morse mutilation" is a term sometimes used to describe a message received in Morse code with errors arising from the transmission process.

Leo Marks, who was in charge of the Special Operations Executive (SOE) code and cipher group during World War II,[27] frequently had to contend with "indecipherables" arising from agents in occupied Europe improperly enciphering. When he began in 1942, about 25% of incoming messages were indecipherable for one reason or another. At that time, only 3% were recovered, but Marks gave lectures on recovery techniques to his team and developed new methods. The

---

[27] The Special Operations Executive was roughly the British equivalent of America's Office of Strategic Services (OSS). They ran risky operations behind enemy lines.

number of solutions quickly rose to a peak of 92% and finally settled at an average of 80%. An individual message required an average of 2,000 attempted decipherments before Marks's team hit upon the real error.[28] It was worth the effort, for the only alternative was to have the agent resend the message, a risky activity with Nazis constantly on the lookout with direction-finding equipment for agents transmitting radio messages. A second attempt at transmission could easily result in capture and torture for the agent.

More recently, a copycat killer nicknamed Zodiac II sent an enciphered message to the *New York Post*. In order to get the correct decipherment, the cryptanalyst must grapple with some typos the killer made. See the online exercises for this cipher.

Sherlock Holmes, of course, was able to break the dancing men cipher completely and compose a message of his own using it (Figure 1.9).

Figure 1.9   Sherlock Holmes's dancing men ciphertext.

Holmes's adversary in this tale had no idea that the message was from Holmes and this deception led to his capture. So, the story raises an issue that we'll visit again in this book. Namely, how can we be sure that the message comes from the person we think it does? Holmes was able to impersonate another, because he learned the secret of the cipher, but in an ideal system this would not be possible.

Many agents were captured and impersonated over the course of World War II. In one instance, when it was suspected that a particular SOE agent had been compromised and that a Nazi was carrying out his end of the radio conversation, a test was devised to ascertain the truth. The suspicious British radio operator signed off on his message to the agent with HH, short for Heil Hitler. This was a standard sort of "goodbye" among the Nazis and the operator abroad responded automatically with HH, giving himself away.[29]

Was Doyle's use of dancing men to represent letters as creative as it might seem? It turns out that these fellows had been protecting messages long before Doyle wrote his story. When confronted with previous uses, he chalked it up to coincidence.[30] Doyle's dancing men are also reminiscent of some figures in a lost script, known as Rongorongo, which was used by the people of Easter Island and, oddly, resembles Indus Valley script, another indecipherable at the moment (Figure 1.10). The idea that there's a connection between the two scripts is not backed by the leading experts, as the two cultures were widely separated, not only in terms of distance but also in time.

Although such topics are only treated in passing here, deciphering lost scripts attracts the interest of a decent number of cryptologists. See the References and Further Reading section at the end of this chapter for papers *Cryptologia* has published on Rongorongo script.

"The Adventure of the Dancing Men" is not the only story in which Doyle's detective encountered cryptology. Another is discussed in Section 1.21.

---

[28] Marks, Leo, *Between Silk and Cyanide: A Codemaker's War, 1941-1945*. The Free Press, New York, pp. 192, 417.
[29] Marks, Leo, *Between Silk and Cyanide: A Codemaker's War, 1941-1945*. The Free Press, New York, pp. 348–349.
[30] Shulman, David, "The Origin of the Dancing Men," *The Baker Street Journal*, (New Series), Vol. 23, No. 1, March 1973, pp. 19–21.

Figure 1.10 A comparison of Rongorongo (right side of each column) and Indus Valley script (left side of each column). (Adapted from Imbelloni, Sr., J., "The Easter Island Script and the Middle-Indus Seals," *The Journal of the Polynesian Society,* Vol. 48, No. 1, whole no. 189, March 1939, pp. 60-69, p. 68 cited here.)

1.9 Frequency Analysis

MASCs are easy to break, as each letter is always enciphered in the same manner when it reappears. Statistics for the language in which the message was written soon reveal the plaintext. In fact, in many cases, we can use these statistics to recognize what language was used, if it is initially unknown. The correct language can often be determined without any deciphering. A sampling of some statistics for English is provided in Table 1.1.

These are only sample statistics, and your results may vary. For example, E usually stands out as much more frequent than any of the other letters, yet an entire novel (E.V. Wright's *Gadsby,* published in 1939) has been written without this letter. For your amusement, the first paragraph of *Gadsby* follows.

Table 1.1 Sample English Statistics

Letter	Relative Frequency (%)	Letter	Relative Frequency (%)
A	8.2	N	6.7
B	1.5	O	7.5
C	2.8	P	1.9
D	4.3	Q	0.1
E	12.7	R	6.0
F	2.2	S	6.3
G	2.0	T	9.0
H	6.1	U	2.8
I	7.0	V	1.0
J	0.2	W	2.4
K	0.8	X	0.2
L	4.0	Y	2.0
M	2.4	Z	0.1

If youth, throughout all history, had had a champion to stand up for it; to show a doubting world that a child can think; and, possibly, do it practically; you wouldn't constantly run across folks today who claim that "a child don't know anything." A child's brain starts functioning at birth; and has amongst its many infant convolutions, thousands of dormant atoms, into which God has put a mystic possibility for noticing an adult's act, and figuring out its purport.

Wright was not alone in meeting this challenge. The French author Georges Perec completed *La Disparation* (1969) without using a single E. Frequency tables vary from language to language, but E is commonly high. Perec's performance was perhaps more impressive, as E is even more frequent in French (17.5%) than in English. Gilbert Adair translated this work without introducing an E. He titled it *A Void* and it begins

Today, by radio, and also on giant hoardings, a rabbi, an admiral notorious for his links to masonry, a trio of cardinals, a trio, too, of insignificant politicians (bought and paid for by a rich and corrupt Anglo-Canadian banking corporation), inform us all of how our country now risks dying of starvation. A rumour, that's my initial thought as I switch off my radio, a rumour or possibly a hoax. Propaganda, I murmur anxiously – as though, just by saying so, I might allay my doubts – typical politicians' propaganda. But public opinion gradually absorbs it as a fact. Individuals start strutting around with stout clubs. 'Food, glorious food!' is a common cry (occasionally sung to Bart's music), with ordinary hard-working folk harassing officials, both local and national, and cursing capitalists and captains of industry. Cops shrink from going out on night shift. In Mâcon a mob storms a municipal building. In Rocadamour ruffians rob a hangar full of foodstuffs, pillaging tons of tuna fish, milk and cocoa, as also a vast quantity of corn – all of it, alas, totally unfit for human consumption. Without fuss or ado, and naturally without any sort of trial, an indignant crowd hangs 26 solicitors on a hastily built scaffold in front of Nancy's law courts (this Nancy is a town, not a woman) and ransacks a local journal, a disgusting right-wing rag that is siding against it. Up and down this land of ours looting has brought docks, shops and farms to a virtual standstill.

Although this paragraph was written without the letter E, it cannot be read without it! By writing 26 using digits instead of letters, Adair avoided that particular E.

Frequency counts used for the purpose of cryptanalysis first appear in Arabic works. The interest was apparently a byproduct of studying the Koran. All sorts of statistics on letters and words used in the Koran had been compiled as part of the intense study this work was subject to. Finally, someone took the next step and applied this data to an enciphered message.

1.10 Biblical Cryptology

In the beginning was the Word, and the Word was encrypted.

Although the Christian world was behind the Islamic world in terms of cryptanalysis, there is some crypto in the Bible. A system known as Atbash enciphered Hebrew by switching the first and last letters of the alphabet, the second and second to last, and so on. With our alphabet, it would look like this

```
A B C D E F G H I J K L M N O P Q R S T U V W X Y Z plaintext
Z Y X W V U T S R Q P O N M L K J I H G F E D C B A ciphertext
```

The name Atbash is derived from the first two pairs of Hebrew letters that are switched: aleph, taw and beth, sin (Figure 1.11).

aleph	beth	gimel	daleth	he	waw	zayin	heth	teth	yod	kaph
א	ב	ג	ד	ה	ו	ז	ח	ט	י	כ
ת	ש	ר	ק	צ	פ	ע	ס	נ	מ	ל
taw	sin shin	resh	qoph	sadhe	pe	ayin	samekh	nun	mem	lamed

Figure 1.11   Atbash substitutions.

This system was used in Jeremiah 25:26 and Jeremiah 51:41 where Sheshach appears as an enciphered version of Babel. Until it was recognized that this encipherment had been used, biblical scholars spent many hours puzzling over where Sheshach was located! In Section 1.5, we examined the Caesar cipher. There is a variant of this system, called a *reverse Caesar cipher*, for which the ciphertext alphabet is written backwards before being shifted. If this is suspected, one may simply transform the ciphertext Atbash style and then break it like a regular Caesar cipher. Unfortunately, much media attention has been focused on a nonexistent Bible Code, where equidistant sequences are alleged to reveal information about events from biblical times through the present and into the future. Rather than waste space on it here, I'll simply reference a paper that debunks this nonsense and point out that such "messages" can be found in any book of sufficient length.[31]

---

[31] Nichols, Randall, "The Bible Code," *Cryptologia*, Vol. 22, No. 2, April 1998, pp. 121–133.

1.11 More Frequencies and Pattern Words

We have seen a frequency table for letters, but these frequencies change if we are interested in particular positions; for example, E is far more frequent as the last letter of a word than the first. Even though E is the most common letter overall, the letter most likely to begin a word in English is not E, but rather T. This is due to the tremendous popularity of the word THE (see Table 1.2).

Table 1.2 Frequency of Letters in Initial and Terminal Positions

Letter	Frequency		Letter	Frequency	
	Initial Position	*Terminal Position*		*Initial Position*	*Terminal Position*
A	123.6	24.7	N	21.6	67.6
B	42.4	0.9	O	70.8	41.4
C	47	3.7	P	40.9	6.4
D	20.6	111	Q	2.2	-
E	27.7	222.9	R	31.2	47.5
F	41.9	49.3	S	69.1	137
G	12.3	25.1	T	181.3	98.1
H	47.5	34.2	U	14.1	2.3
I	59	0.9	V	4.1	-
J	7	-	W	63.3	3.7
K	1.8	8.7	X	0.4	1.4
L	22.4	24.7	Y	7.5	66.2
M	39.2	20.1	Z	0.4	-

Source: Pratt, Fletcher, *Secret and Urgent,* Bobbs-Merrill, New York, 1939, p. 158.

Note: Figures above represent the frequency of occurrence in 1,000 words. V, Q, J, and Z occur so rarely as terminals that their frequencies cannot be expressed in this table.

The letters most commonly doubled are shown in Table 1.3. Frequency tables for all two-letter combinations (known as *digraphs*) can be found in many books and online sources. Figure 1.12 provides such an example. Usually, TH is the most frequent digraph, but for the government telegrams used for Friedman's table EN is the most frequent and TH places fifth.

Pattern word lists are also of value for deciphering without a key. The pattern of the letters in the word may be looked up as in a dictionary. As an example, consider the pattern ABCADEAE. This is meant to indicate that the first, fourth, and seventh letters are all the same. Also, the sixth and eighth letters match. No other matches may be present, as B, C, and D denote distinct letters. There are very few words that fit this form. ARKANSAS, EXPENDED, and EXPENSES are the only possibilities given in the reference cited here.[32]

---

[32] Carlisle, Sheila, *Pattern Words Three Letters to Eight Letters in Length,* Aegean Park Press, Laguna Hills, California, 1986, p. 65. There are more complete lists that include other possibilities.

Table 1.3   Most Common Doubled Letters

Letter	Frequency	Letter	Frequency	Letter	Frequency
LL	19	FF	9	MM	4
EE	14	RR	6	GG	4
SS	15	NN	5	DD	1.5
OO	12	PP	4.5	AA	0.5
TT	9	CC	4	BB	0.25

Source: Pratt, Fletcher, *Secret and Urgent,* Bobbs-Merrill, New York, 1939, p. 259.

Note: Figures above represent the frequency of occurrence in 1,000 words.

If such a rare pattern word appears in a MASC, we can quickly obtain several letters in the cipher alphabet. If the substitutions suggested by ARKANSAS yield impossible words elsewhere, simply try the next choice on the list.

There are websites that allow you to enter a pattern and see all of the words, in a particular language, that fit that pattern. A pair of these are https://design215.com/toolbox/wordpattern.php and https://www.hanginghyena.com/solvers/cryptogram-helper.

It's very easy to manipulate data in electronic form, but this was not how things were done in the old days. An excerpt from an interview with Jack Levine, the creator of several volumes of pattern word books, sheds some light on this.[33]

> **Levine:** Cryptography is my hobby. I enjoy doing it. Years ago there were very few mathematicians working in this area but today there are a good many prestigious mathematicians studying cryptography, in particular algebraic cryptography, which by the way is an expression I invented. My pattern word list, which I produced in the 70s, is now the standard work in this area.
>
> **Burniston:** Tell me about this.
>
> **Levine:** What I did was to take *Webster's Unabridged Dictionary* which has over 500,000 words in it, and copied each word and classified it by its pattern. In other words, if you wanted to know all six-letter words where the first and fourth letters were the same, you could go to my book and find all the words with that pattern quickly.
>
> **Burniston:** Let me get this straight, now. You copied out all the words in *Webster's Unabridged Dictionary?*
>
> **Levine:** Yes, In fact, I started with the second edition and while I was doing this, the third edition came out and I more or less had to start the whole thing again. That was a pain.
>
> **Burniston:** How long did it take you to do this?
>
> **Levine:** About 15 years. I had the word list published by the print shop here on campus [N.C. State] at my own expense and gave the copies to members of the American Cryptogram Association, of which I am a past president. Now because of the very limited number of copies, it has become a valuable item. It also probably ruined my eye sight.

---

[33] *History of the Math Department at NCSU: Jack Levine,* December 31, 1986, interview, https://web.archive.org/web/20160930110613/http://www4.ncsu.edu/~njrose/Special/Bios/Levine.html.

CONFIDENTIAL

TABLE 6–A.—*Frequency distribution of digraphs, based on 50,000 letters of Governmental plaintext telegrams; reduced to 5,000 digraphs*

SECOND LETTER

FIRST LETTER	A	B	C	D	E	F	G	H	I	J	K	L	M	N	O	P	Q	R	S	T	U	V	W	X	Y	Z	Total	Blanks
A	3	6	14	27	1	4	6	2	17	1	2	32	14	64	2	12		44	41	47	13	7	3		12		374	3
B	4				18				2	1		6	1		4			2	1	1	2				7		49	14
C	20		3	1	32	1		14	7		4	5	1	1	41			4	1	14	4		1		1		155	8
D	32	4	4	8	33	8	2	2	27	1		3	5	4	16	5	2	12	13	15	5	3	4		1		209	3
E	35	4	32	60	42	18	4	7	27	1		29	14	111	12	20	12	87	54	37	3	20	7	7	4	1	648	1
F	5		2	1	10	11	1		39			2	1		40	1		9	3	11	3		1		1		141	9
G	7		2	1	14	2	1	20	5	1		2	1	3	6	2		5	3	4	2		1				82	7
H	20	1	3	2	20	5			33			1	2	3	20	1	1	17	4	28	8		1		1		171	7
I	8	2	22	6	13	10	19				2	23	9	75	41	7		27	35	27		25		15		2	368	7
J	1				2										2						2						7	22
K	1		1		6			2				1		1				1									13	19
L	28	3	3	9	37	3	1	1	20			27	2	1	13	3		2	6	8	2	2	2		10		183	5
M	36	6	3	1	26	1		1	9			13			10	8		2	4	2	2				2		126	10
N	26	2	19	52	57	9	27	4	30	1	2	5	5	8	18	3	1	4	24	82	7	3	3		5		397	2
O	7	4	8	12	3	25	2	3	5	1	2	19	25	77	6	25		64	14	19	37	7	8	1	2		376	2
P	14	1	1	1	23	2		3	6			13	4	1	17	11		18	6	8	3	1	1		1		135	6
Q													1					1			15						17	23
R	39	2	9	17	98	6	7	3	30	1	1	5	9	7	28	13		11	31	42	5	5	4		9		382	3
S	24	3	13	5	49	12	2	26	34		1	2	3	4	15	10		5	19	63	11	1	4		1		307	4
T	28	3	6	6	71	7	1	78	45			5	6	7	50	2	1	17	19	19	5		36		41	1	454	4
U	5	3	3	3	11	1	8		5			6	5	21	1	2		31	12	12		1					130	9
V	6				57				12						1						1						77	21
W	12				22		4	13				1		2	19			1	1						1		76	16
X	2		2	1	1	1		1	2				1	1	2			1	1	7							23	13
Y	6	2	4	4	9	11	1	1	3			2	2	6	10	3		4	11	15	1		1				96	7
Z	1				2				1																		4	23
Total	370	46	154	217	657	137	82	170	374	8	14	189	123	397	373	130	17	368	304	462	130	75	77	23	99	4	5,000	
Blanks	1	11	6	7	1	7	12	10	3	18	19	6	6	7	3	8	21	4	4	5	7	15	11	23	10	23		248

CONFIDENTIAL

Figure 1.12 Frequency table for two-letter combinations. (From Friedman, William F. and Lambros D. Callimahos, *Military Cryptanalytics, Part I,* National Security Agency, Washington, DC, 1956, p. 257.)

Pattern word lists were used during World War II, although America's lists weren't as complete as those made later by Levine. Amazingly, I've been unable to find decent lists from before World War II. It seems like such an obvious approach to cracking MASCs must be hundreds of years old, but, if so, where are the lists?

Non-pattern words are also useful to have in lists. These are words that do not use any letter more than once. They are also called *isograms*. The record length seems to be 16 letters. A few examples follow:

uncopyrightables	(16 letters)
uncopyrightable	(15 letters)[34]
dermatoglyphics	(15 letters—the science of fingerprints)
ambidextrously	(14 letters)
thumbscrewing	(13 letters)[35]
sympathized	(11 letters)
pitchforked	(11 letters)
gunpowdery	(10 letters)
blacksmith	(10 letters)
prongbucks	(10 letters)
lumberjack	(10 letters)

Many more words can be added to the shorter word length lists above. How many can you find?

Computer programs that use dictionaries of pattern (and nonpattern) words can rapidly break the simple ciphers discussed in this chapter. But before we start cracking ciphers, let's examine some more tools that may be useful.

1.12 Vowel Recognition Algorithms

To see the basic idea at the heart of various vowel recognition algorithms, consider the following challenge. Construct words that have the following digraphs in them (at the beginning, middle, or end—your choice). Take as much time as you like and then continue reading.

AA	BA
AB	BB
AC	BC
AD	BD
AE	BE
AF	BF
AG	BG
AH	BH
:	:
AZ	BZ

Was this a harder task for the second column of digraphs? I'm sure the second column was more difficult for you. This is because vowels contact more letters than consonants. That is the key characteristic in distinguishing the vowels from the consonants. Hence, your work above shows that A is more likely to be a vowel than B, since you found words for a higher percentage of digraphs in

---

[34] Fourteen- and 15-letter isograms are from Lederer, Richard, *Crazy English*, Pocket Books, New York, 1989, p. 159.
[35] Ten-, 11-, and 13- letter isograms are from http://www.wordways.com/morenice.htm.

the A column than in the B column. We now look at an algorithm that may be applied to mono-alphabetic ciphertexts, as well as unknown scripts.

1.12.1 Sukhotin's Method

1. Count the number of times each letter contacts each of the others and put these values in an n-by-n square (n = alphabet size).
2. Make the diagonal of the square all zeroes.
3. Sum the rows and assume all characters are consonants.
4. Find highest "consonant" row-sum and assume it's a vowel. (Stop if none is positive.)
5. Subtract from the row sum of each consonant twice the number of times that it occurs next to the newly found vowel. Return to Step 4.

As always, an example makes things clearer. Consider the phrase NOW WE'RE RECOGNIZING VOWELS.

Step 1

```
    C E G I L N O R S V W Z
C   0 1 0 0 0 0 0 1 0 0 0 0 0
E   1 0 0 0 1 0 0 3 0 0 2 0
G   0 0 0 0 0 2 1 0 0 0 0 0
I   0 0 0 0 0 2 0 0 0 0 0 2
L   0 1 0 0 0 0 0 0 1 0 0 0
N   0 0 2 2 0 0 1 0 0 0 0 0
O   1 0 1 0 0 1 0 0 0 1 2 0
R   0 3 0 0 0 0 0 0 0 0 0 0
S   0 0 0 0 1 0 0 0 0 0 0 0
V   0 0 0 0 0 0 1 0 0 0 0 0
W   0 2 0 0 0 0 2 0 0 0 0 0
Z   0 0 0 2 0 0 0 0 0 0 0 0
```

Steps 2 and 3

	C	E	G	I	L	N	O	R	S	V	W	Z	Sum	Consonant/Vowel
C	0	1	0	0	0	0	0	1	0	0	0	0	2	C
E	1	0	0	0	1	0	0	3	0	0	2	0	7	V
G	0	0	0	0	0	2	1	0	0	0	0	0	3	C
I	0	0	0	0	0	2	0	0	0	0	0	2	4	C
L	0	1	0	0	0	0	0	0	1	0	0	0	2	C
N	0	0	2	2	0	0	1	0	0	0	0	0	5	C
O	1	0	1	0	0	1	0	0	0	1	2	0	6	C
R	0	3	0	0	0	0	0	0	0	0	0	0	3	C
S	0	0	0	0	1	0	0	0	0	0	0	0	1	C
V	0	0	0	0	0	0	1	0	0	0	0	0	1	C
W	0	2	0	0	0	0	2	0	0	0	0	0	4	C
Z	0	0	0	2	0	0	0	0	0	0	0	0	2	C

Step 4

E looks like a vowel, because it has the highest row sum. We then adjust the row sums.

Step 5

	C	E	G	I	L	N	O	R	S	V	W	Z	Sum	Consonant/Vowel
C	0	1	0	0	0	0	0	1	0	0	0	0	0	C
E	1	0	0	0	1	0	0	3	0	0	2	0	7	V
G	0	0	0	0	0	2	1	0	0	0	0	0	3	C
I	0	0	0	0	0	2	0	0	0	0	0	2	4	C
L	0	1	0	0	0	0	0	0	1	0	0	0	0	C
N	0	0	2	2	0	0	1	0	0	0	0	0	5	C
O	1	0	1	0	0	1	0	0	0	1	2	0	6	V
R	0	3	0	0	0	0	0	0	0	0	0	0	-3	C
S	0	0	0	0	1	0	0	0	0	0	0	0	1	C
V	0	0	0	0	0	0	1	0	0	0	0	0	1	C
W	0	2	0	0	0	0	2	0	0	0	0	0	0	C
Z	0	0	0	2	0	0	0	0	0	0	0	0	2	C

Back to Step 4

Now O looks like a vowel. We adjust the row sums again (Step 5):

	C	E	G	I	L	N	O	R	S	V	W	Z	Sum	Consonant/Vowel
C	0	1	0	0	0	0	0	1	0	0	0	0	-2	C
E	1	0	0	0	1	0	0	3	0	0	2	0	7	V
G	0	0	0	0	0	2	1	0	0	0	0	0	1	C
I	0	0	0	0	0	2	0	0	0	0	0	2	4	C
L	0	1	0	0	0	0	0	0	1	0	0	0	0	C
N	0	0	2	2	0	0	1	0	0	0	0	0	3	C
O	1	0	1	0	0	1	0	0	0	1	2	0	6	V
R	0	3	0	0	0	0	0	0	0	0	0	0	-3	C
S	0	0	0	0	1	0	0	0	0	0	0	0	1	C
V	0	0	0	0	0	0	1	0	0	0	0	0	-1	C
W	0	2	0	0	0	0	2	0	0	0	0	0	-4	C
Z	0	0	0	2	0	0	0	0	0	0	0	0	2	C

After adjusting for O, the next vowel appears to be I:

	C	E	G	I	L	N	O	R	S	V	W	Z	Sum	Consonant/Vowel
C	0	1	0	0	0	0	0	1	0	0	0	0	-2	C
E	1	0	0	0	1	0	0	3	0	0	2	0	7	V
G	0	0	0	0	0	2	1	0	0	0	0	0	1	C
I	0	0	0	0	0	2	0	0	0	0	0	2	4	V
L	0	1	0	0	0	0	0	0	1	0	0	0	0	C
N	0	0	2	2	0	0	1	0	0	0	0	0	-1	C
O	1	0	1	0	0	1	0	0	0	1	2	0	6	V
R	0	3	0	0	0	0	0	0	0	0	0	0	-3	C
S	0	0	0	0	1	0	0	0	0	0	0	0	1	C
V	0	0	0	0	0	0	1	0	0	0	0	0	-1	C
W	0	2	0	0	0	0	2	0	0	0	0	0	-4	C
Z	0	0	0	2	0	0	0	0	0	0	0	0	-2	C

Continuing the process, G and S are declared vowels! The technique is not perfect, but it works much better with a text of greater length. This procedure is well suited for implementation on a computer, so we can often quickly separate the vowels from the consonants in a ciphertext. The most frequent characters are usually E and T and with the help this technique gives us in distinguishing them, we are well on our way to a solution.

1.13 More MASCs

Many popular authors have included ciphers in their works. These include, in addition to those discussed previously, some at a higher level of sophistication by Jules Verne, Dorothy Sayers, Charles Dodgson (better known by his pen name Lewis Carroll), and others whom you'll encounter in the pages to follow. Figure 1.13 displays a dust jacket from one of the many editions of the first book in J.R.R. Tolkien's *Lord of the Rings* trilogy. Tolkien created a rune alphabet, which is often used to conceal secret messages on the covers. Can you crack it with the statistics and techniques discussed in this chapter?

Ozzy Osbourne's *Speak of the Devil* album makes use of Tolkien's rune alphabet on the front and back covers.

For those whose tastes lean more towards classical music, Figure 1.14 provides a cipher "composed" by Wolfgang Amadeus Mozart. Each note simply represents a letter.

Cryptology *is* mathematics in the same sense that music *is* mathematics.—H. Gary Knight[36]

One common MASC sometimes used by children consists solely of non-alphabetic symbols, yet it is easy to remember. It works as shown in Figure 1.15. The symbol drawn is simply a representation of the region in which the letter is found. Decipher the message shown in Figure 1.16 to make sure you have the hang of it.

This system is known as the pigpen cipher, because the letters are separated like pigs in a pen. It is also called the Masonic cipher, because the Society of Freemasons has made use of it. It was used in the Civil Wars in England in the 17th century and even as recently as the U.S. Civil War by prisoners sending messages to friends.[37]

The cipher on the tombstone shown in Figure 1.17 uses a variant of the pigpen cipher. Notice that there are no ciphertext letters resembling the greater than and less than symbols, in any rotation, with or without dots. On the other hand, some pieces have two dots inside. The tombstone can be seen in Trinity Churchyard in New York City. It marks the grave of James Leeson, an officer of a Masonic lodge, who died in 1794.[38]

The same message appears on a flat slab marking the grave of Captain James Lacey (d. 1796) in St. Paul's Churchyard, just a few blocks away; however, this one uses a different key.[39]

---

[36] Knight, H. Gary, "Cryptanalyst's Corner," *Cryptologia*, Vol. 2, No. 1, January 1978, pp. 68–74.

[37] McCormick, Donald, *Love in Code*, Eyre Methuen Ltd, London, UK, 1980, pp. 4–5.

[38] Kruh, Louis, "The Churchyard Ciphers," *Cryptologia*, Vol. 1, No. 4, October 1977, pp. 372–375.

[39] Kruh, Louis, "The Churchyard Ciphers," *Cryptologia*, Vol. 1, No. 4, October 1977, pp. 372–375.

Figure 1.13 Find the message hidden on the cover of this book. (Courtesy of Houghton Mifflin and the Tolkien estate.)

Figure 1.14 A cipher created by Wolfgang Amadeus Mozart—not intended for performance! (Retyped by Nicholas Lyman from McCormick, Donald, *Love in Code,* Eyre Methuen Ltd., London, UK, 1980, p. 49.)

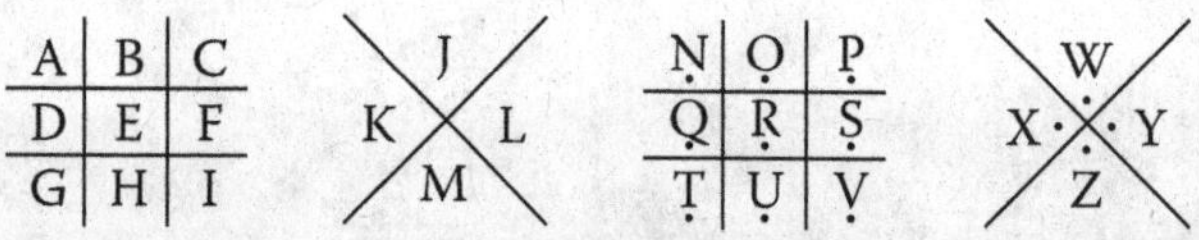

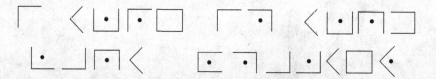

Figure 1.15 A common MASC used by Masons and children.

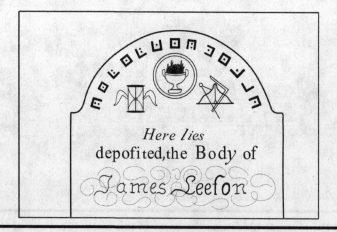

Figure 1.16 A sample message to decipher.

Figure 1.17 Tales from the crypt(ologist)? (Image created by Josh Gross.)

1.14 Cryptanalysis of a MASC

Now that we're seeing ciphers everywhere, it's time for an example of how they can be broken. Suppose we intercept the following ciphertext.

```
JDS CGWMUJNCQV NSIBVMJCBG QGW CJU ZBGUSHMSGZSU DQIS ASSG Q
WCUQUJSN XBN JDS DMTQG NQZS.
```

We can begin our attack by constructing a frequency table for the ciphertext letters (Table 1.4).

Table 1.4 Frequency Table for Sample Ciphertext Letters

Letter	Frequency	Letter	Frequency
A	1	N	5
B	4	O	0
C	5	P	0
D	4	Q	7
E	0	R	0
F	0	S	11
G	7	T	1
H	1	U	6
I	2	V	2
J	6	W	3
K	0	X	1
L	0	Y	0
M	4	Z	3

The letter S sticks out as having the largest frequency; therefore, it's likely that it represents the plaintext letter E. The three letter ciphertext word JDS appears twice. The most common three letter combination (as well as the most common three letter word) is THE. This agrees with our suspicion that S represents E. Is it likely that J represents T? T is very common in English and the ciphertext letter J appears 6 times, so it seems plausible. Writing our guesses above the corresponding ciphertext letters gives:

```
THE     T    E   T       T    E E E H E EE
JDS CGWMUJNCQV NSIBVMJCBG QGW CJU ZBGUSHMSGZSU DQIS ASSG Q

     TE     THE H       E.
WCUQUJSN XBN JDS DMTQG NQZS.
```

There are many possible ways to proceed from here. We have a 12-letter word, ZBGUSHMSGZSU, so let's consider its pattern. It has the form ABCDEFGECAED. At one of the websites referenced earlier, you can type in this pattern and you'll see that there's only one word that fits, namely CONSEQUENCES. Substituting for all of the letters now revealed gives

```
THE  N UST     E  UT ON  N   TS CONSEQUENCES H E   EEN
JDS CGWMUJNCQV NSIBVMJCBG QGW CJU ZBGUSHMSGZSU DQIS ASSG Q

 S STE  O  THE HU N   CE.
WCUQUJSN XBN JDS DMTQG NQZS.
```

The fifth word cannot be anything other than ITS. Placing I above every C quickly leads to more letters and the message is revealed to be:

```
THE INDUSTRIAL REVOLUTION AND ITS CONSEQUENCES HAVE BEEN A
DISASTER FOR THE HUMAN RACE.
```

This is a quote from former mathematician Theodore Kaczynski. The substitutions used to encipher it follow.

```
A B C D E F G H I J K L M N O P Q R S T U V W X Y Z plaintext
Q A Z W S X E D C R F V T G B Y H N U J M I K O L P ciphertext
```

Do you see the pattern in the substitutions now that it is entirely revealed?[40]

Notice that the statistics given in this chapter helped, but they don't match our message perfectly. T was the second most frequent letter in our table, but it's tied for fourth place in the sample cipher. Nevertheless, the frequency was high enough to make it seem like a reasonable substitution. In general, we may make some incorrect guesses in trying to break a cipher. When this happens, simply backtrack and try other guesses!

Now that we've achieved some skill in breaking MASCs, it's time to laugh at those who are not so well informed; may they forever wallow in their ignorance! One fellow, whose name has been lost to history, unwittingly displayed his ignorance when he proudly explained how he had deciphered a message of the type we've been examining.[41]

> From the moment when the note fell into my hands, I never stopped studying from time to time the signs which it bore.... About 15 years more or less passed, until the moment when God (Glory to Him!) did me the favor of permitting me to comprehend these signs, although no one taught them to me...

On a more serious note, Chevalier de Rohan's death was a direct consequence of his inability to decipher such a message.[42] The original message was in French and the ciphertext read

```
mg dulhxcclgu ghj yxuj lm ct ulgc alj
```

---

[40] A hint is provided at the end of this chapter.

[41] Kahn, David, *The Codebreakers*, second edition, Scribner, New York, 1996, p. 99.

[42] For a fuller account, see Pratt, Fletcher, *Secret and Urgent*, Bobbs Merrill, New York, 1939, pp.137–139.

1.15 Ciphers by a Killer and a Composer

The paragraph above ended on a serious note, but on an even *more* serious note, there have been instances of serial killers using ciphers. In some cases, the purpose was to hide incriminating evidence such as details of murders recorded in a journal (e.g. Unabomber and Joseph Edward Duncan), and sometimes simply to taunt the police and the public (e.g. Zodiac, Scorpion, and B.T.K.). Some such ciphers are still unsolved and the killers themselves not yet identified.

The first set of three ciphers that the Zodiac killer sent out were made available to the National Security Agency (NSA), Central Intelligence Agency (CIA), and Federal Bureau of Investigation (FBI), but it was a pair of amateurs that broke them. Donald Harden, a high school history and economics teacher, began working on the ciphers and was later joined by his wife, Bettye Harden, who had no previous experience with cryptology. Nevertheless, she came up with the idea of a probable word search and together they recovered the messages. It took the pair 20 hours. Complications included the presence of five errors. Part one of this cipher is shown in Figure 1.18. How can you tell that it is more complicated than a MASC?

Figure 1.18   A Zodiac killer cipher.

The decipherment for this portion is:

> I like killing people because it is so much fun it is more fun than killing wild game
> in the forest because man is the most dangerous anamal of all to kill something gi

I ended up having a lot to say about another cipher created by the Zodiac killer. It's reproduced in Figure 1.19, but I won't be giving any spoilers here. If you wish to learn more, I encourage you to pursue the references at the end of this chapter.

It seems like a good time to lighten things up again. "Pomp and Circumstance," played at virtually every graduation ceremony, was composed by Edward Elgar (Figure 1.20). Although this is certainly his best known composition, he also composed a cipher message in 1897 that no one has been able to break.[43] It's shown in Figure 1.21.

---

[43] For more information, see Chapter 3 of Bauer, Craig P., *Unsolved! The History and Mystery of the World's Greatest Ciphers from Ancient Egypt to Online Secret Societies*, Princeton University Press, Princeton, New Jersey, 2017.

Figure 1.19 Another Zodiac Killer Cipher.

Figure 1.20 Edward Elgar (https://en.wikipedia.org/wiki/Edward_Elgar).

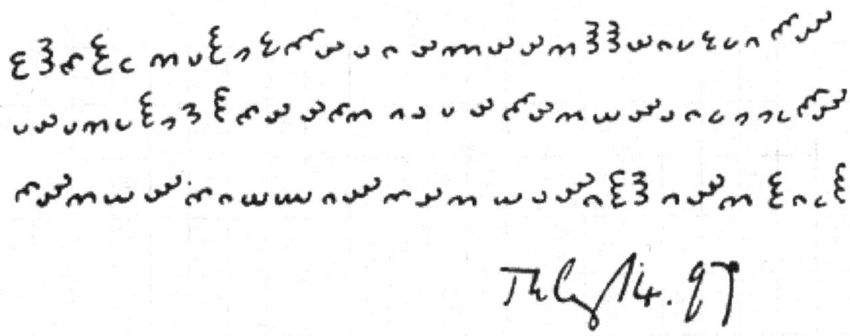

Figure 1.21   An open problem—what does this say? (http://en.wikipedia.org/wiki/File:Dorabella-cipher-image.gif).

1.16 Affine Ciphers

When we introduced the mixed alphabets, the mathematical view of encrypting (via modular arithmetic) seemed to disappear! It appeared again on the other side, cryptanalysis, where we've been using some simple statistics. Returning to Caesar's cipher, we can generalize to bring the mathematical perspective back to enciphering again. Caesar used $C = M + 3$ (mod 26). We made the key arbitrary, giving $C = M + K$ (mod 26). But why not generalize further? We do this by introducing a multiplier, a.

Consider $C = aM + b$ (mod 26). Here our key is $K = (a, b)$, an ordered pair. Using $K = (0, b)$ is not allowed, because every letter would get sent to b and the message would be impossible to decipher—only its length would be known. $K = (1, 3)$ gives a Caesar shift. What about other values? We only get a 1:1 map between plaintext and ciphertext letters when the greatest common divisor of a and 26 is 1. That is, a and 26 must be relatively prime (also called *coprime*).

The various values of a may be investigated individually. All odd values between 1 and 25 inclusive work, with the exception of 13. Euler's totient function, $\varphi(n)$ gives the number of positive integers less than n and relatively prime to n. So we may write $\varphi(26) = 12$. This function is very useful in public key cryptography, as will be seen in Section 14.3. Now, b can take any value in the range 0 to 25, so the keyspace for this cipher is $(12)(26) = 312$. Decipherment is done via the equation $M = a^{-1}(C - b)$, where a^{-1} is the multiplicative inverse of a, that is, the number that gives 1 when multiplied by a (mod 26).

The keyspace for an affine cipher is too small to serve as protection against a brute force attack. Although we are using some mathematics here, the set of possible keys is just a tiny subset of that for the general MASC. Recall that the keyspace for this is 26! = 403,291,461,126,605,635,584,000,000. A mod 26 multiplication table (Table 1.5) is provided for your convenience in using an affine cipher.

Let's encipher a short message using the key (11, 8) to get the hang of this system:

HOW ARE YOU? I'M AFFINE.

We convert the message to numbers (ignoring punctuation) to get:

7, 14, 22, 0, 17, 4, 24, 14, 20, 8, 12, 0, 5, 5, 8, 13, 4

Table 1.5 Mod 26 Multiplication Table

×	1	2	3	4	5	6	7	8	9	10	11	12	13	14	15	16	17	18	19	20	21	22	23	24	25
1	1	2	3	4	5	6	7	8	9	10	11	12	13	14	15	16	17	18	19	20	21	22	23	24	25
2	2	4	6	8	10	12	14	16	18	20	22	24	0	2	4	6	8	10	12	14	16	18	20	22	24
3	3	6	9	12	15	18	21	24	1	4	7	10	13	16	19	22	25	2	5	8	11	14	17	20	23
4	4	8	12	16	20	24	2	6	10	14	18	22	0	4	8	12	16	20	24	2	6	10	14	18	22
5	5	10	15	20	25	4	9	14	19	24	3	8	13	18	23	2	7	12	17	22	1	6	11	16	21
6	6	12	18	24	4	10	16	22	2	8	14	20	0	6	12	18	24	4	10	16	22	2	8	14	20
7	7	14	21	2	9	16	23	4	11	18	25	6	13	20	1	8	15	22	3	10	17	24	5	12	19
8	8	16	24	6	14	22	4	12	20	2	10	18	0	8	16	24	6	14	22	4	12	20	2	10	18
9	9	18	1	10	19	2	11	20	3	12	21	4	13	22	5	14	23	6	15	24	7	16	25	8	17
10	10	20	4	14	24	8	18	2	12	22	6	16	0	10	20	4	14	24	8	18	2	12	22	6	16
11	11	22	7	18	3	14	25	10	21	6	17	2	13	24	9	20	5	16	1	12	23	8	19	4	15
12	12	24	10	22	8	20	6	18	4	16	2	14	0	12	24	10	22	8	20	6	18	4	16	2	14
13	13	0	13	0	13	0	13	0	13	0	13	0	13	0	13	0	13	0	13	0	13	0	13	0	13
14	14	2	16	4	18	6	20	8	22	10	24	12	0	14	2	16	4	18	6	20	8	22	10	24	12
15	15	4	19	8	23	12	1	16	5	20	9	24	13	2	17	6	21	10	25	14	3	18	7	22	11
16	16	6	22	12	2	18	8	24	14	4	20	10	0	16	6	22	12	2	18	8	24	14	4	20	10
17	17	8	25	16	7	24	15	6	23	14	5	22	13	4	21	12	3	20	11	2	19	10	1	18	9
18	18	10	2	20	12	4	22	14	6	24	16	8	0	18	10	2	20	12	4	22	14	6	24	16	8
19	19	12	5	24	17	10	3	22	15	8	1	20	13	6	25	18	11	4	23	16	9	2	21	14	7
20	20	14	8	2	22	16	10	4	24	18	12	6	0	20	14	8	2	22	16	10	4	24	18	12	6
21	21	16	11	6	1	22	17	12	7	2	23	18	13	8	3	24	19	14	9	4	25	20	15	10	5
22	22	18	14	10	6	2	24	20	16	12	8	4	0	22	18	14	10	6	2	24	20	16	12	8	4
23	23	20	17	14	11	8	5	2	25	22	19	16	13	10	7	4	1	24	21	18	15	12	9	6	3
24	24	22	20	18	16	14	12	10	8	6	4	2	0	24	22	20	18	16	14	12	10	8	6	4	2
25	25	24	23	22	21	20	19	18	17	16	15	14	13	12	11	10	9	8	7	6	5	4	3	2	1

Then each cipher letter is obtained by taking $11M + 8$ (modulo 26), where M is the message letter. Make sure to use the mod 26 multiplication table to save time.

$11(7) + 8 = 7$	$11(14) + 8 = 6$	$11(22) + 8 = 16$	$11(0) + 8 = 8$	$11(17) + 8 = 13$
$11(4) + 8 = 0$	$11(24) + 8 = 12$	$11(14) + 8 = 6$	$11(20) + 8 = 20$	$11(8) + 8 = 18$
$11(12) + 8 = 10$	$11(0) + 8 = 8$	$11(5) + 8 = 11$	$11(5) + 8 = 11$	$11(8) + 8 = 18$
$11(13) + 8 = 21$	$11(4) + 8 = 0$			

So our numerical ciphertext is 7, 6, 16, 8, 13, 0, 12, 6, 20, 18, 10, 8, 11, 11, 18, 21, 0. Converting back to letters and replacing punctuation (not good for security!) we have

> HGQ INA MGU? S'K ILLSVA.

To decipher, we need to convert back to numbers and apply the equation $M = a^{-1}(C - b)$. Our choice for a was 11 and the mod 26 multiplication table shows $(11)(19) = 1$, so a^{-1} is 19. Go ahead and apply $M = 19(C - 8)$ to each letter of ciphertext to recover the original message, if you are at all unsure. You can do the subtraction, followed by the multiplication or convert the formula like so:

$$M = 19(C - 8) = 19C - 19(8) = 19C - 22 = 19C + 4$$

and then use the formula $M = 19C + 4$ to decipher in the same manner as we originally enciphered.

Although the affine cipher's keyspace is so small that we don't need to look for anything more sophisticated than a brute-force attack to rapidly break it, I'll point out another weakness. If we're able to obtain a pair of distinct plaintext letters and their ciphertext equivalents (by guessing how the message might begin or end, for example), we can usually recover the key mathematically.

Example 1

Suppose we intercept a message and we guess that it begins DEAR…. If the first two ciphertext letters are RA, we can pair them up with D and E to get

$$R = Da + b, \qquad A = Ea + b$$

Replacing the letters with their numerical equivalents gives

$$17 = 3a + b, \qquad 0 = 4a + b$$

We have several methods we can use to solve this system of equations:

1. Linear Algebra offers several techniques.
2. We can solve for b in one of the equations and substitute for it in the other to get an equation with one unknown, a.
3. We can subtract one equation from the other to eliminate the unknown b.

Let's take approach 3.

$$17 = 3a + b$$
$$- (0 = 4a + b)$$
$$\overline{17 = -a}$$

Since we are working modulo 26, the solution becomes $a = -17 = 9$. Plugging $a = 9$ into $0 = 4a + b$, we get $36 + b = 0$, which is $b = -36 = 16$. We've now completely recovered the key. It is (9, 16).

Example 2

Suppose we intercept a message sent by Cy Deavours and we guess that the last two ciphertext letters arose from his signature: CY. If the ciphertext ended LD, then we have

$$L = aC + b, \quad D = aY + b$$

Replacing the letters with their numerical equivalents gives

$$11 = 2a + b, \quad 3 = 24a + b$$

Let's take approach 3 again to solve this system of equations:

$$3 = 24a + b$$
$$- (11 = 2a + b)$$
$$-8 = 22a$$

Because we're working modulo 26, this result becomes $18 = 22a$. Looking at our mod 26 multiplication table, we see that there are two possible values for a, 2 and 15. Let's see what each equation gives us, when we plug in $a = 2$.

$$3 = 24a + b \rightarrow \quad 3 = 22 + b \rightarrow \quad -19 = b \rightarrow \quad b = 7$$
$$11 = 2a + b \rightarrow \quad 11 = 4 + b \rightarrow \quad 7 = b$$

Now let's try plugging $a = 15$ into each equation:

$$3 = 24a + b \rightarrow \quad 3 = 22 + b \rightarrow \quad -19 = b \rightarrow \quad b = 7$$
$$11 = 2a + b \rightarrow \quad 11 = 4 + b \rightarrow \quad 7 = b$$

So, b is definitely 7, but we cannot determine if a is 2 or 15. In this case, we need another plaintext/ciphertext pair to decide. Of course, because we know it is one or the other, we could try the two key pairs (2, 7) and (2, 15) on the message and see which gives meaningful text.

So why didn't this example work out as nicely as the first? Why didn't we get a unique solution? Our problem began when $18 = 22a$ gave us two choices for the value of a. To get at the heart of the matter, let's back up to the previous step:

$$3 = 24a + b$$
$$- (11 = 2a + b)$$
$$-8 = 22a$$

Now let's make it more general by letting C_1 and C_2 represent the ciphertext values and M_1 and M_2 the plaintext values:

$$C_2 = M_2a + b$$
$$-(C_1 = M_1a + b)$$
$$C_2 - C_1 = (M_2 - M_1)a$$

Thus, we'll fail to have a unique solution whenever the equation $C_2 - C_1 = (M_2 - M_1)a$ fails to have a unique solution. If $M_2 - M_1$ has an inverse modulo 26 (something we can multiply by to get 1 modulo 26), then there will be a unique solution for a, namely $a = (M_2 - M_1)^{-1}(C_2 - C_1)$, where $(M_2 - M_1)^{-1}$ denotes the inverse of $M_2 - M_1$.

In Example 1, $M_2 - M_1$ was 1, which is invertible mod 26. However, for Example 2, $M_2 - M_1$ was 22, which is not invertible mod 26 (there is no number in the mod 26 table that 22 can be multiplied by to get 1).

Before we move on to the next section, a challenge is presented. Can you find the hidden message on the book cover reproduced in Figure 1.22? If not, try again after reading the next few pages.

Figure 1.22   Cryptic science fiction cover art. (Courtesy of DAW Books, www.dawbooks.com.)

1.17 Morse Code and Huffman Coding

Another system in which each letter is consistently swapped out for the same representation is the familiar Morse code. There have actually been two versions, American and international. The international code is shown in Table 1.6.

Table 1.6 International Morse Code

Letter	Code	Letter	Code
A	·_	N	_·
B	_···	O	_ _ _
C	_·_·	P	·_ _·
D	_··	Q	_ _·_
E	·	R	·_·
F	··_·	S	···
G	_ _·	T	_
H	····	U	··_
I	··	V	···_
J	·_ _ _	W	·_ _
K	_·_	X	_··_
L	·_··	Y	_·_ _
M	_ _	Z	_ _··

Although it looks like a substitution cipher, you'll get confused looks if you refer to this system as Morse cipher. Notice that the most common letters have the shortest representations, whereas the rarest letters have the longest. This was done intentionally so that messages could be conveyed more rapidly.

There are also combinations of dots and dashes representing the digits 0 through 9, but as these can be spelled out, they are not strictly necessary. Notice that the most common letters, E and T, have single character representations, while V is represented by four characters. V is easy to remember as the beginning of Beethoven's Fifth Symphony. The allies used V for victory during World War II and made use of Beethoven's Fifth Symphony for propaganda purposes.

Look closely at the back of the coin shown in Figure 1.23. Notice the series of dots and dashes around the perimeter that spell out a message in Morse code. Start reading clockwise from the bottom of the coin just to the left of the "N" in "CENTS" and the message WE WIN WHEN WE WORK WILLINGLY is revealed.[44] There have been much more disturbing uses of Morse code; for example,[45]

> [Jeremiah] Denton is best known for the 1966 North Vietnamese television interview he was forced to give as a prisoner, in which he ingeniously used the opportunity to communicate to American Intelligence. During the interview Denton blinked his eyes in Morse code to spell out the word "*T-O-R-T-U-R-E*" to communicate that his captors were torturing him and his fellow POWs.

---

[44] Anonymous, "DPEPE DPJO," *Cryptologia*, Vol. 1, No. 3, July 1977, pp. 275–277, picture on p. 275.
[45] http://en.wikipedia.org/wiki/Jeremiah_Denton.

Figure 1.23   A Canadian coin with a hidden message (Thanks to Lance Snyder for helping with this image.).

As a historical note, in 1909, the international distress signal (SOS) which in Morse code is … - - - …, was first radioed by Jack Binns, when his ship, the *S.S. Republic* collided with the *S.S. Florida*.

Morse code offers no secrecy, because there isn't a secret key. The substitutions made are available to everyone. In fact, broadcast messages are so easy to intercept, they may as well be sent directly to the enemy. By comparison, it's much more difficult to get messages that are conveyed by a courier. Thus, the telegraph (and radio) made cryptology much more important. If the enemy is going to get copies of your messages, they better be well protected. The convenience of telegraph and radio communication, combined with the usual overconfidence in whatever means of encipherment is used, makes this a very appealing method. The telegraph was first used for military purposes in the Crimean War (1853–1856) and then to a much greater extent in the U.S. Civil War.

It may appear that Morse code only requires two symbols, dots and dashes, but there is a third—the space. If all of the dots and dashes are run together, we cannot tell where one letter ends and the next begins; for example,[46]

```
...  ---  ..-.  ..  .-         decodes to "Sofia"
.  ..-  --.  .  -.  ..  .-      decodes to "Eugenia"
```

There is another, more modern, system for coding that prevents this problem and truly requires just two characters. It's due to David. A. Huffman (Figure 1.24), who came up with the idea while working on his doctorate in computer science at MIT.[47]

---

[46] http://rubyquiz.com/quiz121.html.
[47] Huffman, David A., "A Method for the Construction of Minimum-Redundancy Codes," *Proceedings of the Institute of Radio Engineers*, Vol. 40, No. 9, September 1952, pp. 1098–1101.

Figure 1.24   David A. Huffman (1925–1999). (Courtesy of Don Harris, University of California, Santa Cruz.)

Huffman codes make use of the same idea as Morse code. Instead of representing each character by eight bits, as is standard for computers, common characters are assigned shorter representations while rarer characters receive longer representations (see Table 1.7). The compressed data is then stored along with a key giving the substitutions that were used. This is a simple example of an important area known as *data compression*. High compression rates allow information to be sent more rapidly, as well as take up less space when stored. Zip files are an example. If not zipped, the download time for the file would be longer. Huffman coding is also used to compress images such as JPEGs.

Using the code shown in Table 1.7, MATH would be expressed as 1101001000011001. You may now try to break the 0s and 1s up any way you like, but the only letters you can get are MATH. To see why this is the case, examine the binary graph in Figure 1.25.

To read this graph, start at the top and follow the paths marked by 0s and 1s until you get to a letter. The path you followed is the string of bits that represents that letter in our Huffman code. In Morse code, the letter N (– .) could be split apart to get – and . making TE, but this cannot happen with the letters represented in the graph shown in Figure 1.25, because a particular string of bits that leads to a letter doesn't pass through any other letters on the way. We only labeled letters at the ends of paths. The tree could be extended out more to the right to include the rest of the alphabet, but enough is shown to make the basic idea clear.

The next level of Huffman coding is to replace strings of characters with bit strings. A common word may be reduced to less space than a single character normally requires, while a rarer word

Table 1.7 Huffman Code

Letter	Huffman Code	Letter	Huffman Code
E	000	M	11010
T	001	W	11011
A	0100	F	11100
O	0101	G	111010
I	0110	Y	111011
N	0111	P	111100
S	1000	B	111101
H	1001	V	111110
R	1010	K	1111110
D	10110	J	11111110
L	10111	X	111111110
C	11000	Q	1111111110
U	11001	Z	1111111111

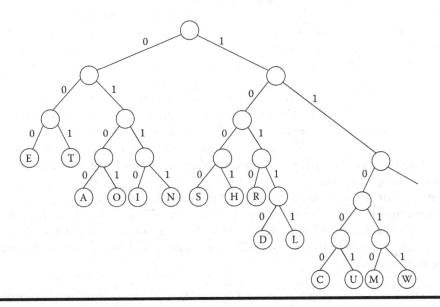

Figure 1.25 Huffman coding graph.

becomes longer after encoding. This method should only be applied to large files. It would not be efficient to encode a short letter in this manner.

Although the terms *encode* and *decode* are used in this context, data compression is not the same as coding theory. Coding theory lengthens messages in an effort to make garbled bits recoverable. It adds redundancy, while data compression seeks to remove it, as does cryptography.

The coding technique described above earned Huffman a place in this book, but he may have done other work that is relevant. A 1955 letter written by John Nash[48] to Major Grosjean of the National Security Agency contains the following passage.[49]

> Recently a conversation with Prof. Huffman here indicated that he has recently been working on a machine with similar objectives. Since he will be consulting for NSA I shall discuss my results with him.

Many of the great mathematicians and computer scientists of the second half of the twentieth century have connections to NSA. I expect that declassification efforts will reveal fascinating stories in the decades to come.

1.18 MASC Miscellanea

In a moment we'll leave the simple substitution cipher behind. It's not that there's nothing else of interest left to be said on the topic. In fact, a great deal has been left out! The simple ciphers of this chapter have been the subject of entire books.[50] However, a great deal must be skipped or the study of classical cryptology will take up thousands of pages.

It should be noted that the Polybius cipher (see Section 1.2) had not really been improved upon until the time of the Renaissance. Other monoalphabetic substitution schemes have the advantage of not doubling the length of the message, but they aren't any harder to crack (assuming mixed alphabets are used in both). The lack of advancement in cryptology during these years has caused many historians to refer to that time period as the "Dark Ages."

One exception is the work of Roger Bacon (c. 1214–1294), author of the first known European book to describe ciphers. During this time period, and indeed through the Renaissance, the art/science that is cryptology was considered magic and its practitioners magicians or, worse yet, in league with devils.[51] There are still a few practitioners today who have done little to dispel this myth (see the Ozzy Osbourne album cover referenced earlier in this chapter). "The Alphabet of the Magi" is reproduced in Figure 1.26.

Cryptology has strong literary roots. Even Chaucer dabbled in the field. We close this section with a sample from his works. Figure 1.27 is one of six ciphers from Chaucer's *The Equatorie of the Planetis*, which is apparently a companion piece to his *Treatise on the Astrolabe* (Chaucer was also an astronomer). It is simply a monoalphabetic substitution cipher, giving simplified instructions for using the equatorie.

---

[48] Yes, this is the John Nash of the book and movie *A Beautiful Mind*. The book is excellent, but the movie made major changes and doesn't, in my opinion, provide an accurate depiction of Nash's life.

[49] Scans of this, and other Nash letters, are available online at http://www.nsa.gov/publicinfo/files/nashletters/nash_letters1.pdf.

[50] See the References and Further Reading at the end of this chapter.

[51] Said of the French cryptanalyst Viète by King Philip II of Spain when writing the Pope in an attempt to get Viète tried in a Cardinal's Court. See Singh, Simon, *The Code Book*, Doubleday, New York, 1999, p. 28

Figure 1.26 Alphabet of the Magi. (From Christian, Paul (pseudonym for Jean Baptiste Pitois). (From *Histoire de la Magie, du Monde Surnaturel et da la Fatalité à travers les Temps et les peuples*, 1870, p. 177.)

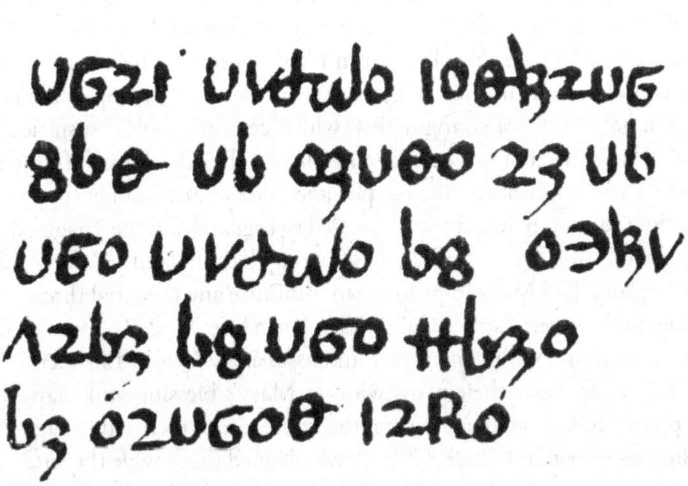

Figure 1.27 Chaucer's cipher. (Courtesy of the David Kahn Collection, National Cryptologic Museum, Fort Meade, Maryland.)

1.19 Nomenclators

From 1400 to 1850, nomenclators were the kings of the European nations' cipher systems. They are basically a combination of a code and a MASC, which may be used to spell out words not provided for in the code portion. The code portions initially just consisted of names, hence, *nomenclator*. Figure 1.28 shows one used by Mary, Queen of Scots.

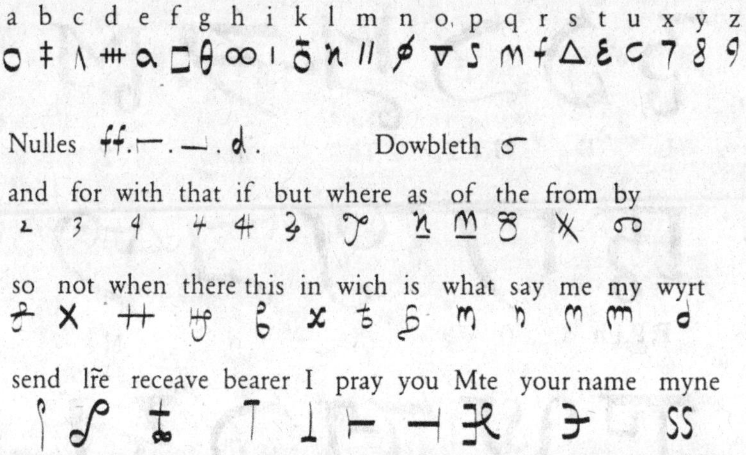

Figure 1.28 Nomenclator used by Mary, Queen of Scots. (From Singh, Simon, *The Code Book*, Doubleday, New York, 1999, p. 38. With permission.)

You can probably guess what happened when Mary's life hung in the balance, dependent on whether or not messages in this cipher could or could not be read by a cryptanalyst without access to the key. Mary was wise to include nulls, but there were no homophones (different symbols representing the same letter, as in the Zodiac ciphers) and far too few words in the code portion.

Mary was crowned queen of Scotland in 1543 at only nine months of age. In 1559, she married Francis, the dauphin of France. It was hoped that this would serve to strengthen the ties between the two Roman Catholic nations, but he died in 1560 and in the meanwhile Scotland was becoming more and more Protestant. Mary then married her cousin Henry Stewart, who caused so much trouble for Scotland that it was planned for him to die by having his house blown up while he was inside. He escaped the explosion only to die of strangulation, which certainly looked suspicious. Mary found a third husband, James Hepburn, but he was exiled in 1567 by the now powerful Protestant population of Scotland, and Mary was imprisoned. She escaped and with an army of her supporters attempted to reclaim her crown, but the attempt failed and so she fled to England and the imagined protection of her cousin Queen Elizabeth I. But, Queen Elizabeth, knowing that England's Catholics considered Mary the true Queen of England, had Mary imprisoned to minimize any potential threat.

Angered by England's persecution of Catholics, Mary's former page, Anthony Babbington, and others put together a plan in 1586 to free Mary and assassinate Queen Elizabeth. The conspirators decided they could not carry out their plans without Mary's blessing and managed to smuggle a message to her in prison. It was enciphered using the nomenclator pictured in Figure 1.29. However, the conspirators didn't realize that Gilbert Gifford, who helped to smuggle the letter (and earlier messages from other supporters of Mary) was a double agent. He turned the messages over to Sir Francis Walsingham, Principal Secretary to Queen Elizabeth. Thus, they were copied before being delivered to Mary, and the cryptanalyst Thomas Phelippes succeeded in breaking them.

Mary responded, supporting the conspiracy, as long as her liberation was before or simultaneous with the assassination, as she feared for her life if the assassination came first. Like the previous message, this one was read by Phelippes and relayed to Walsingham. Both Babbington and Queen Mary were now marked for death, but the other conspirators remained unnamed. To snare the rest, Walsingham had Phelippes add a bit more enciphered text to Mary's response in the style of Mary's own hand (Figure 1.29).

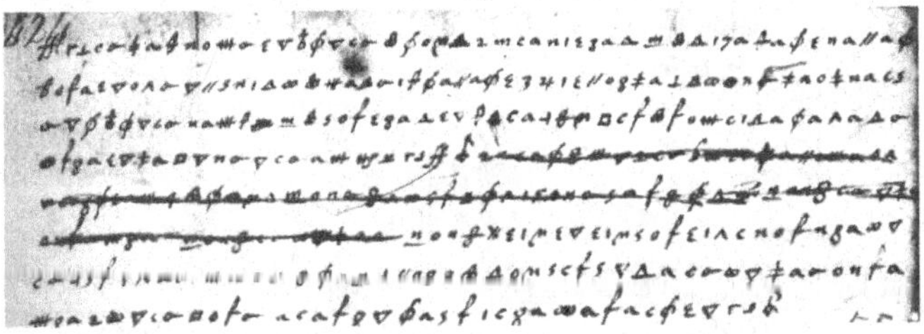

Figure 1.29   A forged ciphertext. (Courtesy of the David Kahn Collection, National Cryptologic Museum, Fort Meade, Maryland.)

The deciphered message reads:

> I would be glad to know the names and qualities of the six gentlemen which are to accomplish the designment; for it may be that I shall be able, upon knowledge of the parties, to give you some further advice necessary to be followed therein, as also from time to time particularly how you proceed: and as soon as you may, for the same purpose, who be already, and how far everyone is privy hereunto.

However, events were such that Babbington was about to leave the country to prepare the overthrow of Queen Elizabeth (he was supported by Philip II of Spain), and it became necessary to arrest him. Following the trial, Mary was beheaded in 1587.[52] But would the plot have succeeded, if Mary had used a stronger cipher? Many more tales can be told concerning nomenclators used in Europe, but we now jump ahead to the American Revolution.

General George Washington, having learned some hard lessons in spycraft (the death of Nathan Hale, for example), went on to protect the identity of spies codenamed Culper Senior and Culper Junior so well that it remained a mystery to historians until 1930! Morton Pennypacker finally uncovered the Culper's true names, when he came upon letters from Robert Townsend in a hand that perfectly matched that of the spy Culper Junior.[53] Townsend was a Quaker who did not want any attention drawn to the work he did on behalf of his country. It was not until the twentieth century that spying lost its stigma and became glamorized. In the days of the American Revolution it was considered dishonorable.

A page from the nomenclator used by the Culper spy ring, operating in occupied New York, is reproduced in Figure 1.30.

The Culper's nomenclator wasn't always used in the best manner. One letter bore the partially enciphered phrase "David Mathews, Mayor of 10."[54] If this message had been intercepted, how long do you think it would have taken a British cryptanalyst to determine that 10 = New York?

[57] Mary's story is entertainingly related in Simon Singh's *The Code Book*, Doubleday, New York, 1999, pp. 1–3, 32–44, from which the above was adapted.

[53] Pennypacker, Morton, *The Two Spies, Nathan Hale and Robert Townsend*, Houghton Mifflin Company, Boston, Massachusetts, 1930.

[54] Pennypacker, Morton, *General Washington's Spies*, Long Island Historical Society, Brooklyn, New York, 1939, facing p. 50.

Figure 1.30 A nomenclator used in the American Revolution. (From *George Washington Papers at the Library of Congress, 1741-1799*: Series 4. General Correspondence. 1697-1799, Talmadge, 1783, Codes.)

Although Culper Junior's identity was eventually revealed, we still don't have a very good idea of his appearance. Pennypacker identified a silhouette (Figure 1.31, left) as Robert Townsend, but it is actually Townsend's brother. The only depiction currently accepted as Robert (Figure 1.31, right) was drawn by his nephew Peter Townsend.

Figure 1.31 **Silhouette incorrectly believed by Pennypacker to be Robert Townsend (left). (Courtesy of the Friends of Raynham Hall Museum.) Drawing of Robert Townsend by his nephew Peter Townsend (right). (Courtesy of the Friends of Raynham Hall Museum.)**

During the Revolutionary War, General Washington only spent $17,617 on espionage activities, and he paid for these activities out of his own pocket! He did later bill Congress, but this sort of budget contrasts very strongly with the situation today. See Chapter 12.

1.20 Cryptanalysis of Nomenclators

Now that we've seen a few examples, let's consider how we might break a nomenclature—one larger and more challenging than those above! Good nomenclatures have tens of thousands of code groups; however, the code portion may have the distinct disadvantage of having the codewords arranged in alphabetical order with corresponding code numbers in numerical order.[55] If this is the case, a nice formula, that is present in many probability texts, allows us to estimate the size of the code.[56]

$$Size \approx \frac{n+1}{n} Max - 1$$

where n is the number of codegroups observed and *Max* is the largest value among the observed codegroups.

---

[55] Two part codes, which avoid this pitfall, are discussed in Section 4.2.

[56] For a derivation see Ghahramani, Saeed, *Fundamentals of Probability*, Prentice-Hall, Upper Saddle River, New Jersey, 1996, pp.146–148. This is not, however, the first place the result appeared or was proven.

For example, if an intercepted message contains 22 codegroups and the largest is 31,672, then the number of entries in the code book is roughly

$$\frac{n+1}{n}Max - 1 = \frac{22+1}{22}(31,672) - 1 \approx 33,111$$

Subtracting the 1 makes no practical difference here, but it was included to be mathematically correct. The formula above has other uses. In the probability text I referenced for the derivation, it is used to estimate the number of enemy tanks based on how many were observed and the highest number seen on them.

Once we know the size of the code, a dictionary of similar size may be consulted. Because it is not likely to be the key, we cannot expect to simply plug in the word in the position of the code number; however, there's a good chance that the first letter, and possibly the second, of that word may be correct. Thus, we can jot down assumptions for the first two letters of each word. This, when combined with context, may allow us to guess some phrases of plaintext.

Table 1.8 shows where words beginning with the given letters stop in a dictionary of 60,000–65,000 words.

1.21 Book Codes

A nomenclator with a large code portion may well employ a code book to list them all, but this is not to be confused with a book code. I'll illustrate the idea of a book code with an example sent by Porlock to Sherlock Holmes in *The Valley of Fear* with confidence that Holmes would be able to read it without the key.[57]

```
534 C2 13 127 36 31 4 17 21 41
DOUGLAS 109 293 5 37 BIRLSTONE
     26 BIRLSTONE 9 47 171
```

Holmes decides that 534 indicates page 534 of some book and C2 indicates column two. The numbers then represent words and a few names are spelled out, because they do not occur on the given page. William Friedman once broke such a cipher without finding the book that was used, but Holmes takes a more direct route. The minimum size of the book being 534 pages and its practicality for use requiring that a copy be available to Holmes, he considers volumes that fit the constraints. The Bible is a possibility broached by Watson, but Holmes comments that he "could hardly name any volume which would be less likely to lie at the elbow of one of Moriarty's associates." Holmes finally turns to the current almanac. It yields nonsense, but the previous year's almanac works and the story is off and running.

Other villains who made use of book codes include Benedict Arnold, in his attempt to betray West Point to the British, and Hannibal Lecter in the novel *Silence of the Lambs*. Arnold's book was a dictionary and Lecter's was *The Joy of Cooking*.

Figure 1.32 shows the book code that William Friedman solved without benefit of the accompanying book, which was only found later. Notice that this particular code doesn't select words, but rather individual letters, to form the ciphertext.

---

[57] Doyle, Arthur Conan, *The Valley of Fear*, Doran, New York, 1915. This story was first serialized in *Strand Magazine* from September 1914 to May 1915. Three months after the New York edition, a British edition appeared.

Table 1.8 Code Dictionary

AA	5	DA	11,646	GL	21,300	LO	30,690	PE	38,121	TE 55,336
AB	207	DE	12,850	GN	21,344	LU	30,850	PH	38,385	TH 55,778
AC	467	DI	13,935	GO	21,592	LY	30,890	PI	38,828	TI 56,036
AD	695	DO	14,210	GR	22,267	MA	31,730	PL	39,245	TO 56,466
AE	741	DR	14,620	GU	22,530	ME	32,362	PO	39,970	TR 57,232
AF	845	DU	14,840	GY	22,588	MI	32,903	PR	41,260	TU 57,432
AG	942	DW	14,855	HA	23,320	MO	33,525	PS	41,320	TW 57,498
AI	1,018	DY	14,900	HE	23,942	MU	33,826	PU	41,740	TY 57,556
AL	1,325	EA	15,000	HI	24,180	MY	33,885	PY	41,815	UB 57,571
AM	1,415	EC	15,075	HO	24,764	NA	34,075	QUA	41,984	UG 57,589
AN	1,957	ED	15,126	HU	24,989	NE	34,387	QUE	42,036	UL 57,639
AP	2,081	EF	15,187	HY	25,190	NI	34,529	QUI	42,159	UM 57,685
AR	2,514	EG	15,225	IC	25,270	NO	34,815	QUO	42,181	UN 59,885
AS	2,737	EI	15,235	ID	25,347	NU	34,928	RA	42,573	UP 59,957
AT	2,860	EL	15,436	IG	25,370	NY	34,946	RE	44,346	UR 60,014
AU	3,014	EM	15,630	IL	25,469	OA	34,970	RH	44,422	US 60,050
AV	3,073	EN	16,030	IM	25,892	OB	35,140	RI	44,712	UT 60,080
AW	3,100	EP	16,145	IN	27,635	OC	35,230	RO	45,207	VA 60,363
AZ	3,135	EQ	16,210	IR	27,822	OD	35,270	RU	45,441	VE 60,692
BA	3,802	ER	16,290	IS	27,868	OF	35,343	SA	46,192	VI 61,113
BE	4,250	ES	16,387	IT	27,910	OG	35,356	SC	46,879	VO 61,277
BI	4,470	ET	16,460	JA	28,046	OI	35,390	SE	47,945	VU 61,307
BL	4,760	EU	16,505	JE	28,135	OL	35,450	SH	48,580	WA 61,830
BO	5,180	EV	16,610	JI	28,168	OM	35,496	SI	49,024	WE 62,133
BR	5,590	EX	17,165	JO	28,290	ON	35,555	SK	49,152	WH 62,472
BU	5,930	EY	17,190	JU	28,434	OO	35,575	SL	49,453	WI 62,800
BY	5,954	FA	17,625	KA	28,500	OP	35,727	SM	49,600	WO 63,079
CA	6,920	FE	17,930	KE	28,583	OR	35,926	SN	49,788	WR 63,175
CE	7,110	FI	18,390	KI	28,752	OS	35,993	SO	50,266	X 63,225
CH	7,788	FL	18,964	KN	28,857	OT	36,018	SP	51,132	YA 63,282
CI	7,970	FO	19,610	KO	28,878	OU	36,159	SQ	51,259	YE 63,345
CL	8,220	FR	20,030	KR	28,893	OV	36,348	ST	52,678	YO 63,397
CO	10,550	FU	20,265	KU	28,910	OW	36,361	SU	53,701	YU 63,409
CR	11,030	GA	20,700	LA	29,457	OX	36,395	SW	53,977	ZE 63,452
CU	11,300	GE	20,950	LE	29,787	OY	36,410	SY	54,206	ZI 63,485
CY	11,380	GI	21,088	LI	30,283	PA	37,226	TA	54,783	ZO 63,542
										ZY 63,561

Source: Mansfield, L.C.S., *The Solution of Codes and Ciphers,* Alexander Maclehose, London, UK, 1936, pp. 154–157.

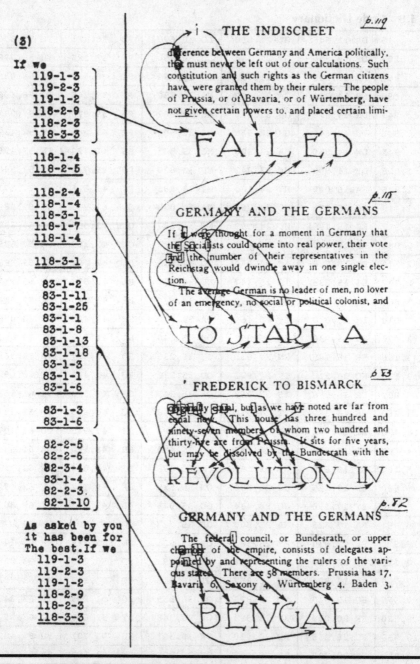

Figure 1.32 A solution to a book code. (Courtesy of the David Kahn Collection, National Cryptologic Museum, Fort Meade, Maryland.)

References and Further Reading

On Ancient Ciphers and Some General References

Balliett, Blue, *Chasing Vermeer*, Scholastic Press, New York, 2004. This is a young adult/children's novel that incorporates a Polybius cipher that mixes letters and numbers by using the following (non-square) rectangle for enciphering

	1	2	3
F	A	M	Y
I	B	N	Z
L	C	O	
N	D	P	
P	E	Q	
T	F	R	
U	G	S	
V	H	T	
W	I	U	
X	J	V	
Y	K	W	
Z	L	X	

Thus, HELLO would become V1 P1 Z1 Z1 L2. The book promotes free thinking and exploration in the spirit of Charles Fort, who is quoted at various points.

Bauer, Craig P., *Discrete Encounters*, CRC/Chapman & Hall, Boca Raton, Florida, 2020. If you like the style of *Secret History: The Story of Cryptology*, you might also like this book. It merges history with the presentation of discrete mathematics, nearly all of which finds applications in cryptology.

Bellovin, Steven, M., *Compression, Correction, Confidentiality, and Comprehension: A Modern Look at Commercial Telegraph Codes*, Department of Computer Science, Columbia University, New York, 2009, available online at http://www.usenix.org/events/sec09/tech/slides/bellovin.pdf. This PowerPoint® presentation on commercial code books provides many entertaining examples, such as this excerpt from *The Theatrical Cipher Code* (1905):

Filacer	An opera company
Filament	Are they willing to appear in tights
Filander	Are you willing to appear in tights
Filar	Ballet girls
Filaria	Burlesque opera
Filature	Burlesque opera company
File	Burlesque people
Filefish	Chorus girl
Filial	Chorus girls

Filially	Chorus girls who are
Filiation	Chorus girls who are shapely and good looking
Filibuster	Chorus girls who are shapely and good looking and can sing
Filicoid	Chorus girls who can sing
Filiform	Chorus man
Filigree	Chorus men
Filing	Chorus men who can sing
Fillet	Chorus people
Fillip	Chorus people who can sing
Filly	Comic opera
Film	Comic Opera Company
Filter	Comic Opera people
Fitering	Desirable chorus girl

Hunt, Arthur S., "A Greek Cryptogram," *Proceedings of the British Academy*, Vol. 15, 1929, pp. 127–134. This easy-to-read paper presents a Greek ciphertext in which the letters were turned half over or modified in other small ways to disguise the writing. No knowledge of Greek is needed to understand this paper, as an English translation of the ciphertext is provided. Unfortunately, an approximate date for the ciphertext examined (The Michigan Cryptographic Papyrus) is not given.

Kahn, David, *The Codebreakers*, Second Edition, Scribner, New York, 1996. Kahn surveys the cryptography of the entire ancient world in Chapter 2 of his classic history.

The following three references deal with controversial decipherments. I'm simply providing the titles; it's up to you to decide whether or not the claims are correct.

Landsverk, Ole G., *Ancient Norse Messages on American Stones*, Norseman Press, Glendale, California, 1969.

Landsverk, Ole G., "Cryptography in Runic Inscriptions," *Cryptologia*, Vol. 8, No. 4, October 1984, pp. 302–319.

Mongé, Alf and Ole G. Landsverk, *Norse Medieval Cryptography in Runic Carvings*, Norseman Press, Glendale, California, 1967.

Reeds, Jim, *Commercial Code Book Database*, Mar 23, 2001, archived at https://web.archive.org/web/20140130084013/http://www.dtc.umn.edu:80/~reedsj/codebooks.txt. This online source provides bibliographic details of 1,745 commercial code books. Here's your checklist collectors!

On Poe

Brigham, Clarence S., "Edgar Allen Poe's Contributions to *Alexander's Weekly Messenger*," *Proceedings of the American Antiquarian Society*, Vol. 52, No. 1, April 1942, pp. 45–125. This paper marks the rediscovery of Poe's columns on cryptography.

Friedman, William F., "Edgar Allan Poe, Cryptographer," *American Literature*, Vol. 8, No. 3, November 1936, pp. 266–280.

Friedman, William F., "Edgar Allan Poe, Cryptographer," *Signal Corps Bulletin*, No. 97, July–September 1937, pp. 41–53; Friedman, William F., "Edgar Allan Poe, Cryptographer (Addendum)," *Signal Corps Bulletin*, No. 98, October–December 1937, pp. 54–75. These items were reprinted in Friedman, William F., editor, *Cryptography and Cryptanalysis Articles*, Vol. 2, Aegean Park Press, Laguna Hills, California, 1976.

Pirie, David, *The Patient's Eyes: The Dark Beginnings of Sherlock Holmes*, St. Martin's Minotaur, New York, 2002. This is a novel that involves some cryptanalysis. Interestingly, the author reproduced Poe's frequency ordering and introduced another error by omitting the letter x.

Silverman, Kenneth, *Edgar A. Poe: Mournful and Never-ending Remembrance*, HarperCollins, New York, 1991.

Wimsatt, Jr., William K., "What Poe Knew about Cryptography," *Publications of the Modern Language Association*, Vol. 58, No. 3, September 1943, pp. 754–779.

Not of Interest

Rosenheim, Shawn James, *The Cryptographic Imagination: Secret Writing from Edgar Poe to the Internet*, Parallax: Re-Visions of Culture and Society, The Johns Hopkins University Press, Baltimore, Maryland, 1997.

For details, see the following review by a professor of computer science at the University of Waterloo:

Shallit, Jeffrey, "Book review of Menezes, van Oorschot, and Vanstone, *Handbook of Applied Cryptography*, and Rosenheim, *The Cryptographic Imagination: Secret Writings from Edgar Poe to the Internet*," *American Mathematical Monthly*, Vol. 106, No. 1, January 1999, pp. 85–88.

On Sherlock Holmes and Cryptology

Many of the references given below propose sources for the dancing men as a cipher. Doyle claimed they were an independent creation and any similarity to previous ciphers was a coincidence. It's strange that some of his fans, who presumably find his stories to be original and imaginative (or they wouldn't *be* fans), cannot credit him with hitting on this idea himself!

Bond, Raymond T., editor, *Famous Stories of Code and Cipher*, Rinehart and Company, New York, 1947. The contents of the paperback edition, Collier Books, New York, 1965, differ from the hardcover by one story, but "The Adventure of the Dancing Men" is present in both, as are the author's interesting introductory comments to the story. See the hardcover pp.136–137 or the paperback, pp. 171–172, for these comments.

Donegall, Lord, *Baker Street and Beyond: Essays on Sherlock Holmes*, Sherlock Holmes Society of London, London, UK, 1993. This book collects essays Donegall wrote for *The New Strand* magazine. If you'd rather see the originals, the relevant pieces are "Baker Street and Beyond (9): A Treaty for Breakfast," Vol. 1, No. 9, August 1962, pp. 1048–1050 and "Baker Street and Beyond (15): Too Hot to Mollies?—Those Phoney Ciphers—Blank for Blank," Vol. 2, No. 2, February 1963, pp. 1717–1720.

Hearn, Otis, "Some Further Speculations upon the Dancing Men," *The Baker Street Journal*, (New Series), Vol. 19, December 1969, pp. 196–202. Hearn is a pseudonym of Walter N. Trenerry.

Kahn, David, *The Codebreakers*, The Macmillan Company, New York, 1967, p. 794–798. I'm citing the first edition here, so that you may see when Kahn weighed in on the issue.

McCormick, Donald, *Love in Code*, Eyre Methuen, London, UK, 1980, p. 5.

Orr, Lyndon, "A Case of Coincidence Relating to Sir A. Conan Doyle," *The Bookman*, Vol. 31, No. 2, April 1910, pp. 178–180, available online at https://archive.org/details/bookman20unkngoog/page/n190/mode/2up. This article was reprinted in *The Baker Street Journal*, (New Series), Vol. 19, No. 4, December 1969, pp. 203–205.

Pattrick, Robert H., "A Study in Crypto-Choreography," *The Baker Street Journal*, (New Series), Vol. 5, No. 4, October 1955, pp. 205–209.

Pratt, Fletcher, "The Secret Message of the Dancing Men," in Smith, Edgar W., editor, *Profile by Gaslight: An Irregular Reader about the Private Life of Sherlock Holmes*, Simon and Schuster, New York, 1944, pp. 274–282.

Schenk, Remsen Ten Eyck, "Holmes, Cryptanalysis and the Dancing Men," *The Baker Street Journal*, (New Series), Vol. 5, No. 2, April 1955, pp. 80–91.

Schorin, Howard R., "Cryptography in the Canon," *The Baker Street Journal*, (New Series), Vol. 13, December 1963, pp. 214–216.

Shulman, David, "Sherlock Holmes: Cryptanalyst," *The Baker Street Journal*, (Old Series), Vol. 3, 1948, pp. 233–237.

Shulman, David, "The Origin of the Dancing Men," *The Baker Street Journal*, (New Series), Vol. 23, No. 1, March 1973, pp. 19–21.

Trappe, Wade and Lawrence C. Washington, *Introduction to Cryptography with Coding Theory*, Prentice Hall, Upper Saddle River, New Jersey, 2002, pp. 26–29. These authors summarize the story (a sort of Cliff's Notes edition) and discuss the typos in various editions.

For more examples of codes and ciphers used in fiction, John Dooley is the person to turn to:

Dooley, John F., "Codes and Ciphers in Fiction: An Overview," *Cryptologia*, Vol. 29, No. 4, October 2005, pp. 290–328. For an updated list, go to https://www.johnfdooley.com/ and follow the link "Crypto Fiction." As of October 7, 2020, this list contains 420 examples.

On RongoRongo script

If you'd like to learn more, these Rongorongo references, in turn, reference many more books and papers.

Fischer, Steven Roger, *Glyphbreaker*, Copernicus, New York, 1997.

Melka, Tomi S., "Structural Observations Regarding the *RongoRongo* Tablet 'Keiti'," *Cryptologia*, Vol. 32, No. 2, January 2008. pp. 155–179.

Melka, Tomi S., "Some Considerations about the *Kohau Rongorongo* Script in the Light of Statistical Analysis of the 'Santiago Staff'," *Cryptologia*, Vol. 33, No. 1, January 2009, pp. 24–73.

Melka, Tomi S., and Robert M. Schoch, "Exploring a Mysterious Tablet from Easter Island: The Issues of Authenticity and Falsifiability in *Rongorongo* Studies," *Cryptologia*, Vol. 44, No. 6, November 2020, pp. 482–544.

Wieczorek, Rafal, "Putative Duplication Glyph in the Rongorongo Script," *Cryptologia*, Vol. 41, No. 1, January 2017, pp. 55–72.

On Arabic Cryptology

Al-Kadi, Ibraham A., "Origins of Cryptology: The Arab Contributions," *Cryptologia*, Vol. 16, No. 2, April 1992, pp. 97–126.

Mrayati, M., Y. Meer Alam and M.H. at-Tayyan, series editors, *Series on Arabic Origins of Cryptology, Volume 1: al-Kindi's Treatise on Cryptanalysis*, KFCRIS (King Faisal Center for Research and Islamic Studies) & KACST (King Abdulaziz City for Science and Technology), Riyadh, 2003.

Mrayati, M., Y. Meer Alam and M.H. at-Tayyan, series editors, *Series on Arabic Origins of Cryptology, Volume 2: Ibn 'Adlān's Treatise al-mu'allaf lil-malik al-Ašraf*, KFCRIS (King Faisal Center for Research and Islamic Studies) & KACST (King Abdulaziz City for Science and Technology), Riyadh, 2003.

Mrayati, M., Y. Meer Alam and M.H. at-Tayyan, series editors, *Series on Arabic Origins of Cryptology, Volume 3: Ibn ad-Duryahim's Treatise on Cryptanalysis*, KFCRIS (King Faisal Center for Research and Islamic Studies) & KACST (King Abdulaziz City for Science and Technology), Riyadh, 2004.

Mrayati, M., Y. Meer Alam and M.H. at-Tayyan, series editors, *Series on Arabic Origins of Cryptology, Volume 4: Ibn Dunaynir's Book: Expositive Chapters on Cryptanalysis*, KFCRIS (King Faisal Center for Research and Islamic Studies) & KACST (King Abdulaziz City for Science and Technology), Riyadh, 2005.

Mrayati, M., Y. Meer Alam and M.H. at-Tayyan, series editors, *Series on Arabic Origins of Cryptology, Volume 5: Three Treatises on Cryptanalysis of Poetry*, KFCRIS (King Faisal Center for Research and Islamic Studies) & KACST (King Abdulaziz City for Science and Technology), Riyadh, 2006.

Schwartz, Kathryn A., "Charting Arabic Cryptology's Evolution," *Cryptologia*, Vol. 33, No. 4, October 2009, pp. 297–305.

On Cryptanalysis

Barker, Wayne G., *Cryptanalysis of the Simple Substitution Cipher with Word Divisions Using Non-Pattern Word Lists*, Aegean Park Press, Laguna Hills, California, 1975. This work also discusses techniques for distinguishing vowels from consonants.

Edwards, D. J. *OCAS – On-line Cryptanalysis Aid system*, MIT Project MAC, TR-27, May 1966. Bruce Schatz wrote, "reported on a SNOBOL-like programming language specially designed for cryptanalysis."

Gaines, Helen F., *Cryptanalysis: A Study of Ciphers and Their Solutions*, corrected from prior printings and augmented with solutions, Dover, New York, 1956, pp. 74ff, 88–92; the first printing appeared in 1939 and was titled *Elementary Cryptanalysis*.

Giredansky, M. B., "Cryptology, the Computer, and Data Privacy," *Computers and Automation*, Vol. 21, No. 4, April 1972, pp. 12–19. This is a survey of automated cryptanalysis, so you can see that this has been investigated publicly for quite some time.

Guy, Jaques B. M., "Vowel Identification: An Old (But Good) Algorithm," *Cryptologia*, Vol. 15, No. 3, July 1991, pp. 258–262.

Mellen. Greg E., "Cryptology, Computers, and Common Sense," in *AFIPS '73: Proceedings of National Computer Conference and Exposition*, Vol. 42, June 4–8, 1973, AFIPS Press, Montvale, New Jersey, pp. 569–579. This is a survey of automated cryptanalysis. AFIPS stands for American Federation of Information Processing Societies.

Moler, Cleve and Donald Morrison, "Singular Value Analysis of Cryptograms," *American Mathematical Monthly*, Vol. 90, No. 2, February 1983, pp. 78–87. This paper uses the singular value decomposition for vowel recognition.

Olson, Edwin, "Robust Dictionary Attack of Short Simple Substitution Ciphers," *Cryptologia*, Vol. 31, No. 4, October 2007, pp. 332–342.

Schatz, Bruce R., "Automated Analysis of Cryptograms," *Cryptologia*, Vol. 1, No. 2, April 1977, pp. 116–142.

Silver, R., "Decryptor," in *MIT Lincoln Laboratory Quarterly Progress Report, Division 5* (Information Processing), December 1959, pp. 57–60.

Sutton, William G., "Modified Sukhotin A Manual Method," from Computer Column in *The Cryptogram*, Vol. 58, No. 5, September–October 1992, pp. 12–14.

Vobbilisetty, Rohit, Fabio Di Troia, Richard M. Low, Corrado Aaron Visaggio, and Mark Stamp, "Classic Cryptanalysis Using Hidden Markov Models," *Cryptologia*, Vol. 41, No. 1, January 2017, pp. 1–28.

Pattern Word Books

Today, you're better off using a website that allows you to search pattern word files. Two of these were referenced in Section 1.11. I'm listing a few of the books to show how much time people devoted to making this powerful cryptanalytic tool available before computers became ubiquitous.

Andree, Richard V., *Pattern & Nonpattern Words of 2 to 6 Letters*, Raja Press,[58] Norman, Oklahoma, 1977. This is described as a byproduct of research carried out by the author and his students at the University of Oklahoma. It used *Webster's Seventh New Collegiate Dictionary* expanded to 152,296 words by adding endings. The words go up to length 35, although there are no entries of length 32 or 34. These numbers are for the total of all volumes. It was compiled using a computer.

Andree, Richard V., *Pattern & Nonpattern Words of 7 & 8 Letters*, Raja Press, Norman, Oklahoma, 1980.

Andree, Richard V., *Pattern & Nonpattern Words of 9 & 10 Letters*, Raja Press, Norman, Oklahoma, 1981.

Carlisle, Sheila, *Pattern Words Three-Letters to Eight-Letters in Length*, Aegean Park Press, Laguna Hills, California, 1986. This work includes about 60,000 words. The author writes, "Without the use of a computer this compilation would have been impossible." Well Sheila, you don't know Jack (Levine)!

---

[58] Raja is the American Cryptogram Association (ACA) nom de plume for the author and his wife Josephine.

Goddard, Eldridge and Thelma Eldridge, *Cryptodyct*, Wagners, Davenport, Iowa, 1976. Upon seeing the title, I thought it must have been written in some foreign language I wasn't familiar with; however, it is in English and is simply a pattern word dictionary of 272 pages for words of length 14 and shorter. This effort took two years and appears to have been privately printed.

Goddard, Eldridge and Thelma Eldridge, *Cryptokyt I, Non-Pattern Nine Letter Word List*, Wagners, Davenport, Iowa, 1977, 28 pages.

Goddard, Eldridge & Thelma, *Cryptokyt II, Non-Pattern Five Letter Word List*, Wagners, Davenport, Iowa, 1977, 64 pages.

Hempfner, Philip and Tania Hempfner, *Pattern Word List For Divided and Undivided Cryptograms*, self-published, 1984. An electronic dictionary was used to generate the 30,000+ entries in this 100-page book, which goes up to the 21-letter word electroencephalograph.

Levine, Jack, *A List of Pattern Words of Lengths Two Through Nine*, self-published, 1971, 384 pages.

Levine, Jack, *A List of Pattern Words of Lengths Ten Through Twelve*, self-published, 1972, 360 pages.

Levine, Jack, *A List of Pattern Words of Lengths Thirteen to Sixteen*, self-published, 1973, 270 pages.

Lynch, Frederick D., Colonel, USAF, Ret., *Pattern-Word List, Volume 1, Containing Words up to 10 Letters in Length*, Aegean Park Press, Laguna Hills, California, 1977, 152 pages. "much of Colonel Lynch's work remains classified," but this work was "compiled manually by the author over a period of many years" using open source material (*Webster's New International Dictionary, third edition*). Words where only a single letter repeats itself once were not included.

On Zodiac

Bauer, Craig P., *Unsolved! The History and Mystery of the World's Greatest Ciphers from Ancient Egypt to Online Secret Societies*, Princeton University Press, Princeton, New Jersey, 2017. See Chapter 4.

Crowley, Kieran, *Sleep My Little Dead: The True Story of the Zodiac Killer*, St. Martin's Paperbacks, New York, 1997. This is about a copycat killer in New York, not the original Zodiac. The author signed my copy "To Craig – What's Your Sign?" It gave me a chill – thanks!

Graysmith, Robert, *ZODIAC*, St. Martin's/Marek, New York, 1986. This is the best book on Zodiac. It's creepy and in 2007 was made into a movie of the same title that is also creepy.[59] It is not to be confused with an extremely low budget film titled *Zodiac Killer*,[60] which appeared in 2005, a year that also saw the release of *The Zodiac*.[61]

Graysmith, Robert, *Zodiac Unmasked: The Identity of American's Most Elusive Serial Killer Revealed*, Berkley Books, New York, 2002.

Hunt for the Zodiac Killer, The. This is a 5-part television series that premiered on History in 2017. I recommend reading chapter 4 of the first reference in this section before viewing it.

Oranchak, David, "Let's Crack Zodiac - Episode 5 - The 340 Is Solved!" December 11, 2020, https://www.youtube.com/watch?v=-1oQLPRE21o&feature=youtu.be. This video was posted when the present book was at the proof stage.

On MASCs

Bamford, James, *Body of Secrets*, Doubleday, New York, 2001. Although this is a book about the National Security Agency, each chapter begins with cryptograms—can you find the correct decipherments?

Fronczak, Maria, "Atbah-Type Ciphers in the Christian Orient and Numerical Rules in the Construction of Christian Substitution Ciphers," *Cryptologia*, Vol. 37, No. 4, October 2013, pp. 338–344.

Huffman, David A., "A Method for the Construction of Minimum-Redundancy Codes," *Proceedings of the Institute of Radio Engineers*, Vol. 40, No. 9, September 1952, pp. 1098–1101.

Kruh, Louis, "The Churchyard Ciphers," *Cryptologia*, Vol. 1, No. 4, October 1977, pp. 372–375.

---

[59] http://www.imdb.com/title/tt0443706/.

[60] http://www.imdb.com/title/tt0469999/.

[61] http://www.imdb.com/title/tt0371739/.

Reeds, Jim, "Solved: The Ciphers in Book III of Trithemius's *Steganographia*," *Cryptologia*, Vol. 22, No. 4, October 1998, pp. 291–317. This paper concerns an old and hidden cipher of Trithemius, only recognized and deciphered hundreds of years after his death. Trithemius's life and his cryptologic work are discussed in Section 2.2, but the paper referenced here can be appreciated now.

On Nomenclators and the Times in Which They Were Used

Budiansky, Stephen, *Her Majesty's Spymaster: Elizabeth I, Sir Francis Walsingham, and the Birth of Modern Espionage*, Viking, New York, 2005.

Dooley, John, "Reviews of Cryptologic Fiction," *Cryptologia*, Vol. 34, No. 1, January 2010, pp. 96–100. One of the books Dooley reviewed in this issue was Barbara Dee's *Solving Zoe* (Margaret K. McElderry Books, 2007). The book includes the letter substitution portion of the Mary Queen of Scotts nomenclator, and Dooley observed that the cipher symbols matched those presented in Fred Wrixon's book, *Codes and Ciphers*, but didn't match those provided by Simon Singh in *The Code Book*. Clearly Singh's version is correct, as it matches a surviving message (reproduced in the present text). Dooley concludes with "Which leads to the question, where did Wrixon's cipher come from?" I love this kind of informed, aggressive reviewing.

Ford, Corey, *A Peculiar Service*, Little, Brown and Company, Boston, Massachusetts, 1965. This is a novelization of the spy activities that took place during the American Revolutionary War. Dialog, of course, had to be invented, but the author carefully points out the few places where conjecture must take the place of established fact in terms of *events*. The result was a very entertaining read that does a great job conveying a feel for the time, the locale, and the personalities.

Groh, Lynn, *The Culper Spy Ring*, The Westminster Press, Philadelphia, Pennsylvania, 1969. This is promoted as a children's book, but outside of its short length, I don't understand why.

Nicholl, Charles, *The Reckoning: The Murder of Christopher Marlowe*, Harcourt Brace & Company, New York, 1992.

Pennypacker, Morton, *The Two Spies, Nathan Hale and Robert Townsend*, Houghton Mifflin Company, Boston, Massachusetts, 1930. This work revealed the identities of the Culper spies for the first time.

Pennypacker, Morton, *General Washington's Spies*, Long Island Historical Society, Brooklyn, New York, 1939. This could be considered a second edition of Pennypacker's 1930 book, but there was so much new content he decided to give it a new title. The book consists mostly of letters, which have great historical value but do not make for an exciting read.

Silbert, Leslie, *The Intelligencer*, Atria Books, New York, 2004. This novel refers to many ciphertexts, but they aren't provided. It is referenced here because the story takes place during the reign of Elizabeth the First and Thomas Phelippes is included as a character.

Singh, Simon, *The Code Book*, Doubleday, New York, 1999. This work is aimed at a general audience and relates the history of codemakers vs. codebreakers in historical context in a lively manner. Singh uses the term "code" in the title in a broader than usual sense and intends it to include ciphers as well. The first chapter covers Mary, Queen of Scots, in much greater detail than I can here.

TURN: Washington's Spies. This television series ran for four seasons from 2014 to 2017. The historical accuracy declined over the years. See https://www.imdb.com/title/tt2543328/ for more.

Solution to Poe's sonnet: If you read along the diagonal (first line/first character, second line/second character, etc.), the name of the woman for whom the sonnet was written is spelled out.

Hint for the pattern in the cipher alphabet from the Kaczynski example: Look at a keyboard.

Chapter 2

Simple Progression to an Unbreakable Cipher

This chapter describes a simple cipher system and proceeds to patch it against attacks until the final result of a theoretically unbreakable cipher is achieved.

2.1 The Vigenère Cipher

In the first chapter we saw how Edgar Allan Poe challenged readers to send him monoalphabetic substitution ciphers to break. For the submission reproduced in Figure 2.1, Poe was not able to offer a solution; however, he was able to demonstrate that the sender did not follow his rules. That is, the cipher is not monoalphabetic. Poe concluded (incorrectly) that it was "a jargon of random characters, having no meaning whatsoever."[1]

```
Ge Jeasgdxv,
    Zij gl mw, laam, xzy zmlwhfzek ejlvdxw
  kwke tx lbr atgh lbmx aanu bai Vsmukkss pwn
  vlwk agh gnumk wdlnzweg jnbxvv oaeg enwb
  zwmgy mo mlw wnbx mw al pnfdcfpkh wzkex
  hssf xkiyahul.  Mk num yexdm wbxy sbc hv
  wyx Phwkgnamcuk?
```

Figure 2.1 A ciphertext that Poe could not solve. (From Winkel, Brian J., *Cryptologia*, Vol. 1, No. 1, January 1977, p. 95; the ciphertext originally appeared in Poe, Edgar Allan, *Alexander's Weekly Messenger*, February 26, 1840.)

---

[1] Poe demonstrated that the cipher was nonsense in his article, "More of the Puzzles," which appeared in *Alexander's Weekly Messenger*, Vol. 4, No. 9, February 26, 1840, p. 2, column 4, and gave the quote provided over a year later (reflecting back) in his article "A Few Words on Secret Writing," which appeared in the July 1841 issue of *Graham's Magazine*, Vol. 19, No. 1, pp. 33–38. Ironically, Poe's "A Few Words on Secret Writing" was the longest of his essays dealing with cryptology.

Jumping ahead to the 1970s, Mark Lyster, an undergraduate in Brian Winkel's cryptology class at Albion College, became curious and attempted a solution. Together, the professor and his student solved it. Brian then challenged *Cryptologia's* readers to attempt their own solutions in the paper referenced with Figure 2.1. In the August 1977 *Scientific American*, Martin Gardner challenged his readers to solve it. You may consider yourself so challenged after reading the material on cryptanalysis that follows in this chapter! A solution was presented in Brian Winkel's article, "Poe Challenge Cipher Solutions," in the October 1977 issue of *Cryptologia* (pp. 318–325). Look at this paper only after making a serious attempt to solve it yourself!

The system behind the Poe cipher was long known as *Le Chiffre Indéchiffrable* ("The Unbreakable Cipher"). Today, it is simply referred to as the Vigenère cipher—we have to make a few improvements before it becomes truly unbreakable!

Figure 2.2   Blaise de Vigenère (1523–1596). (http://fr.wikipedia.org/wiki/Fichier:Vigenere.jpg.)

The main weaknesses of monoalphabetic ciphers are the preservation of letter frequencies (only the symbols representing the letters change) and word patterns, as detailed in the previous chapter. So, a necessary condition for a cipher to be secure is that it be invulnerable to these attacks. The Vigenère cipher accomplishes this by using a variety of substitutions for each letter in the plaintext alphabet. The frequencies of the letters in the ciphertext are thus flattened. Pattern words are also disguised. This is an example of a polyalphabetic substitution cipher. Really, it shouldn't be named after Vigenère (Figure 2.2), but we'll let the history wait for a moment while we take a look at an example of this cipher using the keyword ELVIS, which can be seen running down the first column in the substitution table below.

```
A B C D E F G H I J K L M N O P Q R S T U V W X Y Z plaintext
E F G H I J K L M N O P Q R S T U V W X Y Z A B C D alphabet 1
L M N O P Q R S T U V W X Y Z A B C D E F G H I J K alphabet 2
V W X Y Z A B C D E F G H I J K L M N O P Q R S T U alphabet 3
I J K L M N O P Q R S T U V W X Y Z A B C D E F G H alphabet 4
S T U V W X Y Z A B C D E F G H I J K L M N O P Q R alphabet 5
```

Alphabet 1 is used to encipher the first letter in the message, alphabet 2 is used for enciphering the second letter, and so on. When we get to the sixth letter, we return to alphabet 1. A sample encipherment follows.

```
THANK YOU,  THANK YOU VERY MUCH Plaintext
ELVIS ELV   ISELV ISE LVIS ELVI Key
XSVVC CZP,  BZEYF GGY GZZQ QFXP Ciphertext
```

The words THANK YOU are enciphered in two different ways, depending upon the position relative to the key alphabets. Also, we have doubled letters in the ciphertext, VV and ZZ, where there are no doubled letters in the plaintext. When this system first appeared, there were no cryptanalytic techniques in existence that were any better than simply guessing at the key. In general, longer keys are better. If the key is only a single character, this system reduces to the Caesar cipher.

2.2 History of the Vigenère Cipher

Okay, so who should this system be named after? Well, several men contributed the pieces from which this cipher was built. The first was Leon Battista Alberti (1404–1472) (Figure 2.3). Remember the cipher disk pictured in Figure 1.1? It turns! Alberti suggested turning it every three or four words to prevent an attacker from having a large statistical base from which to crack any particular substitution alphabet. He made this breakthrough to polyalphabeticity in 1466 or 1467; however, he never suggested the use of a keyword or switching alphabets with every letter.

Figure 2.3   Leon Battista Alberti (1404–1472) (http://en.wikipedia.org/wiki/File:Leon_Battista_Alberti2.jpg.)

Figure 2.4 Johannes Trithemius, cryptology's first printed author. (Courtesy of the National Cryptologic Museum, Fort Meade, Maryland.)

The next step was taken by Johannes Trithemius (1462–1516) (Figure 2.4), the author of the first printed book on cryptology, *Polygraphiae* (Figure 2.5). It was written in 1508 and first printed in 1518, after his death. This was actually his second book dealing with cryptology, but the first, *Steganographia*, did not reach printed form until 1606. *Steganographia* had long circulated in manuscript form and had even attracted the attention of the Roman Catholic Church, which placed it on the Index of Prohibited Books. It is now available online at http://www.esotericarchives. com/esoteric.htm#tritem. Most of the cryptographers of Trithemius's era were also alchemists and magicians of a sort. In fact, Trithemius knew the real Dr. Faustus (whom he considered a charlatan) and is said to have been a mentor of Paracelsus and Cornelius Agrippa.[2] According to legend, Trithemius himself was said to have raised the wife of Emperor Maximilian I from the dead.[3]

---

[2] Kahn, David, *The Codebreakers*, second edition, Scribner, New York, 1996, p. 131.
[3] Goodrick-Clarke, Nicholas, *The Western Esoteric Traditions: A Historical Introduction*, Oxford University Press, New York, 2008, p. 52.

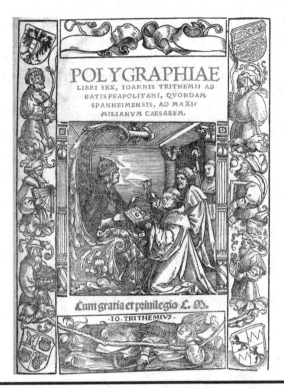

Figure 2.5 Title page of *Polygraphiae* by Trithemius. (Courtesy of the National Cryptologic Museum, Fort Meade, Maryland.)

Polygraphiae contained the first "square table" or "tableau." This is pictured below and simply represents all possible shift ciphers. The first row is the plaintext.

```
A B C D E F G H I J K L M N O P Q R S T U V W X Y Z
B C D E F G H I J K L M N O P Q R S T U V W X Y Z A
C D E F G H I J K L M N O P Q R S T U V W X Y Z A B
D E F G H I J K L M N O P Q R S T U V W X Y Z A B C
E F G H I J K L M N O P Q R S T U V W X Y Z A B C D
F G H I J K L M N O P Q R S T U V W X Y Z A B C D E
G H I J K L M N O P Q R S T U V W X Y Z A B C D E F
H I J K L M N O P Q R S T U V W X Y Z A B C D E F G
I J K L M N O P Q R S T U V W X Y Z A B C D E F G H
J K L M N O P Q R S T U V W X Y Z A B C D E F G H I
K L M N O P Q R S T U V W X Y Z A B C D E F G H I J
L M N O P Q R S T U V W X Y Z A B C D E F G H I J K
M N O P Q R S T U V W X Y Z A B C D E F G H I J K L
N O P Q R S T U V W X Y Z A B C D E F G H I J K L M
O P Q R S T U V W X Y Z A B C D E F G H I J K L M N
P Q R S T U V W X Y Z A B C D E F G H I J K L M N O
Q R S T U V W X Y Z A B C D E F G H I J K L M N O P
R S T U V W X Y Z A B C D E F G H I J K L M N O P Q
S T U V W X Y Z A B C D E F G H I J K L M N O P Q R
T U V W X Y Z A B C D E F G H I J K L M N O P Q R S
U V W X Y Z A B C D E F G H I J K L M N O P Q R S T
V W X Y Z A B C D E F G H I J K L M N O P Q R S T U
W X Y Z A B C D E F G H I J K L M N O P Q R S T U V
X Y Z A B C D E F G H I J K L M N O P Q R S T U V W
Y Z A B C D E F G H I J K L M N O P Q R S T U V W X
Z A B C D E F G H I J K L M N O P Q R S T U V W X Y
```

Trithemius used the alphabets in order, enciphering 24 letters of plaintext with each (his Latin alphabet had 24 letters, which seems to be why he chose this number). He also enciphered by changing the alphabet after each letter, but he always used the alphabets in order. As with Alberti, the idea of using a keyword was not realized. It was finally hit upon by Giovan Battista Bellaso in 1553 in his work *La cifra del Sig. Giovan.*

Figure 2.6 Giovanni Battista Porta (1535–1615) (http://en.wikipedia.org/wiki/File:Giambattista_della_Porta.jpeg).

Now that all of the ideas were finally present, Giovanni Battista Porta[4] (1535–1615) (Figure 2.6) combined them. He used Bellaso's keyword to determine which alphabets to use, but he also mixed the letters within the cipher alphabets, as Alberti had with his cipher disk. It should be noted that the mixed alphabets represent a greater level of security than provided by the straight alphabets of the Vigenère cipher. Porta's work was published as *De Furtivis Literarum Notis* in 1563. This work also included the first digraphic cipher, a topic we shall return to in Section 4.4.

Blaise de Vigenère (1523–1596) published his work in *Traicté des Chiffres* in 1586, by which time the cipher described above already existed. Vigenère was careful to give credit to those who had earned it, yet somehow his name became attached to a system that wasn't his, and his real contribution, the *autokey*, was ignored.[5] We shall also ignore it—for now. As a further example

---

[4] Also of interest is that Porta founded an "Academy of Secrets" in Naples. For more information see Zielinski, Siegfried, "Magic and Experiment: Giovan Battista Della Porta," in Zielinski, Siegfried, editor, *Deep Time of the Media: Toward an Archaeology of Hearing and Seeing by Technical Means*, MIT Press, Cambridge, Massachusetts, pp. 57–100, available online at https://gebseng.com/media_archeology/reading_materials/Zielinsky-deep_time_of_the_media.pdf.

[5] Yet, even this contribution should really be credited to a previous discoverer, Giovan Battista Bellaso, who described it in 1564. See LABRONICUS [ACA pen-name of Augusto Buonafalce], "Historical Tidbits," *The Cryptogram*, Vol. 58, No. 3, May–June 1992, p. 9.

of the involvement of early cryptographers in alchemy and magic, let it be known that *Traicté des Chiffres* contains a recipe for making gold.

The Vigenère cipher was one of the best at the time, especially when using mixed alphabets; nevertheless, there are still various cases of its being cracked. One amusing anecdote involves such a cipher being broken by Casanova (Figure 2.7), who then used his accomplishment as a means to a seduction.[6] He wrote:

> Five or six weeks later, she asked me if I had deciphered the manuscript which had the transmutation procedure. I told her that I had.
>
> "Without the key, sir, excuse me if I believe the thing impossible."
>
> "Do you wish me to name your key, madame?"
>
> "If you please."
>
> I then told her the word, which belonged to no language, and I saw her surprise. She told me that it was impossible, for she believed herself the only possessor of that word which she kept in her memory and which she had never written down.
>
> I could have told her the truth – that the same calculation which had served me for deciphering the manuscript had enabled me to learn the word – but on a caprice it struck me to tell her that a genie had revealed it to me. This false disclosure fettered Madame d'Urfé to me. That day I became the master of her soul, and I abused my power. Every time I think of it, I am distressed and ashamed, and I do penance now in the obligation under which I place myself of telling the truth in writing my memoirs. [I took my leave] bearing with me her soul, her heart, her wits and all the good sense that she had left.

Sadly, Casanova didn't reveal his method of cryptanalysis.

Figure 2.7   Casanova, studly codebreaker.

---

[6] Kahn, David, *The Codebreakers*, second edition, Scribner, 1996, p. 153.

2.3 Cryptanalysis of the Vigenère Cipher

So, how can *we* be like Casanova and break a Vigenère cipher? Knowing the length of the key usually makes breaking this system easy. For example, if the length of the key was 2, then the first, third, fifth, etc., letters would all be enciphered with the same alphabet, as would the second, fourth, sixth, etc., albeit using a different alphabet. Grouping these letters together and looking at the frequencies, one letter usually stands out, namely E. A difficulty may arise; if the message is short, the most frequent letter in a specific group won't always be E. This problem is addressed in the example coming up. For now, let's assume that we can identify E. Because a straight alphabet is used, all the other letters would also be known. Thus, the message is easily broken. If mixed alphabets are used, the decipherment is slower, but still possible with patience, using a frequency table. Making the key longer requires separating the ciphertext into more groups. For example, if the key is of length 5, we would group the letters as follows:

1, 6, 11, 16, …	alphabet 1
2, 7, 12, 17, …	alphabet 2
3, 8, 13, 18, …	alphabet 3
4, 9, 14, 19, …	alphabet 4
5, 10, 15, 20, …	alphabet 5

Notice that each group will be part of an equivalence class mod 5.

So how can we determine the length of the key? Several methods are available. The simplest is to assume that some plaintext portions of the message are repeated in the same alignment with the key. For example, if two computer science professors are communicating, the word COMPUTER may appear repeatedly in the message. If the key is ORANGE, we may have various alignments:

COMPUTER	COMPUTER	COMPUTER
ORANGEOR	RANGEORA	ANGEORAN
COMPUTER	COMPUTER	COMPUTER
NGEORANG	GEORANGE	EORANGEO

It is easy to calculate the probability that one of these alignments will be repeated for various numbers of appearances of the word COMPUTER. Of course, if COMPUTER appears seven or more times in the message, a repeated alignment is guaranteed. This results in eight letters of ciphertext being repeated. The distance between the first letters of these repetitions must be a multiple of the keylength. Many other words may be repeated in the same alignment with the keyword, such as common words like THE and AND. Taking all repeats of say, three letters or more, and looking at the distances between them will suggest a keylength of which all of these distances should be multiples. There may be some repetition arising from different words combining with different alignments of the keyword, but this will usually appear as background noise and the true keylength will be clear. This process is known as the *Kasiski test*.[7] (It might be known as the Babbage test, as Charles Babbage discovered it before Kasiski, but, as was typical for Babbage, he didn't follow through. He never published this particular result.)[8] The Kasiski test will be demonstrated shortly, but first another method of determining the keylength is examined.

---

[7] Friedrich W. Kasiski (1805–1881) was a retired Prussian infantry major, who published his method in *Die Geheimschriften und die Dechiffrir-kunst* in 1863.

[8] For a bit more on this see Singh, Simon, *The Code Book*, Doubleday, New York, 1999, p. 78. For a tremendous amount more see Franksen, Ole Immanuel, *Mr. Babbage's Secret, The Tale of a Cypher and APL*, Prentice-Hall, Englewood Cliffs, New Jersey, 1984.

Figure 2.8   William Friedman (1891–1969). (Courtesy of National Cryptologic Museum, Fort Meade, Maryland.)

A wonderful attack, published in 1920 by William Friedman (Figure 2.8),[9] arises from a calculation called the *index of coincidence* (IC). Simply stated, this is the probability that two randomly chosen letters from a text of length N will be the same.

For both to be A, we take

$$P(\text{first letter is A}) \cdot P(\text{second letter is A}) = \frac{F_A}{N} \cdot \frac{F_A - 1}{N - 1}$$

where F_A denotes the frequency of A.

Because both letters could have been B, or both letters could have been C, etc., we must sum these probabilities over each letter in the alphabet, which then gives

$$\text{IC} = \frac{\sum_{i=A}^{Z} F_i (F_i - 1)}{N(N - 1)}.$$

[9] Friedman, William F., *The Index of Coincidence and Its Applications in Cryptography*, Publication No. 22, Riverbank Laboratories, Geneva, Illinois, 1920.

The use of multiple substitution alphabets in a cipher flattens the frequency distribution for the letters and therefore decreases the chance of two randomly selected ciphertext letters being the same, as compared to the chance for letters in the original plaintext message. Thus, the value of the index of coincidence can be said to measure the flatness of the frequency distribution or, in other words, estimate the number of alphabets in use.

Due to the variation of letter frequencies in normal plaintext, we will not get exactly the same value from the IC every time a keyword of a given length is used; however, the expected value may be calculated for each size keyword. It is provided in the following table. The values depend in part on the length of the text. Separate tables can be constructed for various message lengths. The table below gives values for long messages.

Number of Alphabets (Keyword Length)	Expected Value for Index of Coincidence
1	0.0660
2	0.0520
3	0.0473
4	0.0449
5	0.0435
6	0.0426
7	0.0419
8	0.0414
9	0.0410
10	0.0407

As N gets large, we approach a limiting value of approximately 0.0388. This is the value we'd get for purely random text—that is, text where all letters are equally frequent and thus share a probability of 1/26.

Notice that the difference between expected values is largest when the number of alphabets used is small. We can easily distinguish between one alphabet (a monoalphabetic substitution cipher) and two alphabets, but distinguishing between nine and ten alphabets is difficult.

Suppose the index of coincidence is 0.04085, indicating that nine or ten alphabets have been used. We can investigate further by assuming that the correct keylength is 9 and splitting the ciphertext into nine groups of letters, each of which would have been enciphered by the same alphabet, if our assumption is correct. The first group would contain the letters in positions 1, 10, 19, 28,…, because a key of length 9 forces us to start over with the first alphabet at position 10, and again at positions 19, 28, etc. The second group would contain all letters enciphered with the second alphabet, positions 2, 11, 20, 29, etc. Now applying the IC to group one should indicate (by resulting in a value close to 0.066) if those letters truly did arise from encipherment with the same alphabet. If the value of the IC is closer to 0.038, we lose confidence in a keylength of 9. But the first group is not the only one we should consider. The IC value for this group could be a misleading fluke! Testing all nine groups of letters separately gives a much firmer statistical base upon which to decide if a keylength of 9 is correct. If the nine IC values, considered as a group, are discouraging, we can start over and assume the keylength is 10. Splitting the ciphertext letters into ten groups and computing the IC for each will show if this assumption is better or not. The smaller groups of ciphertext letters, for which these computations are done, are referred to as *decimated alphabets*, even if there aren't exactly ten of them.

A few examples will indicate how reliable the Kasiski and IC tests are. The first is presented below and others are left as exercises. *Caution:* Some books concoct examples where such tests work perfectly, creating the false impression that this is always the case.

The IC equation can be turned around to give the length (L) of the key, when the number of characters (N) in the ciphertext and the ciphertext index of coincidence (IC) are known:

$$L \approx \frac{0.028N}{(IC)(N-1) - 0.038N + 0.066}$$

Although you needn't understand the derivation of the formula above in order to use it, it's easy to demonstrate. Suppose we have a ciphertext consisting of N letters and that the enciphering key has length L. If we randomly pick two letters, what is the probability that they are the same (As *ciphertext*—the plaintext letters they represent needn't match)? The probability that the two letters match is much higher if they were both enciphered with the same alphabet, so we consider two separate cases and combine them for our final answer.

Case 1: The Two Letters Arose from the Same Cipher Alphabet

It doesn't matter which letter we pick first, but the second letter has to come from the same alphabet, so it must be one of the remaining $(N/L) - 1$ letters of this type (The L alphabets will roughly divide the N letters of the text into groups of size N/L, each enciphered differently). Thus, there are $(N/L) - 1$ choices left out of the total remaining $N - 1$ letters. So the probability is

$$\frac{\left(\frac{N}{L}\right) - 1}{N - 1}.$$

But we also want the two letters to be the same. Because we are already within the same alphabet, this probability is simply 0.066, the value of the IC for one alphabet. We multiply these two values together to get

$$\frac{\left(\frac{N}{L} - 1\right)(0.066)}{N - 1}.$$

Case 2: The Two Letters Arose from Different Cipher Alphabets

As with case 1, it doesn't matter which letter we pick first, but now the second letter must come from a different alphabet. Because we already found the probability the second letter came from the same alphabet, we can take the complement to get the probability it came from a different alphabet. We have

$$1 - \left(\frac{\frac{N}{L} - 1}{N - 1} \right).$$

Now that we have two letters from different alphabets, we need to multiply by the probability that they match. This is simply the IC value for random text (or, equivalently, a large number of alphabets), namely 0.038. So, our probability for case 2 is

$$\left(1 - \frac{\frac{N}{L} - 1}{N - 1} \right) (0.038).$$

Combining the two cases, we have

$$\text{IC} \approx \frac{\left(\frac{N}{L} - 1 \right) (0.066)}{N - 1} + \left(1 - \frac{\frac{N}{L} - 1}{N - 1} \right) (0.038).$$

It's now just a matter of doing the algebra to solve for L (see Exercise 23) and obtain the result.

$$L \approx \frac{0.028N}{(IC)(N - 1) - 0.038N + 0.066}$$

Using this equation, you don't need a table of values like the one given above; however, this version of the equation is only intended for ciphers having English plaintexts. For other languages, the constants may vary.

The Kasiski test and the index of coincidence may sound complicated at first, but they are very easy to use. Take a look at the following example to see how simple they make the task of Vigenère cipher cryptanalysis.

Example

```
IZPHY XLZZP SCULA TLNQV FEDEP QYOEB SMMOA AVTSZ VQATL LTZSZ
AKXHO OIZPS MBLLV PZCNE EDBTQ DLMFZ ZFTVZ LHLVP MBUMA VMMXG
FHFEP QFFVX OQTUR SRGDP IFMBU EIGMR AFVOE CBTQF VYOCM FTSCH
ROOAP GVGTS QYRCI MHQZA YHYXG LZPQB FYEOM ZFCKB LWBTQ UIHUY
LRDCD PHPVO QVVPA DBMWS ELOSM PDCMX OFBFT SDTNL VPTSG EANMP
MHKAE PIEFC WMHPO MDRVG OQMPQ BTAEC CNUAJ TNOIR XODBN RAIAF
UPHTK TFIIG EOMHQ FPPAJ BAWSV ITSMI MMFYT SMFDS VHFWQ RQ
```

Several character groups repeat. We need to note their positions to make use of the Kasiski test.

Character Grouping	Starting Positions	Difference Between Starting Positions
IZP	1 and 57	56
HYX	4 and 172	168
EPQ	24 and 104	80
MBU	91 and 123	32
TSM	327 and 335	8

Consider the last column of the table above. All of the values are multiples of 8. This suggests that the key is of length 8. It's possible that the keylength is 4 (or 2), but if this were the case, it's likely that one of the numbers in the difference column would be a multiple of 4 (or 2) but not a multiple of 8.

Calculating the index of coincidence requires more work. We begin by constructing a frequency table (Table 2.1) for the ciphertext letters.

Table 2.1 Frequency Table Used to Calculate Index of Coincidence

Letter	Frequency	Letter	Frequency
A	18	N	7
B	14	O	18
C	12	P	22
D	12	Q	17
E	15	R	10
F	22	S	16
G	9	T	20
H	14	U	8
I	13	V	18
J	2	W	5
K	4	X	7
L	15	Y	9
M	26	Z	14

The numerator of the index of coincidence for this example is then given by

$$
\sum_{i=A}^{Z} F_i(F_i - 1) =
\begin{aligned}
& (18)(17) + (14)(13) + (12)(11) + (12)(11) + (15)(14) + (22)(21) \\
& + (9)(8) + (14)(13) + (13)(12) + (2)(1) + (4)(3) + (15)(14) + (26)(25) \\
& + (7)(6) + (18)(17) + (22)(21) + (17)(16) + (10)(9) + (16)(15) \\
& + (20)(19) + (8)(7) + (18)(17) + (5)(4) + (7)(6) + (9)(8) + (14)(13) \\
& = 5178
\end{aligned}
$$

The index of coincidence is then

$$IC = \frac{\sum_{i=A}^{Z} F_i(F_i - 1)}{N(N-1)} = \frac{5178}{(347)(346)} \approx 0.0431.$$

This value, although not matching an expected value perfectly, suggests that five or six alphabets were used. We can assume the key is of length 5, split the ciphertext letters into groups that would have been enciphered with the same alphabet, and perform the IC calculation for each group. We get the values 0.046, 0.050, 0.041, 0.038, and 0.046. These values are far below 0.066, so it's very unlikely that they are really from the same alphabet. Our assumption that the key is of length 5 must be wrong. Repeating these calculations based on a key of length 6 gives the values 0.050, 0.041, 0.035, 0.048, 0.038, and 0.037. Again, it seems that the key cannot be of this length.

We could try again with four alphabets and then with seven, before moving on to values further from the ones suggested by the IC, but the Kasiski test suggested eight, so let's skip ahead to this value. Splitting the ciphertext into eight separate groups and calculating the IC for each gives 0.105, 0.087, 0.075, 0.087, 0.056, 0.069, 0.065, and 0.046. These values are by far the largest, so we have another test backing up the results of the Kasiski test. Another way in which an IC calculation can be used to support the result of the Kasiski test is described in Exercise 22. We now rewrite the ciphertext in blocks of length 8, so that characters in the same column represent letters enciphered by the same alphabet. We may construct a frequency table for each column (Table 2.2). This had to be done to get the values for the IC given in the paragraph above, but I left out showing the work until it could be seen to lead to a positive conclusion. Take a moment to examine Table 2.2 and the accompanying text before returning here.

For each column, there are 26 possible choices for which letter represents E. In column 1, the maximum value for the sum of the frequencies of E, A, and T is 21 and is obtained when M represents E. Now for column 2, assuming that P represents E yields a sum of only 9. The greatest sum is 11, which is obtained when Q represents E. For column 3, the maximum sum is also obtained when Q represents E. This time the sum is 17. For columns 4 through 8, this technique suggests E is represented by S, V, X, E, and P, respectively. We see that, if these substitutions are correct, E is only the most frequent character in columns 1, 3, 4, and 5, while it is tied for first place in column 7. You are encouraged to investigate other techniques for determining the shift of each alphabet in Exercise 14. The substitutions above imply the keyword (the letters representing A in each alphabet strung together in order) is IMMORTAL. Since this is a real word, we gain some confidence in our solution. Applying this keyword to the ciphertext by subtracting modulo 26 gives:

```
IZPHY XLZZP SCULA TLNQV FEDEP QYOEB SMMOA AVTSZ VQATL LTZSZ
IMMOR TALIM MORTA LIMMO RTALI MMORT ALIMM ORTAL IMMOR TALIM
ANDTH ELORD GODSA IDBEH OLDTH EMANI SBECO NEASO NEOFU STOKN

AKXHO OIZPS MBLLV PZCNE EDBTQ DLMFZ ZFTVZ LHLVP MBUMA VMMXG
MORTA LIMMO RTALI MMORT ALIMM ORTAL IMMOR TALIM MORTA LIMMO
OWGOO DANDE VILAN DNOWL ESTHE PUTFO RTHHI SHAND ANDTA KEALS

FHFEP QFFVX OQTUR SRGDP IFMBU EIGMR AFVOE CBTQF VYOCM FTSCH
RTALI MMORT ALIMM ORTAL IMMOR TALIM MORTA LIMMO RTALI MMORT
OOFTH ETREE OFLIF EANDE ATAND LIVEF OREVE RTHER EFORE THELO

ROOAP GVGTS QYRCI MHQZA YHYXG LZPQB FYEOM ZFCKB LWBTQ UIHUY
ALIMM ORTAL IMMOR TALIM MORTA LIMMO RTALI MMORT ALIMM ORTAL
RDGOD SENTH IMFOR THFRO MTHEG ARDEN OFEDE NTOTI LLTHE GROUN
```

```
LRDCD  PHPVO  QVVPA  DBMWS  ELOSM  PDCMX  OFBFT  SDTNL  VPTSG  EANMP
IMMOR  TALIM  MORTA  LIMMO  RTALI  MMORT  ALIMM  ORTAL  IMMOR  TALIM
DFROM  WHENC  EHEWA  STAKE  NSOHE  DROVE  OUTTH  EMANA  NDHEP  LACED

MHKAE  PIEFC  WMHPO  MDRVG  OQMPQ  BTAEC  CNUAJ  TNOIR  XODBN  RAIAF
MORTA  LIMMO  RTALI  MMORT  ALIMM  ORTAL  IMMOR  TALIM  MORTA  LIMMO
ATTHE  EASTO  FTHEG  ARDEN  OFEDE  NCHER  UBIMS  ANDAF  LAMIN  GSWOR

UPHTK  TFIIG  EOMHQ  FPPAJ  BAWSV  ITSMI  MMFYT  SMFDS  VHFWQ  RQ
RTALI  MMORT  ALIMM  ORTAL  IMMOR  TALIM  MORTA  LIMMO  RTALI  MM
DWHIC  HTURN  EDEVE  RYWAY  TOKEE  PTHEW  AYOFT  HETRE  EOFLI  FE
```

Table 2.2 Frequency Table with Tallies for Each Column

1 2 3 4 5 6 7 8		Column Number							
I Z P H Y X L Z		1	2	3	4	5	6	7	8
Z P S C U L A T									
L N Q V F E D E	A	1	3	5	2	0	2	5	0
P Q Y O E B S M	B	6	0	0	4	0	4	0	0
M O A A V T S Z	C	1	0	0	7	2	0	0	3
V Q A T L L T Z	D	0	0	4	1	2	0	2	2
S Z A K X H O O	E	0	1	0	0	2	4	6	2
I Z P S M B L L	F	0	5	4	5	4	0	3	1
V P Z C N E E D	G	0	0	0	2	1	4	1	1
B T Q D L M F Z	H	1	1	0	3	0	5	5	0
Z F T V Z L H L	I	4	2	0	1	3	1	1	0
V P M B U M A V	J	0	0	0	0	1	0	0	1
M M X G F H F E	K	1	0	0	1	2	0	0	0
P Q F F V X O Q	L	2	0	0	0	2	4	3	4
T U R S R G D P	M	11	3	4	1	2	4	0	1
I F M B U E I G	N	0	2	0	0	1	0	3	1
M R A F V O E C	O	2	2	0	2	0	1	6	5
B T Q F V Y O C	P	2	8	3	0	1	4	0	4
M F T S C H R O	Q	2	3	9	0	0	0	0	3
O A P G V G T S	R	0	4	2	1	1	0	1	1
Q Y R C I M H Q	S	1	0	1	8	0	0	2	4
Z A Y H Y X G L	T	1	4	4	1	1	3	3	2
Z P Q B F Y E O	U	0	1	1	1	4	0	1	0
M Z F C K B L W	V	5	0	0	3	9	0	0	1
B T Q U I H U Y	W	0	0	2	0	1	0	0	2
L R D C D P H P	X	0	0	2	0	1	4	0	0
V O Q V V P A D	Y	0	1	2	0	2	3	0	1
B M W S E L O S	Z	4	4	1	0	1	0	0	4
M P D C M X O F									
B F T S D T N L									
V P T S G E A N									
M P M H K A E P									
I E F C W M H P									
O M D R V G O Q									
M P Q B T A E C									
C N U A J T N O									
I R X O D B N R									
A I A F U P H T									
K T F I I G E O									
M H Q F P P A J									
B A W S V I T S									
M I M M F Y T S									
M F D S V H F W									
Q R Q									

Consider column 1. M is by far the most frequent letter, so we assume that M represents E. This implies that B represents T and I represents A. The frequencies of B and I are high, so this looks like a good fit. For a small sample, E will not always be the most frequent letter. For example, in column 2 the most frequent character is P, yet this implies that E (with frequency 1) is T and L (with frequency 0) is A, values that do not seem likely. A better strategy is to look for the shift that maximizes the sum of the frequencies for E, A, and T, the three most frequent plaintext letters.

The plaintext turns out to be a passage from the Bible about gaining immortality.[10]

The index of coincidence gave too low a value for the number of alphabets, as will often happen when two or more of the "different" alphabets used are, in fact, the same. The repeated M alphabet in the key IMMORTAL is to blame in this instance.

Kasiski's attack worked better in this example, but Friedman's index of coincidence is, in general, the more powerful technique. Friedman would have gained immortality through this work, even if he had done nothing else. It can be applied in many different contexts and can even be used in some cases to distinguish between languages.[11]

Language	IC
English	0.0667
French	0.0778
German	0.0762
Italian	0.0738
Russian	0.0529 (30-letter Cyrillic alphabet)
Spanish	0.0775

There are many other ways to distinguish between languages in a monoalphabetic substitution cipher without deciphering. One such measure, the entropy of a text (discussed more fully later in this book), can even be used to determine (very roughly) the era in which a text was produced. The entropy of a language seems to increase with time, obeying the second law of thermodynamics![12]

The Vigenère cipher has seen extensive use over hundreds of years. It was used by the confederacy in the Civil War and it had long been believed that they only ever used three keys: MANCHESTER BLUFF, COMPLETE VICTORY, and (after General Lee's surrender) COME RETRIBUTION. In 2006, however, Kent Boklan, attempting to break an old confederate message, discovered a fourth key.[13] The small number of keys certainly aided the Union codebreakers! Kasiski's attack was published during this war, but appears to have gone unnoticed by the Confederacy.

Mathematician Charles Dodgson (1832–1898), who wrote *Alice's Adventures in Wonderland* and *Through the Looking Glass* under the pseudonym Lewis Carroll, independently discovered what we call the Vigenère cipher. He wrote that it would be "impossible for any one [sic], ignorant of the key-word, to decipher the message, even with the help of the table."[14]

---

[10] Publishing seems to be the surest path. Some seek immortality through accomplishments in sports, but how many ancient Greek athletes can you name? On the other hand, you can probably name several ancient Greek playwrights and authors of works on mathematics, science, and philosophy.

[11] Kahn, David, *The Codebreakers*, second edition, Scribner, 1996, p. 378.

[12] Bennett, Jr., William Ralph, *Scientific and Engineering Problem-Solving with the Computer*, Prentice-Hall, Englewood Cliffs, New Jersey, 1976. Sections 4.13 and 4.14 are relevant here.

[13] Boklan, Kent, "How I Broke the Confederate Code (137 Years Too Late)," *Cryptologia*, Vol. 30, No. 4, October 2006, pp. 340–345. Boklan later evened things out by breaking an old Union cipher.

[14] The source of this quote is typically cited as "The Alphabet Cipher," which was published in 1868 in a children's magazine. Thus far, I've been unable to obtain further bibliographic information. I have a photocopy of the two-page paper, but even this is no help, as no title, date, author name, or even page numbers appear on the pages. Christie's auctioned off an original as lot 117 of sale 2153. The price realized was $1,000, and the item was described as "Broadsheet on card stock (180 × 123 mm). The table of letters printed on one side and the Explanation on the other." So perhaps it never was in a magazine? See http://www.christies.com/lotfinder/books-manuscripts/dodgson-charles-lutwidge-5280733-details.aspx?from=salesummary&pos=5&intObjectID=5280733&sid=&page=9 for more information.

Even as late as 1917, not everyone had heard that the system had been broken. In that year, *Scientific American* reprinted an article from the *Proceedings of the Engineers' Club of Philadelphia* that proclaimed the system to be new (!) and "impossible of translation."[15] The article also stated, "The ease with which the key may be changed is another point in favor of the adoption of this code by those desiring to transmit important messages without the slightest danger of their messages being read by political or business rivals." However, the mistake was eventually recognized, and a 1921 issue of *Scientific American Monthly* carried an article entitled "The Ciphers of Porta and Vigenère, The Original Undecipherable Code and How to Decipher It" by Otto Holstein.[16]

2.4 *Kryptos*

A more recent example of a Vigenère cipher is in the form of an intriguing sculpture called *Kryptos* (Figure 2.9). Created by James "Jim" Sanborn in 1990, this artwork is located in an outdoor area within the Central Intelligence Agency (CIA). Although the location is not open to the public, it has attracted a great deal of public attention and is even alluded to via its latitude and longitude on the dust jacket of Dan Brown's novel *The Da Vinci Code*.

Figure 2.9 Sanborn's sculpture *Kryptos*. (Courtesy of National Cryptologic Museum, Fort Meade, Maryland.)

---

[15] "A New Cipher Code," *Scientific American Supplement*, Vol. 83, No. 2143, January 27, 1917, p. 61.
[16] Holstein, Otto, "The Ciphers of Porta and Vigenère, The Original Undecipherable Code and How to Decipher It," *Scientific American Monthly*, Vol. 4, No. 4, October 1921, pp. 332–334.

The left half of the sculpture is the ciphertext, which may be divided into two panels. These are referred to as panel 1 (top half of left side) and panel 2 (bottom half of left side). Both panels are reproduced in Figures 2.10 and 2.11, as shown on the CIA's website. Each contains two distinct ciphers. These ciphers are distinguished in the figures here, but not in the original sculpture.

```
E M U F P H Z L R F A X Y U S D J K Z L D K R N S H G N F I V J
Y Q T Q U X Q B Q V Y U V L L T R E V J Y Q T M K Y R D M F D
V F P J U D E E H Z W E T Z Y V G W H K K Q E T G F Q J N C E
G G W H K K ? D Q M C P F Q Z D Q M M I A G P F X H Q R L G
T I M V M Z J A N Q L V K Q E D A G D V F R P J U N G E U N A
Q Z G Z L E C G Y U X U E E N J T B J L B Q C R T B J D F H R R
Y I Z E T K Z E M V D U F K S J H K F W H K U W Q L S Z F T I
H H D D D U V H ? D W K B F U F P W N T D F I Y C U Q Z E R E
E V L D K F E Z M O Q Q J L T T U G S Y Q P F E U N L A V I D X
F L G G T E Z ? F K Z B S F D Q V G O G I P U F X H H D R K F
F H Q N T G P U A E C N U V P D J M Q C L Q U M U N E D F Q
E L Z Z V R R G K F F V O E E X B D M V P N F Q X E Z L G R E
D N Q F M P N Z G L F L P M R J Q Y A L M G N U V P D X V K P
D Q U M E B E D M H D A F M J G Z N U P L G E W J L L A E T G
```

Figure 2.10   Panel 1 of *Kryptos*; a horizontal line has been added to separate cipher 1 from cipher 2. (Adapted from https://www.cia.gov/about-cia/headquarters-tour/kryptos/KryptosPrint.pdf.)

```
E N D Y A H R O H N L S R H E O C P T E O I B I D Y S H N A I A
C H T N R E Y U L D S L L S L L N O H S N O S M R W X M N E
T P R N G A T I H N R A R P E S L N N E L E B L P I I A C A E
W M T W N D I T E E N R A H C T E N E U D R E T N H A E O E
T F O L S E D T I W E N H A E I O Y T E Y Q H E E N C T A Y C R
E I F T B R S P A M H N E W E N A T A M A T E G Y E E R L B
T E E F O A S F I O T U E T U A E O T O A R M A E E R T N R T I
B S E D D N I A A H T T M S T E W P I E R O A G R I E W F E B
A E C T D D H I L C E I H S I T E G O E A O S D D R Y D L O R I T
R K L M L E H A G T D H A R D P N E O H M G F M F E U H E
E C D M R I P F E I M E H N L S S T T R T V D O H W ? O B K R
U O X O G H U L B S O L I F B B W F L R V Q Q P R N G K S S O
T W T Q S J Q S S E K Z Z W A T J K L U D I A W I N F B N Y P
V T T M Z F P K W G D K Z X T J C D I G K U H U A U E K C A R
```

Figure 2.11   Panel 2 of *Kryptos*; a (mostly) horizontal line has been added to separate cipher 3 from cipher 4. (Adapted from https://www.cia.gov/about-cia/headquarters-tour/kryptos/KryptosPrint.pdf.)

The right side of *Kryptos* (panels 3 and 4) provides a clue as to the means of encipherment used on the left side. These panels are reproduced in Figures 2.12 and 2.13, as shown on the CIA's website.

```
A B C D E F G H I J K L M N O P Q R S T U V W X Y Z A B C D
A K R Y P T O S A B C D E F G H I J L M N Q U V W X Z K R Y P
B R Y P T O S A B C D E F G H I J L M N Q U V W X Z K R Y P T
C Y P T O S A B C D E F G H I J L M N Q U V W X Z K R Y P T O
D P T O S A B C D E F G H I J L M N Q U V W X Z K R Y P T O S
E T O S A B C D E F G H I J L M N Q U V W X Z K R Y P T O S A
F O S A B C D E F G H I J L M N Q U V W X Z K R Y P T O S A B
G S A B C D E F G H I J L M N Q U V W X Z K R Y P T O S A B C
H A B C D E F G H I J L M N Q U V W X Z K R Y P T O S A B C D
I B C D E F G H I J L M N Q U V W X Z K R Y P T O S A B C D E
J C D E F G H I J L M N Q U V W X Z K R Y P T O S A B C D E F
K D E F G H I J L M N Q U V W X Z K R Y P T O S A B C D E F G
L E F G H I J L M N Q U V W X Z K R Y P T O S A B C D E F G H
M F G H I J L M N Q U V W X Z K R Y P T O S A B C D E F G H I
```

Figure 2.12 Panel 3 of *Kryptos* provides a clue as to the means of encipherment. (Adapted from https://www.cia.gov/about-cia/headquarters-tour/kryptos/KryptosPrint.pdf.)

```
N G H I J L M N Q U V W X Z K R Y P T O S A B C D E F G H I J
O H I J L M N Q U V W X Z K R Y P T O S A B C D E F G H I J L
P I J L M N Q U V W X Z K R Y P T O S A B C D E F G H I J L M
Q J L M N Q U V W X Z K R Y P T O S A B C D E F G H I J L M N
R L M N Q U V W X Z K R Y P T O S A B C D E F G H I J L M N Q
S M N Q U V W X Z K R Y P T O S A B C D E F G H I J L M N Q U
T N Q U V W X Z K R Y P T O S A B C D E F G H I J L M N Q U V
U Q U V W X Z K R Y P T O S A B C D E F G H I J L M N Q U V W
V U V W X Z K R Y P T O S A B C D E F G H I J L M N Q U V W X
W V W X Z K R Y P T O S A B C D E F G H I J L M N Q U V W X Z
X W X Z K R Y P T O S A B C D E F G H I J L M N Q U V W X Z K
Y X Z K R Y P T O S A B C D E F G H I J L M N Q U V W X Z K R
Z Z K R Y P T O S A B C D E F G H I J L M N Q U V W X Z K R Y
A B C D E F G H I J K L M N O P Q R S T U V W X Y Z A B C D
```

Figure 2.13 Panel 4 of *Kryptos*, in which the clue continues. (Adapted from https://www.cia. gov/about-cia/headquarters-tour/kryptos/KryptosPrint.pdf.). *Note:* The CIA made a mistake in transcribing this panel! Read chapter 9 in my book listed in the References and Further Reading section at the end of this chapter for the details and much more.

Panels 3 and 4 clearly indicate a Vigenère cipher with mixed alphabets.

An encipherment of MATHEMATICS IS THE QUEEN OF THE SCIENCES using the key GAUSS shows how it works. First the plaintext alphabet is written out in a mixed order, beginning with KRYPTOS and then continuing alphabetically with the letters not already listed.

```
K R Y P T O S A B C D E F G H I J L M N Q U V W X Z   plaintext
G H I J L M N Q U V W X Z K R Y P T O S A B C D E F   alphabet 1
A B C D E F G H I J L M N Q U V W X Z K R Y P T O S   alphabet 2
U V W X Z K R Y P T O S A B C D E F G H I J L M N Q   alphabet 3
S A B C D E F G H I J L M N Q U V W X Z K R Y P T O   alphabet 4
S A B C D E F G H I J L M N Q U V W X Z K R Y P T O   alphabet 5
```

Below this, the key GAUSS is written vertically down the left hand side to provide the first letters of our five cipher alphabets. Each of these cipher alphabets is continued from its first letter in the same order as our initial mixed alphabet. Now to encipher, the five alphabets are used in order, as many times as necessary, until the message is at an end. We get:

```
MATHEMATICS IS THE QUEEN OF THE SCIENCES plaintext
GAUSSGAUSSG AU SSG AUSSG AU SSG AUSSGAUS key
OHZQLOHZUIN VR DQX RJLLS FA DQX GTULSJSF ciphertext
```

This sort of cipher is referred to as a *Quagmire III* by members of the American Cryptogram Association (ACA), such as James J. Gillogly. The entire plaintext of *Kryptos* has not yet been recovered, but Gillogly, using computer programs of his own design, deciphered the majority of it in 1999. Several factors served to make his work more difficult:

1. There's no obvious indication that the left side contained four ciphers instead of just one.
2. The mixing of the alphabets was done with a keyword, but not KRYPTOS, as in the clue on the right side.
3. Sanborn intentionally introduced some errors in the ciphers.
4. Only the first two ciphers are Quagmire IIIs. Ciphers 3 and 4 arose from other systems.

Determining the correct keys and recovering the first two messages is left as a challenge to the reader. The solutions can easily be found online. Can you use the techniques detailed in this chapter to meet this challenge? The second part should be easier than the first, since there is more ciphertext to work with.

The third cipher made use of transposition (see Chapter 3) and was also solved by Gillogly. With only a little more than three lines of ciphertext left, Gillogly got stuck. He was unable to break the fourth and final cipher.

After Gillogly's success, the CIA revealed that one of its employees, David Stein, had already deciphered the portions Gillogly recovered back in 1998. Stein's results appeared in a classified publication, which James had no opportunity to see. Not to be bested by CIA, the National Security Agency (NSA) revealed that some of their employees had solved it in 1992, but initially NSA wouldn't provide their names! In 2005, the information was released that it was actually a team at NSA (Ken Miller, Dennis McDaniels, and two others whose identities are still not publicly known).[17] Despite intense attention, the last portion of *Kryptos* has resisted decipherment, at least as far as the general public knows! "People call me an agent of Satan," says artist Sanborn, "because I won't tell my secret."[18]

Sanborn (Figure 2.14) has, at least, revealed valuable clues. On November 20, 2010, he indicated that the letters starting at position 64 of the undeciphered portion, namely NYPVTT, decipher to BERLIN.[19] Surprisingly, this didn't help! No solution came forth! Exactly four years later, on November 20, 2014, Sanborn released another clue. It represented an expansion of his previous clue. He said that NYPVTTMZFPK deciphered to BERLIN CLOCK. Still no solution appeared,

---

[17] http://en.wikipedia.org/wiki/Kryptos.

[18] Levy, Steven, "Mission Impossible: The Code Even the CIA Can't Crack," *Wired*, Vol. 17, No. 5, April 20, 2009.

[19] Scryer, "Kryptos Clue," *The Cryptogram*, Vol. 77, No. 1, January–February 2011, p. 11. Scryer is the American Cryptogram Association (ACA) pen-name used by James J. Gillogly.

Figure 2.14 James Sanborn (1945–) (http://www.pbs.org/wgbh/nova/sciencenow/3411/images/ate-bio-03.jpg.)

and now we have another mini-mystery—why were both clues released on November 20? Does this date have some special significance for Sanborn or *Kryptos*? In 2020, Sanborn provided a third clue: the letters in positions 26 through 34 decipher to NORTHEAST. This clue broke the previous pattern of dates, being given on January 29.[20]

Although matrix encryption is not presented until Chapter 6, I should point out now that some researchers believe that a form of matrix encryption was used to create the still unsolved portion of *Kryptos*. Greg Link, Dante Molle, and I investigated this possibility, but were unable to offer definitive proof one way or the other.[21] There are many ways in which matrices can be used to encipher text and we were only able to use Sanborn's clues to rule out some of the simpler methods.

2.5 Autokeys

Now that we've examined the cryptography, cryptanalysis, and historical uses of the basic Vigenère cipher, let's examine the real contribution that Blaise de Vigenère made to this system, which,

[20] Schwartz, John, and Jonathan Corum, "This Sculpture Holds a Decades-Old Mystery. And Now, Another Clue," *The New York Times*, January 29, 2020, available online at https://www.nytimes.com/interactive/2020/01/29/climate/kryptos-sculpture-final-clue.html.

[21] Our investigation appeared as Bauer, Craig, Gregory Link, and Dante Molle, "James Sanborn's *Kryptos* and the Matrix Encryption Conjecture," *Cryptologia*, Vol. 40, No. 6, 2016, pp. 541–552. The results were later summarized in a broader survey of *Kryptos* as part of chapter 9 (pp. 386–407) of Bauer, Craig P., *Unsolved! The History and Mystery of the World's Greatest Ciphers from Ancient Egypt to Online Secret Societies*, Princeton University Press, Princeton, New Jersey, 2017.

recall, already existed. Vigenère's autokey only used the given key (COMET in the examples below) once. After that single application, the original message (or the ciphertext generated) is used as the key for the rest of the message.

```
S E N D S U P P L I E S …    message
C O M E T S E N D S U P …    key
U S Z H L M T C O A Y H …    ciphertext
```

The ciphertext used as "key":

```
S E N D S U P P L I E S …    message
C O M E T U S Z H L O H …    key
U S Z H L O H O S T S Z …    ciphertext
```

This particular example was presented by Claude Shannon in his classic paper "Communication Theory of Secrecy Systems."[22]

Since the encipherment of each letter depends on previous message or cipher letters, we have a sort of chaining in use here. This idea was applied again when matrix encryption was discovered and is still in use in modern block ciphers. It's examined in greater detail in this book in Section 14.6, which covers various modes of encryption. Using the ciphertext as the key, although sounding like a complication, yields an easily broken cipher!

One risk associated with various autokey methods is that an error in a single position can propagate through the rest of the ciphertext. Observe what happens in our earlier example if the third character is mistakenly enciphered as N instead of Z. An error-free encipherment is reproduced below for comparison.

```
S E N D S U P P L I E S …    message
C O M E T U S N H L O H …    key
U S N H L O H C S T S Z …    ciphertext obtained
U S Z H L O H O S T S Z …    ciphertext desired
```

Continuing on we would find that every fifth ciphertext character after our initial error is also incorrect.

The Vigenère cipher was introduced in this chapter, then broken, and we're about to go on to patch it, creating a stronger system. However, it should be noted that the Vigenère cipher did not get disposed of so quickly in the real world. It had an extremely successful run. We may never again see a system that survives for hundreds of years before successful attacks are discovered.

2.6 The Running Key Cipher and Its Cryptanalysis

Because the major weakness in the Vigenère cipher is the fact that the key repeats at regular intervals, allowing the attacks described above, a natural improvement is to encipher in a similar manner, but with a key that does not repeat. One way to do this is by using the text of a book as the key. This approach is often called a running key (Vigenère) cipher. An example is provided below.

```
BEGIN THE ATTACK AT DAWN    plaintext
ITWAS THE BESTOF TI MESI    key
JXCIF MOI BXKTQP TB PEOV    ciphertext
```

---

[22] Shannon, Claude, "Communication Theory of Secrecy Systems," *The Bell System Technical Journal*, Vol. 28, No. 4, October 1949, pp. 656–715. Shannon noted, "The material in this paper appeared in a confidential report "A Mathematical Theory of Cryptography" dated Sept. 1, 1945, which has now been declassified."

Here the message is enciphered using *A Tale of Two Cities* by Charles Dickens as the key.

The Kasiski attack won't work against this upgrade to the Vigenère cipher, because the key never repeats and Friedman's index of coincidence will only indicate that a large number of cipher alphabets was used. However, that doesn't mean that Friedman was defeated by the running key cipher. In a cover letter introducing his paper "Methods for the Solution of Running-Key Ciphers," he wrote:

> Concerning the possibility of the decipherment of a message or a series of messages enciphered by a running-key, it was said until as recently as three months ago, "It can't be done" or "It is very questionable." It is probably known to you that the U.S. Army Disk in connection with a running-key has been used as a cipher in field service for many years, and is, to the best of our knowledge, in use to-day. I suppose that its long-continued use, and the confidence placed in its safety as a field cipher has been due very probably to the fact that no one has ever taken the trouble to see whether "It could be done." It is altogether probable that the enemy, who has been preparing for war for a long time, has not neglected to look into our field ciphers, and we are inclined to credit him with a knowledge equal to or superior to our own. We have been able to prove that not only is a single short message enciphered by the U. S. Army Disk, or any similar device, easily and quickly deciphered, but that a series of messages sent out in the same key may be deciphered more rapidly than they have been enciphered![23]

Friedman's new attack was based on some very simple mathematics and is now examined.

A list of the probabilities of letters in English is provided once more in Table 2.3 for handy reference.

Table 2.3   Probabilities of Letters in English

Letter	Probability	Letter	Probability
A = 0	0.08167	N = 13	0.06749
B = 1	0.01492	O = 14	0.07507
C = 2	0.02782	P = 15	0.01929
D = 3	0.04253	Q = 16	0.00095
E = 4	0.12702	R = 17	0.05987
F = 5	0.02228	S = 18	0.06327
G = 6	0.02015	T = 19	0.09056
H = 7	0.06094	U = 20	0.02758
I = 8	0.06966	V = 21	0.00978
J = 9	0.00153	W = 22	0.02360
K = 10	0.00772	X = 23	0.00150
L = 11	0.04025	Y = 24	0.01974
M = 12	0.02406	Z = 25	0.00074

Source: Beutelspacher, Albrecht, *Cryptology,* Mathematical Association of America, Washington DC, 1994, p. 10.

---

[23] Friedman, William F., *Methods for the Solution of Running-Key Ciphers*, Publication No. 16, Riverbank Laboratories, Geneva, Illinois, 1918.

Now suppose we see the letter A in a ciphertext arising from some running key. It could have arisen from an A in the message combining with another A from the key or it could have arisen from a B in the message combining with a Z from the key. Which seems more likely to you? The letter A is much more common than B or Z, so the first possibility is more likely. There are other possible combinations that would yield an A in the ciphertext. Table 2.4 lists all plaintext/key combinations along with their probabilities (obtained using Table 2.3).

Note that we must double the probabilities for distinct key and message letters. For example, the pair B and Z, which combine to give A can do so with B in the message and Z in the key *or* with Z in the message and B in the key. Thus, there are two equally probable ways this pair of letters can yield A. However, there is only one way that the letters A and A can combine to yield A. Similarly N and N can only combine in one way to give A. So, we do not double the probability if two of the same letter combine to give the ciphertext letter.

The rankings in Table 2.4 show that a ciphertext A is most likely to arise from combining an H and a T. However, the other pairings will sometimes be correct. Considering the top five possibilities will yield the correct pairings often enough that the remaining solutions can be found in a manner similar to fixing typos. The tables of ranked pairings for each letter of the alphabet are given in Table 2.5:[24]

Table 2.4   Plaintext/Key Combinations and Probabilities

Combination	Probability		Ranking
AA	0.0066699889	= 0.0066699889	3
BZ	0.0000110408 × 2	= 0.0000220816	13
CY	0.0005491668 × 2	= 0.0010983336	9
DX	0.0000637950 × 2	= 0.0001275900	12
EW	0.0029976720 × 2	= 0.0059953440	4
FV	0.0002178984 × 2	= 0.0004357968	10
GU	0.0005557370 × 2	= 0.0011114740	8
HT	0.0055187264 × 2	= 0.0110374528	1
IS	0.0044073882 × 2	= 0.0088147764	2
JR	0.0000916011 × 2	= 0.0001832022	11
KQ	0.0000073340 × 2	= 0.0000146680	14
LP	0.0007764225 × 2	= 0.0015528450	7
MO	0.0018061842 × 2	= 0.0036123684	6
NN	0.0045549001	= 0.0045549001	5

---

[24] Thanks to Adam Reifsneider for writing a computer program to calculate these rankings.

Table 2.5 Ranked Pairings for Each Letter of the Alphabet

A	B	C	D
HT 0.0110374528	IT 0.012616819	OO 0.0056355049	AD 0.0069468502
IS 0.0088147764	NO 0.0101329486	EY 0.0050147496	LS 0.0050932350
AA 0.0066699889	HU 0.0033614504	LR 0.0048195350	OP 0.0028962006
EW 0.0059953440	AB 0.0024370328	AC 0.0045441188	MR 0.0028809444
NN 0.0045549001	DY 0.0016790844	IU 0.0038424456	HW 0.0028763680
MO 0.0036123684	FW 0.0010516160	NP 0.0026037642	KT 0.0013982464
LP 0.0015528450	MP 0.0009282348	HV 0.0011919864	IV 0.0013625496
GU 0.0011114740	KR 0.0009243920	KC 0.0009768888	FY 0.0008796144
CY 0.0010983336	GV 0.0003941340	GW 0.0009510800	BC 0.0008301488
FV 0.0004357968	EX 0.0003810600	JT 0.0002771136	EZ 0.0001879896
JR 0.0001832022	JS 0.0001936062	BB 0.0002226064	NQ 0.0001282310
DX 0.0001275900	LQ 0.0000764750	FX 0.0000668400	JU 0.0000843948
BZ 0.0000220816	CZ 0.0000411736	DZ 0.0000629444	GX 0.0000604500
KQ 0.0000146680		MQ 0.0000457140	

E	F	G	H
AE 0.0207474468	OR 0.0089888818	NT 0.0122237888	OT 0.0135966784
NR 0.0080812526	NS 0.0085401846	OS 0.0094993578	DE 0.0108043212
LT 0.0072900800	MT 0.0043577472	CE 0.0070673928	AH 0.0099539396
IW 0.0032879520	BE 0.0037902768	AG 0.0032913010	NU 0.0037227484
MS 0.0030445524	AF 0.0036392152	IY 0.0027501768	PS 0.0024409566
BD 0.0012690952	HY 0.0024059112	PR 0.0023097846	LW 0.0018998000
GY 0.0007955220	CD 0.0023663692	DD 0.0018088009	CF 0.0012396592
CC 0.0007739524	LU 0.0022201900	MU 0.0013271496	BG 0.0006012760
KU 0.0004258352	IX 0.0002089800	LV 0.0007872900	MV 0.0004706136
PP 0.0003721041	KV 0.0001510032	BF 0.0006648352	QR 0.0001137530
HX 0.0001828200	JW 0.0000722160	KW 0.0003643840	IZ 0.0001030968
OQ 0.0001426330	PQ 0.0000366510	HZ 0.0000901912	JY 0.0000604044
FZ 0.0000329744	GZ 0.0000298220	JX 0.0000045900	KX 0.0000231600
JV 0.0000299268		QQ 0.0000009025	

(Continued)

Table 2.5 (Continued)   Ranked Pairings for Each Letter of the Alphabet

I	J	K	L
EE 0.0161340804	RS 0.0075759498	RT 0.0108436544	EH 0.0154811976
AI 0.0113782644	EF 0.0056600112	DH 0.0051835564	ST 0.0114594624
OU 0.0041408612	CH 0.0033907016	EG 0.0051189060	AL 0.0065744350
RR 0.0035844169	NW 0.0031855280	SS 0.0040030929	DI 0.0059252796
PT 0.0034938048	BI 0.0020786544	CI 0.0038758824	RU 0.0033024292
DF 0.0018951368	DG 0.0017139590	OW 0.0035433040	NY 0.0026645052
BH 0.0018184496	LY 0.0015890700	AK 0.0012609848	PW 0.0009104880
NV 0.0013201044	OV 0.0014683692	MY 0.0009498888	FG 0.0008978840
MW 0.0011356320	PU 0.0010640364	FF 0.0004963984	BK 0.0002303648
CG 0.0011211460	AJ 0.0002499102	PV 0.0003773124	OX 0.0002252100
KY 0.0003047856	QT 0.0001720640	NX 0.0002024700	CJ 0.0000851292
LX 0.0001207500	MX 0.0000721800	LZ 0.0000595700	MZ 0.0000356088
QS 0.0001202130	KZ 0.0000114256	QU 0.0000524020	QV 0.0000185820
JZ 0.0000022644		BJ 0.0000456552	

M	N	O	P
EI 0.0176964264	AN 0.0110238166	AO 0.0122619338	EL 0.0102251100
TT 0.0082011136	TU 0.0049952896	HH 0.0037136836	HI 0.0084901608
AM 0.0039299604	FI 0.0031040496	DL 0.0034236650	TW 0.0042744320
SU 0.0034899732	RW 0.0028258640	SW 0.0029863440	CN 0.0037551436
OY 0.0029637636	GH 0.0024558820	GI 0.0028072980	AP 0.0031508286
FH 0.0027154864	CL 0.0022395100	BN 0.0020139016	RY 0.0023636676
BL 0.0012010600	SV 0.0012375612	EK 0.0019611888	BO 0.0022400888
RV 0.0011710572	PY 0.0007615692	TV 0.0017713536	DM 0.0020465436
CK 0.0004295408	BM 0.0007179504	CM 0.0013386984	UV 0.0005394648
GG 0.0004060225	DK 0.0006566632	UU 0.0007606564	FK 0.0003440032
DJ 0.0001301418	EJ 0.0003886812	RX 0.0001796100	SX 0.0001898100
NZ 0.0000998852	OZ 0.0001111036	FJ 0.0000681768	GJ 0.0000616590
PX 0.0000578700	QX 0.0000028500	QY 0.0000375060	QZ 0.0000014060
QW 0.0000448400		PZ 0.0000285492	

(Continued)

Table 2.5 (Continued)   Ranked Pairings for Each Letter of the Alphabet

Q	R	S	T
EM 0.0061122024	EN 0.0171451596	EO 0.0190707828	AT 0.0147920704
DN 0.0057406994	AR 0.0097791658	AS 0.0103345218	IL 0.0056076300
II 0.0048525156	DO 0.0063854542	HL 0.0049056700	EP 0.0049004316
CO 0.0041768948	TY 0.0035753088	FN 0.0030073544	FO 0.0033451192
SY 0.0024978996	GL 0.0016220750	BR 0.0017865208	CR 0.0033311668
FL 0.0017935400	CP 0.0010732956	DP 0.0016408074	HM 0.0029324328
UW 0.0013017760	FM 0.0010721136	UY 0.0010888584	GN 0.0027198470
BP 0.0005756136	HK 0.0009409136	IK 0.0010755504	BS 0.0018879768
GK 0.0003111160	VW 0.0004616160	GM 0.0009696180	VY 0.0003861144
TX 0.0002716800	IJ 0.0002131596	WW 0.0005569600	DQ 0.0000808070
HJ 0.0001864764	SZ 0.0000936396	TZ 0.0001340288	WX 0.0000708000
AQ 0.0001551730	UX 0.0000827400	CQ 0.0000528580	UZ 0.0000408184
VV 0.0000956484	BQ 0.0000283480	VX 0.0000293400	JK 0.0000236232
RZ 0.0000886076		JJ 0.0000023409	

U	V	W	X
HN 0.0082256812	ER 0.0152093748	ES 0.0160731108	ET 0.0230058624
DR 0.0050925422	IN 0.0094027068	IO 0.0104587524	FS 0.0028193112
AU 0.0045049172	HO 0.0091495316	DT 0.0077030336	IP 0.0026874828
CS 0.0035203428	DS 0.0053817462	AW 0.0038548240	GR 0.0024127610
IM 0.0033520392	CT 0.0050387584	FR 0.0026678072	DU 0.0023459548
GO 0.0030253210	AV 0.0015974652	HP 0.0023510652	LM 0.0019368300
BT 0.0027023104	BU 0.0008229872	LL 0.0016200625	KN 0.0010420456
WY 0.0009317280	GP 0.0007773870	CU 0.0015345512	BW 0.0007042240
FP 0.0008595624	KL 0.0006214600	YY 0.0003896676	CV 0.0005441592
EQ 0.0002413380	JM 0.0000736236	KM 0.0003714864	AX 0.0002450100
JL 0.0001231650	XY 0.0000592200	BV 0.0002918352	JO 0.0002297142
KK 0.0000595984	FQ 0.0000423320	JN 0.0002065194	HQ 0.0001157860
VZ 0.0000144744	WZ 0.0000349280	GQ 0.0000382850	YZ 0.0000292152
XX 0.0000022500		XZ 0.0000022200	

(Continued)

Table 2.5 (Continued)   Ranked Pairings for Each Letter of the Alphabet

Y		Z	
HR	0.0072969556	IR	0.0083410884
EU	0.0070064232	HS	0.0077113476
LN	0.0054329450	LO	0.0060431350
FT	0.0040353536	GT	0.0036495680
AY	0.0032243316	MN	0.0032476188
GS	0.0025497810	EV	0.0024845112
CW	0.0013131040	DW	0.0020074160
KO	0.0011590808	FU	0.0012289648
DV	0.0008318868	BY	0.0005890416
MM	0.0005788836	KP	0.0002978376
IQ	0.0001323540	AZ	0.0001208716
JP	0.0000590274	CX	0.0000834600
BX	0.0000447600	JQ	0.0000029070
ZZ	0.0000005476		

We now take a look at a running key ciphertext to see how Friedman used rankings like those given above to read the original message and key—for example,

L A E K A H B W A G W I P T U K V S G B

The L that starts the ciphertext is likely to have arisen from E + H, S + T, A + L, D + I, or R + U, as these are the top five pairings that yield L. We write these letters under L in a long vertical column, and then write them out again with the ordering reversed in each pair. We do the same with the top five pairings for each of the other letters in the ciphertext. This gives us Table 2.6.

Table 2.6   Sample Running Key Ciphertext

```
L A E K A H B W A G W I P T U K V S G B
E H A R H O I E H N E E A H R E E N I
H T E T T T S T T S E L T N T R O T T
S I N D I D N I I O I A H I D D I A O N
T S R H S E O O S S O I I L R H N S S O
A A L E A A H D A C D O T E A E H H C H
L A T G A H U T A E T U W P U G O L E U
D E I S E N A A E A A R C F C S D F A A
I W W S W U B W W G W R N O S S S N G B
R N M C N P D F N I F P A C I C C B I D
U N S I N S Y R N Y R T P R M I T R Y Y

H T E T T T S T T S E L T N T R O T T
E H A R H O I E H N E E A H R E E N I
T S R H S E O O S S O I I L R H N S S O
S I N D I D N I I O I A H I D D I A O N
L A T G A H U T A E T U W P U G O L E U
A A L E A A H D A C D O T E A E H H C H
I W W S W U B W W G W R N O S S S N G B
D E I S E N A A E A A R C F C S D F A A
U N S I N S Y R N Y R T P R M T T R Y Y
R N M C N P D F N I F P A C I C C B I D
```

Reversing the order of the letter pairs in the second block allows a nice correspondence. If the third letter under the L was actually used in the message or key to generate the L, then the third letter down in the bottom block of text is what it was paired with. You'll soon see how this helps with the cryptanalysis. Letters that pair with themselves to give the desired ciphertext letter will appear twice in our table. This is redundant, but aesthetically pleasing; it keeps all of the columns the same length.

We now focus on the first block of text, the ten rows of letters directly beneath the ciphertext. We try to select a single letter from each column, such that a meaningful message is formed by them, when read across. There may be many, but we go slowly and can easily tell if we are likely to be on the right track. We do this by considering letters in the lower block of text that occupy the same positions as the letters in the message we are attempting to form. If the letters in the bottom block are also forming words, we gain confidence in our solution. This will become clearer as our example continues.

THE is the most common word in English, so we may as well try to start there. We select letters in the top block of text that spell TIIE and see what we get from those positions in the bottom block of text (see Table 2.7).

Table 2.7   Finding THE in the First Block and STA in the Second Block

```
L A E K A H B W A G W I P T U K V S G B
E H A R H O I E H N E E E A H R E E N I
H T E T T T T S T T S E L T N T R O T T
S I N D I D N I I O I A H I D D I A O N
T S R H S E O O S S O I I L R H N S S O
A A L E A A H D A C D O T E A E H H C H
L A T G A H U T A E T U W P U G O L E U
D E I S E N A A E A A R C F C S D F A A
I W W S W U B W W G W R N O S S S N G B
R N M C N P D F N I F P A C I C C B I D
U N S I N S Y R N Y R T P R M I T R Y Y

H T E T T T T S T T S E L T N T R O T T
E H A R H O I E H N E E E A H R E E N I
T S R H S E O O S S O I I L R H N S S O
S I N D I D N I I O I A H I D D I A O N
L A T G A H U T A E T U W P U G O L E U
A A L E A A H D A C D O T E A E H H C H
I W W S W U B W W G W R N O S S S N G B
D E I S E N A A E A A R C F C S D F A A
U N S I N S Y R N Y R T P R M T T R Y Y
R N M C N P D F N I F P A C I C C B I D
```

We get STA, which sounds promising. It could continue as STAY, STATION, STAB, STATISTICIAN, STALINGRAD, STALACTITE, STAPHYLOCOCCUS...—we have many possibilities! However, the top rows of each rectangle contain the letters most likely to be used in the continuations. It's best to look for possibilities there first. The Y in STAY (the first word that came to mind) doesn't even show up in the appropriate column of the bottom rectangle. It is a possibility that STAY is correct, but not the strongest possibility. Take a moment to examine the bottom rectangle for yourself before reading any further. What word do you think is formed?

As the words start to form, we cannot tell which text is the message and which is the key. Hopefully, when we're done we'll be able to distinguish the two.

Okay, did you find the word START? That seems like the best choice. Let's see what it gives us in the corresponding positions of the upper rectangle (Table 2.8).

Table 2.8 Finding THE TH in the First Block and START in the Second Block

```
L A E K A H B W A G W I P T U K V S G B
E H A R H O I E H N E E E A H R E E N I
H T E T T T S T T S E L T N T R O T T
S I N D I D N I I O I A H I D D I A O N
T S R H S E O O S S O I I L R H N S S O
A A L E A A H D A C D O T E A E H H C H
L A T G A H U T A E T U W P U G O L E U
D E I S E N A A E A A R C F C S D F A A
I W W S W U B W W G W R N O S S S N G B
R N M C N P D F N I F P A C I C C B I D
U N S I N S Y R N Y R T P R M I T R Y Y

H T E T T T S T T S E L T N T R O T T
E H A R H O I E H N E E E A H R E E N I
T S R H S E O O S S O I I L R H N S S O
S I N D I D N I I O I A H I D D I A O N
L A T G A H U T A E T U W P U G O L E U
A A L E A A H D A C D O T E A E H H C H
I W W S W U B W W G W R N O S S S N G B
D E I S E N A A E A A R C F C S D F A A
U N S I N S Y R N Y R T P R M T T R Y Y
R N M C N P D F N I F P A C I C C B I D
```

The top rectangle now reads THE TH. This looks okay. It could turn out to be THE THOUGHT IS WHAT COUNTS or THE THREE AMIGOS or THE THREAT OF DEFEAT LOOMS LARGE. To see what happens if we make a wrong turn, let's investigate the result if we guessed the bottom rectangle read STAGE (Table 2.9).

The top text would be THE EW..., which we'd have trouble continuing *or* possibly THEE W..., but THEE seems like an unlikely word, if the message wasn't sent by Shakespeare or Thor.

So continuing on with the texts we've recovered, THE TH and START, we may look at either rectangle, whichever seems easier to continue. I prefer the top rectangle. Examine it yourself and see if your selection matches the one I give in Table 2.10.

THE TH can be extended to THE THOUSAND and the bottom rectangle then yields START THE ATT... This must be START THE ATTACK. It seems that we're getting the message in the bottom block and the key in the top block this time. We try to complete the word ATTACK in the bottom rectangle and check to make sure the top rectangle is still giving something meaningful. But we hit a snag—there's no K to be found where we need one! That's okay. The rectangles list the most likely pairings, but less likely pairings can occur. We simply tack on the K we need, along

Table 2.9 Finding STAGE in the Second Block

```
L A E K A H B W A G W I P T U K V S G B
E H A R H O I E H N E E E A H R E E N I
H T E T T T S T T S E L T N T R O T T
S I N D I D N I I O I A H I D D I A O N
T S R H S E O O S S O I I L R H N S S O
A A L E A A H D A C D O T E A E H H C H
L A T G A H U T A E T U W P U G O L E U
D E I S E N A A E A A R C F C S D F A A
I W W S W U B W W G W R N O S S S N G B
R N M C N P D F N I F P A C I C C B I D
U N S I N S Y R N Y R T P R M I T R Y Y

H T E T T T S T T S E L T N T R O T T
E H A R H O I D N E E E A H R E E N I
T S R H S E O O S S O I I L R H N S S O
S I N D I D N I I O I A H I D D I A O N
L A T G A H U T A E T U W P U G O L E U
A A L E A A H D A C D O T E A E H H C H
I W W S W U B W W G W R N O S S S N G B
D E I S E N A A E A A R C F C S D F A A
U N S I N S Y R N Y R T P R M T T R Y Y
R N M C N P D F N I F P A C I C C B I D
```

Table 2.10 Working from THE TH in the First Block and START in the Second Block

```
L A E K A H B W A G W I P T U K V S G B
E H A R H O I E H N E E E A H R E E N I
H T E T T T S T T S E L T N T R O T T
S I N D I D N I I O I A H I D D I A O N
T S R H S E O O S S O I I L R H N S S O
A A L E A A H D A C D O T E A E H H C H
L A T G A H U T A E T U W P U G O L E U
D E I S E N A A E A A R C F C S D F A A
I W W S W U B W W G W R N O S S S N G B
R N M C N P D F N I F P A C I C C B I D
U N S I N S Y R N Y R T P R M I T R Y Y

H T E T T T S T T S E L T N T R O T T
E H A R H O I E H N E E E A H R E E N I
T S R H S E O O S S O I I L R H N S S O
S I N D I D N I I O I A H I D D I A O N
L A T G A H U T A E T U W P U G O L E U
A A L E A A H D A C D O T E A E H H C H
I W W S W U B W W G W R N O S S S N G B
D E I S E N A A E A A R C F C S D F A A
U N S I N S Y R N Y R T P R M T T R Y Y
R N M C N P D F N I F P A C I C C B I D
```

with (in the top rectangle) the J it must combine with to yield the ciphertext letter T (Table 2.11). We can always add letters in this manner, but really shouldn't unless we are either fairly confident that they are correct or have no other reasonable options.

Table 2.11 Adding J and K as Needed

```
L A E K A H B W A G W I P T U K V S G B
E H A R H O I E H N E E E A H R E E N I
H T E T T T T S T T S E L T N T R O T T
S I N D I D N I I O I A H I D D I A O N
T S R H S E O O S S O I I L R H N S S O
A A L E A A H D A C D O T E A E H H C H
L A T G A H U T A E T U W P U G O L E U
D E I S E N A A E A A R C F C S D F A A
I W W S W U B W W G W R N O S S S N G B
R N M C N P D F N I F P A C I C C B I D
U N S I N S Y R N Y R T P R M I T R Y Y
                                J

H T E T T T T S T T S E L T N T R O T T
E H A R H O I E H N E E E A H R E E N I
T S R H S E O O S S O I I L R H N S S O
S I N D I D N I I O I A H I D D I A O N
L A T G A H U T A E T U W P U G O L E U
A A L E A A H D A C D O T E A E H H C H
I W W S W U B W W G W R N O S S S N G B
D E I S E N A A E A A R C F C S D F A A
U N S I N S Y R N Y R T P R M T T R Y Y
R N M C N P D F N I F P A C I C C B I D
                                K
```

Our two texts now read START THE ATTACK and THE THOUSAND INJ... Once more, please take a moment to look for the continuation of INJ in the top rectangle (to see that it's not so hard to do) before reading on.

Of course, one could proceed in a different direction by trying to extend START THE ATTACK. In general though, it's easier to complete partial words than to find new ones, unless the preceding words are the beginning of a well-known phrase. Speaking of which, there will likely be a reader who has already recognized the source of the key being used here. More in a moment.

Okay, did you extend the top text to THE THOUSAND INJURIES? This makes the bottom text read START THE ATTACK AT NOO. Looking at the bottom text one last time, we extend it by another letter and then do the same in the top rectangle.

We now have the message and the key:

```
THE THOUSAND INJURIES O          key
START THE ATTACK AT NOON         message
```

The key was, by the way, the beginning of Edgar Allan Poe's short story "The Cask of Amontillado."

The example chosen above to illustrate this attack was a little easier than usual. Nine times out of twenty (45%), the pair of letters most likely to yield a particular ciphertext letter did yield that letter. Experiments reveal that this happens less than a third of the time on average. Also, the example only had one ciphertext letter (5% of the cipher) that didn't arise from one of the five most likely pairings. This was a lower percent than usual.

Nevertheless, this is a great attack. Friedman made the attack even quicker by cutting out the columns. This allowed him to move individual columns up and down. When he got a word or phrase going across letters from the top rectangle, he could glance down at the bottom rectangle and also read the corresponding letters there straight across. I used bold text and underlining, because it is easier to present it this way in a book, but the sliding paper works better for classroom demonstrations.

Although Friedman didn't pursue it further, his attack can be expanded. Instead of taking the ciphertext characters one at a time, groups may be considered. For example, suppose the ciphertext begins MOI. Our tables suggest

> M is most likely E + I.
> O is most likely A + O.
> I is most likely E + E.

But these tables don't take context into consideration. The top pairing suggested for O doesn't consider what letters appear before or after it in the ciphertext. Taking the letters MOI as group and using trigraph frequencies to rank the possibilities, we see it is most likely to arise from THE + THE.

Christian N.S. Tate, an undergraduate at the time, and I investigated this new attack. We split the ciphertext into groups of characters and used frequencies for letter combinations of the group size to rank our pairings. For example, if the ciphertext was HYDSPLTGQ, and we were using groups of three characters to launch our attack, we'd split the ciphertext up as HYD SPL TGQ and replace each trigram with the pair most likely to yield it.

The idea needed to be tested and the easiest way to do this was to write computer programs to carry out the calculations and analyze the results. We tested single character analysis (as Friedman had) digraphs, trigraphs, tetragraphs, pentagraphs, and hexagraphs.

I expected that after the results were in, I could compare the new attack to Friedman's and say, "It is a far far better thing I have done." However, this was not the case. The results were only marginally better! We tried to place the results in the best light possible for the write-up. Figuring even the slightest improvement on Friedman's work ought to be worth publishing, we submitted the paper. A kind editor accepted it following some revising.

Fortunately, Alexander Griffing, a student in the Bioinformatics Ph.D. program at North Carolina State University, saw the paper and was able to turn the attack into something truly successful. He did this by not just taking the blocks of ciphertext characters one after the other, but by also considering their overlap when computing the most likely solution.[25] So, when considering the ciphertext HYDSPLTGQ, Griffing's method didn't just look at how HYD, SPL, and TGQ could arise, but rather HYD, YDS, DSP, SPL, PLT, LTG, and TGQ. A graph reproduced from his paper is provided in Figure 2.15. It shows how his method is better than the method Tate and I proposed for every size letter group considered, and dramatically better when pentagraphs and hexagraphs are used.

---

[25] Griffing, Alexander, "Solving the Running Key Cipher with the Viterbi Algorithm," *Cryptologia*, Vol. 30, No. 4, October 2006, pp. 361–367.

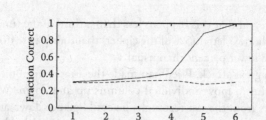

Figure 2.15 Griffing's results (solid line) compared to an earlier attempt by others (dotted line).

With the techniques discussed in this chapter, running key ciphers of any length can be broken; however, extremely short messages are not likely to have unique solutions. The graph in Figure 2.16 shows how the number of potential solutions changes as a function of the message's length. Beyond eight characters, we expect only a single solution.

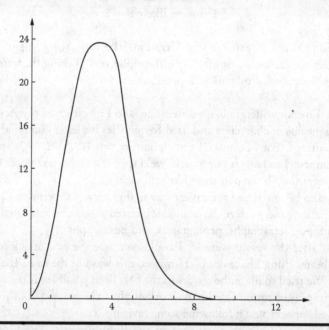

Figure 2.16 Number of spurious decipherments as a function of the message size. (From Deavours, Cipher A., "Unicity Points in Cryptanalysis," *Cryptologia*, Vol. 1, No. 1, January 1977, pp. 46-68, p. 62 cited here.)

2.7 The One-Time Pad or Vernam Cipher

The Vigenère cipher was seen to be easily broken due to a weakness caused by the keyword repeating. Modifying the system so that the key never repeats gives us the running key cipher. Although the new system can't be defeated by the attacks that succeeded against the Vigenère cipher, there are other attacks that work, as we saw. The weakness this time was that both the key and the message are made up of meaningful words. Thus, the next step for those who seek to make ciphers secure seems obvious—use a running key that consists of letters chosen randomly. Such a system resists the attacks already described, and all other attacks. Used properly, it is unbreakable! This is, in fact, the only cipher that is theoretically unbreakable.[26] Edgar Allan Poe didn't know about it when he wrote that "human ingenuity cannot concoct a cipher which human ingenuity cannot resolve." We can forgive him, because this method had not yet been discovered. Despite the way in which the unbreakable system seems to naturally evolve from "patching" the running key cipher, historians long believed[27] that it didn't arise until the years 1917 to 1918, when it was developed by Gilbert Vernam (Figure 2.17) and Major Joseph Mauborgne (Figure 2.18) at AT&T. The form it took then was different from how it is introduced here, but functionally equivalent. We'll return shortly to how

Figure 2.17   **Gilbert Vernam (1890–1960). (http://en.wikipedia.org/wiki/File:Gilbert_Vernam.jpg).**

Figure 2.18   **Joseph Mauborgne (1881–1971) (From *The Signal Corp Bulletin,* October-December, 1937.)**

---

[26] Shannon, Claude, "Communication Theory of Secrecy Systems," *The Bell System Technical Journal*, Vol. 28, No. 4, October 1949, pp. 656–715. Shannon noted, "The material in this paper appeared in a confidential report "A Mathematical Theory of Cryptography" dated Sept. 1, 1945, which has now been declassified."

[27] If you know about Frank Miller, please be patient. I discuss his work at the end of this chapter.

Vernam and Mauborgne described their system. Sometimes it's referred to as a Vernam cipher, but the name *one-time pad* is a bit better because it emphasizes the proper use of the key—only once! If a random key is used for more than one message, it is no longer unbreakable, as we shall see.

The one-time pad could easily have been discovered hundreds of years earlier, as it works in much the same way as a Vigenère cipher or a running key cipher, except the key is random and must be as long as the message. As an example, suppose the one-time pad begins U SNHQ LCIYU and Bob wants to send the message: I LOVE ALICE. Using the pad as the key and adding letters (using their numerical equivalents) mod 26, we have

I LOVE ALICE	plaintext
U SNHQ LCIYU	key
C DBCU LNQAY	ciphertext

If Eve intercepts the message and correctly guesses the key, she recovers the message. However, she has no reason to guess this particular key. If she instead guesses U WBJQ LCIYU, then the message deciphers to I HATE ALICE. Or suppose she tries the key U SNHQ TTYAL. In this case, the message becomes I LOVE SUSAN. Any message ten characters in length will arise from some key. As Eve has no reason to favor one key over another, she receives no information beyond the length of the message. In fact, the length of the ciphertext only provides an upper bound on the length of the message. Padding may have been added to make a very brief message appear longer.

Following the development of an unbreakable cipher, we might well expect that it would quickly be adopted by everyone and all other methods of encryption would vanish. This was not the case! In fact, it wasn't until the early 1920s that this American discovery saw heavy use and it was by the Germans![28] They used it as an extra step for their diplomatic codes. That is, after the message was converted to numerical code groups, the one-time pad, in the form of a list of random digits between 0 and 9, was used to shift each of the ciphertext values. Whenever such an extra step is taken, whether the key is random or not, we refer to the system as an enciphered code. One-time pads were also used by the Office of Strategic Services (OSS),[29] an American group in World War II that evolved into both the CIA and the Green Berets, and heavily by the Soviet Union for diplomatic messages beginning in 1930.[30]

One famous Russian spy caught by the Federal Bureau of Investigation (FBI) in New York City with a one-time pad in 1957 was Rudolf Abel.[31] Five years after his arrest he was traded for Francis Gary Powers, who was shot down while piloting a U-2 spy plane over the Soviet Union. The trade took place on the Glienick Bridge connecting East and West Berlin.

Figure 2.19 is a page from a one-time pad used by a Communist spy in Japan in 1961.[32] One side was probably for enciphering and the other for deciphering.

Although it had been talked about for some time, it was only in 1963, after the Cuban missile crisis, that the "hot line" was set up between Washington, DC, and Moscow (Figure 2.20).

---

[28] This is according to Kahn, David, *The Codebreakers*, second edition, Scribner, New York, 1996, p. 402. However, Alexander "Alastair" Guthrie Denniston recalled German use beginning in 1919 in a memoir (penned in 1944) titled "The Government Code and Cypher School between the Wars," which appeared posthumously in *Intelligence and National Security*, Vol. 1, No. 1, January 1986, pp. 48–70 (p. 54 cited here). Kahn's estimate of 1921–1923 was based on interviews conducted with German cryptographers in 1962.

[29] But not exclusively, as it was just one of many systems they employed. See Kahn, David, *The Codebreakers*, second edition, Scribner, New York, 1996, p. 540 for more.

[30] Kahn, David, *The Codebreakers*, second edition, Scribner, New York, 1996, p. 650.

[31] Kahn, David, *The Codebreakers*, second edition, Scribner, New York, 1996, p. 664.

[32] Kahn, David, *The Codebreakers*, second edition, Scribner, New York, 1996, p. 665.

39892	09897	07361	35736	38309		69801	56628	37254	61467	52308
33571	01448	63458	24848	30238		08098	14542	31851	07595	77970
27135	40220	47079	71707	80633		01536	97896	88209	71480	42063
49941	56035	48846	15111	59324		57188	83556	96509	08657	46851
10051	21816	63253	86240	99495		75643	56639	05326	97662	54705
40048	55040	17710	60896	94366		58493	69423	44744	07023	50651
11512	18996	91403	40539	50135		43896	70213	66610	66808	03001
74168	69956	53870	02897	18192		06724	13542	87558	11061	71468
20349	15133	12850	56853	47799		16904	59833	10280	50670	51183
20883	94649	78587	63065	94545		92600	10425	35061	98370	35554
51802	14552	07608	38392	22224		99718	57838	08540	62986	40799
20348	29842	76282	49048	51771		95196	30638	03983	76992	72652
98905	46438	78295	72769	07178		77170	45854	58100	40649	42651
53669	53304	18152	17691	54117		35868	60370	62207	91750	93298
08658	97627	93221	37250	66427		66368	08297	37727	99832	89892
52053	66220	87679	61332	81960		83742	23755	03930	41515	10297
54208	37131	32366	77519	57374		95762	25255	38703	20509	40545
06587	04827	18084	80288	23274		23049	07190	95129	34875	81629
54419	64469	20538	15087	89185		72724	98390	98735	09156	04417
52776	73748	01537	27259	51549	038	23888	63783	92325	29209	10390	038

Figure 2.19 Page from a one-time pad used by a Communist spy in Japan in 1961. (Courtesy of the David Kahn Collection, National Cryptologic Museum, Fort Meade, Maryland.)

Figure 2.20 The one-time tape is visible on the left side in this picture of the hot line between Washington, DC, and Moscow. (Courtesy of the National Cryptologic Museum, Fort Meade, Maryland.)

Actually, there were two hot lines (a backup is always a good idea), and both were secured with one-time pads. A commercially available system that was keyed by tapes was used.[33]

When Ché Guevara was killed in Bolivia in 1967, he was found to be carrying a one-time pad.[34] A message Ché sent to Fidel Castro months earlier, which used a Polybius cipher to convert

[33] Kahn, David, *The Codebreakers*, second edition, Scribner, New York, 1996, pp. 715–716.
[34] Polmar, Norman, and Thomas B. Allen, *Spy Book: The Encyclopedia of Espionage*, Random House, New York, 1997, p. 413.

the text to numbers prior to applying a numerical one-time pad, was later decoded by Barbara Harris and David Kahn using Ché's calculation sheet.[35]

2.8 Breaking the Unbreakable

Although other historic uses for the one-time pad have been documented, its use has been far from universal. The reason for this is that it presents a serious problem in terms of key distribution. If millions of enciphered messages will be sent over the course of a war, then millions of pages of key will be needed. Generating truly random sequences is hard, and keeping large volumes of such material secret while getting them to the people who need them is also very difficult. If you think these problems are not so tough after all, consider some of the failed implementations of the system described below.

During World War II, the German Foreign Office made use of one-time pads for its most important messages, but the pads were generated by machines, which cannot create truly random sequences. In the winter of 1944–1945, the U.S. Army's Signal Security Agency was able to break these messages.[36]

Another way the system can fail is by using the same key twice. To see how this can be broken let M_1 and M_2 denote two distinct messages that are sent with the same page of key K. Our two ciphertexts will then be

$$C_1 = M_1 + K \text{ and } C_2 = M_2 + K$$

Anyone who intercepts both ciphertexts can calculate their difference:

$$C_1 - C_2 = (M_1 + K) - (M_2 + K) = M_1 + K - M_2 - K = M_1 - M_2$$

So the eavesdropper can create a text that is the difference of two meaningful messages. It makes no important difference that the meaningful messages were combined with subtraction instead of addition. A table similar to Table 2.5 can be used to break it in the manner of a running key cipher. One must be careful in constructing the table, however, for in this instance we need to rank pairs of letters whose differences, not their sums, yield the desired letter. After the table is made, the approach used in Section 3.6 can be taken to recover the messages. Thus, using a one-time pad twice is equivalent to using a running key. And this does sometimes happen.

During World War II, one of the Soviet diplomatic cipher systems was an enciphered code. After the message was converted to code groups, random numbers from a one-time pad would be added so repeated codegroups would appear different. This is a great system. However, early 1942 was a very tough time for the Soviet Union (Germany invaded Russia in June 1941), and for a few months they printed over 35,000 one-time pad pages twice. We've already seen how such a page used twice may be broken, but the American cryptanalysts didn't have it so easy! The pages were bound into books in different orders, so identical pages might be used at quite different times, even years apart, and by different people. Most of the duplicate pages were used between 1942 and 1944, but some were as late as 1948.[37]

If the ciphers were strictly one-time pads, with no underlying code, those using the same key could be paired using Friedman's index of coincidence. To do so, simply take a pair of ciphertexts C_1 and C_2 and consider $C_1 - C_2$. If they used the same key, this would be the difference of

[35] James, Daniel, *Ché Guevara*, Stein and Day, New York, 1969.
[36] Erskine, Ralph, "Enigma's Security: What the Germans Really Knew," in Erskine, Ralph and Michael Smith, *Action this Day*, Bantam Press, London, UK, 2001, pp 370–385, p. 372 cited here.
[37] Phillips, Cecil James, "What Made Venona Possible?" in Benson, Robert Louis and Michael Warner, editors, *Venona: Soviet Espionage and the American Response, 1939-1957*, NSA/CIA, Washington, DC, 1996, p. xv.

meaningful messages. In such differences, we expect identical pairs of letters to align about 6.6% of the time (using English as an example). So, we should have about 6.6% of the difference be $A = 0$. If C_1 and C_2 did not arise from the same one-time pad key, the difference $C_1 - C_2$ would essentially be random and for such text the expected probability of $A = 0$ is only about 3.8%. But in the case of the Soviet ciphers, the underlying code made the pairing and deciphering processes more difficult. Despite this extra obstacle, portions of over 2,900 messages were eventually read.[38] In late 1953, after years of effort, it was discovered that a copy of a partially burned codebook found in April 1945[39] had been used for messages for 1942 and most of 1943.[40] However, by this time, the main breakthrough had already been made, the hard way.[41]

The intelligence derived from this material was first codenamed Jade, but then changed to Bride, Drug, and finally Venona, the name it's referred to by historians today. Many of the decipherments weren't found until the 1960s and 1970s.[42] Declassified documents containing the plaintexts of those messages were released beginning in July 1995.

Because all of the security lies in keeping the key secret, it must be easily hidden and easy to destroy if the agent is compromised; otherwise, we have another way the system can fail.

Figure 2.21   A one-time pad, very much like those used by Cold War spies. (Copyright Dirk Rijmenants, 2009; https://users.telenet.be/d.rijmenants/en/onetimepad.htm.)

A one-time pad, like the one pictured in Figure 2.21, was found in the possession of Helen and Peter Kroger, two spies for the Soviets caught in England in 1961. Both Americans, their

---

[38] Benson, Robert Louis and Michael Warner, editors, *Venona: Soviet Espionage and the American Response, 1939-1957*, NSA/CIA, Washington, DC, 1996, p. viii.

[39] It was originally found by Finnish troops, who overran the Soviet consulate in Finland in 1941. The Germans then got the book from the Finns, and finally, in May 1945, Americans found a copy in a German signals intelligence archive in Saxony, Germany. See Haynes, John Earl and Harvey Klehr, *Venona: Decoding Soviet Espionage in America*, Yale University Press, New Haven, Connecticut, 1999, p. 33.

[40] Haynes, John Earl and Harvey Klehr, *Venona: Decoding Soviet Espionage in America*, Yale University Press, New Haven, Connecticut, 1999, p. 33.

[41] Benson, Robert L., *The Venona Story*, Center for Cryptologic History, National Security Agency, Fort George G. Meade, Maryland, 2001, available online at https://www.nsa.gov/Portals/70/documents/about/cryptologic-heritage/historical-figures-publications/publications/coldwar/venona_story.pdf.

[42] Benson, Robert Louis and Michael Warner, editors, *Venona: Soviet Espionage and the American Response, 1939-1957*, NSA/CIA, Washington, DC, 1996, p. xxx.

true names were Morris and Lona Cohen. They had done spy work in the United States, but fled the country following the arrest of Julius Rosenberg. After their capture, they were sentenced to 20 years in prison, but 8 years later they were traded to the Soviets. A Soviet newspaper stated, "Thanks to Cohen, designers of the Soviet atomic bomb got piles of technical documentation straight from the secret laboratories in Los Alamos."[43]

Our initial example used letters for the key, but the pad depicted above used numbers. Suppose we use a random number generator to get a string of values, each between 0 and 9. We then shift each letter of a message by the digits, one at a time. Here's an example:

```
IS THIS SECURE?          message
74 9201 658937           key
PW CJIT YJKDUL           ciphertext
```

The answer to the question posed by the message is no! If there are only ten possible shifts for each letter, then there are only ten possible decipherments for each ciphertext character. For example, the initial ciphertext letter P couldn't possible represent plaintext A, as that would require the first digit of the key to have been 15.

In order to securely implement a numerical one-time pad, the message must first be converted to numbers. Using a numerical code like the Russian spies did is fine, as is first applying the Polybius cipher, like Ché Guevara.

Such ciphers remained popular with spies during the cold war, even though the field continued to advance with machine encryption. There are a number of reasons for this. High among them is the fact that they can be created with pencil and paper; the spy needn't carry around a cipher machine, which would tend to be incriminating!

2.9 Faking Randomness

Randomness is essential for the one-time pad, but generating randomness is quite difficult, so in addition to the difficulty of generating the large volumes of key needed, we have the difficulty involved in generating any of it! Solar radiation and other natural phenomena have been used, in addition to artificial sources, to generate keys; however, artificial sources cannot be truly random. We refer to them as *pseudorandom number generators*. If we use one of these, which is often more convenient, we have what is known as a *stream cipher*. These systems are discussed in Chapter 19.

As was mentioned earlier, the manner in which the one-time pad was presented here is not how it was implemented by Vernam. His story began in 1917 when he devised a machine to perform encryption by making use of tape. A later version of his mechanical solution is pictured in Figure 2.22.

Vernam's implementation was for messages sent by wire or radio telegraphy, so it didn't act on our 26-letter alphabet or on numbers, but rather involved punching the message character by character into a long piece of paper tape. This was done using the existing five-unit printing telegraph code (Figure 2.23), where each symbol is represented by an arrangement of holes punched in the paper.

With up to five holes for each symbol, $2^5 = 32$ distinct characters could be represented. Only 26 of these possibilities are needed for letters. The remaining six represented space, carriage return, line feed, figure shift, letter shift, and the blank or idle signal.

---

[43] *Komsomolskaya Pravda*, quoted here from Polmar, Norman and Thomas B. Allen, *Spy Book: The Encyclopedia of Espionage*, Random House, New York, 1997, p. 128.

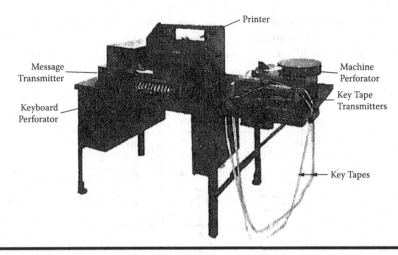

Figure 2.22 Cipher printing telegraph machine. (From Vernam, Gilbert S., "Cipher printing telegraph systems: For secret wire and radio telegraphic communications," *Journal of the American Institute of Electrical Engineers*, Vol. 45, No. 2, February 1926, pp. 109-115, p. 109 cited here.)

The key tape looks like the message tape except that the sequence of characters represented is random. Vernam suggested it be generated in advance by "working the keyboard at random." Each pair of message and key characters is combined by the machine to yield a cipher character. This is done using a rule like for the Vigenère tableau, but with 32 shifted alphabets, instead of 26. On the recipient's end, the cipher tape is combined with a duplicate of the key tape to reclaim the original message, which is then automatically printed in letter form.

Vernam intended for the tape to be used in loops, but this repetition made it equivalent to a Vigenère cipher. An engineer, Lyman F. Morehouse, had the idea of using *two* loops of tape, with one of them being one character longer than the other. The characters produced by combining the pairs of characters (one from each tape loop) would be used as the key. Although he knew such a sequence could not be truly random, he did get much longer key segments (the product of the two lengths) than he would if each tape had been used individually (the sum of the two lengths).[44]

Mauborgne's contribution was to recognize, in 1918, as the Vernam cipher was evolving, that the system would be completely unbreakable if the key was random and never repeated. Years later, Vernam promoted his device in a paper by describing how impractical it would be to implement this unbreakable cipher by hand:[45]

> This method, if carried out manually, is slow and laborious and liable to errors. If errors occur, such as the omission of one or more letters, the messages are difficult for the recipient to decipher. Certain difficulties would also be involved in preparing, copying and guarding long random keys. The difficulties with this system are such as to make it unsuitable for general use, unless mechanical methods are used.

---

[44] He also suggested that the two lengths could differ by some other amount, as long as that amount was not a factor of the length of either tape. If this doesn't sound quite right to you, good! There's a much better way to restrict the lengths of the two tapes if we want to gain a long period.

[45] Vernam, Gilbert S., "Cipher Printing Telegraph Systems: For Secret Wire and Radio Telegraphic Communications," *Journal of the American Institute of Electrical Engineers*, Vol. 45, No. 2, February 1926, pp. 109–115, p. 113 cited here.

Vernam also noted in this paper, "This cipher was demonstrated before the delegates to the Preliminary International Communications Conference in October, 1920."[46] In today's world a government employee would not be demonstrating the latest in encryption technology at an international conference, or publishing it in an open journal!

NUMBER OF SIGNAL	START ELEMENT	1	2	3	4	5	STOP ELEMENT	AMERICAN TELETYPE COMMERCIAL KEYBOARD	
1	•	•	•				•	A	-
2	•	•			•	•	•	B	?
3	•		•	•	•		•	C	:
4	•	•			•		•	D	$
5	•	•					•	E	3
6	•	•		•	•		•	F	!
7	•		•		•	•	•	G	&
8	•			•		•	•	H	£
9	•		•	•			•	I	8
10	•	•	•		•		•	J	'
11	•	•	•	•	•		•	K	(
12	•		•			•	•	L	)
13	•			•	•	•	•	M	.
14	•			•	•		•	N	,
15	•				•	•	•	O	9
16	•		•	•		•	•	P	0
17	•	•	•	•		•	•	Q	1
18	•		•		•		•	R	4
19	•	•		•			•	S	BELL
20	•					•	•	T	5
21	•	•	•	•			•	U	7
22	•		•	•	•	•	•	V	;
23	•	•	•			•	•	W	2
24	•	•		•	•	•	•	X	/
25	•	•		•		•	•	Y	6
26	•	•				•	•	Z	"
27				•			•	CARRIAGE RETURN	
28			•				•	LINE FEED	
29		•	•	•	•	•	•	LETTERS	
30		•	•		•	•	•	FIGURES	
31				•			•	SPACE	
32							•	BLANK	

Figure 2.23 Five-unit printing telegraph code. This is sometimes referred to as a *Baudot code,* after its French inventor J.M.E. Baudot, who is also the source for the term *baud*. (Image drawn by and courtesy of Sam Hallas.)

---

[46] Vernam, Gilbert. S., "Cipher Printing Telegraph Systems: For Secret Wire and Radio Telegraphic Communications," *Journal of the American Institute of Electrical Engineers*, Vol. 45, No. 2, February 1926, pp. 109–115, p. 115 cited here.

2.10 An Unsolved Cipher from 1915

Another example of a cipher that was created by Mauborgne, in 1915, currently has no known decipherment. It appears in Table 2.12.[47]

Table 2.12 Cipher Created by Mauborgne in 1915

```
PMVEB DWXZA XKKHQ RNFMJ VATAD YRJON FGRKD TSVWF TCRWC
RLKRW ZCNBC FCONW FNOEZ QLEJB HUVLY OPFIN ZMHWC RZULG
BGXLA GLZCZ GWXAH RITNW ZCQYR KFWVL CYGZE NQRNI JFEPS
RWCZV TIZAQ LVEYI QVZMO RWQHL CBWZL HBPEF PROVE ZFWGZ
RWLJG RANKZ ECVAW TRLBW URVSP KXWFR DOHAR RSRJJ NFJRT
AXIJU RCRCP EVPGR ORAXA EFIQV QNIRV CNMTE LKHDC RXISG
RGNLE RAFXO VBOBU CUXGT UEVBR ZSZSO RZIHE FVWCN OBPED
ZGRAN IFIZD MFZEZ OVCUS DPRJII IIVCRO IPCIP WHUKP NHKTV
IVONS TNADX UNQDY PERRB PNSOR ZCLRE MLZKR YZNMN PJMQB
RMJZL IKEFV CDRRN RHENC TKAXZ ESKDR GZCXD SQFGD CXSTE
ZCZNI GFHGN ESUNR LYKDA AVAVX QYVEQ FMWET ZODJY RMLZJ
QOBQ
```

If the enciphered text in Table 2.12 arose from a one-time pad, we cannot read it without the key; however, it is believed that Mauborgne didn't yet have the idea of the one-time pad in 1915. On the other hand, the known systems from this year (or earlier) shouldn't be too hard to crack with modern attacks and technology. So, why don't we have a plaintext yet? My best guess is that it used a wheel cipher of the sort described in Section 4.2.[48]

2.11 OTPs and the SOE

Leo Marks (Figure 2.24), who worked for Britain's Special Operations Executive (SOE) during World War II, independently discovered the alphabetical version of the OTP, calling it a *letter one-time pad* or LOP for short, but he was informed by Commander Dudley Smith that "letter one-time pads have been working very successfully for quite a long time."[49] The SOE did eventually make use of it for contact with their agents, who had been tasked by Winston Churchill to "set Europe ablaze." The keys were issued on silk that would be less suspicious sewn into clothing than the lump produced by a pad of paper. As with paper keys, used portions of the silk keys could be cut or torn off and destroyed. Some silks were printed with the keys in invisible ink. Of course, a strong case had to be made for using silk in this manner, as there were wartime shortages and silk was also needed for parachutes. Prior to Marks's rediscovery of the letter one-time pad system, the SOE made use of transposition ciphers. These are discussed in the next chapter.

---

[47] Kruh, Louis, "A 77-Year Old Challenge Cipher," *Cryptologia*, Vol. 17, No. 2, April 1993, pp. 172–174.

[48] See Bauer, Craig P., *Unsolved! The History and Mystery of the World's Greatest Ciphers from Ancient Egypt to Online Secret Societies*, Princeton University Press, Princeton, New Jersey, 2017, Chapter 8, for a much deeper look at this unsolved cipher.

[49] Marks, Leo, *Between Silk and Cyanide: A Codemaker's War, 1941-1945*, The Free Press, New York, 1998, p. 250.

Figure 2.24 Leo Marks holding a silk one-time pad from Sara Krulwich/*The New York Times*/Redux Pictures. With permission.)

2.12 History Rewritten!

At the beginning of the discussion of the one-time pad, I wrote that "historians long believed that it didn't arise until the years 1917 to 1918, when it was developed by Gilbert Vernam and Major Joseph Mauborgne at AT&T." In 2011, history was rewritten by Steven M. Bellovin, a computer science professor and a collector of code books. Bellovin's hobby had previously intersected with his professional life when he gave an entertaining lecture titled "Compression, Correction, Confidentiality, and Comprehension, A Modern Look at Commercial Telegraph Codes" at the 2009 Cryptologic History Symposium sponsored by the National Security Agency's Center for Cryptologic History.[50] But he struck gold without even trying when filling some leisure hours at the Library of Congress (and indulging his hobby) by examining some of the code books kept there. One of these, dated 1882 and written by Frank Miller, contained a description of a one-time pad![51]

---

[50] The slides from this lecture are available online at http://www.usenix.org/events/sec09/tech/slides/bellovin.pdf.

[51] Miller, Frank, *Telegraphic Code to Insure Privacy and Secrecy in the Transmission of Telegrams*, Charles M. Cornwell, New York, 1882.

In that instant, Bellovin had a publishable result. It could've been the easiest paper he'd ever written. In fact, I saw a comment posted online that was critical of stumbling over a codebook being considered scholarly research. Adacrypt wrote "'Trainspotting' old links to defunct cryptography is hardly to be called crypto research."[52] David Eather responded with "And like you say it is really nothing... so it is strange you couldn't do it for yourself."[53] I mention this exchange to emphasize the point that chance discoveries are usually made by people who are looking and not by people who are busy criticizing others. There's plenty to do in Washington, DC. Bellovin didn't have to spend his time at the Library of Congress looking at code books. Because he was looking, though, he was much more likely to make such a discovery. And, having made the discovery he brought the full force of his research abilities to bear on it. Not anyone could have done this. In fact, Bellovin crafted a 20-page paper with 78 references! It impressed the editor of *Cryptologia* enough to earn it the position of lead article in the July 2011 issue. It was also covered in *The New York Times*.[54] Finds like Bellovin's are sometimes among the earned rewards of a life spent passionately engaged in research. A paper by a pair of retired NSA historians detailing two more examples of this sort of reward, as well as giving tips on how to increase the likelihood of such finds is Hanyok, Robert J. and Betsy Rohaly Smoot, "Sources and Methods: Contingency and its Role in Researching Records of Cryptologic History – A Discussion and Some Lessons to Apply for Future Research," *Cryptologia*, Vol. 44, No. 6, November 2020.

References and Further Reading

On the Vigenère Cipher

Barr, Thomas H. and Andrew J. Simoson, "Twisting the Keyword Length from a Vigenère Cipher," *Cryptologia*, Vol. 39, No. 4, October 2015, pp. 335–341.

Bauer, Craig P., *Unsolved! The History and Mystery of the World's Greatest Ciphers from Ancient Egypt to Online Secret Societies*, Princeton University Press, Princeton, New Jersey, 2017. A more detailed examination of *Kryptos* than can be given here appears in Chapter 9 (pp. 386–407).

Berntsen, Matthew C., *Automating the Cracking of Simple Ciphers*, A Thesis Presented to the Faculty of Bucknell University in Partial Fulfillment of the Requirements for the Degree of Bachelor of Science with Honors in Computer Science, Bucknell University, Lewisburg, Pennsylvania, April 19, 2005, available online at http://mattberntsen.net/code/thesis/MatthewBerntsenBUThesis.pdf.

Boklan, Kent D., "How I Broke the Confederate Code (137 Years Too Late)," *Cryptologia*, Vol. 30, No. 4, October 2006, pp. 340–345.

Boklan, Kent D., "How I Deciphered a Robert E. Lee Letter – And a Note on the Power of Context in Short Polyalphabetic Ciphers," *Cryptologia*, Vol. 40, No. 5, September 2016, pp. 406–410.

Bowers, William Maxwell, "Decipherment of the Casanova Cryptogram," *Casanova Gleanings*, Vol. 14, 1971, pp. 11–16.

Brawley, J. V. and Jack Levine, "Equivalences of Vigenère Systems," *Cryptologia*, Vol. 1, No. 4, October 1977, pp. 338–361. This paper dresses up the Vigenère system by using the notation and lingo of abstract algebra; it also generalizes the cipher system.

---

[52] sci.crypt, Frank Miller Invented the OTP in 1882, https://groups.google.com/forum/#!topic/sci.crypt/RJATDxgg6BQ, December 1, 2011.

[53] sci.crypt, Frank Miller Invented the OTP in 1882, https://groups.google.com/forum/#!topic/sci.crypt/RJATDxgg6BQ, December 1, 2011.

[54] Markoff, John, "Codebook Shows an Encryption Form Dates Back to Telegraphs," *The New York Times*, July 25, 2011, available online at http://www.nytimes.com/2011/07/26/science/26code.html.

Dunin, Elonka, Elonka's *Kryptos* Page, http://elonka.com/kryptos/. This page, focused on *Kryptos*, is a great resource for anyone wanting to learn more about the mysterious sculpture and its creator.

Friedman, William F., "Jacques Casanova de Seingalt, Cryptologist," *Casanova Gleanings*, Vol. 4, 1961, pp. 1–12.

Gardner, Martin, "Mathematical Games, A New Kind of Cipher that Would Take Millions of Years to Break," *Scientific American*, Vol. 237, No. 2, August 1977, pp.120–124. This important paper will be revisited in Section 15.4. For now, the relevant part is inset on page 124 of this article and concerns the decipherment of the Vigenère cipher sent to Poe.

Grošek, Otokar, Eugen Antal, and Tomáš Fabšič, "Remarks on Breaking the Vigenère Autokey Cipher," *Cryptologia*, Vol. 43, No. 6, November 2019, pp. 486–496.

Hamilton, Michael and Bill Yankosky, "The Vigenère Cipher with the TI-83," *Mathematics and Computer Education*, Vol. 38, No. 1, Winter 2004. Hamilton was an undergraduate at North Carolina Wesleyan College when this paper was written.

Kaeding, Thomas, "Slippery Hill-climbing Technique for Ciphertext-only Cryptanalysis of Periodic Polyalphabetic Substitution Ciphers," *Cryptologia*, Vol. 44, No. 3, May 2020, pp. 205–222.

Levy, Steven, "Mission Impossible: The Code Even the CIA Can't Crack," *Wired*, Vol. 17, No. 5, April 20, 2009, http://www.wired.com/science/discoveries/magazine/17-05/ff_kryptos.

Lipson, Stanley H. and Francine Abeles, "The Key-Vowel Cipher of Charles L. Dodgson," *Cryptologia*, Vol. 15, No. 1, January 1991, pp. 18–24. This paper describes a cipher invented by the famed author of *Alice in Wonderland* (under the pseudonym Lewis Carroll) that turns out to be of the Vigenère type with nulls inserted systematically.

McCloy, Helen, *Panic*, William Morrow, New York, 1944. This is a novel. The author thought she found a nice solution to the problem of having sufficiently mixed alphabets and a hard-to-guess key without requiring the users to write any of it down; that is, the key and alphabets are easy to generate when needed. The work is of no value to cryptographers but might interest literature buffs. The cipher seems to have been the motivation for the novel.

Park, Seongmin, Juneyeun Kim, Kookrae Cho, and Dae Hyun Yum, "Finding the Key Length of a Vigenère Cipher: How to Improve the Twist Algorithm," *Cryptologia*, Vol. 44, No. 3, May 2020, pp. 197–204.

Schwartz, John and Jonathan Corum, "This Sculpture Holds a Decades-Old Mystery. And Now, Another Clue," *The New York Times*, January 29, 2020, available online at https://www.nytimes.com/interactive/2020/01/29/climate/kryptos-sculpture-final-clue.html.

Scryer, "The Kryptos Sculpture Cipher: A Partial Solution," *The Cryptogram*, Vol. 65, No. 5, September–October 1999, pp. 1–7. Scryer is the ACA pen name used by James J. Gillogly.

Scryer, "Kryptos Clue," *The Cryptogram*, Vol. 77, No. 1, January–February 2011, p. 11. Scryer is the ACA pen name used by James J. Gillogly.

Tuckerman, Bryant, *A Study of the Vigenère-Vernam Single and Multiple Loop Enciphering Systems*, IBM Research Report RC-2879, T. J. Watson Research Center, Yorktown Heights, New York, May 14, 1970. This 115-page report shows such systems to be insecure.

de Vigenère, Blaise, *Traicté des Chiffres, ou, Secretes Manieres D'escrire*, Abel l'Angelier, Paris, 1586.

Vigenère, *Cryptool – Online*, https://www.cryptool.org/en/cto/ciphers/vigenere. This website allows users to encipher and decipher using Vigenère. It's part of a much larger (and still growing) site that cover many ciphers and includes online cryptanalysis programs.

Winkel, Brian J., "Casanova and the Beaufort Cipher," *Cryptologia*, Vol. 2, No. 2, April 1978, pp. 161–163.

On Running Key Ciphers

Bauer, Craig and Christian N. S. Tate, "A Statistical Attack on the Running Key Cipher," *Cryptologia*, Vol. 26, No. 4, October 2002, pp. 274–282.

Bauer, Craig and Elliott Gottloeb, "Results of an Automated Attack on the Running Key Cipher," *Cryptologia*, Vol. 29, No. 3, July 2005, pp. 248–254. This paper described a computer attack on the running key cipher that used large files of English words to find all combinations of words from a message and a key that would yield the ciphertext. The solutions were not ranked by probability or checked for grammatical correctness. Due to the latter omission, a great number of potential solutions arose.

Friedman, William F., *Methods for the Solution of Running-Key Ciphers*, Publication No. 16, Riverbank Laboratories, Geneva, Illinois, 1918. Friedman showed that the U.S. Army's field cipher was insecure in this paper, even for short messages. This was reprinted together with other Friedman papers in Friedman, William, F. *The Riverbank Publications*, Vol. 1, Aegean Park Press, Laguna Hills, California, 1979. As the original printing only consisted of 400 copies, I suggest looking for the reprint instead.

Griffing, Alexander, "Solving XOR Plaintext Strings with the Viterbi Algorithm," *Cryptologia*, Vol. 30, No. 3, July 2006, pp. 257–265. This paper attacks running key ciphers where word spacing is preserved in the message and the key.

Griffing, Alexander, "Solving the Running Key Cipher with the Viterbi Algorithm," *Cryptologia*, Vol. 30, No. 4, October 2006, pp. 361–367. This paper dramatically improves upon the results in Bauer and Tate and in Bauer and Gottleib, to the point that these papers ought to be burned.

On One-Time Pads

Note: There is some overlap between papers on generating one-time pads and papers on random number generators. Papers that fall in the overlap are only referenced in this book in Chapter 19, which is on stream ciphers. Stream ciphers serve as approximations of the one-time pad without having the problems associated with true one-time pads.

Anon., "Automatic Code Messages," in "Science News" section of *Science*, New Series, Vol. 63, No. 1625, February 19, 1926, pp. x, xii.

Anon., "A Secret-Code Message Machine," *The Literary Digest*, Vol. 89, No. 3, Whole No. 1878, April 17, 1926, p. 22. This article, after an introductory paragraph, reproduces text from *Science Service's Daily News Bulletin*: "The new machine was described by G. S. Vernam, engineer of the American Telegraph and Telephone Company, who stated that it had been developed for the use of the Signal Corps of the U.S. Army during the war, but until recently it had been kept secret." Kept secret? Why? It's not like the Signal Corps was actually *using it* or anything...

Bellovin, Steven M., "Frank Miller: Inventor of the One-Time Pad," *Cryptologia*, Vol. 35, No. 3, July 2011, pp. 203–222.

Benson, Robert L., *The Venona Story*, Center for Cryptologic History, National Security Agency, Fort George G. Meade, Maryland, 2001, available online at https://www.nsa.gov/Portals/70/documents/about/cryptologic-heritage/historical-figures-publications/publications/coldwar/venona_story.pdf.

Benson, Robert Louis and Michael Warner, editors, *Venona: Soviet Espionage and the American Response, 1939–1957*, NSA/CIA, Washington DC, 1996. The bulk of this book is reproductions of formerly classified documents. The preface is nice, but the rest is dry. Although the reproductions are of value to historians, casual readers will prefer the book by John Earl Haynes and Harvey Klehr referenced below.

Bury, Jan, "Breaking Unbreakable Ciphers: The Asen Georgiyev Spy Case," *Cryptologia*, Vol. 33, No. 1, 2009, pp. 74–88.

Bury, Jan, "From the Archives: Breaking OTP Ciphers," *Cryptologia*, Vol. 35, No. 2, April 2011, p. 176–188.

Filby, P. William, "Floradora and a Unique Break into One-Time Pad Ciphers," *Intelligence and National Security*, Vol. 10, No. 3, July 1995, pp. 408–422.

Foster, Caxton C., "Drawbacks of the One-Time Pad," *Cryptologia*, Vol. 21, No. 4, October 1997, pp. 350–352. This paper briefly addresses the matter of determining the random sequence used as the key. If it is not truly random, then the cipher ceases to be unbreakable. Algorithms run on traditional computers are never truly random. To get true random numbers, one needs a quantum computer.

Haynes, John Earl and Harvey Klehr, *Venona: Decoding Soviet Espionage in America*, Yale University Press, New Haven, Connecticut, 1999. Although focused on the history, this book has a chapter ("Breaking the Code") that gives more detail on the cryptology than other works.

Marks, Leo, *Between Silk and Cyanide: a Codemaker's War, 1941–1945*. The Free Press, New York, 1998. Marks, a cryptographer for the Special Operations Executive (SOE), writes about his experiences in a very entertaining manner. (After the war, but before this volume, he was a screenwriter, so he can write!) The back cover sports blurbs from David Kahn and Martin Scorsese.

Mauborgne, Ben P., *Military Foundling*, Dorrance and Company, Philadelphia, Pennsylvania, 1974. A mix of fact and fiction, this novel is of interest to us mainly for its dedication:

> This book is respectfully dedicated to the memory of my illustrious, talented and versatile father, Major General Joseph O. Mauborgne, chief signal officer of the United States Army from 1937 to 1941; scientist, inventor, cryptographer, portrait painter, etcher, fine violin maker and author.
>
> Military history has recorded that he was the first person to establish two-way wireless communication between the ground and an airplane in flight; that he invented an unbreakable cipher; and that he was "directly responsible" for probably the greatest feat of cryptanalysis in history – the breaking of the Japanese PURPLE code – more than a year prior to the sneak attack on Pearl Harbor.

Miller, Frank, *Telegraphic Code to Insure Privacy and Secrecy in the Transmission of Telegrams*, Charles M. Cornwell, New York, 1882.

Philips, Cecil, "The American Solution of a German One-Time-Pad Cryptographic System," *Cryptologia*, Vol. 24, No. 4, October 2000, pp. 324–332.

Redacted, "A New Approach to the One-Time Pad," *NSA Technical Journal*, Vol. 19, No. 3, Summer 1974. The title of this paper was released by the National Security Agency as part of a much redacted index to this journal. In fact, the author's name was redacted. But we do know it comes somewhere between Gurin, Jacob and Jacobs, Walter. Any guesses?

Rubin, Frank, "One-Time Pad Cryptography," *Cryptologia*, Vol. 20, No. 4, October 1996, pp. 359–364. This paper attempts to make the one-time pad more practical.

Shannon, Claude, "Communication Theory of Secrecy Systems," *The Bell System Technical Journal*, Vol. 28, No. 4, October 1949, pp. 656–715. Shannon shows that the one-time pad is unbreakable and anything that is unbreakable must be a one-time pad. Despite his having found this result over 70 years ago, one needn't look hard for other cipher systems that are billed as being unbreakable.

Vernam, Gilbert S., "Cipher Printing Telegraph Systems: For Secret Wire and Radio Telegraphic Communications," *Journal of the American Institute of Electrical Engineers*, Vol. 45, February 1926, pp. 109–115.

Vinge, Vernor, *A Fire Upon the Deep*, St. Martin's Press, New York, 1993. A one-time pad is used in this science fiction novel, but it must first be pieced together by three starship captains.

Yardley, Herbert O., "Are We Giving Away Our State Secrets?," *Liberty*, Vol. 8, December 19, 1931, pp. 8–13. Yardley argued that America ought to be making use of the one-time pad.

Chapter 3

Transposition Ciphers

The 21st century will see transposition regain its true importance.

– **Friedrich L. Bauer**[1]

3.1 Simple Rearrangements and Columnar Transposition

Transposition ciphers represent a class separate from the substitution ciphers we have been examining.

Imagine taking a novel and writing all the As first, followed by all the Bs and so on. None of the letters would be altered; only their positions would change. Nonetheless, this would be a very difficult cipher to crack. It is not very useful, as decipherment would not be unique. Rearranging letters often produces several possible phrases with even a small number of letters. Take, for example, EILV. This could be VEIL, EVIL, VILE, or LIVE. Such ciphers are easy to spot, as the frequencies of the individual letters are not altered by encryption. What varies from one transposition cipher to another is the systematic way of scrambling and unscrambling the letters. We begin with a few very simple examples.

3.1.1 Rail-Fence Transposition

Example 1

> ANYONE WHO LOOKS AT US THE WRONG WAY TWICE WILL SURELY DIE.[2]

We simply write the text moving back and forth in a zig-zag fashion from the top line to the bottom line

```
A Y N W O O K A U T E R N W Y W C W L S R L D E
  N O E H L O S T S H W O G A T I E I L U E Y I
```

and then read across the top line first to get the ciphertext:

> AYNWO OKAUT ERNWY WCWLS RLDEN OEHLO STSHW OGATI EILUE YI

---

[1] Bauer, Friedrich L., *Decrypted Secrets: Methods and Maxims of Cryptology*, second edition, Springer, Berlin, Germany, 2000, p. 100.

[2] Quote from a soldier in the Middle East, as heard on the nightly news.

The "fence" needn't be limited to two tiers. We could encipher the same message as follows.

```
A     W     K     T     N     W     L     L
  N  EH  OS  SH  OG  TI  IL  EY
    YN  OO  AU  ER  WY  CW  SR  DE
      O     L     T     W     A     E     U     I
```

to get the ciphertext

AWKTN WLLNE HOSSH OGTII LEYYN OOAUE RWYCW SRDEO LTWAE UI

3.1.2 Rectangular Transposition

We can also write the text in a rectangle in a particular manner and read it off in another manner.

Example 2

ATTACK DAMASCUS AT DAWN.

We write the message in the form of a rectangle (any dimensions can be used), filling in by rows from top to bottom, but we get our ciphertext by pulling the text out by columns from left to right

```
ATTACK
DAMASC    →    ADUWT ASNTM AKAAT ECSDT KCAW
USATDA
WNKETW
```

Note: The last four letters in the message are only there to complete the rectangle. It is common to see the letter X used for this purpose, but it is better to use more frequent letters, making the cryptanalyst's job a bit harder.

The rectangle can have any dimension. If a cryptanalyst suspects this manner of encipherment, the number of possible cases to consider depends on the number of factors of the ciphertext's length. For example, consider the following intercept:

YLAOH TEROO YNNEO WLNUW FGSLH ERCHO UTIIS DAIRN AKPMH
NPSTR ECAWO AOITT HNCNM LLSHA SU

With 72 letters of ciphertext, the enciphering rectangle could be 2×36, 3×24, 4×18, 6×12, 8×9, 9×8, 12×6, 18×4, 24×3, or 36×2. If forced to look at each of these possibilities individually, it would be wise to start with the dimensions closest to a square and work out, but we have a nice technique to eliminate this tedium.

A probable word search often works nicely and because every message has some sort of context from which to guess a crib, this is fair. Suppose we can guess the word WHIP appears in the message. Reproducing the ciphertext with the appropriate letters underlined and boldfaced reveals a pattern. (We only need to consider characters between the first W and the last P).

YLAOH TEROO YNNEO **W**LNU**W** FGSL**H** ER**C**HO UT**II**S DAIRN
AK**P**MH N**P**STR ECAWO AOITT HNCNM LLSHA SU

Looking at the position of each of these letters and the distances between them:

W_1 16
W_2 20

H_1 25	$H_1 - W_1 = 9$	$H_1 - W_2 = 5$
H_2 29	$H_2 - W_1 = 13$	$H_2 - W_2 = 9$

I_1 33	$I_1 - H_1 = 8$	$I_1 - H_2 = 4$
I_2 34	$I_2 - H_1 = 9$	$I_2 - H_2 = 5$
I_3 38	$I_3 - H_1 = 13$	$I_3 - H_2 = 9$

P_1 43	$P_1 - I_1 = 10$	$P_1 - I_2 = 9$	$P_1 - I_3 = 5$
P_2 47	$P_2 - I_1 = 14$	$P_2 - I_2 = 13$	$P_2 - I_3 = 9$

The only number that shows up as a distance between each pair of letters in WHIP is 9. It may look like 13 works, but chaining the letters to get through the probable word is not possible, as we would have to use two different Hs. Reproducing the table and boldfacing the 9s for convenience, we have:

W_1 16
W_2 20

H_1 25	**$H_1 - W_1 = 9$**	$H_1 - W_2 = 5$
H_2 29	$H_2 - W_1 = 13$	**$H_2 - W_2 = 9$**

I_1 33	$I_1 - H_1 = 8$	$I_1 - H_2 = 4$
I_2 34	**$I_2 - H_1 = 9$**	$I_2 - H_2 = 5$
I_3 38	$I_3 - H_1 = 13$	**$I_3 - H_2 = 9$**

P_1 43	$P_1 - I_1 = 10$	**$P_1 - I_2 = 9$**	$P_1 - I_3 = 5$
P_2 47	$P_2 - I_1 = 14$	$P_2 - I_2 = 13$	**$P_2 - I_3 = 9$**

If we start with W_1 to form WHIP, we then have to use H_1, as it is the only H 9 units away from W_1. This eliminates the ambiguity in the I section. Because we had to use H_1, we must also use I_2 and, it follows from there, P_1. Thus, we have a consistent solution. The 9s indicate that nine rows were used.

Similarly, if we start with W_2 to form WHIP, we then have to use H_2, as it is the only H 9 units away from W_2. This eliminates the ambiguity in the I section. Because we had to use H_2, we must also use I_3 and, it follows from there, P_2. Thus, we have another consistent solution. Apparently the word WHIP appeared twice in this message. Again, the 9s indicate that nine rows were used. To decipher, we write the ciphertext as columns and read the message out in rows.

```
Y O U C A N O N
L Y W H I P A M
A N F O R S O L
O N G U N T I L
H E S T A R T S
T O L I K E T H
E W H I P C H A
R L E S M A N S
O N R D H W C U
```

Message: YOU CAN ONLY WHIP A MAN FOR SO LONG UNTIL HE STARTS TO LIKE THE WHIP – CHARLES MANSON

A few random letters were used to round out the block. This approach will work when the probable word appears on a single line of the enciphering block.

3.1.3 More Transposition Paths

Many paths are possible. We do not have to fill our rectangle row by row and read out column by column to encipher. One may read off on the diagonals or by spiraling in or out or any other pattern that eventually gets each letter.[3] To decipher, simply plug the letters into the rectangle in the order they were taken off and read in the order they were inscribed. In general, the number of transposition ciphers operating on blocks of size n is given by $n!$ Hence, the keyspace is very large for large n. In practice, few of these are represented by some easily memorized route. Most would jump around wildly. One very common technique for scrambling a block of text is known as *columnar transposition*.

Example 3

The number represented by e can serve as the key to encipher the message

```
THIS CONSTANT WAS THE COMBINATION TO THE
SAFES AT LOS ALAMOS DURING WORLD WAR II
```
[4]

We write out our message in rows of length 10 and then label the columns with the digits of the number represented by e (leaving out digits that repeat). Because $e \approx 2.7182818284590452353 6\ldots$, we have

```
2 7 1 8 4 5 9 0 3 6
T H I S C O N S T A
N T W A S T H E C O
M B I N A T I O N T
O T H E S A F E S A
T L O S A L A M O S
D U R I N G W O R L
D W A R I I T A N R
```

Reading out the columns in the order indicated by the key, we get the ciphertext

```
IWIHORA TNMOTDD TCNSORN CSASANI OTTALGI
AOTASLR HTBTLUW SANESIR NHIFAWT SEOEMOA
```

The key needn't be given numerically. It could be a word such as VALIDATE.

Example 4

Using the keyword VALIDATE to encipher SECRECY IS THE BEGINNING OF TYRANNY, we have

```
V A L I D A T E
S E C R E C Y I
S T H E B E G I
N N I N G O F T
Y R A N N Y E T
```

---

[3] In Dan Brown's otherwise awful novel *The Lost Symbol*, magic squares are used for transposition. For a review of the book see Dooley, John, "Reviews of Cryptologic Fiction," *Cryptologia*, Vol. 34, No. 2, April, 2010, pp. 180–185. The review in question is on pp. 183–185.

[4] Feynman, Richard, *"Surely You're Joking Mr. Feynman!"* W. W. Norton & Company, New York, 1985. A key or password that can be guessed is a poor choice. See Section 16.5 of the present book for examples of poorly chosen passwords.

We'd like to remove the text by columns, taking them in alphabetical rather than numerical order, but there are two As. No problem. Start with the first A, then move on to the second A, and then take the rest of the columns in alphabetical order. Our ciphertext is

ETNR CEOY EBGN IITT RENN CHIA YGFE SSNY.

In the examples above, we'd be wise to express the final ciphertexts in groups of five letters (the standard) so as not to reveal the number of rows in our rectangle of text.

3.2 Cryptanalysis of Columnar Transposition

So, how can columnar transposition ciphers be broken? By using high-tech graph paper and scissors! An example will make this clear. Suppose we suspect columnar transposition for the following intercepted ciphertext.

```
HAESE UTIER KHKHT ERIPB SADPA IREVH HUIOU TELTO RTHFR TTSTV
RETLO AGHRY STASE UUEUT SYPEI AEIRU CEDNY ABETH LOBVT ALBDO
HTTYT BOLOE EAEFN TTMAT TOTOT I
```

A quick count reveals 126 characters. This number factors as $2 \times 3 \times 3 \times 7$, so there are several possibilities for the dimensions of the enciphering rectangle:

$2 \times 63, 3 \times 42, 6 \times 21, 7 \times 18, 9 \times 14, 14 \times 9, 18 \times 7, 21 \times 6, 42 \times 3, 63 \times 2$

Guessing the wrong dimensions isn't the end of the world. If you've ever encountered a "Prove or Disprove" question on a test or homework set, you know how this goes. Try to prove it for a while, and if you can't then look for a counterexample. Or, if you started by looking for a counterexample and can't find one, then consider that maybe it's true and try to prove it. Thus, we pick somewhere to start, and if we can't get a decipherment then we simply try some other set of dimensions. If a team is working on the intercept, each member could take a different case. So, trying 9 rows and 14 columns, we read our ciphertext off in columns to get

```
H R P E E T O S P E L O O M
A K B V L T A E E D O H E A
E H S H T S G U I N B T E T
S K A H O T H U A Y V T A T
E H D U R V R E E A T Y E O
U T P I T R Y U I B A T F T
T E A O H E S T R E L B N O
I R I U F T T S U T B O T T
E I R T R L A Y C H D L T I
```

Of course, we cannot read the message yet because the columns were not inscribed in the proper order. If our guess at the dimensions of the rectangle is correct, rearranging the columns in the right way will yield the message. Trying all possible rearrangements would keep us busy though, because with 14 columns, we have $14! = 87{,}178{,}291{,}200$ possibilities.

Instead of using brute force, a better way to find the right order for the columns is to try to form a common word in one row and see if it yields anything reasonable for the other rows. For example, because the first row has T, H, and E, we may put those letters together to get the most common word in English. We actually have three Es in the first row, so we start by joining the first to the T and H and see what we get. Then we try the second E and finally the third E. Comparing the results, we have

```
THE          THE          THE
TAV          TAL          TAD
SEH          SET          SEN
TSH          TSO          TSY
VEU          VER          VEA
RUI          RUT          RUB
ETO          ETH          ETE
TIU          TIF          TIT
LET          LER          LEH
```

Take a close look and decide which option looks the best to you before reading on. We could easily rank the possibilities by summing the frequencies of the trigraphs produced in each case and seeing which scores the highest—that is, this process is easy to automate.

Without examining frequencies, I'd opt for the middle choice. The second to last row of the first choice has TIU, which seems unlikely (although I can think of a possibility: anTIUnion). For the third choice, the second position has TAD, which doesn't suggest a lot of possibilities, other than sTADium and amisTAD. Now we attempt to build our rectangle out by joining other columns to the left or right hand sides one at a time.

```
THER THEP THEE THEO THES THEP THEE THEL THEO THEO THEM
TALK TALB TALV TALA TALE TALE TALD TALO TALH TALE TALA
SETH SETS SETH SETG SETU SETI SETN SETB SETT SETE SETT
TSOK TSOA TSOH TSOH TSOU TSOA TSOY TSOV TSOT TSOA TSOT
VERH VERD VERU VERR VERE VERE VERA VERT VERY VERE VERO
RUTT RUTP RUTI RUTY RUTU RUTI RUTB RUTA RUTT RUTF RUTT
ETHE ETHA ETHO ETHS ETHT ETHR ETHE ETHL ETHB ETHN ETHO
TIFR TIFI TIFU TIFT TIFS TIFU TIFT TIFB TIFO TIFT TIFT
LERI LERR LERT LERA LERY LERC LERH LERD LERL LERT LERI
```

Which of the 11 possibilities above looks the best to you? Decide before continuing. Also, it should be noted that it could be none of the above! Perhaps the first row ends with THE. In that case, the other columns would all have to be joined to the left-hand side of the portion of the rectangle we have recreated in the previous step.

Did you like the first option best? If so, you may have a word or pair of words in mind to complete the fragment that one of the rows presents. Feel free to copy the columns and try rearranging them before and after the four columns we have to see if the words you have in mind form impossible combinations elsewhere or lead to other words leaping to mind and a quick decipherment. My solution follows, but you'll get more out of trying it for yourself.

Taking the first of the options above, the TIFR makes me a bit nervous. ANTIFRENCH? ANTIFRUGAL? But the TSOK seems likely to be IT'S OKAY or THAT'S OKAY, while the other solution options don't seem to have much to offer in that position. There are three columns that have an A in the necessary position to continue OKAY. Each of them is considered below, followed by the only column having a Y, where it's needed for OKAY.

```
THERPE THERPE THEROE
TALKBD TALKED TALKED
SETHSN SETHIN SETHEN
TSOKAY TSOKAY TSOKAY
VERHDA VERHEA VERHEA
RUTTPB RUTTIB RUTTFB
ETHEAE ETHERE ETHENR
TIFRIT TIFRUT TIFRTT
LERIRH LERICH LERITH
```

Which possibility looks the best to you? The first choice can almost certainly be eliminated because of the TALKBD in the second row and the third seems unlikely due to TIFRTT in row nine. Going with the second possibility, and working on the left hand side, we see that there is no I available in the necessary position to form IT'S OKAY, but we do have two columns with an A in the position needed to make THAT'S OKAY. We try each.

```
OTHERPE   PTHERPE
ETALKED   BTALKED
ESETHIN   SSETHIN
ATSOKAY   ATSOKAY
EVERHEA   DVERHEA
FRUTTIB   PRUTTIB
NETHERE   AETHERE
TTIFRUT   ITIFRUT
TLERICH   RLERICH
```

Consider the fifth rows. Which looks better EVERHEA or DVERHEA? It seems that the first possibility is the better choice. Other rows support this selection. To continue forming THAT'S OKAY, we have two choices for a column with a T in the necessary position and two choices for a column with an H in the necessary position. Thus, there are four possibilities, altogether. We compare these below.

```
OOOTHERPE   OEOTHERPE   MOOTHERPE   MEOTHERPE
HAETALKED   HVETALKED   AAETALKED   AVETALKED
TGESETHIN   THESETHIN   TGESETHIN   THESETHIN
THATSOKAY   THATSOKAY   THATSOKAY   THATSOKAY
YREVERHEA   YUEVERHEA   OREVERHEA   OUEVERHEA
TYFRUTTIB   TIFRUTTIB   TYFRUTTIB   TIFRUTTIB
BSNETHERE   BONETHERE   OSNETHERE   OONETHERE
OTTTIFRUT   OUTTIFRUT   TTTTIFRUT   TUTTIFRUT
LATLERICH   LTTLERICH   IATLERICH   ITTLERICH
```

The first choice is comical with three consecutive Os in the first row. The second isn't any better, and the third only makes sense if the message was composed by a cow. Thus, we go with the fourth.

```
MEOTHERPE
AVETALKED
THESETHIN
THATSOKAY
OUEVERHEA
TIFRUTTIB
OONETHERE
TUTTIFRUT
ITTLERICH
```

The deciphering should be moving along faster now. Can you guess what letter comes just before AVETALKED or OUEVERHEA or ITTLERICH? The strangest rows are even beginning to make sense now. TIFRUTTIB and TUTTIFRUT may look odd by themselves, but taken together, anyone familiar with early rock and roll music ought to recognize them.

Also, as the unused columns dwindle, the placements become easier, because there are fewer possibilities. Thus, at this point, you should have no trouble completing the rectangle to get

```
SOMEOTHERPEOPL
EHAVETALKEDABO
UTTHESETHINGSB
UTTHATSOKAYHAV
EYOUEVERHEARDT
UTTIFRUTTIBYPA
TBOONETHERESAL
SOTUTTIFRUTTIB
YLITTLERICHARD
```

The final message, with punctuation inserted, is

```
Some other people have talked about these things, but that's
okay. Have you heard Tutti Frutti by Pat Boone? There's also
Tutti Frutti by Little Richard.
```

Although the ciphertext was obtained by columnar transposition using a keyword, we recovered the message without learning the keyword. It was actually a name: Joseph A. Gallian. The quote is his, as well, and was delivered at the start of one of his popular talks.

The attack shown above can be easily converted to a computer program. Begin with any column and then generate a score (based on digraph frequencies) for each row placed to its left or right. The highest score stays and the process is repeated with the remaining columns until none is left and the message can be read.

3.3 Historic Uses

Columnar transposition has seen extensive use. A few examples are given below and are not intended to provide anything near a comprehensive list.

In the 1920s, transposition ciphers were used by the Irish Republican Army. Tom Mahon came upon some of these in an archive, and his search for someone who could break them ultimately led to James J. Gillogly (Figure 3.1) of *Kryptos* fame. James's story of cryptanalysis is conveyed in a thorough and interesting manner in *Decoding the IRA*.[5]

The IRA message in Figure 3.2 was sent in 1927, but dated February 24, 1924 to confuse the police in case they got hold of it. Consider the middle ciphertext (the others are left as exercises):

```
96: UEIMS NRFCO OBISE IOMRO POTNE NANRT HLYME PPROM TERSI
HEELT NBOFO LUMDT TWOAO ENUUE RMDIO SRILA SSYHP PRSGI IOSIT B
```

The 96 tells us that there are that many letters in the ciphertext. It serves as a useful check that nothing was accidentally dropped or repeated. To decipher, we need a rectangle whose dimensions

---

[5] Mahon, Tom and James J. Gillogly, *Decoding the IRA*, Mercier Press, Cork, Ireland, 2008.

Figure 3.1 James J. Gillogly. (Photo taken in Egypt, courtesy of Gillogly.)

```
C. S.
                                        24th. February '24
Y. 18.
             DESTROY  WHEN  READ.  .
To/
    O.C. No3 Area, Britain.


I.   113: UHTAO, EURSI, YSOIO, ONTOG, OSHNY, DMERS, OSRAS, NOMEO,
MRYUR, TRRRF, CNTYR, NIRIH, IUSNR, TNENF, UMYOA, SIREO, TOIME,
IPEFR, WIAOT, TRHDT, AOTNP, TOCOA, NMB.

     96: UEIMS, NRFCO, OBISE, IOMRO, POTNE, NANRT, HLYOL, PPROM,
TERSI, HEELT, NBOFO, LUMDT, TWOAO, ENUUL, RHDIC, SKILA, SSYHP,
PRSGI, IOSIT, B.

     84: ETHEU, OIKLD, IYTTE, UOTWI, HBRUA, EYTHY, DHHOA, SESRR,
NIFEO, ITNNS, ESROS, OISIE, ERBTL, TTTSG, OTSRA, OTACC, CPAU.

2.   Bearer of this despatch will bring back the £ 20 which you are
to refund.  This is the cash which was advanced in connection with
sailors.

3.   I was very sorry I could not go over last week.  I hoped to, but
could not get away.


                                  CHIEF  OF  STAFF.
```

Figure 3.2 IRA ciphertexts from 1927. (From Mahon, Thomas and James J. Gillogly, *Decoding the IRA,* Mercier Press, Cork, Ireland, 2008, p. 11. With permission)

multiply together to yield 96. Although we might not try the right dimensions the first time, we eventually get around to considering 8 rows and 12 columns.

```
 1  2  3  4  5  6  7  8  9 10 11 12
 U  C  O  E  Y  T  L  U  O  D  S  G
 E  O  M  N  M  E  T  M  E  I  S  I
 I  O  R  A  E  R  N  D  N  C  Y  I
 M  B  O  N  P  S  B  T  U  S  H  O
 S  I  P  R  P  I  O  T  U  R  P  S
 N  S  O  T  R  H  F  W  E  I  P  I
 R  E  T  H  O  E  O  O  R  L  R  T
 F  I  N  L  M  E  L  A  M  A  S  B
```

Cutting the columns out and rearranging, soon yields the result below.

```
 2  3  1  7 10  5  9  8 12  4  6 11
 C  O  U  L  D     Y  O  U  G     E  T  S
 O  M  E  T  I     M  E  M  I     N  E  S
 O  R  I  N  C     E  N  D  I     A  R  Y
 B  O  M  B  S     P  U  T  O     N  S  H
 I  P  S  O  R     P  U  T  S     R  I  P
 S  O  N  F  I     R  E  W  I     T  H  P
 E  T  R  O  L     O  R  O  T     H  E  R
 I  N  F  L  A     M  M  A  B     L  E  S
```

The plaintext, with one typo, may then be read off as

COULD YOU GET SOME TIME MINES OR INCENDIARY BOMBS PUT ON SHIPS OR PUT SRIPS ON FIRE WITH PETROL OR OTHER INFLAMMABLES

Some IRA transposition ciphers were made harder to crack by inserting columns of nulls.

During World War II, Hong Kong was attacked by the Japanese at about the same time as Pearl Harbor. Royal Air Force (RAF) Pilot Donald Hill was captured and kept a journal of his experiences between December 7, 1941 and March 31, 1942, even though this was forbidden by the Ministry of War in London, who feared intelligence might be relayed to the enemy through such writings. Hill fooled his Japanese captors by disguising his entries as mathematical tables (he converted the letters to numbers prior to transposing). The story of Donald Hill and his love, as well as the cipher, its cryptanalysis (by Philip Aston) and the story the plaintext revealed are all detailed in Andro Linklater's *The Code of Love*.[6] A page of Hill's cipher is reproduced in Figure 3.3.

Various forms of transposition were used during World War II by Britain's Special Operations Executive (SOE), in addition to the one-time pads mentioned in Section 2.11. The best form, double transposition is described shortly.

More recently, investigators found a tremendous amount of enciphered text in the Unabomber's shack. The enciphering algorithm included numerical substitutions for letters, punctuation, common combinations of letters, and some words, followed by extensive transposition. Could a genius mathematician, such as the Unabomber, have come up with a pencil and paper cipher that could withstand the scrutiny of our nation's best cryptanalysts? We may never know, as the key to the system was also found in the shack. FBI cryptanalyst Jeanne Anderson detailed the system in a 2015 paper.[7] The original ciphers were sold at a government auction, along with many other items that were once property of the Unabomber. The funds raised went to his victims and their families.[8]

---

[6] Linklater, Andro, *The Code of Love*, Doubleday, New York, 2001.

[7] Anderson, Jeanne, "Kaczynski's Ciphers," *Cryptologia*, Vol. 39, No. 3, July, 2015, pp. 203–209.

[8] Hoober, Sam, "Items in Unabomber Auction Net More than $200,000," *NewsyType.com*, June 3, 2011, http://www.newsytype.com/7120-unabomber-auction/.

Figure 3.3 A page from the enciphered diary of POW Donald Hill. (Thanks to Phillip Aston, the mathematician who cracked the diary, for providing this image.)

3.4 Anagrams

If the transposition key is very long and random (not arising from valid words) compared to the length of the message, this system can be difficult to break. In particular, if the length of the transposition key equals the length of the message, the cryptanalyst is essentially playing Scrabble with a large number of tiles and may be able to form several meaningful solutions with no statistical reason for favoring one over another. A rearrangement of letters is also known as an *anagram*.[9] Both Galileo and Newton concealed discoveries through anagramming; however, they did not scramble their messages in a systematic way, so they could not be recovered as easily as in the example above. William Friedman also used an anagram to state his opinion on the Voynich manuscript.

Example 5

Galileo Galilei

```
Haec immature a me iam frustra leguntur O. Y.
```

Giuliano de Medici received word of a discovery made by Galileo in the anagram form given above. As you can see, Galileo constructed his anagram such that another sentence was formed. However, his new sentence didn't use all the letters of the original. He had an O and Y left over and simply placed them at the end. The translation is "These unripe things are now read by me in vain." It was meant to disguise

```
cynthiae figures aemulatur mater amorum
```

which translates as "The mother of love [Venus] imitates the phases of Cynthia [the moon]." This was revealed by Galileo on January 1, 1611.

---

[9] Some reserve this term for when the letters of a word are rearranged to make another word. I use it more generally to indicate any rearrangement.

Example 6

Christian Huygens

$a^7c^5d^1e^5g^1h^1i^7l^4m^2n^9o^4p^2q^1r^2s^1t^5u^5$

In this alphabetized transposition, the exponents indicate how many of each letter appeared in the original message. The decipherment is

annulo cingitur tenui plano, nusquam cohaerente, ad
eclipticam inclinato

which translates as "[Saturn] is girdled by a thin flat ring, nowhere touching, inclined to the ecliptic."

Example 7

Isaac Newton

$a^7c^2d^2e^{14}f^2i^7l^3m^1n^8o^4q^3r^2s^4t^8v^{12}x^1$

Concerned with establishing priority, Newton included this alphabetized transposition in his second letter to Leibniz (1677). It deciphers to

Data aequatione quodcumque fluentes quantitates involvente,
fluxiones invenire et vice versa.

which translates as "From a given equation with an arbitrary number of fluentes to find the fluxiones, and vice versa."

Example 8

William F. Friedman

I put no trust in anagrammatic acrostic ciphers, for they are
of little real value-a waste-and may prove nothing-Finis.

In reference to the Voynich Manuscript (an enciphered manuscript of well over 200 pages that no one has been able to crack), Friedman wrote that he "has had for a number of years a new theory to account for its mysteries. But not being fully prepared to present his theory in plain language, and following the precedents of three more illustrious predecessors, he wishes to record in brief the substance of his theory:" His theory followed in anagram form (see above), with the rearrangement, like Galileo's, making sense; however, he topped Galileo by not having any letters left over. Three (incorrect) solutions were found by others for this anagram:[10]

William F. Friedman in a feature article arranges to use
cryptanalysis to prove he got at that Voynich Manuscript. No?

This is a trap, not a trot. Actually I can see no apt way
of unraveling the rare Voynich Manuscript. For me, defeat
is grim.

To arrive at a solution of the Voynich Manuscript, try
these general tactics: a song, a punt, a prayer. William F.
Friedman.

---

[10] Zimansky, Curt A., "William F. Friedman and the Voynich Manuscript," *Philological Quarterly*, Vol. 49, No. 4, 1970, pp. 433–442. This was reprinted in Brumbaugh, Robert S., editor, *The Most Mysterious Manuscript*, Southern Illinois University Press, Carbondale and Edwardsville, Illinois, 1978, pp. 99–108.

The correct solution was revealed by Friedman in 1970 as

```
The Voynich MSS was an early attempt to construct an
artificial or universal language of the a priori type.
- Friedman
```

With multiple possible solutions for such anagrams, the best that the cryptanalyst can hope for is to intercept a second message of the same length that uses the same transposition key. He or she can then anagram them simultaneously, as if solving a columnar transposition cipher that used only two rows. It is very unlikely that there will be more than one arrangement that yields a meaningful result for both intercepts. In general, the number of characters of ciphertext needed to expect there to be just one possible solution is known as the unity distance. Claude Shannon calculated the unicity distance for transposition systems with period d to be $1.7d$.[11]

3.5 Double Transposition

As before, when we encounter an attack on a cipher system, we can think about how to block it and thus create a stronger cipher system. In this case, the weakness is that when we try to form words, we have many rows to work with and can evaluate them as a group as being more or less likely than other possible alignments. To block this, we could use a method known as *double transposition.* This is demonstrated below.

Example 9

We'll encipher the following using double transposition with the key FRIEDRICH NIETZSCHE.

```
YES SOMETHING INVULNERABLE UNBURIABLE IS WITH ME SOMETHING
THAT WOULD REND ROCKS ASUNDER IT IS CALLED MY WILL SILENTLY
DOES IT PROCEED AND UNCHANGED THROUGHOUT THE YEARS
```

We start off exactly as we did for columnar transposition:

```
F R I E D R I C H N I E T Z S C H E
Y E S S O M E T H I N G I N V U L N
E R A B L E U N B U R I A B L E I S
W I T H M E S O M E T H I N G T H A
T W O U L D R E N D R O C K S A S U
N D E R I T I S C A L L E D M Y W I
L L S I L E N T L Y D O E S I T P R
O C E E D A N D U N C H A N G E D T
H R O U G H O U T T H E Y E A R S N
```

but we place the ciphertext under the key

```
F R I E D R I C H N I E T Z S C H E
T N O E S T D U U E T A Y T E R O L
M L I L D G S B H U R I E U G I H O
L O H E N S A U I R T N Y E W T N L
O H H B M N C L U T L I H S W P D S
S A T O E S E O E U S R I N N A N R
T R L D C H I U E D A Y N T E R I W
D L C R M E E D T E A H V L G S M I
G A I A I C E E A Y N B N K D S N E
```

---

[11] Shannon, Claude, "Communication Theory of Secrecy Systems," *The Bell System Technical Journal*, Vol. 28, No. 4, October 1949, pp. 656–715. Page 695 cited here.

and transpose once again to get

```
UBULO UDERI TPARS SSDNM ECMIE LEBOD RAAIN IRYHB LOLSR WIETM
LOSTD GUHIU EETAO HNDNI MNOIH HTLCI DSACE IEETR TLSAA NEURT
UDEYN LOHAR LATGS NSHEC EGWWN EGDYE YHINV NTUES NTLK
```

Another sort of double transposition was used in the second half of the 19th century by the anarchist enemies of the czars. This Nihilist cipher, rather than performing columnar transposition twice, separately transposed columns and rows. Other Nihilist ciphers based on substitution were in use as well.

William Friedman described double transposition as a "very excellent" method;[12] however, he did mention special cases in which it could fail. The biggest concern is that a careless encipherer will forget to perform the second transposition. In this event, an intercepted message can be easily read and will then provide the key for other messages that were correctly transposed twice. Other special cases where solutions may be obtained include the following:

1. Interception of two messages having the same length.
2. A single message enciphered using a rectangle that is also a square.
3. A non-square rectangle that is completely filled.

Friedman wrote this in 1923, before high speed computer attacks were feasible. A dictionary attack on the keyword could now yield a solution even if the message does not satisfy any of the three special conditions he mentioned.

However, we did not have to wait for the digital age, as a general attack was (secretly) published in 1934.[13] The author was Solomon Kullback, whom you will hear more about in Chapter 8.

One way to square the number of keys that would need to be checked using a dictionary attack is to use two different words, one for the first transposition and another for the second. Although the composition of two transpositions is a transposition, the "composite word" is not likely to be in the dictionary. An attack for this improved version was also presented in Kullback's paper.

Several years later, Britain's Special Operations Executive (SOE) would use single and double transposition with their agents in occupied Europe. Leo Marks tried to replace this system with one-time pads, but the result was that both were then used, although not for the same messages! On the other side (of the pond and the war), German operatives in Latin America used columnar transposition until the spring of 1941.[14]

3.6 Word Transposition

While transposition is most commonly used on letters (or bits for computerized encryption), it can be done at the level of words. During the U.S. Civil War, the Union enciphered much of its

---

[12] Friedman, William F., *Elements of Cryptanalysis*, Aegean Park Press, Laguna Hills, California, 1976, p. 103. This is an easily available reprint edition of the May 1923 first edition, which was marked FOR OFFICIAL USE ONLY and published by the Government Printing Office for the War Department.

[13] Kullback, Solomon, *General Solution for the Double Transposition Cipher*, published by the Government Printing Office for the War Department, Washington, DC, 1934. This was eventually declassified by the National Security Agency and then quickly reprinted by Aegean Park Press, Laguna Hills, California, in 1980.

[14] Bratzel, John F. and Leslie B. Rout, Jr., "Abwehr Ciphers in Latin America," *Cryptologia*, Vol 7, No 2, April 1983, pp. 132–144.

communications in this manner. As a sample ciphertext, consider the following June 1, 1863 dispatch from Abraham Lincoln.[15]

```
GUARD ADAM THEM THEY AT WAYLAND BROWN FOR KISSING VENUS
CORESPONDENTS AT NEPTUNE ARE OFF NELLY TURNING UP CAN GET
WHY DETAINED TRIBUNE AND TIMES RICHARDSON THE ARE ASCERTAIN
AND YOU FILLS BELLY THIS IF DETAINED PLEASE ODOR OF LUDLOW
COMMISSIONER
```

GUARD indicates the size of the rectangle and what path to follow for the transposition. In this case, to decipher, the words should be filled in by going up the first column, down the second, up the fifth, down the fourth, and up the third. After GUARD, every eighth word is a null, and is therefore ignored.[16] We get

FOR	VENUS	LUDLOW	RICHARDSON	AND
BROWN	CORRESPONDENTS	OF	THE	TRIBUNE
WAYLAND	AT	ODOR	ARE	DETAINED
AT	NEPTUNE	PLEASE	ASCERTAIN	WHY
THEY	ARE	DETAINED	AND	GET
THEM	OFF	IF	YOU	CAN
ADAM	NELLY	THIS	FILLS	UP

If transposition were the only protection, we'd be able to read the message now; however, the Union used an extra level of protection—code words:

VENUS = colonel
WAYLAND = captured
ODOR = Vicksburg
NEPTUNE = Richmond
ADAM = President of the U.S.
NELLY = 4:30 pm

Applying the code words (and removing the last words THIS FILLS UP, which are more nulls used to fill out the block above) yields the original message:

```
For Colonel Ludlow,

Richardson and Brown, correspondents of the Tribune, captured
at Vicksburg, are detained at Richmond. Please ascertain why
they are detained and get them off if you can.
                              The President, 4:30 pm
```

This system completely stymied the Confederacy. It's been claimed often in the cryptologic literature that the Confederacy even resorted to printing some intercepted ciphertexts in southern newspapers, along with solicitations for help! Despite this being a frequently repeated claim, all of my attempts to find the actual solicitations only led to others seeking the same! Finally, in April 2012, I found an article that I believe solves the mystery of the missing Confederate ads. It was a piece by Albert J. Myer of the Signal Corps titled "The Cypher of the Signal Corps." It ran in the October 7, 1865 edition of *Army Navy Journal*, p. 99. I know this is the wrong side, and after the war, but read on! The piece is reproduced below.

---

[15] Kahn, David, *The Codebreakers*, second edition, Scribner, New York, 1996, p. 215.

[16] That is, we ignore KISSING, TURNING, TIMES, BELLY, and COMMISSIONER.

The Cypher of the Signal Corps.

An article which has appeared in a western paper has attracted some attention from the fact that its author was at one time employed in the War Department and by its reckless statements. Among those is that the principal utility of the Signal Corps has been to catch and read the message of Rebel signal officers, just as they have caught and read ours, "for it should be understood that our signals and their's were substantially the same, and that no system of visible signals has yet been invented which cannot be deciphered by an expert."

The following message is enciphered with the simple apparatus of the Signal Corps:

```
CLBHBQHBAG        &YFSINGYBINGS       AMPCT-KTION
MZYPXOTSXB.       INGU&PSDZSYN        VTELYTIONTQJY
WKINGLQPM&        OEINGHFOY           FILOUSPN
ISGTIONEAHCS      RSAVJOSXCYJ         QJAG
```

It is held, first, that no expert not of the Signal Corps, who is now, or has been, in the employ of the War Department, or of any Army of the United States during the war, can interpret this message at all; and, second, that no expert in the United States, not of the Signal Corps, can interpret it with less than three days' labor.

To compensate any responsible endeavor the editor of the Army and Navy Journal will pay the sum of fifty dollars to the first successful interpreter, to be determined by himself.

This cypher can be wholly changed at the wave of a flag in twenty seconds' time. It can be more difficult. It is plainer in print than it appears in signals. A second message need not resemble this. It will benefit the service to know the rules by which this message may be interpreted, and no one will be more willingly assured that it can be than the writer.

A.J.M.

As the article quoted above appeared soon after the end of the Civil War, someone seeing it could easily confuse the date slightly, when recalling the article years later, and believe it ran during the war. Furthermore, recalling that it asked the reader to break a Union cipher, the person recalling the ad might well think it must have been in a southern newspaper. Through the haze of memory, years after the fact, it would all seem very reasonable. This I propose is the origin of the myth of the Rebel newspaper ads.

3.6.1 Civil War Reenactors

In 2010, Kent Boklan and Ali Assarpour, found a Union cipher, like the transposition example detailed above, for which no solution was extant. They went on to solve it, which balanced things for the lead author, who had previously broken an unsolved cipher from the Confederacy (see Section 2.3).[17]

3.7 Transposition Devices

In Section 1.2 we examined the skytale, an ancient Greek device to perform transposition encryption. Another device known as the *Cardano grille* may also be used for this purpose. Before detailing its use, we take a brief look at the life of Girolamo Cardano (Figure 3.4).

---

[17] Boklan, Kent D. and Ali Assarpour, "How We Broken the Union Code (148 Years Too Late)," *Cryptologia*, Vol. 34, No. 3, July 2010, pp. 200–210.

Figure 3.4 **"I was so exceptionally clever that everyone marveled at my extraordinary skill."** **(Girolamo Cardano, 1501–1576.) (From University of Pennsylvania, Rare Book & Manuscript Library, Elzevier Collection, Elz D 786. Thanks to Regan Kladstrup for help with this image.)**

Cardano is best remembered for being the first to publish (in *Ars Magna*, 1545) the solution to the cubic equation. However, a controversy ensued immediately, as he had obtained the formula from Tartaglia, after much harassment and a promise to keep it secret. Cardano is also credited with authoring the first book on probability[18] (*Liber de Ludo Aleae*) and making the first explicit use of complex numbers in a calculation (*Ars Magna* again). On the personal side, things didn't go as well. In 1560, Cardano's son Giambattista, was jailed and executed following his conviction for killing his wife. Another son is alleged to have had his ears cut off by Cardano for some offense!

Cardano himself was jailed in 1570. He was charged with heresy for casting the horoscope of Jesus Christ and writing a book that praised Nero.[19] His punishment was much less severe than what others faced in the hands of the Inquisition; he spent 77 days in prison and more under house arrest. Although he was banned from publishing (and even writing!) for a time, he authored 100 works over the course of his life, some of which consisted of more than one "book."

Cardano had been plagued by health problems throughout his life. His list includes catarrh, indigestion, congenital heart palpitations, hemorrhoids, gout, rupture, bladder trouble, insomnia, plague, carbuncles, tertian fever, colic, and poor circulation, plus various other ailments. One is tempted to add hypochondria. He seemed to delight in recounting his troubles: "My struggles, my worries, my bitter grief, my errors, my insomnia, intestinal troubles, asthma, skin ailments and even phtheiriasis, the weak ways of my grandson, the sins of my own son… not to mention my daughter's barrenness, the drawn out struggle with the College of Physicians, the constant intrigues, the slanders, poor health, no true friends" and "so many plots against me, so many tricks to trip me up, the thieving of my maids, drunken coachmen, the whole dishonest, cowardly, traitorous, arrogant crew that it has been my misfortune to deal with."[20]

---

[18] Yet he didn't know as much as he thought he did about the laws of chance, for in 1533 he was forced to pawn his wife's jewelry and some of his furniture to pay gambling debts.

[19] If you're curious, the horoscope was reprinted in Shumaker, Wayne, *Renaissance Curiosa*, Medieval & Renaissance Texts & Studies Vol. 8, Center for Medieval & Renaissance Studies, Binghamton, New York, 1982, pp. 53–90. An introduction is included.

[20] Muir, Jane, *Of Men and Mathematics: The Story of the Great Mathematicians*, Dodd, Mead and Company, New York, 1965, p. 45.

Initially Cardano's grille was used steganographically—that is, to conceal the presence of the message. It consisted of a piece of paper (usually heavy to withstand repeated use) with rectangular holes cut out at various locations. The encipherer placed this grille atop the paper and wrote his message in the holes. The holes could be big enough for entire words or just individual letters. After removing the grille, he would then attempt to fill in other words around the real message to create a cover text that he hoped would fool any interceptor into thinking it was the real message. This last step can be tricky and awkward phrasings and handwriting that doesn't seem to flow naturally can result, and tip off the interceptor to the fact that a grille was used. Nevertheless, grilles saw use. One with a single large hole, in the shape of an hourglass, was used in the American Revolution.

As described thus far, this is not a transposition scheme. The words or letters remain in their original order. However, a slight twist turns this into a transposition device, called a *turning grille*. To see how this works, consider the ciphertext shown in Figure 3.5.

	T	F	C	H	O	P	M		
L	A	T	P	E	N		D	A	R
A	E	B	H	E	D	E	E	I	T
		L	F	O	I		I	R	T
C	Y	S	T		O		E	I	
N	D	S		T	A	O	R	T	I
H	G	E	T	N	I			F	H
	P	V	I	D	E	C	O	I	W
S	T	J	E	A	R	R	A	D	E
N	O	R	Y			T	O		A

Figure 3.5   Original ciphertext.

By itself, the above looks like a crossword puzzle gone bad, or perhaps a word search puzzle, but see what happens when we slide the grille in Figure 3.6 over it.

Figure 3.6   THE ABILITY TO DESTROY A

A message begins to take shape, with word spacing preserved, (THE ABILITY TO DESTROY A), but it seems to be incomplete. We rotate our grille 90° clockwise (from your perspective, not the clock's!) and place it down again to observe more of the message (Figure 3.7).

Figure 3.7 PLANET IS INSIGNIFICANT

Our message continues (PLANET IS INSIGNIFICANT), but still doesn't make much sense, although it has meaning. We rotate our grille another 90° clockwise (Figure 3.8) to get more of the message (COMPARED TO THE POWER O).

Figure 3.8 COMPARED TO THE POWER O

We turn the grille yet another 90° clockwise (last time - Figure 3.9) to get the final part of the message (F THE FORCE – DARTH VADER).

Figure 3.9 F THE FORCE – DARTH VADER

The full message is now revealed to be a quote from the Dark Lord of the Sith:

THE ABILITY TO DESTROY A PLANET IS INSIGNIFICANT
COMPARED TO THE POWER OF THE FORCE - DARTH VADER

A close look at the original ciphertext shows there are four letters that were not used. Punching one more hole in the grille would allow us to make use of those four extra positions, if we needed them. Instead, these were filled in with nulls. Actually the four letters used can be anagrammed to continue the theme of the message.

Turning grilles were used as recently as World War I by the Germans, although only for a period of four months. French cryptanalysts learned to break these, and the Germans moved on to a better system, which will be examined in Section 5.2.

Most modern ciphers use both substitution and transposition. Some of them are detailed in the second half of this volume.

References and Further Reading

Anderson, Jeanne, "Kaczynski's Ciphers," *Cryptologia*, Vol. 39, No. 3, July 2015, pp. 203–209.

Barker, Wayne, *Cryptanalysis of the Double Transposition Cipher*, Aegean Park Press, Laguna Hills, California, 1996.

Bean, Richard W., George Lasry, and Frode Weierud, "Eavesdropping on the Biafra-Lisbon Link - Breaking Historical Ciphers from the Biafran War," *Cryptologia*, to appear. This paper presents a successful attack on a real-world variant of columnar transposition.

Bratzel, John F. and Leslie B. Rout, Jr., "Abwehr Ciphers in Latin America," *Cryptologia*, Vol. 7, No. 2, April 1983, pp. 132–144. This paper details the ciphers used by German operatives in Latin America during World War II.

Carroll, John M. and Lynda E. Robbins, "Computer Cryptanalysis of Product Ciphers," *Cryptologia*, Vol. 13, No. 4, October 1989, pp. 303–326. By product ciphers, the authors mean ciphers that combine two techniques, such as substitution and transposition. We'll learn more about such systems in Section 5.2.

Dimovski, Aleksandar, and Danilo Gligoroski, "Attacks on the Transposition Ciphers Using Optimization Heuristics," in *Proceeding of the 38th International Scientific Conference on Information, Communications and Energy Systems and Technologies (ICEST 2003)*, held in Sofia, Bulgaria, Heron Press, Birmingham, UK, 2003, pp. 322–324, available online at http://www.icestconf.org/wp-content/uploads/2016/proceedings/icest_2003.pdf.

> The abstract reads:
>
> In this paper three optimization heuristics are presented which can be utilized in attacks on the transposition cipher. These heuristics are simulated annealing, genetic algorithm and tabu search. We will show that each of these heuristics provides effective automated techniques for the cryptanalysis of the ciphertext. The property which make this cipher vulnerable, is that it is not sophisticated enough to hide the inherent properties or statistics of the language of the plaintext.

Eyraud, Charles, *Precis de Cryptographie Moderne*, Editions Raoul Tari, Paris, 1953. An attack on double transposition is presented here.

Friedman, William F., *Formula for the Solution of Geometrical Transposition Ciphers*, Riverbank Laboratories Publication No. 19, Geneva, Illinois, 1918.

Friedman, William F. and Elizebeth S. Friedman, "Acrostics, Anagrams, and Chaucer," *Philological Quarterly*, Vol. 38, No. 1, January 1959, pp. 1–20.

Giddy, Jonathan P. and Reihaneh Safavi-Naini, "Automated Cryptanalysis of Transposition Ciphers," *Computer Journal*, Vol. 37, No. 5, 1994, pp. 429–436.

The abstract reads:
   In this paper we use simulated annealing for automatic cryptanalysis of transposition ciphers. Transposition ciphers are a class of ciphers that in conjunction with substitution ciphers form the basis of all modern symmetric algorithms. In transposition ciphers, a plaintext block is encrypted into a ciphertext block using a fixed permutation. We formulate cryptanalysis of the transposition cipher as a combinatorial optimization problem, and use simulated annealing to find the global minimum of a cost function which is a distance measure between a possible decipherment of the given ciphertext and a sample of plaintext language. The success of the algorithm depends on the ratio of the length of ciphertext to the size of the block. For lower ratios there are cases that the plaintext cannot be correctly found. This is the expected behaviour of all cryptanalysis methods. However, in this case, examining the output of the algorithm provides valuable 'clues' for guiding the cryptanalysis. In summary, simulated annealing greatly facilities cryptanalysis of transposition ciphers and provides a potentially powerful method for analyzing more sophisticated ciphers.

Kullback, Solomon, *General Solution for the Double Transposition Cipher*, published by the Government Printing Office for the War Department, Washington, DC, 1934. This was eventually declassified by the National Security Agency and then quickly reprinted by Aegean Park Press, Laguna Hills, California, in 1980.
Lasry, George, Nils Kopal, and Arno Wacker, "Solving the Double Transposition Challenge with a Divide-and Conquer Approach," *Cryptologia*, Vol. 38, No. 3, July 2014, pp. 197–214.
Lasry, George, Nils Kopal, and Arno Wacker, "Cryptanalysis of Columnar Transposition Cipher with Long Keys," *Cryptologia*, Vol. 40, No. 4, July 2016, pp. 374–398.
Leighton, Albert C., "Some Examples of Historical Cryptanalysis," *Historia Mathematica*, Vol. 4, No. 3, August 1977, pp. 319–337. This paper includes, among others, a Union transposition cipher from the U.S. Civil War.
Leighton, Albert C., "The Statesman Who Could Not Read His Own Mail," *Cryptologia*, Vol. 17, No. 4, October 1993, pp. 395–402. In this paper, Leighton presents how he cracked a columnar transposition cipher from 1678.
Michell, Douglas W., ""Rubik's Cube" as a Transposition Device," *Cryptologia*, Vol. 16, No. 3, July 1992, pp. 250–256. Although the keyspace makes this cipher sound impressive, a sample ciphertext I generated was broken overnight by a Brett Grothouse, a student of mine who was also a cube enthusiast. Recall that a large keyspace is a necessary condition for security, but not a sufficient condition. It was recently shown that any scrambling of Rubik's Cube can be solved in 20 moves or less.[21]
Ritter, Terry, "Transposition Cipher with Pseudo-random Shuffling: The Dynamic Transposition Combiner," *Cryptologia*, Vol. 15, No. 1, January 1991, pp. 1–17.
Zimansky, Curt A., "Editor's Note: William F. Friedman and the Voynich Manuscript," *Philological Quarterly*, Vol. 49, No. 4, October 1970, pp. 433–442. This was reprinted in Brumbaugh, Robert S., editor, *The Most Mysterious Manuscript*, Southern Illinois University Press, Carbondale and Edwardsville, Illinois, 1978, pp. 99–108.

---

[21] Fildes, Jonathan, "Rubik's Cube Quest for Speedy Solution Comes to an End," *BBC News*, August 11, 2010, available online at http://www.bbc.co.uk/news/technology-10929159.

Chapter 4

Shakespeare, Jefferson, and JFK

In this chapter, we examine a controversy involving the works of William Shakespeare, the contributions of Thomas Jefferson, and a critical moment in the life of John F. Kennedy.

4.1 Shakespeare vs. Bacon

Figure 4.1   Sir Francis Bacon (1561–1626). (http://en.wikipedia.org/wiki/File:Francis_Bacon_2.jpg.)

Sir Francis Bacon (Figure 4.1) is best known as a philosopher and advocate of applied science and the scientific method, which he called the *New Instrument*. His views became more influential following his death. In particular, he provided inspiration to the men who founded the Royal

Society. Bacon earned a place in these pages because he also developed a binary cipher—that is, a cipher in which only two distinct symbols are needed to convey the message. An updated example of his biliteral cipher follows.

A = aaaaa		N = abbab
B = aaaab		O = abbba
C = aaaba		P = abbbb
D = aaabb		Q = baaaa
E = aabaa		R = baaab
F = aabab		S = baaba
G = aabba		T = baabb
H = aabbb		U = babaa
I = abaaa		V = babab
J = abaab		W = babba
K = ababa		X = babbb
L = ababb		Y = bbaaa
M = abbaa		Z = bbaab

One could use this to encipher a message, sending the 25-letter string aabbb aabaa ababb ababb abbba to say "hello," but this is a particularly inefficient way to do a monoalphabetic substitution! The strength in this cipher lies in its invisibility. Let a be represented by normal text characters and let b be represented by boldface characters. Now observe the message hidden behind the text that follows.

> Joe **will help in** the **heist.** He's a good **man**
> and **he** knows a **lot about bank** security.

```
Joewi llhel pinth eheis t.He'sa goodm anand hekno wsalo
aaabb abbba abbab baabb b aa bb baaab babaa baaba baabb
D     O     N     T         T     R     U     S     T
```

```
tabou tbank secur ity.
abaab abbba aabaa
J     O     E
```

Less obvious means of distinguishing a from b may be employed. For example, two fonts that differ very slightly may be employed. As long as they can be distinguished by some means, the hidden message can be recovered. This simple system was eventually used to bolster arguments that Shakespeare's plays were actually written by Bacon, but such claims pre-date the alleged cryptologic evidence.

According to Fletcher Pratt, the idea that Bacon was the true author of Shakespeare's plays was first put forth by Horace Walpole in *Historic Doubts*. Pratt also reported that Walpole claimed Julius Caesar never existed.[1] Contrarians who seek out *Historic Doubts* will be disappointed, as neither claim is actually present. It would be interesting to know how Pratt made this error.

Ignatius Donnelly (Figure 4.2), a politician from Minnesota, wrestled with the authorship controversy for 998 pages in his work *The Great Cryptogram: Francis Bacon's Cipher in the So-Called Shakespeare Plays* published in 1888. He also wrote a book on Atlantis. His evidence for Bacon's

---

[1] Pratt, Fletcher, *Secret and Urgent*, Bobbs Merrill, New York, 1939, p. 85.

Figure 4.2   Ignatius Donnelly (1831–1901) (http://nla.gov.au/pub/nlanews/apr01/donelly.html).

authorship involved a convoluted numerical scheme with a great deal of flexibility. Such flexibility in determining the plaintext caused most to react with great skepticism. In fact, in the same year that Donnelly's book appeared, Joseph Gilpin Pyle authored a parody *The Little Cryptogram*, which in only 29 pages used Donnelly's technique to generate messages of his own that could not possibly have been intended by Bacon. Although Pyle's approach doesn't provide a rigorous proof that Donnelly was wrong, it is very satisfying.

In general, we may draw all sorts of conclusions, depending on what sort of evidence we are willing to accept. For example, let's take a look at Psalm 46. The words from the psalm have been numbered from beginning and end to position 46.

```
  1 2 3   4    5    6
God is our refuge and strength,

7 8    9    10 11 12
a very present help in trouble.

   13    14 15 16 17
Therefore will not we fear,

  18   19 20 21   22     23
though the earth be removed, and

  24   25    26    27 28
though the mountains be carried

  29  30 31  32 33 34
into the midst of the sea;
```

35   36   37   38   39
Though the waters thereof roar

40 41   42   43   44
and be troubled, though the

45   46
mountains **shake** with the
swelling thereof. Selah.
There is a river, the streams
whereof shall make glad the city of
God, the holy place of the
tabernacles of the most High.
God is in the midst of her; she
shall not be moved: God shall help her,
and that right early.
The Heathen raged, the kingdoms
were moved: he uttered his voice,
the earth melted.
The Lord of hosts is with us;
the God of Jacob is our refuge. Selah.
Come, behold the works of the Lord,
what desolations he hath made in the earth.
He maketh wars to cease unto
the end of the earth; he breaketh the bow,

46 45 44
and cutteth the **spear** in sunder;

43   42   41   40   39 38 37
he burneth the chariot in the fire.

36 35   34 33   32 31 30 29
Be still, and know that I am God:

28 27 26 25   24   23   22
I will be exalted among the heathen,

21 20 19 18   17 16 15
I will be exalted in the earth.

14 13   12 11   10 9   8
The Lord of hosts is with us;

7   6 5 4 3 2 1
The God of Jacob is our refuge. Selah.

The words we end on are **shake** and **spear**! From this, we may conclude that Shakespeare
was the author! Or perhaps, less dramatically, we may conclude that a slight change was made in

translation to honor his 46th birthday (the King James Version was used). Or perhaps it is all a coincidence.

Elizabeth Gallup was the first to publish an anti-Shakespeare theory using the cipher introduced in this chapter. Her 1899 work was titled *The Bi-literal Cypher of Francis Bacon*. In it she claimed that different-looking symbols for the same letters used in printing Shakespeare's plays actually represented two distinct symbols and spelled out messages in Bacon's biliteral cipher to indicate that he was the true author. Bacon revealed his system of concealing messages in 1623, the year that the first folio was published, so the timing is right, but that is about all.

The evidence against Gallup's case is convincing. At the time of the printing of the first folio, various old typefaces were commonly mixed, and broken characters were used along with better copies. Thus, the characters seem to form a continuum between the new and the old, rather than two distinct forms. Also, Gallup had Bacon use words that did not exist at the time his hidden message was allegedly printed.[2] Over a half-century after Gallup's book appeared, William and Elizebeth Friedman authored a book examining the controversy.[3] Their research ended with the conclusion that the Bacon supporters were mistaken. The uncondensed version of the book (1955) won the Folger Shakespeare Library Literature Prize of $1,000.[4]

The Friedmans got much more out of this controversy than a published book and a prize, as it was actually how they met. They both worked for Riverbank Laboratories, run by eccentric millionaire George Fabyan, in Geneva, Illinois, just outside Chicago. The research areas included acoustics, chemistry, cryptology (only with the aim of glorifying Bacon as the author of those famous plays—Gallup worked there), and genetics. William Friedman was hired as a geneticist, but he helped the dozen plus workers in the cryptology section with his skill at enlarging photographs of texts believed to contain hidden messages. It could be said that William fell in love twice at Riverbank Labs. In addition to meeting his wife-to-be Elizebeth,[5] who worked in the cryptology section, he also began researching valid cryptologic topics for Fabyan. When a cryptologist hears someone refer to "The Riverbank Publications," Friedman's cryptologic publications are what come to mind, even though other works were put out by the Lab's press. As America headed into World War I, Friedman's cryptologic education proved valuable. Remember, America still didn't have a standing cryptologic agency. Like spies, codemakers and codebreakers went back to other work at the end of each of America's wars.

Broken type in Shakespeare first folios may not reveal a hidden message, but something can be learned from type in general. Penn State biology professor S. Blair Hedges found a way to estimate the print dates of various editions of books by comparing the cracks in the wood blocks used to print the illustrations. These cracks appear at a continuous rate. Other processes acting on copper plates allow dating for images printed in that manner, as well.[6]

Contrarians attempting to show that Shakespeare's plays were really written by Sir Francis Bacon seem to have been gradually replaced by contrarians attempting to show that Shakespeare's

---

[2] Pratt, Fletcher, *Secret and Urgent*, Bobbs Merrill, New York, 1939, p. 91.

[3] Friedman, William F. and Elizebeth S. Friedman, *The Shakespearean Ciphers Examined*, Cambridge University Press, Cambridge, UK, 1957.

[4] Kahn, David, *The Codebreakers*, second edition, Scribner, New York, 1996, p. 879.

[5] They were married in May 1917.

[6] Marino, Gigi, "The Biologist as Bibliosleuth," *Research Penn State*, Vol. 27, No. 1, Fall 2007, pp. 13–15.

plays were really written by Edward de Vere, although there are several other names bandied about. The arguments made today for de Vere are much the same as those of their predecessors, and are not taken seriously by professional cryptologists.

On the bright side, progress is still being made in the study of Shakespeare himself. In 2009, for the first time, a contemporary portrait of Shakespeare was publicly revealed. It was made in 1610 and is reproduced in Figure 4.3. For generations it passed down through the Cobbe family, in a house outside of Dublin, without anyone realizing who the painting depicted. Finally, someone noticed the resemblance and some top experts agree that it is actually William Shakespeare.

Figure 4.3   A contemporary portrait of Shakespeare. (http://en.wikipedia.org/wiki/File:Cobbe_portrait_of_Shakespeare.jpg).

4.2 Thomas Jefferson: President, Cryptographer

The cryptographic work of Thomas Jefferson (Figure 4.4) includes a cipher he (re)created that was previously described by Sir Francis Bacon. This system was reinvented repeatedly and was still in use up to the middle of World War II. But, before we look at it, some of Jefferson's other cryptographic work is detailed.

For the Lewis and Clark Expedition, Jefferson instructed Lewis to "communicate to us, seasonable at intervals, a copy of your journal, notes and observations, of every kind, putting into cipher whatever might do injury if betrayed." Jefferson had the Vigenère cipher in mind, but it was never used.[7] It seems reasonable to assume that Jefferson chose this system because he was aware of weaknesses in simpler systems.

Some knowledge of cryptanalysis is also demonstrated by a nomenclator Jefferson created in 1785. Part of the code portion of this nomenclator is reproduced in Figure 4.5.

Notice that Jefferson didn't number the words consecutively. He was aware of attacks on codes constructed in that manner (see Section 1.16) and made sure his own couldn't be defeated by

---

[7] https://web.archive.org/web/20021002043241/http://www.loc.gov/exhibits/lewisandclark/preview.html.

Figure 4.4 Thomas Jefferson.

N	O	P	Qu	S	T	U V	X
N *64*	O *527*	P *941*	Qu *451*	S *7.*	T *965*	U V *822*	X *1685*
na *502*	oa *746*	pa *490*	qua *103*	sa *821*	ta *1187*	va *990*	xy *1475*
nal *115*	oach *559*	paid *405*	quan *614*	safe *075*	tact *1260*	vail *1011*	Y *1247*
name *717*	oad *217*	pam *942*	quarter *607*	sal *1296*	tain *190*	val *1193*	ya *1605*
nince *909*	oal *104*	pal *555*	que *666*	same *970*	take *475*	van *786*	yea *1550*
nancy *1028*	oam *1690*	pan *581*	queen *604*	satis *343*	tal *838*	vap *971*	year *1577*
nant *41*	oam *1562*	paper *207*	quer *1155*	saturday *1400*	tance *515*	uce *1116*	yesterday *1547*
nar *64*	oar *1015*	par *879*	quest *1586*	sc *152*	tar *1266*	ved *783*	yet *1681*
narrow *166*	oard *634*	paris *1662*	question *799*	scr *557*	tation *721*	ven *911*	yeild *147*
nat *34*	oat *455*	parliament *..*	qui *1655*	se *479*	tax *578*	vent *1278*	yo *1245*
nation *127*	ob *370*	part *1375*	quin *1202*	sea *754*	tch *313*	vention *77*	yoke *1468*
native *181*	object *296*	particular *..*	quire *655*	second *831*	te *947*	ver *977*	yond *1677*
natur *839*	objection *70*	party *951*	quo *398*	secret *1050*	ted *766*	versailles *..*	york *1461*
navigation	obligation *..*	pas *678*	quota *713*	sect *1218*	temp *607*	very *1151*	you *1543*
navy *901*	oblige *109*	past *209*	R *1105*	sel *710*	ten *599*	vest *996*	your *1570*
nay *901*	observation *..*	pat *490*	ra *562*	self *440*	tend *500*		Z *1506*
nd *111*	observe *91*	pay *268*	race *144*	sem *255*	tenth *634*	ure *774*	zeal *1566*
ne *169*	obstacle *471*	pe *977*	rag *270*	sen *777*	tention *1172*	uf *1013*	&c *1677*
nea *1111*	obstruct *85*	pea *614*	rai *856*	sent *120*	ter *881*	vienna *1589*	0 *101*
near *099*	obtain *100*	peace *1370*	ral *370*	september *..*	test *700*	view *1376*	1 *476*
nearly *500*	oc *291*	pec *1131*	ram *747*	ser *933*	text *977*	vil *1024*	2 *739*
nece:sary *96*	occasion *96*	peculiar *..*	ran *1355*	serve *509*	th *417*	vin *1191*	3 *577*
necessit *..*	occur *601*	ped *1027*	rap *1540*	set *347*	than *974*	ving *1165*	4 *980*
ned *379*	ock *196*	pen *866*	rare *865*	seven *1048*	that *698*	vir *1394*	5 *858*
need *121*	october *316*	pensylvania *..*	ras *148*	seventeen *..*	the *812*	virginia *..*	6 *271*
neg *119*	od *1040*	people *136*	rash *281*	seventh *446*	their *1096*	virtu *1298*	7 *511*
negative *706*	ode *943*	per *576*	rat *613*	seventy *535*	them *412*	virtue *1182*	8 *643*
negociat *79*	oe *1276*	perhaps *..*	rate *502*	sever *172*	then *781*	vis *1204*	9 *461*
negociation	of *1552*	person *111*	rather *622*	sh *242*	thence *833*	ult *1583*	
neighbors	offer *1031*	pet *107*	ration *1134*	shal *701*	ther *794*	ume *1044*	
neither *823*	oft *1518*	ph *749*	rch *1694*	she *1070*	there *596*	un *1495*	
nier *489*	often *1698*	phe *1635*	rd *1330*	shi *1238*	therefore *..*	under *..*	
nefs *746*	og *1004*	philadelphia *..*	re *130*	ship *1411*	these *749*	united *1589*	
		phy *..*	rea *465*	sho *1550*	they *1173*	united states *..*	

Figure 4.5 A two-part code by Thomas Jefferson. (Courtesy of the David Kahn Collection, National Cryptologic Museum, Fort Meade, Maryland.)

them. We now come to his most famous discovery (Figure 4.6). The older wheel cipher pictured on the right is described below.[8]

> This enciphering and deciphering device was acquired from West Virginia by NSA in the early 1980s. It was first thought to have been a model of the "Jefferson cipher wheel," so called because Thomas Jefferson described a similar device in his writings. We believe it to be the oldest extant device in the world, but the connection with Jefferson is unproven. Such devices are known to have been described by writers as

---

[8] http://www.nsa.gov/museum/wheel.html.

Figure 4.6 A pair of wheel ciphers. (Left, https://web.archive.org/web/20130601064810/http://ilord. com/m94.html, courtesy of Robert Lord; right, https://web.archive.org/web/20031011031452/http:// www.nsa.gov/museum/wheel.html).

early as Francis Bacon in 1605 and may have been fairly common among the arcane "black chambers" of European governments. This cipher wheel was evidently for use with the French language, which was the world's diplomatic language up through World War I. How it came to be in West Virginia is unknown.

Jefferson, and several others, independently invented an enciphering device like the ones pictured in Figure 4.6. For this reason, it is sometimes referred to as the "Thomas Jefferson cipher wheel" or "Thomas Jefferson wheel cipher." To encipher using the wheel cipher, simply turn the individual wheels to form the desired message across one of the lines of letters. Copy any of the other lines to get the ciphertext. Deciphering is just as easy. To do this, form the ciphertext along one line of the wheel and then search the other lines for a meaningful text.

The wheel cipher pictured on the left in Figure 4.6 has 25 wheels. Each wheel has the alphabet ordered differently around the edge (notice the distinct letters appearing above the four Rs). The key is given by the order in which the wheels are placed on the shaft. Hence, the 25-wheel model has a keyspace almost as big as a monoalphabetic substitution cipher. It is, however, much more difficult to break. Jefferson's version had 36 wheels.[9]

Others following Jefferson also came up with the idea independently. Major Etienne Bazeries proposed such a device with 20 disks in 1891 for the French Ministry of War (which turned it down).[10] Captain Parker Hitt came up with the idea in 1914 in the strip-cipher variant.[11] Here, vertical slips of paper bearing scrambled alphabets are held in place horizontally by a backing that allows vertical motion (Figure 4.7).

Moving the strips up and down is equivalent to turning the wheels on Jefferson's device. In this format, it is necessary to have two copies of the shuffled alphabet on each strip. Otherwise, when attempting to read a given row off the device, one or more letters might be missing due to strips being shifted too far up or down. If it weren't for this repetition in the alphabets, joining the ends of each strip would turn the device into a wheel cipher. Hitt's device became, in cylinder form,

---

[9] Salomon, David, *Data Privacy and Security*, Springer, New York, 2003, p. 82.
[10] Kahn, David, *The Codebreakers*, second edition, Scribner, New York, 1996, p. 247.
[11] Kahn, David, *The Codebreakers*, second edition, Scribner, New York, 1996, p. 493.

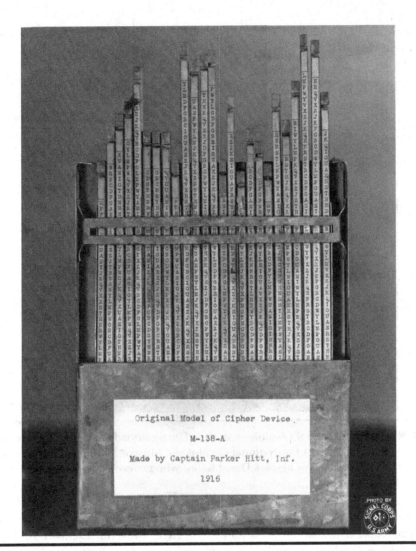

Figure 4.7   A device equivalent to a wheel cipher. (Courtesy of the National Cryptologic Museum, Fort Meade, Maryland.)

the U.S. Army's field cipher in 1922. In this form, it is known as the M-94, and was used until the middle of World War II.[12] The Navy adopted the device in 1928, naming it CSP 488, and the Coast Guard was using it by 1939 under the name CSP 493.[13] The U.S. Navy still had a version

[12] Mellen, Greg and Lloyd Greenwood, "The Cryptology of Multiplex Systems," *Cryptologia*, Vol. 1. No. 1, January 1977, pp. 4–16, p. 13 cited here

[13] Gaddy, David W., "The Cylinder-Cipher," *Cryptologia*, Vol. 19, No. 4, October 1995, pp. 385–391, p. 386 cited here. Note that the dates of adoption given for the various service branches vary from author to author! For example, in Weller, Robert, "Rear Admiral Joseph N. Wenger USN (Ret) and the Naval Cryptologic Museum," *Cryptologia*, Vol. 8, No. 3, July 1984, pp. 208–234 these wheel ciphers were delivered to the Navy in December 1926 and use by the Coast Guard began "about 1935." p. 214 cited here.

of this cipher in use in the mid-1960s![14] A wheel cipher is shown in Figure 4.8 with an operator to give a sense of scale.

Figure 4.8   Former high school cryptology student Dustin Rhoades gives us a sense of scale as he examines a wheel cipher in the National Cryptologic Museum library. A poster in the background seems to show that this pleases David Kahn, who generously donated his own cryptologic library to the museum.

The wheel cipher is an example of a multiplex system. This simply means that the user is able to choose from more than one ciphertext for each message. The term is actually an abbreviation coined by William Friedman for *multiple possible ciphertexts*. In this case, we have 25 choices, although the line or two directly beneath the message was sometimes forbidden. An advantage of a multiplex system is that identical plaintext portions of a message needn't generate identical portions of ciphertext.

4.3 Wheel Cipher Cryptanalysis

Suppose we capture an M-94 and then intercept a message. We might turn the wheels to form the ciphertext on one row and then eagerly look at each of the other 25 rows only to be disappointed. Clearly, the key (the order of the disks) has changed. If the sender errs with stereotyped beginnings or has also sent the message in another (compromised) system, we have a crib and can attempt to

---

[14] Mellen, Greg and Lloyd Greenwood, "The Cryptology of Multiplex Systems," *Cryptologia*, Vol. 1. No. 1, January 1977, pp. 4–16, p. 5 cited here.

determine the new key. There are 25! possible orderings, so we need an approach more sophisticated than brute force. The alphabets for the U.S. Navy wheel cipher were as follows.[15]

```
 1  BCEJIVDTGFZRHALWKXPQYUNSMO
 2  CADEHIZFJKTMOPUQXWBLVYSRGN
 3  DGZKPYESNUOAJXMHRTCVBWLFQI
 4  EIBCDGJLFHMKRWQTVUANOPYZXS
 5  FRYOMNACTBDWZPQIUHLJKXEGSV
 6  GJIYTKPWXSVUEDCOFNQARMBLZH
 7  HNFUZMSXKEPCQIGVTOYWLRAJDB
 8  IWVXRZTPHOCQGSBJEYUDMFKANL
 9  JXRSFHYGVDQPBLIMOAKZNTCWUE
10  KDAFLJHOCGEBTMNRSQVPXZIYWU
11  LEGIJBKUZARTSOHNPFXMWQDVCY
12  MYUVWLCQSTXHNFAZGDRBJEOIPK
13  NMJHAEXBLIGDKCRFYPWSZOQUVT
14  OLTWGANZUVJEFYDKHSMXQIPBRC
15  PVXRNQUIYZSJATWBDLGCEHFOKM
16  QTSEOPIDMNFXWUKYJVHGBLZCAR
17  RKWPUTQEBXLNYVFCIMZHSAGDOJ
18  SONMQUVAWRYGCEZLBKDFIJXHTP
19  TSMZKXWVRYUFIGJDABEOPCHNLQ
20  UPKGSCFJOWAYDHVELZNRTBMQIX
21  VFLQYSORPMHZUKXACGJIDNTEBW
22  WHOLBDMKEQNIXRTUZJFYCSVPAG
23  XZPTVOBMQCWSLJYGNEIUFDRKHA
24  YQHACRLNDPBOVZSXWITEGKUMJF
25  ZUQNXWRYALIVPBESMCOKHGJTFD
```

One possible means of breaking it, if modern technology is allowed, is to use a probable word
search. Suppose we believe the word MONEY appears in the message. There are $_{25}P_5 = (25)(24)(23)$
$(22)(21) = 6,375,600$ possibilities as to which of the five wheels were used to encipher this word. A
computer can examine, for each possibility, the various ciphertexts that would result. If one of them
matches part of the ciphertext that has been intercepted, we may know the order of five of the wheels
(a coincidence is also possible, as the word MONEY may not be present in the message). This attack
assumes that we know the order of the letters on each wheel and only the ordering of the wheels on
the shaft is unknown. Once we know the order of a few of the wheels, the calculations to determine
the rest become less time-consuming. The various wheels can be tried on the end of the one containing
the crib such that the ciphertext is continued on the appropriate line, while looking to see if the line
of plaintext continues to make sense. If there's a given ciphertext of length 25 or greater for which we
know, or can guess, the plaintext, we can recover the order of the wheels with just pencil and paper.

Example 1

Perhaps a commonly sent message is NOTHING TO REPORT AT THIS TIME. If we suspect
the ciphertext YTWML GHWGO PVRPE SDKTA QDVJO represents this message, we pair the
two up and examine the distance between each pair of plaintext/ciphertext letters for each of the
25 disks. Table 4.1 shows the result.

---

[15] This is according to Salomon, David, *Data Privacy and Security*, Springer, New York, 2003, p. 84. Elsewhere, other alphabets have been stated as being in use.

Table 4.1 Distance between Pairs of Plaintext/Ciphertext Letters for 25 Disks

| | N | O | T | H | I | N | G | T | O | R | E | P | O | R | T | A | T | T | H | I | S | T | I | M | E |
	Y	T	W	M	L	G	H	W	G	O	P	V	R	P	E	S	D	K	T	A	Q	D	V	J	O
1	24	08	08	12	10	12	04	08	09	14	16	13	12	07	21	10	25	09	21	09	22	25	01	05	23
2	22	24	07	07	14	25	06	07	12	15	10	07	11	16	19	21	18	25	06	22	19	18	15	23	09
3	23	07	04	25	23	19	14	04	17	20	24	15	06	14	15	22	09	12	02	12	17	08	20	24	04
4	03	21	24	01	06	12	04	24	11	08	21	21	18	09	11	07	15	22	06	17	15	15	15	22	20
5	23	05	03	13	03	18	20	03	20	02	17	12	24	12	14	18	02	12	17	17	16	02	10	15	07
6	12	15	03	22	21	09	25	03	11	21	20	04	05	12	08	16	09	01	05	17	09	09	08	06	03
7	17	25	03	05	07	13	12	03	23	22	01	05	04	15	19	10	08	18	16	09	06	08	02	18	08
8	19	23	21	12	25	14	22	21	03	05	17	21	21	03	10	16	13	16	24	23	24	13	02	21	19
9	12	05	02	10	25	13	24	02	09	14	12	23	12	09	04	12	14	23	16	03	07	14	20	11	17
10	09	05	12	07	08	21	23	12	02	18	09	25	08	04	24	14	15	14	06	06	01	15	22	18	23
11	10	24	09	05	23	13	12	09	15	03	15	07	23	06	16	03	11	21	23	06	09	11	20	11	12
12	15	13	21	15	08	04	21	21	20	04	03	05	22	06	12	20	08	16	24	17	25	08	06	20	01
13	16	04	19	24	25	10	19	19	15	07	12	07	19	03	06	15	12	13	22	21	03	12	15	01	16
14	07	02	08	02	06	24	12	08	04	02	11	13	24	24	09	12	12	13	12	10	03	12	14	18	15
15	04	16	01	04	10	14	03	01	21	20	06	01	06	23	07	24	03	11	18	05	21	03	20	12	03
16	06	23	11	16	15	10	25	11	15	05	02	12	21	06	02	04	06	13	09	18	24	06	11	08	01
17	01	07	23	24	20	11	23	23	24	24	22	10	02	03	02	25	18	22	12	05	12	18	23	08	17
18	08	23	10	06	21	09	12	10	10	18	12	07	08	16	15	19	20	19	01	13	04	20	12	18	14
19	12	07	06	06	12	16	09	06	20	11	02	13	15	12	18	11	15	04	04	04	02	15	21	12	01
20	19	12	15	09	18	11	10	15	21	15	12	13	11	08	21	20	18	08	07	12	19	18	16	11	19
21	09	16	03	25	09	22	19	03	11	25	11	18	01	01	01	16	24	17	12	22	24	24	07	09	09
22	09	12	12	05	18	15	02	12	23	15	15	25	11	10	20	24	17	19	13	13	14	17	11	11	01
23	24	24	07	09	20	25	09	07	10	09	11	02	17	06	14	12	18	20	05	07	23	18	12	06	14
24	19	07	24	21	15	13	08	24	09	06	16	03	20	04	01	11	16	03	16	12	13	16	21	01	18
25	04	05	08	22	25	18	25	08	03	12	24	25	14	06	17	07	02	22	03	24	13	02	01	06	04

Because each ciphertext character is the same fixed distance on the wheel it arose from, when compared to the message letter it represents, we need to find a numerical value that appears in every row and column.

Column 1 doesn't contain 2, 5, 11, 13, 14, 18, 20, 21, and 25, so these may be eliminated as possible shifts. The possibilities that remain are 1, 3, 4, 6, 7, 8, 9, 10, 12, 15, 16, 17, 19, 22, 23, and 24. But column 2 doesn't contain 1, 3, 6, 9, 10, 17, 19, or 22, so our list is quickly reduced to just 4, 7, 8, 12, 15, 16, 23, and 24. Column 3 doesn't contain a 16, so we are then left with 4, 7, 8, 12, 15, 23, and 24. Column 4 eliminates 8 and 23, leaving 4, 7, 12, 15, and 24. Column 5 reduces the choices to 7, 12, and 15. There's no 7 in column 6, so we now know the shift is either 12 or 15. Things now start to move slower! Every column contains both a 12 and a 15 until we get to column 18, which lacks the 15. Finally (with seven columns to spare!) we conclude the shift was by 12.

Locating all of the 12s in Table 4.1 will help us to find the order of the wheels (Table 4.2).

Table 4.2 shows that N is followed 12 places later by Y on wheels 6, 9, and 19, but we don't know which of these it is. The best strategy is to not worry about it for now. Moving on to the fifth letter in the message, we see that I is followed 12 places later by L on wheel 19. Thus, wheel 19 must be in position 5 on the shaft of the wheel cipher. Similarly, wheels 2, 25, 12, 17, and 11 must be in positions 9, 10, 15, 21, and 25, respectively. We may label these determinations like so

```
        19               2 25                12               17         11
N O T H  I  N G T O R E P O R  T  A T T H I S T I M  E
Y T W M  L  G H W G O P V R P  E  S D K T A Q D V J  O
```

We now take these wheels "off the table" since their positions in the key have been determined (Table 4.3).

Table 4.2 Locating All of the 12s in the Table

	N	O	T	H	I	N	G	T	O	R	E	P	O	R	T	A	T	T	H	I	S	T	I	M	E
	Y	T	W	M	L	G	H	W	G	O	P	V	R	P	E	S	D	K	T	A	Q	D	V	J	O
1	24	08	08	12	10	12	04	08	09	14	16	13	12	07	21	10	25	09	21	09	22	25	01	05	23
2	22	24	07	07	14	25	06	07	12	15	10	07	11	16	19	21	18	25	06	22	19	18	15	23	09
3	23	07	04	25	23	19	14	04	17	20	24	15	06	14	15	22	09	12	02	12	17	08	20	24	04
4	03	21	24	01	06	12	04	24	11	08	21	21	18	09	11	07	15	22	06	17	15	15	15	22	20
5	23	05	03	13	03	18	20	03	20	02	17	12	24	12	14	18	02	12	17	17	16	02	10	15	07
6	12	15	03	22	21	09	25	03	11	21	20	04	05	12	08	16	09	01	05	17	09	09	08	06	03
7	17	25	03	05	07	13	12	03	23	22	01	05	04	15	19	10	08	18	16	09	06	08	02	18	08
8	19	23	21	12	25	14	22	21	03	05	17	21	21	03	10	16	13	16	24	23	24	13	02	21	19
9	12	05	02	10	25	13	24	02	09	14	12	23	12	09	04	12	14	23	16	03	07	14	20	11	17
10	09	05	12	07	08	21	23	12	02	18	09	25	08	04	24	14	15	14	06	06	01	15	22	18	23
11	10	24	09	05	23	13	12	09	15	03	15	07	23	06	16	03	11	21	23	06	09	11	20	11	12
12	15	13	21	15	08	04	21	21	20	04	03	05	22	06	12	20	08	16	24	17	25	08	06	20	01
13	16	04	19	24	25	10	19	19	15	07	12	07	19	03	06	15	12	13	22	21	03	12	15	01	16
14	07	02	08	02	06	24	12	08	04	02	11	13	24	24	09	12	12	13	12	10	03	12	14	18	15
15	04	16	01	04	10	14	03	01	21	20	06	01	06	23	07	24	03	11	18	05	21	03	20	12	03
16	06	23	11	16	15	10	25	11	15	05	02	12	21	06	02	04	06	13	09	18	24	06	11	08	01
17	01	07	23	24	20	11	23	23	24	24	22	10	02	03	02	25	18	22	12	05	12	18	23	08	17
18	08	23	10	06	21	09	12	10	10	18	12	07	08	16	15	19	20	19	01	13	04	20	12	18	14
19	12	07	06	06	12	16	09	06	20	11	02	13	15	12	18	11	15	04	04	04	02	15	21	12	01
20	19	12	15	09	18	11	10	15	21	15	12	13	11	08	21	20	18	08	07	12	19	18	16	11	19
21	09	16	03	25	09	22	19	03	11	25	11	18	01	01	01	16	24	17	12	22	24	24	07	09	09
22	09	12	12	05	18	15	02	12	23	15	15	25	11	10	20	24	17	19	13	13	14	17	11	11	01
23	24	24	07	09	20	25	09	07	10	09	11	02	17	06	14	12	18	20	05	07	23	18	12	06	14
24	19	07	24	21	15	13	08	24	09	06	16	03	20	04	01	11	16	03	16	12	13	16	21	01	18
25	04	05	08	22	25	18	25	08	03	12	24	25	14	06	17	07	02	22	03	24	13	02	01	06	04

Table 4.3 Taking Wheels off the Table

	N	O	T	H	I	N	G	T	O	R	E	P	O	R	T	A	T	T	H	I	S	T	I	M	E
	Y	T	W	M	L	G	H	W	G	O	P	V	R	P	E	S	D	K	T	A	Q	D	V	J	O
1	24	08	08	12	10	12	04	08	09	14	16	13	12	07	21	10	25	09	21	09	22	25	01	05	23
3	23	07	04	25	23	19	14	04	17	20	24	15	06	14	15	22	09	12	02	12	17	08	20	24	04
4	03	21	24	01	06	12	04	24	11	08	21	21	18	09	11	07	15	22	06	17	15	15	15	22	20
5	23	05	03	13	03	18	20	03	20	02	17	12	24	12	14	18	02	12	17	17	16	02	10	15	07
6	12	15	03	22	21	09	25	03	11	21	20	04	05	12	08	16	09	01	05	17	09	09	08	06	03
7	17	25	03	05	07	13	12	03	23	22	01	05	04	15	19	10	08	18	16	09	06	08	02	18	08
8	19	23	21	12	25	14	22	21	03	05	17	21	21	03	10	16	13	16	24	23	24	13	02	21	19
9	12	05	02	10	25	13	24	02	09	14	12	23	12	09	04	12	14	23	16	03	07	14	20	11	17
10	09	05	12	07	08	21	23	12	02	18	09	25	08	04	24	14	15	14	06	06	01	15	22	18	23
13	16	04	19	24	25	10	19	19	15	07	12	07	19	03	06	15	12	13	22	21	03	12	15	01	16
14	07	02	08	02	06	24	12	08	04	02	11	13	24	24	09	12	12	13	12	10	03	12	14	18	15
15	04	16	01	04	10	14	03	01	21	20	06	01	06	23	07	24	03	11	18	05	21	03	20	12	03
16	06	23	11	16	15	10	25	11	15	05	02	12	21	06	02	04	06	13	09	18	24	06	11	08	01
18	08	23	10	06	21	09	12	10	10	18	12	07	08	16	15	19	20	19	01	13	04	20	12	18	14
20	19	12	15	09	18	11	10	15	21	15	12	13	11	08	21	20	18	08	07	12	19	18	16	11	19
21	09	16	03	25	09	22	19	03	11	25	11	18	01	01	01	16	24	17	12	22	24	24	07	09	09
22	09	12	12	05	18	15	02	12	23	15	15	25	11	10	20	24	17	19	13	13	14	17	11	11	01
23	24	24	07	09	20	25	09	07	10	09	11	02	17	06	14	12	18	20	05	07	23	18	12	06	14
24	19	07	24	21	15	13	08	24	09	06	16	03	20	04	01	11	16	03	16	12	13	16	21	01	18

Notice that wheel 4 only moves one character of the message to the appropriate ciphertext character—namely, the letter in position 6. Although there are other wheels that move the letter in position 6 to where it needs to go, it must be wheel 4 that actually does so. This is because wheel 4 must be used somewhere, and it doesn't work anywhere else. We may therefore take wheel 4 off the table and remove the underlining and boldfacing that indicated the other possible wheels for position 6. We do the same (by following the same reasoning) for wheels 7, 8, 15, 16, 21, and 24. This is reflected in the updated key below and in Table 4.4.

```
        8 19  4  7       2 25      16       12          21 24 17       15 11
N O T H I N G T O R E P O R T A T T H I S T I M E
Y T W M L G H W G O P V R P E S D K T A Q D V J O
```

Table 4.4 Taking More Wheels off the Table

	N	O	T	H	I	N	G	T	O	R	E	P	O	R	T	A	T	T	H	I	S	T	I	M	E
	Y	T	W	M	L	G	H	W	G	O	P	V	R	P	E	S	D	K	T	A	Q	D	V	J	O
1	24	08	08	12	10	12	04	08	09	14	16	13	**12**	07	21	10	25	09	21	09	22	25	01	05	23
3	23	07	04	25	23	19	14	04	17	20	24	15	06	14	15	22	09	**12**	02	12	17	08	20	24	04
5	23	05	03	13	03	18	20	03	20	02	17	12	24	**12**	14	18	02	**12**	17	17	16	02	10	15	07
6	**12**	15	03	22	21	09	25	03	11	21	20	04	05	**12**	08	16	09	01	05	17	09	09	08	06	03
9	**12**	05	02	10	25	13	24	02	09	14	**12**	23	**12**	09	04	**12**	14	23	16	03	07	14	20	11	17
10	09	05	**12**	07	08	21	23	**12**	02	18	09	25	08	04	24	14	15	14	06	06	01	15	22	18	23
13	16	04	19	24	25	10	19	19	15	07	**12**	07	19	03	06	15	**12**	13	22	21	03	**12**	15	01	16
14	07	02	08	02	06	24	12	08	04	02	11	13	24	24	09	**12**	**12**	13	12	10	03	**12**	14	18	15
18	08	23	10	06	21	09	12	10	10	18	**12**	07	08	16	15	19	20	19	01	13	04	20	**12**	18	14
20	19	**12**	15	09	18	11	10	15	21	15	**12**	13	11	08	21	20	18	08	07	12	19	18	16	11	19
22	09	**12**	**12**	05	18	15	02	**12**	23	15	15	25	11	10	20	24	17	19	13	13	14	17	11	11	01
23	24	24	07	09	20	25	09	07	10	09	11	02	17	06	14	**12**	18	20	05	07	23	18	**12**	06	14

With some of the underlining and boldfacing removed in the previous step, we see that we can apply the same argument again. Wheels 1 and 3 must be in positions 13 and 18, respectively. We now update our key and table (Table 4.5).

```
        8 19  4  7       2 25      16  1     12       3 21 24 17       15 11
N O T H I N G T O R E P O R T A T T H I S T I M E
Y T W M L G H W G O P V R P E S D K T A Q D V J O
```

Another bit of underlining and boldfacing removed, as a consequence of the previous step, reveals wheel 5 must be in position14. We update again to get the following key and Table 4.6.

```
        8 19  4  7       2 25      16  1  5 12       3 21 24 17       15 11
N O T H I N G T O R E P O R T A T T H I S T I M E
Y T W M L G H W G O P V R P E S D K T A Q D V J O
```

Table 4.5 Updated Table

	N	O	T	H	I	N	G	T	O	R	E	P	O	R	T	A	T	T	H	I	S	T	I	M	E
	Y	T	W	M	L	G	H	W	G	O	P	V	R	P	E	S	D	K	T	A	Q	D	V	J	O
5	23	05	03	13	03	18	20	03	20	02	17	12	24	**12**	14	18	02	12	17	17	16	02	10	15	07
6	**12**	15	03	22	21	09	25	03	11	21	20	04	05	**12**	08	16	09	01	05	17	09	09	08	06	03
9	**12**	05	02	10	25	13	24	02	09	14	**12**	23	12	09	04	**12**	14	23	16	03	07	14	20	11	17
10	09	05	**12**	07	08	21	23	**12**	02	18	09	25	08	04	24	14	15	14	06	06	01	15	22	18	23
13	16	04	19	24	25	10	19	19	15	07	**12**	07	19	03	06	15	**12**	13	22	21	03	**12**	15	01	16
14	07	02	08	02	06	24	12	08	04	02	11	13	24	24	09	**12**	**12**	13	12	10	03	**12**	14	18	15
18	08	23	10	06	21	09	12	10	10	18	**12**	07	08	16	15	19	20	19	01	13	04	20	**12**	18	14
20	19	**12**	15	09	18	11	10	15	21	15	**12**	13	11	08	21	20	18	08	07	12	19	18	16	11	19
22	09	**12**	**12**	05	18	15	02	**12**	23	15	15	25	11	10	20	24	17	19	13	13	14	17	11	11	01
23	24	24	07	09	20	25	09	07	10	09	11	02	17	06	14	**12**	18	20	05	07	23	18	**12**	06	14

Table 4.6 Updated Table

	N	O	T	H	I	N	G	T	O	R	E	P	O	R	T	A	T	T	H	I	S	T	I	M	E
	Y	T	W	M	L	G	H	W	G	O	P	V	R	P	E	S	D	K	T	A	Q	D	V	J	O
6	**12**	15	03	22	21	09	25	03	11	21	20	04	05	12	08	16	09	01	05	17	09	09	08	06	03
9	**12**	05	02	10	25	13	24	02	09	14	**12**	23	12	09	04	**12**	14	23	16	03	07	14	20	11	17
10	09	05	**12**	07	08	21	23	**12**	02	18	09	25	08	04	24	14	15	14	06	06	01	15	22	18	23
13	16	04	19	24	25	10	19	19	15	07	**12**	07	19	03	06	15	**12**	13	22	21	03	**12**	15	01	16
14	07	02	08	02	06	24	12	08	04	02	11	13	24	24	09	**12**	**12**	13	12	10	03	**12**	14	18	15
18	08	23	10	06	21	09	12	10	10	18	**12**	07	08	16	15	19	20	19	01	13	04	20	**12**	18	14
20	19	**12**	15	09	18	11	10	15	21	15	**12**	13	11	08	21	20	18	08	07	12	19	18	16	11	19
22	09	**12**	**12**	05	18	15	02	**12**	23	15	15	25	11	10	20	24	17	19	13	13	14	17	11	11	01
23	24	24	07	09	20	25	09	07	10	09	11	02	17	06	14	**12**	18	20	05	07	23	18	**12**	06	14

This reveals that wheel 6 must be in position 1. Again, we update to get the following key and Table 4.7.

```
6          8 19 4 7       2 25     16 1 5 12          3 21 24 17       15 11
N O T H I N G T O R E P O R T A T T H I S T I M E
Y T W M L G H W G O P V R P E S D K T A Q D V J O
```

Table 4.7 Updated Table

	N	O	T	H	I	N	G	T	O	R	E	P	O	R	T	A	T	T	H	I	S	T	I	M	E
	Y	T	W	M	L	G	H	W	G	O	P	V	R	P	E	S	D	K	T	A	Q	D	V	J	O
9	12	05	02	10	25	13	24	02	09	14	**12**	23	12	09	04	**12**	14	23	16	03	07	14	20	11	17
10	09	05	**12**	07	08	21	23	**12**	02	18	09	25	08	04	24	14	15	14	06	06	01	15	22	18	23
13	16	04	19	24	25	10	19	19	15	07	**12**	07	19	03	06	15	**12**	13	22	21	03	**12**	15	01	16
14	07	02	08	02	06	24	12	08	04	02	11	13	24	24	09	**12**	**12**	13	12	10	03	**12**	14	18	15
18	08	23	10	06	21	09	12	10	10	18	**12**	07	08	16	15	19	20	19	01	13	04	20	**12**	18	14
20	19	**12**	15	09	18	11	10	15	21	15	**12**	13	11	08	21	20	18	08	07	12	19	18	16	11	19
22	09	**12**	**12**	05	18	15	02	**12**	23	15	15	25	11	10	20	24	17	19	13	13	14	17	11	11	01
23	24	24	07	09	20	25	09	07	10	09	11	02	17	06	14	**12**	18	20	05	07	23	18	**12**	06	14

Positions 17 and 22 must be wheels 13 and 14 (in one order or another), so the other underlined and boldfaced options for these wheels no longer need to be considered (Table 4.8).

In the same manner, positions 3 and 8 must be wheels 10 and 22 (in one order or another), so the other underlined and boldfaced option for wheel 22 no longer needs to be considered (Table 4.9).

The updated table now reveals that position 2 must be wheel 20. We update our key (below) and the table (Table 4.10 on p. 146) again.

```
6 20       8 19 4 7       2 25     16 1 5 12          3 21 24 17       15 11
N O T H I N G T O R E P O R T A T T H I S T I M E
Y T W M L G H W G O P V R P E S D K T A Q D V J O
```

Notice that Table 4.10 uses three shades of highlighting and underlining/boxing for the undetermined possibilities that remain. This is because we cannot continue as we've been going. To brute force a solution at this stage would seem to require 128 configurations of the wheel cipher (two possibilities for each of seven unknowns). A little reasoning will reduce this but we cannot narrow it down to a single possibility based on the information we have. With more pairs of plaintext and ciphertext we would likely be able to do so, but we don't have this.

Table 4.8 Updated Table

N	O	T	H	I	N	G	T	O	R	E	P	O	R	T	A	T	T	H	I	S	T	I	M	E
Y	T	W	M	L	G	H	W	G	O	P	V	R	P	E	S	D	K	T	A	Q	D	V	J	O

| 9 | 12 | 05 | 02 | 10 | 25 | 13 | 24 | 02 | 09 | 14 | **12** | 23 | 12 | 09 | 04 | **12** | 14 | 23 | 16 | 03 | 07 | 14 | 20 | 11 | 17 |
| 10 | 09 | 05 | **12** | 07 | 08 | 21 | 23 | **12** | 02 | 18 | 09 | 25 | 08 | 04 | 24 | 14 | 15 | 14 | 06 | 06 | 01 | 15 | 22 | 18 | 23 |

| 13 | 16 | 04 | 19 | 24 | 25 | 10 | 19 | 19 | 15 | 07 | 12 | 07 | 19 | 03 | 06 | 15 | **12** | 13 | 22 | 21 | 03 | **12** | 15 | 01 | 16 |
| 14 | 07 | 02 | 08 | 02 | 06 | 24 | 12 | 08 | 04 | 02 | 11 | 13 | 24 | 24 | 09 | 12 | **12** | 13 | 12 | 10 | 03 | **12** | 14 | 18 | 15 |

| 18 | 08 | 23 | 10 | 06 | 21 | 09 | 12 | 10 | 10 | 18 | **12** | 07 | 08 | 16 | 15 | 19 | 20 | 19 | 01 | 13 | 04 | 20 | **12** | 18 | 14 |

| 20 | 19 | **12** | 15 | 09 | 18 | 11 | 10 | 15 | 21 | 15 | **12** | 13 | 11 | 08 | 21 | 20 | 18 | 08 | 07 | 12 | 19 | 18 | 16 | 11 | 19 |

| 22 | 09 | **12** | **12** | 05 | 18 | 15 | 02 | **12** | 23 | 15 | 15 | 25 | 11 | 10 | 20 | 24 | 17 | 19 | 13 | 13 | 14 | 17 | 11 | 11 | 01 |
| 23 | 24 | 24 | 07 | 09 | 20 | 25 | 09 | 07 | 10 | 09 | 11 | 02 | 17 | 06 | 14 | **12** | 18 | 20 | 05 | 07 | 23 | 18 | **12** | 06 | 14 |

Table 4.9 Updated Table

N	O	T	H	I	N	G	T	O	R	E	P	O	R	T	A	T	T	H	I	S	T	I	M	E
Y	T	W	M	L	G	H	W	G	O	P	V	R	P	E	S	D	K	T	A	Q	D	V	J	O

| 9 | 12 | 05 | 02 | 10 | 25 | 13 | 24 | 02 | 09 | 14 | **12** | 23 | 12 | 09 | 04 | **12** | 14 | 23 | 16 | 03 | 07 | 14 | 20 | 11 | 17 |
| 10 | 09 | 05 | **12** | 07 | 08 | 21 | 23 | **12** | 02 | 18 | 09 | 25 | 08 | 04 | 24 | 14 | 15 | 14 | 06 | 06 | 01 | 15 | 22 | 18 | 23 |

| 13 | 16 | 04 | 19 | 24 | 25 | 10 | 19 | 19 | 15 | 07 | 12 | 07 | 19 | 03 | 06 | 15 | **12** | 13 | 22 | 21 | 03 | **12** | 15 | 01 | 16 |
| 14 | 07 | 02 | 08 | 02 | 06 | 24 | 12 | 08 | 04 | 02 | 11 | 13 | 24 | 24 | 09 | 12 | **12** | 13 | 12 | 10 | 03 | **12** | 14 | 18 | 15 |

| 18 | 08 | 23 | 10 | 06 | 21 | 09 | 12 | 10 | 10 | 18 | **12** | 07 | 08 | 16 | 15 | 19 | 20 | 19 | 01 | 13 | 04 | 20 | **12** | 18 | 14 |

| 20 | 19 | **12** | 15 | 09 | 18 | 11 | 10 | 15 | 21 | 15 | **12** | 13 | 11 | 08 | 21 | 20 | 18 | 08 | 07 | 12 | 19 | 18 | 16 | 11 | 19 |

| 22 | 09 | 12 | **12** | 05 | 18 | 15 | 02 | **12** | 23 | 15 | 15 | 25 | 11 | 10 | 20 | 24 | 17 | 19 | 13 | 13 | 14 | 17 | 11 | 11 | 01 |
| 23 | 24 | 24 | 07 | 09 | 20 | 25 | 09 | 07 | 10 | 09 | 11 | 02 | 17 | 06 | 14 | **12** | 18 | 20 | 05 | 07 | 23 | 18 | **12** | 06 | 14 |

Table 4.10 Updated Table

	N	O	T	H	I	N	G	T	O	R	E	P	O	R	T	A	T	T	H	I	S	T	I	M	E
	Y	T	W	M	L	G	H	W	G	O	P	V	R	P	E	S	D	K	T	A	Q	D	V	J	O
9	12	05	02	10	25	13	24	02	09	14	12	23	12	09	04	12	14	23	16	03	07	14	20	11	17
10	09	05	12	07	08	21	23	12	02	18	09	25	08	04	24	14	15	14	06	06	01	15	22	18	23
13	16	04	19	24	25	10	19	19	15	07	12	07	19	03	06	15	12	13	22	21	03	12	15	01	16
14	07	02	08	02	06	24	12	08	04	02	11	13	24	24	09	12	12	13	12	10	03	12	14	18	15
18	08	23	10	06	21	09	12	10	10	18	12	07	08	16	15	19	20	19	01	13	04	20	12	18	14
22	09	12	12	05	18	15	02	12	23	15	15	25	11	10	20	24	17	19	13	13	14	17	11	11	01
23	24	24	07	09	20	25	09	07	10	09	11	02	17	06	14	12	18	20	05	07	23	18	12	06	14

Consider the four lightly shaded values. Positions 3 and 8 are occupied by wheels 10 and 22, in one order or the other. This represents two possibilities, not the four it might seem to be at first glance, because a particular wheel cannot be in two positions at once. Similarly, the four underlined/boxed values give us two possibilities altogether. For the six darkly shaded values, assigning wheel 9 to either position forces wheels 18 and 23 to particular positions. Thus, there are only two ways to assign those three wheels. Assignments for the various shaded and underlined/boxed values are all independent.

Thus, the total number of possibilities left to check (and these must be checked by hand) is $(2)(2)(2) = 8$. These possibilities all convert the given message to the given ciphertext, but only one is likely to correctly decipher the next message that is received.

The attack presented here relied on knowing some plaintext and the corresponding ciphertext, as well as the order of the alphabet on each wheel. Only the key was unknown. There are more sophisticated attacks that do not demand as much. See the paper "The Cryptology of Multiplex Systems. Part 2: Simulation and Cryptanalysis" by Greg Mellen and Lloyd Greenwood in the References and Further Reading section, if you would like to learn more about these attacks.

Another weakness with this cipher is that a letter can never be enciphered as itself. If we have a phrase that we believe appears in the message, this weakness can sometimes help us decide where.

Although primarily known for his "invention" of the wheel cipher, it is interesting to note that Thomas Jefferson (1743–1826) also wrote the Declaration of Independence, served as the third president of the United States, and founded the University of Virginia. One might expect that a figure as important as Jefferson would have been so closely examined that there is no room left for original research; however, this is not the case. In the winter of 2007, mathematician Lawren Smithline learned from a neighbor, who was working on a project to collect and publish all of

Jefferson's letters and papers that several were written in code or cipher. In June, the neighbor mentioned that one of these letters, from Robert Patterson to Jefferson, included a cipher or code portion that couldn't be read. The letter discussed cryptography and the unreadable passage was a sample ciphertext that Patterson thought couldn't be broken. Lawren got a copy of the letter, which was dated December 19, 1801, and went to work. It was a columnar transposition cipher with nulls and Lawren was able to solve it. The plaintext turned out to be the preamble to the Declaration of Independence.[16] There are two lessons we can take away from this. First, don't assume there's nothing new to be discovered, just because a topic is old or already much studied. Second, be social. Because Lawren talked with a neighbor, both benefited. You may be amazed at how often you profit from letting people know your interests.

In mathematics, we begin with a small number of assumptions that we cannot prove and then try to prove everything else in terms of them. We call these assumptions *axioms* or *postulates* and ideally they would seem "obviously true" although no proof of them can be given. Jefferson must have been in a mathematical mindset when he began his greatest piece of writing with "We hold these truths to be self-evident…"

4.4 The Playfair Cipher

In 19th-century London, it wasn't unusual for lovers or would-be lovers to encipher their personal communications and pay to place them in *The Times*. Known collectively as "Agony Columns," these notes most commonly used monoalphabetic substitution ciphers. Such simple ciphers provided a bit of sport for Baron Lyon Playfair and Charles Wheatstone.[17]

> On Sundays we usually walked together and used to amuse ourselves by deciphering the cipher advertisements in *The Times*. An Oxford student who was in a reading party at Perth, was so sure of his cipher that he kept up a correspondence with a young lady in London. This we had no difficulty in reading. At last he proposed an elopement. Wheatstone inserted an advertisement in *The Times*, a remonstrance to the young lady in the same cipher, and the last letter was, 'Dear Charles, write me no more, our cipher is discovered.'
>
> **—Lyon Playfair**

The cipher system named after Playfair (although he is not the creator of it) is more sophisticated than those that typically appeared in the papers. It's an example of a digraphic substitution cipher, which simply means that the letters are substituted for two at a time.

Before the Playfair Cipher, digraphic ciphers required the users to keep copies of the key written out, because they were clumsy and not easy to remember, as Porta's example (believed to be the first) demonstrates in Figure 4.9.

We'll use this table to encipher REVOLT. First we split the message into two-letter groups: RE VO LT. To encipher the first group, RE, we find R in the alphabet that runs across the top of the table and then move down that column until we come to the row that has E on the right hand

---

[16] See Smithline, Lawren M., "A Cipher to Thomas Jefferson: A Collection of Decryption Techniques and the Analysis of Various Texts Combine in the Breaking of a 200-year-old Code," *American Scientist*, Vol. 97, No. 2, March-April 2009, pp. 142–149.

[17] McCormick, Donald, *Love in Code*, Eyre Methuen Ltd., London, UK, 1980, p. 84.

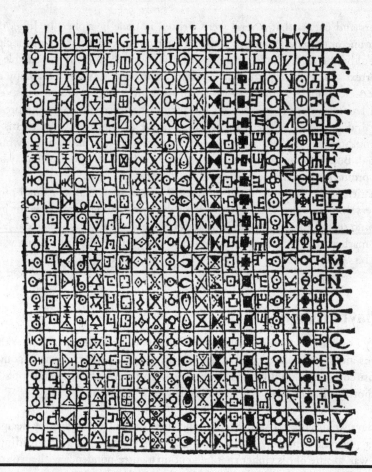

Figure 4.9 A digraphic cipher created by Porta. (Courtesy of the David Kahn Collection, National Cryptologic Museum, Fort Meade, Maryland.)

side. The symbol in this position, Ψ, takes the place of RE. In the same manner, VO becomes ✦ and LT becomes ⬡. Thus, our complete ciphertext is Ψ ✦ ⬡.

The unicity point for a random digraphic substitution cipher, like the one pictured above, is 1,460.61.[18] That is, we'd need about 1,461 characters of ciphertext in order to be able to expect a unique solution.

The Playfair Cipher was invented by Charles Wheatstone (Figure 4.10), who described it in 1854.[19] As Wheatstone and Playfair were both British (and friends), this did not lead to a huge controversy like the Newton–Leibniz feud over the discovery of calculus. Wheatstone also invented a telegraphic system before Samuel Morse, so he lost at least two naming opportunities! On the

---

[18] Deavours, Cipher A., "Unicity Points in Cryptanalysis," *Cryptologia*, Vol. 1, No. 1, January 1977, pp. 46–68.

[19] Kahn, David, *The Codebreakers*, second edition, Scribner, New York, 1996, p. 198.

Figure 4.10 Charles Wheatstone (1802–1875). (http://en.wikipedia.org/wiki/File:Wheatstone_Charles_drawing_1868.jpg).

plus side for Wheatstone, Wadsworth invented a cipher that became known as the Wheatstone cipher.[20] We'll now examine how the Playfair cipher works.

Example 2

To start, we fill a rectangle with the alphabet. I and J are once again (see Polybius) equated:

```
A   B   C   D     E
F   G   H   I&J   K
L   M   N   O     P
Q   R   S   T     U
V   W   X   Y     Z
```

Given the message

    LIFE IS SHORT AND HARD - LIKE A BODYBUILDING ELF.[21]

we begin by breaking it into pairs:

    LI FE IS SH OR TA ND HA RD LI KE AB OD YB UI LD IN GD WA RF

To encipher the first pair, LI, we find those letters in the square above. We can then find two more letters, F and O, to get the four corners of a rectangle.

```
A   B   C   D     E
F   G   H   I&J   K
L   M   N   O     P
Q   R   S   T     U
V   W   X   Y     Z
```

We take these two new corners as our ciphertext pair. But should we take them in the order FO or OF? It was arbitrarily decided that the letter to appear first in the ciphertext pair should be the one

---

[20] Clark, Ronald, *The Man Who Broke Purple*, Little, Brown and Company, Boston, Massachusetts, 1977, pp. 57–58.
[21] From "Lift Your Head Up High (And Blow Your Brains Out)," by Bloodhound Gang.

in the same row as the first plaintext letter. Making note of this first encryption and continuing in the same manner we have

$$
\begin{aligned}
\text{LI} &\to \text{OF} \\
\text{FE} &\to \text{KA} \\
\text{IS} &\to \text{HT} \\
\text{SH} &\to \text{??}
\end{aligned}
$$

Here we have a problem. S and H appear in the same column, so we cannot "make a rectangle" by finding two other letters as we did for the previous pairs. We need a new rule for this special case: if both letters appear in the same column, encipher them with the letters that appear directly beneath each. We then have

$$
\text{SH} \to \text{XN}
$$

If one of the letters was in the last row, we'd circle back to the top of the column to find its encipherment. Now we continue with the other pairs.

$$
\begin{aligned}
\text{OR} &\to \text{MT} \\
\text{TA} &\to \text{QD} \\
\text{ND} &\to \text{OC} \\
\text{HA} &\to \text{FC} \\
\text{RD} &\to \text{TB} \\
\text{LI} &\to \text{OF} \\
\text{KE} &\to \text{PK} \\
\text{AB} &\to \text{??}
\end{aligned}
$$

Another problem! A and B appear in the same row. Again, we cannot form a rectangle. In this case, we simply take the letters directly to the right of each of the plaintext letters We get

$$
\text{AB} \to \text{BC}
$$

If one of the letters was in the last column, we'd circle back to the start of the row to find its encipherment. Our rules now allow us to finish the encryption:

$$
\begin{aligned}
\text{OD} &\to \text{TI} \\
\text{YB} &\to \text{WD} \\
\text{UI} &\to \text{TK} \\
\text{LD} &\to \text{OA} \\
\text{IN} &\to \text{HO} \\
\text{GD} &\to \text{IB} \\
\text{WA} &\to \text{VB} \\
\text{RF} &\to \text{QG}
\end{aligned}
$$

Thus, our ciphertext is

OFKAH TXNMT QDOCF CTBOF PKBCT IWDTK OAHOI BVBQG.

Although it did not arise with this message, there is an ambiguous case. What do we do when a plaintext pair consists of two of the same letter? Do we shift down (because they are in the same column) or shift to the right (because they are in the same row)?

The solution is to avoid this situation! An X is to be inserted between doubled letters prior to encipherment to break them up. Because X is a rare letter, it will not cause any confusion. A recipient who, after deciphering, sees an X between two Ls or two Os, for example, would simple remove the X. The example above was just for instructional purposes. The alphabet in the grid would normally be scrambled (perhaps using an easy to remember keyword).

The first recorded solution of the Playfair cipher was by Joseph O. Mauborgne in 1914. At this time, Playfair was the field cipher for the British. There are reports that this cipher was used in the Boer War (1899–1902),[22] but the example below is more recent. Imagine that, as an Australian coastwatcher, you're the intended recipient of the following Playfair cipher sent on August 2, 1943, in the midst of the war in the Pacific (Figure 4.11).

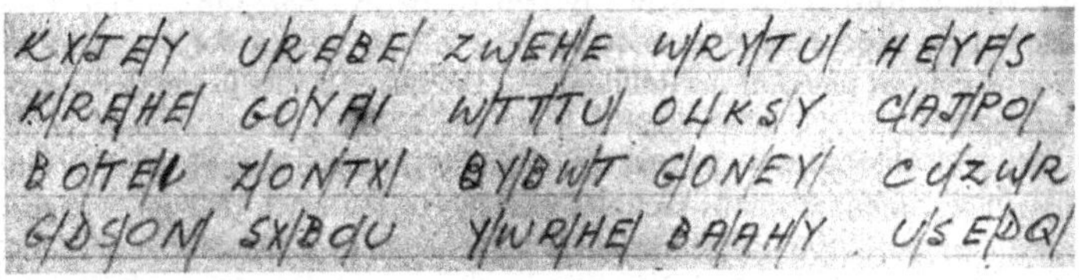

Figure 4.11 Playfair message sent during World War II in the Pacific Theater. (Courtesy of the David Kahn Collection, National Cryptologic Museum, Fort Meade, Maryland.)

The ciphertext, which is typically sent in groups of five letters, has already been split into groups of size two. At about the middle of the second line you notice a doubled letter, TT. You fear the message has been garbled or, perhaps, isn't a Playfair cipher, after all.

In any case, the key is ROYAL NEW ZEALAND NAVY, so you form the following square:

```
R O Y A L
N E W Z D
V B C F G
H I K M P
Q S T U X
```

You begin deciphering. (recall I and J are not distinguished here)

KX → PT	JE → BO	YU → AT	RE → ON	BE → EO
ZW → WE	EH → NI	EW → NE	RY → LO	TU → ST
HE → IN	YF → AC	SK → TI	RE → ON	HE → IN
GO → BL	YF → AC	IW → KE	TT → ??	TU → ST
OL → RA	KS → IT	YC → TW	AJ → OM	PO → IL
BO → ES	TE → SW	IZ → ME	ON → RE	TX → SU
BY → CO	BW → CE	TG → XC	ON → RE	EY → WO
CU → FT	ZW → WE	RG → LV	DS → EX	ON → RE
SX → QU	BO → ES	UY → TA	WR → NY	HE → IN
BA → FO	AH → MR	YU → AT	SE → IO	DQ → NX

Putting it all together, you get

PTBOATONEOWENINELOSTINACTIONINBLACKE??STRAITTWOMILESSW
MERESUCOCEXCREWOFTWEL VEXREQUESTANYINFOMRATIONX

The mystery ciphertext pair TT, deciphered as ?? temporarily, is easy to determine in the context of the plaintext BLACKE??STRAIT. This must be BLACKETT STRAIT. The TT was left as is, not even enciphered![23] After inserting word spacing you get

PT BOAT ONE OWE NINE LOST IN ACTION IN BLACKETT STRAIT TWO MILES
SW MERESU COCE X CREW OF TWELVE X REQUEST ANY INFORMATION X

---

[22] Kahn, David, *The Codebreakers*, second edition, Scribner, New York, 1996, p. 202.

[23] Although an obvious weakness, the Playfair cipher was actually sometimes used this way, as the present example shows!

There's an error, but again, context makes it easy to fix. You produce the final message.

```
PT BOAT ONE ONE NINE LOST IN ACTION IN BLACKETT STRAIT TWO MILES
SW MERESU COVE X CREW OF TWELVE X REQUEST ANY INFORMATION X
```

The message is describing John F. Kennedy's patrol torpedo boat, which had been sliced in half by a Japanese destroyer that had rammed it. More messages will follow and eventually allow the crew, which had swum ashore, to be rescued from the behind enemy lines. Perhaps years later you will recall how the failure of the Japanese to read this (and other messages) may have saved the life of a future American president.

> On dividing the unknown substitution into groups of two letters each, examine the groups and see if any group consists of a repetition of the same letter, as SS. If so, the cipher is not a Playfair.
>
> —J. O. Mauborgne[24]

Although Mauborgne was one of the (re)discoverers of the only unbreakable cipher, his advice above wasn't correct this time. Ciphers are often used improperly by individuals in highly stressful situations. Also, a letter could repeat accidentally due to Morse mutilation.

4.5 Playfair Cryptanalysis

The unicity point for a Playfair cipher is 22.69 letters, so a message longer than this should have a unique solution.[25] Sir George Aston issued the following 30-letter Playfair as a challenge.[26]

BUFDA GNPOX IHOQY TKVQM PMBYD AAEQZ

Alf Mongé solved it (by hand) in the following manner.[27] Splitting the ciphertext into pairs and numbering the pairs for easy reference, we have:

```
 1   2   3   4   5   6   7   8   9   10  11  12  13  14  15
 BU  FD  AG  NP  OX  IH  OQ  YT  KV  QM  PM  BY  DA  AE  QZ
```

Indicating the pairs OQ and QM in positions 7 and 10, Mongé pointed out that O and Q are close to each other in a straight alphabet, as are Q and M. Looking for two other high frequency digraphs with letters that are close to each other in the alphabet and have a letter in common between the pairs, Mongé came up with NO and OU. (He did not say how many other possibilities he tried first!) The proposed ciphertext/plaintext pairings would arise from the following square.

---

[24] Mauborgne, Joseph O., *An Advanced Problem in Cryptography and its Solution*, second edition, Army Service Schools Press, Fort Leavenworth, Kansas, 1918.

[25] Deavours, Cipher A., "Unicity Points in Cryptanalysis," *Cryptologia*, Vol. 1, No. 1, January 1977, pp. 46–68.

[26] Aston, George, *Secret Service*, Faber & Faber, London, England, 1933.

[27] Mongé, Alf, "Solution of a Playfair Cipher," *Signal Corps Bulletin*, No. 93, November–December 1936, reprinted in Friedman, William F., *Cryptography and Cryptanalysis Articles*, Vol. 1, Aegean Park Press, Laguna Hills, California, 1976 and in Winkel, Brian J., "A Tribute to Alf Mongé," *Cryptologia*, Vol. 2, No. 2, April 1978, pp. 178–185.

```
    1  2  3  4  5
    6  7  8  9 10
   11 12 13 14 15
    M  N  O  Q  U
    V  W  X  Y  Z
```

Thus, Mongé determined 40% of the square already! Returning to the ciphertext, he filled in as much as he could, indicating multiple possibilities where they existed and were not too numerous.

```
1   2   3   4   5   6   7   8   9  10  11  12  13  14  15
BU  FD  AG  NP  OX  IH  OQ  YT  KV  QM  PM  BY  DA  AE  QZ
--  --  --  --  -O  --  NO  --  --  OU  --  --  --  --  UY
M       M               V   W       N   Q
N       U               W   X       O   V
O       Q               X   Y       Q   W
Q       U               Z   Z       U   X
                        Q   M           Z
```

Which letters do you think would make the best choices for positions 8 and 9? Think about it for a minute before reading the answer below!

Mongé selected W and Y to form the words NOW and YOU, but positions 8 and 9 represent pairs of letters, so there must be a two-letter word connecting NOW and YOU in the plaintext. Making these partial substitutions, T must occur in position 2, 7, or 12 of the enciphering square and K must be in position 4, 9, or 14. Mongé assumed K didn't occur in the key, forcing it to be in position 14. He then had the following partially recovered square to work with:

```
    1  T?  3  4  5
    6  T?  8  9 10
   11  T? 13  K 15
    M   N  O  Q  U
    V   W  X  Y  Z
```

Position 15 must be L, so the square quickly becomes:

```
    1  T?  3  4  5
    6  T?  8  9 10
   11  T? 13  K  L
    M   N  O  Q  U
    V   W  X  Y  Z
```

Moving back to the ciphertext/plaintext again gives:

```
1   2   3   4   5   6   7   8   9  10  11  12  13  14  15
BU  FD  AG  NP  OX  IH  OQ  YT  KV  QM  PM  BY  DA  AE  QZ
--  --  --  --  -O  --  NO  W-  -Y  OU  --  --  --  --  UY
M       M                           N   Q
N       O                           O   V
O       Q                           Q   W
Q       U                           U   X
L       T                               Z
```

Mongé then focused on the letters T and K from ciphertext groups 8 and 9. If T was in position 12 of the square, then the keyword would be at least 12 letters long and consist of A, B, C, D, E, F, G, IJ, P, R, S, and T. Mongé rejected this as unlikely, so T was in either position 2 or 7 of the square.

If the keyword was less than 11 letters long, then three of the letters A, B, C, D, E, F, G, H, and IJ would have to appear in positions 11, 12, and 13 of the square.

Mongé noticed that H and IJ cannot appear in position 11 of the square, as there are not enough letters between them and K to fill in positions 12 and 13. Thus, position 11 must be A, B, C, D, E, F, or G. Mongé simply tried each possibility and found that only one worked. For example, placing A in position 11, causes ciphertext block 9 to decipher to AY, which, in context, gives a plaintext of NOW –A YOU. There is no letter that can be placed in front of the plaintext A that makes sense. Similarly, all but one of the other possibilities fizzle out.

Placing F in square 11 makes the ciphertext block 9 decipher to FY, so that the plaintext contains the phrase NOW –F YOU, which may sound vulgar until it is recalled that "–" represents an unknown letter. It is then easy to see that the plaintext must be NOW IF YOU. Thus, it is also revealed that I must be in position 4 or 9 of the square. Once F is placed in position 11 and I is forced in position 4 or 9 of the partially recovered square, positions 12 and 13 can be nothing but G and H. We now have

```
1   T?  3   I?  5
6   T?  8   I?  10
F   G   H   K   L
M   N   O   Q   U
V   W   X   Y   Z
```

Continuing to work back and forth between the square and the ciphertext, Mongé wrote

```
1    2    3    4    5    6    7    8    9    10   11   12   13   14   15
BU   FD   AG   NP   OX   IH   OQ   YT   KV   QM   PM   BY   DA   AE   QZ
--   --   --   --   HO   -K   NO   W-   -Y   OU   --   --   --   --   UY
```

and saw that the phrase NOW IF YOU was really KNOW IF YOU.

If the attacker can recover the keyword, the solution is immediately obtained. Mongé's work thus far indicates that IJ, P, R, S, and T must be part of the key. He expected more than a single vowel in the key, and so supposed either A or E or both were part of the key. That then left B, C, and D, as (perhaps) not part of the key. Mongé therefore placed them in the square as follows

```
1   T?  3   I   5
6   T?  B   C   D
F   G   H   K   L
M   N   O   Q   U
V   W   X   Y   Z
```

This has the added benefit (if correct) of eliminating the ambiguity over the location of I. This conjecture may be tested against the ciphertext as follows:

```
1    2    3    4    5    6    7    8    9    10   11   12   13   14   15
BU   FD   AG   NP   OX   IH   OQ   YT   KV   QM   PM   BY   DA   AE   QZ
DO   L-   --   --   HO   -K   NO   W-   -Y   OU   --   CX   --   --   UY
```

The decipherment of group 12 as CX may be discouraging at first, but we recall that doubled plaintext letters are broken up with an X if they are to be enciphered together. Because C can be doubled in English words, we continue on, now following up on group 13 representing C–. This hypothesis suggests A belongs in position 7 of the square.

1	T	3	I	5
6	A	B	C	D
F	G	H	K	L
M	N	O	Q	U
V	W	X	Y	Z

This assignment also eliminates the ambiguity concerning the position of T in the square.

Mongé was then able to determine the keyword, which consisted of the letters E, I, P, R, S, T, but in his explanation continued the analysis by looking at the ciphertext again. Feel free to take a moment to determine the key before reading the rest of Mongé's explanation! The ciphertext/plaintext pairings now become[28]

1	2	3	4	5	6	7	8	9	10	11	12	13	14	15
BU	FD	AG	NP	OX	IH	OQ	YT	KV	QM	PM	BY	DA	AE	QZ
DO	L-	TA	--	HO	-K	NO	WI	FY	OU	--	CX	C-	--	UY
	LP		MA		PK							CP		
	LR		MT		RK							CR		
	LS		OT		SK							CS		
	LE		UT		EK							CE		

Group 2 must be LE, which then forces E to take position 6 in the square, which, in turn, makes ciphertext groups 13 and 14, CE and ED, respectively. Thus, the message ends with --CXCEEDUY. Recalling that the X is only in the plaintext to split up the pair CC, the ending of the message should read --CCEEDUY. At this point, the attacker either may guess SUCCEED followed by two meaningless letters so that the ciphertext could be evenly split into groups of five characters or he may look at the recovered square again, which is now almost complete, testing the very few remaining for a meaningful plaintext. With either technique, the square and message are both quickly revealed as

S	T	R	I	P
E	A	B	C	D
F	G	H	K	L
M	N	O	Q	U
V	W	X	Y	Z

and

BU	FD	AG	NP	OX	IH	OQ	YT	KV	QM	PM	BY	DA	AE	QZ
DO	LE	TA	UT	HO	RK	NO	WI	FY	OU	SU	CX	CE	ED	UY

DO LET AUTHOR KNOW IF YOU SUCCEED

There were several places in the above solution where guesses or assumptions were made. They all proved correct, but it wouldn't have been a disaster if one or more were wrong. We'd simply generate some impossible plaintext and then backtrack to try another guess. It's no different from the backtracking that is typically needed when attempting to solve a monoalphabetic substitution cipher by hand.

---

[28] Mongé left out the possibilities for ciphertext blocks 11 and 14 as being too numerous.

Mongé didn't indicate how many incorrect guesses he may have made, but as the challenge appeared in 1933 and Mongé's solution appeared in 1936, there is a bound of a few years on how long this could possibly have taken. It's unlikely, however, that this is a tight bound! William Friedman had written, "The author once had a student [Mongé] who 'specialized' in Playfair ciphers and became so adept that he could solve messages containing as few as 50-60 letters within 30 minutes."[29]

4.5.1 Computer Cryptanalysis

In 2008, a paper by Michael Cowan described a new attack against short (80–120 letters) Playfair ciphers.[30] This attack has nothing in common with Mongé's approach. In fact, it would be completely impractical to try to implement Cowan's attack by hand; however, with the benefit of the computer, it is a very efficient approach. It's important to note that Cowan's attack doesn't assume the key is based on a word. The order of the letters in the enciphering square may be random. Cowan's approach was to use simulated annealing, which is a modification of hill climbing.

In hill climbing, we start by guessing at a solution (be it a substitution alphabet for a monoalphabetic substitution cipher or a key for a Playfair cipher). Then we make a change to the guess (switch a few letters around, for example). The original guess and the new slightly changed guess are compared. Some method of scoring assigns a value to each, based on how close they are to readable messages in whatever language is expected. Scoring can be done, for example, by summing the frequencies of the individual letters or digraphs. We keep whichever guess, the original or the modified, scores higher and discard the other. Then we make another small change and compare again. This process is repeated thousands of times, which is why it is not practical to do by hand. The idea is that the scores continue to climb until we get to the top, where the correct solution is found.

The analogy of physically climbing a real hill allows us to see how this method can fail. Suppose we seek to get to the highest point in our neighborhood. Taking random steps and only backtracking if we do not ascend in that particular direction seems like a good idea, but we could end up at the top of a small hill from which we can see a higher peak but cannot get there, as a step in any direction will take us downhill. In mathematical lingo, we have found a local (or relative) max, but not the global (or absolute) max.

Simulated annealing provides an opportunity to escape from local maxima and make it to the global maximum by only moving in the uphill direction (to a higher scoring guess) with a certain probability. That is, after scoring two guesses, we might move to the lower scoring guess 40% of the time. This percentage is known as the *temperature* of the process. The temperature is lowered slowly over the course of tens of thousands of modifications. The name *simulated annealing* makes an analogy with the annealing process in metallurgy in which a metal is heated to a specific temperature and then slowly cooled to make it softer.

Cowan's changes to the key, for the purpose of comparing the resulting scores, consisted of a mix of row swaps, column swaps, and individual letter swaps, as well as the occasional flip of the square around the NE–SW axis. For scoring, he found tetragraph frequencies worked best.

---

[29] Friedman, William F., *Military Cryptanalysis, Part I*, Aegean Park Press, Laguna Hills, California, 1996, p. 97, taken here from Winkel, Brian J., "A Tribute to Alf Mongé," *Cryptologia*, Vol. 2, No. 2, April 1978, pp. 178–185.

[30] Cowan, Michael J., "Breaking Short Playfair Ciphers with the Simulated Annealing Algorithm," *Cryptologia*, Vol. 32, No. 1, January 2008, pp. 71–83.

Cowan gives much more detail in his paper. It seems that his approach is reliable, but runtime may vary greatly depending on the particular cipher being examined and the initial guess. His solving times for particular ciphers (averaged over many initial guesses) ranged from about 6 seconds to a little more than a half hour.

After the idea of replacing characters two at a time is contemplated, a good mathematician ought to quickly think of a generalization. Why not replace the characters three at a time (trigraphic substitution) or four at a time, or *n* at a time? A nice mathematical way of doing this (using matrices) is described in Section 6.1.

4.6 Kerckhoffs's Rules

I have pointed out how the method of encipherment can be determined for many cipher systems by an examination of sample ciphertext. In general, this is not necessary. It's usually assumed that the method is known. The security of a cipher must lie in the secrecy of the key. You cannot hide the algorithm (see K2 below). This basic tenet of cryptography goes back to Auguste Kerckhoffs (Figure 4.12). In his *La Cryptographie Millitaire* (1883), he stated six rules that, with the change of only a word or two, are still valid today:[31]

Figure 4.12   Auguste Kerckhoffs (1835–1903). (https://en.wikipedia.org/wiki/Auguste_Kerckhoffs.)

K1. The system should be, if not theoretically unbreakable, unbreakable in practice.

K2. Compromise of the system should not inconvenience the correspondents.

K3. The method for choosing the particular member (key) of the cryptographic system to be used should be easy to memorize and change.

K4. Ciphertext should be transmittable by telegraph.

---

[31] Taken here from Konheim, Alan G., *Cryptography, A Primer*, John Wiley & Sons, New York, 1981, p. 7.

K5. The apparatus should be portable.

K6. Use of the system should not require a long list of rules or mental strain.

Item K6 was echoed by Claude Shannon years later in his paper "Communication Theory of Secrecy Systems" with a justification: "Enciphering and deciphering should, of course, be as simple as possible. If they are done manually, complexity leads to loss of time, errors, etc. If done mechanically, complexity leads to large expensive machines."[32] Shannon also shortened K2 to "the enemy knows the system," which is sometimes referred to as *Shannon's maxim*.

Revealing the details of the system is actually a good way to make sure it's secure. If the world's best cryptanalysts cannot crack it, you have an ad campaign that money can't buy. Despite all of this, some modern purveyors of cryptosystems still try to keep their algorithms secret. An example that will be examined in greater detail in Section 19.5 is RC4, sold by RSA Data Security, Inc. Despite the effort to maintain secrecy, the algorithm appeared on the cypherpunks mailing list.[33]

References and Further Reading

On Bacon's Biliteral Cipher (and Some Bad Ideas It Inspired)

Bacon, Francis, *Of the Proficience and Advancement of Learning, Divine and Humane*, Henrie Tomes, London, 1605. Bacon's biliteral cipher is only alluded to here.

Bacon, Francis, *De Dignitate et Augmentis Scientarum*, 1623. Bacon describes his cipher in detail here. An English translation first appeared in 1640.

Donnelly, Ignatius, *The Great Cryptogram*, R. S. Peale and Company, Chicago, Illinois, 1888.

Donnelly, Ignatius, *The Cipher in the Plays and on the Tombstone*, Verulam Publishing, Minneapolis, Minnesota, 1899. Donnelly didn't give up!

Friedman, William F., "Shakespeare, Secret Intelligence, and Statecraft," *Proceedings of the American Philosophical Society*, Vol. 106, No. 5, October 11, 1962, pp. 401–411.

Friedman, William F. and Friedman, Elizebeth S., *The Shakespearean Ciphers Examined*, Cambridge University Press, Cambridge, UK, 1957.

Friedman, William, and Elizebeth S. Friedman, "Afterpiece," *Philological Quarterly*, Vol. 41, No. 1, January 1962, pp. 359–361.

Gallup, Elizabeth Wells, *The Bi-literal Cypher of Francis Bacon*, Howard Publishing Company, Detroit, Michigan, 1899.

Hedges, S. Blair, "A Method for Dating Early Books and Prints Using Image Analysis," *Proceedings of the Royal Society A: Mathematical, Physical, and Engineering Sciences*, Vol. 462, No. 2076, December 8, 2006, pp. 3555–3573. This paper was described in laymen's terms in Marino (2007).

Howe, Norma, *Blue Avenger Cracks the Code*, Henry Holt and Company, New York, 2000. This is a young adult novel that incorporates Bacon's cipher into the text by giving an explanation and using it to relate messages. I enjoyed it even though I disagree with the arguments it makes for attributing authorship of the plays to Edward de Vere (apparently the present favorite of the contrarians). Encouraging youth to think for themselves and not to fear being different is present as a nice theme, and the authorship question does not play an irritatingly large role, in my opinion. Although the book is a sequel, it stands very well on its own.

Jenkins, Sally, "Waiting for William," *The Washington Post Magazine*, August 30, 2009, pp. 8–15, 25–28. This article reports on the newly discovered Shakespeare portrait. I've assumed it's authentic, but the experts aren't unanimous in accepting it.

---

[32] Shannon, Claude, "Communication Theory of Secrecy Systems," *The Bell System Technical Journal*, Vol. 28, No. 4, October 1949, pp. 656–715. Shannon noted, "The material in this paper appeared in a confidential report, 'A Mathematical Theory of Cryptography,' dated Sept. 1, 1945, which has now been declassified."

[33] See http://www.cypherpunks.to/ for more information on the cypherpunks.

Marino, Gigi, "The Biologist as Bibliosleuth," *Research Penn State*, Vol. 27, No. 1, Fall 2007, pp. 13–15.

Pyle, Joseph Gilpin, *The Little Cryptogram*, The Pioneer Press Co., St. Paul, Minnesota, 1888, 29 pages. This is a spoof of Donnelly's 998-page book *The Great Cryptogram*.

Schmeh, Klaus, "The Pathology of Cryptology – A Current Survey," *Cryptologia*, Vol. 36, No. 1, January 2012, pp. 14–45. Schmeh recommends Pyle's approach to investigation of alleged hidden messages: If the technique yields messages in similar items, selected at random, or different messages from the original source, then it is likely to be an invalid technique.

Stoker, Bram, *Mystery of the Sea*, Doubleday and Company, New York, 1902. Stoker used the biliteral cipher extensively in this novel—not to hide a message within its text, but rather for two characters in the novel to communicate with each other. The two symbols needed for the cipher are manifested in a great variety of ways, not limited to print.

Walpole, Horace, *Historic Doubts on the Life and Reign of King Richard the Third*, J. Dodsley, London, UK, 1768, reprinted by Rowman & Littlefield, Totowa, New Jersey, 1974. The claims Pratt says Walpole makes are not to be found in here!

Zimansky, Curt A., "Editor's Note: William F. Friedman and the Voynich Manuscript," *Philological Quarterly*, Vol. 49, No. 2, October 1970, pp. 433–443. The last two pages reproduce text masking messages via Bacon's biliteral cipher. This paper was reprinted in Brumbaugh, Robert S., editor, *The Most Mysterious Manuscript*, Southern Illinois University Press, Carbondale and Edwardsville, Illinois, 1978, pp. 99–108 with notes on pp. 158–159.

On Wheel Ciphers

Bazeries, Étienne, *Les Chiffres Secrets Dévoilés*, Charpentier-Fasquelle, Paris, France, 1901.

Bedini, Silvio A., *Thomas Jefferson Statesman of Science*, Macmillan, New York, 1990. Although this biography contains only a few paragraphs dealing with cryptology, it does focus on Jefferson's scientific interests and accomplishments.

de Viaris, Gaëtan, *L'art de Chiffrer et Déchiffrer les Dépêches Secretes*, Gauthier-Villars, Paris, France, 1893. The attack described by de Viaris makes the same assumption as the example in this chapter.

Friedman, William F., *Several Machine Ciphers and Methods for their Solution*, Publication No. 20, Riverbank Laboratories, Geneva, Illinois, 1918. Friedman showed attacks on the wheel cipher in part III of this paper. This paper was reprinted together with other Friedman papers in Friedman, William F., *The Riverbank Publications*, Vol. 2, Aegean Park Press, Laguna Hills, California, 1979. As the original printing only consisted of 400 copies, I suggest looking for the reprint instead.

Gaddy, David W., "The Cylinder-Cipher," *Cryptologia*, Vol. 19, No. 4, October 1995, pp. 385–391. Gaddy argues that the wheel cipher was probably not an independent invention of Jefferson, but rather that he got the idea from an already existing wheel or description.

Kruh, Louis, "The Cryptograph that was Invented Three Times," *The Retired Officer*, April 1971, pp. 20–21.

Kruh, Louis, "The Cryptograph that was Invented Three Times," *An Cosantoir: The Irish Defense Journal*, Vol. 32, No. 1–4, January–April, 1972, pp. 21–24. This is a reprint of Kruh's piece from *The Retired Officer*.

Kruh, Louis, "The Evolution of Communications Security Devices," *The Army Communicator*, Vol. 5, No. 1, Winter 1980, pp. 48–54.

Kruh, Louis, "The Genesis of the Jefferson/Bazeries Cipher Device," *Cryptologia*, Vol. 5, No. 4, October 1981, pp. 193–208.

Mellen, Greg and Lloyd Greenwood, "The Cryptology of Multiplex Systems," *Cryptologia*, Vol. 1, No. 1, January 1977, pp. 4–16. This is an interesting introduction and overview of wheel cipher/strip cipher systems. The cryptanalysis is done in the sequel, referenced below.

Mellen, Greg and Lloyd Greenwood, "The Cryptology of Multiplex Systems. Part 2: Simulation and Cryptanalysis," *Cryptologia*, Vol. 1. No. 2, April 1977, pp. 150–165. A program in FORTRAN V to simulate the M-94 is described. Cryptanalysis for three cases is examined: (1) known alphabets and known crib; (2) unknown alphabets and known crib ("A crib of 1000–1500 characters is desirable. Shorter cribs of several hundred letters can be used but prolong the effort."); and (3) unknown alphabets and unknown crib. The authors noted, "The general method for this case was originated by the Marquis de Viaris in 1893 [15] and elaborated upon by Friedman [16]." In the reference section at the end of this paper, we see that [15] refers to David Kahn's *The Codebreakers*, pp. 247–249, but [16] is followed by blank space. Perhaps this work by Friedman was classified at the time and couldn't be cited!

Rohrbach, Hans, "Report on the Decipherment of the American Strip Cipher O-2 by the German Foreign Office (Marburg 1945)," *Cryptologia*, Vol. 3, No. 1, January 1979. Rohrbach was one of the German codebreakers who cracked this cipher during World War II. Following a preface, his 1945 report on how this was done is reprinted.

Smithline, Lawren M., "A Cipher to Thomas Jefferson: A Collection of Decryption Techniques and the Analysis of Various Texts Combine in the Breaking of a 200-year-old Code," *American Scientist*, Vol. 97, No. 2, March–April 2009, pp. 142–149.

Smoot, Betsy Rohaly, "Parker Hitt's First Cylinder Device and the Genesis of U.S. Army Cylinder and Strip Devices," *Cryptologia*, Vol. 39, No. 4, October 2015, pp. 315–321.

For 29 years (116 issues), *Cryptologia* almost never repeated a cover. When it was decided to settle on a single cover, only changing the dates each time, the image that won was of a wheel cipher (Figure 4.13). This is fitting, as a wheel cipher cover marked the journal's debut.

Figure 4.13 *Cryptologia's* new look.

On the Playfair Cipher

Cowan, Michael J., "Breaking Short Playfair Ciphers with the Simulated Annealing Algorithm," *Cryptologia*, Vol. 32, No. 1, January 2008, pp. 71–83.

Gillogly, James J. and Larry Harnisch, "Cryptograms from the Crypt," *Cryptologia*, Vol. 20, No. 4, October 1996, pp. 325–329.

Ibbotson, Peter, "Sayers and Ciphers," *Cryptologia*, Vol. 25, No. 2, April 2001, pp. 81–87. This article is on the ciphers of Dorothy Sayers.

Knight, H. Gary, "Cryptanalysts' Corner," *Cryptologia*, Vol. 4, No 3, July 1980, pp. 177–180.

Mauborgne, Joseph O., *An Advanced Problem in Cryptography and its Solution*, second edition, Army Service Schools Press, Fort Leavenworth, Kansas, 1918. This pamphlet shows how to break Playfair ciphers.

Mitchell, Douglas W., "A Polygraphic Substitution Cipher Based on Multiple Interlocking Applications of Playfair," *Cryptologia*, Vol. 9, No. 2, April 1985, pp. 131–139.

Rhew, Benjamin, *Cryptanalyzing the Playfair Cipher Using Evolutionary Algorithms*, December 9, 2003, 8 pages, http://web.mst.edu/~tauritzd/courses/ec/fl2003/project/Rhew.pdf.

Stumpel, Jan, *Fast Playfair Programs*, https://web.archive.org/web/20090802015059/http://www.jw-stumpel.nl:80/playfair.html. Cowan's program was based on Stumpel's platform.

Winkel, Brian J., "A Tribute to Alf Mongé," *Cryptologia*, Vol. 2, No. 2, April 1978, pp. 178–185.

Chapter 5

World War I and
Herbert O. Yardley

In World War I, all of the ciphers previously discussed in this book saw use, even the weakest ones! We focus here on the best systems, ADFGX and ADFGVX. The fascinating life of Herbert O. Yardley is also covered, but first we look at a coded telegram that had a huge effect on the war.

5.1 The Zimmermann Telegram

It is a common misconception that the sinking of the *Lusitania* got the United States into World War I. A simple checking of dates casts serious doubt on this idea. The Germans sank the *Lusitania* on May 7, 1915, and the United States didn't declare war until April 6, 1917. Now compare the latter date to the revelation of what has become known as the Zimmermann Telegram, which was given to the press (at President Wilson's request) on March 1, 1917.

The famous Zimmermann Telegram is pictured in Figure 5.1. It was sent to Felix von Eckhardt, the German ambassador in Mexico, by Arthur Zimmerman (Director of the German Ministry of Foreign Affairs). It was written using a code, rather than a cipher; that is, the basic unit of substitution was the word. For example, `alliance` was written as `12137` and `Japan` was written as `52262`. The Germans normally included an extra step in which the numbers were enciphered (i.e., an enciphered code), but the process was skipped for this message. After decoding and translating, the telegram reads:[1]

> We intend to begin on the first of February unrestricted submarine warfare. We shall endeavor in spite of this to keep the United States of America neutral. In the event of this not succeeding, we make Mexico a proposal or alliance on the following basis: make war together, make peace together, generous financial support and an understanding on our part that Mexico is to reconquer the lost territory in Texas, New Mexico, and Arizona. The settlement in detail is left to you. You will inform the

---

[1] http://www.nara.gov/education/teaching/zimmermann/zimmerma.html.

President of the above most secretly as soon as the outbreak of war with the United States of America is certain and add the suggestion that he should, on his own initiative, invite Japan to immediate adherence and at the same time mediate between Japan and ourselves. Please call the President's attention to the fact that the ruthless employment of our submarines now offers the prospect of compelling England in a few months to make peace.

—Signed, Zimmermann.

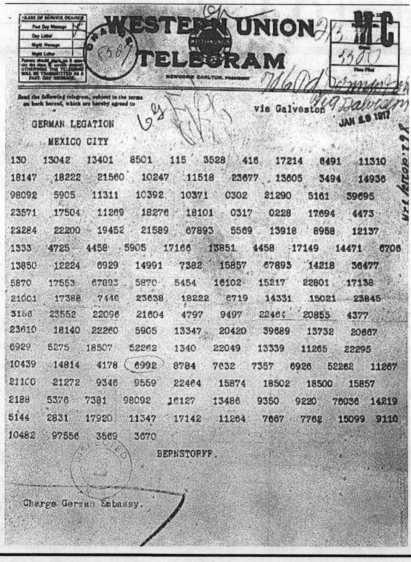

Figure 5.1 The Zimmermann telegram. (http://www.nara.gov/education/teaching/ zimmermann/bernstor.jpg.)

Mexico is hardly regarded as a military threat to the United States today, but 1917 was only one year after the punitive expedition of American troops into Mexico. Bearing this is mind helps us to see how Mexico might have reacted positively to German overtures. Herbert O. Yardley noted, "Mexico was openly pro-German. Our own spies who had been sent into Mexico reported that hundreds of German reservists who fled across the border at the declaration of war were recruiting and drilling Mexican troops."[2]

The British intercepted a copy of the telegram and broke the code. The message, along with the sinking of the *Laconia* (two years after the *Lusitania*), prompted America to join the war on the side of England. If America had not joined the war, Germany may have won. The British faced a challenging problem following their decoding of the telegram. How could they share it with America and (1) not tip the Germans off to the fact that their code had been broken, and (2) convince President Wilson that the telegram was real? Problem 2 ended up being solved by Zimmermann himself, when he admitted on March 3 that the telegram was genuine, crushing theories that it was a British invention designed to gain the badly needed military strength of the United States.[3]

The telegram didn't arrive like an email. It passed through Washington, where it was decoded and put into an older code, as the ultimate destination didn't have the codebook it was originally sent in. The British were able to obtain the second version of the telegram that was received in Mexico. It is this version, after decoding, that they shared with President Wilson. It differed slightly from the original. The Germans recognized these differences and, instead of realizing their code was broken, assumed there must have been a traitor, or a flaw in the security protocol, in Mexico.

Although this is usually the only Zimmermann telegram mentioned in cryptology books, another enciphered message of interest was sent earlier, on January 26, 1915:[4]

> For Military Attaché: You can obtain particulars as to persons suitable for carrying on sabotage in the U.S. and Canada from the following persons: one, Joseph MacGarrity, Philadelphia; two, John P. Keating, Michigan Avenue, Chicago; three, Jeremiah O'Leary, 16 Park Row, New York. One and two are absolutely reliable and discreet. Number three is reliable but not always discreet. These persons were indicated by Sir Roger Casement. In the U.S. sabotage can be carried out in every kind of factory for supplying munitions of war. Railway embankments and bridges must not be touched. Embassy must in no circumstances be compromised. Similar precautions must be taken in regard to Irish pro-German propaganda.
>
> **—Signed, Zimmermann.**

A few words should be written on the British cryptologists at this point. First, they were ahead of the Germans, who didn't even have any cryptanalysts on the western front for the first two years of the war![5] But compared to today's gigantic cryptologic agencies, they were very few in number.

---

2   Yardley, Herbert O., *The American Black Chamber*, Espionage/Intelligence Library, Ballantine Books, New York, 1981, p. 90.

3   Kippenhahn, Rudolf, *Code Breaking: A History and Exploration*, The Overlook Press, New York, 1999, p. 65.

4   Sayers, Michael and Albert E. Kahn, *Sabotage! The Secret War Against America*, Harper & Brothers Publishers, New York, 1942, p. 8. A pair of pictures is provided on page 9.

5   Kahn, David, *The Codebreakers*, second edition, Scribner, New York, 1996, p. 313.

A total of 50 or so cryptanalysts worked in Room 40 of the Old Admiralty Building, where they recovered about 15,000 encoded or enciphered messages between October 1914 and February 1919.[6] Imagine yourself in a classroom with 50 of your peers. How many messages would your group be able to crack? Although the cryptanalysts that were recruited were carefully chosen, to be of very high intelligence, and many possessed fluency in foreign languages, they initially knew less about cryptanalysis than anyone who has read this far. So, the comparison is fair, and it's a good thing those in Room 40 were quick learners.

These cryptanalysts received help on a few occasions in the form of recovered German code books. One came as a gift from the Russians. On August 26, 1914, the German light cruiser *Magdeburg* became stuck in shallow water at Odensholm (now Osmussaar) in the Baltic Sea. This was Russian territory and their troops were able to recover the German Navy's main code book from the wreck, despite attempts by the Germans to destroy everything. The Russians then passed it on to the British, who were the stronger naval power and could use it to great advantage. Indeed, the Germans kept this particular code in use for years![7]

World War I is referred to as "The Chemists' War" due to the major role of chemical warfare, and World War II is called "The Physicists' War" because of the atomic bomb. It has been claimed that, if it occurs, World War III will be "The Mathematicians' War" (if anyone is left to talk about it). Just imagine a cyberattack that renders all of the enemies' computer systems useless and shuts down all enemy communications.

5.2 ADFGX: A New Kind of Cipher

Although the chemists had the lead role in World War I, there were new cryptologic developments, as well. The German systems that became known as ADFGX and ADFGVX provide an example; however, these new ciphers, soon to be described, were not the only ones the Germans employed. Over the course of the war they used a wide range of ciphers that even included monoalphabetic substitution![8]

It's always a good idea to introduce new ciphers shortly before major offensives and this is what the Germans did. A little more than two weeks before General Ludendorff (Figure 5.2) launched their March 21, 1918 attack ADFGX hit the wires and airwaves, and on June 1, the system was modified to become ADFGVX.[9] The cipher got its name from the fact that these were the only letters that appeared in the ciphertext. They were specifically chosen for their ease in being distinguished from one another when transmitted in Morse code. A lot of thought and testing went into these ciphers. The designer, Fritz Nebel, had 60 cryptanalysts try to crack the system prior to its deployment.[10]

Nebel's new system, although exotic sounding, is simply a combination of two ciphers we have already seen, namely Polybius (Section 1.2) and columnar transposition (Section 3.1).[11] When

---

6   Kahn, David, *The Codebreakers*, second edition, Scribner, New York, 1996, pp. 275 and 273.

7   Rislakki, Jukka, "Searching for Cryptology's Great Wreck," *Cryptologia*, Vol. 31, No. 3, July 2007, pp. 263–267.

8   Kahn, David, *The Codebreakers*, second edition, Scribner, New York, 1996, p. 307.

9   Kahn, David, *The Codebreakers*, second edition, Scribner, New York, 1996, p. 344.

10 Norman, Bruce, "The ADFGVX Men," *The Sunday Times Magazine*, August 11, 1974, pp. 8–15, p. 11 cited here.

11 This is not the first time that substitution and transposition were combined. Some earlier instances are pointed out in the References and Further Reading list at the end of this chapter.

Figure 5.2 General Erich Ludendorff directed Germany's spring offensive under the protection of a new and seemingly secure cipher system. (http://en.wikipedia.org/wiki/Erich_Ludendorff.)

two enciphering algorithms are combined, we refer to it as a *superencipherment*. Here's a Polybius square that was used:[12]

	A	D	F	G	V	X
A	c	o	8	x	f	4
D	m	k	3	a	z	9
F	n	w	1	0	j	d
G	5	s	i	y	h	u
V	p	l	v	b	6	r
X	e	q	7	t	2	g

If our message is GOTT MIT UNS,[13] our first step is to convert it to

XX AD XG XG DA GF XG GX FA GD

Next we use the second part of the key, a word that determines in what order the columns are to be read out for the transposition portion of the encipherment:

3	2	4	1
X	X	A	D
X	G	X	G
D	A	G	F
X	G	G	X
F	A	G	D

The ciphertext is then read off as DGFXD XGAGA XXDXF AXGGG. In actual use, the messages were typically much longer, as was the transposition key. Although the length of the key varied, it

---

[12] Kahn, David, *The Codebreakers*, second edition, Scribner, New York, 1996, p. 345.

[13] This translates to "God is With Us" and appeared on the belt buckles of some German troops in both World War I and World War II. If true, it looks like God is 0 for 2 in world wars.

was common to have 20 values. This is an example of a *fractionating cipher*, so-called because the original message letters are replaced by pairs that become split in the transposition step.

5.3 Cryptanalysis of ADFGX

The French were saved by Georges Painvin (Figure 5.3), who exerted a tremendous effort over nearly three months, first breaking ADFGX, which used a 5-by-5 Polybius square, and then ADFGVX, losing 33 pounds (15 kg) in the process but finally obtaining solutions revealing where the next attack was to be.[14] A former American diplomat, J. Rives Childs, observed that,[15]

> His masterly solutions of German ciphers caused him to become known as "artisan of the victory" over the Germans when Paris might have fallen but for the knowledge gained of German intentions by Painvin of where they would strike.

Figure 5.3 French cryptanalyst Georges Painvin. (Courtesy of the David Kahn Collection at the National Cryptologic Museum.)

---

[14] Kahn, David, *The Codebreakers*, second edition, Scribner, New York, 1996, p. 347.
[15] Childs, J. Rives, "My Recollections of G.2 A.6," *Cryptologia*, Vol. 2, No. 3, July 1978, pp. 201–214, p. 206 quoted here.

Painvin's solution didn't allow all ADFGVX ciphers to be read, but did crack some special cases. A general solution was found only after the war ended. We examine a special case below, but some of the comments can be applied more generally. If the following paragraphs are too abstract, feel free to skip ahead to the example.

To attack an ADFGVX cipher, all we need to worry about is how to unravel the transposition portion. Once this is done, we're left with a Polybius cipher that's very easy to break. The first step in unraveling the transposition portion is determining how many columns were used. We'll assume that the intercepted message has no "short columns." That is, the message forms a perfect rectangle. This will make things somewhat easier. As a first step, we determine if the number of columns is even or odd. This can be done by comparing two sets of frequencies. To see this, consider the generic examples below.

Expressing the pairs of ciphertext letters representing each message letter as BE (B stands for "beginning" and E stands for "end"), our rectangle, prior to transposition will take one of two forms, depending on whether the number of columns is even or odd.

Even # of Columns	Odd # of Columns
B E B E B E … B E	B E B E B E … B E B
B E B E B E … B E	E B E B E B … E B E
B E B E B E … B E	B E B E B E … B E B
B E B E B E … B E	E B E B E B … E B E
⋮ ⋮ ⋮	⋮ ⋮ ⋮ ⋮
B E B E B E … B E	(form of last row depends on number of rows)

For the even case, after transposing columns, each column will still be all Bs or all Es. For the odd case, after transposing columns, each column will still alternate Bs and Es.

Unless the placement of the letters in the Polybius square is carefully done to avoid it, the frequencies of the individual letters A, D, F, G, V, and X will differ as beginning characters and end characters in the ciphertext pairs. This allows a cryptanalyst to determine, using the patterns above, if the number of columns is even or odd. The manner in which this is done is now described.

Given a message of n characters, we divide by a number we feel is an upper bound for the number of columns, c, used. The result, n/c, will be a lower bound on the number of rows. Suppose n/c is 18, for example, then the first 18 letters in the ciphertext are all from the same column. It could be a column of all Bs, all Es, or half and half. Take the characters in the odd positions and construct a frequency distribution. These characters *must* all be of the same type (all Bs or all Es), whether the number of columns is even or odd. Now take the characters in the even positions and construct a frequency distribution.

Again, these characters *must* all be of the same type, B or E. Now compare the two frequency distributions. If they look similar, then all characters in that column are of the same type, so the number of columns must be even. If they look dissimilar, an odd number of columns must have been used.

If we decide that the frequency distributions match, then the number of columns is even. We can then plug the ciphertext into rectangles representing each possible number of columns, 2, 4, 6, …22, 24, 26, 28,… (the extreme ends are not likely). For each case, we may then calculate frequency distributions for each column. For the correct case, the distributions should fall into two distinct groups, each containing the same number of columns. A similar approach is used to determine the number of columns when it is known to be odd (See Exercise 7).

Once the number of columns is known, the ciphertext may be written out in the appropriate size rectangle. In order to undo the transposition, we first use the distinct frequency distributions to label each column as either a B or an E (in the odd case, this label would merely indicate which type of character begins the column). At this stage there is no way of knowing which is which, but that's not a problem! We simply label an arbitrary column as a B, then label the other columns with similar frequency distributions as Bs, and finally label the rest as Es.

If the first column we labeled was labeled correctly, all is well. If it was actually an E column, that's okay too, as the only change that makes is to index the entries in the Polybius square by column and row, rather than by row and column.

We may then pair B columns with E columns such that the resulting ciphertext letters have a frequency distribution that resembles the suspected plaintext language. Once the letters have been recreated by joining Bs to Es, we should be able to pick out some high frequency plaintext letters such as E and T. This will help us to order the paired columns correctly, especially if we have a crib or can find the appearances of common words such as THE and AND. When the columns are all correctly ordered, the rest is easy—just a monoalphabetic substitution cipher without word spacing. The example that follows should make this attack much clearer.

Example 1

We'll attack a message enciphered with the original version, ADFGX, so that we needn't be concerned with the frequencies of the various numbers, after we unravel the transposition. We'll assume that the message is in English and that I and J are enciphered by the same pair, as a 5-by-5 grid only allows for 25 plaintext characters. Our ciphertext is

```
AXDXD XDDDX DXXDD DXXDD DXXDG DXGXX XDFGA AGGAF FGGFA AAFFA
ADGGF GFFAD FAFAD FGGAF DFDXD XDFFX AXDXG FGFGX DXGXX DXFAD
XGFDA AFADF FFGGA DFGDF FADFA GAAFF GAAGG XFFDF GGDFG FDFFF
GAFDA FAFAF GAFAA FAFFX DXFXF GDDGX DFFFG XDFXX XDFFX ADAFA
FDXFX FGADD GGDDA AXXXX FFGXX FDXXD FXFGD DFFFD DXDDA DDXDD
GXAFD DXXXX DGGDF XXXXF XXDDD AGGDA FAAGF GGGFA GFGAG FFXAG
FFFGF FXXFX AFXDG DXXXD XXXXD XAADF FXDDF GXGDX XFXXX AGGXD
AFFAX FGFAX XXXAD FFDFD DFDXD XFFXX XDXDA GDGFX XGDFA FGXFG
DDXXX XXGXF XFXXF AXGXF DXDDD AXDDD XFXFD XAFDG XFGGA AAAGF
GAAAA FGAGA AGAAA FDGAF DAGAA GGFDF FGGGG GGAGG AFGAA GFFFG
FGAFF DFAFA GGAGA FGAAD AGGGF GFGFG FFAGA GGAAF AAAGD GGXGF
GGAFF AGAFG AAAAF GDAAG DGFGF FGGXX DDXFD FXXXG GAXXX GGDDG
FFGXD XGDGX FXXGA AGAFG ADAGG FXFGG GAAGA FFGFD DAAAA DGAFF
AFGDA ADFGD FAAGG AFAAG FGGGG FFGDG
```

We have 680 characters and we assume that no more than 30 columns were used, so there must be at least 22 characters in each column ($680/30 \approx 22.67$). We take the first 22 characters, AXDXD XDDDX DXXDD DXXDD DX, and find the frequency distribution for the characters in the odd positions:

$A = 1, D = 8, F = 0, G = 0, X = 2$

and in the even positions:

$A = 0, D = 4, F = 0, G = 0, X = 7$

Experience helps us to decide whether the two distributions are similar or dissimilar. The marked difference in the frequency of X might incline us to the latter, but F and G have identical frequencies, and A is as close as possible without being identical. With three out of five letters matching so closely, we conclude that the distributions are the same.

Under the assumption that the rectangle is completely filled in, we can also examine the last 22 characters, GD FAAGG AFAAG FGGGG FFGDG, to see if our conclusion is reinforced. For characters in odd positions, we have:

$$A = 2, D = 1, F = 4, G = 4, X = 0$$

and in the even positions:

$$A = 3, D = 1, F = 1, G = 6, X = 0$$

The values for D and X match exactly, the values for A only differ by one, and the values for G differ by two, so our conclusion gains further support. Also, observe how markedly both of these distributions (no Xs!) differ from the first 22 characters. It seems that the columns these letters represent cannot both be of the same type (B or E).

So, we have an even number of columns, and that number must divide 680. Our choices are 2, 4, 8, 10, 20, 34, 68, 170, 340, or 680. We already assumed that no more than 30 columns were used, so our list quickly shrinks to 2, 4, 8, 10, 20. The smaller values seem unlikely, so we test 10 and 20.

10 Columns												
Column	Letters	A	D	F	G	X	Column	Letters	A	D	F	G	X
1	1-68	12	18	12	12	14	6	341-408	7	13	15	9	24
2	69-136	13	13	18	12	12	7	409-476	21	11	12	12	12
3	137-204	11	12	27	8	10	8	477-544	19	3	19	27	0
4	205-272	5	20	12	8	23	9	545-612	10	10	13	21	14
5	273-340	10	12	16	14	16	10	613-680	21	8	15	22	2

Columns 8 and 10 stand out as having very few Xs, but we need to split the columns into two groups, Bs and Es, so each group must contain five columns. What other three columns resemble these? The next lowest frequencies for X are 10, 12, and 12—quite a jump!

20 Columns												
Column	Letters	A	D	F	G	X	Column	Letters	A	D	F	G	X
1	1-34	1	15	1	3	14	11	341-374	5	6	9	3	11
2	35-68	11	3	11	9	0	12	375-408	2	7	6	6	13
3	69-102	3	8	7	5	11	13	409-442	3	9	8	2	12
4	103-136	10	5	11	7	1	14	443-476	18	2	4	10	0
5	137-170	8	4	16	5	1	15	477-510	7	2	12	13	0
6	171-204	3	8	11	3	9	16	511-544	12	1	7	14	0
7	205-238	3	8	8	5	10	17	545-578	9	3	8	13	1
8	239-272	2	12	4	3	13	18	579-612	1	7	5	8	13
9	273-306	7	4	11	11	1	19	613-646	12	4	6	10	2
10	307-340	3	8	5	3	15	20	647-680	9	4	9	12	0

From this table, it's easy to split the columns into two groups with distinct frequency distributions. The frequency of X, by itself, clearly distinguishes them. Thus, we conclude that 20 columns were used. Our two distinct groups are

Group 1: Columns 1, 3, 6, 7, 8, 10, 11, 12, 13, 18.
Group 2: Columns 2, 4, 5, 9, 14, 15, 16, 17, 19, 20.

We must now pair them together to represent the plaintext letter. Our work thus far fails to indicate whether Group 1 columns are beginnings or ends of pairs. Happily, it doesn't matter. As mentioned prior to this example, reversing the order of the pairs arising from the Polybius cipher will simply correspond to someone having misused the table—writing first the column header, then the row header, instead of vice versa. As long as all pairs are switched, switching doesn't matter. So, we'll assume that the high frequency X group provides the beginnings. To determine which Group 2 column completes each of the Group 1 beginnings, Painvin, and the American cryptanalysts who examined the problem in the years to follow, simply looked at the frequency distributions for the various possibilities and selected the ones that looked the most like the language of the message. We'd prefer a more objective method, but the obvious approaches don't produce great results. Two approaches are examined below.

Although not discovered until after World War I, the index of coincidence seems like it should be a good measure. If a potential pairing of columns yields a value near 0.066, we favor it over pairings yielding other values. The complete results are given below, with the correct pairings underlined and boldfaced.

		End Column									
		2	4	5	9	14	15	16	17	19	20
	1	0.0909	0.0802	0.1087	0.0891	0.1462	0.1052	**0.1034**	0.0856	0.0963	0.0873
	3	0.0517	0.0481	0.0749	**0.0624**	0.0731	0.0517	0.0784	0.0481	0.0481	0.0446
	6	0.0535	0.0463	**0.0766**	0.0553	0.0766	0.0713	0.0731	0.0517	0.0446	0.0588
Start	7	0.0570	**0.0446**	0.0606	0.0606	0.0660	0.0642	0.0677	0.0535	0.0517	0.0535
Column	8	**0.0731**	0.0553	0.0856	0.0695	0.0998	0.0820	0.0784	0.0677	0.0980	0.0624
	10	0.0713	0.0606	0.0749	0.0588	0.1248	**0.0802**	0.1230	0.0570	0.0588	0.0695
	11	0.0624	0.0535	0.0517	0.0446	0.0873	0.0695	0.0802	**0.0463**	0.0624	0.0677
	12	0.0606	0.0713	0.0677	0.0660	0.0731	0.0695	0.0731	0.0624	0.0570	**0.0695**
	13	0.0570	0.0535	0.0660	0.0606	0.0873	0.0624	0.1141	0.0588	**0.0535**	0.0570
	18	0.0713	0.0624	0.0570	0.0499	**0.1016**	0.0713	0.0677	0.0606	0.0660	0.0749

The correct values range from 0.0446 to 0.1034; thus, this test is not as useful as we might expect. Another obvious approach is to examine, for each possible pairing, the frequency table and see how it compares to that of normal English. To do this, we order the frequencies for each pairing and the regular alphabet, then compare the most frequent in each group, the second most frequent in each group, and so on. To attach a number to this, we sum the squares of the differences between observed and expected frequencies. This yields the table below. Once again, values for correct pairings are underlined and boldfaced.

					End Column					
		2	4	5	9	14	15	16	17	19	20
	1	0.0208	0.0140	0.0310	0.0187	0.0556	0.0277	**0.0285**	0.0167	0.0245	0.0191
	3	0.0039	0.0029	0.0123	**0.0064**	0.0105	0.0039	0.0161	0.0029	0.0072	0.0040
	6	0.0061	0.0045	**0.0115**	0.0056	0.0149	0.0109	0.0105	0.0046	0.0033	0.0055
	7	0.0050	**0.0040**	0.0068	0.0076	0.0078	0.0084	0.0099	0.0070	0.0046	0.0054
Start Column	8	**0.0112**	0.0069	0.0162	0.0084	0.0269	0.0175	0.0135	0.0105	0.0254	0.0064
	10	0.0119	0.0076	0.0140	0.0052	0.0388	**0.0127**	0.0438	0.0053	0.0074	0.0084
	11	0.0071	0.0045	0.0035	0.0028	0.0192	0.0092	0.0144	**0.0045**	0.0083	0.0103
	12	0.0073	0.0109	0.0087	0.0100	0.0105	0.0115	0.0105	0.0063	0.0061	**0.0119**
	13	0.0047	0.0038	0.0092	0.0058	0.0174	0.0084	0.0347	0.0055	**0.0046**	0.0062
	18	0.0103	0.0063	0.0057	0.0049	**0.0253**	0.0090	0.0089	0.0075	0.0092	0.0135

Looking at the fourth row (headed with 7) we see that the smallest value represents the correct pairing. Sadly, this is the only row for which this happens! Thus, this approach also fails to readily pair the columns.

As was mentioned before, Painvin, and later American cryptanalysts who approached this problem, didn't use either of these measures. They simply looked at the frequency distributions for possible pairings and determined by sight which were most likely. Being that there are 10! ways to pair the columns when 20 columns are used, this must have taken them a great deal of time. Surely this was the most difficult step in solving ADFGX and ADFGVX.

With today's technology, we can consider all 10! possibilities. Each possibility then gives 10 columns (each consisting of two letters per row), which may be arranged in 10! ways. The correct arrangement then represents a monoalphabetic substitution cipher without word spacing, which may easily be solved with technology or by hand.

We continue our attack, assuming that the correct pairings have been determined, probably after tremendous trial and error. The pairings are

$$1 \leftrightarrow 16$$
$$3 \leftrightarrow 9$$
$$6 \leftrightarrow 5$$
$$7 \leftrightarrow 4$$
$$8 \leftrightarrow 2$$
$$10 \leftrightarrow 15$$
$$11 \leftrightarrow 17$$
$$12 \leftrightarrow 20$$
$$13 \leftrightarrow 19$$
$$18 \leftrightarrow 14$$

We must now find the proper order for these ten pairs of columns and solve the Polybius cipher (without word divisions) that they provide. There are 10! = 3,628,800 ways to arrange pairs of columns, so we could brute force a solution with a computer.

If we prefer to stick to World War I-era technology, we can use the frequency distribution for all of the pairs to guess at some letters and piece together columns by making words. We have

AA = 6	DA = 27	FA = 19	GA = 15	XA = 36
AD = 3	DD = 7	FD = 8	GD = 6	XD = 8
AF = 12	DF = 22	FF = 23	GF = 7	XF = 31
AG = 5	DG = 31	FG = 12	GG = 13	XG = 43
AX = 0	DX = 1	FX = 2	GX = 0	XX = 3

Based on these frequencies, we conjecture that XG = E and XA = T. Substituting these values in wherever they appear, we have

1↔16	3↔9	6↔5	7↔4	8↔2	10↔15	11↔17	12↔20	13↔19	18↔14
AG	AD	DF	XF	FA	E	XD	DG	XX	E
E	FD	XF	FD	DA	XF	FG	T	FG	E
DA	DD	FD	GA	DG	FD	E	FF	T	DA
E	FA	XF	AA	E	XF	XX	FF	FA	DA
DA	DG	FG	DF	DA	AF	E	T	E	T
XF	E	GG	DA	DF	FG	AF	XF	T	FA
DG	DD	DD	GD	AF	E	GG	E	FF	DG
DA	T	DF	GF	DG	DG	GG	DD	AG	FF
DA	DF	GG	DF	DG	GG	T	T	T	E
XD	FA	XF	DF	XF	DG	DF	DA	GD	T
DA	FA	DD	AG	DA	E	AF	AD	T	T
E	E	FF	AG	DA	T	FA	GF	FG	GA
E	AF	FF	T	GA	E	FG	DG	DG	GA
DG	E	FF	XD	XF	DG	AA	GD	XF	AF
DF	DG	GG	XF	AF	T	XF	FF	DX	E
DG	E	T	E	FA	XF	FG	T	DF	T
XF	GF	DF	FD	DA	E	GA	T	DG	E
E	FA	FD	FF	DD	T	FA	GG	AG	GA
DF	GG	T	GF	E	DA	AA	DG	E	GA
DG	FF	XF	T	E	E	T	FA	DA	DG
DF	GG	T	XD	XF	AF	XF	AF	DA	DA
XF	T	DF	FF	E	AF	E	FA	DG	GA
T	DG	FA	DA	DF	DF	XD	GA	T	FA
DG	XF	FF	E	GF	FG	AA	E	FF	FF
GA	GF	E	T	GA	FF	DA	FF	XF	GD
DG	XX	AA	DA	DD	E	FG	GG	FG	E
E	T	DF	FF	FF	DA	FD	DG	DF	DA
GA	DG	AA	XF	T	DF	DG	DG	XD	XF
T	XF	FA	FG	XF	FF	FF	E	AD	GD
XF	FF	AF	GA	T	GD	DG	XF	FA	DA
T	AF	FA	DA	XD	XF	DF	XF	DA	GG
DA	DG	DF	DG	FF	GA	FF	E	GA	T
FA	XF	XF	FG	E	DF	DG	XD	T	FA
GG	GF	FX	FX	E	T	E	GG	FD	E

We'll use the high frequency of the trigraph THE to order the columns. We start with column 1↔16. Looking down at position 12 and 18, we find Es that match the positions of Ts in column 10↔15. We expect that there is a column that appears between these two that has a plaintext H in positions 12 and 18. We look for a column such that the pairs in positions 12 and 18 match and come up with two possibilities 11↔17 and 18↔14. They can be considered separately.

Case 1			Case 2		
10↔15	11↔17	1↔16	10↔15	18↔14	1↔16
E	XD	AG	**E**	**E**	AG
XF	FG	**E**	XF	**E**	**E**
FD	**E**	DA	FD	DA	DA
XF	XX	**E**	XF	DA	**E**
AF	**E**	DA	AF	**T**	DA
FG	AF	XF	FG	FA	XF
E	GG	DG	**E**	DG	DG
DG	GG	DA	DG	FF	DA
GG	**T**	DA	GG	**E**	DA
DG	DF	XD	DG	**T**	XD
E	AF	DA	**E**	**T**	DA
T	FA	**E**	**T**	GA	**E**
E	FG	**E**	**E**	GA	**E**
DG	AA	DG	DG	AF	DG
T	XF	DF	**T**	**E**	DF
XF	FG	DG	XF	**T**	DG
E	GA	XF	**E**	**E**	XF
T	FA	**E**	**T**	GA	**E**
DA	AA	DF	DA	GA	DF
E	**T**	DG	**E**	DG	DG
AF	XF	DF	AF	DA	DF
AF	**E**	XF	AF	GA	XF
DF	XD	**T**	DF	FA	**T**
FG	AA	DG	FG	FF	DG
FF	DA	GA	FF	GD	GA
E	FG	DG	**E**	**E**	DG
DA	FD	**E**	DA	DA	**E**
DF	DG	GA	DF	XF	GA
FF	FF	**T**	FF	GD	**T**
GD	DG	XF	GD	DA	XF
XF	DF	**T**	XF	GG	**T**
GA	FF	DA	GA	**T**	DA
DF	DG	FA	DF	FA	FA
T	**E**	GG	**T**	**E**	GG
If Case 1 is correct, then FA = H.			If Case 2 is correct, then GA = H.		

At the moment, we don't have enough information to decide, which is correct, so we begin to pursue each possibility further. For Case 1, we look for a column that can appear at the beginning of the chain we're building such that a T would be joined to an E in columns 10↔15. We have four columns that do this: 7↔4, 12↔20, 13↔19, and 18↔14. Initially, 7↔4 looks best, because it pairs two Ts and Es, whereas the others only pair one T and E. However, there is no column that can be placed between 7↔4 and 10↔15 to place an H between each T and E. Recall, H is identified as FA in Case 1, so it is not enough to find a column that has the same letter pair in each of these two positions; we also want that pair to be FA. (12↔20 offers an FA in one of those positions, but then the other T–E would have some other letter in the middle, which is possible…) Now suppose 12↔20 leads into Case 1. Then we'd want an FA in position 17 of some other column (to fill in the H), and we don't have it anywhere. Moving on to 13↔19, we'd want an FA in position 11. We are given this by 3↔9. The last case is the same: For 18↔14, we'd want an FA in position 11 and we are given this by 3↔9. So, there are three reasonable possibilities, with the last two being slightly favored, because they don't require a non-H to appear between T and E anywhere. We examine these last two possibilities as subcases of Case 1.

Case 1a					Case 1b				
13↔19	3↔9	10↔15	11↔17	1↔16	18↔14	3↔9	10↔15	11↔17	1↔16
XX	AD	**E**	XD	AG	**E**	AD	**E**	XD	AG
FG	FD	XF	FG	**E**	**E**	FD	XF	FG	**E**
T	DD	FD	**E**	DA	DA	DD	FD	**E**	DA
FA	FA	XF	XX	**E**	DA	FA	XF	XX	**E**
E	DG	AF	**E**	DA	**T**	DG	AF	**E**	DA
T	**E**	FG	AF	XF	FA	**E**	FG	AF	XF
FF	DD	**E**	GG	DG	DG	DD	**E**	GG	DG
AG	**T**	DG	GG	DA	FF	**T**	DG	GG	DA
T	DF	GG	**T**	DA	**E**	DF	GG	**T**	DA
GD	FA	DG	DF	XD	**T**	FA	DG	DF	XD
T	FA	**E**	AF	DA	**T**	FA	**E**	AF	DA
FG	**E**	**T**	FA	**E**	GA	**E**	**T**	FA	**E**
DG	AF	**E**	FG	**E**	GA	AF	**E**	FG	**E**
XF	**E**	DG	AA	DG	AF	**E**	DG	AA	DG
DX	DG	**T**	XF	DF	**E**	DG	**T**	XF	DF
DF	**E**	XF	FG	DG	**T**	**E**	XF	FG	DG
DG	GF	**E**	GA	XF	**E**	GF	**E**	GA	XF
AG	FA	**T**	FA	**E**	GA	FA	**T**	FA	**E**
E	GG	DA	AA	DF	GA	GG	DA	AA	DF
DA	FF	**E**	**T**	DG	DG	FF	**E**	**T**	DG
DA	GG	AF	XF	DF	DA	GG	AF	XF	DF
DG	**T**	AF	**E**	XF	GA	**T**	AF	**E**	XF
T	DG	DF	XD	**T**	FA	DG	DF	XD	**T**
FF	XF	FG	AA	DG	FF	XF	FG	AA	DG
XF	GF	FF	DA	GA	GD	GF	FF	DA	GA
FG	XX	**E**	FG	DG	**E**	XX	**E**	FG	DG
DF	**T**	DA	FD	**E**	DA	**T**	DA	FD	**E**
XD	DG	DF	DG	GA	XF	DG	DF	DG	GA
AD	XF	FF	FF	**T**	GD	XF	FF	FF	**T**
FA	FF	GD	DG	XF	DA	FF	GD	DG	XF
DA	AF	XF	DF	**T**	GG	AF	XF	DF	**T**
GA	DG	GA	FF	DA	**T**	DG	GA	FF	DA
T	XF	DF	DG	FA	FA	XF	DF	DG	FA
FD	GF	**T**	**E**	GG	**E**	GF	**T**	**E**	GG

In Case 1a, T is followed by FA three times, but in Case 1b, T is followed by FA four times. We thus conclude that Case 1b looks better. Now we attempt to expand out Case 1b further.

Consider position 6. We have an FA (presumed H) followed by an E in the next column. We therefore look for a column with a T in position 6 to complete another THE. Column 13↔19 is the only one that works. We now have

13↔19	18↔14	3↔9	10↔15	11↔17	1↔16
XX	**E**	AD	**E**	XD	AG
FG	**E**	FD	XF	FG	**E**
T	DA	DD	FD	**E**	DA
FA	DA	FA	XF	XX	**E**
E	**T**	DG	AF	**E**	DA
T	FA	**E**	FG	AF	XF
FF	DG	DD	**E**	GG	DG
AG	FF	**T**	DG	GG	DA
T	**E**	DF	GG	**T**	DA
GD	**T**	FA	DG	DF	XD
T	**T**	FA	**E**	AF	DA
FG	GA	**E**	**T**	FA	**E**
DG	GA	AF	**E**	FG	**E**
XF	AF	**E**	DG	AA	DG
DX	**E**	DG	**T**	XF	DF
DF	**T**	**E**	XF	FG	DG
DG	**E**	GF	**E**	GA	XF
AG	GA	FA	**T**	FA	**E**
E	GA	GG	DA	AA	DF
DA	DG	FF	**E**	**T**	DG
DA	DA	GG	AF	XF	DF
DG	GA	**T**	AF	**E**	XF
T	FA	DG	DF	XD	**T**
FF	FF	XF	FG	AA	DG
XF	GD	GF	FF	DA	GA
FG	**E**	XX	**E**	FG	DG
DF	DA	**T**	DA	FD	**E**
XD	XF	DG	DF	DG	GA
AD	GD	XF	FF	FF	**T**
FA	DA	FF	GD	DG	XF
DA	GG	AF	XF	DF	**T**
GA	**T**	DG	GA	FF	DA
T	FA	XF	DF	DG	FA
FD	**E**	GF	**T**	**E**	GG

We could continue to work the front of the key, but none of the matches at this stage can be made with great confidence, so we move to the end. Notice how 1↔16 has Ts in positions 23, 29, and 31. Column 6↔5 provides a wonderful match with FAs (presumed Hs) in all of these positions. We place this in our partially reconstructed key.

13↔19	18↔14	3↔9	10↔15	11↔17	1↔16	6↔5
XX	**E**	AD	**E**	XD	AG	DF
FG	**E**	FD	XF	FG	**E**	XF
T	DA	DD	FD	**E**	DA	FD
FA	DA	FA	XF	XX	**E**	XF
E	**T**	DG	AF	**E**	DA	FG
T	FA	**E**	FG	AF	XF	GG
FF	DG	DD	**E**	GG	DG	DD
AG	FF	**T**	DG	GG	DA	DF
T	**E**	DF	GG	**T**	DA	GG
GD	**T**	FA	DG	DF	XD	XF
T	**T**	FA	**E**	AF	DA	DD
FG	GA	**E**	**T**	FA	**E**	FF
DG	GA	AF	**E**	FG	**E**	FF
XF	AF	**E**	DG	AA	DG	FF
DX	**E**	DG	**T**	XF	DF	GG
DF	**T**	**E**	XF	FG	DG	**T**
DG	**E**	GF	**E**	GA	XF	DF
AG	GA	FA	**T**	FA	**E**	FD
E	GA	GG	DA	AA	DF	**T**
DA	DG	FF	**E**	**T**	DG	XF
DA	DA	GG	AF	XF	DF	**T**
DG	GA	**T**	AF	**E**	XF	DF
T	FA	DG	DF	XD	**T**	FA
FF	FF	XF	FG	AA	DG	FF
XF	GD	GF	FF	DA	GA	**E**
FG	**E**	XX	**E**	FG	DG	AA
DF	DA	**T**	DA	FD	**E**	DF
XD	XF	DG	DF	DG	GA	AA
AD	GD	XF	FF	FF	**T**	FA
FA	DA	FF	GD	DG	XF	AF
DA	GG	AF	XF	DF	**T**	FA
GA	**T**	DG	GA	FF	DA	DF
T	FA	XF	DF	DG	FA	XF
FD	**E**	GF	**T**	**E**	GG	FX

We could continue trying to place the three remaining columns in our partially recovered key, but there is not a very strong reason to make another placement at this stage, so instead we consider these remaining columns as a group. There are six ways to order three objects and we quickly notice that two of these orderings, 12↔20 8↔2 7↔4 and 7↔4 8↔2 12↔20, are such that THE appears (positions 16 and 20, respectively). See below.

12↔20	8↔2	7↔4	vs.	7↔4	8↔2	12↔20
DG	FA	XF		XF	DG	FA
T	DA	FD		FD	T	DA
FF	DG	GA		GA	FF	DG
FF	E	AA		AA	FF	E
T	DA	DF		DF	T	DA
XF	DF	DA		DA	XF	DF
E	AF	GD		GD	E	AF
DD	DG	GF		GF	DD	DG
T	DG	DF		DF	T	DG
DA	XF	DF		DF	DA	XF
AD	DA	AG		AG	AD	DA
GF	DA	AG		AG	GF	DA
DG	GA	T		T	DG	GA
GD	XF	XD		XD	GD	XF
FF	AF	XF		XF	FF	AF
T	FA	E		E	T	FA
T	DA	FD		FD	T	DA
GG	DD	FF		FF	GG	DD
DG	E	GF		GF	DG	E
FA	E	T		T	FA	E
AF	XF	XD		XD	AF	XF
FA	E	FF		FF	FA	E
GA	DF	DA		DA	GA	DF
E	GF	E		E	E	GF
FF	GA	T		T	FF	GA
GG	DD	DA		DA	GG	DD
DG	FF	FF		FF	DG	FF
DG	T	XF		XF	DG	T
E	XF	FG		FG	E	XF
XF	T	GA		GA	XF	T
XF	XD	DA		DA	XF	XD
E	FF	DG		DG	E	FF
XD	E	FG		FG	XD	E
GG	E	FX		FX	GG	E

We could look more closely at these two possibilities in isolation, but we can also see how they fit onto the key we've partially assembled; for example, placing 12↔20 8↔2 7↔4 at the start of our partial key gives us some nice results.

12↔20	8↔2	7↔4	13↔19	18↔14	3↔9	10↔15	11↔17	1↔16	6↔5
DG	FA	XF	XX	**E**	AD	**E**	XD	AG	DF
T	DA	FD	FG	**E**	FD	XF	FG	**E**	XF
FF	DG	GA	**T**	DA	DD	FD	**E**	DA	FD
FF	**E**	AA	FA	DA	FA	XF	XX	**E**	XF
T	DA	DF	**E**	**T**	DG	AF	**E**	DA	FG
XF	DF	DA	**T**	FA	**E**	FG	AF	XF	GG
E	AF	GD	FF	DG	DD	**E**	GG	DG	DD
DD	DG	GF	AG	FF	**T**	DG	GG	DA	DF
T	DG	DF	**T**	**E**	DF	GG	**T**	DA	GG
DA	XF	DF	GD	**T**	FA	DG	DF	XD	XF
AD	DA	AG	**T**	**T**	FA	**E**	AF	DA	DD
GF	DA	AG	FG	GA	**E**	**T**	FA	**E**	FF
DG	GA	**T**	DG	GA	AF	**E**	FG	**E**	FF
GD	XF	XD	XF	AF	**E**	DG	AA	DG	FF
FF	AF	XF	DX	**E**	DG	**T**	XF	DF	GG
T	FA	**E**	DF	**T**	**E**	XF	FG	DG	**T**
T	DA	FD	DG	**E**	GF	**E**	GA	XF	DF
GG	DD	FF	AG	GA	FA	**T**	FA	**E**	FD
DG	**E**	GF	**E**	GA	GG	DA	AA	DF	**T**
FA	**E**	**T**	DA	DG	FF	**E**	**T**	DG	XF
AF	XF	XD	DA	DA	GG	AF	XF	DF	**T**
FA	**E**	FF	DG	GA	**T**	AF	**E**	XF	DF
GA	DF	DA	**T**	FA	DG	DF	XD	**T**	FA
E	GF	**E**	FF	FF	XF	FG	AA	DG	FF
FF	GA	**T**	XF	GD	GF	FF	DA	GA	**E**
GG	DD	DA	FG	**E**	XX	**E**	FG	DG	AA
DG	FF	FF	DF	DA	**T**	DA	FD	**E**	DF
DG	**T**	XF	XD	XF	DG	DF	DG	GA	AA
E	XF	FG	AD	GD	XF	FF	FF	**T**	FA
XF	**T**	GA	FA	DA	FF	GD	DG	XF	AF
XF	XD	DA	DA	GG	AF	XF	DF	**T**	FA
E	FF	DG	GA	**T**	DG	GA	FF	DA	DF
XD	**E**	FG	**T**	FA	XF	DF	DG	FA	XF
GG	**E**	FX	FD	**E**	GF	**T**	**E**	GG	FX

The last column has a T in position 19. When reading the plaintext out of this rectangle, the end of this row is continued at the start of row 20, where we have HE. This indicates a nice placement of the three remaining columns.

Placing 12↔20 8↔2 7↔4 at the end instead doesn't yield such a nice result, nor does placing 7↔4 8↔2 12↔20 at either end. Therefore, we assume that we now have the proper ordering of the columns and that we have correctly identified T, H, and E. Recall that this was all part of Case 1b. If we had reached an impasse, we would have backed up to consider Case 1a or even Case 2, but because things seem to be working out, there's no need.

We fill in all 19 Hs to get the partial plaintext below:

```
DG   H  XF  XX   E  AD   E  XD  AG  DF
 T  DA  FD  FG   E  FD  XF  FG   E  XF
FF  DG  GA   T  DA  DD  FD   E  DA  FD
FF   E  AA   H  DA   H  XF  XX   E  XF
 T  DA  DF   E   T  DG  AF   E  DA  FG
XF  DF  DA   T   H   E  FG  AF  XF  GG
 E  AF  GD  FF  DG  DD   E  GG  DG  DD
DD  DG  GF  AG  FF   T  DG  GG  DA  DF
 T  DG  DF   T   E  DF  GG   T  DA  GG
DA  XF  DF  GD   T   H  DG  DF  XD  XF
AD  DA  AG   T   T   H   E  AF  DA  DD
GF  DA  AG  FG  GA   E   T   H   E  FF
DG  GA   T  DG  GA  AF   E  FG   E  FF
GD  XF  XD  XF  AF   E  DG  AA  DG  FF
FF  AF  XF  DX   E  DG   T  XF  DF  GG
 T   H   E  DF   T   E  XF  FG  DG   T
 T  DA  FD  DG   E  GF   E  GA  XF  DF
GG  DD  FF  AG  GA   H   T   H   E  FD
DG   E  GF   E  GA  GG  DA  AA  DF   T
 H   E   T  DA  DG  FF   E   T  DG  XF
AF  XF  XD  DA  DA  GG  AF  XF  DF   T
 H   E  FF  DG  GA   T  AF   E  XF  DF
GA  DF  DA   T   H  DG  DF  XD   T   H
 E  GF   E  FF  FF  XF  FG  AA  DG  FF
FF  GA   T  XF  GD  GF  FF  DA  GA   E
GG  DD  DA  FG   E  XX   E  FG  DG  AA
DG  FF  FF  DF  DA   T  DA  FD   E  DF
DG   T  XF  XD  XF  DG  DF  DG  GA  AA
 E  XF  FG  AD  GD  XF  FF  FF   T   H
XF   T  GA   H  DA  FF  GD  DG  XF  AF
XF  XD  DA  DA  GG  AF  XF  DF   T   H
 E  FF  DG  GA   T  DG  GA  FF  DA  DF
XD   E  FG   T   H  XF  DF  DG   H  XF
GG   E  FX  FD   E  GF   T   E  GG  FX
```

There are many ways to pursue the decipherment from here. The problem has been reduced to a Polybius cipher without word spacing. A vowel recognition algorithm (as seen in Section 1.11) could be applied and, perhaps after a few guesses the As, Os, and Is could all be filled. From there, it ought to be easy to guess a few words. The solving process would then quickly accelerate. Other approaches are possible. Ultimately, it's up to you, as the final steps are left as an exercise.

The approach used in this example won't work so nicely (although it can be patched), if the rectangle is not completely filled in. However, if the length of the transposition key is 20, then 5% of the messages should, by chance, complete a rectangle. Without knowing the length of the key, the attack could be tried on every message, and once the key is found, the other messages may be easily broken. With the large number of messages that flow in every war, 5% will give cryptanalysts a lot to work with.

It wasn't until 1966 that the creator of ADFGVX, Fritz Nebel, learned that his system had been cracked. He commented:[16]

> I personally would have preferred the second stage to have had a double transposition rather than a single one, but, in discussion with the radiotelegraphy and decipherment chiefs, the idea was rejected. The result was a compromise between technical and tactical considerations. Double transposition would have been more secure but too slow and too difficult in practice.

Nebel met Painvin in 1968. He described it as "the enemies of yesterday meeting as the friends of today." At this meeting, Painvin recalled, "I told him, if he had his way and they'd used a double transposition I'd never have been able to make the break."[17]

Following previous wars, America's spies and codebreakers went back to the lives they led before their time of service, but America was now headed into an era of permanent institutions to handle such activities. In this regard, America was behind the Europeans, and there would still be some bumps in the road, but the nation was on its way to establishing an intelligence empire.

William Friedman would play an important role as he transitioned from his work at Riverbank Laboratories to the government. We'll come back to him soon, but for now we'll take a look at the contributions and controversies of Herbert O. Yardley.

Figure 5.4   Herbert O. Yardley (1889–1958). (http://www.nsa.gov/about/_images/pg_hi_res/yardley.jpg.)

5.4 Herbert O. Yardley

We now come to perhaps the most colorful figure in the cryptologic world. Herbert O. Yardley (Figure 5.4) could be called the Han Solo of cryptology. He's often described as a gambler, a drinker, and a womanizer, and debate is still ongoing concerning his loyalty to the United States. Although both did important cryptanalytic work for the United States, Yardley and Friedman strike a strong contrast in almost every category. Friedman was very neat, dapper even. Yardley was often scruffy and would sometimes water the lawn in his underwear. At the time of his death he didn't own a tie;

---

[16] Norman, Bruce, "The ADFGVX Men," *The Sunday Times Magazine*, August 11, 1974, pp. 8–15, p. 11 cited here.

[17] Norman, Bruce, "The ADFGVX Men," *The Sunday Times Magazine*, August 11, 1974, pp. 8–15, p. 15 cited here.

one was given to his widow for Yardley to be buried in.[18] Yardley was anti-Semitic, whereas Friedman was Jewish.[19] Friedman had a long, apparently happy marriage, but one gets the impression that he didn't have a great deal of confidence with women. Yardley, on the other hand, was not at all shy. He even bragged about knowing his way around a Chinese whorehouse and hosted orgies for visiting journalists and diplomats while he was in China.[20] Although it's not often a good indicator, even their handwriting represents their distinct personalities (Figures 5.5 and 5.6).

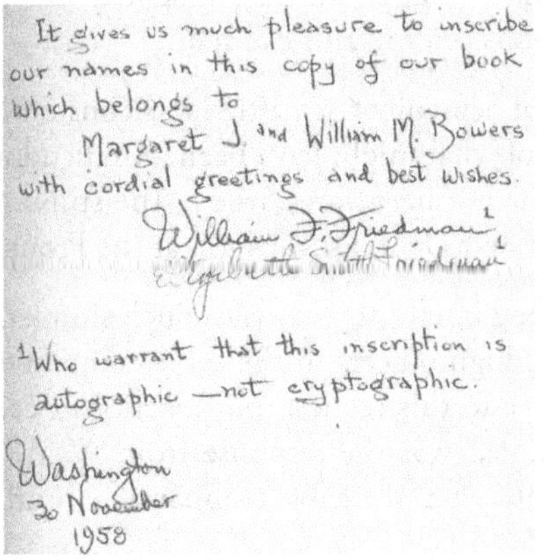

Figure 5.5   The precise penmanship of the Friedmans. (from the collection of the author.)

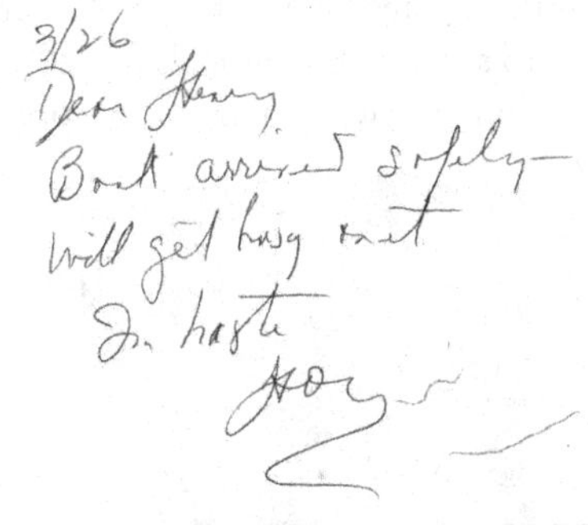

Figure 5.6   Yardley's sloppy scrawl. (from the collection of the author.)

Yardley started out in a noncryptographic position at the State Department; he was just a telegraph operator and code clerk, but having access to coded messages, including ones addressed to President Wilson, he made attempts to break them. He eventually produced a report "Solution of American Diplomatic Codes" for his boss. His successes made a strong case for the need for an improved American cryptographic bureau, or black chamber, and his skills at self-promotion catapulted him into a new position as chief of this new organization under the War Department, formally called the Cipher Bureau, in 1917. Black chambers had played an important role in European history, but such cryptanalytic units were new in the United States. In 18 months, Yardley's team (Military Intelligence Section 8, or MI-8) had read almost 11,000 messages in 579 cryptographic systems. This is even more amazing

---

18   Kahn, David, *The Reader of Gentlemen's Mail*, Yale University Press, New Haven, Connecticut, 2004, p. 288.
19   Kahn, David, *The Reader of Gentlemen's Mail*, Yale University Press, New Haven, Connecticut, 2004, pp. 88 and 146.
20   Kahn, David, *The Reader of Gentlemen's Mail*, Yale University Press, New Haven, Connecticut, 2004, p. 196.

when one considers that the team initially consisted of only Yardley himself and two clerks. At its peak in November 1918, the team consisted of 18 officers, 24 civilians, and 109 typists.[21]

Of course, protecting America's messages was also an important aspect of Yardley's organization. Messages traveling to and from Europe via the transatlantic cable could be obtained by the Germans by having their submarines place long cables (*hundreds* of feet long) next to the transatlantic cable to pick up the messages by induction. Yardley's team overhauled our codes and ciphers so that such interceptions wouldn't matter. Yardley lamented the results of the Germans reading our World War I communications prior to this overhaul:

> The American offensive of September 12, 1918, was considered a triumph, but it represents only a small part of what might have been a tremendous story in the annals of warfare, had the Germans not been forewarned. The stubborn trust placed in inadequate code and cipher systems had taken its toll at the Front.[22]

Yardley had a method for gaining intercepts that was much simpler than induction. He or a State department official approached high officers in the cable companies and asked for the messages. It was illegal for anyone to agree to this request, but reactions weren't all negative. "The government can have anything it wants," was the response from W. E. Roosevelt of the All-America Cable Company.[23] At one point, when the cable companies cut off Yardley's supply of messages, he regained them with bribes.[24]

The Cipher Bureau's work wasn't strictly limited to codes and ciphers. They also dealt with secret inks and shorthand systems. The ability to detect messages hidden by the use of secret inks exposed German spy networks in the United States, but such work could be dangerous, even in times of peace. In an experiment with secret ink chemicals in 1933, Yardley cut his palm on a piece of glass and an infection led to one of his fingers having to be amputated (Figure 5.7).[25]

Figure 5.7 Yardley's injured right hand. (Courtesy of the David Kahn Collection, National Cryptologic Museum, Fort Meade, Maryland.)

[21] Lewand, Robert Edward, *Cryptological Mathematics*, MAA, Washington, DC, 2000, p. 42. Kahn put the total at 165. Perhaps the few extra didn't fit into any of the categories listed here. See p. xvii of the foreword to Yardley, Herbert O., *The American Black Chamber*, Espionage/Intelligence Library, Ballantine Books, New York, 1981.

[22] Yardley, Herbert O., *The American Black Chamber*, Espionage/Intelligence Library, Ballantine Books, New York, 1981, p. 19.

[23] Kahn, David, *The Reader of Gentlemen's Mail*, Yale University Press, New Haven, Connecticut, 2004, p. 58.

[24] Kahn, David, *The Reader of Gentlemen's Mail*, Yale University Press, New Haven, Connecticut, 2004, p. 84.

[25] Kahn, David, *The Reader of Gentlemen's Mail*, Yale University Press, New Haven, Connecticut, 2004, pp. 146–147.

The chamber might have been disbanded when World War I ended, but Yardley convinced his superiors to keep it open. This represented a tremendous change for America. It is the first time that the American codebreakers didn't return to their prior lives following the end of a war. Yardley explained, "As all the Great Powers maintained such a Cipher Bureau, the United States in self-defense must do likewise."[26]

5.5 Peacetime Victory and a Tell-All Book

Yardley's peacetime Cipher Bureau relocated to New York.[27] His first mission was to break the ciphers of the Japanese, even though he didn't know Japanese. Despite this obstacle, the mission was successful. Yardley labeled the first Japanese code he broke in 1919 as Ja. J stood for Japan and the a acted like a subscript. As new Japanese codes came into being, and were broken, Yardley named them Jb, Jc, etc.

On November 12, 1921, the Washington Disarmament Conference began with the aim of settling disputes in the Far East. One issue was the tonnage ratios for the American, British, and Japanese navies. The United States favored a ratio of 10:10:6 for The United States, Britain, and Japan, respectively. That is, the Americans and Brits could have equal tonnage, but the Japanese Navy could not exceed 60% of that figure. The Japanese favored a 10:10:7 ratio. The messages between the negotiators and their bosses were encoded with what Yardley termed Jp. The most important message sent in this system is reproduced below, as Yardley's team decoded it.[28]

> From Tokio
> To Washington.
>
> , Conference No. 13. November 28, 1921.
>
> SECRET.
>   Referring to your conference cablegram No. 74, we are of your opinion that *it is necessary to avoid any clash with Great Britain and America, particularly America, in regard to the armament limitation question.* You will to the utmost maintain a middle attitude and *redouble your efforts to carry out our policy. In case of inevitable necessity you will work to establish your second proposal of 10 to 6.5.* If, in spite of your utmost efforts, *it becomes necessary* in view of the situation and in the interests of general policy *to fall back on your proposal No. 3,* you will *endeavor to limit* the power of concentration and maneuver of the Pacific *by a guarantee to reduce or at least to maintain the status quo of Pacific defenses* and to make an adequate reservation which will make clear that [this is] our intention in agreeing to a 10 to 6 ratio.
>   *No. 4 is to be avoided as far as possible.*

---

[26] Yardley, Herbert O., *The American Black Chamber*, Espionage/Intelligence Library, Ballantine Books, New York, 1981, p. 133.

[27] The funding from the State Department couldn't be used within DC; thus, a move was required. See Yardley, Herbert O., *The American Black Chamber*, Espionage/Intelligence Library, Ballantine Books, New York, 1981, p. 156.

[28] Yardley, Herbert O., *The American Black Chamber*, Espionage/Intelligence Library, Ballantine Books, New York, 1981, p. 208.

Knowing how far Japan could be pushed concerning ship tonnage ratios allowed the United States to do so with full confidence, merely waiting for the Japanese to give in. The Japanese finally accepted the 10:10:6 ratio on December 10, 1921.

This was the greatest peacetime success of the Cipher Bureau, but they did attack many other systems and eventually cracked the ciphers of 20 different countries.[29] With the election of Herbert Hoover in 1928, however, the politics of decipherment changed. Hoover's secretary of state, Henry L. Stimson, was actually offended when decrypts were provided to him. He later summed up his feelings with the famous quote, "Gentlemen do not read each other's mail." Stimson withdrew the Cipher Bureau's funding and it was formally shut down on October 31, 1929.[30]

Of course, such organizations have a way of staying around whether they're wanted or not. Usually they just change names. In this case, William Friedman took possession of Yardley's files and records and the work continued under the Signal Intelligence Service (SIS),[31] part of the Army Signal Corps. Apparently Stimson was not aware of *this* group! Yardley was offered a position with SIS, but the salary was low, and he was expected to refuse, as he did.

Out of work, in the aftermath of the stock market crash, and strapped for cash, Yardley decided to write about his adventures in codebreaking. On June 1, 1931, Yardley's book, *The American Black Chamber*, was released. He stated in the foreword, "Now that the Black Chamber has been destroyed there is no valid reason for withholding its secrets."[32] The book sold 17,931 copies, which was a remarkable number for the time.[33] An unauthorized Japanese edition was even more popular. American officials denied that the Cipher Bureau existed, but privately sought to prosecute Yardley for treason. The nations whose ciphers had been broken were now aware of the fact and could be expected to change systems. It's often claimed that, as a result of Yardley's disclosures, the Japanese changed their ciphers and eventually began to make use of a tough machine cipher that they called "type B cipher machine." The Americans called it Purple and this was the system in use at the time of the attack on Pearl Harbor. However, David Kahn makes a good case for Yardley's revelations having done no harm! Graphs of the number of cryptanalytic solutions to Japanese codes and ciphers (and those of foreign nations in general) show no dip following the publication of *The American Black Chamber* (Figure 5.8).

---

[29] From the foreword (p. xi) to Yardley, Herbert O., *The American Black Chamber*, Espionage/Intelligence Library, Ballantine Books, New York, 1981. On page 222 of this book, Yardley lists the countries as "Argentina, Brazil, Chile, China, Costa Rica, Cuba, England, France, Germany, Japan, Liberia, Mexico, Nicaragua, Panama, Peru, Russia, San Salvador, Santo Domingo, Soviet Union and Spain."

[30] From the foreword (p. xii) to Yardley, Herbert O., *The American Black Chamber*, Espionage/Intelligence Library, Ballantine Books, New York, 1981. Years later, in 1944, Secretary of State Edward P. Stettinius acted similarly and had the Office of Strategic Services (OSS) return Soviet cryptographic documents (purchased by the OSS from Finnish codebreakers in 1944) to the Soviet Embassy! See Benson, Robert Louis and Michael Warner, editors, *Venona: Soviet Espionage and the American Response, 1939-1957*, NSA/CIA, Washington, DC, 1996, p. xviii and p. 59.

[31] The size of the SIS at various times is given in Foerstel, Herbert N., *Secret Science: Federal Control of American Science and Technology*, Praeger, Westport, Connecticut, 1993, p. 103.

[32] Yardley, Herbert O., *The American Black Chamber*, Espionage/Intelligence Library, Ballantine Books, New York, 1981, p. xvii.

[33] From the foreword (p. xiii) to Yardley, Herbert O., *The American Black Chamber*, Espionage/Intelligence Library, Ballantine Books, New York, 1981.

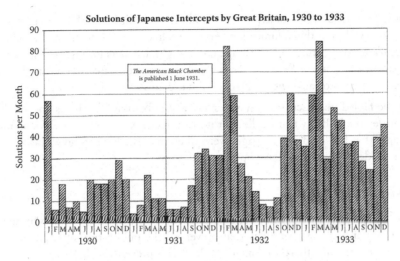

Solutions of Japanese Intercepts by Great Britain, 1930 to 1933

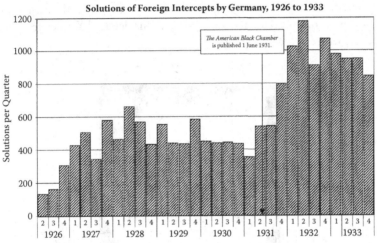

Solutions of Foreign Intercepts by Germany, 1926 to 1933

Figure 5.8 **Graphs charting cryptanalytic successes show no dip from Yardley's book. (Courtesy of the David Kahn Collection, National Cryptologic Museum, Fort Meade, Maryland.)**

Frank Rowlett, who played a key role in breaking Japanese ciphers during World War II, actually thought Yardley's book helped U.S. codebreakers![34]

5.6 The Case of the Seized Manuscript

Despite his close brush with being charged with treason, Yardley prepared another manuscript, *Japanese Diplomatic Secrets*. Thomas E. Dewey, an assistant in the U.S. Attorney's office at the time, made the decision to seize it.[35] The U.S. government did just this in February 1933. The text of 1,000+ pages wasn't declassified until March 2, 1979.[36] It was later offered for sale as a pdf on a CD put out by Aegean Park Press. The excitement historians felt from being able to finally read this work quickly dissipated as

---

[34] Kahn, David, *The Reader of Gentlemen's Mail*, Yale University Press, New Haven, Connecticut, 2004, p. 136.
[35] Foerstel, Herbert N., *Secret Science: Federal Control of American Science and Technology*, Praeger, Westport, Connecticut, 1993, p. 101.
[36] Foerstel, Herbert N., *Secret Science: Federal Control of American Science and Technology*, Praeger, Westport, Connecticut, 1993, p. 101.

they actually began to read. Kahn described it as "suffocatingly dull."[37] Yardley's prior book was written in a lively exciting style, but this volume didn't even seem like it had been written by the same person. In fact, it hadn't. Yardley didn't enjoy writing and used ghost writers for most of his work. Such was the case for *Japanese Diplomatic Secrets*.[38] Although anyone can purchase this book at the present time, there is a new mystery associated with it. David Kahn explains:

> I saw the fabled manuscript many years ago at the National Archives and used it in my Yardley book. But when I called for it again recently, the half-dozen or so manila envelopes that had held its 970 pages were all empty. The file had no withdrawal slip. I have no idea where the manuscript may be.[39]

5.7 Cashing In, Again

In June of 1933, in reaction to Yardley's efforts to increase his income through writing (with ghost writers), the United States passed a law that prevented the publication of any material that had once been prepared in any official diplomatic code.[40] This prevented *The American Black Chamber* from being reprinted, although it eventually saw light again (legally). Meanwhile, the first printing and bootleg copies continued to circulate. None of this deterred Yardley from trying to profit further from his expertise as a codebreaker.

Figure 5.9 An advertisement promoting Yardley's novel *The Blonde Countess*. (From Hannah, T. M., The many lives of Herbert O. Yardley, *Cryptologic Spectrum*, Vol. 11, No. 4, 1981, p. 14.)

---

[37] From the introduction (p. xiv) to Yardley, Herbert O., *The American Black Chamber*, Espionage/Intelligence Library, Ballantine Books, New York, 1981.

[38] The real author was Marie Stuart Klooz; for more information, see https://web.archive.org/web/20111130025234/http://www.intelligence-history.org/jih/reviews-1-2.html.

[39] Kahn, David, email to the author, July 13, 2010. Note: the manuscript is well over 1,000 pages with the appendices. Kahn's figure of 970 is for the text proper.

[40] From the forward (p. xiv) to Yardley, Herbert O., *The American Black Chamber*, Espionage/Intelligence Library, Ballantine Books, New York, 1981.

The publicity generated by the government's attack on *The American Black Chamber* was used by Yardley's publisher for promotional purposes. The small print at the top of the ad reproduced in Figure 5.9 instructs "Be sure your Congressman's in Washington—*Then dip this sheet in water.*" Doing so revealed the hidden message

> The only author ever Gagged by an Act of Congress resorts to fiction in his novel about the American Black Chamber. "The Blonde Countess," by Major Herbert O. Yardley, published by Longmans, Green and Company, 114 Fifth Avenue, New York.

Yardley wrote several other books, including the fictional spy/adventure novels *The Blonde Countess*, *Red Sun of Nippon* (Figure 5.10), and *Crows Are Black Everywhere*. These books were not as controversial as his first two. Yardley's last book sold over 100,000 copies. It was titled *The Education of a Poker Player*.[41]

Figure 5.10   One of Yardley's novels. (Book cover by Paul Bartlett Taylor.)

---

[41] Lewand, Robert Edward, *Cryptological Mathematics*, MAA, Washington, DC, 2000, p. 44.

Other attempts to cash in on his fame included a radio program, "Stories of the Black Chamber," which aired in 1935, and two movies: *Rendezvous* (Metro-Goldwyn-Mayer, 1935) was an adaptation of Yardley's novel *The Blonde Countess*, and *Cipher Bureau* followed in 1938 (see Figure 5.11).

Figure 5.11   Can Hollywood make frequency counts exciting? (Thanks to René Stein, former National Cryptologic Museum librarian, for finding this poster for me and to Nicholas Altland for photographing it.)

In addition to making money with books about cryptology, Yardley wanted to continue doing cryptologic work. From September 1938 to July 1940, he attacked Japanese codes and ciphers for the Chinese. Years after Yardley's death, James Bamford visited with his widow and was thrilled to learn that a manuscript by Yardley detailing his cryptanalytic work in China had been gathering dust in a closet. Bamford penned an introduction and got the work published as *The Chinese Black Chamber*.[42]

From June 1941 through November 1941, Yardley worked as a codebreaker in Canada, where he cracked transposition ciphers used by German spies in South America.[43] There are rumors (you might even say evidence) of Yardley having once again landed a role with American intelligence, after his time in Canada, but like so much else having to do with Yardley, we don't have proof one way or the other. The greatest Yardley mystery is detailed below.

5.8 Herbert O. Yardley—Traitor?

Yardley was always looking for ways to supplement his income. In 1967, Ladislas Farago accused him of including treason as a source of income back in 1928. The relevant paragraphs from Farago's book follow.[44]

---

[42] Yardley, Herbert O., *The Chinese Black Chamber: An Adventure in Espionage*, Houghton Mifflin, Boston, 1983.

[43] Kahn, David, *The Reader of Gentlemen's Mail*, Yale University Press, New Haven, Connecticut, 2004, p. 206.

[44] Farago, Ladislas, *The Broken Seal: The Story of "Operation Magic" and the Pearl Harbor Disaster*, Random House, New York, 1967, pp. 57–58.

At 1661 Crescent Place, an elegant little graystone house off Connecticut Avenue, he [Yardley] was received by Setsuzo Sawada, counselor of the Japanese embassy. Yardley went to the heart of the matter at once. He introduced himself as the United States government's senior cryptologist, briefly sketched his background, then told Sawada that he was prepared to sell his country's most closely guarded secret – for $10,000 in cash.

The offer was so staggering that it aroused Sawada's suspicions. According to his first report to Tokyo, in which he described this strange encounter, he had said to Yardley: "But you're making a lot of money in your job! Why are you willing to sell your country?"

"Simple, sir," Yardley replied, according to Sawada. "It just so happens that I need *more* money."

This was an unparalleled opportunity and Sawada acted quickly to make the most of it. When his report reached Tokyo the two most important Japanese officials in cryptography were sent to Washington, under assumed names on diplomatic passports, to examine Yardley's proposition and advise Sawada. One of them was Captain Kingo Inouye of the Imperial Navy, on loan to the Foreign Ministry to organize a Code Research Group within its Cable Section. The other was Naoshi Ozeki, chief cryptographer of the Foreign Ministry.

A deal was made, but not without some haggling. Contrary to the popular belief that the Japanese had unlimited funds for such transactions, the Foreign Ministry operated on a very tight budget from secret funds. Sawada countered Yardley's demand by offering him $3000 at first, and then $5000. For a while Yardley refused to lower his price, but an agreement was finally reached. Yardley received $7000, with the understanding that he would be paid more if he decided to continue to work for the Japanese.

It was an amazing bargain at any price. In return for their money, the Japanese obtained all the secrets of the "black chamber" – Yardley's methodology in breaking their codes, copies of his work sheets, and his solutions of other codes as well, including those of the British Foreign Office, which they were especially anxious to get. Moreover, Yardley agreed to cut back his work on Japanese messages.

A convincing amount of detail is provided; however, we must be careful to not confuse good writing with good history.

John F. Dooley went back to the original source Farago cited and pursued other avenues of investigation, as well. Ultimately, he concluded that Yardley was innocent of this particular crime. Farago's work is, in general, not very reliable. He was not a careful researcher. Indeed, Dooley found that some of the details in the paragraphs reproduced above cannot be true[45] and others are not backed by the material Farago referenced. The reader is encouraged to examine the literature, especially Dooley's paper, and come to his or her own conclusions. While I side with Dooley on this issue, I'm sure the debate will continue. Just as English professors will always have the sanity or insanity of Hamlet to contemplate, we'll continue to have the guilt or innocence of Yardley to argue about.

---

[45] Dooley wrote, "Farago says "1661 Crescent Place, an elegant little graystone house off Connecticut Avenue" but such a house does not currently exist. The closest current building to this address is 1661 Crescent Place NW, which is a six story apartment building between 16th and 17th Streets NW and about 5 or 6 blocks from Connecticut Avenue."

You might think that Yardley's writings alone would forever place him on some sort of "enemy of the state" list, but his previous work was eventually recognized by his placement in NSA's Hall of Honor. He was also buried with military honors in Arlington National Cemetery.[46]

Often when looking for one thing, something else of interest turns up. Such was the case with Dooley's research into Yardley. He had the following telegram translated and presented it in his paper for the first time.

Telegram Section

Confidential #48

Date: March 10th, 1925
From: Isaburo Yoshida, Acting Ambassador to the US
To: Kijuro Shidehara, Minister of Foreign Affairs
Re: Telegram Codes

Mr. W. Friedman, an American, from Cornell University seems very skilled in breaking codes; for he was engaged in breaking codes at the war in Europe (i.e., WWI), and he is now working for the US Army. When he came to see me recently, he mentioned that the US Army had no difficulty breaking codes. In order to prevent this, we have no choice but change codes very frequently. I am sending this note for your information.

So even if Dooley managed to resolve Yardley's alleged treason, another mystery has arisen: What was Friedman up to?

5.9 Censorship

Japanese Diplomatic Secrets has been referred to as the *only* manuscript ever seized by the U.S. government. This is not correct. The oral history of Captain Joseph Rochefort, who was involved with Navy cryptologic activities during World War II, was impounded by the National Security Agency (NSA).[47] In another case, the U.S. Government bought a manuscript to block its publication. *War Secrets of the Ether*, by Wilhelm F. Flicke, described Germany's interception and cryptanalysis of messages from 1919 to 1945, and included how phone conversations between Roosevelt and Churchill were unscrambled. The book passed from the Army Security Agency to the National Security Agency and was classified "Restricted."[48]

The National Security Agency considered the "purchase and hide" method to stop David Kahn's *The Codebreakers* from appearing. They even considered "clandestine service applications" against Kahn, which certainly sounds sinister. James Bamford interpreted this as possibly meaning "anything from physical surveillance to a black-bag job [theft]." The "surreptitious entry" into Kahn's home that was considered leaves less room for interpretation. In the end, NSA settled for getting Kahn to delete three paragraphs; however, an endnote that made it through, allowed anybody willing to trace sources to see much of what was removed. The deleted material appeared years later, exactly as written by Kahn, in James Bamford's *The Puzzle Palace*.[49]

---

[46] From the Introduction (p. xvi) to Yardley, Herbert O., *The American Black Chamber*, Espionage/Intelligence Library, Ballantine Books, New York, 1981.

[47] Lewin, Ronald, *The American Magic*, Farrar Straus Giroux, New York, 1982, p. 139.

[48] Clark, Ronald, *The Man Who Broke Purple*, Little, Brown and Company, Boston, Massachusetts, 1977, p. 209.

[49] Bamford, James, *The Puzzle Palace*, Penguin Books, New York, 1983, pp. 168–173.

By 1942, Stimson, who had shut down the Cipher Bureau, felt differently about how gentlemen behaved. He now heartily approved of reading others' mail.[50] As Secretary of War, Stimson wrote to the American Library Association (ALA):[51]

> It has been brought to the attention of the War Department that member libraries of the American Library Association have received numerous requests for books dealing with explosives, secret inks, and ciphers.
>
> It is requested that these books be removed from circulation and that these libraries be directed to furnish their local office of the Federal Bureau of Investigation with the names of persons making requests for same.

Stimson's letter closed with the following paragraph:

> This document contains information affecting the national defense of the United States within the meaning of the espionage Act, U.S.C. 50; 31 and 32. Its transmission or the revelation of its contents in any manner to an unauthorized person is prohibited by law.

Librarians who thought Stimson's actions worthless due to a number of popular books on cryptology being available at bookstores could not make the debate public. This approach has been used in recent years in the form of the Federal Bureau of Investigation's National Security Letters.

Despite the gag order, word got out, as evidenced by a piece of short fiction by Anthony Boucher titled "QL69.C9." This story, copyright 1943, describes such a program of librarian cooperation with the FBI, although it is simply part of the backdrop and not the point of the tale.

Happily, Stimson's censorship ended a few months after the war:

> Since hostilities have ceased, it is agreed that the necessity for limiting the circulation and use of these types of books no longer exists and that the libraries which participated in this program should be notified to this effect.[52]

It appears that the FBI didn't catch anyone with bad intentions attempting to borrow books on cryptology, but this official censorship was nothing new.[53]

> In 1918, the US War Department told the American Library Association to remove a number of pacifist and "disturbing" books, including Ambrose Bierce's *Can Such Things Be?* from camp libraries, a directive which was taken to also apply to the homefront.

Bierce had passed away by this time, but those still living took a risk if they chose to openly oppose the draft.

> During World War I, the US government jailed those who were distributing anti-draft pamphlets like this one.[54] Schenck, the publisher of the pamphlet, was convicted, and his conviction was upheld by the Supreme Court in 1919. (This decision was the source of the well-known "fire in a theatre" quote.)[55]

---

50 Of course, the senders could hardly be considered gentlemen at this time!
51 Letter from Henry L. Stimson to Carl H. Milam, August 13, 1942, War department file, Record Series 89/2/6 (War Service Committee), Box 2, ALA Archives, University of Illinois at Urbana-Champaign.
52 Letter from Henry L. Stimson to Carl H. Milam, September 21, 1945, War department file, Record Series 89/2/6 (War Service Committee), Box 2, ALA Archives, University of Illinois at Urbana-Champaign.
53 http://onlinebooks.library.upenn.edu/banned-books.html.
54 The website this paragraph is taken from is hyperlinked at this point to https://web.archive.org/web/20140922041848/ http://faculty-web.at.northwestern.edu:80/commstud/freespeech/cont/cases/schenck/pamphlet.html.
55 http://onlinebooks.library.upenn.edu/banned-books.html.

And there were political censorship demands made by the government during the Cold War, as well.[56]

> In the 1950s, according to Walter Harding, Senator Joseph McCarthy had overseas libraries run by the United States Information Service pull an anthology of American literature from the shelves because it included Thoreau's *Civil Disobedience*.

The Food and Drug Administration (FDA) has engaged in book censorship by claiming that they are used as labeling for banned foods and drugs. For example, the FDA seized a large quantity of the book *Folk Medicine* by C. D. Jarvis, MD., claiming it was used to promote the sale of honey and vinegar. In November 1964, the U.S. Court of Appeals ruled that the seizure was improper. The court also ruled against the FDA when they seized a book that promoted molasses, claiming that it was shipped with molasses and therefore constituted labeling. Another FDA seizure was *Calories Don't Count* by Dr. Herman Taller.[57]

The British government has also cracked down on material deemed inappropriate for public consumption. They seized the manuscript *GCHQ: The Negative Asset* by Jock Kane under the Official Secrets Act in 1984. It was never published, but James Bamford was able to obtain a copy of the manuscript before the seizure and incorporate some of it into his own book, *Body of Secrets*.[58]

In addition to these examples, there has been a great deal of censorship in America targeted at works that some find offensive because of sexual content. A somewhat typical example is provided by James Joyce's *Ulysses*. It was declared obscene and barred from the United States for 15 years. U.S. Postal Authorities even seized copies of it in 1918 and 1930. The ban was finally lifted in 1933. The Modern Library recently chose *Ulysses* as the best novel of the 20th century.[59] This example is typical, as many of the works once considered lewd have since become modern classics. The Catholic Church's Index of Prohibited Books, which included many classic scientific works (as well as more risqué titles) over the centuries, was not abolished until 1966.

While this section is by no means meant to provide a thorough survey, another example will help to illustrate the breadth of censorship in 20th-century America.

> In 1915, Margaret Sanger's husband was jailed for distributing her *Family Limitation*, which described and advocated various methods of contraception. Sanger herself had fled the country to avoid prosecution, but would return in 1916 to start the American Birth Control League, which eventually merged with other groups to form Planned Parenthood.[60]

Later in the 20th-century, various methods of contraception were taught in many public schools, including the junior high school I attended. Other portions of my formal education, such as skimming a few of William Shakespeare's plays, would also have been controversial in generations past. Some individuals even went as far as to blame Shakespeare's violent plays for inspiring the assassination of Abraham Lincoln.[61] There have been many censored versions of these plays published

---

[56] http://onlinebooks.library.upenn.edu/banned-books.html.

[57] Garrison, Omar V., *Spy Government: The Emerging Police State in America*, Lyle Stuart, New York, 1967, pp. 145–149.

[58] Bamford, James, *Body of Secrets*, Doubleday, New York, 2001, p. 645.

[59] See http://onlinebooks.library.upenn.edu/banned-books.html, which provides many more examples.

[60] http://onlinebooks.library.upenn.edu/banned-books.html.

[61] For example, according to the assassin's friend and fellow actor John M. Barron, "the characters he [John Wilkes Booth] assumed, all breathing death to tyrants, impelled him to do the deed." I found this quote in Shapiro, James, *Shakespeare in a Divided America: What His Plays Tell Us About Our Past and Future*, Penguin Press, New York, 2020, p. 247, which cites Alford, Terry, *Fortune's Fool: The Life of John Wilkes Booth*, Oxford University Press, New York, 2015, p. 246.

over the centuries. An example is *The Family Shakespeare* edited by Harriet and Thomas Bowdler (1807 and 1818) which removed profanity, sexual content, and violence. This led to a new verb being added to the English language. To "bowdlerize" basically means to ruin a work of literature by removing profanity, sexual content, and violence.

So many of today's classics appear in old lists of censored works that one might reasonably expect to be able to predict the future curriculum by looking at what is presently banned.

References and Further Reading

On World War I Codes and Ciphers

Anon., "Strategic Use of Communications During the World War," *Cryptologia*, Vol. 16, No. 4, October 1992, pp. 320–326.

Beesly, Patrick, *Room 40: British Naval Intelligence 1914–18*, Hamish Hamilton, London, UK, 1982.

Brunswick, Benoît, *Dictionnaire pour la Correspondance télégraphique secrète*, précédé d'instructions détaillées et suivi de la convention télégraphique internationale conclue à Paris le 17 Mai 1865, Vve Berger-Levrault et fils, Paris, France, 1867. This work proposes a superenciperment of a code (substitution) followed by a shuffling (transposition) of the four-digit code groups. Thanks to Paolo Bonavoglia for pointing this reference out to me.

Childs, J. Rives, *The History and Principles of German Military Ciphers, 1914-1918*, Paris, France, 1919.

Childs, J. Rives, *General Solution of the ADFGVX Cipher System*, War Department, Office of the Chief Signal Officer, United States Government Printing Office, Washington, DC, 1934. This volume was reprinted in 1999 by Aegean Park Press, Laguna Hills, California, along with Konheim's paper and actual World War I intercepts.

Childs, J. Rives, *German Military Ciphers From February to November 1918*, War Department, Office of the Chief Signal Officer, United States Government Printing Office, Washington, DC, 1935.

Ewing, Alfred W., *The Man of Room 40: The Life of Sir Alfred Ewing*, Hutchinson, London, UK, 1939. This is a biography of Sir Alfred Ewing written by his son. Ewing directed what soon became known as Room 40.

Ferris, John, editor, *The British Army and Signals Intelligence During the First World War*, Publications of the Army Records Society, Vol. 8, Alan Sutton, Wolfeboro Falls, New Hampshire, 1992.

Freeman, Peter, "The Zimmermann Telegram Revisited: A Reconciliation of the Primary Sources," *Cryptologia*, Vol. 30, No. 2, April 2006, pp. 98–150. Freeman was the historian at GCHQ (the British cryptologic agency).

Friedman, William F., *Military Cryptanalysis*, Vol. 4, U.S. Government Printing Office, Washington DC, pp. 97–143. A general solution for ADFGVX is discussed.

Friedman, William F. and Charles Mendelsohn, *The Zimmermann Telegram of January 16, 1917 and Its Cryptographic Background*, Aegean Park Press, Laguna Hills, California, 1976.

Friedman, William F., *Solving German Codes in World War I*, Aegean Park Press, Laguna Hills, California, 1976.

Hinrichs, Ernest H., *Listening In: Intercepting Trench Communications in World War I*, White Mane Books, Shippensburg, Pennsylvania, 1996.

Hitt, Parker, *Manual for the Solution of Military Ciphers*, Army Service Schools Press, Fort Leavenworth, Kansas, 1916, second edition 1918, reprinted by Aegean Park Press, Laguna Hills, California, 1976, available online, but not in a nice format, at https://archive.org/details/manualforthesolu48871gut. This was the first book on cryptology to be published in America, although it was preceded by articles and pamphlets. In it, Hitt discussed substitution and transposition being used together, prior to Nebel's invention of ADFGX and ADFGVX. A pair of relevant passages from the first edition are reproduced below.

It is evident that a message can be enciphered by any transposition method, and the result enciphered again by any substitution method, or vice versa. But this takes time and leads to errors in the work, so that, if such a process is employed, the substitution and transposition ciphers used are likely to be very simple ones which can be operated with fair rapidity.

It is very rare to find both complicated transposition and substitution methods used in combination. If one is complicated, the other will usually be very simple; and ordinarily both are simple, the sender depending on the combination of the two to attain indecipherability. It is evident how futile this idea is.

Hoy, Hugh Cleland, *40 O.B. or How the War Was Won*, Hutchinson & Co., London, UK, 1932. "How the War Was Won" books typically explain how the author won it. But not in this case! Hoy was never in Room 40. The *O.B.* in the title indicates that Room 40 was in the *Old Buildings* (of the Admiralty).

James, Admiral Sir William, *The Code Breakers of Room 40: The Story of Admiral Sir William Hall, Genius of British Counter-Intelligence*, St. Martin's Press, New York, 1956.

Kelly, Saul, "Room 47: The Persian Prelude to the Zimmermann Telegram," *Cryptologia*, Vol. 37, No. 1, January 2013, pp. 11–50.

Knight, H. Gary, "Cryptanalysts' Corner," *Cryptologia*, Vol. 4, No. 4, October 1980, pp. 208–212. Knight describes a cipher similar to ADFGVX that was used in Christopher New's 1979 novel *Goodbye Chairman Mao* and presents several ciphertexts of his own in the system for readers to solve. These ciphers are substantially easier than ADFGVX because the Polybius square pairs are not split in the transposition stage.

Konheim, Alan, "Cryptanalysis of ADFGVX Encipherment Systems," in Blakley, George Robert "Bob" and David Chaum, editors, *Advances in Cryptology: Proceedings of CRYPTO 84*, Lecture Notes in Computer Science, Vol. 196, Springer, Berlin, Germany, 1985, pp. 339–341. This is an "Extended Abstract" rather than a full paper. A full-length paper by Konheim on this topic appears as appendix A in Childs, J. Rives, *General Solution of the ADFGVX Cipher System*, as reprinted in 1999 by Aegean Park Press, Laguna Hills, California.

Langie, Andre, *How I Solved Russian and German Cryptograms During World War I*, Imprimerie T. Geneux, Lausanne, 1944, translated from German into English by Bradford Hardie, El Paso, Texas, 1964.

Lasry, George, Ingo Niebel, Nils Kopal, and Arno Wacker, "Deciphering ADFGVX Messages from the Eastern Front of World War I," *Cryptologia*, Vol. 41, No. 2, March 2017, pp. 101–136.

Lerville, Edmond, "The Radiogram of Victory (La Radiogramme de la Victoire)," *La Liason des Transmissions*, April 1969, pp. 16–23, translated from French to English by Steven M. Taylor.

Lerville, Edmond, "The Cipher: A Face-to-Face Confrontation After 50 Years," *L'Armee*, May 1969, pp. 36–53, translated from the French "Le Chiffre «face à face» cinquante ans après" to English by Steven M. Taylor.

Mendelsohn, Charles, *Studies in German Diplomatic Codes Employed During the World War*, War Department, Office of the Chief Signal Officer, United States Government Printing Office, Washington, DC, 1937.

Mendelsohn, Charles, *An Encipherment of the German Diplomatic Code 7500*, War Department, Office of the Chief Signal Officer, United States Government Printing Office, Washington, 1938. This is a supplement to the item listed above.

Norman, Bruce, "The ADFGVX Men," *The Sunday Times Magazine*, August 11, 1974, pp. 8–15. For this piece, Norman interviewed both Fritz Nebel, the German creator of the ADFGVX cipher, and Georges Painvin, the Frenchman who cracked it.

Ollier, Alexandre, *La Cryptographie Militaire avant la guerre de 1914*, Lavauzelle, Panazol, 2002.

Partenio, Pietro, unpublished booklet, 1606. This booklet was only used by students in Partenio's cryptography course. A copy is in the Venice State Archives. Partenio combined substitution and transposition by using a two-letter nomenclator (each group was encrypted with two letters) followed by a shuffling of the pairs according to a phrase to be kept in memory. His sample phrase was "*Lex tua meditatio mea uoluntas tua …*" Thanks to Paolo Bonavoglia for pointing this reference out to me. He is planning to publish it. The transcription is still in progress as of this writing.

Pergent, Jacques, "Une figure extraordinaire du chiffre français de 1914 à 1918: le capitaine Georges Painvin," *Armée et Défense*, Vol. 47, No. 4, April 1968, pp. 4–8.

Rislakki, Jukka, "Searching for Cryptology's Great Wreck," *Cryptologia*, Vol. 31, No. 3, July 2007, pp. 263–267. This paper summarizes the Russian capture of the German Navy's main code book from the wreck of the *Magdeburg* and David Kahn's excursion to the site 92 years later. This is what historians of cryptology do on vacation!

Samuels, Martin, "Ludwig Föppl: A Bavarian cryptanalyst on the Western front," *Cryptologia*, Vol. 40, No. 4, July 2016, pp. 355–373.

Tuchman, Barbara W., *The Zimmermann Telegram*, The Macmillan Company, New York, 1970. This book has gone through many editions, and this is not the first.

von zur Gathen, Joachim, "Zimmermann Telegram: The Original Draft," *Cryptologia*, Vol. 31, No. 1, January 2007, pp. 2–37.

Books and Articles by Yardley (and/or His Ghostwriter)

Yardley, Herbert O., *Universal Trade Code*, Code Compiling Co., New York, 1921. This one doesn't make for an exciting read! Yardley's first book was a commercial code book.

Yardley, Herbert O., "Secret Inks," *Saturday Evening Post*, April 4, 1931, pp. 3–5, 140–142, 145.

Yardley, Herbert O., "Codes," *Saturday Evening Post*, April 18, 1931, pp. 16–17, 141–142.

Yardley, Herbert O., "Ciphers," *Saturday Evening Post*, May 9, 1931, pp. 35, 144–146, 148–149.

Yardley, Herbert O., *The American Black Chamber*, Bobbs-Merrill, Indianapolis, Indiana, 1931. Unauthorized translations were made in Japanese and Chinese.

Yardley, Herbert O., "How They Captured the German Spy, Mme. Maria de Victorica, Told at Last," *Every Week Magazine*, July 12, 1931.

Yardley, Herbert O., "Secrets of America's Black Chamber," *Every Week Magazine*, July 26, 1931.

Yardley, Herbert O., "Double-Crossing America: A glimpse of the International Intrigue that Goes on behind Uncle Sam's Back," *Liberty*, October 10, 1931, pp. 38–42.

Yardley, Herbert O., "Cryptograms and Their Solution," *Saturday Evening Post*, November 21, 1931, pp. 21, 63–65.

Yardley, Herbert O., "Are We Giving Away Our State Secrets?" *Liberty*, December 19, 1931, pp. 8–13.

Yardley, Herbert O., *Yardleygrams*, The Bobbs-Merrill Company, Indianapolis, Indiana, 1932. This was ghost written by Clem Koukol.

Yardley, Herbert O., *Ciphergrams*, Hutchinson & Co., London, 1932. This is the British version of *Yardleygrams*, with a new title but still ghostwritten by Clem Koukol.

Yardley, Herbert O., "The Beautiful Secret Agent," *Liberty*. December 30, 1933, pp. 30–35.

Yardley, Herbert O., "Spies Inside Our Gates," *Sunday* [Washington] *Star Magazine*, April 8, 1934, pp. 1–2.

Yardley, Herbert O., "Spies Inside Our Gates," *New York Herald Tribune Magazine*, April 8, 1934. This article is the same as the one above.

Yardley, Herbert O., "H-27, The Blonde Woman from Antwerp," *Liberty*. April 21, 1934, pp. 22–29.

Yardley, Herbert O., *The Blonde Countess*, Longmans, Green and Co., New York, 1934. This novel, primarily written by Carl Grabo, was made into the movie *Rendezvous*,[62] released by Metro-Goldwyn-Mayer in 1935.

Yardley, Herbert O., *Red Sun of Nippon*, Longmans, Green and Co., New York, 1934. Carl Grabo was the main author.

Yardley, Herbert O. with Carl Grabo, *Crows Are Black Everywhere*, G. P. Putnam's Sons, New York, 1945. Again, Carl Grabo worked on this behind the scenes.

Yardley, Herbert O., *The Education of a Poker Player*, Simon and Schuster, New York, 1957.

Yardley, Herbert O., *The Chinese Black Chamber An Adventure in Espionage*, Houghton Mifflin, Boston, Massachusetts, 1983. The dustjacket includes, "Because of State Department disapproval, this manuscript has lain hidden for forty-two years."

Yardley, Herbert O., "From the Archives: The Achievements of the Cipher Bureau (MI-8) During the First World War," *Cryptologia*, Vol. 8, No. 1, January 1984, pp. 62–74.

About Yardley

Anonymous, "Yardley Sold Papers to Japanese," *Surveillant*, Vol. 2, No. 4, January/February 1992, p. 99

Denniston, Robin, "Yardley's Diplomatic Secrets," *Cryptologia*, Vol. 18, No. 2, April 1994, pp. 81–127.

---

[62] https://en.wikipedia.org/wiki/Rendezvous_(1935_film).

Dooley, John F., "Was Herbert O. Yardley a Traitor?" *Cryptologia*, Vol. 35, No. 1, January 2011, pp. 1–15.

Farago, Ladislas, *The Broken Seal: The Story of "Operation Magic" and the Pearl Harbor Disaster*, Random House, New York, 1967, pp. 56–58.

Hannah, Theodore M., "The Many Lives of Herbert O. Yardley," *Cryptologic Spectrum*, Vol. 11, No. 4, Fall 1981, pp. 5–29, available online at http://www.nsa.gov/public_info/_files/cryptologic_spectrum/many_lives.pdf.

Kahn, David, "Nuggets from the Archive: Yardley Tries Again," *Cryptologia*, Vol. 2, No. 2, April 1978, pp. 139–143.

Kahn, David, *The Reader of Gentlemen's Mail*, Yale University Press, New Haven, Connecticut, 2004.

Nedved, Gregory J., "Herbert O. Yardley Revisited: What Does the New Evidence Say?" *Cryptologia*, to appear.

Turchen, Lesta VanDerWert, "Herbert Osborne Yardley and American Cryptography," Master's Thesis University of South Dakota, Vermillion, South Dakota, May 1969. Turchen made the following comment concerning Farago's book: "Obviously biased against Mr. Yardley *The Broken Seal* could have been more clearly documented and more judicious in its conclusions." (p. 95).

Addendum

Long before meeting his World War I adversary Fritz Nebel, Painvin met his American counterpart Herbert Yardley. An image from their meeting (Figure 5.12) is one of many treasures preserved by the National Cryptologic Museum adjacent to Fort Meade, Maryland.

Figure 5.12 Georges Painvin and Herbert O. Yardley. (Courtesy of the National Cryptologic Museum, Fort Meade, Maryland.)

Chapter 6

Matrix Encryption

Polygraphic substitution ciphers replace characters in groups, rather than one at a time. In Section 4.4, we examined the Playfair cipher, which falls into this category, because the characters are substituted for in pairs. We now turn to a more mathematically sophisticated example, matrix encryption.

6.1 Levine and Hill

Lester Hill (Figure 6.1) is almost always given credit for discovering matrix encryption, but as David Kahn pointed out in his definitive history of the subject, Jack Levine's work in this area preceded Hill's. Levine (Figure 6.2) also pointed this out in a 1958 paper.[1] Nevertheless, matrix encryption is typically referred to as the Hill cipher. Levine explained how this happened:[2]

> Sometime during 1923-1924 while I was a high-school student I managed to construct a cipher system which would encipher <u>two</u> <u>completely</u> <u>independent</u> <u>messages</u> so that one message could be deciphered without disclosing that a second message was still hidden (I also did for three messages). Not long thereafter (end of 1924) _Flynn's Weekly_ magazine started a cipher column conducted by M. E. Ohaver (Sanyam of A.C.A.), a most excellent series appearing every few weeks. Readers were encouraged to submit their systems which Ohaver then explained in a later issue. So I sent him my system mentioned above and it appeared in the Oct. 22, 1926 issue with its explanation in the Nov. 13, 1926 issue. Ohaver gave a very good explanation and some other interesting remarks [...] In 1929 Hill's first article appeared in the Math. Association Monthly Journal recognized he had a very general system which could encipher plaintext units of any length very easily, but the basic principle was the same as my 2-message system

---

[1] Levine, Jack, "Variable Matrix Substitution in Algebraic Cryptography," _American Mathematical Monthly_, Vol. 65, No. 3, March 1958, pp. 170–179.

[2] The Jack Levine Papers, 1716--994, North Carolina State University, MC 308.1.7, Correspondence 1981–1991, Various, Levine, Jack, letter to Louis Kruh, July 24, 1989.

(or 3-message). I had some correspondence with Hill and I think told him of my youthful efforts. All this to explain why I believe my system was the precursor to his very general mathematical formulation (I also used equations).

Figure 6.1 Lester Hill (1890–1961). (Courtesy of the David Kahn Collection, National Cryptologic Museum, Fort Meade, Maryland.)

Figure 6.2 Jack Levine (1907–2005). (Courtesy of the Jack Levine Archive at North Carolina State University.)

Flynn's Weekly was a pulp magazine that consisted mainly of detective fiction. Agatha Christie had a short story in the same November 13, 1926 issue in which Levine's system was explained.

This was not the best place to publish a mathematical idea, but Levine was still a teenager at the time.

Hill's description came three years later, but it appeared in *American Mathematical Monthly*, so it isn't hard to see why the system was named after Hill.[3] The *American Mathematical Monthly* paper provided a complete explanation of matrix encryption with examples, whereas Levine's letter didn't actually reveal his method, although it can be inferred. And, although Levine used algebra, he wasn't quite doing the same thing as Hill, at least not publicly.

Levine went on to earn a doctorate in mathematics from Princeton University, serve his country in a cryptologic capacity (in the Army), and author many more papers on cryptology, some of which are described in the pages to follow. Hill, on the other hand, only published two papers dealing directly with cryptology. Hill served in the Navy during World War I, but not in a cryptologic capacity. In later decades, he did do some cryptologic work for the Navy, but it appears to have been unsolicited and not considered valuable.[4] In any case, the military work of both Levine and Hill was not known to the academic community.

6.2 How Matrix Encryption Works

Matrix encryption can most easily be explained by example. Consider Oscar Wilde's quote "THE BEST WAY TO DEAL WITH TEMPTATION IS TO YIELD TO IT." We first replace each letter with its numerical equivalent:

```
A  B  C  D  E  F  G  H  I  J  K   L   M   N   O   P   Q   R   S   T   U   V   W   X   Y   Z
0  1  2  3  4  5  6  7  8  9  10  11  12  13  14  15  16  17  18  19  20  21  22  23  24  25
```

This gives

```
19 7 4 1 4 18 19 22 0 24 19 14 3 4 0 11 22 8 19 7 19 4 12 15
19 0 19 8 14 13 8 18 19 14 24 8 4 11 3 19 14 8 19.
```

The key we select in this system is an invertible matrix (modulo 26) such as $\begin{pmatrix} 6 & 11 \\ 3 & 5 \end{pmatrix}$. To enci-

pher, we simply multiply this matrix by the numerical version of the plaintext in pieces of length two and reduce the result modulo 26; for example,

$$\begin{pmatrix} 6 & 11 \\ 3 & 5 \end{pmatrix}\begin{pmatrix} 19 \\ 7 \end{pmatrix} = \begin{pmatrix} 6 \cdot 19 + 11 \cdot 7 \\ 3 \cdot 19 + 5 \cdot 7 \end{pmatrix} = \begin{pmatrix} 191 \\ 92 \end{pmatrix} = \begin{pmatrix} 9 \\ 14 \end{pmatrix} \text{ (modulo 26)}$$

---

[3] Hill, Lester, S., "Cryptography in an Algebraic Alphabet," *American Mathematical Monthly*, Vol. 36, No. 6, June–July 1929, pp. 306–312.

[4] For details, see Christensen, Chris, "Lester Hill Revisited," *Cryptologia* Vol. 38, No. 4, October 2014, pp. 293–332.

gives the first two ciphertext values as 9 and 14 or, in alphabetic form, J O. Continuing in this manner we have

$$\begin{pmatrix} 6 & 11 \\ 3 & 5 \end{pmatrix}\begin{pmatrix} 4 \\ 1 \end{pmatrix} = \begin{pmatrix} 9 \\ 17 \end{pmatrix} \qquad \begin{pmatrix} 6 & 11 \\ 3 & 5 \end{pmatrix}\begin{pmatrix} 4 \\ 18 \end{pmatrix} = \begin{pmatrix} 14 \\ 24 \end{pmatrix} \qquad \begin{pmatrix} 6 & 11 \\ 3 & 5 \end{pmatrix}\begin{pmatrix} 19 \\ 22 \end{pmatrix} = \begin{pmatrix} 18 \\ 11 \end{pmatrix}$$

$$\begin{pmatrix} 6 & 11 \\ 3 & 5 \end{pmatrix}\begin{pmatrix} 0 \\ 24 \end{pmatrix} = \begin{pmatrix} 4 \\ 16 \end{pmatrix} \qquad \begin{pmatrix} 6 & 11 \\ 3 & 5 \end{pmatrix}\begin{pmatrix} 19 \\ 14 \end{pmatrix} = \begin{pmatrix} 8 \\ 23 \end{pmatrix} \qquad \begin{pmatrix} 6 & 11 \\ 3 & 5 \end{pmatrix}\begin{pmatrix} 3 \\ 4 \end{pmatrix} = \begin{pmatrix} 10 \\ 3 \end{pmatrix}$$

$$\begin{pmatrix} 6 & 11 \\ 3 & 5 \end{pmatrix}\begin{pmatrix} 0 \\ 11 \end{pmatrix} = \begin{pmatrix} 17 \\ 3 \end{pmatrix} \qquad \begin{pmatrix} 6 & 11 \\ 3 & 5 \end{pmatrix}\begin{pmatrix} 22 \\ 8 \end{pmatrix} = \begin{pmatrix} 12 \\ 2 \end{pmatrix} \qquad \begin{pmatrix} 6 & 11 \\ 3 & 5 \end{pmatrix}\begin{pmatrix} 19 \\ 7 \end{pmatrix} = \begin{pmatrix} 9 \\ 14 \end{pmatrix}$$

$$\begin{pmatrix} 6 & 11 \\ 3 & 5 \end{pmatrix}\begin{pmatrix} 19 \\ 4 \end{pmatrix} = \begin{pmatrix} 2 \\ 25 \end{pmatrix} \qquad \begin{pmatrix} 6 & 11 \\ 3 & 5 \end{pmatrix}\begin{pmatrix} 12 \\ 15 \end{pmatrix} = \begin{pmatrix} 3 \\ 7 \end{pmatrix} \qquad \begin{pmatrix} 6 & 11 \\ 3 & 5 \end{pmatrix}\begin{pmatrix} 19 \\ 0 \end{pmatrix} = \begin{pmatrix} 10 \\ 5 \end{pmatrix}$$

$$\begin{pmatrix} 6 & 11 \\ 3 & 5 \end{pmatrix}\begin{pmatrix} 19 \\ 8 \end{pmatrix} = \begin{pmatrix} 20 \\ 19 \end{pmatrix} \qquad \begin{pmatrix} 6 & 11 \\ 3 & 5 \end{pmatrix}\begin{pmatrix} 14 \\ 13 \end{pmatrix} = \begin{pmatrix} 19 \\ 3 \end{pmatrix} \qquad \begin{pmatrix} 6 & 11 \\ 3 & 5 \end{pmatrix}\begin{pmatrix} 8 \\ 18 \end{pmatrix} = \begin{pmatrix} 12 \\ 10 \end{pmatrix}$$

$$\begin{pmatrix} 6 & 11 \\ 3 & 5 \end{pmatrix}\begin{pmatrix} 19 \\ 14 \end{pmatrix} = \begin{pmatrix} 8 \\ 23 \end{pmatrix} \qquad \begin{pmatrix} 6 & 11 \\ 3 & 5 \end{pmatrix}\begin{pmatrix} 24 \\ 8 \end{pmatrix} = \begin{pmatrix} 24 \\ 8 \end{pmatrix} \qquad \begin{pmatrix} 6 & 11 \\ 3 & 5 \end{pmatrix}\begin{pmatrix} 4 \\ 11 \end{pmatrix} = \begin{pmatrix} 15 \\ 15 \end{pmatrix}$$

$$\begin{pmatrix} 6 & 11 \\ 3 & 5 \end{pmatrix}\begin{pmatrix} 3 \\ 19 \end{pmatrix} = \begin{pmatrix} 19 \\ 0 \end{pmatrix} \qquad \begin{pmatrix} 6 & 11 \\ 3 & 5 \end{pmatrix}\begin{pmatrix} 14 \\ 8 \end{pmatrix} = \begin{pmatrix} 16 \\ 4 \end{pmatrix} \qquad \begin{pmatrix} 6 & 11 \\ 3 & 5 \end{pmatrix}\begin{pmatrix} 19 \\ 23 \end{pmatrix} = \begin{pmatrix} 3 \\ 16 \end{pmatrix}$$

If you're reading very closely, you noticed that the 23 used in the last matrix multiplication wasn't part of the original message. The message had an odd number of characters, so it was necessary to add one more in order to encipher everything. I added an X, which has a numerical value of 23.

The ciphertext is

9 14 9 17 14 24 18 11 4 16 8 23 10 3 17 3 12 2 9 14 2 25 3 7 10 5 20 19 19 3 12 10 8 23 24 8 15 15 19 0 16 4 3 16

which may be converted back to letters to get

JOJ ROYS LEQ IX KDRD MCJO CZDHKFUTTD MK IX YIPPT AQ ED Q

Deciphering is done in the same manner, except that we need to use the inverse of the original matrix. Sometimes the enciphering matrix is chosen to be one that is self-inverse. In that case, deciphering is identical to enciphering. However, a heavy price is paid for this convenience, because limiting oneself to such keys drastically reduces the keyspace. In our example, the deciphering matrix is given by $\begin{pmatrix} 7 & 21 \\ 1 & 24 \end{pmatrix}$. Applying this matrix to the ciphertext in order to recover the original message is left as an exercise for the reader. You may wish to refer to the modulo 26 multiplication table in Section 1.16 when using this system.

A comparison of the plaintext and ciphertext shows an advantage of this system.

THE BEST WAY TO DEAL WITH TEMPTATION IS TO YIELD TO IT X
JOJ ROYS LEQ IX KDRD MCJO CZDHKFUTTD MK IX YIPPT AQ ED Q

The first time TO appeared it was enciphered as IX. The same encipherment happened the second time TO appeared, but for the third appearance it was enciphered as AQ. This is because TO was enciphered using the matrix the first and second time it appeared, but for the third time T and O were split apart and ST and OI were the plaintext pairs to be enciphered. Hence, the same word may be enciphered in two different ways, depending on how it is split up when the message is broken into pairs of letters. The Playfair cipher also had this feature.

The details of how to test if a matrix is invertible and, if so, calculate its inverse follow below, so that you can create your own keys for matrix encryption. A natural starting point is the following definition.

The **determinant** of a 2 × 2 matrix $M = \begin{pmatrix} a & b \\ c & d \end{pmatrix}$ is given by $ad - bc$. It is often denoted by det(M).

The determinant offers a quick test to determine if the matrix is invertible.

Theorem:

A matrix M is invertible if and only if its determinant is invertible.

If the entries of the matrix are real numbers, then the matrix is invertible if and only if its determinant is not zero. This is because every nonzero real number is invertible. However, for matrix encryption the entries of the matrix are not real numbers. They are integers modulo n. In the Example I provided above, $n = 26$. That means we are only allowed to use the numbers 0, 1, 2, 3, 4, 5, 6, 7, 8, 9, 10, 11, 12, 13, 14, 15, 16, 17, 18, 19, 20, 21, 22, 23, 24, and 25. Which of these numbers are invertible?

Theorem:

A number a is invertible modulo n if and only if gcd(a, n) = 1.

That is, a number a is invertible modulo n if and only if the greatest common divisor of a and n is 1. Another way of saying gcd(a, n) = 1 is "a and n are relatively prime." A mod 26 multiplication table was given in Section 1.16. You can use it to confirm that in the case of $n = 26$, the invertible values are 1, 3, 5, 7, 9, 11, 15, 17, 19, 21, 23, and 25. This is simply all of the odd values with the exception of 13. You can quickly find the inverse of each of these numbers in the table. Simply look for the number you have to multiply the given number by to get 1.

Example:

If $M = \begin{pmatrix} 6 & 11 \\ 3 & 5 \end{pmatrix}$, then det(M) = 6·5 − 11·3 = 30 − 33 = −3 = 23 (mod 26).

Because 23 is invertible (mod 26), the matrix M is invertible.

Theorem:

If $M = \begin{pmatrix} a & b \\ c & d \end{pmatrix}$ is an invertible matrix, then $M^{-1} = (\det(M))^{-1} \begin{pmatrix} d & -b \\ -c & a \end{pmatrix}$.

You can see that the formula won't work if the determinant of M is not invertible, because the formula begins with the inverse of the determinant. Applying this formula to the matrix used in the encryption example at the beginning of this section, we have

$$M^{-1} = (23)^{-1} \begin{pmatrix} 5 & -11 \\ -3 & 6 \end{pmatrix} = 17 \begin{pmatrix} 5 & 15 \\ 23 & 6 \end{pmatrix} = \begin{pmatrix} 17 \cdot 5 & 17 \cdot 15 \\ 17 \cdot 23 & 17 \cdot 6 \end{pmatrix} = \begin{pmatrix} 7 & 21 \\ 1 & 24 \end{pmatrix}.$$

So, if we use $M = \begin{pmatrix} 6 & 11 \\ 3 & 5 \end{pmatrix}$ to encipher, then we can use $M^{-1} = \begin{pmatrix} 7 & 21 \\ 1 & 24 \end{pmatrix}$ to decipher, as was claimed.

There are other methods that can be used to calculate the inverse of a matrix. You can find them detailed in linear algebra textbooks. Some of these generalize easily to larger matrices, which is great because square (invertible) matrices of any dimension may be used for encryption. A 5×5 matrix would allow us to encipher characters in groups of five, and a particular word could be enciphered in as many as five different ways depending on its position in the message. There are other advantages to using larger matrices, such as larger keyspaces.

6.3 Levine's Attacks

A brute-force attack on 2×2 matrix encryption will quickly yield the plaintext, as there are only 157,248 possible keys. Indeed, Levine pursued this approach back in 1961 for the special case in which an involutory (self-inverse) matrix was used.[5] There are 736 of these especially convenient keys and even primitive 1961 computer technology was up to the task, although a human was needed to look at the results and determine which provided a meaningful message.[6]

The keyspace grows rapidly with the size of the matrix. For 3×3 matrix encryption there are 1,634,038,189,056 possible keys and for the 4×4 case we have 12,303,585,972,327,392,870,400 invertible matrices (modulo 26) to consider. The references at the end of this chapter include a paper showing how these values are calculated. When larger matrices are used, cryptanalysts need an attack significantly better than brute-force, if they hope to recover the message.

If we are able to guess the beginning of a message (or any other part), the combination of known plaintext and ciphertext makes it easy to recover the matrix and break the rest of the cipher.[7] This is referred to as a *known-plaintext attack*. However, we often have a word or phrase that we believe appears in the message, although we don't know where. Such suspected plaintexts are called *cribs*.

In another 1961 paper, Levine demonstrated how cribs can be applied to break matrix encryption.[8] The mathematics needed to understand this attack consists of two very basic facts:

1. The product of two integers is odd if and only if both of the integers are odd.
2. The sum of two integers is odd if and only if exactly one of the integers is odd.

To see how these facts become useful, suppose we are enciphering with a matrix of the form $\begin{pmatrix} even & odd \\ odd & even \end{pmatrix}$. There are four distinct possibilities for the form of the plaintext pair that is enciphered at each step, namely

$$\begin{pmatrix} even \\ even \end{pmatrix}, \begin{pmatrix} even \\ odd \end{pmatrix}, \begin{pmatrix} odd \\ even \end{pmatrix}, \begin{pmatrix} odd \\ odd \end{pmatrix}$$

[5] Levine, Jack, "Some Applications of High-Speed Computers to the Case $n = 2$ of Algebraic Cryptography," *Mathematics of Computation*, Vol. 15, No. 75, July 1961, pp. 254–260.

[6] Levine gave the incorrect value 740 in his paper. A list of the number of self-inverse matrices for various moduli is given at http://oeis.org/A066907.

[7] The number of plaintext/ciphertext pairs needed to uniquely determine the matrix depends on the size of the matrix and another factor. You are asked to determine this for 3×3 matrix encryption in one of the online exercises for this chapter.

[8] Levine, Jack, "Some Elementary Cryptanalysis of Algebraic Cryptography," *American Mathematical Monthly*, Vol. 68, No. 5, May 1961, pp. 411–418.

Labeling these four forms as 0, 1, 2, and 3, we see (by applying the facts above in the context of matrix multiplication) that the enciphering matrix sends form 0 to another pair of letters having form 0. In the case of form 1, enciphering yields a pair of letters having form 2. A nice symmetry exists, as a plaintext pair of form 2 will be sent to a ciphertext pair of form 1. Finally, a plaintext pair of form 3 will be sent to a ciphertext pair of form 3.

These pairings can be tersely expressed as

$$0 \leftrightarrow 0, \ 1 \leftrightarrow 2, \ 3 \leftrightarrow 3$$

If the enciphering matrix has a form other than $\begin{pmatrix} even & odd \\ odd & even \end{pmatrix}$, these pairings may differ, but the pairings are always constant for a given matrix and we always have $0 \leftrightarrow 0$. If the matrix is self-inverse, the pairings will be reciprocal, otherwise some of the arrows could be one-way.

Now consider the following ciphertext obtained by enciphering with a 2×2 self-inverse matrix of unknown form:

```
CRSFS HLTWB WCSBG RKBCI PMQEM FOUSC
PESHS GPDVF RTWCX FJDPJ MISHE W
```

If this message was sent from one mathematician to another, we might guess that the word MATHEMATICS appears somewhere in the plaintext. Word spacing has not been preserved in the ciphertext, so we cannot immediately locate the word's position. However, the word can appear in only two fundamentally different positions with respect to the enciphering matrix. We have either:

1. MA TH EM AT IC Sx

or

2. xM AT HE MA TI CS

We disregard the pairs with an x (representing an unknown plaintext letter) in them. The remaining pairs have the forms (0, 3, 0, 1, 0) and (1, 2, 0, 2, 0), respectively. Finding the forms of the ciphertext pairs yields:

```
1  1  1  3  1  0  1  1  1  0  2  0  1  0  0  2  1  0  3  3  3
CR SF SH LT WB WC SB GR KB CI PM QE MF OU SC PE SH SG PD VF RT

0  3  3  3  0  1  0
WC XF JD PJ MI SH EW
```

Suppose MATHEMATICS lined up with the enciphering matrix in the manner of the first possibility listed, (0, 3, 0, 1, 0). Could the message begin with this word? If it did, this would imply the plaintext/ciphertext pairing $0 \leftrightarrow 1$, while we know $0 \leftrightarrow 0$ for all matrices. Therefore, if MATHEMATICS appears in the message, it cannot be at the very beginning. Lining up with the first zero in the ciphertext gives:

```
       0 3 0 1 0
1 1 1 3 1 0 1 1 1 0 2 0 1 0 0 2 1 0 3 3 3 0 3 3 3 0 1 0
```

This is not possible, because the middle pairing is $0 \leftrightarrow 1$. Also, if $3 \leftrightarrow 1$, we must have $1 \leftrightarrow 3$, but the above alignment has $1 \leftrightarrow 1$. We now line up the first 0 in our proposed plaintext with the second zero in the ciphertext.

```
            0 3 0 1 0
1 1 1 3 1 0 1 1 1 0 2 0 1 0 0 2 1 0 3 3 3 0 3 3 3 0 1 0
```

The pairings given by this alignment are consistent. This is, in fact, the only consistent alignment for when MATHEMATICS is broken up as MA TH EM AT IC Sx and enciphered. However, as explained earlier, MATHEMATICS could be broken up as xM AT HE MA TI CS and then enciphered. In this case, the form pattern is (1, 2, 0, 2, 0). Because this pattern has two zeros separated by a single value we check alignments in positions where the ciphertext also has this form:

```
            1 2 0 2 0
1 1 1 3 1 0 1 1 1 0 2 0 1 0 0 2 1 0 3 3 3 0 3 3 3 0 1 0
```

This alignment cannot be correct, because it sends the first 2 to 1 and the second 2 to 2, which is inconsistent.

```
          1 2 0 2 0
1 1 1 3 1 0 1 1 1 0 2 0 1 0 0 2 1 0 3 3 3 0 3 3 3 0 1 0
```

This alignment is rejected, because it sends the 1 to 0.

```
                                    1 2 0 2 0
1 1 1 3 1 0 1 1 1 0 2 0 1 0 0 2 1 0 3 3 3 0 3 3 3 0 1 0
```

This alignment is also impossible, as 1 cannot be paired with 3, if 2 is also paired with 3.

Sometimes there will be several consistent alignments. The fact that the ciphertext was short made this less likely for the example above. Usually having more ciphertext makes the cipher easier to break, but for this attack it creates more work! Our only consistent alignment

```
            0 3 0 1 0
1 1 1 3 1 0 1 1 1 0 2 0 1 0 0 2 1 0 3 3 3 0 3 3 3 0 1 0
```

tells us that MA TH EM AT IC Sx was enciphered as CI PM QE MF OU.

Representing our unknown enciphering matrix as $\begin{pmatrix} a & b \\ c & d \end{pmatrix}$, we get the following equations:

$$\text{MA} \to \text{CI} \implies \begin{pmatrix} a & b \\ c & d \end{pmatrix}\begin{pmatrix} 12 \\ 0 \end{pmatrix} = \begin{pmatrix} 2 \\ 8 \end{pmatrix} \implies 12a = 2 \quad \text{and} \quad 12c = 8$$

$$\text{TH} \to \text{PM} \implies \begin{pmatrix} a & b \\ c & d \end{pmatrix}\begin{pmatrix} 19 \\ 7 \end{pmatrix} = \begin{pmatrix} 15 \\ 12 \end{pmatrix} \implies 19a + 7b = 15 \quad \text{and} \quad 19c + 7d = 12$$

$$\text{EM} \to \text{QE} \implies \begin{pmatrix} a & b \\ c & d \end{pmatrix}\begin{pmatrix} 4 \\ 12 \end{pmatrix} = \begin{pmatrix} 16 \\ 4 \end{pmatrix} \implies 4a + 12b = 16 \quad \text{and} \quad 4c + 12d = 4$$

$$\text{AT} \to \text{MF} \implies \begin{pmatrix} a & b \\ c & d \end{pmatrix}\begin{pmatrix} 0 \\ 19 \end{pmatrix} = \begin{pmatrix} 12 \\ 5 \end{pmatrix} \implies 19b = 12 \quad \text{and} \quad 19d = 5$$

$$\text{IC} \to \text{OU} \implies \begin{pmatrix} a & b \\ c & d \end{pmatrix}\begin{pmatrix} 8 \\ 2 \end{pmatrix} = \begin{pmatrix} 14 \\ 20 \end{pmatrix} \implies 8a + 2b = 14 \quad \text{and} \quad 8c + 2d = 20$$

Sx → SC is comparatively less useful, because we don't know what x is.

If you flip back to the mod 26 multiplication table in Section 1.16, you'll see that $12a = 2$ has two solutions (because 12 is not invertible modulo 26). So, either $a = 11$ or $a = 24$. Similarly, $12c = 8$ implies $c = 5$ or $c = 18$. The equations in the fourth plaintext/ciphertext pairing are nicer. Because 19 has an inverse modulo 26, we have a unique solution for each equation. We get $b = 2$ and $d = 3$. Substituting these values into the second pair of equations gives $19a + 14 = 15$ and $19c + 21 = 12$. These equations have the unique solution $a = 11$ and $c = 5$. Thus the matrix used to encipher was $\begin{pmatrix} 11 & 2 \\ 5 & 3 \end{pmatrix}$. The inverse of this matrix is $\begin{pmatrix} 25 & 18 \\ 19 & 11 \end{pmatrix}$. Applying this matrix to each pair of ciphertext letters reveals the message to be

STUDENTS WHO MAJOR IN MATHEMATICS OFTEN MINOR IN COMPUTER SCIENCE

A little time could have been saved by calculating the deciphering matrix first—that is, by finding the matrix that sends the known ciphertext to the known plaintext. The steps would be the same as above; only the numbers would change. It should also be noted that the equations generated by the crib may be solved by placing them in a matrix and row reducing, rather than investigating them in the manner detailed above. For larger matrices, with more unknowns to solve for, the more systematic approach of row reduction would be preferred.

A good cipher must resist crib attacks. An intercepted message always has some context that allows an attacker to guess cribs. If one particular crib doesn't lead to a solution, the attacker can start over with another crib. Modern ciphers are even expected to resist chosen plaintext attacks, where the attacker gets to pick any plaintext he or she likes and receive the corresponding cipher-text. For a matrix encryption system using a 2 × 2 matrix, the chosen plaintext pairs BA and AB would be excellent choices. The ciphertext that results, upon being converted back to numerical values would be the entries in rows 1 and 2 of the matrix, respectively. One could work these pairs into a meaningful message like so:

ABBA RELEASED THEIR FIRST ALBUM IN NINETEEN SEVENTY-THREE.

For a 3 × 3 matrix, the ideal chosen plaintext triplets would be BAA, ABA, and AAB. Casually fitting these into an innocent sounding message would be a little harder.

6.4 Bauer and Millward's Attack

In 2007, a paper by the author of this text (Figure 6.3) and Katherine Millward, an undergraduate mathematics major, detailed an attack on matrix encryption that doesn't require a crib.[9] The attack is made possible by a simple observation: given a ciphertext, we may recover the rows of the deciphering matrix individually, instead of searching for the complete matrix. To see this, consider the ciphertext $C_1C_2C_3C_4C_5C_6C_7C_8$, ... generated by a 2 × 2 matrix. If we denote the unknown

---

9 Bauer, Craig and Katherine Millward, "Cracking Matrix Encryption Row by Row," *Cryptologia*, Vol. 31, No. 1, January 2007, pp. 76–83. The following pages are adapted from this paper and updated with material from sequels by other authors.

Figure 6.3 Craig P. Bauer. (Photograph by Mike Adams and used with permission.)

deciphering matrix as $\begin{pmatrix} a & b \\ c & d \end{pmatrix}$ and guess that a = 7 and b = 3, applying this row will then give the

plaintext equivalents for positions 1, 3, 5, 7, ... We cannot tell at a glance if our guess is correct, as we could if we made guesses for both rows and had a complete potential decipherment, but we can analyze the letters we do obtain statistically to see if they seem reasonable. We do this by comparing the frequencies of the letters recovered (every other letter of the complete plaintext) with the frequencies of the characters in normal English. The choice for *a* and *b* that yields the best match provides our most likely values for the first row of the matrix.

Basically, if our guess for row 1 of the deciphering matrix yields common letters such as E, T, and A, it is considered a good guess; however, if we get rare letters like Z and Q, it is a bad guess. All we need is a way to assign scores to each guess based on the letters generated, so that the process can be automated.

Repeating this procedure for the second row will suggest the same values as the most likely. For a matrix to be invertible, the rows must be distinct, so we need to consider a few of the most likely solutions for the rows and try them in different orders to recover the matrix. For example, if the rows yielding the best match to normal English letter frequencies are (4 13) and (9 6), the deciphering matrix could be $\begin{pmatrix} 4 & 13 \\ 9 & 6 \end{pmatrix}$ or $\begin{pmatrix} 9 & 6 \\ 4 & 13 \end{pmatrix}$. Applying each of these matrices to the ciphertext would quickly reveal which is

correct. However, for some ciphertexts, the two rows that seem to be the most likely are not correct! In those cases we have to consider a larger number of potential rows, again trying them in various orders within the matrix until we finally recover a meaningful message.

So how can we score each potential row to determine which are most likely to be correct? A natural way of ranking them is to award each a point value equal to the sum of the probabilities of the letters it generates when applied to the ciphertext. However, this approach to scoring was not as successful as we had hoped and we actually achieved better results with a less refined approach!

We settled on the following scheme for awarding points to a potential row:

2 points for each A, E, T, N, O
1 point for each H, I, L, R, S

1/2 point for each F, G, P

0 points for each M, B, C, D, U, V, W, Y

−1 points for each Q, J, K, X, Z

In order to test this attack, we needed a selection of ciphertexts. To create these we took a list of books, namely *Modern Library's 100 Best English-Language Novels of the Twentieth Century*,[10] and arbitrarily took the first 100 letters of each of the top 25 novels as our plaintext messages. We then used a computer program to generate invertible matrices, which were used to encipher these messages. Separate programs examined the attack against 2 × 2, 3 × 3, and 4 × 4 matrices. While working on the programs, we discovered that our attack for the special case of a 3 × 3 matrix was previously and independently put forth online by Mark Wutka of the Crypto Forum.[11] We contacted him by email and were encouraged to continue with our investigation.

Our first program investigated the case in which a 2 × 2 matrix was used. All possible rows were considered, and it was discovered that the highest scoring rows were often impossible! For example, the row (13 0) sends every pair of ciphertext letters where the first letter is even to 0 = A, because we are performing all arithmetic modulo 26. The letter A, being so frequent in normal English, was worth 2 points in our scoring scheme, so this row scored very high. However, any matrix containing this row would have a determinant that is a multiple of 13 and therefore not invertible modulo 26. Hence, this row could not possibly arise as part of the deciphering matrix. Similarly, (0 13) and (13 13) cannot be rows in an invertible matrix. Also, all rows of the form (*even even*) would be impossible, as well, because such a row would make the determinant an even number and therefore not invertible modulo 26. In terms of coding, it was easiest to investigate possibilities for the row (*a b*) by using nested loops where a and b each run from 0 to 25. The impossible rows described above were assigned scores, but were ignored when displaying the results.

The results (possible matrix rows and their scores) were then ordered by score. The rows with the highest scores should be tried first; some combination of them is very likely to yield the correct deciphering matrix.

The results for our investigation of the 2 × 2 case are summarized graphically in Figure 6.4. The graph shows an increasing number of ciphertexts being correctly deciphered as we go deeper into our list of possible matrix rows ordered by the scores they generate.

Examining some particular results, in detail, will make the graph clearer. For the fourth ciphertext we attacked, the top scoring rows were (7 25) with a score of 62 and (14 21) with a score of 57.5. The correct deciphering matrix was $\begin{pmatrix} 7 & 25 \\ 14 & 21 \end{pmatrix}$, so our attack worked wonderfully.

For the seventh ciphertext, the top scoring rows were (19 16) with a score of 63 and (22 19) with a score of 59.5. The correct deciphering matrix in this case was $\begin{pmatrix} 22 & 19 \\ 19 & 16 \end{pmatrix}$, so the rows appeared in a different order, but we still had the correct rows as our top two possibilities. We found that, for

[10] *Modern Library's 100 Best English-Language Novels of the Twentieth Century*, https://web.archive.org/web/20150910153230/http://home.comcast.net/~http://home.comcast.net/~netaylor1/modlibfiction.html.

[11] Wutka, Mark, *The Crypto Forum*, http://s13.invisionfree.com/Crypto/index.php?showtopic=80.

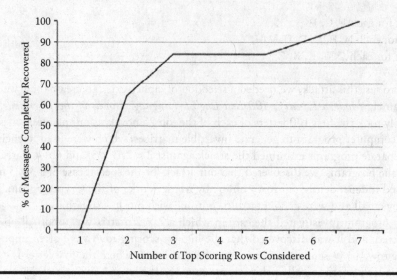

Figure 6.4 **Recovery rate of keys as a function of the number of high scoring rows considered for the 2 × 2 case.**

64% of the ciphertexts we attacked, the two most likely rows could be used to obtain the correct deciphering matrix, as described above. Hence, in Figure 6.4, we have the point (2, 64).

For the second ciphertext we attacked, the two most likely rows (highest scoring) did not yield the correct deciphering matrix. The list of highest scoring rows began (13 22), (0 9), (13 9), (0 7), (13 7), (4 15), … The correct deciphering matrix was $\begin{pmatrix} 4 & 15 \\ 13 & 22 \end{pmatrix}$, which used the rows in positions 1 and 6 of our ordered list. So, while the correct rows usually appeared as the two most likely, we sometimes had to go further down the list. When considering the six highest scoring rows, we found the correct two rows for the deciphering matrix to be among them 92% of the time. Hence, the graph in Figure 6.4 includes the point (6, 92).

Considering the top seven rows, by our ranking scheme, caught the correct two 100% of the time. Selecting two rows from the seven most likely to form a matrix allows for $_7P_2 = 42$ possibilities. It was necessary to use a permutation in this calculation, because the order of the rows matters. The 42 distinct matrices obtained as possibilities in this worst-case scenario (we can usually find our answer within a smaller set) can still be found and checked far more rapidly than the set of all 157,248 invertible 2 × 2 matrices. Even considering the overhead of calculating scores for $26^2 = 676$ rows, we still have a great savings.

We continued on to apply our attack to the same set of messages enciphered with 3 × 3 matrices. Once again, our list of ranked rows greatly reduced the number of matrices that needed to be considered. Our attack found the correct three rows over half the time among those scoring in the top 17. This leaves more possibilities to consider than in the 2 × 2 case, yet testing $_{17}P_3 = 4,080$ matrices once again represents a tremendous savings over a brute force attack on the set of all 1,634,038,189,056 possibilities. In this case, the overhead consists in calculating scores for $26^3 = 17,576$ rows. The correct rows were among the top 76 in 88% of the trials, but we had to consider the top 394 scorers before correctly identifying all 25 matrices. Still, $_{394}P_3 = 60,698,064$ represents only about 0.0037% of the full brute-force search.

The last case we investigated was when a 4 × 4 matrix had been applied to yield the ciphertext. The results continue the trend established by the smaller cases. We achieve success in 52% of the cases with a relatively small number of rows considered from our ranking (namely 1,469), continue on to get 88% within the top 7,372, and then had to go all the way out to the top 24,541 rows to get 100% of the test messages correctly deciphered. Thus, the trend is that a larger and larger number of rows must be considered as the dimension of the matrix grows, yet as a percentage of the total keyspace, which, for the 4 × 4 case, is 12,303,585,972,327,392,870,400, our attack is seen to be improving in terms of efficiency over brute-force. It should also be noted that, because the messages were kept at a fixed length of 100 characters, as the enciphering matrix grew from 2 × 2 to 4 × 4, the number of characters on which the rankings of the rows is determined diminished from 50 to 25. It is likely that the results would be improved, if the sample ciphertexts were longer.

The scoring scheme we used was never claimed to be optimal. Indeed, in 2009, Dae Hyun Yum and Pil Joong Lee, a pair of Korean researchers, found a better scoring method, improving the efficiency of the attack.[12] They called their scoring scheme "simplified multinomial" or SM for short and directly compared it to our scoring scheme (Bauer-Millward, or BM for short) in Table 6.1.

Table 6.1 Bauer-Millward vs. Simplified Multinomial Scoring Schemes

2×2			3×3			4×4		
Candidate Rows Considered	Recovered Message		Candidate Rows Considered	Recovered Message		Candidate Rows Considered	Recovered Message	
	BM	SM		BM	SM		BM	SM
2	72	90	2	4	18	4	0	0
4	92	100	4	35	60	16	10	6
8	99	100	8	58	76	64	24	22
10	100	100	16	73	90	128	41	32
			32	85	98	512	58	68
			50	90	100	2048	77	84
			100	98	100	8192	94	97
			234	100	100	16702	96	100
						42832	100	100

Yum and Lee also generalized the attack to work on Hill ciphers where the numerical representations of the letters do not simply follow the alphabet.

Further improvements were made by Elizabethtown College professors Tom Leap (computer science) and Tim McDevitt (mathematics), working with a pair of undergraduate applied mathematics majors, Kayla Novak and Nicolette Siermine. They refined the scoring scheme by first doing preliminary scoring with the index of coincidence. An important observation they made was that if a row yields a low score for the index of coincidence, then it is safe to not only reject that row, but to also reject all multiples of it, where the multiplier is relatively prime to the modulus, because they would give the same IC value. This leads to a savings of a factor of

[12] Yum, Dae Hyun and Pil Joong Lee, "Cracking Hill Ciphers with Goodness-of-Fit Statistics," *Cryptologia*, Vol. 33, No. 4, October 2009, pp. 335–342.

φ(L), where L is the size alphabet being used.[13] They followed this calculation (for the highest scoring rows) with a check of a goodness-of-fit statistic, like Yum and Lee used. For the best of these possibilities, they used plaintext digraphs statistics to determine the correct deciphering matrix.[14]

The Elizabethtown College team confirmed that their attack works for matrices all the way up to 8 × 8, for which the attack took 4.8 hours on an ordinary quad core desktop computer. The ciphertext in this instance was 1,416 characters long and arose from a message using a 27-letter alphabet that included the space needed to separate words. For smaller matrices, attacks were made on shorter ciphertexts. The team noted, "For short texts with small matrices, the multinomial approach of Yum and Lee may be best, but for larger matrices or longer texts, our method seems to be a substantial improvement."[15]

Two years later, Tim McDevitt, working with a new pair of coauthors, Jessica Lehr (an actuarial science major), and Ting Gu (a computer science professor) put forth an even better attack. This time the team was able to crack 8 × 8 matrix encryption, over a 29-character alphabet, in seconds. The attack allowed them to succeed all the way up to the 14 × 14 case, with an average runtime just under four hours.[16]

In 2020, George Teşeleanu, a Romanian mathematician and computer scientist, broadened the attack to other modes of matrix encryption.[17] These modes are covered in Section 13.6.

The Hill cipher is important because of the explicit connection it makes between algebra and cryptography. It is not known to have been important in use. In fact, Jack Levine enjoyed working on it because he didn't have to worry about intersecting classified work. That is not to say that it wasn't used during World War II. Indeed, it was used by the American military to encipher radio call signs in that war [18] and in the Korean War.[19] There are also rumors of it having been used in Vietnam, where jungle conditions sometimes prevented more secure machine systems from being implemented successfully.

6.5 More Stories Left to Tell

We close this chapter with a quote that shows historians have yet to write the last word on Levine's contributions. Eventually, declassification efforts will reveal at least one more story.

---

[13] The function φ was previously seen in Section 1.16 of this book and will be seen again later.

[14] Leap, Tom, Tim McDevitt, Kayla Novak, and Nicolette Siermine, "Further Improvements to the Bauer-Millward Attack on the Hill Cipher," *Cryptologia*, Vol. 40, No. 5, September 2016, pp. 452–468.

[15] Leap, Tom, Tim McDevitt, Kayla Novak, and Nicolette Siermine, "Further Improvements to the Bauer-Millward Attack on the Hill Cipher," *Cryptologia*, Vol. 40, No. 5, September 2016, pp. 452–468.

[16] McDevitt, Tim, Jessica Lehr, and Ting Gu, "A Parallel Time-Memory Tradeoff Attack on the Hill Cipher," *Cryptologia*, Vol. 42, No. 5, September 2018, pp. 408–426. The authors noted that the results were "generated on a Supermicro 828-14 Server with 4× AMD Operation 6376 Sixteen Core 2.30 GHz processors. It has sixteen 8 GB DIMM RAM and a 120 GB SSD drive, and the programming was done in Java."

[17] Teşeleanu, George, "Cracking Matrix Modes of Operation with Goodness-of-Fit Statistics," in Megyesi, Beáta, editor, *Proceedings of the 3rd International Conference on Historical Cryptology, HistoCrypt 2020*, pp. 135–145.

[18] See Kahn, David, *The Codebreakers*, second edition, Scribner, New York, 1996, p. 408 and Bauer, Friedrich L., *Decrypted Secrets: Methods and Maxims of Cryptology*, second edition, Springer, Berlin, Germany, 2007, p. 85.

[19] Burke, Colin B., "It wasn't All Magic: The Early Struggle to Automate Cryptanalysis, 1930s-1960s," United States Cryptologic History, Special Series, Vol. 6, Center for Cryptologic History, National Security Agency, Fort George G. Meade, MD, 2002, p. 254, available online at http://cryptome.org/2013/06/NSA-WasntAllMagic_2002.pdf and https://fas.org/irp/nsa/automate.pdf.

Actually I worked as a civilian in the Signal Intelligence Service at the beginning of the war but then during the war I was in the Army doing the exact same job, for less pay. I rose to the rank of Technical Sergeant. Much of the work I did there is still classified. By the way, I have never told anyone this, but I was awarded the Legion of Merit for my work. The citation did not say why—again because of the classified material.[20]

—Jack Levine

References and Further Reading

Bauer, Craig and Millward, Katherine, "Cracking Matrix Encryption Row by Row," *Cryptologia*, Vol. 31, No. 1, January 2007, pp. 76–83.

Bauer, Craig, Gregory Link, and Dante Molle, "James Sanborn's *Kryptos* and the Matrix Encryption Conjecture," *Cryptologia*, Vol. 40, No. 6, November 2016, pp. 541–552.

Brawley, Joel V., "In Memory of Jack Levine (1907-2005)," *Cryptologia*, Vol. 30, No. 2, April 2006, pp. 83–97.

Brawley, Joel V. and Jack Levine, "Equivalence Classes of Linear Mappings with Applications to Algebraic Cryptography, I," *Duke Mathematical Journal*, Vol. 39, No. 1, 1972, pp. 121–142.

Brawley, Joel V. and Jack Levine, "Equivalence Classes of Involutory Mappings," *Duke Mathematical Journal*, Vol. 39, No. 2, 1972, pp. 211–217.

Buck, Friederich Johann, *Mathematischer Beweiß: daß die Algebra zur Entdeckung einiger verborgener Schriften bequem angewendet werden könne*, Königsberg, 1772. This is the first known work on algebraic cryptology, which eventually blossomed with matrix encryption. It predates the work of Charles Babbage, who is often given credit as the first to model encryption with an equation.

Christensen, Chris, David Joyner, and Jenna Torres, "Lester Hill's Error-Detecting Code," *Cryptologia*, Vol. 36, No. 2, April 2012, pp. 88–103.

Christensen Chris, "Lester Hill Revisited," *Cryptologia*, Vol. 38, No. 4, October 2014, pp. 293–332.

Flannery, Sarah (with David Flannery), *In Code: A Mathematical Journey*, Profile Books, London, UK, 2000. This is a very enjoyable autobiographical tale of a teenage girl's interest in mathematics and success in developing a new form of public key cryptography using matrices. The coauthor is Sarah's father and a professor of mathematics.

Greenfield, Gary R., "Yet another Matrix Cryptosystem," *Cryptologia* Vol. 18, No. 1, January 1994, pp. 41–51. The title says it all! There has been a lot of work done in this area.

Hill, *Cryptool—Online*, https://www.cryptool.org/en/cto/ciphers/hill. This website allows users to encipher and decipher using 2×2 or 3×3 matrix encryption. It's part of a much larger (and still growing) site that cover many ciphers and includes online cryptanalysis programs.

Hill, Lester, S., "Cryptography in an Algebraic Alphabet," *American Mathematical Monthly*, Vol. 36, No. 6, June–July 1929, pp. 306–312.

Hill, Lester, S., "Concerning Certain Linear Transformations Apparatus of Cryptography," *American Mathematical Monthly*, Vol. 38, No. 3, March 1931, pp. 135–154.

Leap, Tom, Tim McDevitt, Kayla Novak, and Nicolette Siermine, "Further improvements to the Bauer-Millward attack on the Hill cipher," *Cryptologia*, Vol. 40, No. 5, September 2016, pp. 452–468.

Levine, Jack, "Variable Matrix Substitution in Algebraic Cryptography," *American Mathematical Monthly*, Vol. 65, No. 3, March 1958, pp. 170–179.

---

[20] *History of the Math Department at NCSU: Jack Levine*, December 31, 1986, interview, https://web.archive.org/web/20160930110613/~http://www4.ncsu.edu/~njrose/Special/Bios/Levine.html.

Levine, Jack, "Some Further Methods in Algebraic Cryptography," *Journal of the Elisha Mitchell Scientific Society*, Vol. 74, 1958, pp. 110–113.

Levine, Jack, "Some Elementary Cryptanalysis of Algebraic Cryptography," *American Mathematical Monthly*, Vol. 68, No. 5, May 1961, pp. 411–418.

Levine, Jack, "Some Applications of High-Speed Computers to the Case $n = 2$ of Algebraic Cryptography," *Mathematics of Computation*, Vol. 15, No. 75, July 1961, pp. 254–260.

Levine, Jack, "Cryptographic Slide Rules," *Mathematics Magazine*, Vol. 34, No. 6, September-October 1961, pp. 322–328. Levine presented a device for performing matrix encryption in this paper. Also see Figures 6.5 and 6.6.

Levine, Jack, "On the Construction of Involutory Matrices," *American Mathematical Monthly*, Vol. 69, No. 4, April, 1962, pp. 267–272. Levine provides a technique for generating matrices that are self-inverse in this paper.

Noninvertible matrices can be used for encryption as the following four papers demonstrate.

Levine, Jack and Robert E. Hartwig, "Applications of the Drazin Inverse to the Hill Cryptographic System Part I," *Cryptologia*, Vol. 4, No. 2, April 1980, pp. 71–85.

Levine, Jack and Robert E. Hartwig, "Applications of the Drazin Inverse to the Hill Cryptographic System Part II," *Cryptologia*, Vol. 4, No. 3, July 1980, pp. 150–168.

Levine, Jack and Robert E. Hartwig, "Applications of the Drazin Inverse to the Hill Cryptographic System Part III," *Cryptologia*, Vol. 5, No 2, April 1981, pp. 67–77.

Levine, Jack and Robert E. Hartwig, "Applications of the Drazin Inverse to the Hill Cryptographic System Part IV," *Cryptologia*, Vol. 5, No. 4, October 1981, pp. 213–228.

Levine, Jack and Richard Chandler, "The Hill Cryptographic System with Unknown Cipher Alphabet but Known Plaintext," *Cryptologia*, Vol. 13, No. 1, January 1989, pp. 1–28.

McDevitt, Tim, Jessica Lehr, and Ting Gu, "A Parallel Time-Memory Tradeoff Attack on the Hill Cipher," *Cryptologia*, Vol. 42, No. 5, September 2018, pp. 408–426.

Ohaver, M. E., "Solving Cipher Secrets," *Flynn's Weekly*, October 22, 1926, p. 798. M. E. Ohaver, is one of the pseudonyms used by Kendell Foster Crossen (1910–1981). Problem No. 6, on page 798 of this article, is the one posed by Jack Levine.

Ohaver, M. E., "Solving Cipher Secrets," *Flynn's Weekly*, November 13, 1926, pp. 794–800. This is the column in which an explanation of Levine's system, from the October 22, 1926 issue, appeared (see pages 799–800). Levine believed it laid the foundation for matrix encryption.

Overbey, Jeffrey, William Traves, and Jerzy Wojdylo, "On the Keyspace of the Hill Cipher," *Cryptologia*, Vol. 29, No. 1, January 2005, pp. 59–72, available online at https://web.archive.org/web/20050910055747/http://jeff.actilon.com/keyspace-final.pdf this paper presents formulas yielding the size of the keyspace for matrix encryption with arbitrary dimension and moduli. Some of this material may also be found (more tersely) in Friedrich L. Bauer's *Decrypted Secrets*, 1st edition, Springer, Berlin, Germany, 1997, p. 81. Second, third, and fourth editions of the latter have since been released.

Teşeleanu, George, "Cracking Matrix Modes of Operation with Goodness-of-Fit Statistics," in Megyesi, Beáta, editor, *Proceedings of the 3rd International Conference on Historical Cryptology, HistoCrypt 2020*, pp. 135–145.

Thilaka, B. and K. Rajalakshni, "An Extension of Hill Cipher Using Generalized Inverses and m^{th} Residue Modulo m," *Cryptologia*, Vol. 29, No. 4, October 2005, pp. 367–376.

Wutka, Mark, The Crypto Forum, http://s13.invisionfree.com/Crypto/index.php?showtopic=80. This link is now broken and the page was not archived by Wayback Machine, but I'm including it, because I want to continue pointing out Wutka's priority on what is sometimes called the "Bauer-Millward attack."

Yum, Dae Hyun and Pil Joong Lee, "Cracking Hill Ciphers with Goodness-of-Fit Statistics," *Cryptologia*, Vol. 33, No. 4, October 2009, pp. 335–342.

There have been many other papers in recent years in less specialized journals that attempt to describe stronger variants of matrix encryption; however, the cryptographic community is, in general, skeptical. One modification described in some of the papers referenced above foreshadows the various modes of encryption used for modern cipher systems. This is discussed in Section 13.6.

The Jack Levine archive at North Carolina State University is home to cipher wheels for performing matrix encryption. They are shown in Figures 6.5 and 6.6.

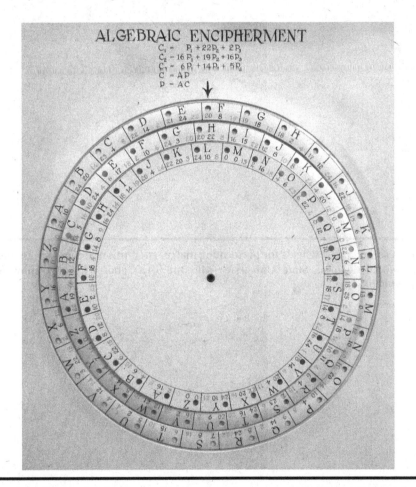

Figure 6.5 A cipher wheels for performing matrix encryption. (From Jack Levine Papers, 1716–1994, North Carolina State University, MC 308.5.1, General Cryptography, Algebraic Encipherment Wheels.)

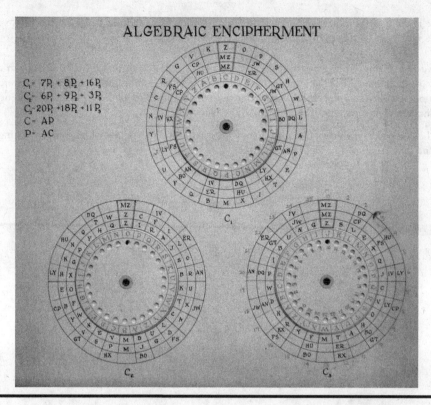

Figure 6.6 A set of cipher wheels for performing matrix encryption. (From Jack Levine Papers, 1716–1994, North Carolina State University, MC 308.5.1, General Cryptography, Algebraic Encipherment Wheels.)

Chapter 7

World War II: The Enigma of Germany

Three may keep a secret if two of them are dead.

—Ben Franklin

The quote above is often used to debunk conspiracy theories, but as this chapter will show, Ben was completely wrong. Thousands can, and have, kept secrets for decades.

7.1 Rise of the Machines

As World War I was ending, a new device for enciphering messages came into being—the *rotor*. It was independently invented by four men in four countries over the years 1917 to 1923.

The rotor pictured in Figure 7.1 has 26 electrical contacts on each side. Each contact represents a letter, and 26 wires inside the rotor perform a substitution as the letters pass from one side of the rotor, through the wires, and out to different positions on the other side. Thus, what was once done by hand could now be done by machine. Also, as we shall see, rotors can be easily combined to offer stronger encryption than was previously available.

Edward Hugh Hebern sketched the first rotor system in 1917, and had a prototype machine in 1918. In 1921, Hebern Electric Code became the first company in America incorporated to offer cipher machines for sale. Although he sold some machines to the U.S. Navy, business didn't go well.[1] Hugo Alexander Koch, in the Netherlands, had the idea in 1919, but outside of filing a patent he did nothing with it. He assigned the patent rights to Arthur Scherbius in 1927.[2]

Scherbius originally used rotors to only encipher the digits 0 through 9 but later increased the number of contacts and wires to handle 26 letters. It is through Scherbius that the first Enigma machines came on the market in 1923. But, as with Hebern, business was bad.[3]

---

[1] Kahn, David, *The Codebreakers*, second edition, Scribner, New York, 1996, pp. 415–420.
[2] Kahn, David, *The Codebreakers*, second edition, Scribner, New York, 1996, p. 420.
[3] Kahn, David, *The Codebreakers*, second edition, Scribner, New York, 1996, pp. 420–422.

Figure 7.1   The two sides of an Enigma rotor.

Eventually, with Hitler's rise to power and the re-arming of Germany, Enigma machines were mass produced. Scherbius was probably dead by this time, and it is not known if anyone else profited from these sales. In any case, the Nazis did not nationalize the business.[4] The majority of this chapter is focused on the Enigma, but the last of the four rotor inventors should be mentioned before moving on.

Arvid Gerhard Damm filed a patent in Sweden for a cipher machine with rotors only a few days after Koch. Damm died just as the company he began to market cipher machines started to take off. Boris Hagelin took over and in 1940 came to America, where he was eventually able to sell machines to the U.S. Army.[5] The M-209 created by Hagelin and used by America is pictured in Figures 7.2 and 7.3. It was not the most secure machine America had in use during World War II, but it was lightweight and easy to use.

Hagelin earned millions and returned to Sweden in 1944. The Cold War paved the way for millions more to be made, and Hagelin relocated his business to Switzerland, so the Swedish government couldn't take over in the name of national defense. In Switzerland, the company became Crypto Aktiengesellschaft, or Crypto AG for short.[6] The Swiss reputation for neutrality worked to Hagelin's advantage, as many nations felt safe purchasing cipher machines from them. This may have been to the United States' advantage, as well. Hagelin's story is continued in Section 12.7, but for now we return to Scherbius's Enigma.

The Enigma machine existed first in a commercial version. It was modified (in small ways) for military use and adopted by the German Navy around 1926 and by the German Army on July 15, 1928.[7] The Luftwaffe also came to use it. Altogether, about 40,000 of these machines were used

---

4 Email from Frode Weierud to the author, November 22, 2010.
5 Kahn, David, *The Codebreakers*, second edition, Scribner, New York, 1996, pp. 422–427.
6 Kahn, David, *The Codebreakers*, second edition, Scribner, New York, 1996, p. 432.
7 Kahn, David, "The Significance of Codebreaking and Intelligence in Allied Strategy and Tactics," *Cryptologia*, Vol. 1, No. 3, July 1977, pp. 209–222, p. 211 cited here.

Figure 7.2 Former National Cryptologic Museum curator Patrick Weadon with an M-209B. The M-209B is a later version of the M-209, but the differences between the two are very minor.

Figure 7.3 A closer look at an M-209B. (This machine is in the collection of the National Cryptologic Museum and was photographed by the author.)

by the Nazis during World War II.[8] A description of the military version follows, referencing the commercial version only when necessary.

7.2 How Enigma Works

Figure 7.4   An Enigma machine.

When a key was pressed on the Enigma machine (Figure 7.4), one of the lights above the keys would switch on. This represented the ciphertext letter. As will be seen, a letter could never be enciphered as itself. This is a weakness. Another weakness is that, at a given setting, if pressing X lights Y, then pressing Y lights X. This reciprocity allows the same setting that was used to encipher the message to be used for deciphering. It is analogous to using an involutory (self-inverse) matrix for the Hill cipher, although this machine operates in a completely different manner. We will now examine the electrical path connecting the depressed key to the lighted ciphertext letter. Each key is linked to a pair of holes in the *Steckerbrett* (literally "stick-board" or "plugboard")[9] pictured in Figure 7.5. Cables may connect these in pairs, performing the first substitution. Initially, six cables were used.

---

[8] Erskine, Ralph, "Enigma's Security: What the Germans Really Knew," in Erskine, Ralph and Michael Smith, editors, *Action this Day*, Bantam Press, London, UK, 2001, pp. 370–385, p. 370 cited here.

[9] The plugboard didn't exist for early (commercial) versions of the Enigma. It was introduced in 1928.

Figure 7.5 **An Enigma plugboard with a cable connecting F and M.**

Later, the number of cables used could be anywhere from 0 to 13, inclusive. Because the letters are connected in pairs, 13 is the limit for a 26-letter alphabet. The German alphabet has a few extra letters, namely ä, ö, ü, and ß, which are rendered as ae, oe, ue, and ss, respectively, for this machine.

After the plugboard substitution, we may follow the path in the schematic diagram provided in Figure 7.6 to the rightmost rotor. There are 26 wires internal to the machine connecting the plugboard to the rotor system. This offers the opportunity for another scrambling (more on this later). The rotors each make another substitution.

Figure 7.7 shows a disassembled rotor. Each side has 26 electrical contacts, one representing each letter. The wires inside connect these, yielding the substitution.

Initially, there were only three distinct rotors that could be used, although they could be placed in the machine in any order. Their inner wirings differed from those on the commercial version of the Enigma and were kept secret. They were as follows:

Rotor I
Input: ABCDEFGHIJKLMNOPQRSTUVWXYZ
Output: EKMFLGDQVZNTOWYHXUSPAIBRCJ

Rotor II
Input: ABCDEFGHIJKLMNOPQRSTUVWXYZ
Output: AJDKSIRUXBLHWTMCQGZNPYFVOE

Rotor III
Input: ABCDEFGHIJKLMNOPQRSTUVWXYZ
Output: BDFHJLCPRTXVZNYEIWGAKMUSQO

The above might seem like the tersest possible notation for the action of the rotors, but this isn't so! Such permutations (as mathematicians usually refer to them) can be written even more briefly. Consider Rotor I. It sends A to E, E to L, L to T, T to P, P to H, H to Q, Q to X, X to R, R to U, and U to A, which is where I started this chain. We may write this as (AELTPHQXRU), leaving off the last A with the understanding that when we reach the end, we loop back to the start. Now, this is very nice, but it doesn't include all of the letters. So, we pick a letter that was missed and repeat the process to get (BKNW). Putting these together yields (AELTPHQXRU)(BKNW), which still doesn't include all 26 letters, so again we start with one that was missed and form another *cycle*, as these groups of letters are called. Eventually we get

Rotor I = (AELTPHQXRU) (BKNW) (CMOY) (DFG) (IV) (JZ) (S)

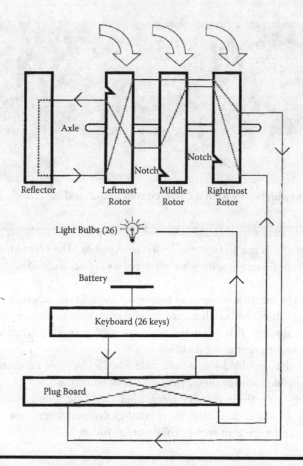

Figure 7.6 From keyboard to bulb—the enciphering path of an Enigma. (From Miller, R., *The Cryptographic Mathematics of Enigma*, Center for Cryptologic History, Fort Meade, Maryland, 2001, p. 2.)

Figure 7.7 A disassembled Enigma rotor. (Courtesy of René Stein, National Cryptologic Museum.)

Repeating this process for the other rotors gives

Rotor II = (A) (BJ) (CDKLHUP) (ESZ) (FIXVYOMW) (GR) (NT) (Q)
Rotor III = (ABDHPEJT) (CFLVMZOYQIRWUKXSG) (N)

The most popular single area in which mathematics doctorates are awarded in American universities is abstract algebra. An important part of abstract algebra is group theory, which we'll see more of later. For now, I'll point out that every group that has been studied (or ever can be) is equivalent to some set of permutations. Thus, as you may imagine, these permutations have been subject to intense study. There were, in fact, theorems waiting to be exploited in favor of the cryptanalysts and to the great disadvantage of the Nazis.

Figure 7.8   Reflector B and the rotor assembly.

In Figure 7.8, we see three rotors side by side. Each rotor performs a substitution. After passing through all three rotors, the electrical impulse passes to the reflector, pictured up close in Figure 7.9. The reflector makes another substitution.

In our cyclic permutation notation, the reflector first used by the German military, called *reflector A*, was (AE)(BJ)(CM)(DZ)(FL)(GY)(HX)(IV)(KW)(NR)(OQ)(PU)(ST). A differently wired *reflector B* was introduced later. Notice that all of the cycles in the permutation for reflector A are in pairs (2-cycles) and no two 2-cycles have a letter in common. When cycles have nothing in common, they are said to be *disjoint*.

After passing through the reflector, the electrical impulse passes back through the three rotors (along a different path than the first time), through the wire connecting it to the plugboard, through the plugboard itself, and finally to one of the lights above the keypad.

The composition of monoalphabetic substitutions is a monoalphabetic substitution, so, if Enigma worked exactly as described above, it would not result in a strong encryption. What makes the machine special is that the first rotor turns by one position every time a key is pressed, changing the substitution that will be performed. The first two rotors have notches that cause the next rotor to turn by one position for every 26 turns they make themselves.

Thus, it *appears* that after each of the three rotors has turned all of the way around—that is, after (26)(26)(26) = 17,576 letters have been typed—the machine returns to its original rotor setting. In this way, the Enigma would work like a Vigenère Cipher with 17,576 independently mixed

Figure 7.9   View of a reflector after removing the rotors.

alphabets. Actually, although many authors have made this mistake,[10] the period is not 17,576, but rather a little smaller. Stephen Budiansky is one of the authors who got this right. In his excellent history of World War II cryptanalysis, *Battle of Wits*, he explained:[11]

> There are 26 × 26 × 26, or 17,576, different possible combinations of the three rotor settings. However, in actual operation of the Enigma, the turnover mechanism causes a "double stepping" to occur in the middle rotor: each time the middle rotor advances to the position where it will trigger a turnover of the left rotor, it then immediately advances again (along with the left rotor) as the next letter is typed in. If, for example, the turnover occurs between E and F on the middle rotor and between V and W on the right rotor, then an actual rotor sequence would be as follows:
>
> $$\begin{matrix} ADU \\ ADV \\ AEW \\ BFX \\ BFY \\ BFZ \\ BFA \end{matrix}$$
>
> Thus the key length of the normal Enigma is actually 26 × 25 × 26, or 16,900. When rotors with multiple turnover notches were later introduced, the key length was shortened even further.

---

[10] Even Gordon Welchman, a mathematician at Bletchley Park who worked on breaking Enigma, made this error when he wrote about his work decades later! See Welchman, Gordon, *The Hut Six Story*, McGraw-Hill Book Company, New York, p. 45 footnote.

[11] Budiansky Stephen, *Battle of Wits: The Complete Story of Codebreaking in World War II*, The Free Press, New York, pp. 370–371.

7.3 Calculating the Keyspace

As we've seen in previous chapters, a large keyspace is a necessary, but not sufficient, condition for a cipher to be secure. What is the keyspace of Enigma? The literature offers many different values. There's room for different answers, depending on whether we look at how the machine was used or how it could have been used; however, mistakes have been made in some previous calculations. The problem is attacked here by following the encryption path from keyboard to light bulb. We start our count with the first substitution, performed by the plugboard.

If p cables are used, there are $\binom{26}{2p}$ ways to choose the letters to be connected. *How* these let-

ters are connected still remains to be examined. Plugging a cable into one of the letters allows $(2p - 1)$ choices for the other end. Now that two holes have been plugged, inserting another cable leaves $(2p - 3)$ choices for its other end, and so forth. The total number of possible connections (once the letters to be involved are determined) is

$$(2p-1)(2p-3)(2p-5)\dots(1)$$

Thus, the number of ways the plugboard may be wired is

$$\binom{26}{2p}(2p-1)(2p-3)(2p-5)\dots(1) = \left(\frac{26!}{(26-2p)!(2p)!}\right)\left(\frac{(2p)!}{p!(2^p)}\right)$$

$$= \frac{26!}{(26-2p)!\,p!\,2^p}$$

But the above result is only for when p cables are used. Because p is a variable somewhere between 0 and 13 inclusive, the total number of possible plugboard settings is

$$\sum_{p=0}^{13}\frac{26!}{(26-2p)!\,p!\,2^p} = 532{,}985{,}208{,}200{,}576$$

Originally, exactly six cables were used. So, in calculating the keyspace, some authors use the figure for six cables, instead of the much larger summation value given above. Later on, the number of cables used varied.

The next factor is how the internal wiring connects the plugboard to the rotor assembly. There are 26! ways to do this. We now come to the first rotor. There are 26! ways for a rotor to be wired. We assume that the users will want to be able to reorder the rotors in the machine to get different encipherments. Therefore, it makes sense for the rotors to be wired differently. If they are all wired identically, reordering would have no effect. So, if each rotor is wired differently, we have (26!) (26! – 1) (26! – 2) possibilities for the wirings.

This brings us to some common errors. It is tempting to insert other factors at this point having to do with the possible orderings of the three rotors and the 26 positions each individual rotor can be rotated to, but these have already been accounted for. To see this, imagine setting a rotor in place and then turning it by one position. This is no different from having inserted another rotor that is wired this way, and we already counted all 26! ways that the first rotor can be wired.

Similarly, having accounted for the possible wirings of each of the three rotors, rearranging them simply counts these possibilities again. We do not want this duplication. To make a simple analogy, consider three mailboxes arranged side by side. If we have five letters, and can place only one in each box, then there are five choices for the first box, four choices for the second box, and three choices for the third box. We get a total of $5 \times 4 \times 3 = 60$ possibilities. We do not then consider rearranging the order of the letters in the boxes. For the Enigma, distinctly wired rotors take the place of letters in this example and we have 26! of them instead of just five, but the argument against counting rearrangements is unchanged.

The locations of the notches (ring setting) on the two fastest rotors (right and middle) determine when the next wheel turns, so they must be considered as part of the key. This gives us another factor of $(26)(26) = 676$. The contact points of the reflector were wired together in pairs, so the number of possible wirings is the same as for a plugboard with 13 cables, namely 7,905,853,580,625. To summarize, we have the following enumerations:

Plugboard Settings:	532,985,208,200,576
Wiring from Plugboard to Rotors	403,291,461,126, 605,635,584,000,000
Wiring of the Rotors:	65,592, 937,459,144,468,297, 405,473,480,371,753, 615,896,841,298,988, 710,328,553,805,190, 043,271,168,000,000
Notch Position of Rotors:	676
Reflector Wiring:	7,905,853,580,625

The keyspace is obtained by multiplying all of the numbers above together to get 753,506,019,827, 465,601,628,054,269,182,006,024,455,361,232,867,996,259,038,139,284,671,620,842,209,198, 855,035,390,656,499,576,744,406,240,169,347,894,791,372,800,000,000,000,000.

This is ridiculously larger than is necessary to prevent a brute force attack.

7.4 Cryptanalysis Part 1: Recovering the Rotor Wirings

A large keyspace is a necessary condition for security, but it is not sufficient. Much of the rest of this chapter is devoted to how Enigma was broken.

The Polish mathematicians shown in Figure 7.10 were the first to break Enigma. Poles were also the first to publicly reveal the secret, decades later, and the first to issue a postage stamp commemorating the work (Figure 7.11). We'll take a historical look at how they did it and also discuss contributions made by the French, British, Americans, and even a German.

While in prison for an attempt to overthrow the government following World War I, Adolph Hitler penned *Mein Kampf*, which included his claim of Germany's need for more *Lebensraum* ("living space") in the east. When President Paul von Hindenburg appointed Hitler chancellor of Germany in January 1933, the Poles were well aware of the danger. Fortunately, they had previously acquired a commercial version of an Enigma machine and had succeeded in breaking Enigma messages before the war began. However, the task was highly nontrivial. The military version of the Enigma differed from the commercial version in small, but important, ways, such as

Figure 7.10 Marian Rejewski (c. 1932), Jerzy Różycki, and Henryk Zygalski. (Rejewski photograph Creative Commons Attribution-Share Alike 2.5 Generic license, Obtained from Marian Rejewski's daughter and published in commons under CC-BY-SA with her permission.)

Figure 7.11 A first day cover for the first stamp honoring codebreaking.

the wiring of the rotors. Enigma messages were intercepted by the Poles from July 15, 1928, when they first went on the air (with Army messages), but the only progress that was made in the first few years was the creation of a mathematical model of the machine. Later writers have continued the use of the Pole's notation, which follows:

S = plugboard permutation (the S stands for *Steckerbrett*, German for "stickerboard")
N = rightmost rotor permutation (this is the fast turning rotor)
M = middle rotor permutation
L = leftmost rotor permutation
R = reflector permutation
H = permutation representing the internal wiring from the plugboard to the entry point for the set of rotors

For the commercial Enigma, H = (AJPRDLZFMYSKQ)(BW)(CUGNXTE)(HOI)(V), which seems to have been randomly chosen. But, if we switch back to our old-school style notation, we see the pattern clearly!

H:

Input:	ABCDEFGHIJKLMNOPQRSTUVWXYZ
Output:	JWULCMNOHPQZYXIRADKEGVBTSF

H^{-1}:

Input:	ABCDEFGHIJKLMNOPQRSTUVWXYZ
Output:	QWERTZUIOASDFGHJKPYXCVBNML

Does it look more familiar now? H^{-1} almost matches our American QWERTY keyboards. In fact, it matched the Enigma keyboard perfectly. Of course, the World War II cryptanalysts didn't have to look at different notations to see this pattern. The Poles had a commercial Enigma and could simply observe, without using any notation at all, that the keys were connected, in order, to the rotor system input. I chose to present it in this manner to illustrate that notation needs to fit the circumstance and we can't simply say one form is always superior.

The values of the reflector, called R now, and the three rotors, N, M, and L, were given earlier in this chapter. Of course, the rotors could be used in any order, so we cannot simply say N = Rotor I, for example. The Poles, however, didn't know any of these permutations at first. They had to derive them mathematically. Hence, all of the letters above represented unknowns initially.

We need one more permutation to show the change that occurs every time the fast rotor advances one position. Fortunately, this one is known and is given by

$$P = (ABCDEFGHIJKLMNOPQRSTUVWXYZ).$$

In order to encipher a message, the user set the machine up to the "daily key," say HLD, by turning the rotors so these are the three letters on top.[12] He then selected another "session key" (randomly, if he was following orders…), say EBW. Typing the session key twice might result in the ciphertext GDOMEH. This would be sent at the start of the message. The intended recipient, who knew the daily key and had set his Enigma to it, typed in GDOMEH to get out EBWEBW. Now back to the encipherer. After he typed the session key twice, he reset his Enigma to the session key and typed out the message. The intended recipient, having used the daily key setting to recover the session key, also reset the machine to the session key and typed the rest of the ciphertext to recover the original message. Clearly, the session key needn't be typed twice. This redundancy was intentionally introduced to ensure the session key was received correctly. It worked in that regard, but it also proved to be a weakness.

An example will illustrate how Marian Rejewski exploited the repeated session key.[13] Consider the following beginnings for ciphertexts using various session keys, but sent on the same day (thus using the same daily key for the first six letters):[14]

---

[12] If the rotors used displayed numbers, rather than letters, the user would set them using the correspondence A = 1, B = 2, etc.

[13] The method that follows has been adapted from Rejewski's own explanation.

[14] From Bauer, Friedrich L., *Decrypted Secrets: Methods and Maxims of Cryptology*, second edition, Springer, New York, 2000, p. 390.

AUQ AMN	IND JHU	PVJ FEG	SJM SPO	WTM RAO
BNH CHL	JWF MIC	QGA LYB	SJM SPO	WTM RAO
BCT CGJ	JWF MIC	QGA LYB	SJM SPO	WTM RAO
CIK BZT	KHB XJV	RJL WPX	SUG SMF	WKI RKK
DDB VDV	KHB XJV	RJL WPX	SUG SMF	XRS GNM
EJP IPS	LDR HDE	RJL WPX	TMN EBY	XRS GNM
FBR KLE	LDR HDE	RJL WPX	TMN EBY	XOI GUK
GPB ZSV	MAW UXP	RFC WQQ	TAA EXB	XYW GCP
HNO THD	MAW UXP	SYX SCW	USE NWH	YPC OSQ
HNO THD	NXD QTU	SYX SCW	VII PZK	YPC OSQ
HXV TTI	NXD QTU	SYX SCW	VII PZK	ZZY YRA
IKG JKF	NLU QFZ	SYX SCW	VQZ PVR	ZEF YOC
IKG JKF	OBU DLZ	CYX SCW	VQZ PVR	ZSJ YWG

It will be convenient to let A, B, C, D, E, and F denote the permutations that the Enigma setting yielding these messages imposes upon the first, second, third, fourth, fifth, and sixth letters typed.

The first encrypted session key is AUQ AMN. Now looking at the first and fourth ciphertext letters (which result from enciphering the same plaintext letter), we see that permutations A and D both send the first letter of the session key to the ciphertext letter A. The first letter of the session key is unknown, but we may denote it by α. We have $A(\alpha) = A$ and $D(\alpha) = A$. Because the Enigma is self-inverse, we also have $A(A) = \alpha$ and $D(A) = \alpha$. Thus, if we form the composition of the permutations, AD (read left to right—that is, perform A first, then D), we see that this new permutation will send the letter A to A.

Thus, the permutation AD begins with (A). Continuing this argument with the second and fourth encrypted session keys, BNH CHL and CIK BZT, we see that the permutation AD sends B to C, and also sends C to B, so we now have AD = (A) (BC)...

Examining more session keys, we next get the longer cycles, (DVPFKXGZYO) and (EIJMUNQLHT). Finally, we end the permutation with (RW) and (S). Putting it all together, with the cycles in order of decreasing size, we have

$$AD = (DVPFKXGZYO) (EIJMUNQLHT) (BC) (RW) (A) (S)$$

Repeating this approach for the second and fifth letters and then once more for the third and sixth letters we get two more permutations.

$$BE = (BLFQVEOUM) (HJPSWIZRN) (AXT) (CGY) (D) (K)$$
$$CF = (ABVIKTJGFCQNY) (DUZREHLXWPSMO)$$

Notice that the cycles making up AD have the lengths 10, 10, 2, 2, 1, and 1. For BE the cycle lengths are 9, 9, 3, 3, 1, and 1. And for CF we have 13 and 13. The exact results for the lengths of the cycles depended upon the daily key, but if a cycle of a given length is present, then another cycle of that same length will also appear.[15] Rejewski referred to the pattern in the cycle lengths as the *characteristic structure*, or more briefly the *characteristic* for a given day.

It would be more useful to know A, B, C, D, E, and F, instead of the products above, but we cannot get them as easily. In fact, we will need to determine A, B, C, D, E, and F by factoring the

---

[15] This is a consequence of the Enigma being self-inverse. The permutations A, B, C, D, E, and F must therefore all be self-inverse and, hence, consist of disjoint 2-cycles. Any product of permutations that consists of disjoint 2-cycles will have a cycle structure with all lengths appearing in pairs.

more readily available product permutations AD, BE and CF. Fortunately, there's a nice formula for doing so. That is, knowing the product AD we can quickly find A and D and similarly for BE and CF. However, unlike factoring an integer as a product of primes, the factorizations here won't be unique. This non-uniqueness of factorization makes extra work for the cryptanalyst! For instance, if XY = (AB)(CD), factorizations are given by

$$X = (AD)(BC) \text{ and } Y = (BD)(AC)$$

and

$$X = (AC)(BD) \text{ and } Y = (BC)(AD).$$

For larger pairs of cycles, there are more possible factorizations. In general,[16] if XY = $(x_1x_3x_5\ldots x_{2n-1})$ $(y_{2n}\ldots y_6y_4y_2)$ then we can factor it as

$$X = (x_1y_2)(x_3y_4)(x_5y_6)\ldots(x_{2n-1}y_{2n}) \text{ and } Y = (y_2x_3)(y_4x_5)(y_6x_7)\ldots(y_{2n}x_1).$$

Expressing the cycle $(x_1x_3x_5\ldots x_{2n-1})$ as $(x_3x_5\ldots x_{2n-1}\, x_1)$ and following the same rule gives a different factorization. Because a cycle of length n can be expressed in n different ways, we'll get n distinct factorings. We can factor pairs of disjoint cycles having the same length independently and then piece all such results together to get our "overall" factoring. This will be done below. When Rejewski explained his work, he skipped over much of what follows by writing

> We assume that thanks to the theorem on the product of transpositions, combined with a knowledge of encipherers' habits, we know separately the permutations A through F.

It is hoped that what the following pages lose in terms of terseness, compared to Rejewski's explanation, is made up for in clarity.

We start by factoring AD = (DVPFKXGZYO) (EIJMUNQLHT) (BC) (RW) (A) (S) by making repeated use of our factoring rule, for each pair of cycles of equal length, with the example to guide us.

AD includes the 1-cycles (A) and (S) ⇒

(AS) is part of A and (SA) is part of D.

AD includes the 2-cycles (BC) and (RW) ⇒

(BR)(CW) is part of A and (RC)(WB) is part of D

or

(BW)(CR) is part of A and (WC)(RB) is part of D.

AD includes the 10-cycles (DVPFKXGZYO) and (EIJMUNQLHT) ⇒

(DT)(VH)(PL)(FQ)(KN)(XU)(GM)(ZJ)(YI)(OE) is part of A and
(TV)(HP)(LF)(QK)(NX)(UG)(MZ)(JY)(IO)(ED) is part of D

or

(DE)(VT)(PH)(FL)(KQ)(XN)(GU)(ZM)(YJ)(OI) is part of A and
(EV)(TP)(HF)(LK)(QX)(NG)(UZ)(MY)(JO)(ID) is part of D.

---

[16] It should be noted that this result and other theorems from abstract algebra used in this chapter existed before they were needed for cryptanalysis. The Poles didn't have to invent or discover the mathematics; they only had to know it and apply it.

or

    (DI)(VE)(PT)(FH)(KL)(XQ)(GN)(ZU)(YM)(OJ) is part of A and
    (IV)(EP)(TF)(HK)(LX)(QG)(NZ)(UY)(MO)(JD) is part of D.

or

    (DJ)(VI)(PE)(FT)(KH)(XL)(GQ)(ZN)(YU)(OM) is part of A and
    (JV)(IP)(EF)(TK)(HX)(LG)(QZ)(NY)(UO)(MD) is part of D.

or

    (DM)(VJ)(PI)(FE)(KT)(XH)(GL)(ZQ)(YN)(OU) is part of A and
    (MV)(JP)(IF)(EK)(TX)(HG)(LZ)(QY)(NO)(UD) is part of D.

or

    (DU)(VM)(PJ)(FI)(KE)(XT)(GH)(ZL)(YQ)(ON) is part of A and
    (UV)(MP)(JF)(IK)(EX)(TG)(HZ)(LY)(QO)(ND) is part of D.

or

    (DN)(VU)(PM)(FJ)(KI)(XE)(GT)(ZH)(YL)(OQ) is part of A and
    (NV)(UP)(MF)(JK)(IX)(EG)(TZ)(HY)(LO)(QD) is part of D.

or

    (DQ)(VN)(PU)(FM)(KJ)(XI)(GE)(ZT)(YH)(OL) is part of A and
    (QV)(NP)(UF)(MK)(JX)(IG)(EZ)(TY)(HO)(LD) is part of D.

or

    (DL)(VQ)(PN)(FU)(KM)(XJ)(GI)(ZE)(YT)(OH) is part of A and
    (LV)(QP)(NF)(UK)(MX)(JG)(IZ)(EY)(TO)(HD) is part of D.

or

    (DH)(VL)(PQ)(FN)(KU)(XM)(GJ)(ZI)(YE)(OT) is part of A and
    (HV)(LP)(QF)(NK)(UX)(MG)(JZ)(IY)(EO)(TD) is part of D.

Mixing and matching the choices we have for the decomposition of the 10-cycle and the 2-cycle to go with our only option for the 1-cycle, yields 20 possible decompositions for AD.

This is a bit misleading, because each of the 20 possibilities, if taken one by one, can only reveal the first letter in each of the three-letter session keys. To recover the three letters of the session key, we need to simultaneously try a solution for A, a solution for B, and a solution for C. Thus, to make this demonstration complete, we must list the possibilities for B and C. E and F follow from the same work, but aren't needed for this step.

We found previously that BE = (BLFQVEOUM)(HJPSWIZRN)(AXT)(CGY)(D)(K).
BE includes the 1-cycles (D) and (K) ⇒

    (DK) is part of B and (KD) is part of E.

BE includes the 3-cycles (AXT) and (CGY) ⇒

   (AY)(XG)(TC) is part of B and (GT)(YX)(CA) is part of E

or

   (AG)(XC)(TY) is part of B and (CT)(GX)(YA) is part of E

or

   (AC)(XY)(TG) is part of B and (YT)(CX)(GA) is part of E

BE includes the 9-cycles (BLFQVEOUM)(HJPSWIZRN) ⇒

   (BN)(LR)(FZ)(QI)(VW)(ES)(OP)(UJ)(MH) is part of B and
   (NL)(RF)(ZQ)(IV)(WE)(SO)(PU)(JM)(HB) is part of E

or

   (BH)(LN)(FR)(QZ)(VI)(EW)(OS)(UP)(MJ) is part of B and
   (HL)(NF)(RQ)(ZV)(IE)(WO)(SU)(PM)(JB) is part of E

or

   (BJ)(LH)(FN)(QR)(VZ)(EI)(OW)(US)(MP) is part of B and
   (JL)(HF)(NQ)(RV)(ZE)(IO)(WU)(SM)(PB) is part of E

or

   (BP)(LJ)(FH)(QN)(VR)(EZ)(OI)(UW)(MS) is part of B and
   (PL)(JF)(HQ)(NV)(RE)(ZO)(IU)(WM)(SB) is part of E

or

   (BS)(LP)(FJ)(QH)(VN)(ER)(OZ)(UI)(MW) is part of B and
   (SL)(PF)(JQ)(HV)(NE)(RO)(ZU)(IM)(WB) is part of E

or

   (BW)(LS)(FP)(QJ)(VH)(EN)(OR)(UZ)(MI) is part of B and
   (WL)(SF)(PQ)(JV)(HE)(NO)(RU)(ZM)(IB) is part of E

or

   (BI)(LW)(FS)(QP)(VJ)(EH)(ON)(UR)(MZ) is part of B and
   (IL)(WF)(SQ)(PV)(JE)(HO)(NU)(RM)(ZB) is part of E

or

   (BZ)(LI)(FW)(QS)(VP)(EJ)(OH)(UN)(MR) is part of B and
   (ZL)(IF)(WQ)(SV)(PE)(JO)(HU)(NM)(RB) is part of E

or

(BR)(LZ)(FI)(QW)(VS)(EP)(OJ)(UH)(MN) is part of B and
(RL)(ZF)(IQ)(WV)(SE)(PO)(JU)(HM)(NB) is part of E

Mixing and matching the choices we have for the decomposition of the 9-cycle and the 3-cycle to go with our only option for the 1-cycle, yields 27 possible decompositions for BE.

We also have CF = (ABVIKTJGFCQNY)(DUZREHLXWPSMO).
CF includes only 13-cycles, so we have one of the following:

C = (AO)(BM)(VS)(IP)(KW)(TX)(JL)(GH)(FE)(CR)(QZ)(NU)(YD)
F = (OB)(MV)(SI)(PK)(WT)(XJ)(LG)(HF)(EC)(RQ)(ZN)(UY)(DA)

or

C = (AD)(BO)(VM)(IS)(KP)(TW)(JX)(GL)(FH)(CE)(QR)(NZ)(YU)
F = (DB)(OV)(MI)(SK)(PT)(WJ)(XG)(LF)(HC)(EQ)(RN)(ZY)(UA)

or

C = (AU)(BD)(VO)(IM)(KS)(TP)(JW)(GX)(FL)(CH)(QE)(NR)(YZ)
F = (UB)(DV)(OI)(MK)(ST)(PJ)(WG)(XF)(LC)(HQ)(EN)(RY)(ZA)

or

C = (AZ)(BU)(VD)(IO)(KM)(TS)(JP)(GW)(FX)(CL)(QH)(NE)(YR)
F = (ZB)(UV)(DI)(OK)(MT)(SJ)(PG)(WF)(XC)(LQ)(HN)(EY)(RA)

or

C = (AR)(BZ)(VU)(ID)(KO)(TM)(JS)(GP)(FW)(CX)(QL)(NH)(YE)
F = (RB)(ZV)(UI)(DK)(OT)(MJ)(SG)(PF)(WC)(XQ)(LN)(HY)(EA)

or

C = (AE)(BR)(VZ)(IU)(KD)(TO)(JM)(GS)(FP)(CW)(QX)(NL)(YH)
F = (EB)(RV)(ZI)(UK)(DT)(OJ)(MG)(SF)(PC)(WQ)(XN)(LY)(HA)

or

C = (AH)(BE)(VR)(IZ)(KU)(TD)(JO)(GM)(FS)(CP)(QW)(NX)(YL)
F = (HB)(EV)(RI)(ZK)(UT)(DJ)(OG)(MF)(SC)(PQ)(WN)(XY)(LA)

or

C = (AL)(BH)(VE)(IR)(KZ)(TU)(JD)(GO)(FM)(CS)(QP)(NW)(YX)
F = (LB)(HV)(EI)(RK)(ZT)(UJ)(DG)(OF)(MC)(SQ)(PN)(WY)(XA)

or

C = (AX)(BL)(VH)(IE)(KR)(TZ)(JU)(GD)(FO)(CM)(QS)(NP)(YW)
F = (XB)(LV)(HI)(EK)(RT)(ZJ)(UG)(DF)(OC)(MQ)(SN)(PY)(WA)

or

C = (AW)(BX)(VL)(IH)(KE)(TR)(JZ)(GU)(FD)(CO)(QM)(NS)(YP)
F = (WB)(XV)(LI)(HK)(ET)(RJ)(ZG)(UF)(DC)(OQ)(MN)(SY)(PA)

or

C = (AP)(BW)(VX)(IL)(KH)(TE)(JR)(GZ)(FU)(CD)(QO)(NM)(YS)
F = (PB)(WV)(XI)(LK)(HT)(EJ)(RG)(ZF)(UC)(DQ)(ON)(MY)(SA)

or

C = (AS)(BP)(VW)(IX)(KL)(TH)(JE)(GR)(FZ)(CU)(QD)(NO)(YM)
F = (SB)(PV)(WI)(XK)(LT)(HJ)(EG)(RF)(ZC)(UQ)(DN)(OY)(MA)

or

C = (AM)(BS)(VP)(IW)(KX)(TL)(JH)(GE)(FR)(CZ)(QU)(ND)(YO)
F = (MB)(SV)(PI)(WK)(XT)(LJ)(HG)(EF)(RC)(ZQ)(UN)(DY)(OA)

That is, there are 13 possible decompositions for CF.

We had 20 choices for A and D, 27 choices for B and E, and 13 choices for C and F. Thus, altogether, we have $(20)(27)(13) = 7,020$ possibilities for these six permutations. How can we tell which of these is right? Recall how simple Rejewski made it sound. He wrote:

> We assume that thanks to the theorem on the product of transpositions, combined with a knowledge of encipherers' habits, we know separately the permutations A through F.

I think you can figure out what this means by yourself. To help you do so, I will simplify things a bit. Instead of showing you the 7,020 sets of session keys that result from the 7,020 different factorizations, I will present just two of them. One is correct and one is not. Look at them closely and try to determine which set of session keys was actually used. Here's the first factorization to consider (option 1):

A = (AS)(BR)(CW)(DT)(VH)(PL)(FQ)(KN)(XU)(GM)(ZJ)(YI)(OE)
B = (DK)(AY)(XG)(TC)(BN)(LR)(FZ)(QI)(VW)(ES)(OP)(UJ)(MH)
C = (AO)(BM)(VS)(IP)(KW)(TX)(JL)(GH)(FE)(CR)(QZ)(NU)(YD)
D = (SA)(RC)(WB)(TV)(HP)(LF)(QK)(NX)(UG)(MZ)(JY)(IO)(ED)
E = (KD)(GT)(YX)(CA)(NL)(RF)(ZQ)(IV)(WE)(SO)(PU)(JM)(HB)
F = (OB)(MV)(SI)(PK)(WT)(XJ)(LG)(HF)(EC)(RQ)(ZN)(UY)(DA)

We will use this information to decipher all of the session keys. The substitutions performed by D, E, and F are redundant and therefore not needed, but they do serve as a way to check that our previous work was correct. The enciphered session keys are reproduced below for convenience:[17]

---

[17] From Bauer, Friedrich L., *Decrypted Secrets: Methods and Maxims of Cryptology*, second edition, Springer, New York, 2000, p. 390.

AUQ	AMN	IND	JHU	PVJ	FEG	SJM	SPO	WTM	RAO
BNH	CHL	JWF	MIC	QGA	LYB	SJM	SPO	WTM	RAO
BCT	CGJ	JWF	MIC	QGA	LYB	SJM	SPO	WTM	RAO
CIK	BZT	KHB	XJV	RJL	WPX	SUG	SMF	WKI	RKK
DDB	VDV	KHB	XJV	RJL	WPX	SUG	SMF	XRS	GNM
EJP	IPS	LDR	HDE	RJL	WPX	TMN	EBY	XRS	GNM
FBR	KLE	LDR	HDE	RJL	WPX	TMN	EBY	XOI	GUK
GPB	ZSV	MAW	UXP	RFC	WQQ	TAA	EXB	XYW	GCP
HNO	THD	MAW	UXP	SYX	SCW	USE	NWH	YPC	OSQ
HNO	THD	NXD	QTU	SYX	SCW	VII	PZK	YPC	OSQ
HXV	TTI	NXD	QTU	SYX	SCW	VII	PZK	ZZY	YRA
IKG	JKF	NLU	QFZ	SYX	SCW	VQZ	PVR	ZEF	YOC
IKG	JKF	OBU	DLZ	CYX	CCW	VQZ	PVR	ZSJ	YWG

To recover the first enciphered session key, AUQ AMN, we begin with the first character, A. Because it is in the first position, it was enciphered by permutation A. Permutation A contains the swap (AS), so the ciphertext letter A becomes plaintext S. The second cipher letter is U, so we look for U in permutation B and find the swap (UJ). Thus, U deciphers to J. Finally, we look up Q in permutation C. We find the swap (QZ), so ciphertext Q deciphers to Z. We've now recovered the session key SJZ. The three remaining ciphertext letters, AMN, serve to check our result. We look for A, M, and N in permutations D, E, and F, respectively, and the swaps once again give us SJZ. Thus, we are confident that, if the permutations are correct, the first session key is SJZ.

Deciphering the entire list of session keys with our proposed permutations, A through F, gives

SJZ	SJZ	YBY	YBY	LWL	LWL	AUB	AUB	CCB	CCB
RBG	RBG	ZVE	ZVE	FXO	FXO	AUB	AUB	CCB	CCB
RTX	RTX	ZVE	ZVE	FXO	FXO	AUB	AUB	CCB	CCB
WQW	WQW	NMM	NMM	BUJ	BUJ	AJH	AJH	CDP	CDP
TKM	TKM	NMM	NMM	BUJ	BUJ	AJH	AJH	ULV	ULV
OUI	OUI	PKC	PKC	BUJ	BUJ	DHU	DHU	ULV	ULV
QNC	QNC	PKC	PKC	BUJ	BUJ	DHU	DHU	UPP	UPP
MOM	MOM	GYK	GYK	BZR	BZR	DYO	DYO	UAK	UAK
VBA	VBA	GYK	GYK	AAT	AAT	XEF	XEF	IOR	IOR
VBA	VBA	KGY	KGY	AAT	AAT	HQP	HQP	IOR	IOR
VGS	VGS	KGY	KGY	AAT	AAT	HQP	HQP	JFD	JFD
YDH	YDH	KRN	KRN	AAT	AAT	HIQ	HIQ	JSE	JSE
YDH	YDH	ENN	ENN	AAT	AAT	HIQ	HIQ	JEL	JEL

This is the first of the two potential solutions you are asked to consider.

Now, consider another possible factorization (option 2):

$$A = (AS)(BR)(CW)(DI)(VE)(PT)(FH)(KL)(XQ)(GN)(ZU)(YM)(OJ)$$
$$B = (DK)(AY)(XG)(TC)(BJ)(LH)(FN)(QR)(VZ)(EI)(OW)(US)(MP)$$
$$C = (AX)(BL)(VH)(IE)(KR)(TZ)(JU)(GD)(FO)(CM)(QS)(NP)(YW)$$
$$D = (SA)(RC)(WB)(IV)(EP)(TF)(HK)(LX)(QG)(NZ)(UY)(MO)(JD)$$
$$E = (KD)(GT)(YX)(CA)(JL)(HF)(NQ)(RV)(ZE)(IO)(WU)(SM)(PB)$$
$$F = (XB)(LV)(HI)(EK)(RT)(ZJ)(UG)(DF)(OC)(MQ)(SN)(PY)(WA)$$

With these selections, the session keys decipher to

SSS	SSS	DFG	DFG	TZU	TZU	ABC	ABC	CCC	CCC
RFV	RFV	OOO	OOO	XXX	XXX	ABC	ABC	CCC	CCC
RTZ	RTZ	OOO	OOO	XXX	XXX	ABC	ABC	CCC	CCC
WER	WER	LLL	LLL	BBB	BBB	ASD	ASD	CDE	CDE
IKL	IKL	LLL	LLL	BBB	BBB	ASD	ASD	QQQ	QQQ
VBN	VBN	KKK	KKK	BBB	BBB	PPP	PPP	QQQ	QQQ
HJK	HJK	KKK	KKK	BBB	BBB	PPP	PPP	QWE	QWE
NML	NML	YYY	YYY	BNM	BNM	PYX	PYX	QAY	QAY
FFF	FFF	YYY	YYY	AAA	AAA	ZUI	ZUI	MMM	MMM
FFF	FFF	GGG	GGG	AAA	AAA	EEE	EEE	MMM	MMM
FGH	FGH	GGG	GGG	AAA	AAA	EEE	EEE	UVW	UVW
DDD	DDD	GHJ	GHJ	AAA	AAA	ERT	ERT	UIO	UIO
DDD	DDD	JJJ	JJJ	AAA	AAA	ERT	ERT	UUU	UUU

Before you continue reading, stop and ponder the lists of session keys arising from factorization option 1 and option 2. Which of these were more likely to be chosen by the Nazis using the machines? The answer follows below.

Recall that the Nazis operating these machines were told to select their session keys randomly. Well, despite what some war criminals claimed at the Nuremburg trials, Nazis didn't always follow orders. The second set of session keys is the correct one, and these keys are far from random. For example, there are many triplets, such as AAA and KKK.

Looking at the layout of an Enigma keyboard reveals the inspiration for other keys.

```
    Q W E R T Z U I O
     A S D F G H J K
     P Y X C V B N M L
```

Of the 65 keys, 39 use a triplet, 18 read along a row of keys, 3 read down a diagonal on the keyboard (RFV, IKL, QAY), and 5 use three consecutive letters (one of which, CDE, happens to also be up a diagonal on the keyboard). Every single session key had some pattern! By contrast, in option 1 we can't find any of these patterns.

This is what Rejewski meant by "knowledge of encipherers' habits" The patterns he looked for did not have to be of the type shown above. Enigma operators sometimes used a person's initials, letters connected to a girlfriend's name, or something else that could be guessed. Taking advantage of this human tendency toward the nonrandom is often referred to as using the *psychological method*.

With this method, the session keys from the correct factorization will stand out from a large group of possibilities just as clearly as from the pair presented above. Having to consider 7,020 possibilities may sound painful, but the correct answer will be found, on average, after 3,510 attempts (half the total), and the problem is perfect for parallel processing. Of course, during World War II, this meant many people working simultaneously on separate possibilities. This is why cryptanalytic groups often had a large number of clerks! With 100 people working on this problem, a single person would have only 35 possibilities to check on average. Also, to speed things up, one could move on to the next possibility if, say, the first five session keys all fail to fit any of the expected forms. But due to security concerns, the Poles didn't adopt the parallel processing approach. It should be noted that the number of factorization possibilities varies depending on the characteristic for the set of messages. Some days there were more possibilities to investigate and some days fewer.

Rejewski presented the correct solution in a manner that looks a bit different, but recall that disjoint cycles commute and 2-cycles may be written in the form (XY) or (YX). Rejewski's representation of permutations A through F follows:

$$A = (AS)(BR)(CW)(DI)(EV)(FH)(GN)(JO)(KL)(MY)(PT)(QX)(UZ)$$
$$B = (AY)(BJ)(CT)(DK)(EI)(FN)(GX)(HL)(MP)(OW)(QR)(SU)(VZ)$$
$$C = (AX)(BL)(CM)(DG)(EI)(FO)(HV)(JU)(KR)(NP)(QS)(TZ)(WY)$$
$$D = (AS)(BW)(CR)(DJ)(EP)(FT)(GQ)(HK)(IV)(LX)(MO)(NZ)(UY)$$
$$E = (AC)(BP)(DK)(EZ)(FH)(GT)(IO)(JL)(MS)(NQ)(RV)(UW)(XY)$$
$$F = (AW)(BX)(CO)(DF)(EK)(GU)(HI)(JZ)(LV)(MQ)(NS)(PY)(RT)$$

If only the rightmost rotor turns while enciphering the session key, then we have the following.[18]

$$A = SHPNP^{-1}MLRL^{-1}M^{-1}PN^{-1}P^{-1}H^{-1}S^{-1}$$
$$B = SHP^2NP^{-2}MLRL^{-1}M^{-1}P^2N^{-1}P^{-2}H^{-1}S^{-1}$$
$$C = SHP^3NP^{-3}MLRL^{-1}M^{-1}P^3N^{-1}P^{-3}H^{-1}S^{-1}$$
$$D = SHP^4NP^{-4}MLRL^{-1}M^{-1}P^4N^{-1}P^{-4}H^{-1}S^{-1}$$
$$E = SHP^5NP^{-5}MLRL^{-1}M^{-1}P^5N^{-1}P^{-5}H^{-1}S^{-1}$$
$$F = SHP^6NP^{-6}MLRL^{-1}M^{-1}P^6N^{-1}P^{-6}H^{-1}S^{-1}$$

Recalling our notations, the first equation, read from left to right, shows the composition of the permutations S (plugboard), H (wiring from plugboard to rotor assembly), P (representing the turning of the first rotor), N (the fast rotor), P^{-1} (we then need to undo the turning effect), M (the middle rotor), L (the final rotor), R (the reflector), and then out through the same permutations again, but in reverse order and direction (hence the inverse notation). Now, if the middle rotor turns, the above won't hold true, but this would only happen with probability 5/26.

Because $MLRL^{-1}M^{-1}$ appears in the center of all of the equations, we can simplify matters by setting this equal to Q. We then have

$$A = SHPNP^{-1}QPN^{-1}P^{-1}H^{-1}S^{-1}$$
$$B = SHP^2NP^{-2}QP^2N^{-1}P^{-2}H^{-1}S^{-1}$$
$$C = SHP^3NP^{-3}QP^3N^{-1}P^{-3}H^{-1}S^{-1}$$
$$D = SHP^4NP^{-4}QP^4N^{-1}P^{-4}H^{-1}S^{-1}$$
$$E = SHP^5NP^{-5}QP5N^{-1}P^{-5}H^{-1}S^{-1}$$
$$F = SHP^6NP^{-6}QP^6N^{-1}P^{-6}H^{-1}S^{-1}$$

---

[18] Rejewski, in explaining his work, sometimes wrote the first equation without the necessary P^{-1} in position 5 and P in position 11. It seems he didn't want to confuse the reader with too many details. Later on he includes it (when he's being more detailed). Christensen followed this style in his excellent paper.

Because there are six equations and only four unknowns (S, H, N, and Q), we ought to be able to find a solution.

Figure 7.12 Hans Thilo Schmidt. (From the David Kahn Collection, National Cryptologic Museum, Fort Meade, Maryland.)

The identification of S (the plugboard setting) came about by noncryptanalytic means. A German, Hans Thilo Schmidt (Figure 7.12), sold it, along with other Enigma information, to the French on November 8, 1931. In December of the following year, the Frenchman Captain Gustav Bertrand (later to become a General) passed Schmidt's information on to the Poles.[19]

H was guessed to be the same as the commercial Enigma and A, B, C, D, E, and F were determined above. Thus, to simplify things, we multiply both sides of each equation on the left by $H^{-1}S^{-1}$ and on the right by SH to get

$$H^{-1}S^{-1}ASH = PNP^{-1}QPN^{-1}P^{-1}$$
$$H^{-1}S^{-1}BSH = P^2NP^{-2}QP^2N^{-1}P^{-2}$$
$$H^{-1}S^{-1}CSH = P^3NP^{-3}QP^3N^{-1}P^{-3}$$
$$H^{-1}S^{-1}DSH = P^4NP^{-4}QP^4N^{-1}P^{-4}$$
$$H^{-1}S^{-1}ESH = P^5NP^{-5}QP5N^{-1}P^{-5}$$
$$H^{-1}S^{-1}FSH = P^6NP^{-6}QP^6N^{-1}P^{-6}$$

All of the permutations on the left side of the above equalities were believed to be known.

---

[19] Rejewski, Marian, "Mathematical Solution of the Enigma Cipher," *Cryptologia*, Vol. 6, No. 1, January 1982, pp. 1–18, p. 8 cited here.

A quick definition: If we start with a permutation A and use another permutation P to form the product $P^{-1}AP$, we say that A has been transformed by the action of P.

Now take the six equations above and transform both sides by the action of P, P^2, P^3, P^4, P^5, and P^6, respectively. We label the results with new letters for convenience:

$$U = P^{-1}H^{-1}S^{-1}ASHP \qquad U = NP^{-1}QPN^{-}$$
$$V = P^{-2}H^{-1}S^{-1}BSHP^2 \qquad V = NP^{-2}QP^2N^{-1}$$
$$W = P^{-3}H^{-1}S^{-1}CSHP^3 \qquad W = NP^{-3}QP^3N^{-1}$$
$$X = P^{-4}H^{-1}S^{-1}DSHP^4 \qquad X = NP^{-4}QP^4N^{-1}$$
$$Y = P^{-5}H^{-1}S^{-1}ESHP^5 \qquad Y = NP^{-5}QP^5N^{-1}$$
$$Z = P^{-6}H^{-1}S^{-1}FSHP^6 \qquad Z = NP^{-6}QP^6N^{-1}$$

Note: The left-hand sides are known.

We now use the second column of equations to take some products.

$$UV = (NP^{-1}QPN^{-1})(NP^{-2}QP^2N^{-1})$$
$$VW = (NP^{-2}QP^2N^{-1})(NP^{-3}QP^3N^{-1})$$
$$WX = (NP^{-3}QP^3N^{-1})(NP^{-4}QP^4N^{-1})$$
$$XY = (NP^{-4}QP^4N^{-1})(NP^{-5}QP^5N^{-1})$$
$$YZ = (NP^{-5}QP^5N^{-1})(NP^{-6}QP^6N^{-1})$$

Removing the parenthesis, the $N^{-1}N$ in the middle drops out of each equation. We then combine the powers of P that appear next to one another and insert a new pair of parenthesis (to stress a portion the equations all share) and get the following.

$$UV = NP^{-1}(QP^{-1}QP)PN^{-1}$$
$$VW = NP^{-2}(QP^{-1}QP)P^2N^{-1}$$
$$WX = NP^{-3}(QP^{-1}QP)P^3N^{-1}$$
$$XY = NP^{-4}(QP^{-1}QP)P^4N^{-1}$$
$$YZ = NP^{-5}(QP^{-1}QP)P^5N^{-1}$$

Now take the first four equations above and transform both sides of each by the action of NPN^{-1}:

$$NP^{-1}N^{-1}(UV)NPN^{-1} = NP^{-1}N^{-1}NP^{-1}(QP^{-1}QP)PN^{-1}NPN^{-1}$$
$$NP^{-1}N^{-1}(VW)NPN^{-1} = NP^{-1}N^{-1}NP^{-2}(QP^{-1}QP)P^2N^{-1}NPN^{-1}$$
$$NP^{-1}N^{-1}(WX)NPN^{-1} = NP^{-1}N^{-1}NP^{-3}(QP^{-1}QP)P^3N^{-1}NPN^{-1}$$
$$NP^{-1}N^{-1}(XY)NPN^{-1} = NP^{-1}N^{-1}NP^{-4}(QP^{-1}QP)P^4N^{-1}NPN^{-1}$$

Simplifying the right-hand sides yields

$$NP^{-1}N^{-1}(UV)NPN^{-1} = NP^{-2}(QP^{-1}QP)P^2N^{-1}$$
$$NP^{-1}N^{-1}(VW)NPN^{-1} = NP^{-3}(QP^{-1}QP)P^3N^{-1}$$
$$NP^{-1}N^{-1}(WX)NPN^{-1} = NP^{-4}(QP^{-1}QP)P^4N^{-1}$$
$$NP^{-1}N^{-1}(XY)NPN^{-1} = NP^{-5}(QP^{-1}QP)P^5N^{-1}$$

Recognizing the right-hand sides, we have

$$NP^{-1}N^{-1}(UV)NPN^{-1} = VW$$
$$NP^{-1}N^{-1}(VW)NPN^{-1} = WX$$
$$NP^{-1}N^{-1}(WX)NPN^{-1} = XY$$
$$NP^{-1}N^{-1}(XY)NPN^{-1} = YZ$$

Now we have four equations and the only unknown is N (remember, P represents the advancement of the fast rotor).

Observe that $NP^{-1}N^{-1}$ is the inverse of the permutation NPN^{-1}. Thus, the first of the equations above has the form $T^{-1}(UV)T = VW$ and the others are similar. When we have this situation, we say UV and VW are *conjugate permutations*, and it is very easy to find a permutation T that transforms UV into VW.

To do so, we simply write VW under UV and pretend that the top line is the plaintext and the bottom line is the ciphertext. We then express this "encryption" in disjoint cycle notation. For example,

$$UV = (AEPFTYBSNIKOD) (RHCGZMUVQWLJX)$$
$$VW = (AKJCEVZYDLWNU) (SMTFHQIBXOPGR)$$

gives

$$T = (A) (EKWONDUILPJGFCT) (YVBZHMQXRS)$$

The only problem is that UV could be written, switching the order of the 13-cycles, as

$$UV = (RHCGZMUVQWLJX) (AEPFTYBSNIKOD)$$

and this will give a different result. That is, the solution to the problem of finding a permutation T such that $T^{-1}(UV)T = VW$ is not unique. Indeed, beginning one of the 13-cycles that make up UV with a different letter also changes the solution.

So, the first equation will give us dozens of possibilities for $NP^{-1}N^{-1}$ (recall that T was taking the place of $NP^{-1}N^{-1}$ in the discussion above). Exactly how many solutions exist depends on the cycle structure of UV.

In any case, each of the equations below will offer various possibilities for $NP^{-1}N^{-1}$.

$$NP^{-1}N^{-1}(UV)NPN^{-1} = VW$$
$$NP^{-1}N^{-1}(VW)NPN^{-1} = WX$$
$$NP^{-1}N^{-1}(WX)NPN^{-1} = XY$$
$$NP^{-1}N^{-1}(XY)NPN^{-1} = YZ$$

However, there will only be one possibility suggested repeatedly—this is the one we take. We won't even need all four equations. We can simply find all solutions given by the first two equations above and take the one that arises twice.

Continuing with our example, we have

$$UV = (\text{AEPFTYBSNIKOD}) \ (\text{RHCGZMUVQWLJX})$$
$$VW = (\text{AKJCEVZYDLWNU}) \ (\text{SMTFHQIBXOPGR})$$
$$WX = (\text{AQVLOIKGNWBMC}) \ (\text{PUZFTJRYEHXDS})$$

So we write VW under UV all possible ways ($13 \times 13 \times 2 = 338$ of them!)[20] and see what permutations they give, and we also write WX under VW all possible ways, and then look for a match. For example,

$$VW = (\text{AKJCEVZYDLWNU}) \ (\text{SMTFHQIBXOPGR})$$
$$WX = (\text{AQVLOIKGNWBMC}) \ (\text{PUZFTJRYEHXDS})$$

gives

$$(\text{A}) \ (\text{KQJVIRSPXEOHTZ}) \ (\text{CLWBYGDNMU}) \ (\text{F})$$

The only match tells us we have $NPN^{-1} = (\text{AYURICXQMGOVSKEDZPLFWTNJHB})$.

Now, we can use this new equation to get N. It's the same sort of problem we addressed above. Because, $P = (\text{ABCDEFGHIJKLMNOPQRSTUVWXYZ})$, we could simply place the alphabet underneath NPN^{-1}, pretend it's an encryption, and proceed to write out the disjoint cycle form for it. Because we can start P with any of the 26 letters, 26 distinct solutions will arise.

Alternatively, if we shuffle the top permutation to get it alphabetized, and apply the same reordering to the bottom permutation, we'll get the same results, but in a different form.

```
AYURICXQMGOVSKEDZPLFWTNJHB          ABCDEFGHIJKLMNOPQRSTUVWXYZ
                            ⇒
ABCDEFGHIJKLMNOPQRSTUVWXYZ          AZFPOTJYEXNSIWKRHDMVCLUGBP

AYURICXQMGOVSKEDZPLFWTNJHB          ABCDEFGHIJKLMNOPQRSTUVWXYZ
                            ⇒
BCDEFGHIJKLMNOPQRSTUVWXYZA          BAGQPUKZFYOTJXLSIENWDMVHCR

AYURICXQMGOVSKEDZPLFWTNJHB          ABCDEFGHIJKLMNOPQRSTUVWXYZ
                            ⇒
CDEFGHIJKLMNOPQRSTUVWXYZAB          CBHRQVLAGZPUKYMTJFOXENWIDS
                            ⋮
AYURICXQMGOVSKEDZPLFWTNJHB          ABCDEFGHIJKLMNOPQRSTUVWXYZ
                            ⇒
ZABCDEFGHIJKLMNOPQRSTUVWXY          ZYEONSIXDWMRHVJQGCLUBKTFAP
```

One of the 26 possibilities indicated above will be N, the wiring of the fast rotor.

---

[20] The factors of 13 come from the fact that either 13-cycle can be expressed beginning with any of its 13 letters. The factor of 2 comes from the choice as to which of the two 13-cycles we write first.

The attack described above was based on Rejewski's description. He didn't use real data, so, although the approach works, the rotor possibilities we end up with don't match any that were actually used.

In any case, the work we just went through only provides the wiring for one rotor—the rightmost and fastest rotor. The Germans, however, placed the rotors in the machine in different orders, every three months, as part of the key, and the settings supplied by Schmidt straddled two quarters. Happily, two different rotors fell in the rightmost position in these two quarters and both were recovered by the method detailed above.

There was some more work involved, as the wiring of the third rotor and the reflector still had to be recovered. But the above gives the flavor of the work. It should also be noted that we cannot narrow down the possibilities for each rotor individually. We must look at them as a group to decide which are correct.[21]

After everything was determined, and the Poles were expecting to be able to read messages, only gibberish appeared! Something was wrong. Finally, Rejewski turned to the wiring from the plugboard to the rotor entry points. Perhaps it wasn't the same as in the commercial Enigma the Poles had acquired. So, what complex mathematical machinery did Rejewski apply to this new problem? None—he guessed. Maybe H was simply

Input: ABCDEFGHIJKLMNOPQRSTUVWXYZ
Output: ABCDEFGHIJKLMNOPQRSTUVWXYZ

He tried it and it worked; Enigma had been broken.

There's a much simpler way of explaining how the Poles were able to learn all of the details of Enigma. Simply cite the unreliable source, *A Man Called Intrepid*:

> The new Enigmas were being delivered to frontier units, and in early 1939 a military truck containing one was ambushed. Polish agents staged an accident in which fire destroyed the evidence. German investigators assumed that some charred bits of coils, springs, and rotors were the remains of the real Enigma.[22]

The author does mention the star of this chapter, sort of; he refers to "Mademoiselle Marian Rejewski." Richard A. Woytak admired Stevenson's "thereby managing, with masterful economy of expression, to get wrong both Rejewski's sex and marital status."[23]

That was a lot easier than wrapping your head around the mathematics wasn't it? Everyone knows you can't believe everything you read on the internet, but I really don't see how it's any different from the print world.

Thus, the Poles, having been given the daily keys, were able to reconstruct the Enigma machine. Eventually the keys expired and the Poles had to face the opposite problem: Having the machine, how could the daily keys be recovered?

---

[21] Rejewski, Marian, translated by Christopher Kasparek, "Mathematical Solution of the Enigma Cipher," *Cryptologia*, Vol. 6, No. 1, pp. 1–18, p. 11 states "But those details may only be established following the basic reconstruction of the connections in all the rotors."

[22] Stevenson, William, *A Man Called Intrepid*, Harcourt Brace Jovanovich, New York, 1976, p. 49. Or see the paperback edition, Ballantine Books, New York, 1977, p. 53.

[23] Rejewski, Marian, "Remarks on Appendix 1 to British Intelligence in the Second World War by F. H. Hinsley," *Cryptologia*, Vol. 6, No.1, January 1982, pp. 75–83. This piece was translated into English by Christopher Kasparek and contains a prefatory note by Richard A. Woytak.

7.5 Cryptanalysis Part 2: Recovering the Daily Keys

We now examine the problem of recovering the keys (rotor order, settings, ring settings, and plugboard). With Schmidt's keys expired, the Poles needed to determine all of these key details, but on the bright side, they had the equivalent of military Enigmas (which they made) to use. Also, they discovered an important pattern, which is detailed below.

In Section 7.4, we saw that for one particular day (and thus a particular daily key) we were able to determine:

$$AD = (DVPFKXGZYO)\ (EIJMUNQLHT)\ (BC)\ (RW)\ (A)\ (S)$$
$$BE = (BLFGVEOUM)\ (HJPSWIZRN)\ (AXT)\ (CGY)\ (D)\ (K)$$
$$CF = (ABVIKTJGFCQNY)\ (DUZREHLXWPSMO)$$

Thus,

AD had the cycle structure 10, 10, 2, 2, 1, 1.
BE had the cycle structure 9, 9, 3, 3, 1, 1.
CF had the cycle structure 13, 13.

The Poles observed that as the settings on Enigma were changed from day to day, this disjoint cycle structure also changed. That is, the cycle structure is determined by the order of the rotors and their initial positions. The ring settings do not affect it and can therefore be ignored. If the Poles could build a catalog showing the correspondence between the rotor settings and the disjoint cycle structures, then the latter, when recovered from a set of intercepted messages, would tell them how to set up their bootleg Enigmas to decipher the intercepts. But how large would this catalog be? Would its creation even be possible?

There are 6 ways to order the three rotors and $26^3 = 17,576$ ways to select a daily key to determine their initial positions. Thus, the total number of possibilities comes to $(6)(17,576) = 105,456$. A catalog with this many entries would take some time to create, but it could be done. However, there is also the plugboard. We saw in Section 7.3 that there are 532,985,208,200,576 possible plugboard settings. Having a factor of this size in the calculation of the catalog size would make constructing it impossible. This bring us to what is sometimes called "The Theorem that Won the War." It can be stated tersely as:

Conjugate permutations have the same disjoint cycle structure.

Recall that if P is a permutation and C is some other permutation, then P and the product $C^{-1}PC$ are conjugate permutations. That is, we get such a pair by multiplying the original on one side by a permutation C and on the other side by the inverse of that permutation, C^{-1}. The fact that this does not change the cycle structure can be stated as follows.

If P and C are permutations and $P(\alpha) = \beta$, then the permutation $C^{-1}PC$ sends $C(\alpha)$ to $C(\beta)$. Hence, P and $C^{-1}PC$ have the same disjoint cycle structure.[24]

---

[24] A proof of this theorem is provided in Rejewski, Marian, *Memories of my work at the Cipher Bureau of the General Staff Second Department 1930–1945*, Adam Mickiewicz University, Poznań, Poland, 2011.

The plugboard, represented by S, acts on the rest of the Enigma encryption by conjugation. We can see this by looking again at the mathematical model the Poles built of Enigma. A set of parenthesis is included around the letters representing permutations caused by non-plugboard components.

$$A = S(HPNP^{-1}MLRL^{-1}M^{-1}PN^{-1}P^{-1}H^{-1})S^{-1}$$
$$B = S(HP^2NP^{-2}MLRL^{-1}M^{-1}P^2N^{-1}P^{-2}H^{-1})S^{-1}$$
$$C = S(HP^3NP^{-3}MLRL^{-1}M^{-1}P^3N^{-1}P^{-3}H^{-1})S^{-1}$$
$$D = S(HP^4NP^{-4}MLRL^{-1}M^{-1}P^4N^{-1}P^{-4}H^{-1})S^{-1}$$
$$E = S(HP^5NP^{-5}MLRL^{-1}M^{-1}P^5N^{-1}P^{-5}H^{-1})S^{-1}$$
$$F = S(HP^6NP^{-6}MLRL^{-1}M^{-1}P^6N^{-1}P^{-6}H^{-1})S^{-1}$$

We have S on one side and S^{-1} on the other.[25] Thus, the plugboard doesn't alter cycle structures, no matter how it is wired. It will change what letters are involved in each cycle, but not the lengths of the cycles. So, the plugboard can be ignored when building the catalog! The catalog only needs to contain the 105,456 entries corresponding to the order and initial positions of the three rotors.

To create the catalog, the Poles could have set their new bootleg military Enigma to a particular key and enciphered a session key twice, noting the result, then reset the key and enciphered another session key twice, and so on. After obtaining dozens of these enciphered session keys, Rejewski's method, described in Section 7.4, could be applied to yield the permutations AD, BE, and CF, and thus determine their cycle structure; however, to do this 105,456 times would be very time consuming.

Instead, to create the desired catalog, the Poles made a machine called a *cyclometer*, depicted in Figure 7.13.

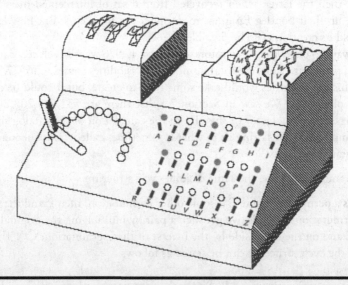

Figure 7.13 The Polish cyclometer. (Illustration by Dan Meredith from Christensen, Chris, *Mathematics Magazine,* **Vol. 80, No. 4, 2007, p. 260.)**

---

[25] *Note*: It doesn't matter if we have S on the left and S^{-1} on the right or S^{-1} on the left and S on the right. All that matters is that the permutations we have on either side are inverses of each other. The labeling is arbitrary anyway.

This time-saving device consisted of two rotor sets (with reflectors). As you may have already guessed, one represents A and the other D (or B and E, C and F, whichever permutation is being investigated at that moment). A charge can be applied to any of the letters and the current will flow through the set of rotors representing permutation A and then through the rotors representing permutation D; the letter that comes out will be illuminated, but this is not the end. The charge continues through A and D again and may light another letter, if the cycle is not yet complete. The handle on the left side of the front of the cyclometer is to control the amount of current.[26] If the amount of current needed to clearly light the bulbs for all of the letters in a large cycle were used for a very short cycle, it might burn out the filaments. Hence, the operator should start out low, and increase the current only if it looks like enough bulbs are lighting to avoid damage.

Naturally it took some time to construct the cyclometer, but it allowed the catalog to be built much more quickly than by applying the method described in Section 7.4. The original catalog no longer exists, but Alex Kuhl, an undergraduate at Northern Kentucky University at the time, reconstructed it.[27] Naturally, he used a personal computer rather than doing it by hand with a cyclometer. Although the cyclometer method was quicker than a mathematical analysis, it still took over a year for the Poles to complete their catalog. I find it amazing that more people were not assigned to this extremely important work in order to speed its completion. It was a task extremely well suited to what we would now call parallel processing. Twice as many workers could have completed the job in half the time, and so on for three, four, etc. times as many workers. Once the cyclometers were made, anyone could have been trained to work on a portion of the catalog. It didn't require any skills other than attention to detail. In fact, it must have been very monotonous work. But, even though it was tremendously important, the Poles weren't even working on the catalog full time; they were also reconstructing Enigma keys by another method each day. Kozaczuk described what it was like when cyclometer work was being done:[28]

> This was a tedious, time-consuming job and, on account of the work's secrecy, the mathematicians could not delegate it to the Cipher Bureau's technical personnel. In their haste the men would scrape their fingers raw and bloody.

When completed, the catalog did not offer a one-to-one correspondence, for there were 105,456 ways to order and position the rotors, but only 21,230 different disjoint cycle structures. A recovered disjoint cycle structure would lead to, on average, 5 possibilities for the rotors. However, averages can be misleading. On average, humans have one testicle and one ovary each. Perhaps the average of 5 rotor settings is also unrepresentative of a typical result. What does the catalog

---

[26] The technical term for such a control is *rheostat*.

[27] Kuhl, Alex, "Rejewski's Catalog," *Cryptologia*, Vol. 31, No. 4, October 2007, pp. 326–332.

[28] Kozaczuk, Wladyslaw, *Enigma: How the German Machine Cipher Was Broken, and How It Was Read by the Allies in World War Two*, edited and translated from the original 1979 Polish version by Christopher Kasparek, University Publications of America, Inc., Frederick, Maryland, 1984, p. 29.

actually look like? Kuhl's reconstruction revealed exactly how the 105,465 rotor settings map to the 21,230 different disjoint cycle structures. Some of his results are provided below.[29]

- The good news is that 11,466 of the disjoint cycle structures have unique rotor settings that give rise to them. Thus, over half the time the cycle structure, when checked in the catalog, would immediately yield the rotor settings. Over 92% of the disjoint cycle structures correspond to 10 or fewer possible rotor settings.
- The bad news is that there are disjoint cycle structures that give far more possibilities (See Table 7.1).

It's not known what the Poles did on the worst days, but such days were rare, and recovery of the daily key typically took only 10 to 20 minutes.

Nevertheless, the Poles' work was not ended! Rejewski lamented:

> Unfortunately, on November 2, 1937, when the card catalogue was ready, the Germans exchanged the reversing drum[30] that they had been using, which they designated by the letter A, for another drum, a B drum, and consequently, we had to do the whole job over again, after first reconstructing the connections in drum B, of course.[31]

Still, a catalog with 105,456 entries can be generated by hand, even twice.

It should be pointed out that the catalog doesn't reveal how the plugboard is wired. The catalog only helps find the order and positions of the rotors. Once the correct rotor setting is determined, we still won't get perfect plaintext out unless no plugboard cables were in use. The more cables in use, the worse our result will be; however, unless 13 cables were used, some correct plaintext letters will be revealed. Recall that, originally, exactly six cables were used. This, coupled with the use of cribs, and the fact that no letter can be enciphered as itself, allows the plugboard to be reconstructed.

7.6 After the Break

The story of how Enigma was broken is far from over at this point. Just as the Poles faced a setback following the introduction of a new reflector, many other changes would follow. On September 15, 1938, the method in which the session key was sent changed. The Poles fought back with new techniques: Zygalski sheets, and bomba, the latter of which had roots in the cyclometer.

On December 15, 1938, the Germans introduced two new rotors. The method detailed above allowed Rejewski to recover their wirings, but the recovery of the daily keys became ten times more difficult (60 possible rotor orderings in place of 6). On July 24, 1939, the Poles shared their Enigma results with the French and British.[32]

The Nazis invaded Poland on September 1, 1939, and the cryptanalysts fled to France. Soon France too was invaded. The Poles remained in unoccupied France (while it still existed), but fled to England once all of France was occupied. The entire time, they had been sending recovered

---

[29] Kuhl, Alex, "Rejewski's Catalog," *Cryptologia*, Vol. 31, No. 4, October 2007, pp. 326–332, pp. 329–330 cited here.

[30] I've been referring to this as the reflector.

[31] Rejewski, Marian, "How the Polish Mathematicians Broke Enigma," Appendix D, in Kozaczuk, Wladyslaw, *Enigma: How the German Machine Cipher Was Broken, and How It Was Read by the Allies in World War Two*, Arms & Armour Press, London, UK, 1984, pp. 246–271, p. 264 cited here.

[32] According to Kahn's chronology. In contrast, the date of July 25 is given in Welchman, Gordon, *The Hut Six Story*, McGraw-Hill, New York, 1982, p. 16, and in Rejewski, Marian, "Mathematical Solution of the Enigma Cipher," *Cryptologia*, Vol. 6, No. 1, January 1982, pp. 1–18, p. 17 cited here.

Table 7.1   Most Frequent Cycle Structures

Disjoint Cycle Structure (AD)(BE)(CF)	*Number of Occurrences*
(13 13)(13 13)(13 13)	1771
(12 12 1 1)(13 13)(13 13)	898
(13 13)(13 13)(12 12 1 1)	866
(13 13)(12 12 1 1) (13 13)	854
(11 11 2 2)(13 13)(13 13)	509
(13 13)(12 12 1 1)(12 12 1 1)	494
(13 13)(13 13)(11 11 2 2)	480
(12 12 1 1)(13 13)(12 12 1 1)	479
(13 13)(11 11 2 2)(13 13)	469
(12 12 1 1)(12 12 1 1)(13 13)	466
(13 13)(10 10 3 3)(13 13)	370
(13 13)(13 13)(10 10 3 3)	360
(10 10 3 3)(13 13)(13 13)	358
(13 13)(13 13)(9 9 4 4)	315
(9 9 4 4)(13 13)(13 13)	307

Source: Kuhl, Alex, Rejewski's Catalog, *Cryptologia*, Vol. 31, No. 4, October 2007, p. 329. With permission.

Enigma keys to England. Once there, however, in the words of David Kahn, "The British showed their gratitude by excluding them from any further contact with codebreaking."[33]

7.7 Alan Turing and Bletchley Park

England's cryptanalytic efforts during World War II were centered at Bletchley Park, halfway between Oxford and Cambridge. But before discussing the work that went on there, we take a look at the early life of England's most famous codebreaker, Alan Turing (Figure 7.14).

In 1926, the year of the general strike, Turing entered Sherborne School. He had to cycle 60 miles to get there. He would eventually become an athlete of almost Olympic caliber.[34] However, he found it difficult to fit in at the school. Conventional schooling is inappropriate for the most original thinkers. Turing's headmaster wrote:[35]

> If he is to stay at Public School, he must aim at becoming educated. If he is to be solely a Scientific Specialist, he is wasting his time at a Public School.

---

[33] Kahn, David, "The Significance of Codebreaking and Intelligence in Allied Strategy and Tactics," *Cryptologia*, Vol. 1, No. 3, July 1977, pp. 209–222.

[34] http://www-groups.dcs.st-and.ac.uk/~history/Mathematicians/Turing.html.

[35] Hodges, Andrew, *Alan Turing: The Enigma*, Simon and Schuster, New York, 1983, p. 26.

Figure 7.14 Alan Turing (1912–1954).

This says more about the failings of public schools than about Turing. The Headmaster seemed to think education meant becoming "familiar with the ideas of authority and obedience, of cooperation and loyalty, of putting the house and the school above your personal desires."[36] He was later to complain of Turing, who did not buy into this, "He should have more *esprit de corps*."[37] The spirit at Sherborne was perhaps summarized by Turing's form-master for the fall of 1927: "This room smells of mathematics! Go out and fetch a disinfectant spray!" Turing later observed that[38]

---

[36] Hodges, Andrew, *Alan Turing: The Enigma*, Simon and Schuster, New York, 1983, p. 22.
[37] Hodges, Andrew, *Alan Turing: The Enigma*, Simon and Schuster, New York, 1983, p. 24.
[38] Hodges, Andrew, *Alan Turing: The Enigma*, Simon and Schuster, New York, 1983, p. 381.

The great thing about a public school education is that afterwards, however miserable you are, you know it can never be quite so bad again.

Turing educated himself by reading Einstein's papers on relativity and Eddington's *The Nature of the Physical World*. He seems to have encountered one decent teacher before graduating and moving on to King's College, Cambridge, in 1931. This teacher wrote:[39]

All that I can claim is that my deliberate policy of leaving him largely to his own devices and standing by to assist when necessary, allowed his natural mathematical genius to progress uninhibited.

In 1933, Turing joined the Anti-War Council, telling his mother in a letter, "Its programme is principally to organize strikes amongst munitions and chemical workers when government intends to go to war. It gets up a guarantee fund to support the workers who strike."[40] The group also protested films such as *Our Fighting Navy*, which Turing called "blatant militarist propaganda."[41]

Turing graduated from King's College in 1934. In 1935, he attended a course that dealt with a result found by the Austrian logician Kurt Gödel and an open question that traced back to the German mathematician David Hilbert. Gödel had shown, in 1931, that in any axiomatic system sufficient to do arithmetic, there will always be statements that can be neither proven nor disproven. That is, such statements are independent of the axioms. This is very disappointing and it is known as Gödel's incompleteness theorem, because it shows that mathematics will always be incomplete in a sense. The question of decidability asked if there was a way, ideally an efficient way, to identify which statements fall into this unfortunate category. That is, can we decide whether or not a given statement is provable (in the system under consideration)? This was known as the decision problem, although more commonly referred to by its German name, the *Entscheidungsproblem*. It represented a generalization of Hilbert's 10th problem, from a famous talk he gave in 1900, in which he described the most important problems in mathematics for the coming century.

Turing began working on the *Entscheidungsproblem* at this time, but his dissertation was on another topic, the reasons behind why so many phenomena follow a Gaussian distribution. He proved the central limit theorem (seen in every college-level course in probability), but later learned that his proof was not the first. Although another proof had been published slightly earlier, Turing discovered his independently.

Turing then took on the decision problem in his landmark paper "On Computable Numbers, with an Application to the *Entscheidungsproblem*." It was submitted in 1936 and published in 1937. Like Gödel's incompleteness theorem, the answer was disappointing. Turing proved that there can be no general process for determining if a given statement is provable or not. And, again, Turing's proof was not the first. Alonzo Church had established this result shortly before him, with a different manner of proof.[42]

However, there is a silver lining. Turing's proof differed from Church's. It included a description of what is now known as a *Turing machine*. The theoretical machine read symbols from a tape

---

[39] Hodges, Andrew, *Alan Turing: The Enigma*, Simon and Schuster, New York, 1983, p. 32.

[40] Hodges, Andrew, *Alan Turing: The Enigma*, Simon and Schuster, New York, 1983, p. 71.

[41] Hodges, Andrew, *Alan Turing: The Enigma*, Simon and Schuster, New York, 1983, p. 87.

[42] Church's paper saw print in 1936. The citation is Church, Alonzo, "An Unsolvable Problem of Elementary Number Theory," *American Journal of Mathematics*, Vol. 58, No. 2, April 1936, pp. 345–363. Also see Church, Alonzo, "A Note on the *Entscheidungsproblem*," *The Journal of Symbolic Logic*, Vol. 1, No. 1, March 1936, pp. 40–41.

and could also delete or write symbols on the tape as well. A computable number was defined to be a real number whose decimal expansion could be produced by a Turing machine starting with a blank tape. Because only countably many real numbers are computable and there are uncountably many real numbers, there exists a real number that is not computable. This argument was made possible by Georg Cantor's work on transfinite numbers. Turing described a number which is not computable, remarking that this seemed to be a paradox, because he had apparently described in finite terms, a number that cannot be described in finite terms. The answer was that it is impossible to decide, using another Turing machine, whether a Turing machine with a given table of instructions will output an infinite sequence of numbers.[43] The Turing machine provides a theoretical foundation for modern computers. Thus, it's one of the key steps leading to the information age.

Turing also discussed a universal machine in his 1936 paper. It is a machine:[44]

> ... which can be made to do the work of any special-purpose machine, that is to say to carry out any piece of computing, if a tape bearing suitable "instructions" is inserted into it.

Turing then began studying at Princeton. *Systems of Logic Based on Ordinals* (1939) was the main result of this work.

Figure 7.15 The mansion at Bletchley Park. (Creative Commons Attribution-Share Alike 4.0 International license, by Wikipedia user DeFacto, https://en.wikipedia.org/wiki/File:Bletchley_Park_Mansion.jpg)

When war was declared in 1939, Turing moved to the Government Code and Cypher[45] School (GCCS) at Bletchley Park (Figure 7.15).[46] Bletchley eventually grew to employ about 10,000 people. The vast majority were women.[47] His work there in breaking some of the German ciphers,

[43] http://www-groups.dcs.st-and.ac.uk/~history/Mathematicians/Turing.html.

[44] Newman, Maxwell Herman Alexander, "Alan Mathison Turing, 1912-1954," *Biographical Memoirs of Fellows of the Royal Society of London*, Vol. 1, November 1955, pp. 253–263, pp. 257–258 quoted here, available online at https://royalsocietypublishing.org/doi/pdf/10.1098/rsbm.1955.0019.

[45] This is how the British spell Cipher.

[46] Also known as Station X, as it was the tenth site acquired by MI-6 for its wartime operations.

[47] Smith Michael, *The Emperor's Codes: The Breaking of Japan's Secret Ciphers*, Penguin Books, New York, 2002, p. 2.

provided the Allies with information that saved many lives. It has been estimated that this shortened the war by about two years.[48] It was also a lot of fun. Turing remarked,[49]

> Before the war my work was in logic and my hobby was cryptanalysis and now it is the other way round.

The cryptanalysts at Bletchley Park had been working on cracking Enigma prior to learning of the successes of the Polish mathematicians. They had been stymied by the wiring from the plugboard to the rotor assembly, never having considered that it simply took each letter to itself. Despite his earlier impressive academic work, not even Turing considered this possibility. His colleague and fellow mathematician, Gordon Welchman, was furious upon learning how simple the answer was.

Turing redesigned the Polish bomba, and then Welchman made further improvements. The machine was now known as a *bombe*. This seems to have simply been a modification of the Polish name, but nobody knows why the Poles chose the name in the first place. According to one of several stories, it was because it ticked while in operation, like a time bomb.[50] The ticking stopped when a potential solution arose.

Stories of Turing's eccentricities circulated at Bletchley; for example, beginning each June, he would wear a gas mask while bicycling. This was to keep the pollen out; he suffered from hay fever. The bicycle was also unusual. Periodically a bent spoke would touch a particular link and action would have to be taken to prevent the chain from coming off. Turing kept track of how many times the bicycle's wheel had turned and would stop the bike and reset the chain, before it came off.[51]

Still, Turing and the other eccentrics at Bletchley were very good at what they did. By 1942, the deciphered messages totaled around 50,000 per month;[52] however, in February 1942, the German Navy put a four-rotor Enigma into use. This caused the decipherments to come to an immediate halt and the results were bloody.

Kahn convincingly illustrated the importance of Enigma decipherments in the Atlantic naval war by comparing sinkings of Allied ships in the second half of 1941, when Enigma messages were being read, with the second half of 1942, when the messages went unsolved. The respective figures for tons sunk are 600,000 vs. 2,600,000.[53] Kahn also puts a human face on these figures:[54]

> And each of the nearly 500 ships sunk in those six months meant more freezing deaths in the middle of the ocean, more widows, more fatherless children, less food for some toddler, less ammunition for some soldier, less fuel for some plane—and the prospect of prolonging these miseries.

Reading Enigma messages also allowed the allies to sink Axis convoys in greater numbers. Rommel was greatly handicapped in Africa by the lack of much needed gasoline due to these sinkings. At

[48] Smith Michael, *The Emperor's Codes: The Breaking of Japan's Secret Ciphers*, Penguin Books, New York, 2002, p. 3.

[49] Hodges, Andrew, *Alan Turing: The Enigma*, Simon and Schuster, New York, 1983, pp. 214–215.

[50] There are several different stories given to explain why the machines were called bombes. There is no consensus as to which is correct.

[51] Hodges, Andrew, *Alan Turing: The Enigma*, Simon and Schuster, New York, 1983, p. 209.

[52] Hodges, Andrew, *Alan Turing: The Enigma*, Simon and Schuster, New York, 1983, p. 237.

[53] Kahn, David, *Seizing the Enigma*, Houghton Mifflin Company, Boston, Massachusetts, 1991, pp. 216–217.

[54] Kahn, David, *Seizing the Enigma*, Houghton Mifflin Company, Boston, Massachusetts, 1991, p. 217.

one point he sent a sarcastic message of thanks to Field-Marshal Kesselring for the few barrels that washed ashore from wrecks of a convoy.[55]

7.8 The Lorenz Cipher and Colossus

Enigma wasn't the only German cipher machine. Others, called Lorenz SZ 40 and SZ 42, came into use in 1943 (Figures 7.16 and 7.17).[56] The messages these carried were even higher level than Enigma traffic. The British referred to both the Lorenz machines and their ciphertexts as *Tunny*, in accordance with their convention of using names of fish for various ciphers. In America, this particular fish is called "tuna."

Figure 7.16   Former National Cryptologic Museum curator Patrick Weadon with a Lorenz machine. This machine dwarfs the M-209 and is also significantly larger than the Enigma.

For a description of how the Lorenz SZ machines operated and how the British cryptanalysts broke them, the reader is referred to items listed in the References and Further Reading portion of this chapter. Due to its great historical importance, I will mention that the cryptanalysis of these ciphers marked a key moment in the history of computer science.

Colossus, shown in Figure 7.18, was constructed by Tommy Flowers, a telegraph engineer, to defeat the Lorenz ciphers. It saw action by D-Day, June 6, 1944 and is now considered the first programmable electronic computer. It was long believed that ENIAC was first, as the existence of Colossus remained classified until the 1970s! A Colossus has since been rebuilt, by a team led by Tony Sale, and is on display at Bletchley Park.[57]

---

[55] Winterbotham, Frederick William, *The Ultra Secret*, Harper & Row, New York, 1974, p. 82 (p. 122 of the paperback edition).

[56] The "SZ" in the name is short for *Schlüsselzusatz*, which translates to "cipher attachment." These machines were attached to Lorenz teleprinters.

[57] See http://www.codesandciphers.org.uk/lorenz/rebuild.htm for more information.

Figure 7.17 Lorenz SZ 42 cipher machine. (http://en.wikipedia.org/wiki/File:Lorenz-SZ42-2. jpg)

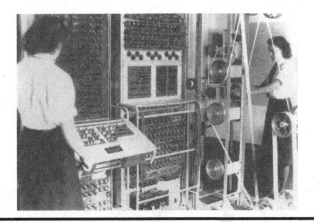

Figure 7.18 Nazi ciphers being broken by the first programmable computer, Colossus. (https:// en.wikipedia.org/wiki/Colossus_computer#/media/File:Colossus.jpg.)

Bletchley also has a bombe that was rebuilt by a team led by John Harper. Again, no originals survived in England to the present day. One U.S. Navy bombe, built in Dayton, Ohio, was preserved and is now on display at the National Cryptologic Museum.

7.9 What If Enigma Had Never Been Broken?

So, how great of a difference did all of the allied codebreaking make?

> I feel that intelligence was a vital factor in the Allied victory – I think that without it we might not have won, after all.

> **—Harold Deutsch[58]**

---

[58] Kahn, David, "The Significance of Codebreaking and Intelligence in Allied Strategy and Tactics," *Cryptologia*, Vol. 1, No. 3, July 1977, pp. 209–222, p. 221 cited here.

There are other experts who share Deutsch's opinion. Among experts who believe we would have won without the cryptanalysts, the consensus is that the victory would have arrived at least a year later, perhaps two. Certainly, a tremendous number of lives were saved by the work of the cryptanalysts.

These cryptanalytic heroes had to wait in silence for decades before their contributions could be revealed. Eventually some honors were bestowed upon them. We saw the Polish postage stamp earlier in this chapter. Other countries eventually issued special stamps, as well. In addition to a monument in Poland (Figure 7.19), there are cryptologic museums in England and America. This is good. We should pay our respects in this manner, but this is not the reward of the cryptanalysts. Look closely the next time you see a crowd of people at a sporting event or at a shopping center and know that many of them would never have existed, if the war had lasted a year or two longer. How many of our parents, grandparents, and great-grandparents made it back from the war, but wouldn't have, if it had gone on longer? How many of us have forefathers who weren't called to duty because they were a little bit too young? How many more hospital beds would be filled with the horribly wounded? How many more Holocaust victims would there have been? The reward of the cryptanalysts is pride in knowing their work made an incredible difference in the lives of those around them.

Figure 7.19 Monument in Poznan, Poland, honoring Rejewski, Różycki, and Zygalski. (From Grajek, Marek, "Monument *in Memoriam* of Marian Rejewski, Jerzy Różycki and Henryk Zygalski Unveiled in Poznań," *Cryptologia*, Vol. 32, No. 2, April 2008, pp. 101-103, p. 103 cited here.)

7.10 Endings and New Beginnings

At the end of World War II, the Allies hoped to capture German cryptologic equipment as well as the codemakers and codebreakers themselves. The results turned out to be better than they could have anticipated. The men in question had buried some equipment, fearing the possibility of capture by Soviet troops; however, upon seeing that the Allied troops arrived first, they quickly revealed the location of the equipment and began digging. It was literally buried treasure. The Germans demonstrated how the recovered equipment could be used to crack Soviet ciphers. Thus, the United States began the Cold War with a great cryptologic advantage. This story remained secret even longer than the breaking of Enigma. It was revealed at last in Thomas Parrish's *The Ultra Americans* in 1986.[59]

The fact that Enigma was broken was called the Ultra secret and it remained a secret for about 30 years following the war. One reason for maintaining secrecy long after the Nazis were defeated was to keep alive the illusion that machine ciphers were unbreakable. The techniques used could continue to be applied to those same machines, now in the hands of smaller nations. Also, some of the cryptanalytic techniques used on Enigma can be generalized to other machines. These machines, once made in huge numbers are, today, valuable collectors' items. Of the original 40,000 Enigmas, only 400 some are known to still exist.[60] Several may be seen (and used!) at the National Cryptologic Museum.[61]

This chapter opened with a quote from Ben Franklin. The chapter itself is the best counterexample I have. Over 10,000 people were in on the conspiracy of silence concerning the breaking of Enigma and they maintained their silence for decades. A common reaction to conspiracy theories is "There's no way they could keep something like that secret." Yes, there is.

After the war, Turing had more time to devote to exercise. He was a member of Walton Athletic Club and won their three-mile and ten-mile championship in record time.[62] He was unable to try out for the Olympic team due to an injury. The marathon that year was won by an Argentinian with a time 17 minutes faster than what Turing could do.[63] How did he learn to run so fast? Well, in grade school:[64]

> …he hated and feared the gym class and the afternoon games. The boys played hockey in winter, and Alan later claimed that it was the necessity of avoiding the ball that had taught him to run fast.

In 1950, Turing's paper "Computing Machinery and Intelligence" was published in *Mind*. The question as to whether or not a machine could be considered intelligent was posed in this paper.

---

[59] Parrish, Thomas, *The Ultra Americans*, Stein and Day, Briarcliff Manor, New York, 1986. A trade paperback edition of this book published in 1991 by Scarborough House, Chelsea, Michigan, bears the title The *American Codebreakers*.

[60] The best list of extant Enigmas is available online at http://enigmamuseum.com/dhlist.xls.

[61] More information on the National Cryptologic Museum can be found at http://www.nsa.gov/about/cryptologic_heritage/museum/.

[62] http://www-groups.dcs.st-and.ac.uk/~history/Mathematicians/Turing.html.

[63] Hodges, Andrew, *Alan Turing: The Enigma*, Simon and Schuster, New York, 1983, p. 386.

[64] Hodges, Andrew, *Alan Turing: The Enigma*, Simon and Schuster, New York, 1983, p. 11.

The Turing test was introduced, in which questions are posed and the examiner is to decide whether a person or a machine is answering him.[65] Turing wrote[66]

> …I believe that at the end of the century the use of words and general educated opinion will have altered so much that one will be able to speak of machines thinking without expecting to be contradicted.

These ideas must have seemed absurd to many in 1950 when "it was not unknown for computer users to be sweltering in 90° F heat, and banging the racks with a hammer to detect loose valves."[67]

In 1952, Turing, attempting to evade blackmail, gave the police details of a homosexual affair he had. He was charged with "gross indecency" for violating the British homosexuality statutes. On the advice of his counsel, he decided to not offer a defense and pled guilty. He likely made a defense that he saw nothing wrong in his actions, but not to the judge. Turing was found guilty and given a choice of prison or a year of estrogen treatment (as a subcutaneous implant). He took the estrogen treatment. After this incident, some speculate that Turing was viewed as a security risk. His clearance was withdrawn and his foreign colleagues were investigated. Hodge's biography of Turing indicates that Turing was fairly open about his sexuality, although Turing's nephew, Sir Dermot Turing, states that Alan was only open about it in safe company, with those who were very close to him. Alan's older brother John had no idea for 40 years.[68] It is possible that some of the powers that be knew Turing was gay, but chose to ignore the fact, because he was so useful in the war effort.

Turing died in 1954 of potassium cyanide poisoning. He had been conducting electrolysis experiments and cyanide was found on a half-eaten apple beside him. It was ruled a suicide, but Turing's mother maintained that it was an accident.[69]

In September 2009, Gordon Brown, the Prime Minister of England, offered a formal apology to Turing. While this pleased many, mathematics professor Steve Kennedy had a different reaction:[70]

> I didn't feel elated and I wondered if there was something wrong with me. Oh sure, I recognized that this was a good and necessary step, but I couldn't help but feel that it was not proportionate. The British government, in the name of the British people, tortured this good and decent man (and thousands of others) because they disapproved of his sexual habits. Now, half a century later, they offer only words of regret. Maybe I'd feel better if Gordon Brown vowed not to rest until gay marriage was legal in Britain. Of course in Britain today the legal status of gays is a thousand times better than in the US, so maybe Brown could work on educating America? How about passing a heavy tax—the Turing Tariff—on all computing hardware and software imported into the UK from countries, like the US, that still discriminate against homosexuals by banning gay marriage? The proceeds of the tariff could be donated, in the name of Alan Mathison Turing, to the leading gay rights organizations in the exporting country.

---

[65] Philip K. Dick made use of this test in his novel *Do Androids Dream of Electric Sheep?*, which was later made into the film *Bladerunner*. The book is, of course, superior. There are also three humorous installments of the comic strip "Dilbert" that deal with Turing tests. They can be viewed online at http://search.dilbert.com/comic/Turing%20Test.

[66] http://www-groups.dcs.st-and.ac.uk/~history/Quotations/Turing.html.

[67] Hodges, Andrew, *Alan Turing: The Enigma*, Simon and Schuster, New York, 1983, p. 402.

[68] Turing, Sir Dermot, email to the author, June 22, 2018.

[69] http://www-groups.dcs.st-and.ac.uk/~history/Mathematicians/Turing.html.

[70] Kennedy, Steve, "A Politician's Apology," *Math Horizons*, Vol. 17, No. 2, November 2009, p. 34.

Kennedy knew his suggestions wouldn't be adopted, but his point was well made. Actions speak louder than words.

I'll try to lighten the mood a little bit by closing this chapter with an anecdote. This and several other humorous incidents are recounted in Neal Stephenson's *Cryptonomicon*. Although it is a work of fiction, many real events have been incorporated into the book. Alan Turing is one of the characters in the book and it is obvious that Stephenson read Hodge's biography of Turing and that it had a big influence on parts of the novel.

At one point during the war, Turing enrolled in the infantry section of the Home Guard. He had to complete forms to do this. One question asked, "Do you understand that by enrolling in the Home Guard you place yourself liable to military law?" Turing saw no advantage in writing yes, so he wrote no. Of course, nobody ever looked closely at the form. He was taught how to shoot, became very good at it, and then no longer having any use for the Home Guard, moved on to other things. He was eventually summoned to explain his absence. Turing explained that he only joined to learn how to shoot and that he was not interested in anything else. He did not want, for example, to attend parades. The conversation progressed as follows:

> "But it is not up to you whether to attend parades or not. When you are called on parade, it is your duty as a soldier to attend."
>
> "But I am not a soldier."
>
> "What do you mean, you are not a soldier! You are under military law!"
>
> "You know, I rather thought this sort of situation could arise. I don't know I am under military law. If you look at my form you will see that I protected myself against this situation."

He was right and nothing could be done.

References and Further Reading

Note: Cryptologia has published hundreds of articles on Enigma. Only a few of these are listed below.

Bertrand, Gustave, *Enigma*, Plon, Paris, 1973. This is a pre-Winterbotham revelation that the allies broke Nazi ciphers during World War II. It's written in French, from the French perspective.

Brown, Anthony Cave, *Bodyguard of Lies*, Harper & Row, New York, 1975.

Budiansky Stephen, *Battle of Wits: The Complete Story of Codebreaking in World War II*, The Free Press, New York, 2000

Christensen, Chris, "Polish Mathematicians Finding Patterns in Enigma Messages," *Mathematics Magazine*, Vol. 80, No. 4, October 2007, pp. 247–273. This paper won the Carl B. Allendoerfer Award in 2008.

Clayton, Mike, "Letter Repeats in Enigma Ciphertext Produced by Same-Letter Keying," *Cryptologia*, Vol. 43, No. 5, September 2019, pp. 438–457.

Copeland, B. Jack, editor, *The Essential Turing*, Clarendon Press, Oxford, 2004.

Davies, Donald W., "The Lorenz Cipher Machine SZ42," *Cryptologia*, Vol. 19, No. 1, January 1995, pp. 39–61, reprinted in Deavours, Cipher A., David Kahn, Louis Kruh, Greg Mellen, and Brian J. Winkel, editors, *Selections from Cryptologia: History, People, and Technology*, Artech House, Boston, Massachusetts, 1998, pp. 517–539.

Evans, N. E., "Air Intelligence and the Coventry Raid," *Royal United Service Institution Journal* (The RUSI Journal), Vol. 121, No. 3, 1976, pp. 66–74.

Girard, Daniel J., "Breaking *"Tirpitz"*: Cryptanalysis of the Japanese-German Joint Naval Cipher," *Cryptologia*, Vol. 40, No. 5, September 2016, pp. 428–451.

Grajek, Marek, "Monument in Memoriam of Marian Rejewski, Jerzy Różycki and Henryk Zygalski Unveiled in Poznań," *Cryptologia*, Vol. 32, No. 2, April 2008, pp. 101–103.

Grey, Christopher, "From the Archives: Colonel Butler's Satire of Bletchley Park," *Cryptologia*, Vol. 38, No. 3, July 2014, pp. 266–275.

Hinsley, Francis Harry and Alan Stripp, editors, *Codebreakers: The Inside Story of Bletchley Park*, Oxford University Press, Oxford, 1993.

Hodges, Andrew, *Alan Turing: The Enigma*, Simon and Schuster, New York, 1983. Hodges, able to relate to Turing on many levels, is an ideal biographer. Like Turing, Hodges is a mathematician, a homosexual, an atheist, and British. More mathematical detail is provided in this biography than one could fairly expect from anything written by a non-mathematician. If you want to learn more about Turing's life and work, start here. The 2014 film *The Imitation Game*[71] was based on this book, but it introduced many errors that Hodges would never have made.

Kahn, David, *Seizing the Enigma*, Houghton Mifflin Company, Boston 1991.

Kahn, David, "An Enigma Chronology," *Cryptologia*, Vol. 17, No. 3, July 1993, pp. 237–246. This is a very handy reference for anyone wishing to keep the dates and details straight when writing on or speaking about Enigma.

Kenyon, David and Frode Weierud, "Enigma G: The Counter Enigma," *Cryptologia*, Vol. 44, No. 5, September 2020, pp. 385–420.

Körner, Thomas William, *The Pleasure of Counting*, Cambridge University Press, Cambridge, UK, 1996. Part IV of this book takes a look at Enigma cryptanalysis.

Kuhl, Alex, "Rejewski's Catalog," *Cryptologia*, Vol. 31, No. 4, October 2007, pp. 326–331. This paper won one of *Cryptologia's* undergraduate paper competitions.

Lasry, George, Nils Kopal, and Arno Wacker, "Ciphertext-only Cryptanalysis of Hagelin M-209 Pins and Lugs," *Cryptologia*, Vol. 40, No. 2, March 2016, pp. 141–176.

Lasry, George, Nils Kopal, and Arno Wacker, "Automated Known-Plaintext Cryptanalysis of Short Hagelin M-209 Messages," *Cryptologia*, Vol. 40, No. 1, pp. 49–69.

Lasry, George, Nils Kopal, and Arno Wacker, "Ciphertext-only Cryptanalysis of Short Hagelin M-209 Ciphertexts," *Cryptologia*, Vol. 42, No. 6, November 2018, pp. 485–513.

Lasry, George, Nils Kopal, and Arno Wacker, "Cryptanalysis of Enigma Double Indicators with Hill Climbing," *Cryptologia*, Vol. 43, No. 4, July 2019, pp. 267–292.

List, David and John Gallehawk, "Revelation for Cilli's," *Cryptologia*, Vol. 38, No. 3, July 2014, pp. 248–265.

Marks, Philip, "Enigma Wiring Data: Interpreting Allied Conventions from World War II," *Cryptologia*, Vol. 39, No. 1, January 2015, pp. 25–65.

Marks, Philip, "Mr. Twinn's bombes," *Cryptologia*, Vol. 42, No. 1, January 2018, pp. 1–80.

Miller, Ray, *The Cryptographic Mathematics of Enigma*, revised edition, Center for Cryptologic History, National Security Agency, Fort George G. Meade, Maryland, 2019. This is available at no charge at the National Cryptologic Museum in print form, as well as online at https://tinyurl.com/y635qumc.

Muggeridge, Malcolm, *Chronicles of wasted Time. Chronicle 2: The Infernal Grove*, Collins, London, 1973. This is a pre-Winterbotham revelation that the allies broke Nazi ciphers during World War II.

Ostwald, Olaf and Frode Weierud, "History and Modern Cryptanalysis of Enigma's Pluggable Reflector," *Cryptologia*, Vol. 40, No. 1, January 2016, pp. 70–91.

Ostwald, Olaf and Frode Weierud, "Modern Breaking of Enigma Ciphertexts," *Cryptologia*, Vol. 41, No. 5, September 2017, pp. 395–421.

Parrish, Thomas, *The Ultra Americans, Stein and Day*, Briarcliff Manor, New York, 1986. A trade paperback edition of this book published in 1991 by Scarborough House, Chelsea, Michigan, bears the title The *American Codebreakers*.

Randell Brian, *The Colossus*, Technical Report Series No. 90, Computing Laboratory, University of Newcastle upon Tyne, 1976.

Randell, Brian, "Colossus: Godfather of the Computer," *New Scientist*, Vol. 73, No. 1038, February 10, 1977, pp. 346–348.

Rejewski, Marian, "How Polish Mathematicians Deciphered the Enigma," *Annals of the History of Computing*, Vol. 3, No. 3, July 1981, pp. 213–234.

---

[71] https://www.imdb.com/title/tt2084970/?ref_=fn_al_tt_1.

Rejewski, Marian, "Mathematical Solution of the Enigma Cipher," *Cryptologia*, Vol. 6, No. 1, January 1982, pp. 1–18.

Rejewski, Marian, *Memories of My Work at the Cipher Bureau of the General Staff Second Department 1930-1945*, Adam Mickiewicz University, Poznań, Poland, 2011.

Sebag-Montefiore, Hugh, *Enigma: The Battle for the Code*, The Folio Society, London, UK, 2005. This book was first published by Weidenfeld & Nicolson, London, UK, 2000.

Stevenson, William, *A Man Called Intrepid*, Harcourt Brace Jovanovich, New York, 1976. A paperback edition is from Ballantine Books, New York, 1977.

Sullivan, Geoff, Geoff's Crypto page, http://www.hut-six.co.uk/. This page has links to emulators for various cipher machines, including several versions of the Enigma.

Teuscher, Christof, editor, *Alan Turing: Life and Legacy of a Great Thinker*, Springer, New York, 2004.

Thimbleby, Harold, "Human Factors and Missed Solutions to Enigma Design Weaknesses," *Cryptologia*, Vol. 40, No. 2, March 2016, pp. 177–202.

Turing, Sara, *Alan M. Turing*, W. Heffer & Sons, Ltd., Cambridge, UK, 1959.

Turing, Sara, *Alan M. Turing: Centenary Edition*, Cambridge University Press, Cambridge, UK, 2012. This special edition of the long out-of-print title listed above commemorates what would have been Alan Turing's 100th birthday in 2012. It includes a new foreword by Martin Davis and a never-before-published memoir by John F. Turing, Alan's older brother.

Vázquez, Manuel and Paz Jiménez-Seral, "Recovering the Military Enigma Using Permutations–Filling in the Details of Rejewski's Solution," *Cryptologia*, Vol. 42, No. 2, March 2018, pp. 106–134.

Weierud, Frode and Sandy Zabell, "German Mathematicians and Cryptology in WWII," *Cryptologia*, Vol. 44, No. 2, March 2020, pp. 97–171.

Welchman, Gordon, *The Hut Six Story*, McGraw-Hill, New York, 1982.

Wik, Anders, "Enigma Z30 Retrieved," *Cryptologia*, Vol. 40, No. 3, May 2016, pp. 215–220.

Winterbotham, Frederick William, *The Ultra Secret*, Harper & Row, New York, 1974.

> Despite what you might read elsewhere, this was not the first public revelation that the allies had broken Nazi ciphers during World War II. Winterbotham claimed that Ultra revealed that the Germans would be bombing Coventry, but Churchill declined an evacuation order for fear that the Germans would take it as a sign that the Brits had inside information, perhaps from a compromised cipher! If the Germans replaced Enigma or modified it in such a way as to shut out the codebreakers, the loss would far exceed the damage in terms of lives and material sacrificed at Coventry. This claim was supported by Anthony Cave Brown (*Bodyguard of Lies*) and William Stevenson (*A Man Called Intrepid*) but has not stood up to the scrutiny of other historians.[72] Brown knew that Enigma had been broken before Winterbotham revealed it, but he did not get his own book out until after *The Ultra Secret*.[73]

Wright, John, "Rejewski's Test Message as a Crib," *Cryptologia*, Vol. 40, No. 1, January 2016, pp. 92–106.

Wright, John, "A Recursive Solution for Turing's H-M Factor," *Cryptologia*, Vol. 40, No. 4, July 2016, pp. 327–347.

Wright, John, "The Turing Bombe *Victory* and the First Naval Enigma Decrypts," *Cryptologia*, Vol. 41, No. 4, July 2017, pp. 295–328.

Wright, John, "Rejewski's Equations: Solving for the Entry Permutation," *Cryptologia*, Vol. 42, No. 3, May 2018, pp. 222–226.

Video

There are many videos, aimed at a general audience, describing World War II codebreaking. Just one of these is singled out here, for it contains information on the American construction of bombes (Figure 7.20) in Dayton, Ohio, an important topic mentioned only very briefly in this book. The focus of the video is on

---

[72] See Evans, N. E., "Air Intelligence and the Coventry Raid," *Royal United Service Institution Journal*, September 1976, pp. 66–73 for an early refutation.

[73] See Parish, Thomas, *The Ultra Americans*, Stein and Day, Briarcliff Manor, New York, 1986, p. 287.

Figure 7.20 René Stein, former National Cryptologic Museum librarian, in front of an American-made bombe that is preserved in the museum's collection.

Joe Desch, who was responsible for much of the success in Dayton. Decades after carrying out his wartime work, Desch was inducted into NSA's Hall of Honor.

Dayton Codebreakers, The Dayton Codebreakers Project, 2006, 56 minutes. See http://www.daytoncode-breakers.org/, a website maintained by Desch's daughter, Debbie Anderson, for much more information. Ordering instructions for the DVD can be found at https://daytoncodebreakers.org/video/dvds/.

Equipment

An Enigma machine made with modern technology (Figure 7.21) shows the positions of the simulated rotors digitally. These are available for sale in kit form—some assembly required!

Figure 7.21 Enigma machine made with modern technology. (Simons, Marc and Paul Reuvers, Crypto Museum, http://www.cryptomuseum.com/kits/.)

Chapter 8

Cryptologic War against Japan

This chapter examines how American cryptanalysts broke Japanese diplomatic ciphers during World War II, while some of the United States' own communications were protected by being "enciphered" using the natural language of the Navajo, as spoken by the Indians themselves, with some code words mixed in. But before we get to these topics, we take a look at a potential case of steganography.

8.1 Forewarning of Pearl Harbor?

The November 22, 1941 issue of *The New Yorker* had a strange advertisement that appeared repeatedly (on 14 different pages!). It is reproduced in Figure 8.1. Look closely. Do you notice anything interesting?

Figure 8.1   Advertisement appearing in *The New Yorker*, November 22, 1941.

Flipping ahead to page 86 of the magazine, we find the rest of the ad (Figure 8.2).

Figure 8.2   Continuation of advertisement appearing in *The New Yorker,* November 22, 1941.

Beneath this second image is the following text and image (see also Figure 8.3).

> We hope you'll never have to spend a long winter's night in an air-raid shelter, but we were just thinking… it's only common sense to be prepared. If you're not too busy between now and Christmas, why not sit down and plan a list of the things you'll want to have on hand. …Canned goods, of course, and candles, Sterno, bottled water, sugar, coffee or tea, brandy, and plenty of cigarettes, sweaters and blankets, books or magazines, vitamin capsules… and though it's no time, really, to be thinking of what's fashionable, we bet that most of your friends will remember to include those intriguing dice and chops which make Chicago's favorite game.

THE
DEADLY DOUBLE

**$2.50 at leading Sporting Goods
and Department Stores Everywhere**

Figure 8.3   Image appearing below advertisement appearing in *The New Yorker,* November 22, 1941.

Did you notice that the dice show 12-07 (December 7)? Was it meant to serve as a warning? Perhaps for Japanese living in America? Federal Bureau of Investigation (FBI) agent Robert L. Shivers wondered the same thing and on January 2, 1942 sent a radiogram to FBI Director J. Edgar Hoover posing the question. Other inquiries followed from government employees, as well as private citizens. The investigation that followed revealed that it was simply a coincidence.[1]

8.2 Friedman's Team Assembles

Sometimes it's easy to see hidden messages where none exist, but in the years leading up to Pearl Harbor, both American and British cryptanalysts had been working hard to crack the very real ciphers produced by the new machines that the Japanese were now using. The story of these machines and the American team that defeated them begins in 1930.

On April 1 of that year, mathematician Frank Rowlett reported to Washington, DC, for his first day on the job as a "junior cryptanalyst," despite having no idea what a cryptanalyst was. He was William Friedman's first hire in that capacity for the Signals Intelligence Section of the Army Signal Corps. Friedman gave him books on cryptology in German and French to read, with the help of dictionaries, while waiting for two other new hires, Abraham Sinkov and Solomon Kullback, to arrive later in the month. The job continued to be unlike any other, as Rowlett found himself studying cryptology in a third-floor vault, almost half of which was filled with filing cabinets he was not allowed to open (Major Crawford told him, "I'll have you shot next morning at sunrise if I catch you near those file cabinets."[2]). Friedman later gave Rowlett some English-language materials to read, including the Riverbank publications. Between them, the three mathematicians had some knowledge of German, French, and Spanish, but no mathematician with expertise in Japanese could be found. Friedman finally hired John B. Hurt, who was not a mathematician, but had an impressive command of the Japanese language, despite never having been there.

One morning in late June, Friedman led the three mathematicians to a second-floor room with a steel door like the one on the vault they worked in, secured by a combination lock. After opening it, Friedman pulled out a key and unlocked another door behind it. Finally he lit a match so he could find the pull cord for the ceiling light in what was then revealed to be a windowless room approximately 25 feet square and nearly filled completely with filing cabinets. Friedman then said theatrically, "Welcome, gentlemen to the secret archives of the American Black Chamber." The three mathematicians had no idea what he was talking about.[3] This room marked the new location of Herbert Yardley's defunct Cipher Bureau. Henry Stimson thought he shut it down, but it had merely been placed under new management, as it were. Yardley's work on Japanese codes and ciphers was now at the disposal of the Army cryptanalysts.

[1] Kruh, Louis, "The Deadly Double Advertisements - Pearl Harbor Warning or Coincidence?" *Cryptologia*, Vol. 3, No. 3, July 1979, pp. 166–171.

[2] Rowlett, Frank B., The *Story of Magic: Memoirs of an American Cryptologic Pioneer*, Aegean Park Press, Laguna Hills, California, 1989, p. 17.

[3] Rowlett, Frank B., The *Story of Magic: Memoirs of an American Cryptologic Pioneer*, Aegean Park Press, Laguna Hills, California, 1989, pp. 6–35.

8.3 Cryptanalysis of Red, a Japanese Diplomatic Cipher

Around the time Rowlett, Sinkov, Kullback, and Hurt began working for Friedman, the Japanese Foreign Office put *Angooki Taipu A* (type A cipher machine)[4] into use. The American codebreakers called it "Red."[5] Other Japanese ciphers received color code names, as well. Rather than explain how it worked, we play the role of the cryptanalyst who has received an intercept in this new system. To make things a bit easier, the underlying message will be in English—our example is actually a training exercise the British used during World War II.[6]

Ciphertext:

```
EBJHE  VAWRA  UPVXO  VVIHB  AGEWK  IKDYU  CJEKB  POEKY  GRYLU
PPOPS  YWATC  CEQAL  CULMY  CERTY  NETMB  IXOCY  TYYPR  QYGHA
HZCAZ  GTISI  YMENA  RYRSI  ZEMFY  NOMAW  AHYRY  QYVAB  YMYZQ
YJIIX  PDYMI  FLOTA  VAINI  AQJYX  RODVU  HACET  EBQEH  AGZPD
UPLTY  OKKYG  OFUTL  NOLLI  WQAVE  MZIZS  HVADA  VCUME  XKEHN
PODMA  EFIMA  ZCERV  KYXAV  DCAQY  RLUAK  ORYDT  YJIMA  CUGOF
GUMYC  SHUUR  AEBUN  WIGKX  YATEQ  CEVOX  DPEMT  SUCDA  RPRYP
PEBTU  ZIQDE  DWLBY  RIVEM  VUSKK  EZGAD  OGNNQ  FEIHK  ITVYU
CYBTE  LWIBC  IROQX  XIZIU  GUHOA  VPAYK  LIZVU  ZBUWY  ZHWYL
YHETG  IANWZ  YXFSU  ZIDXN  CUCSH  EFAHR  OWUNA  JKOLO  QAZRO
QKBPY  GIKEB  EGWWO  LRYPI  RZZIB  YUHEF  IHICE  TOPRY  ZYDAU
BIDSU  GYNOI  NQKYX  EFOGT  EOSZI  PDUQG  UZUZU  LVUUI  YLZEN
OWAAG  WLIDP  UYXHK  OHZYI  BKYFN
```

In this ciphertext, A, E, I, O, U, and Y make up 39% of the total and their distribution in the ciphertext resembles that of vowels in plaintext. Hence, it appears that the encryption process takes vowels to vowels and consonants to consonants, but both the frequencies of individual letters and the index of coincidence indicate that it is not simply a monoalphabetic substitution cipher. After some further study, we find what appear to be isomorphs (identical plaintexts enciphered differently)[7]:

```
VXOVVIHBAGEWKIKD      SKKEZGADOGNN      BYRIVEM     QXXIZIUG
LNOLLIWQAVEMZIZS      GWWOLRYPIRZZ      PYGIKEB     RZZIBYUH
PRYPPEBTUZIQDEDW
```

---

[4] Also known as *91-shiki-obun In-ji-ki*, which translates to "alphabetical typewriter 91." The 91 refers to the year it was first developed, 2591 by the Japanese reckoning, 1931 for the Americans.

[5] Originally, it was referred to in conversation and official reports as "A Machine," but Friedman realized this was poor security, as it was so close to the Japanese name for the device. The color code name was settled on after much discussion, and the other colors of the spectrum were assigned to other machines, including Enigma, Kryha, and Hagelin. (From Kruh, Louis, "Reminiscences of a Master Cryptologist," *Cryptologia*, Vol. 4, No. 1, January 1980, pp. 45–50, p. 49 cited here.) Also, the binders in which information on these messages were kept were red.

[6] Found, along with the solution I provide, in Deavours, Cipher and Louis Kruh, *Machine Cryptography and Modern Cryptanalysis*, Artech House, Inc., Dedham, Massachusetts, 1985, pp. 213–215.

[7] In general, two strings of characters are isomorphic if one can be transformed into the other via a monoalphabetic substitution. Isomorph attacks have been used against various other cipher systems, including the Hebern machine. See Deavours, Cipher A., "Analysis of the Hebern Cryptograph Using Isomorphs," *Cryptologia*, Vol. 1, No. 2, April 1977, pp. 167–185.

We list the consonants, because they appear to be enciphered separately from the group of vowels:

BCDFGHJKLMNPQRSTVWXZ (20 consonants)

Looking at the longest set of three isomorphs, we make some interesting observations.

    1.   LNOLLIWQAVEMZIZS
            PRYPPEBTUZIQDEDW

From any consonant in the top line to the one directly beneath it is distance 3 in our consonant alphabet. This is not likely to be a coincidence!

    2.   VXOVVIHBAGEWKIKD
            LNOLLIWQAVEMZIZS

For this pair, the distance is always 12. We start, for example, at V and move forward through the consonants W, X, and Z, and then start back at the beginning of the alphabet and continue until we arrive at L. Again, this is not likely to be a coincidence!

This attack, which should remind you of Kasiski's attack on the Vigenère cipher (but using isomorphs instead of identical ciphertext segments), indicates that the basic consonant substitution alphabet is simply shifted to create the various encipherment possibilities. Thus, our substitution table should look something like this:

	BCDFGHJKLMNPQRSTVWXZ	**Plaintext**
1.	BCDFGHJKLMNPQRSTVWXZ	**Cipher Alphabet 1**
2.	CDFGHJKLMNPQRSTVWXZB	**Cipher Alphabet 2**
3.	DFGHJKLMNPQRSTVWXZBC	⋮
4.	FGHJKLMNPQRSTVWXZBCD	
5.	GHJKLMNPQRSTVWXZBCDF	
6.	HJKLMNPQRSTVWXZBCDFG	
7.	JKLMNPQRSTVWXZBCDFGH	
8.	KLMNPQRSTVWXZBCDFGHJ	
9.	LMNPQRSTVWXZBCDFGHJK	
10.	MNPQRSTVWXZBCDFGHJKL	
11.	NPQRSTVWXZBCDFGHJKLM	
12.	PQRSTVWXZBCDFGHJKLMN	
13.	QRSTVWXZBCDFGHJKLMNP	
14.	RSTVWXZBCDFGHJKLMNPQ	
15.	STVWXZBCDFGHJKLMNPQR	
16.	TVWXZBCDFGHJKLMNPQRS	
17.	VWXZBCDFGHJKLMNPQRST	
18.	WXZBCDFGHJKLMNPQRSTV	
19.	XZBCDFGHJKLMNPQRSTVW	
20.	ZBCDFGHJKLMNPQRSTVWX	

However, our results above don't specify that we start with what is labeled Cipher Alphabet 1. We may start anywhere in this table. We can simply try each of the 20 possible start positions for the first line, as if we were breaking a Caesar shift cipher by brute force. We delete the artificial spacing in groups of five to save space and progress through our 20 cipher alphabets one at a time with each letter of text.

Starting cipher:

```
    EBJHEVAWRAUPVXOVVIHBAGEWKIKDYUCJEKBPOEKYGRYLU
 1  -ZGD-P-NH--BGH-CB-KC-F-SF-CT--PT-SJV--M-GQ-H-
 2  -XFC-N-MG--ZFG-BZ-JB-D-RD-BS--NS-RHT--L-FP-G-
 3  -WDB-M-LF--XDF-ZX-HZ-C-QC-ZR--MR-QGS--K-DN-F-
 4  -VCZ-L-KD--WCD-XW-GX-B-PB-XQ--LQ-PFR--J-CM-D-
 5  -TBX-K-JC--VBC-WV-FW-Z-NZ-WP--KP-NDQ--H-BL-C-
 6  -SZW-J-HB--TZB-VT-DV-X-MX-VN--JN-MCP--G-ZK-B-
 7  -RXV-H-GZ--SXZ-TS-CT-W-LW-TM--HM-LBN-F-XJ-Z-
 8  -QWT-G-FX--RWX-SR-BS-V-KV-SL--GL-KZM--D-WH-X-
 9  -PVS-F-DW--QVW-RQ-ZR-T-JT-RK--FK-JXL--C-VG-W-
10  -NTR-D-CV--PTV-QP-XQ-S-HS-QJ--DJ-HWK--B-TF-V-
11  -MSQ-C-BT--NST-PN-WP-R-GR-PH--CH-GVJ--Z-SD-T-
12  -LRP-B-ZS--MRS-NM-VN-Q-FQ-NG--BG-FTH--X-RC-S-
13  -KQN-Z-XR--LQR-ML-TM-P-DP-MF--ZF-DSG--W-QB-R-
14  -JPM-X-WQ--KPQ-LK-SL-N-CN-LD--XD-CRF--V-PZ-Q-
15  -HNL-W-VP--JNP-KJ-RK-M-BM-KC--WC-BQD--T-NX-P-
16  -GMK-V-TN--HMN-JH-QJ-L-ZL-JB--VB-ZPC--S-MW-N-
17  -FLJ-T-SM--GLM-HG-PH-K-XK-HZ--TZ-XNB--R-LV-M-
18  -DKH-S-RL--FKL-GF-NG-J-WJ-GX--SX-WMZ--Q-KT-L-
19  -CJG-R-QK--DJK-FD-MF-H-VH-FW--RW-VLX--P-JS-K-
20  -BHF-Q-PJ--CHJ-DC-LD-G-TG-DV--QV-TKW--N-HR-J-
```

Now there's a small surprise—we can't simply read across any of the lines, as we could normally do if one of them were a plaintext only missing the vowels! Double checking reveals no errors were made. Line 10 starts out promising, but then fizzles out. Taking a closer look at line 10, along with nearby lines shows that the word INTRODUCTION seems to begin on line 10, but break off and continue on line 11.

```
 7  -RXV-H-GZ--SXZ-TS-CT-W-LW-TM--HM-LBN-F-XJ-Z-
 8  -QWT-G-FX--RWX-SR-BS-V-KV-SL--GL-KZM-D-WH-X-
 9  -PVS-F-DW--QVW-RQ-ZR-T-JT-RK--FK-JXL-C-VG-W-
10  -NTR-D-CV--PTV-QP-XQ-S-HS-QJ--DJ-HWK-B-TF-V-
11  -MSQ-C-BT--NST-PN-WP-R-GR-PH--CH-GVJ-Z-SD-T-
12  -LRP-B-ZS--MRS-NM-VN-Q-FQ-NG--BG-FTH-X-RC-S-
13  -KQN-Z-XR--LQR-ML-TM-P-DP-MF--ZF-DSG-W-QB-R-
```

Another shift occurs later from line 11 to line 12. These shifts are referred to as a *stepping action*. The mechanics of the Red cipher, which will be detailed momentarily, makes how this happens clearer.

Filling in vowels and word breaks is now easy:

INTRODUCTION STOP NEW PARAGRAPH EACH OF THE EXERCISE(S)...

The word STOP shows the message to be in the style of telegraph traffic. We were even able to complete the last word, which requires one more letter of ciphertext than was provided. One can now go back and see that the vowel substitutions were made using the mixed order alphabet AOEUYI with various shifts, like the consonant alphabet.

Enciphering vowels as vowels and consonants as consonants is clearly a disadvantage. The motivation for the Japanese to do this was apparently purely economic, as cable services charged a lower rate for transmitting text that could be pronounced.

The Japanese word for "and" is *oyobi*, which has the unusual pattern vowel, vowel, vowel, consonant, vowel; thus, with vowels enciphered as vowels and consonants enciphered as consonants, the cryptanalysts could easily locate this word in a ciphertext and then note whatever it revealed of the encryption process.

The British cracked Red in November 1934 and built a machine to simulate it in August 1935. U.S. Army cryptanalysts broke it in late 1936 and took two more years to make a simulator of their own.[8] The key insight for the Americans was made by Frank Rowlett, working with Solomon Kullback.

In the example above, our familiar 26-letter alphabet was used. Although the actual intercepts were in Japanese, this was the case for those ciphertexts as well. Because the traditional manner of writing Japanese is not well suited for electronic transmission, the words were first written out using the Latin alphabet,[9] then enciphered, and finally transmitted.

One needn't know how the machine actually accomplishes the encipherment in order to break such messages. In fact, there is more than one possible way to realize Red encryption electromechanically. The Japanese did it with two half-rotors, driven by a gearwheel. Typically, the rotor unit moved forward by one position for each letter, thus advancing through the alphabets, but there were pins that could be removed, causing the rotor unit to advance one position more at that point. This is the mechanism responsible for the stepping action we observed in our example above. Two adjacent pins may be removed to cause an advance of three positions. The American version of Red, used to recover plaintexts for intercepted messages, did so with two rotors.[10]

The sample ciphertext analyzed above was simplified by ignoring another component of Red: the plugboard. This had the effect of performing a monoalphabetic substitution on both the plaintext and cipher alphabets in our Vigenère tableaus. The plugboard connections changed daily. This "daily sequence" as it came to be known had to be recovered as part of the cryptanalytic process. As with Enigma, changes to this machine and how it was used were made over its time of service. The above is only intended to convey the general idea.

8.3.1 Orange

The Americans used the code name Orange for a variant of Red used as the Japanese naval attaché machine. This machine enciphered *kana* syllables, rather than *romaji* letters. The first breaks for the Americans came in February 1936.[11] Lieutenant Jack S. Holtwick, Jr. of the U.S. Navy made a machine that broke this naval variant.[12] The British also attacked this system and had results even earlier. Hugh Foss and Oliver Strachey broke it for the British in November 1934.[13] The

---

[8] Smith, Michael, *The Emperor's Codes: The Breaking of Japan's Secret Ciphers*, Penguin Books, New York, 2002, pp. 46–47.

[9] The Japanese called this *romaji*. It is still used.

[10] Deavours, Cipher and Louis Kruh, *Machine Cryptography and Modern Cryptanalysis*, Artech House, Inc., Dedham, Massachusetts, 1985, The American Red machine is pictured on pages 216–217.

[11] Smith, Michael, *The Emperor's Codes: The Breaking of Japan's Secret Ciphers*, Penguin Books, New York, 2002, p. 35.

[12] Kahn, David, *The Codebreakers*, second edition, Scribner, New York, pp. 20 and 437. Also see Deavours, Cipher and Louis Kruh, *Machine Cryptography and Modern Cryptanalysis*, Artech House, Inc., Dedham, Massachusetts, 1985, p. 11.

[13] Smith, Michael, *The Emperor's Codes: The Breaking of Japan's Secret Ciphers*, Penguin Books, New York, 2002, pp. 34–35.

successes were only temporary, however. The Japanese would eventually switch over to a more secure machine, *Angooki Taipu B* (type B cipher machine).[14]

8.4 Purple—How It Works

The American cryptanalysts continued their use of colors as code names for the various cipher machines, but by the time Japan introduced *Angooki Taipu B*, the seven colors of the rainbow (Red, Orange, Yellow, Green, Blue, Indigo, Violet)[15] had all been used. The new machine cipher became known as Purple, which seemed the most appropriate choice at the time, but more colors would soon be needed. Japan's Purple cipher was a machine, so in this sense it was similar to both Red and Enigma. However, the inner workings of Purple are completely different from both, as you will see. So, how exactly did Purple work? Figure 8.4 will help to clarify the description that follows.

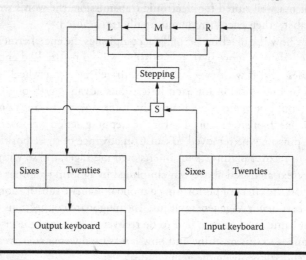

Figure 8.4   Purple schematics.

The characters typed are immediately permuted by a plugboard, which then separates them into two groups of unequal size—the sixes[16] and the twenties. If a letter is among the sixes, it follows one of 25 paths through S, each of which is a permutation (among these six letters). The paths are cycled through in order (controlled by a telephone switch, *not a rotor*, as in Red and Enigma). The result is then fed through another plugboard to get the output.

If the letter typed does not become one of the sixes, it follows a path through R, M, and L before returning to the plugboard on the left side of the schematic. R, M, and L each contain 25 permutations of the alphabet, but these are not simply cycled through in order; rather, a stepping mechanism determines the rotation. The stepping mechanism takes its cue from S. The details follow, but we first note that R, M, and L will differ in that one will switch slowly, another quickly and the third at an intermediate (medium) value. This is in regards to how long it takes to cycle

---

[14] Also known as *97-shiki-obun In-ji-ki*, which translates to "alphabetical typewriter 97." The 97 refers to the year it was first developed, 2597 by the Japanese reckoning, 1937 for the Americans.

[15] Of course, this division of the rainbow into seven colors in arbitrary. That particular value was chosen by Isaac Newton to create a nice symmetry with the seven-note musical scale (A, B, C, D, E, F, G).

[16] Unlike Red, the sixes in Purple weren't necessarily vowels; they could be any six letters and they changed daily. Actually, this innovation was one of the changes made in Red usage before its demise.

through all 25 permutations, not the speed of a single switch, which is the same for all. The speeds for the switches are part of the key.

If we label the permutations for S, R, M, and L as 0 through 24, we then have:

If S is in position 23 and M is in position 24, the slow switch advances.
If S is in position 24, the medium switch advances.
In all other cases, the fast switch advances.

The period for Purple is a bit shorter than for the Enigma, which cycled through 16,900 permutations before returning to the first. Purple cycled through $25^3 = 15,625$. Of course, the sixes for Purple had a much shorter cycle of 25.

As with the Enigma, the plugboard settings were determined in advance and the same for all users on a particular day; however, the Japanese limited themselves to 1,000 different connections (out of a possible 26!; the connections needn't be reciprocal, as with Enigma).[17] Like Enigma, part of the Purple key for any given message was to be determined at random and sent as a header along with the ciphertext. This was a five-digit number, selected initially from a list of 120 (later on the list offered 240 choices), which stood for the initial positions of each switch and the stepping switch. There was no mathematical relation between the five-digit number and the setting it stood for. It was simply a code number. For example, 13579 meant the sixes switch starts at position 3 and the twenties switches start at positions 24, 8, and 25, while the 20-switch motion (assignment of speeds slow, medium, and fast to particular switches) is 3-2-1.[18] Of the 120 five digit codes, half consisted only of odd digits and half only of even digits. It was a great mistake for the Japanese to artificially restrict themselves to this tiny fraction of the initial settings. With 25 possible settings for each of the twenties and the sixes, we have $25^4 = 390,625$ possibilities.

The Japanese did make some wise choices in how Purple was used. The first was enciphering the five digit key using an additive. Thus, messages using the same key would not bear identical indicators. Another good decision was encoding messages, prior to running them through Purple, but they did this with the commercially available Phillips Code—not the best choice![19]

Because a large keyspace is essential for security, we should address the overall keyspace for Purple. A variety of values may be given, depending on how we perform the calculation. Determining the keyspace of Purple requires that several subjective choices be made; for example, should one use all possible plugboard configurations as a factor, or limit it to the 1,000 that were actually used? Should all possible initial positions of the switches be considered or only the 120 (later 240) the Japanese allowed? Should we count all possible internal wirings? The cryptanalysts had to determine these, but the Japanese couldn't change them once the machines were assembled, so one could say they weren't part of the key. Thus, it's not surprising that a range of values for the keyspace can be found in the literature. Stephen J. Kelley offers a figure of 1.8×10^{138}, which is much larger than for the three rotor Enigma.[20] Mark Stamp, more conservative in assigning values to various factors, came up with $2^{198} \approx 4 \times 10^{59}$ as the keyspace.[21] Quite a difference! However,

---

[17] Smith, Michael, *The Emperor's Codes: The Breaking of Japan's Secret Ciphers*, Penguin Books, New York, 2002, p. 68.

[18] Freeman, Wes, Geoff Sullivan, and Frode Weierud, "Purple Revealed: Simulation and Computer-Aided Cryptanalysis of *Angooki Taipu B*," *Cryptologia*, Vol. 27, No. 1, January 2003, pp. 1–43, p. 38 cited here.

[19] Smith, Michael, *The Emperor's Codes: The Breaking of Japan's Secret Ciphers*, Penguin Books, New York, 2002, p. 68.

[20] Kelley, Stephen J., *Big Machines*, Aegean Park Press, Laguna Hills, California, 2001, p. 178

[21] Stamp, Mark and Richard M. Low, *Applied Cryptanalysis: Breaking Ciphers in the Real World*, John Wiley & Sons, Hoboken, New Jersey, 2007, pp. 44–45.

even this lower value was sufficiently large to avoid a brute force attack. So, the keyspace was definitely large enough. This was not why Purple ultimately proved insecure.

8.5 Purple Cryptanalysis

As with many classical ciphers, a frequency analysis of Purple ciphertexts provides useful information to get the cryptanalysis started. The six letters that are separated from the rest will each have a frequency about equal to the average of that group, because, over the course of the 25 substitutions, each letter is about equally likely to be replaced by any letter in the sixes. In the same manner, every letter in the group of twenty will be substituted for, on average, equally often by each letter in the twenties. Thus, every letter in this group will also have about the same frequency in the ciphertext.

As an experiment, I placed one Scrabble tile for each letter in a container, shook it, and then selected six. The six, in alphabetical order were A, E, J, L, V, and Z; thus, the set happened to contain both the most frequent and the rarest English letters. According to the frequency table from Section 1.9, the average frequency of the letters in this group is 4.3666. For the remaining 20 letters, the average frequency is 3.7, for a difference of 0.666.

This is how we distinguish the sixes from the twenties. After calculating the frequencies, we either split off the six most frequent or the six least frequent letters to be the sixes (whichever set seems to stand out most strongly from the other frequencies) and assume these are, in fact, the sixes. The remaining letters should be the twenties.

Rowlett's team had the advantage of having broken Red prior to attacking Purple, so the 6-20 split was familiar. The sixes being easier, with a period of only 25, were recovered first. The twenties were more difficult, but a number of factors worked to the cryptanalysts' advantage.

Switching from Red to Purple was a good move for the Japanese, but they made the move in the worst possible way—slowly! Ideally, there would be an abrupt change, with everyone shifting to the new çipher at, for example, the stroke of midnight on a particular date. The slow change made by the Japanese over many months meant that many messages sent out in Purple were also sent out in Red to locations that hadn't yet made the transition. Because Red was already broken, this allowed the cryptanalysts to compare those plaintexts with the corresponding ciphertexts in the Purple system. In other words, they had cribs in abundance! Messages that weren't also sent in Red were cribbed with stereotyped phrases and the numbering of messages used by the Japanese. Finally, the Japanese sometimes sent enciphered copies of U.S. State Department communiqués for which the cryptanalysts had no trouble obtaining plaintexts.[22]

Leo Rosen, an electrical engineering major from the Massachusetts Institute of Technology (MIT) and part of the cryptanalytic team, made an important discovery in September 1939. While looking at an electrical supply catalog, he recognized that telephone stepping switches could have been used to accomplish the substitutions. There was no commercially available cipher machine that worked in this manner, so it was not an obvious solution!

Even with Rosen's insight and the help inadvertently supplied by the Japanese, recovering the twenties with their large period of $25^3 = 15,625$ would have been difficult; however, Genevieve Grotjan recognized that there was a relationship to be found at shorter intervals of just 25 letters. Some illustrations of reconstructed substitution alphabets make her finding clearer.

---

[22] Smith, Michael, *The Emperor's Codes: The Breaking of Japan's Secret Ciphers*, Penguin Books, New York, 2002, pp. 70–71.

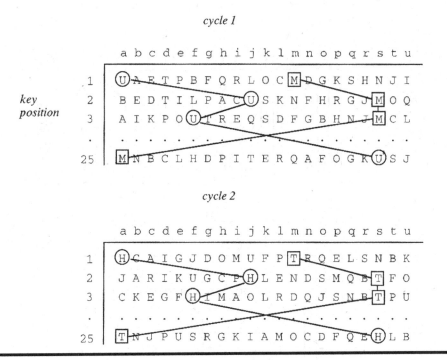

Figure 8.5 A pattern in the Purple alphabets. (From Budiansky, Stephen, *Battle of Wits: The Complete Story of Codebreaking In World War II*, **Free Press, New York, 2000, p. 354. With permission.)**

The example in Figure 8.5 shows one of the patterns in the Purple alphabets. Suppose the first alphabet enciphers A as U, then when we come to the 26th alphabet (i.e., alphabet 1 of cycle 2), A is enciphered as H. Then whatever is enciphered as U in alphabets 2 through 25 will be enciphered as H in alphabets 2 through 25 of cycle 2. The same goes for all other letters. So, knowing alphabets 1 through 25 of cycle 1 and the first alphabet of cycle 2 reveals the remaining 24 alphabets of cycle 2. In real cryptanalysis, the process wouldn't be so orderly. The cycles would get filled in like a jigsaw puzzle with some values in each cycle known from cribs allowing corresponding values to be filled in other cycles.

Another pattern Grotjan recognized (Figure 8.6) is even easier to grasp and exploit for cryptanalytic purposes. The columns in the substitution table for cycle 2 are identical to those of cycle 1, but reordered. Which pattern held for a particular message depended on the location of the slow switch during that message's encipherment. Grotjan first found a pattern on September 20, 1940, and Purple was completely broken on September 27, 1940. Thus ended 18 months of intense work. Images of Rowlett, Rosen, and Grotjan are provided in Figure 8.7.

Following this feat, Friedman had a nervous breakdown that required several months off and Leo Rosen led the construction of a "Purple analog" to simulate the Japanese machine (Figure 8.8).

A later American version is shown in Figure 8.9. These recreations were actually better than the originals! They used brass for some contact points that, in the Japanese version, were made of copper, which would wear with use and then tend to produce garbled text.[23]

---

[23] Lewin, Ronald, *The American Magic: Codes, Ciphers and the Defeat of Japan*, Farrar Straus Giroux, New York, p. 40 footnote.

cycle 1

	a	b	c	d	e	f	g	h	i	j	k	l	m	n	o	p	q	r	s	t	u
1	U	A	E	T	P	B	F	Q	R	L	O	C	M	D	G	K	S	H	N	J	I
2	B	E	D	T	I	L	P	A	C	U	S	K	N	F	H	R	G	J	M	O	Q
3	A	I	K	P	O	U	T	R	E	Q	S	D	F	G	B	H	N	J	M	C	L
.	.	.	.	.	.	.	.	.	.	.	.	.	.	.	.	.	.	.	.	.	.
25	M	N	B	C	L	H	D	P	I	T	E	R	Q	A	F	O	G	K	U	S	J

key position

cycle 2

	a	b	c	d	e	f	g	h	i	j	k	l	m	n	o	p	q	r	s	t	u
1	A	P	R	I	J	E	L	D	F	U	T	S	O	Q	M	K	B	N	G	C	H
2	E	I	C	Q	O	D	U	F	P	B	T	G	S	A	N	R	L	M	H	K	J
3	I	O	E	L	C	K	Q	G	T	A	P	N	S	R	F	H	U	M	B	D	J
.	.	.	.	.	.	.	.	.	.	.	.	.	.	.	.	.	.	.	.	.	.
25	N	L	I	J	S	B	T	A	D	M	C	G	E	P	Q	O	H	U	F	R	K

Figure 8.6 Another pattern in the Purple alphabets. (From Budiansky, Stephen, *Battle of Wits: The Complete Story of Codebreaking In World War II*, Free Press, New York, 2000, p. 355. With permission.)

Figure 8.7 Frank Rowlett (http://www.fas.org/irp/agency/inscom/journal/98-oct-dec/article6_1.jpg); Leo Rosen (courtesy of René Stein, former National Cryptologic Museum Librarian); and Genevieve Grotjan (http://www.nsa.gov/about/cryptologic_heritage/women/honorees/feinstein.shtml).

Figure 8.8 **A Purple analog built in 1940 without seeing the original. (Courtesy of the National Security Agency, https://web.archive.org/web/20160325223400/http://www.nsa.gov/about/_images/pg_hi_res/purple_analog.jpg.)**

Figure 8.9 **René Stein, former National Cryptologic Museum librarian, stands by the back of a Purple Analog from 1944.**

8.6 Practical Magic

The collection of deciphered Japanese intercepts was codenamed Magic. Magic was in operation before Pearl Harbor was attacked by Admiral Yamamoto's forces. A question that has been long debated is whether or not such decipherments should have given us advance warning of the attack. The consensus view is that they didn't provide a detailed warning. There was no Purple message saying that Pearl Harbor, specifically, would be attacked, or even that an attack was imminent. Still, there was enough, when the intercepts were taken in the context of the times, to generate a general warning that would include Pearl Harbor. Recall, Purple was a diplomatic cipher, not military. Perfect understanding of the Japanese Naval code could have provided a timely warning, but that particular code, JN-25, was only partially broken at that time. David Kahn examined the Pearl Harbor issue in great detail in the first chapter of *The Codebreakers* and the reader is referred there for more information. Cryptanalytic intelligence proved extremely valuable in the years following Pearl Harbor. A few key episodes are detailed, in chronological order, in the following paragraphs.

The Americans eventually had success in breaking the Japanese naval codes JN-25 as well as subsequent codes such as JN-25a and JN-25b. A JN-25b message decoded on May 14, 1942 provided warning of a large invasion force heading to AF, where AF represented a still unbroken portion of the message. Although there was no definite proof, Navy cryptanalyst Joseph Rochefort believed that AF stood for Midway Island. He was supported in this by Admiral Nimitz, but back in Washington, DC, the Aleutian Islands were believed to be the target. To test his conjecture, Rochefort had a message sent in plain language[24] from Midway saying that their desalination plant had broken. Following this, a Japanese message encoded in JN-25b was soon intercepted that included the unknown code group AF. After plugging in the known code groups and translating, the message stated that AF was short of water. Thus, the conjecture was proven.[25] Yamamoto was indeed planning to invade Midway, and when he did on June 4, the U.S. Navy was ready. The battle of Midway proved to be the turning point of the Pacific war. Prior to Midway, the United States had never had a victory, and following Midway, she never saw defeat.

Yamamoto suffered a much more personal defeat when a message encoded with the latest version of JN-25, and sent on April 13, 1943, provided the American codebreakers with his itinerary for the next five days. This itinerary brought him close to the combat zone and Nimitz made the decision to attempt to shoot down his plane. It was risky, for if the Japanese realized how the U.S. Navy knew where to find Yamamoto, their naval codes would be promptly changed. On April 18, Major John Mitchell led a formation of 16 P-38 Lightnings, which included four ace pilots. They carried out their mission successfully. Because of the tremendous importance of Admiral Yamamoto to the Japanese, David Kahn described his assassination as "the equivalent of a major victory."[26]

The decoded message that resulted in Yamamoto's death was translated by Marine Corps Captain Alva Lasswell. Despite being a farmboy from Arkansas with only an eighth-grade education, Lasswell, as a Marine, traveled to Japan, where he mastered the language. Tom "Captain T" Hunnicutt commemorated Lasswell's accomplishments by dubbing him "The Sigint Sniper" in a song of the same title.[27]

Reading Purple kept the Allies informed of German plans as well as Japanese. Ōshima Hiroshi, the Japanese military attaché in Berlin, sent detailed messages concerning Nazi plans back to Tokyo. Carl Boyd devoted a book to the study of intelligence about The Nazis obtained from Hiroshi.[28] A few paragraphs are reproduced below from messages Hiroshi sent on November 10, 1943 (nearly six months before the June 6, 1944, D-Day invasion) enciphered with Purple.[29]

> All of the German fortifications on the French coast are very close to the shore and it is quite clear that the Germans plan to smash any enemy attempt to land as close to the edge of the water as possible.
>
> The Strait of Dover area is given first place in the German Army's fortification scheme and troop dispositions, and Normandy and the Brittany peninsula come next. Other parts of the coast are regarded with less significance. Although the possibility

---

[24] In some accounts, the message was enciphered, but in a weak system, that it was assumed the Japanese could break.

[25] Kahn, David, *The Codebreakers*, second edition, Scribner, New York, 1996, p. 569.

[26] Kahn, David, *The Codebreakers*, second edition, Scribner, New York, 1996, pp. 595–601.

[27] "Sigint Sniper (The Yamamoto Shoot Down)," The Hunnicutt Collection, White Swan Records, ASCAP Music, 2009.

[28] Boyd, Carl, *Hitler's Japanese Confidant: General Ōshima Hiroshi and Magic Intelligence, 1941–1945*, University of Kansas Press, Lawrence, Kansas, 1993.

[29] All excerpts reproduced from Boyd, pp.186–189.

of an Anglo-American landing on the Iberian Peninsula, followed by a crossing of the Pyrenees, is not altogether ruled out, no special defences have been constructed for this area. Instead, mobile forces are held in reserve at Narbonne and other strategic points, and these are prepared to hold the mountain passes in the Pyrenees should an emergency arise.

The coastal defense divisions are distributed as follows:

    a. Netherlands Defense Army (covering the Netherlands down to the mouth of the Rhine)-4 divisions.
    b. 15th Army (covering the area extending from the mouth of the Rhine to West of Le Havre)-9 divisions.
    c. 7th Army (extending thence to the southern bank of the Loire)-8 divisions.
    d. 1st Army (extending thence to the Spanish border)-4 divisions.
    e. 19th Army (covering the French Mediterranean coast)-6 divisions.

With messages containing details like these, Hiroshi unknowingly became a great friend of the allies!

It should be stressed again that the American cryptanalysts did their work without seeing any parts of the Japanese cipher machines until after the war. Because it was a purely mathematical analysis, many regard this work as even more impressive than the cryptanalysis of Enigma.

Pictured in Figure 8.10 is the largest of three surviving pieces of a Purple machine. It was found in the destroyed Japanese Embassy in Berlin in 1945. It may now be seen at the National

Figure 8.10 Part of a Japanese Purple machine. (Courtesy of the National Security Agency, http://www.nsa.gov/about/_images/pg_hi_res/purple_switch.jpg.)

Cryptologic Museum adjacent to Fort Meade, Maryland. In the photograph behind the Purple fragment, Hiroshi may be seen shaking hands with Hitler.

President Truman awarded William Friedman the Medal of Merit, which was the highest Presidential civilian award. Friedman retired in 1955 and died on November 12, 1969. He's buried in Arlington National Cemetery.[30] Frank Rowlett received the National Security Medal from President Johnson.

It has been estimated that cryptanalysis saved a year of war in the Pacific.

—**David Kahn**[31]

Just as in the Enigma chapter, there is much more to this story of cryptanalysis. The Japanese also had a Green machine, which was "a rather strangely constructed version of the commercial Enigma machine,"[32] and variants of Purple, codenamed Coral and Jade.

There was overlap between the American and British cryptanalytic efforts, but a basic division of labor emerged with the British attacking the codes and ciphers of the Germans, while the Americans focused on those of the Japanese. The two nations shared their results in what began as an uneasy relationship between intelligence agencies. This is discussed in greater detail in Section 12.10.

8.7 Code Talkers

Were it not for the Navajos, the marines would never have taken Iwo Jima!

—**Major Howard M. Conner**[33]

It's been claimed that the Japanese would be the world's worst foreign language students if it weren't for the Americans. So perhaps it isn't surprising that the Americans have repeatedly used foreign languages as codes and that they had their most famous success against the Japanese. The story of such codes goes back to the U.S. Civil War, when the North used Hungarians to befuddle the South.

In the last month of World War I, Native Americans began serving as code talkers, but as in the Civil War, this language code did not play a major role. Only eight Choctaws were initially used as radio operators. They were in Company D, 141st Infantry, under Captain E. W. Horner in Northern France.[34] Before World War I ended, the number of Choctaw code talkers grew to fifteen.[35] They were able to use their own language to communicate openly without fear of the enemy understanding. The effort was successful. As Colonel A. W. Bloor of the 142nd Infantry put it, "There was hardly one chance in a million that Fritz would be able to translate these dialects."[36]

---

[30] http://www.sans.org/infosecFAQ/history/friedman.htm.

[31] Kahn, David, *The Codebreakers*, second edition, Scribner, New York, 1996, p. xi.

[32] Deavours, Cipher and Louis Kruh, *Machine Cryptography and Modern Cryptanalysis*, Artech House, Inc., Dedham, Massachusetts, 1985, p. 212.

[33] Paul, Doris A., *The Navajo Code Talkers*, Dorrance Publishing Co., Inc., Pittsburgh, Pennsylvania, 1973, p. 73.

[34] Singh, Simon, *The Code Book*, Doubleday, New York, 1999, pp. 194–195.

[35] Meadows, William C., *The Comanche Code Talkers of World War II*, University of Texas Press, Austin, Texas, 2002, p. 18.

[36] Meadows, William C., *The Comanche Code Talkers of World War II*, University of Texas Press, Austin, Texas, 2002, p. 20.

Other Native American tribes who contributed to the code war in World War I, with their native tongues, included the Comanche and Sioux.[37]

Some problems arose from the lack of necessary military terms in the language. A few Indian words were applied to these terms, but they were few in number and did not cover everything that was needed. Marine TSgt. Philip Johnston found a way around this difficulty in time for World War II.[38] By adopting code words for terms not provided for in the Indian language, and spelling out other needed words or names or locations that arose and weren't in the code, the code talkers could have even greater success. Johnston, the son of a missionary, had lived on a Navajo reservation for 22 years beginning at age 4; thus, he learned the language, which he recognized as being very difficult.

Johnston's reasons for suggesting use of Navajos weren't solely due to his own experience. He pointed out that, despite their low literacy rate compared with other tribes, the sheer size of the Navajo Nation, at nearly 50,000 people (more than twice the size of any other tribe at that time), would make it easier to recruit the desired numbers. Another advantage of using Navajo was expressed by Major General Clayton B. Vogel.

> Mr. Johnston stated that the Navajo is the only tribe in the United States that has not been infested with German students during the past twenty years. These Germans, studying the various tribal dialects under the guise of art students, anthropologists, etc., have undoubtedly attained a good working knowledge of all tribal dialects except Navajo. For this reason the Navajo is the only tribe available offering complete security for the type of work under consideration.[39]

But would the idea work? The Germans weren't the only enemy who had attempted to study the Indian languages between the wars. Some Japanese had been employed by the Indian Affairs Bureau. And why would the Navajo, who faced countless agonies at the hands of the white men, including attempted genocide, be willing to help? This is an obvious question that many modern authors attempt to answer. I'll let the Navajos answer for themselves. The following is a resolution passed unanimously by the Navajo Tribal Council at Window Rock on June 3, 1940:

> Whereas, the Navajo Tribal Council and the 50,000 people we represent, cannot fail to recognize the crisis now facing the world in the threat of foreign invasion and the destruction of the great liberties and benefits which we enjoy on the reservation, and
>
> Whereas, there exists no purer concentration of Americanism than among the first Americans, and
>
> Whereas, it has become common practice to attempt national destruction through the sowing of seeds of treachery among minority groups such as ours, and
>
> Whereas, we hereby serve notice that any un-American movement among our people will be resented and dealt with severely, and
>
> Now, Therefore, we resolve that the Navajo Indians stand ready as they did in 1918, to aid and defend our Government and its institutions against all subversive and armed conflict and pledge our loyalty to the system which recognizes minority rights and a way of life that has placed us among the greatest people of our race.[40]

---

[37] Meadows, William C., *The Comanche Code Talkers of World War II*, University of Texas Press, Austin, Texas, 2002, p. 29.

[38] Johnston came up with his idea in late December 1941 (after Pearl Harbor).

[39] Paul, Doris A., *The Navajo Code Talkers*, Dorrance Publishing Co., Inc., Pittsburgh, Pennsylvania, 1973, p. 157.

[40] Paul, Doris A., *The Navajo Code Talkers*, Dorrance Publishing Co., Inc., Pittsburgh, Pennsylvania, 1973, pp. 2–3.

Eventually, 540 Navajo served as Marines, of which 420 were code talkers.[41]

The lack of Navajo words for needed military terms was remedied by the creation of easily remembered code words. A tank would be referred to as a TORTOISE, an easy code word to remember because of the tortoise's hard shell. Planes became BIRDS, and so on. The basic idea was Johnston's, but the Navajo came up with the actual code words themselves.

Words that were not part of the code, such as proper names and locations, or anything else that might arise, could be spelled out. Initially the alphabet consisted of one Navajo word for each letter, but as the enemy might catch on when words with easily recognized patterns of letters, such as GUADALCANAL, were spelled out, alternate representations of frequent letters were soon introduced.[42] The expanded alphabet is reproduced in Table 8.1 along with a handful of the 411 code words. The complete list of code words may be found in various books on the Navajo Code Talkers, as well as online at https://web.archive.org/web/20130329065820/http://www.history.navy.mil/faqs/faq61-4.htm, which was the source used here.

Even with the initial code (prior to its expansion), tests showed that Navajo who were not familiar with the code words couldn't decipher the messages.[43] An important feature of using a natural language was increased speed. There was no lengthy process of looking up code groups in order to recover the original messages. A significant savings in time (minutes instead of hours) yielded a combat advantage to the American troops. The expanded code was even faster, as there were fewer delays due to having to spell out words not in the Navajo language or the code. Also, both the original and expanded code caused fewer errors than traditional codes. It did cause some confusion though, for allies not in on the secret. When the Navajo first hit the combat airwaves in Guadalcanal, some of the other American troops thought it was the Japanese broadcasting.

With regard to the Navajo role at Iwo Jima (Figure 8.11), Major Conner had this to say:[44]

> The entire operation was directed by Navajo code. Our corps command post was on a battleship from which orders went to the three division command posts on the beachhead, and on down to the lower echelons. I was signal officer of the Fifth Division. During the first forty-eight hours, while we were landing and consolidating our shore positions, I had six Navajo radio nets operating around the clock. In that period alone they sent and received over eight hundred messages without an error.

The "without an error" portion of the quote above is not something that was taken for granted in World War II-era coded transmissions. When Leo Marks began his cryptographic work for Britain's Special Operations Executive (SOE) in 1942, about 25% of incoming messages from their agents couldn't be read for one reason or another.

---

[41] Paul, Doris A., *The Navajo Code Talkers*, Dorrance Publishing Co., Inc., Pittsburgh, Pennsylvania, 1973, p. 117.

[42] This idea was due to Captain Stilwell, a cryptographer. See Paul, Doris A., *The Navajo Code Talkers*, Dorrance Publishing Co., Inc., Pittsburgh, Pennsylvania, 1973, p. 38.

[43] Paul, Doris A., *The Navajo Code Talkers*, Dorrance Publishing Co., Inc., Pittsburgh, Pennsylvania, 1973, p. 30.

[44] Paul, Doris A., *The Navajo Code Talkers*, Dorrance Publishing Co., Inc., Pittsburgh, Pennsylvania, 1973, p. 73.

Table 8.1 Navajo Code Talkers' Dictionary. (Revised 15 June 1945, and Declassified under Department of Defense Directive 5200.9)

Letter	Navajo Word	Translation	Letter	Navajo Word	Translation
A	WOL-LA-CHEE	Ant	K	KLIZZIE-YAZZIE	Kid
A	BE-LA-SANA	Apple	L	DIBEH-YAZZIE	Lamb
A	TSE-NILL	Axe	L	AH-JAD	Leg
B	NA-HASH-CHID	Badger	L	NASH-DOIE-TSO	Lion
B	SHUSH	Bear	M	TSIN-TLITI	Match
B	TOISH-JEH	Barrel	M	BE-TAS-TNI	Mirror
C	MOASI	Cat	M	NA-AS-TSO-SI	Mouse
C	TLA-GIN	Coal	N	TSAH	Needle
C	BA-GOSHI	Cow	N	A-CHIN	Nose
D	BE	Deer	O	A-KHA	Oil
D	CHINDI	Devil	O	TLO-CHIN	Onion
D	LHA-CHA-EH	Dog	O	NE-AHS-JAH	Owl
E	AH-JAH	Ear	P	CLA-GI-AIH	Pant
E	DZEH	Elk	P	BI-SO-DIH	Pig
E	AH-NAH	Eye	P	NE-ZHONI	Pretty
F	CHUO	Fir	Q	CA-YEILTH	Quiver
F	TSA-E-DONIN-EE	Fly	R	GAH	Rabbit
F	MA-E	Fox	R	DAH-NES-TSA	Ram
G	AH-TAD	Girl	R	AH-LOSZ	Rice
G	KLIZZIE	Goat	S	DIBEH	Sheep
G	JEHA	Gum	S	KLESH	Snake
H	TSE-GAH	Hair	T	D-AH	Tea
H	CHA	Hat	T	A-WOH	Tooth
H	LIN	Horse	T	THAN-ZIE	Turkey
I	TKIN	Ice	U	SHI-DA	Uncle
I	YEH-HES	Itch	U	NO-DA-IH	Ute
I	A-CHI	Intestine	V	A-KEH-DI-GLINI	Victor
J	TKELE-CHO-G	Jackass	W	GLOE-IH	Weasel
J	AH-YA-TSINNE	Jaw	X	AL-NA-AS-DZOH	Cross
J	YIL-DOI	Jerk	Y	TSAH-AS-ZIH	Yucca
K	JAD-HO-LONI	Kettle	Z	BESH-DO-TLIZ	Zinc
K	BA-AH-NE-DI-TININ	Key			

(Continued)

Table 8.1 (Continued)   Navajo Code Talkers' Dictionary. (Revised 15 June 1945, and Declassified under Department of Defense Directive 5200.9)

Countries	Navajo Word	Translation
Africa	ZHIN-NI	Blackies
Alaska[1]	BEH-HGA	With winter
America	NE-HE-MAH	Our mother
Australia	CHA-YES-DESI	Rolled hat
Britain	TOH-TA	Between waters
China	CEH-YEHS-BESI	Braided hair
France	DA-GHA-HI	Beard
Germany	BESH-BE-CHA-HE	Iron hat
Iceland	TKIN-KE-YAH	Ice land
India	AH-LE-GAI	White clothes
Italy	DOH-HA-CHI-YALI-TCHI	Stutter[2]
Japan	BEH-NA-ALI-TSOSIE	Slant eye
Philippine	KE-YAH-DA-NA-LHE	Floating island
Russia	SILA-GOL-CHI-IH	Red army
South America	SHA-DE-AH-NE-HI-MAH	South our mother
Spain	DEBA-DE-NIH	Sheep pain
Airplanes	*Navajo Word*	*Translation*
Planes	WO-TAH-DE-NE-IH	Air Force
Dive bomber	GINI	Chicken hawk
Torpedo plane	TAS-CHIZZIE	Swallow
Obs. plane	NE-AS-JAH	Owl
Fighter plane	DA-HE-TIH-HI	Humming bird
Bomber plane	JAY-SHO	Buzzard

(Continued)

[1] I know. This was the category it was placed under in the code. Don't blame me.
[2] The Navajos were unable to think of an appropriate, easy to remember, code word for Italy. Finally, one mentioned that he knew an Italian who stuttered....

Table 8.1 (Continued)   Navajo Code Talkers' Dictionary. (Revised 15 June 1945, and Declassified under Department of Defense Directive 5200.9)

Patrol plane	GA-GIH	Crow
Transport	ATSAH	Eagle
Ships	*Navajo Word*	*Translation*
Ships	TOH-DINEH-IH	Sea force
Battleship	LO-TSO	Whale
Aircraft	TSIDI-MOFFA-YE-HI	Bird carrier
Submarine	BESH-LO	Iron fish
Mine sweeper	CHA	Beaver
Destroyer	CA-LO	Shark
Transport	DINEH-NAY-YE-HI	Man carrier
Cruiser	LO-TSO-YAZZIE	Small whale
Mosquito boat	TSE-E	Mosquito

Figure 8.11   The U.S. Marine Cemetery on Iwo Jima shows the price of victory, a price that would've been even higher without the Navajo; Mount Suribachi, site of the famous flag raising, is in the background (https://web.archive.org/web/20130306120012/http://history.navy.mil/library/online/battleiwojima.htm).

The security surrounding the use of the Navajo as code talkers was very poor. Several accounts appeared in the media before the war's end. Without leaks like these, however, the program might never have existed! Johnston explained how he came up with his idea:

> [O]ne day, a newspaper story caught my eye. An armored division on practice maneuvers in Louisiana had tried out a unique idea for secret communication. Among the enlisted personnel were several Indians from one tribe. Their language might possibly

offer a solution for the oldest problem in military operations – sending a message that no enemy could possibly understand.[45]

William C. Meadows has tentatively identified this "newspaper story" with a piece from the November 1941 issue of *The Masterkey for Indian Lore and History*, from which the relevant paragraphs are reproduced below.[46]

> The classic World War I trick of using Indians speaking their own languages as "code" transmitters, is again being used in the Army, this time during the great maneuvers in the South, says *Science Service*. Three units of the 32nd Division have small groups of Indians from Wisconsin and Michigan tribes, who receive instructions in English, put them on the air in a tongue intelligible only to their listening fellow-tribesmen, who in turn retranslate the message into English at the receiving end.
>
> The Indians themselves have had to overcome certain language difficulties, for there are no words in their primitive languages for many of the necessary military terms. In one of the groups, ingenious use was made of the fact that infantry, cavalry, and artillery wear hat cords and other insignia of blue, yellow, and red, respectively. The Indian word for "blue" thus comes to mean infantry, "yellow" means cavalry, and "red" means artillery. The Indian term for "turtle" signifies a tank.

The article went on to state that 17 [Comanche] Indians had been trained.

Recall that Johnston preferred Navajo, in part, because it hadn't been studied by the Germans. Yet, despite the Germans' study of the other dialects, the Comanche code talkers, used by the U.S. Army for the D-day landing at Normandy and after, sent and received messages that the Germans failed to crack.[47] The Comanche were recruited to serve as code talkers about 16 months before the Navajo, but have attracted much less attention, in large part because of their much smaller numbers. Although 17 were trained, only 14 actually served in Europe.[48] Like the Navajo did later, the Comanche created their own code words (nearly 250 of them) and the result was that non-code talking Comanche couldn't understand the messages. In contrast to the Navajo, there was no attempt to keep the Comanche code talkers secret, which is ironic considering how few people are presently aware of them, compared to the Navajo![49]

Despite poor security (see the items in the reference section that appeared during the war!), the World War II code talkers were highly successful. Although it is difficult to measure the impact of any single component in a war, due to the many other variable factors, at least one statistic does support their impact being substantial. American pilots faced a 53% fatality rate prior to the introduction of the Navajo code talkers, a number that dropped afterwards to less than 7%.[50]

---

[45] Johnston, Philip, "Indian Jargon Won Our Battles," *The Masterkey for Indian Lore and History*, Vol. 38, No. 4, October–December, 1964, pp. 130–137, p. 131 quoted here.

[46] Meadows, William C., *The Comanche Code Talkers of World War II*, University of Texas Press, Austin, Texas, 2002, p. 75.

[47] Meadows, William C., *The Comanche Code Talkers of World War II*, University of Texas Press, Austin, Texas, 2002, p. xv.

[48] Meadows, William C., *The Comanche Code Talkers of World War II*, University of Texas Press, Austin, Texas, 2002, p. 80.

[49] Meadows, William C., *The Comanche Code Talkers of World War II*, University of Texas Press, Austin, Texas, 2002, pp. 108–109.

[50] McClain, Sally, *Navajo Weapon: The Navajo Code Talkers*, Rio Nuevo Publishers, Tucson, Arizona, 2001, p. 118.

In the two world wars, the U.S. military used Native Americans from at least 19 different tribes who spoke their natural languages with or without the addition of code words.[51] Of these, the Hopi also deserve to be singled out for mixing in code words with their natural language, just as the Navajo, Comanche, and Choctaw did. It is believed that most of the tribes did not make this step. The Hopi first hit the airwaves in the Marshall Islands, then New Caledonia and Leyte.[52] They only numbered 11. As far as is known, the largest code talking group after the Navajo was only 19 strong, and it did not make use of code words. For most of the groups, very little is known, but the presence of Navajo in such comparatively large numbers helped to ensure that their story would be told. Despite honors being bestowed upon various code talkers following the official declassification of the not-so-secret program in 1968,[53] the veterans hadn't always been treated fairly. One of the code talkers complained to Philip Johnston in a letter dated June 6, 1946:

> The situation out in the Navajoland is very bad and we as vets of World War II are doing everything we can to aid our poor people. We went to Hell and back for what? For the people back here in America to tell us we can't vote!! Can't do this! Can't do that!, because you don't pay taxes and are not citizens!! We did not say we were not citizens when we volunteered for service against the ruthless and treacherous enemies, the Japs and Germans! Why?[54]

Indians in Arizona and New Mexico weren't allowed to vote until 1948.

8.8 Code Talkers in Hollywood

The code talkers have captured the imagination of Hollywood. For example, the opening credits of Season 2, Episode 25 (May 19, 1995), of *The X-Files* which featured a Navajo code talker, had the Navajo words "EL 'AANIGOO 'AHOOT'E" take the place of the usual "The Truth is Out There."[55] In Season 1, Episode 2 (October 3, 1998), of *Highlander: The Raven*, Nick and Amanda found themselves thrown together again by the murder of a Navajo code talker.

In the 2002 film *Windtalkers*, the Navajo code talkers are given "bodyguards," who were actually under orders to shoot the code talkers in situations in which capture was likely. This was to prevent the enemy from being able to use torture to force secrets of the code from a prisoner. Cryptologic historian David Hatch, has claimed that no such order was given, and that the assignment was to protect the code talker from fellow soldiers who may never have seen a Navajo before and might mistake him for the enemy. This was a serious problem; several code talkers were "captured" by their own troops, one on two separate occasions![56] Another, Harry Tsosie, was killed by friendly fire, although this seems to have had more to do with his moving in the trench, while instructions were for everybody to sit tight, than with his physical appearance.[57] True or not, the claim in *Windtalkers*

---

[51] Meadows, William C., *The Comanche Code Talkers of World War II*, University of Texas Press, Austin, Texas, 2002, p. xv.

[52] Meadows, William C., *The Comanche Code Talkers of World War II*, University of Texas Press, Austin, Texas, 2002, p. 68.

[53] Navajo and other code talkers served in the Korean and Vietnam wars, so despite the many leaks, it was still officially secret.

[54] Paul, Doris A., *The Navajo Code Talkers*, Dorrance Publishing Co., Inc., Pittsburgh, Pennsylvania, 1973, p. 111.

[55] The episode was titled "Anasazi."

[56] Paul, Doris A., *The Navajo Code Talkers*, Dorrance Publishing Co., Inc., Pittsburgh, Pennsylvania, 1973, p. 85.

[57] McClain, Sally, *Navajo Weapon: The Navajo Code Talkers*, Rio Nuevo Publishers, Tucson, Arizona, 2001, p. 104. Ten other Navajo code talkers died in World War II.

appeared before the film was made and was not simply a creation of the script writer. It was presented in Deanne Durrett's *Unsung Heroes of World War II: The Story of the Navajo Code Talkers*.[58]

In one case of mistaken identity, non-Navajo Marines, who had advanced to a location previously held by Japanese, were being bombarded by artillery from their fellow troops and when they tried to call off the attack, it continued! The Japanese had so often imitated Americans on the airwaves that these real Americans were thought to be fakes. Finally headquarters asked, "Do you have a Navajo?" The Japanese couldn't imitate the Navajo, and when one responded, the salvo ceased.[59]

Joe Kieyoomia, a Navajo who was not a code talker, was captured early in the war by the Japanese. At first, despite his denials, they thought he was a Japanese-American. Eventually, when they realized an Indian language was being used as a code, they came to believe he was in fact Navajo, but they didn't believe he couldn't understand the coded messages. Joe, who had already survived the Bataan Death March, now faced more torture, but there was nothing he could tell them. In all, he spent 1,240 days as a POW before being freed after the end of the war.[60]

Figure 8.12   A code talker action figure. (Courtesy of Chris Christensen.)

The action figure shown in Figure 8.12, with voice supplied by the real Navajo code Talker Sam Billison, was first released in 1999, and was described by a vendor as follows:[61]

---

[58] Durrett, Deanne, *Unsung Heroes of World War II: The Story of the Navajo Code Talkers*, Facts On File, New York, 1998, p. 77. Thanks to Katie Montgomery for introducing me to this source.

[59] Paul, Doris A., *The Navajo Code Talkers*, Dorrance Publishing Co., Inc., Pittsburgh, Pennsylvania, 1973, p. 66.

[60] McClain, Sally, *Navajo Weapon: The Navajo Code Talkers*, Rio Nuevo Publishers, Tucson, Arizona, 2001, pp. 119–121.

[61] http://www.southernwestindian.com/prod/G-I-Joe-Navajo-Code-Talker.cfm. This link is now broken and was not archived by Internet Archive Wayback Machine,

G I Joe Navajo Code Talker

"Request Air Support!" "Attack by Machine Gun!" This dynamic, talking G I Joe speaks seven different phrases - in both Navajo and English! The complete equipment list includes a camouflage-covered helmet, web belt, hand phone set, backpack radio, shirt, pants, boots, and M-1 rifle. Also included in this Deluxe Edition (not available in any store or catalog) is a handsome 3″ Embroidered Iron-On Patch featuring the famous silhouette of the Marines hoisting the flag on Iwo Jima and proudly identifying you as a "Junior Navajo Code Talker." And to help you master the Code, we've also included a sturdy, laminated list of over 200 authentic Code Words (stamped "Top Secret—Confidential") actually used by the original Navajo Code Talkers, including the English word, the Navajo equivalent, and the literal Navajo translation! Did you know the literal translation for "tank destroyer" is "tortoise killer"? Now you can write secret messages to your friends in Navajo Code! (G I Joe 11″ Tall)

Price: $39.00

8.9 Use of Languages as Oral Codes

The discussion in this chapter has focused on the United States' use of code talkers, but Canada also used code talkers in World War I and the British used Latin for this purpose in the Boer War.[62] Latin was also used by the ball player/spy Moe Berg as a secret language on the baseball diamond for passing messages between himself and the second baseman when he played for Princeton University.[63] It seems likely that there are many other examples of languages used as oral codes. See Figure 8.13 for a bit of cryptologic humor on this topic.

Figure 8.13   Cryptologic humor (http://xkcd.com/257/).

[62] Meadows, William C., *The Comanche Code Talkers of World War II*, University of Texas Press, Austin, Texas, 2002, p. 5.

[63] Dawidoff, Nicholas, *The Catcher Was a Spy: The Mysterious Life of Moe Berg*, Pantheon Books, New York, 1994, p. 34.

References and Further Reading

On Japanese Codes and Ciphers

Boyd, Carl, *Hitler's Japanese Confidant: General Ōshima Hiroshi and Magic Intelligence, 1941–1945*, University of Press of Kansas, Lawrence, Kansas, 1993.

Budiansky, Stephen, *Battle of Wits: The Complete Story of Codebreaking in World War II*, The Free Press, New York, 2000.

Carlson, Elliot, *Joe Rochefort's War: The Odyssey of the Codebreaker Who Outwitted Yamamoto at Midway*, Naval Institute Press, Annapolis, Maryland, 2011.

Clark, Ronald, *The Man Who Broke Purple: The Life Of Colonel William F. Friedman, Who Deciphered The Japanese Code In World War II, Little Brown and Company*, Boston, Massachusetts, 1977. Rowlett and others could justifiably become angry at the title of this book. It was hardly a one-man show. It now appears that Rowlett, in fact, deserves more credit than Friedman; however, Rowlett did make some gracious comments:

> Clark's biography of Friedman does not do him justice. He deserves better. The book should have been written by someone who knew more of what went on after the Signal Intelligence Service was formed. The early years of Friedman's career were excellently characterized but the biography was very weak during the period after 1930.[64]

> A good deal of misinformation has been written about U.S. cryptologic work and, unfortunately, succeeding writers such as Clark pick up that kind of incorrect material with the result errors are perpetuated and are eventually accepted as facts.[65]

Clarke, Brigadier General Carter W., with introductory material from the editors of *Cryptologia*, "From the Archives: Account of Gen. George C. Marshall's Request of Gov. Thomas E. Dewey," *Cryptologia*, Vol. 7, No. 2, April 1983, pp. 119–128. The secrecy of the successful cryptanalysis of Purple wasn't maintained nearly as well as the Ultra secret. Dewey, a political opponent of Roosevelt, could have used his awareness of this success to claim that Roosevelt should have anticipated the Pearl Harbor attack; however, General Marshall was able to convince Dewey to sacrifice his most potent political weapon to the greater good—keeping a secret that would continue to save lives and shorten the war.

Currier, Prescott, "My "Purple" Trip to England in 1941," *Cryptologia*, Vol. 20, No. 3, July 1996, pp. 193–201.

Deavours, Cipher and Louis Kruh, *Machine Cryptography and Modern Cryptanalysis*, Artech House, Inc., Dedham, Massachusetts, 1985.

Freeman, Wes, Geoff Sullivan, and Frode Weierud, "Purple Revealed: Simulation and Computer-Aided Cryptanalysis of *Angooki Taipu B*," *Cryptologia*, Vol. 27, No. 1, January 2003, pp. 1–43. In this paper, the authors provide a level of detail sufficient for implementing Purple, as well as a modern attack.

Jacobsen, Philip H., "Radio Silence of the Pearl Harbor Strike Force Confirmed Again: The Saga of Secret Message Serial (SMS) Numbers," *Cryptologia*, Vol. 31, No. 3, July 2007, pp. 223–232.

Kahn, David, *The Codebreakers*, second edition, Scribner, 1996. The first chapter concerns Pearl Harbor, and more information on the various World War II-era codes and ciphers used by the Japanese can be found elsewhere in the book.

Kahn, David, "Pearl Harbor and the Inadequacy of Cryptanalysis," *Cryptologia*, Vol. 15, No. 4, October 1991, pp. 273–294. Pages 293–294 are devoted to Genevieve Grotjan, whom Kahn had interviewed.

Kelley, Stephen J., *Big Machines*, Aegean Park Press, Laguna Hills, California, 2001. This book focuses on Enigma, Purple (and its predecessors), and SIGABA, the top American machine of World War II, one that was never broken (See Chapter 9 of the present book).

---

[64] Kruh, Louis, "Reminiscences of a Master Cryptologist," *Cryptologia*, Vol, 4, No. 1, January 1980, pp. 45–50.
[65] Kruh, Louis, "Reminiscences of a Master Cryptologist," *Cryptologia*, Vol, 4, No. 1, January 1980, pp. 45–50.

Kruh, Louis, "The Deadly Double Advertisements - Pearl Harbor Warning or Coincidence?" *Cryptologia*, Vol. 3, No. 3, July 1979, pp.166–171.

Kruh, Louis, "Reminiscences of a Master Cryptologist," *Cryptologia*, Vol. 4, No. 1, January 1980, pp. 45–50. The following are some quotes from Frank Rowlett, from this article:

> The successful cryptanalysis of the Japanese Purple system was accomplished by a team of Army cryptanalysts. At first, it was a joint Army-Navy project, but after a few months the Navy withdrew its cryptanalytic resources to apply them to the Japanese naval systems. The Navy did however, continue to provide some intercept coverage of diplomatic traffic.

> The Chief Signal Officer, General Joseph O. Mauborgne, was personally interested in the Purple effort and supported our work to the fullest degree possible. He liked to refer to us as his magicians and called the translations of the messages we produced by the name "magic".

> Friedman played a signal role in the selection and assignment of personnel and participated in the analytical work on a part-time basis.

Lewin, Ronald, *The American Magic: Codes, Ciphers and the Defeat of Japan*, Farrar Straus Giroux, New York, 1982.

Parker, Frederick D., "The Unsolved Messages of Pearl Harbor," *Cryptologia*, Vol. 15, No. 4, October 1991, pp. 295–313.

Rowlett, Frank B., The *Story of Magic: Memoirs of an American Cryptologic Pioneer*, Aegean Park Press, Laguna Hills, California, 1989. Not only was Rowlett there, but he also writes well! This book does the best job of capturing the atmosphere of World War II-era codebreaking in America.

Smith, Michael, *The Emperor's Codes: The Breaking of Japan's Secret Ciphers*, Penguin Books, New York, 2002. Smith describes the British successes against Japanese codes and ciphers, pointing out that they were the first to crack a Japanese diplomatic cipher machine (Orange, see pp. 34–35) and JN-25 (see pp. 5, 59–60).

Stamp, Mark and Richard M. Low, *Applied Cryptanalysis: Breaking Ciphers in the Real World*, John Wiley & Sons, Hoboken, New Jersey, 2007.

Tucker, Dundas P., edited and annotated by Greg Mellen, "Rhapsody in Purple: A New History of Pearl Harbor – Part I," *Cryptologia*, Vol. 6, No. 3, July 1982, pp. 193–228.

Weierud, Frode, The PURPLE Machine *97-shiki-obun In-ji-ki Angooki Taipu B*, http://cryptocellar.org/simula/purple/index.html. An online Purple simulator can be found at this website.

On Code Talkers

Aaseng, Nathan, *Navajo Code Talkers*, Thomas Allen & Son, Markham, Ontario, Canada, 1992. This is a young adult book.

Anon., "Comanches Again Called for Army Code Service," *New York Times*, December 13, 1940, p. 16.

Anon., "DOD Hails Indian Code Talkers," *Sea Services Weekly*, November 27, 1992, pp. 9–10.

Anon., "Pentagon Honors Navajos, Code Nobody Could Break," *Arizona Republic*, September 18, 1992, p. A9.

Anon., "Played Joke on the Huns," *The American Indian Magazine*, Vol. 7, No. 2, 1919, p. 101. This article revealed the role the Sioux played in World War I with their native language. It is quoted in Meadows, William C., *The Comanche Code Talkers of World War II*, University of Texas Press, Austin, Texas, 2002, p. 30.

Bianchi, Chuck, *The Code Talkers*, Pinnacle Books, New York, 1990. This is a novel.

Bixler, Margaret, *Winds of Freedom: The Story of the Navajo Code Talkers of World War II*, Two Bytes Publishing Company, Darien, Connecticut, June 1992.

Bruchac, Joseph, *Codetalker: A Novel About the Navajo Marines of World War Two*, Dial Books, New York, 2005.

Davis, Jr., Goode, "Proud Tradition of the Marines' Navajo Code Talkers: They Fought With Words–Words No Japanese Could Fathom," *Marine Corps League*, Vol. 46, No. 1, Spring 1990, pp. 16–26.

Donovan, Bill, "Navajo Code Talkers Made History Without Knowing It," *Arizona Republic*, August 14, 1992, p. B6.

Durrett, Deanne, *Unsung Heroes of World War II: The Story of the Navajo Code Talkers*, Facts On File, New York, 1998

Gyi, Maung, "The Unbreakable Language Code in the Pacific Theatre of World War II," *ETC: A Review of General Semantics*, Vol. 39, No. 1, Spring 1982, pp. 8–15.

Hafford, William E., The Navajo Code Talkers, *Arizona Highways* Vol. 65, No. 2, February 1989, pp. 36–45.

Huffman, Stephen, "The Navajo Code Talkers: A Cryptologic and Linguistic Perspective," *Cryptologia*, Vol. 24 No. 4, October 2000, pp. 289–320.

Johnston, Philip, "Indian Jargon Won Our Battles," *The Masterkey for Indian Lore and History*, Vol. 38, No. 4, October–December 1964, pp.130–137.

Kahn, David, "From the Archives: Codetalkers Not Wanted," *Cryptologia*, Vol. 29, No. 1, January 2005, pp. 76–87.

Kawano, Kenji. *Warriors: Navajo Code Talkers*, Northland Pub. Co., Flagstaff, Arizona, 1990.

King, Jodi A., "DOD Dedicates Code Talkers Display," *Pentagram*, September 24, 1992, p. A3.

Langille, Vernon, "Indian War Call," *Leatherneck*, Vol. 31, No. 3, March 1948, pp. 37–40.

Levine, Captain Lincoln A., "Amazing Code Machine That Sent Messages Safely to U.S. Army in War Baffles Experts: War Tricks That Puzzled Germans," *New York American*, November 13, 1921. America's use of Choctaw code talkers in World War I was described in this article, which is quoted in Meadows, William C., *The Comanche Code Talkers of World War II*, University of Texas Press, Austin, Texas, 2002, p. 23–24.

Marder, Murrey, "Navajo Code Talkers," *Marine Corps Gazette*, September 1945, pp. 10–11.

McClain, Sally, *Navajo Weapon*, Books Beyond Borders, Inc., Boulder, Colorado, 1994.

McCoy, Ron, "Navajo Code Talkers of World War II: Indian Marines Befuddled the Enemy," *American West*, Vol. 18, No. 6, November/December 1981, pp. 67–73, 75.

Meadows, William C., *The Comanche Code Talkers of World War II*, University of Texas Press, Austin, Texas, 2002. This thorough and scholarly account also contains very useful appendices with data concerning all of the Native American tribes, identified thus far, that served as code talkers in World War I or World War II.

Paul, Doris A., *The Navajo Code Talkers*, Dorrance Publishing Co., Inc., Pittsburgh, Pennsylvania, 1973.

Price, Willson H., "I Was a Top-Secret Human Being During World War 2," *National Enquirer*, February 4, 1973. In general, National Enquirer is not a reliable source!

Shepherdson, Nancy, "America's Secret Weapon," *Boy's Life*, November 1997, p. 45.

Stewart, James, "The Navajo at War," *Arizona Highways*, June 1943, pp. 22–23. This article (published while the war was still on!) included the following passage:

> [T]he U.S. Marine Corps has organized a special Navajo signal unit for combat communications service… Its members were trained in signal work using the Navajo language as a code, adapting a scheme tried with considerable success during World War I.

Whether or not the Japanese saw this before the war ended is unknown. In any case, they did catch on to the fact that some of the conversations they couldn't understand were being carried out in Navajo.

Thomas, Jr., Robert McG, "Carl Gorman, Code Talker in World War II, Dies at 90," *The New York Times*, February 1, 1998, p. 27.

United States Congress, "Codetalkers Recognition: Not Just the Navajos," *Cryptologia*, Vol. 26, No. 4, October 2002, pp. 241–256. This article provides the text of the Code Talkers Recognition Act.

U.S. Marine Corps, *Navajo Dictionary*, June 15, 1945.

Watson, Bruce, "Navajo Code Talkers: A Few Good Men," *Smithsonian*, Vol. 24, No. 5, August 1993, pp. 34–40, 42–43.

Wilson, William, "Code Talkers," *American History*, February 1997, pp.16–20, 66–67.

Bibliography

A bibliography that includes unpublished (archival) sources can be found at https://web.archive.org/web/20130306115918/http://www.history.navy.mil/faqs/faq12-1.htm.

Videography

Chibitty, Charles, Bob Craig, and Brad Agnew, *American Indian Code Talkers* [VHS], Center for Tribal Studies, College of Social and Behavioral Sciences, Northeastern Oklahoma State University, Tahlequah, Oklahoma, 1998.

Chibitty, Charles, Dwayne Noble, Eric Noble, and Jeff Eskew, *Recollections of Charles Chibitty—The Last Comanche Code Talker* [VHS], Hidden Path Productions, Mannford, Oklahoma, 42 minutes, 2000.

Hayer, Brandi, Hinz Cory, and Matt Wandzel, Dine College, and Winona State University, *Samuel Tso: Code Talker, 5th Marine Division* [DVD], Dine College, Tsaile, Arizona, and Winona State University, Winona, Minnesota, 2009.

Meadows, William C., *Comanche Code Talkers of World War II* [VHS], August 4, 1995. This is a videotaped interview with Comanche code talkers Roderick Red Elk and Charles Chibitty. Meadows was both the host and producer. A copy is available in the Western History Collections of the University of Oklahoma.

NAPBC, *In Search of History: The Navajo Code Talkers* [VHS], History Channel and Native American Public Broadcasting Consortium, Lincoln, Nebraska, 50 minutes, 2006 (originally broadcast in 1998).

Red-Horse, Valerie, Director, *True Whispers, The Story of the Navajo Code Talkers* [DVD], PBS Home Video, ~60 minutes, 2007 (originally broadcast 2002).

Sam, David, Patty Talahongva, and Craig Baumann, *The Power of Words: Native Languages as Weapons of War* [DVD], National Museum of the American Indian, Smithsonian Institution, Washington, DC, 2006.

Tully, Brendan W., Director, *Navajo Code Talkers: The Epic Story* [VHS], Tully Entertainment, 55 minutes, 1994.

Wright, Mike, *Code Talkers Decoration Ceremony, Oklahoma State Capitol, November 3, 1989* [VHS], Oral History Collections, Oklahoma Historical Society, Oklahoma City, Oklahoma, 1989.

Chapter 9

SIGABA:
World War II Defense

It seems that most writers are concerned with the breaking of other nation's ciphers. Isn't it more important and even more of a feat to make your own systems secure against foreign cryptanalysts?

—Frank Rowlett[1]

Figure 9.1 Frank Rowlett. (Courtesy of the National Cryptologic Museum, Fort Meade, Maryland.)

9.1 The Mother of Invention

Machine ciphers were vulnerable. Frank Rowlett (Figure 9.1) and his colleagues knew this, but Rowlett was not tasked with inventing a superior machine; William Friedman thought that he had already created one, namely the M-134. In 1934, Friedman assigned Rowlett the job of simply

---

[1] Quoted in Kruh, Louis, "Reminiscences of a Master Cryptologist," *Cryptologia*, Vol. 4, No. 1, January 1980, pp. 45–50, p. 49 cited here.

creating the paper tape keys (Figure 9.2) that needed to be fed through Friedman's device.[2] The key on the tape would control which rotor(s) turned at each step, thus avoiding the regularity of the turning in other machines, such as Enigma. Fortunately, creating the key tape was a horrible job, made even worse for Rowlett by the fact that Friedman told him to spend half of his time at it, while the other half was to be devoted to his continued training, which he much preferred.[3]

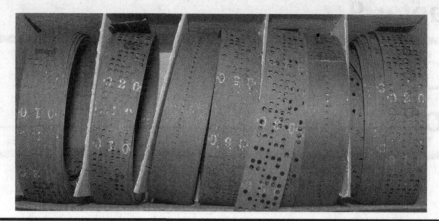

Figure 9.2   M-134 paper key tape. (Courtesy of the National Cryptologic Museum, Fort Meade, Maryland.)

Friedman showed Rowlett how to operate the equipment to make the tape and observed him making a test run. He then suggested Rowlett make several more test runs and left the room Rowlett related what came next:

> After he departed, I continued as he had proposed. I decided that I would duplicate the test run I had made under his supervision to see if the keys prepared on two separate runs were identical as they should be. When I finished the second run and compared the two keys, I found several points of discrepancy. I decided that I would make another attempt to duplicate the first run. When it was finished and I compared it with the two previous runs, I found all three to be different. And when I tried two more duplicate runs, I found that I got different results for each. At this point I decided that I had better consult with Friedman.
>
> When I showed Friedman the results I had obtained, he came with me to the equipment room. When he tried to produce a duplicate of the test run I had made, he also obtained different results. We spent until lunchtime trying to get satisfactory results, but with only moderate success.
>
> My first day's experience with the equipment was only a preview of the succeeding days. The equipment operated erratically, and frequently I had to dismantle a piece in order to locate the trouble. After pursuing this course for some time, I was finally able to make several runs with identical results. By the end of the first month I had completed only a small portion of the compilation task that I had been assigned.

2   Mucklow, Timothy J., *SIGABA/ECM II: A Beautiful Idea*, Center for Cryptologic History, National Security Agency, Fort George G. Meade, Maryland, 2015, p. 7.

3   Rowlett, Frank B., *The Story of Magic: Memoirs of an American Cryptologic Pioneer*, Aegean Park Press, Laguna Hills, California, 1999, p. 92.

Friedman seemed to be disappointed with my progress. He spent hours with me, trying to determine why the results I had been obtaining were so unsatisfactory. At first he seemed to think that I was at fault, but after operating the equipment himself on a number of runs, he reluctantly admitted that the equipment was operating unreliably. Finally, in desperation, he told me to continue with the preparation of the keying materials while he undertook the procurement of more reliable equipment.

By this time I was fully fed up with the assignment. No matter how hard I tried, the equipment kept performing erratically and I had to reject over three-quarters of the keys that I had prepared. There was no relief in sight, and I soon began to feel that I would be spending at least several months on a most unrewarding assignment.[4]

Rowlett summarized his frustration, writing "My morale had never before been so low. [...] I was stuck with what I considered to be an impossible assignment, I had been given inadequate equipment and support, and my supervisor [Friedman] seemed anything but sympathetic."[5] He naturally thought that there had to be a better way! Eventually he came up with the idea of a different sort of cipher machine, one that retained the seemingly random turning of the rotors caused by the tape, but did so without actually needing a paper tape key.

Rowlett took his idea for a better cipher machine to Friedman and was basically told to forget about it and get back to work. Nevertheless, he persisted, and after many attempts, over some 6–10 months, which finally included a threat to quit and a threat to go over Friedman's head with the matter, Rowlett got Friedman to take his idea seriously.[6] The idea was shared with the Navy, who produced a prototype, which included some modifications of their own (replacing a plugboard with another set of ten-pin rotors).[7] A full description of the final machine is given later in this chapter, but in the meanwhile, let's consider a pair of historical "What ifs?"

1. What if Friedman had tasked someone with a less powerful intellect than Rowlett to create the paper tape key? Would the new and improved tapeless cipher machine have been invented in time for World War II? In this instance at least, it seems that assigning someone a tedious task for which he was overqualified paid off!

2. What if Rowlett had not persisted with his idea? If Friedman's machine went into service using the paper tape keys, how many hours of labor would have to be devoted to tape creation? Who would carry out this work? What work would he or she be diverted from to do so? What impact would this have on World War II? Also, would the paper tape be distributed successfully and function properly in all cases? If not, which messages would fail to go through and what would the effect of this be?

---

4 Rowlett, Frank B., *The Story of Magic: Memoirs of an American Cryptologic Pioneer*, Aegean Park Press, Laguna Hills, California, 1999, pp. 92–93.

5 Rowlett, Frank B., *The Story of Magic: Memoirs of an American Cryptologic Pioneer*, Aegean Park Press, Laguna Hills, California, 1999, p. 94.

6 Mucklow, Timothy J., *SIGABA/ECM II: A Beautiful Idea*, Center for Cryptologic History, National Security Agency, Fort George G. Meade, Maryland, 2015, p. 9, which cites Frank B. Rowlett, Oral History Interview 1974, OH-1974-01, Part B, 45c, Center for Cryptologic History, National Security Agency, Fort George G. Meade, Maryland. Also see Rowlett, Frank B., *The Story of Magic: Memoirs of an American Cryptologic Pioneer*, Aegean Park Press, Laguna Hills, California, 1999, p. 96.

7 Mucklow, Timothy J., *SIGABA/ECM II: A Beautiful Idea*, Center for Cryptologic History, National Security Agency, Fort George G. Meade, Maryland, 2015, p. 15.

The new and improved cipher machine was dubbed SIGABA by the Army, while the Navy called it ECM (Electric Cipher Machine) II or CSP-888/889. A modified Navy version was known as the CSP-2900. The machines were first sent into the field in June 1941 and, before being replaced with another device, years after World War II, a closely accounted for 10,060 machines saw use.[8] These SIGABAs enciphered the most important messages, while lesser secrets were run through weaker machines such as the M-209. The Germans were often able to take advantage of operational mistakes by M-209 operators, such as sending messages in depth (encrypted with the same key), and recover the messages. However, this process typically took seven to ten days, by which time the information might well be worthless. Between such cryptanalysis and captured keys, about 10% of the M-209 traffic was compromised.[9] By contrast, SIGABA was never cracked.

9.2 Making the Rotors

Rowlett is not the only hero in this chapter. Someone had to come up with the idea of SIGABA, but that was not enough. The 10,060 machines had to actually be built. Because SIGABA was a rotor-based machine, an extremely important detail in each machine was the wiring of the rotors. If just one wire was soldered incorrectly, that machine would be useless, with potentially disastrous consequences in the field. So, before getting into the details of how exactly SIGABA worked, it's worth taking a closer look at the rotors and how they were manufactured. Figure 9.3 shows a SIGABA rotor before the wires were soldered to connect the letters on its opposite sides.

Originally, male shipyard electricians did all of the soldering work. The average production rate for these men was seven rotors per day.[10] Later, women got a turn to try their hands at this task (Figure 9.4). On the Navy side it was WAVES (Women Accepted for Volunteer Emergency, and for the Army WACs (Women Army Corps).[11] They worked far faster than the men:

> In 1943 the average WAVE managed to solder the connections for fourteen wheels per day. One actually assembled a record twenty-two wheels on her shift.[12]

This rate was attained without sacrificing accuracy:

> Midway through the war, the Navy women alone had wired more than 150,000 rotor wheels. Remarkably, there was not a single configuration error and only one instance where a wheel had been mislabeled![13]

---

[8] Mucklow, Timothy J., *SIGABA/ECM II: A Beautiful Idea*, Center for Cryptologic History, National Security Agency, Fort George G. Meade, Maryland, 2015, pp. 19–20.

[9] Simons, Marc and Paul Reuvers, "M-209," *Crypto Museum*, https://www.cryptomuseum.com/crypto/hagelin/m209/index.htm.

[10] Mucklow, Timothy J., *SIGABA/ECM II: A Beautiful Idea*, Center for Cryptologic History, National Security Agency, Fort George G. Meade, Maryland, 2015, p. 20.

[11] Mucklow, Timothy J., *SIGABA/ECM II: A Beautiful Idea*, Center for Cryptologic History, National Security Agency, Fort George G. Meade, Maryland, 2015, p. 20.

[12] Mucklow, Timothy J., *SIGABA/ECM II: A Beautiful Idea*, Center for Cryptologic History, National Security Agency, Fort George G. Meade, Maryland, 2015, p. 20.

[13] Mucklow, Timothy J., *SIGABA/ECM II: A Beautiful Idea*, Center for Cryptologic History, National Security Agency, Fort George G. Meade, Maryland, 2015, p. 20, which cites Ratcliff, Rebecca Ann, *Delusions of Intelligence: Enigma, Ultra, and the End of Secure Ciphers*, Cambridge University Press, New York, 2006, p. 81 and Safford, Captain Laurance, *History of Invention and Development of the Mark II ECM*, SRH-360, United States Navy OP-20-S-5, Office of the Chief of Naval Operations, Washington, DC, October 30, 1943, p. 52. This history is available at NARA (National Archives and Records Administration) RG 457, Box 1124, College Park, Maryland. Note: SRH stands for Special Research Histories.

Figure 9.3   A SIGABA rotor prior to being wired. (Courtesy of the National Cryptologic Museum, Fort Meade, Maryland.)

Figure 9.4   A pair of women at work on SIGABA rotors. (Courtesy of the National Cryptologic Museum, Fort Meade, Maryland.)

Altogether, over 450,000 wheels were made.[14]

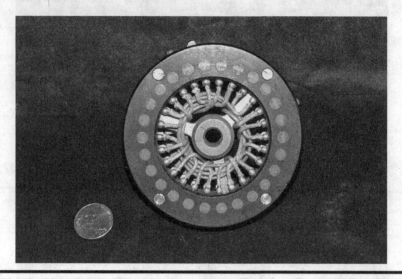

Figure 9.5   A completely wired SIGABA rotor. (Courtesy of the National Cryptologic Museum, Fort Meade, Maryland.)

A completely wired SIGABA rotor is shown in Figure 9.5, along with a nickel to give a sense of scale. Looking at the wires makes me think of knitting. It has been speculated that one of the reasons the women out-performed the men at the task of rotor wiring is that they tended to have greater prior experience with activities such as knitting, crocheting, sewing, embroidery, cross-stitching, etc., and that these skills transferred over. The women's typically smaller hands may also have been advantageous to carrying out such precise small-scale work. It has also been suggested that the women exhibited greater patience, becoming frustrated less quickly than males.

Prior to America entering World War II and the women's involvement, Rear Admiral Leigh Noyes, Director of Naval Communications, foresaw the contributions they could make. He sent a letter to Ada Comstock, the President of Radcliffe, Harvard University's women's college, asking that she raise the possibility of extra-curricular training of some seniors in naval cryptanalytic work. He wrote, "In the event of total war, … women will be needed for this work, and they can do it probably better than men."[15] In the instance of constructing rotors, at least, he was right! And the women contributed in many other ways. Their work is detailed in Liza Mundy's excellent book *Code Girls*.[16] Following the appearance of this book, Mundy accepted the opportunity to serve as the 11th Scholar-in-Residence in the National Security Agency's Center for Cryptologic History. This guarantees that there will be sequel to *Code Girls*, for that is part of Mundy's contractual obligation in the SiR role. I am looking forward to it!

---

[14] Mucklow, Timothy J., *SIGABA/ECM II: A Beautiful Idea*, Center for Cryptologic History, National Security Agency, Fort George G. Meade, Maryland, 2015, p. 20.

[15] Bauer, Craig, "The Cryptologic Contributions of Dr. Donald Menzel," *Cryptologia*, Vol. 30, No. 4, 2006, pp. 306–339, p. 306 cited here. The original document is: Leigh Noyes to Ada Comstock, letter, September 25, 1941, Papers of President Comstock, Radcliffe Archives, Radcliffe Institute for Advanced Study, Harvard University.

[16] Mundy, Liza, *Code Girls: the untold story of the American women code breakers of World War II*, Hachette Books, New York, 2017.

Figure 9.6   The 15-rotor heart of SIGABA. (Courtesy of the National Cryptologic Museum, Fort Meade, Maryland.)

While the Enigma machines used by the Nazis had 3, and later 4, rotors in use at a time, SIGABA employed 15, placed in three banks of 5 rotors each. They can be seen in Figure 9.6. The five smaller rotors on the left-hand side are called *index rotors*. The five rotors in the middle are called *stepping control rotors*, or simply *control rotors*. The right-most rotors are called the *alphabet rotors* or *cipher rotors*. It is with this last batch of rotors that the explanation of SIGABAs functioning begins in the next section.

9.3 Anatomy of a Success

Figure 9.7 shows how the cipher rotors of SIGABA encipher a letter. Like Enigma rotors, there are 26 contacts on each side. A plaintext letter begins its journey through these rotors from the left-hand side. Each of the five rotors makes a substitution, as shown by the lines internal to each

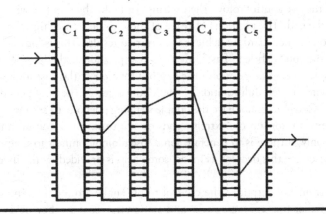

Figure 9.7   Cipher rotors.

rotor, and the enciphered letter comes out on the right-hand side. In contrast to Enigma, there is no plugboard and no reflector. The letter to be enciphered makes only one trip through the cipher rotors. When the enciphered message is received by the intended recipient, he sets the rotors in the same position, but runs the ciphertext letter through them in the opposite direction, starting at the right-hand side and receiving the plaintext letter out on the left-hand side. The lack of a reflector requires the user to carefully select the correct direction, depending on whether the message is being enciphered or deciphered. This can be done with the turn of a switch on the machine.

The cipher rotors are where the action happens, but the other two banks of rotors are what make SIGABA secure. These are the rotors responsible for making the cipher rotors turn in a very irregular manner. Remember, the predictable way in which the rotors of Enigma turned was one of its major weaknesses. To see how the irregularity is introduced, we examine the last two banks of rotors one at a time, starting with the control rotors (Figure 9.8).

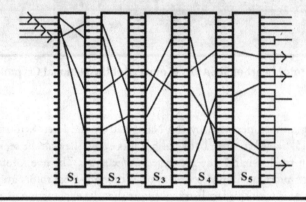

Figure 9.8   Control rotors.

While the cipher rotors only have current passing through one wire at a time, determined by which letter is being enciphered, there are four live wires for the control rotors at each step. These are indicated with arrows at the top left of Figure 9.8. Note that it is always these specific wires that are live, regardless of what letter is being enciphered. After passing through all five control rotors, some of the live wires may meet up on the right-hand side. As Figure 9.8 shows, many of the output wires, following the fifth control rotor, are bundled. In the specific instance shown, the four live wires entering this rotor bank exited in just three live wires, indicated by arrows on the right. Depending on the positions of the five control rotors, their wirings can take the four live wires to anywhere from 1 to four wires at the end. The positions of three of these rotors can change, as the user enters the plaintext message on the keyboard, to achieve these varied results. This is indicated in Figure 9.9.

In Figure 9.9, the rotor labeled "Fast" advances one position with every letter of the message. The Medium rotor advances one position for every full rotation of the Fast rotor and the Slow rotor advances one position for every full rotation of the Medium rotor. The total period for these rotors is thus $26^3 = 17,576$. SIGABA does not have the double stepping phenomenon seen in Enigma (see the end of section 7.2). The two rotors on the extreme ends do not turn. These control rotors do not do any enciphering. Their only purpose is to generate some (pseudo)randomness to determine which cipher rotors will turn, but before this decision is made, some help is provided by the index rotors shown in Figure 9.10.

The live wires from the output of the control rotors snake around and enter the index rotors. These rotors are smaller (ten contacts each) and stationary. That is, they never turn. Depending on how many wires coming out of the control rotors are live, there could be input to anywhere

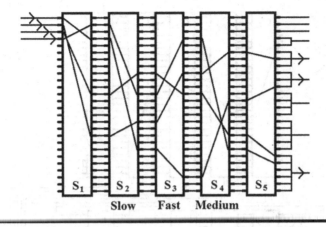

S_1　S_2　S_3　S_4　S_5

Slow　Fast　Medium

Figure 9.9　Control rotors' rotation.

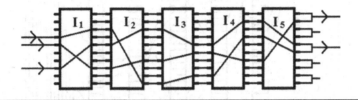

I_1　I_2　I_3　I_4　I_5

Figure 9.10　Index rotors.

from 1 to 4 of the contacts on the left-hand side of the first index rotor. If more than one wire is live, there is a possibility for a smaller number of live wires coming out of the bundling that occurs after all five index rotors have been traversed. In general, the result will be somewhere between one and four live wires at the very end, although it will never be more than the number of live wires entering the index rotor bank. That is, the number of live wires may decrease, but can never increase. Also, there will always be at least one live wire at the end. Figure 9.11 pieces all of the above together and shows how the cipher rotors are made to turn.

Figure 9.11 shows four live wires entering the bank of control rotors (in the middle of the diagram). Three live wires exit this rotor bank and snake down to enter the index rotors. Two live wires exit the index rotor bank and pass their current on to a pair of cipher rotors, C_1 and C_3, making them turn one position each. Every time a letter is typed on the SIGABA keyboard, a control rotor turns and the current starting at the control rotors follows a different path through those rotors and the index rotors, leading to a selection of anywhere from one to four cipher rotors to be turned. Thus, the cipher rotors turn in a manner that is very difficult to predict!

The paper tape that so frustrated Rowlett is not needed. However, after each letter's cipher equivalent is determined, it is printed on a narrow paper tape. This is yet another difference between SIGABA and Enigma. For Enigma, a bulb would be illuminated to indicate the enciphered letter. By automatically printing the letter, SIGABA allowed encryption to be carried out more rapidly. SIGABA could actually encipher at a rate of 60 words per minute, if the operator could type that fast![17] However, the SIGABA user's manual, *Crypto-Operating Instructions for Converter M-134-C,*

---

[17] Mucklow, Timothy J., *SIGABA/ECM II: A Beautiful Idea*, Center for Cryptologic History, National Security Agency, Fort George G. Meade, Maryland, 2015, p. 16.

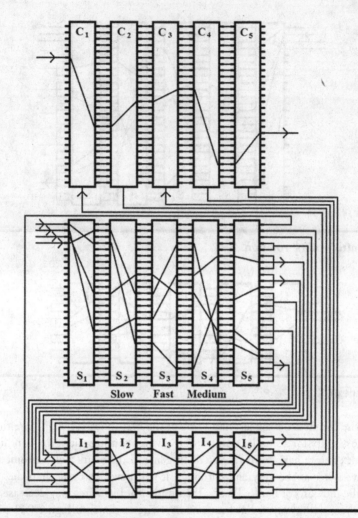

Figure 9.11   The entire encryption process.

warned that the device "should be operated at a maximum speed of 45 to 50 words per minute... if this speed is exceeded, characters may fail to print."[18]

For many cipher machines of this era, the intended recipient would have to figure out where to insert word breaks to make the recovered message readable. In contrast to such machines, SIGABA was implemented in a way that conveyed the spacing along with the message. Because there are only 26 wires in each Cipher Rotor it might seem that this is impossible, but a 27th symbol really isn't needed. The space-preserving method worked like so:

- Prior to enciphering, SIGABA converts every Z to an X. Real Xs are left unchanged. Basically, Zs and Xs are combined in a single character denoted by X.
- Spaces are converted to (the now available!) Zs.

---

[18] Mucklow, Timothy J., *SIGABA/ECM II: A Beautiful Idea*, Center for Cryptologic History, National Security Agency, Fort George G. Meade, Maryland, 2015, p. 33.

Example[19]

```
ZERO ONE TWO THREE FOUR FIVE SIX
```
is converted by SIGABA to

```
XEROZONEZTWOZTHREEZFOURZFIVEZSIX
```

Using a particular key, this enciphers to

```
IEQDEMOKGJEYGOKWBXAIPKRHWARZODWG
```

and then deciphers to

```
XERO ONE TWO THREE FOUR FIVE SIX
```

It is left to the decipherer to recognize that the first X should be taken as a Z, while the last X is an X. Because Z is the rarest letter in the English alphabet, there won't be many Xs that need to be changed to Zs. In any case, context should allow them to be recognized easily.

9.4 SIGABA Production

While it was found that women produced SIGABA rotors more efficiently than men, production of the machines involved both men and women. Figure 9.12 shows some men at work, as well as a female (on the far left).

Figure 9.12 Part of a SIGABA factory. (Courtesy of the National Cryptologic Museum, Fort Meade, Maryland.)

---

[19] Taken from Stamp, Mark and Wing On Chan, "SIGABA: Cryptanalysis of the Full Keyspace," *Cryptologia*, Vol. 31, No. 3, July 2007, pp. 201–222.

Figure 9.13   SIGABA. (from CryptoMuseum, cryptomuseum.com, under license CM500576.)

A completed machine, weighing in at 100 pounds, is shown in Figure 9.13.

SIGABAs didn't simply set on desks 24-7. They had to be stored securely. Because the security cabinets (i.e. safes) weighed 800 pounds, the secured machines had to be transported via "tactical communications vehicles." As a further precaution, thermite emergency destruction devices usually traveled with them. These devices could reduce the top secret components of a SIGBA to molten metal in just 97 seconds. This protection was foregone on ships due to the potential for a larger than desired fire.[20]

The history of cryptology spans thousands of years, so covering it in a single volume means that many stories cannot be told in as much detail as I would desire. In the case of SIGABA, a closer look shows that there were many versions between Rowlett's great initial insight and the machines that helped win World War II. A sequence of prototypes was manufactured from late 1936 to January 1941 becoming gradually smaller, lighter, and faster, as well as less susceptible to failure due to heat, humidity, and vibration.[21]

9.5 Keyspace and Modern Cryptanalysis

An important question to ask about any cipher is "What is the keyspace?" As with Enigma, several answers can be put forth in response to this question for SIGABA.

---

[20] See Mucklow, Timothy J., *SIGABA/ECM II: A Beautiful Idea*, Center for Cryptologic History, National Security Agency, Fort George G. Meade, Maryland, 2015, p. 39, which cites Safford, Captain Laurance, *History of Invention and Development of the Mark II ECM*, SRH-360, United States Navy OP-20-S-5, Office of the Chief of Naval Operations, Washington, DC, October 30, 1943, p. 61. This history is available at NARA (National Archives and Records Administration) RG 457, Box 1124, College Park, Maryland. *Note*: SRH stands for Special Research Histories.

[21] Mucklow, Timothy J., *SIGABA/ECM II: A Beautiful Idea*, Center for Cryptologic History, National Security Agency, Fort George G. Meade, Maryland, 2015, p. 16.

If the enemy has no idea how any of the SIGABA rotors are wired, there are 26! choices for each of the 10 large rotors and 10! Choices for each of the smaller index rotors. Thus, there would seem to be about $(26!)^{10}(10!)^5 \approx 7.2 \times 10^{298}$ total possibilities. For modern encryption algorithms, the keysize is usually stated in bits. Because $7.2 \times 10^{298} \approx 2^{992.8}$, we see that the keyspace is about 993 bits. I wrote "about" because it is easy to nitpick this number by pointing out that it is unlikely for a system to use two rotors that have the exact same wiring or a rotor whose wiring is the identity (although the internal wiring from the military Enigma's plugboard to its rotor assembly was the identity!). A more serious nitpick arises from the fact that the index rotors are stationary. Thus, a collection of 5 of them is equivalent to some differently wired single rotor. Hence, once should replace the factor $(10!)^5$ in the calculation above with $10!$.[22] This reduces the keyspace to $\approx 4.13 \times 10^{272} \approx 2^{905.6}$.

If an enemy was able to learn the wiring of all of SIGABA's rotors through a spy, a double agent, blackmail, surreptitious entry, etc., then there are far fewer possible keys to consider, but the number is still immense. To calculate it, we first note that the 10 large rotors can be placed in the machine in 10! different orders. However, each could be placed in a given position right-side-up or upside-down! These orientations were referred to as "forward" or "reverse." Thus, we have another factor of 2^{10}. Once each rotor is placed in the machine, in whatever orientation, there are 26 choices as to how far along in its rotation it is started. This gives another factor of 26^{10} when all 10 large rotors are considered. Altogether then, the large rotors may be set in $(10!)(2^{10})(26^{10})$ ways. Similarly, the 5 small index rotors may be ordered in 5! ways, inserted in normal or reverse position[23] (a factor of 2^5) and set to any of 10 initial positions each (a factor of 10^5). The grand total is $(10!)(2^{10})(26^{10})\,(5!)(2^5)(10^5) \approx 2.0 \times 10^{32}$. This is approximately $2^{107.3}$, so SIGABA could be said to have about a 107 bit key, if the wirings of all of the rotors are known.

The above calculation shows the number of potential keys, given only the limitations imposed by the wirings of the available rotors. However, there were other limitations imposed by the procedures that dictated how the machine was actually used. For example, the SIGABA manual instructs for the settings of the control rotors to be sent in plaintext (!) as a message indicator, along with the enciphered message. This obviously reduces the keyspace for an enemy who knows what a message indicator means. On the other hand, communications between Roosevelt and Churchill were not carried out in this manner. For all users, the bundling of outputs from the index rotors has the effect that different orderings of the index rotors can produce identical results, reducing the effective keyspace.[24]

Mark Stamp and Wing On Chan calculated the keyspace available for Roosevelt and Churchill to be about $2^{95.6}$, and that achieved by following the manual as to message indicators as about $2^{48.4}$. They pointed out that while this smaller keyspace could be brute-forced at the time of their writing (2007), it "would have been unassailable using 1940s technology, provided no shortcut attack was available." They attacked the more impressive keyspace of $2^{95.6}$, assuming 100 characters of known plaintext, and found that they could achieve success 82% of the time with a total workload of only $2^{84.5}$. While they conceded that this was "far from practical," anything better than brute-force is

---

[22] This was pointed out in Stamp, Mark and Wing On Chan, "SIGABA: Cryptanalysis of the Full Keyspace," *Cryptologia*, Vol. 31, No. 3, July 2007, pp. 201–222.

[23] Although the reverse position *could* be utilized for the index rotors, it never actually was during World War II (see Mucklow, Timothy J., *SIGABA/ECM II: A Beautiful Idea*, Center for Cryptologic History, National Security Agency, Fort George G. Meade, Maryland, 2015, p. 29), so it would be reasonable to eliminate the factor of 2^5 in calculating the keyspace.

[24] Stamp, Mark and Wing On Chan, "SIGABA: Cryptanalysis of the Full Keyspace," *Cryptologia*, Vol. 31, No. 3, July 2007, pp. 201–222.

considered an attack, as it shows the cipher to have less than its apparent strength. They also commented that "it is certainly possible to improve on the attack presented here."[25]

The next attack to be published came from George Lasry in 2019. It also required some known plaintext, but only needed $2^{60.2}$ steps.[26]

9.6 Missing or Captured Machines?

> In late autumn 1944, an intelligence report about the Japanese capturing SIGABA shocked and disheartened readers until they learned that the "SIGABA" in question was a village in New Guinea.[27]

There was another SIGABA scare late in the war. On February 3, 1945, two U.S. Army sergeants in Colmar, France left their truck, which contained a SIGABA, unguarded as they entered a brothel. When they returned, the truck was gone. A frantic search was begun by counterintelligence, but only the trailer that had been attached to the truck was located. The SIGABA was still missing. General Eisenhower made locating the machine an extremely high priority, but weeks went by with no leads. U.S. and French counterintelligence agents formed a joint squad to try to find the SIGABA, General Fay B. Prickett became involved, inquiries were made with Swiss spies, and General Charles de Gaulle was even consulted to see if the French might have taken the device to learn how to strengthen their own cryptographic efforts![28]

Eventually, a tip from some French source led to a pair of safes lying in the mud after apparently having been dumped into the Giessen river from a bridge upstream. This was fantastic progress, but there was a third safe. Where was it? Men searched the banks and divers checked under the water in vain. In desperation, the river was dammed and a bulldozer dredged the bottom. Thus, days went by with no certainty that the efforts would be rewarded. Finally, on March 20, a reflection from the sun revealed the last safe to be mired in the mud at a spot previously underwater.[29]

After the recovery, the French explained that one of their military chauffeurs had lost his truck and simply "borrowed" the one with the SIGABA as a replacement. He ditched the safes (pushing them off a bridge into the Giessen) because he didn't want to be accused of stealing them![30] Prior to the recovery and this explanation coming forth, Eisenhower had no way of knowing whether SIGABA had been in the hands of the enemy or not. In the meanwhile, top level communications

[25] Stamp, Mark and Wing On Chan, "SIGABA: Cryptanalysis of the Full Keyspace," *Cryptologia*, Vol. 31, No. 3, July 2007, pp. 201–222.

[26] Lasry, George, "A Practical Meet-in-the-Middle Attack on SIGABA," in Schmeh, Klaus and Eugen Antal, editors, *Proceedings of the 2nd International Conference on Historical Cryptology, HistoCrypt 2019*, Mons, Belgium, June 23-26, 2019, Linköping University Electronic Press, Linköping, Sweden, pp. 41–49, available online at http://www.ep.liu.se/ecp/158/005/ecp19158005.pdf.

[27] *ULTRA and the Army Air Forces in World War II: An Interview with Associate Justice of the U.S. Supreme Court Lewis F. Powell, Jr.*, edited with an introduction and essay by Diane T. Putney, Office of Air Force History, United States Air Force, Washington DC, 1987, p. 96, available online at https://tinyurl.com/ydcdw78b.

[28] Kahn, David, *The Codebreakers*, second edition Scribner, New York, 1996, pp. 510–512.

[29] Kahn, David, *The Codebreakers*, second edition Scribner, New York, 1996, pp. 510–512.

[30] Kahn, David, *The Codebreakers*, second edition Scribner, New York, 1996, pp. 510–512.

had to continue. To be on the safe side, Eisenhower ordered the production of a new, differently wired, set of 15 rotors for all 10,060 SIGABAs.[31]

9.7 The End of SIGABA

SIGABAs were used extensively during the Korean War at higher echelons, but nothing lasts forever (Figure 9.14).[32]

> SIGABA and its temporary successor SIGROD were slowly replaced in the 1950s by the TSEC/KL-7 (ADONIS/POLLUX). The new cipher machine was an electronic-mechanical hybrid that employed a programmable cipher rotors/bezel assembly (eight rotors/thirty-six pins), cams, and vacuum tube technology along with a novel re-flexing principle. It was phased out of the U.S. military inventory in the early 1980s.[33]

Figure 9.14   A SIGABA, no longer needed, rests in a locked case. (Courtesy of the National Cryptologic Museum, Fort Meade, Maryland.)

Historians of cryptology long thought that the reason SIGABA went out of use was that it was no longer fast enough. However, this wasn't exactly true. The real reason is that the machine was unbreakable for the time period and it was feared that, if use continued, the Soviets might

---

[31] Mucklow, Timothy J., *SIGABA/ECM II: A Beautiful Idea*, Center for Cryptologic History, National Security Agency, Fort George G. Meade, Maryland, 2015, p, 30.

[32] Mucklow, Timothy J., *SIGABA/ECM II: A Beautiful Idea*, Center for Cryptologic History, National Security Agency, Fort George G. Meade, Maryland, 2015, p. 26.

[33] Mucklow, Timothy J., *SIGABA/ECM II: A Beautiful Idea*, Center for Cryptologic History, National Security Agency, Fort George G. Meade, Maryland, 2015, p. 41.

manage to learn the basic operating principles and create their own version.[34] So, it was decided to mothball SIGABA. If the cold war turned hot, however, it would quickly be brought out again with confidence that it was still secure. Luckily, the nukes haven't started flying (as of this writing), and, in the meanwhile, SIGABA has truly become too slow. Thus, the only machines brought out of mothballs were for the purpose of being placed in museums (Figure 9.15). One is on display at The National Cryptologic Museum and others (on loan from this museum) can be found at the National Museum of the U.S. Air Force at Wright-Patterson Air Force Base, The Dr. Dennis F. Casey Heritage Center on Joint Base San Antonio, and The National Archives of Australia.[35]

Figure 9.15 Frank Rowlett shows off a SIGABA to Admiral Bobby Ray Inman (Director NSA) and Ann Caracristi (NSA's first female Deputy Director). (Courtesy of the National Cryptologic Museum, Fort Meade, Maryland.)[36]

On January 16, 2001 a patent was finally granted for SIGABA (U.S. Patent 6,175,625 B1), following the filing from December 15, 1944.[37]

---

[34] Mucklow, Timothy J., *SIGABA/ECM II: A Beautiful Idea*, Center for Cryptologic History, National Security Agency, Fort George G. Meade, Maryland, 2015, pp. 26 and 41.

[35] Thanks to Robert Simpson, National Cryptologic Museum librarian, for providing this list.

[36] A similar image at https://www.nsa.gov/Resources/Everyone/Digital-Media-Center/Image-Galleries/Historical/NSA-60th-1970s/igphoto/2002138805/About-Us/EEO-Diversity/Employee-Resource-Groups/, identifies the people named here, but not the third man.

[37] Safford, Laurance F. and Donald W. Seller, *Control Circuits for Electric Coding Machines*, Patent No. 6,175,625 B1, United States Patent and Trademark Office, January 16, 2001, available online at https://tinyurl.com/ybmehc3h.

References and Further Reading

Army Security Agency, *History of Converter M-134-C*, Vol. 1, SRH-359, Army Security Agency, Washington, DC, no date, available online at https://tinyurl.com/yd6xy3w9. Note: SRH stands for Special Research Histories.

Bauer, Craig, "The Cryptologic Contributions of Dr. Donald Menzel," *Cryptologia*, Vol. 30, No. 4, 2006, pp. 306–339.

Budiansky, Stephen. *Battle of Wits: The Complete Story of Codebreaking in World War II*, The Free Press, New York, 2000.

Chan, Wing On, *Cryptanalysis of SIGABA*, master's thesis, San Jose State University, San Jose, California, May 2007, available online at https://web.archive.org/web/20161005080518/http://cs.sjsu.edu/faculty/stamp/students/Sigaba298report.pdf.

Johnson, Thomas M., "Search for the Stolen Sigaba," *Army*, Vol. 12, February 1962, pp. 50–55.

Kahn, David, *The Codebreakers*, second edition, Scribner, New York, 1996.

Kelley, Stephen J., *Big Machines: Cipher Machines of World War II*, Aegean Park Press, Laguna Hills, California, 2001.

Kwong, Heather Ellie, *Cryptanalysis of the Sigaba Cipher*, master's thesis, San Jose State University, San Jose, California, December 2008, available online at https://scholarworks.sjsu.edu/cgi/viewcontent.cgi?referer=https://www.google.com/&httpsredir=1&article=4626&context=etd_theses.

Lasry, George, "A Practical Meet-in-the-Middle Attack on SIGABA," in Schmeh, Klaus and Eugen Antal, editors, *Proceedings of the 2nd International Conference on Historical Cryptology, HistoCrypt 2019*, Mons, Belgium, June 23–26, 2019, Linköping University Electronic Press, Linköping, Sweden, pp. 41–49, available online at http://www.ep.liu.se/ecp/158/005/ecp19158005.pdf.

Lee, Michael, *Cryptanalysis of the Sigaba*, master's thesis, University of California, Santa Barbara, June 2003, available online at http://ucsb.curby.net/broadcast/thesis/thesis.pdf.

Mucklow, Timothy and LeeAnn Tallman. "SIGABA/ECM II: A Beautiful Idea." *Cryptologic Quarterly (CQ)*, Vol. 30, 2011, pp. 3–25.

Mucklow, Timothy J., *SIGABA/ECM II: A Beautiful Idea*, Center for Cryptologic History, National Security Agency, Fort George G. Meade, MD, 2015, available online at https://tinyurl.com/yaqc29en.

Mundy, Liza, *Code Girls: The Untold Story of the American Women Code Breakers of World War II*, Hachette Books, New York, 2017.

Pekelney, Rich, "Electronic Cipher Machine (ECM) Mark II," *San Francisco Maritime National Park Association*, http://www.maritime.org/ecm2.htm, 2010.

Rowlett, Frank B., *The Story of Magic: Memoirs of an American Cryptologic Pioneer*, Aegean Park Press, Laguna Hills, California, 1999.

Safford, Captain Laurance. *History of Invention and Development of the Mark II ECM*, SRH-360, United States Navy OP-20-S-5, Office of the Chief of Naval Operations, Washington, DC, October 30, 1943. This history is available at NARA (National Archives and Records Administration) RG 457, Box 1124, College Park, Maryland. Note: SRH stands for Special Research Histories.

Safford, Laurance F. and Donald W. Seller, *Control Circuits for Electric Coding Machines*, Patent No. 6,175,625 B1, United States Patent and Trademark Office, January 16, 2001, available online at https://tinyurl.com/ybmehc3h. Thus is the SIGABA patent.

Savard, John J. G. and Richard S. Pekelney, "The ECM Mark II: Design, History and Cryptology," *Cryptologia*, Vol. 23, No. 3, July 1999, pp. 211–228.

Stamp, Mark and Wing On Chan, "SIGABA: Cryptanalysis of the Full Keyspace," *Cryptologia*, Vol. 31, No. 3, July 2007, pp. 201–222.

Stamp, Mark and Richard M. Low, *Applied Cryptanalysis*, Wiley, Hoboken, New Jersey, 2007.

Sullivan, Geoff, "The ECM Mark II: Some Observations on the Rotor Stepping," *Cryptologia*, Vol. 26, No. 2, April 2002, pp. 97–100.

Wilcox, Jennifer. *Sharing the Burden: Women in Cryptology during World War II*, Center for Cryptologic History, National Security Agency Ft. George G. Meade, Maryland, 1998.

Chapter 10

Enciphering Speech[1]

Mathematical ideas seem to inevitably find applications that were undreamed of when they were originally discovered. This chapter details how modular arithmetic and logarithms helped the Allies win World War II.

10.1 Early Voice Encryption

Voice encryption, also known as ciphony, goes back as far as the 1920s, when AT&T put an analog system into use. During this decade, inverters swapped high tones with low tones, and vice versa. Expressing it more mathematically, the frequency p of each component is replaced with $s - p$, where s is the frequency of a carrier wave. The equation reveals a major weakness with this form of encryption. Namely, tones near the middle are hardly changed. So, that dull professor you remember not too fondly wouldn't be able to speak securely using an inverter, if his tone of choice was near the middle (Figure 10.1).

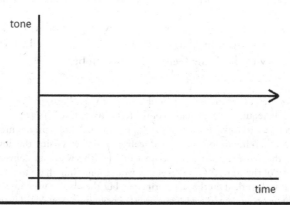

Figure 10.1   Tone as a function of time for some professors.

---

[1]   This chapter originally appeared in a slightly different form as Bauer, Craig, "How Modular Arithmetic Helped Win World War II," *Cryptologic Quarterly* (*CQ*), 2015-01, Vol. 34, No. 1, pp. 43-57, Center for Cryptologic History, National Security Agency, Fort George G. Meade, Maryland.

Actually, nobody could speak securely using an inverter. This system protected only against casual eavesdropping and could be easily inverted back by determined amateurs. There was no key as such, and inverters are not hard to build. In some cases, the devices were not even needed. With practice it is possible to understand much inverted speech, even if it isn't that old professor of yours speaking.

AT&T and RCA offered a slightly more sophisticated scheme in 1937. Known as the A-3 Scrambler, this system split the speech into five channels (aka subbands), each of which could be inverted, and shuffled them before transmitting. However, this was still weak, and it was implemented in an especially weak manner. Because there are only $5! = 120$ ways to reorder the 5 subbands and $2^5 = 32$ ways to decide which (if any) of the subbands will be inverted, we have a total of $(120)(32) = 3,840$ ways to scramble the speech. Thus, the key space is way too small. If the attacker knows how the system works, he or she could simply try all of the possibilities. Even worse, many of these keys failed to garble the speech sufficiently to prevent portions of it from remaining understandable. Worst of all, of the 11 keys deemed suitable for use, only 6 were actually used! They were applied in a cycle of 36 steps, each lasting 20 seconds, for a full period of 12 minutes.[2]

Hence, like the inverters of the 1920s, the A-3 Scrambler was understood to offer "privacy, not security." A good analogy is the privacy locks on interior doors of homes. If someone walks up to a home bathroom that is in use, and the lock prevents the doorknob from turning, he'll think, "Oh, someone's in there," and walk away. Privacy is protected. However, there's no real security. Someone intent on entering that bathroom will not be stopped by the lock. In the same manner, a scrambler would protect someone on a party line,[3] but could not be expected to protect national secrets against foreign adversaries.

When President Franklin D. Roosevelt and Prime Minister Winston Churchill spoke on the phone, they needed real security, not just privacy, yet they initially used the A-3 Scrambler! It was solved by the Germans by September 1941, after only a few months' work.[4] As the following quotes show, allies on both sides of the Atlantic were aware of the problem.

> The security device has not yet been invented which is of any protection whatever against the skilled engineers who are employed by the enemy to record every word of every conversation made.—British Foreign Office Memorandum, June 1942[5]

> In addition, this equipment furnishes a very low degree of security, and we know definitely that the enemy can break the system with almost no effort.—Colonel Frank McCarthy, Secretary to the Army General Staff, October 1943[6]

---

[2] Kahn, David, *The Codebreakers*, second edition, Scribner, New York, 1996, p. 554.

[3] Younger readers will likely require an explanation of the term "party line." As a first step, imagine a house with phones that actually connect to jacks in the walls (i.e., landlines). A boy upstairs might pick up the phone in his room and hear his dad talking to someone. He'd realize his dad was using the downstairs phone and hang up. All of the phones in the house were wired via a common line. This would be convenient for conference calls, but inconvenient the rest of the time. A family member would sometimes have to wait his turn, when wanting to make a call. "Party lines" worked on the same principle, but the phones were in different homes. That is, in the old days, you might be on a party line with one or more neighbors. You could listen in on their calls, if you desired, but would hopefully respect their privacy and hang up when you discovered the line was in use.

[4] Kahn, David, *The Codebreakers*, second edition, Scribner, New York, 1996, pp. 555-556.

[5] British Foreign Office memorandum FO/371/32346. Taken here from Hodges, Andrew, *Alan Turing: The Enigma* Simon & Schuster, New York, 1983, p. 236.

[6] From a letter to Harry Hopkins, assistant to President Roosevelt. Taken here from Mehl, Donald E., *The Green Hornet*, self-published, 1997, p. 5.

Given that the Americans and the British knew that the system they were using for voice encryption offered no security, it's natural to ask why they didn't use something better. The answer is that securing speech with encryption is much more difficult than encrypting text. There are several reasons why this is so, but one of the most important is redundancy. Redundancy in speech allows us to comprehend it through music, background noise, bad connections, mumbling, other people speaking, etc. Text is at least 50% redundant (in other words, removing half of the letters from a given paragraph does not typically prevent it from being reconstructed—see Section 11.3 for more details), but speech is much more redundant and it is hard to disguise because of this.

Speech that is scrambled in the manner of the A-3 Scrambler can be reconstructed using a sound spectrograph, which simply involves plotting the tones and reassembling them like a jigsaw puzzle. So, although splitting the voice into more channels would increase the number of possible keys, the attacker could simply reassemble what amounts to a jigsaw puzzle with more pieces. A successful voice encryption system would have to operate in a fundamentally different manner than inverting and shuffling.

10.2 The Cost of Insecurity

There was a very high cost associated with the lack of a secure voice system. Shortly before the Japanese attack on Pearl Harbor, American cryptanalysts broke a message sent in the Japanese diplomatic cipher known as Purple. It revealed that Japan would be breaking off diplomatic relations with the United States. In the context of the times, this meant war. General Marshall knew he needed to alert forces at Pearl Harbor to be prepared for a possible attack, but, not trusting the A-3 Scrambler, he refused to use the telephone. If the Japanese were listening in, they would learn that their diplomatic cipher had been broken, and would likely change it. The United States would thus lose the benefit of the intelligence those messages provided. The result was that the message was sent by slower means and didn't arrive until after the attack.[7]

10.3 SIGSALY—A Solution from the Past Applied to Speech

Fortunately, the simpler problem of enciphering text had been mastered—a perfect system had been found, namely the one-time pad (see Section 2.7)—and it was possible to create an analog of it for voice. The new device would add random values to the sound wave. It's a method completely different from inverting and reordering subbands. It's the story of SIGSALY.

The following are equivalent:

1. SIGSALY
2. RC-220-T-1
3. The Green Hornet
4. Project X-61753
5. Project X (the atomic bomb was Project Y)
6. X-Ray
7. Special Customer

Proof—see the literature.

---

[7] For a reasoned argument that it would have made little difference if the warning had arrived in time, see Christensen, Chris, "Review of two collections of essays about Alan Turing," *Cryptologia*, Vol. 44, No. 1, 2020, pp. 82-86.

As indicated above, SIGSALY, the ciphony system that would replace the A-3 Scrambler for Roosevelt and Churchill (and others), had many different names. This is an indication of its importance. The sixth name may be seen on the cover of a formerly classified directory for the system (Figure 10.2). The cover is certainly attention grabbing, but the contents are quite dry by comparison.

Figure 10.2   Is this how we should market texts?

Before getting into the details of how SIGSALY worked, a picture is presented (Figure 10.3). Upon first seeing this image, I asked, "So where in the room is SIGSALY?" I wasn't sure which item I should be looking at. The answer was, "It *is* the room!" The result of the quest for secure voice communication led to a 55-ton system that took up 2,500 square feet. In fact, the image only shows *part* of SIGSALY. It literally filled a house. Some reflection makes sense of why the project didn't turn out a more compact device.

Necessity is the mother of invention, so it's not surprising that the need to keep voice communications secure from Nazi cryptanalysts is what finally motivated the design of a secure system. But this impetus also meant that no time could be wasted. The designers didn't have the luxury of taking a decade to make a system of utmost elegance. Instead, they based it on earlier technology that could be readily obtained, saving much time. The heart of the system was a *vocoder*, which is a portmanteau of *voice coder*. The original intent of such devices was to digitize speech so that it might be sent on undersea phone cables using less bandwidth, thus reducing costs. Due to the aforementioned high redundancy of human speech, compression down to 10 percent of the original was found to be possible, while still allowing the original meaning to be recovered.[8] For SIGSALY, the compression was a bonus. The important thing was to digitize the voice, so that a

---

[8] Tompkins, Dave, *How to Wreck a Nice Beach*, Stopsmiling Books, Chicago, Illinois, 2010, p. 23.

Figure 10.3 A view of SIGSALY. (from http://www.cryptologicfoundation.org/content/A-Museum-Like-No-Other/COMSEC.shtml.)

random digital key could be added to it in the manner of the one-time pad. Off-the-shelf vocoder technology took up much space!

For those interested in hearing how early vocoders transformed speech, a recording of a Bell Labs vocoder from 1936 may be heard at http://www.complex.com/music/2010/08/the-50-greatest-vocoder-songs/bell-telephone-laboratory.

Middle-aged readers might find the sound reminds them of the Cylons in the original (1970s) *Battlestar Galactica* TV series. Indeed, this sound effect was produced using a vocoder.[9] Decades earlier, Secretary of War Henry Stimson had remarked of a vocoder, "It made a curious kind of robot voice."[10]

This brings us to an interesting point. Vocoders sound cool. For this reason, many musicians have used them. Dave Tompkins, a hip-hop journalist, aware of the use of vocoders in voice encryption and music, wrote a very entertaining book that examines both applications. The front cover of this book appears in Figure 10.4. The title of Tompkins's book arose from the manner in which vocoders were tested. Various phrases would be passed through the vocoders, and listeners, ignorant of what they were supposed to hear, would try to determine the messages. In one instance, the phrase "How to recognize speech" was misheard as "How to wreck a nice beach." Clearly that vocoder was not suitable to military applications in which a slight misunderstanding could have a calamitous effect.

---

[9] A Cylon from a 1977 episode of *Battlestar Galactica* may be heard at http://www.youtube.com/watch?v=0ccKPSVQcFk&feature=endscreen&NR=1.

[10] Tompkins, Dave, *How to Wreck a Nice Beach*, Stopsmiling Books, Chicago, Illinois, 2010, p. 63.

Figure 10.4 For a book with cryptologic content, Tompkins's work contains a record-shattering amount of profanity. (Courtesy of Dave Tompkins).

The diverse applications of the vocoder, detailed in Tompkins's book, are represented by Figures 10.5 and 10.6.

The vocoder used by SIGSALY broke the speech into ten channels (from 150 Hz to 2950 Hz), and another channel represented pitch. Some sources describe the pitch as being represented by a pair of channels. Both points of view can be considered accurate, as will be made clear shortly. Each channel was 25 Hz, so the total bandwidth (with two pitch channels) was $(12)(25) = 300$ Hz. Ultimately, the communications were sent at VHF. The digitization of each channel was done on a senary scale; that is, the amplitude of each signal was represented on a scale from 0 to 5, inclusive. A binary scale was tried initially, but such rough approximation of amplitudes didn't allow for an understandable reconstruction of the voice on the receiving end.[11] For some reason the pitch had to be measured even more precisely, on a scale from 0 to 35. Because such a scale can be represented by a pair of numbers between 0 and 5, pitch may be regarded as consisting of two channels.

---

[11] Hodges, Andrew, *Alan Turing: The Enigma*, Simon & Schuster, New York, 1983, p. 246.

Figure 10.5 These men knew nothing about the future use of vocoders by musicians. (Courtesy of the National Cryptologic Museum, Fort Meade, Maryland).

Figure 10.6 Musicians, represented here by Michael Jonzun (and a Roland SVC vocoder), knew nothing of the use of vocoders by the military. (Courtesy of Dave Tompkins and Michael Jonzun.)

Before we get to modular arithmetic, the mathematical star of this tale, we examine how logarithms contributed to winning the war. When discretizing sound, it seems reasonable to represent the amplitude using a linear scale, but the human ear doesn't work in this fashion. Instead, the ear distinguishes amplitudes at lower amplitudes more finely. Thus, if we wish to ease the ability of the ear to reconstruct the sound from a compressed form, measuring the amplitude on a logarithmic scale is a wiser choice. This allows for greater discernment at lower amplitudes. Thus, the difference in amplitude between signals represented by 0 and 1 (in our senary scale) is much smaller than the difference in amplitude between signals represented by 4 and 5. This technique goes by the technical name *logarithmic companding*, where *companding* is itself a compression of *compressing* and *expanding*.[12] The concept described above will already have been familiar to all readers. Who hasn't used heard of the (logarithmic) decibel scale for measuring sound intensity?

Having discretized the speech, we're ready to add the random key. With both the speech and the key taking values between 0 and 5, the sum will always fall between 0 and 10. SIGSALY, however, performed the addition modulo 6, so that the final result remained between 0 and 5, as represented in Figure 10.7.

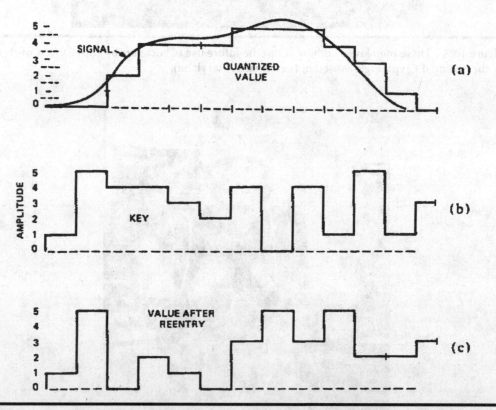

Figure 10.7 The mod 6 addition of the key was referred to as "reentry" by the creators of SIGSALY. (From Boone, James V. and Peterson, R. R., *The Start of the Digital Revolution: SIGSALY Secure Digital Voice Communications in World War II*, Center for Cryptologic History, National Security Agency, Fort Meade, Maryland, July 2000, p. 19.)

---

12 The pitch channel, however, wasn't companded.

Why was the addition of the key done in this complicated manner? Why not just add without the mod 6 step? Three reasons are given below.

1. The mod 6 step was Harry Nyquist's idea.[13] Students of information theory will recognize this name and, for them, it certainly lends a stamp of authority to support the inclusion of this step. But an argument from authority is not a proof! Fortunately, we have two more reasons.
2. If we don't perform the mod 6 step, then a cipher level of 0 can arise only from both message and key being 0. So, whenever a 0 is the output, an interceptor will know a portion of the signal. Similarly, a great cipher level of 10 can only arise from both message and key being 5. Hence, without the mod 6 step, an interceptor would be able to immediately identify $2/36 \approx 5.5\%$ of the signal from the simple analysis above.
3. Simply adding the key without the mod step would result in random increases in amplitude, which may be described as hearing the message over the background noise of the key. Are you able to understand a friend talking despite the white noise produced by an air-conditioner or chainsaw in the background?

SIGSALY enciphered every channel in this manner using a separate random key for each. A simplified schematic for the overall encryption process is provided in Figure 10.8.

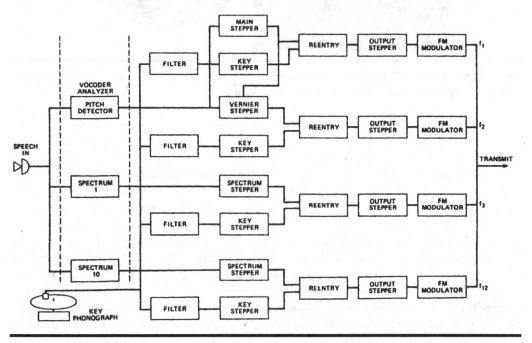

Figure 10.8 An incredibly simplified schematic of a SIGSALY transmit terminal. (Courtesy of the National Cryptologic Museum, Fort Meade, Maryland.)

Figure 10.8 shows the speech entering the system on the left-hand side and getting broken down into a pitch channel (pitch detector) and ten voice channels (spectrum 1 through spectrum 10). There are steps, not discussed here, both before and after the mod 6 (reentry) takes place.

---

[13] Mehl, Donald E., *The Green Hornet*, self-published, 1997, p. 38.

The "missing steps" are of greater interest to engineers than mathematicians, and can be found in Donald E. Mehl's book *The Green Hornet*.[14]

At this point I'd like to draw your attention to the lower left-hand corner of Figure 10.8. The "key phonograph" is exactly what it sounds and looks like. The source of the key that needed to be combined with each channel was simply a record (see Figure 10.9). The one-time key for voice encryption was codenamed SIGGRUV. As with text, the key was added to encipher and subtracted to decipher. Taking the form of a record, a built-in safety mechanism caused communication to cease if the key stopped. Otherwise, the speaker would suddenly be broadcasting in the clear.

Figure 10.9   A SIGSALY turntable and record, with a modern CD for scale. (Courtesy of the National Cryptologic Museum, Fort Meade, Maryland.)

The digitized speech was sampled 50 times per second, so to separately encipher all of the channels, the record had to be simultaneously playing twelve tones at different frequencies, and these tones had to change every fiftieth of a second. It's natural to ask why the sampling rate was 50 times per second and not higher or lower. The fundamental unit of speech, known as a phoneme, has a duration of about a fiftieth of a second, so the sampling rate is just high enough to allow it to be captured. A higher sampling rate is not needed to make the digitized voice comprehensible and would worsen the synchronization problem—the record at the receiving terminal, used to subtract the key, must be synchronized with the incoming message, if there is to be any hope of recovering it! While we're on the topic of synchronization, it should be mentioned that the records contained tones for purposes other than encryption. For example, a tone at one particular frequency was used for fine-tuning the synchronization.

Ideally the keys would be random, a condition simulated for SIGGRUV by recording thermal noise backward. None of these records would become classic tunes, but the military was content with one-hit wonders. Indeed, the system would become vulnerable if the same record were ever replayed. Although not labeled as such, the implicit warning was "Don't Play it Again, Uncle Sam!," and the records were destroyed after use.

Vinyl aficionados may have noticed that the record in Figure 10.9 is unexpectedly large in comparison to the CD. SIGSALY's records measured sixteen inches and could be played from start to finish in twelve minutes. Over 1,500 of these key sets were made.[15]

---

[14] Mehl, Donald E., *The Green Hornet*, self-published, 1997.
[15] Tompkins, Dave, *How to Wreck a Nice Beach*, Stopsmiling Books, Chicago, Illinois, 2010, p. 68.

10.4 Plan B

Once the SIGSALY installations were in place, all that was necessary for communication was that each location have the same record. Initially spares were made, but as confidence was gained, only two copies of each record were made. Still, there was a Plan B.

Figure 10.10 looks like a locker room, but it is simply SIGSALY's back-up key, codenamed SIGBUSE. If for some reason the records couldn't be used for keying purposes, SIGBUSE could generate a pseudorandom key mechanically.

Figure 10.10   SIGSALY's back-up key SIGBUSE. (Courtesy of the National Cryptologic Museum, Fort Meade, Maryland.)

Because SIGSALY would link Roosevelt and Churchill, the Americans and the British needed to be satisfied that it was secure. The British had the added concern that the operating teams, which would consist of Americans, even in London, would hear everything. Thus, in January 1943, the British sent their top cryptanalyst, Alan Turing, to America to evaluate the system. After much debate, probably reaching President Roosevelt,[16] Turing was allowed access to details of the closely guarded secret project.

Turing helped by suggesting improvements to the SIGBUSE key, and he reported to the British, "If the equipment is to be operated solely by U.S. personnel it will be impossible to prevent them listening in if they so desire." In reality, the Americans were often so focused on their jobs they had no idea what was actually said.

---

[16] We have no proof, but Mehl, Donald E., *The Green Hornet*, self-published, 1997, p. 69; Hodges, Andrew. *Alan Turing: The Enigma*. Simon & Schuster, New York, 1983, p. 245; and Tompkins, Dave, *How to Wreck a Nice Beach*, Stopsmiling Books, Chicago, Illinois, 2010, p. 59, all believe the matter reached Roosevelt. In any case, Secretary of War Stimson resolved it.

Turing's examination of SIGSALY inspired him to create his own (completely different) system, Delilah. Turing's report on Delilah appeared publicly for the first time in the October 2012 issue of *Cryptologia*.[17]

Ultimately, SIGBUSE turned out to be wasted space. The records never failed, so the alternate key was never used. A more critical part of SIGSALY was the air-conditioning system. It is shown in Figure 10.11. A voice encryption system that fills a house requires a cooling system on the same scale!

Figure 10.11 SIGSALY's air conditioning system. Donald Mehl appears on the right in this photo. (Courtesy of the National Cryptologic Museum, Fort Meade, Maryland.)

10.5 SIGSALY in Action

In November 1942 an experimental station was installed in New York, and in July 1943 a final version was activated linking Washington, DC, and London. This marked the first transmission of digital speech and the first practical "Pulse Code Modulation."[18] Although the bandwidth-compressing vocoder was described earlier in this chapter as preexisting technology (which it was), it had not become practical enough for use. Eventually, SIGSALY installations made it to Algiers, Berlin, Brisbane, Frankfurt, Guam, Hawaii, Manila, Oakland, OL-31 (on a barge), Paris, and Tokyo.[19]

---

[17] Turing, Alan M. and Donald Bayley, "Report on speech secrecy system DELILAH, a Technical Description Compiled by A. M. Turing and Lieutenant D. Bayley REME, 1945–1946," *Cryptologia*, Vol. 36, No. 4, October 2012, pp. 295–340.

[18] This refers to the digitization process.

[19] Mehl, Donald E., *The Green Hornet*, self-published, 1997, p. 86.

Figure 10.12   Another view of SIGSALY. (Courtesy of the David Kahn Collection, National Cryptologic Museum, Fort Meade, Maryland.)

Figure 10.12 provides another view of a SIGSALY installation. In this one, a phone is clearly visible, but this is not what the caller would be using. The phone you see was used by a member of the operating team to make sure synchronization was being maintained.

A separate room existed to allow the user(s) to converse in a more comfortable condition (Figure 10.13).

Figure 10.13   SIGSALY users—fighting the Germans and Japanese... and loving it! (Courtesy of the National Cryptologic Museum, Fort Meade, Maryland.)[20]

---

[20] An alternate caption for this image is "SIGSALY: Your digital pal who's fun to be with!"

10.6 SIGSALY Retires

SIGSALY received an honorable discharge, having never been broken. The Germans didn't even recognize it as enciphered speech. They thought it was just noise or perhaps a teletype signal. The sound they heard was similar to the buzz in the introduction of the *Green Hornet* radio show of that era. Although they might not have been familiar with the program, Americans certainly were, and this is why the system was sometimes referred to as the Green Hornet.

Although we now recognize SIGSALY as a complete success, back during World War II, General Douglas MacArthur didn't trust it! Happily, others did, and the rewards of the instant communication it provided were reaped. Given its success, it's natural to ask why it wasn't kept in use longer. There were several reasons:

1. It weighed 55 tons and had a seventy-ton shipping weight.
2. It took up 2,500 square feet.
3. It cost $250,000–$1,000,000+ per installation.
4. It converted 30 kilowatts of power into 1 milliwatt of low-quality speech.[21]
5. The deciphered speech sounded like Donald Duck.[22]

Technological developments, over the decades that followed, rapidly diminished the space needed for secure voice encryption. Still, JFK's system (Figure 10.14) looked decidedly less cool.

Newsday *Center Section*

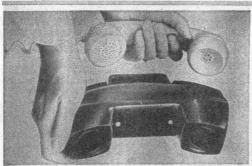

Guarding the privacy of communication of President Kennedy, as well as the security of high level diplomatic conversations around the world is the telephone scrambler, which electronically jumbles speech. The device's strange appearance, as shown above in a portable model, is matched only by the unusual role it has played and is playing in world affairs.

THE SOUND OF SECRECY

By David Kahn   radiotelephones in his official car is   broken simply by repeated listening. The
                equipped with a scrambler. The White   so-called New Zealand device, which com-

Figure 10.14 President Kennedy's voice encryption system (Courtesy of the David Kahn Collection, National Cryptologic Museum, Fort Meade, Maryland.)

The system in Figure 10.14 looks like something Maxwell Smart of the TV series *Get Smart* might have used. What would the next step be, three phones?

---

[21] Hodges, Andrew, *Alan Turing: The Enigma*, Simon & Schuster, New York, 1983, p. 247.

[22] General Eisenhower complained that it made his wife sound like an old woman. The system was optimized for male voices, and as a result, deciphered female voices sounded worse.

Long since retired, SIGSALY was finally declassified in 1976. This allowed patents, applied for decades earlier, to finally be granted. Three recipients, also long since retired (from Bell Telephone Laboratories Inc.), were the engineers Robert C. Nathes, Ralph K. Potter, and P. W. Blye.[23]

A mock-up of a portion of SIGSALY (Figure 10.15) may be seen today at the National Cryptologic Museum adjacent to Ft. Meade, Maryland. This museum also has an excellent library that includes the David Kahn Collection.[24] Kahn is widely regarded as cryptology's greatest historian and, prior to his donation, his collection was the largest in private hands.

Figure 10.15   The National Cryptologic Museum's SIGSALY mock-up, which has been scaled down since this photo was taken.

In Section 10.2, we saw the consequences that may be faced when a nation is without a secure voice encryption system. We close with a reminder of the advantage gained when a nation does possess such a system.

10.7 Voice vs. Text

Text systems take longer to encipher and decipher than voice systems. The situation was far worse during the precomputer era of World War II. Then, an enciphered message might take an hour to reach readable form. Sometimes this was too long to wait! The instant communication voice encryption allows can make a tremendous difference when speed is of the essence. In Section 8.7, the solution provided by another voice system, the Navajo code talkers, was detailed. The rapid communication made possible by these men allowed for equally rapid and coordinated movement of troops, in response to changing conditions. This was an advantage the Japanese did not possess. But code talkers couldn't be used forever, while digital voice encryption has been continuously improved up to the present day.

---

[23] Jones, Stacy V., "From WWII era 'Green Hornet' Patent Awarded," *The New York Times*, July 3, 1976, p. 27, found at the National Cryptologic Museum, David Kahn Collection, Folder 12-7.

[24] Hamer, David, "The David Kahn Collection at NSA's National Cryptologic Museum," *Cryptologia*, Vol. 35, No. 2, April 2011, pp. 110-113.

References and Further Reading

Bauer, Craig, "Alan Turing and Voice Encryption: A Play in Three Acts," in Cooper, S. Barry and Jan van Leeuwen, editors, *Alan Turing: His Work and Impact*, Elsevier, 2013.

Bauer, Craig, "How Modular Arithmetic Helped Win World War II," *Cryptologic Quarterly (CQ)*, 2015-01, Vol. 34, No. 1, pp. 43–57, Center for Cryptologic History, National Security Agency, Fort George G. Meade, Maryland. The current chapter is a slight modification of this paper.

Bauer, Craig, "The Early History of Voice Encryption," in Floyd, Juliet and Alisa Bokulich, editors, *Philosophical Explorations of the Legacy of Alan Turing, Turing 100*, Boston Studies in the Philosophy and History of Science, Vol. 324, Springer, Cham, Switzerland, 2017, pp. 159–187.

Boone, James V. and R. R. Peterson. *The Start of the Digital Revolution: SIGSALY Secure Digital Voice Communications in World War II*, Center for Cryptologic History, National Security Agency, Fort George G. Meade, Maryland, July 2000, available online at https://www.nsa.gov/Portals/70/documents/about/cryptologic-heritage/historical-figures-publications/publications/wwii/sigsaly.pdf.

Hodges, Andrew. *Alan Turing: The Enigma*, Simon & Schuster, New York, 1983.

Kahn, David, *The Codebreakers*, second edition, Scribner, New York, 1996.

Mehl, Donald E. *The Green Hornet*, self-published, 1997.

Paul, Jon D., "Re-creating the Sigsaly Quantizer: This 1943 analog-to-digital converter gave the allies an unbreakable scrambler," *IEEE Spectrum*, Vol. 56, No. 2, February 2019, pp. 16–17, available online at https://ieeexplore.ieee.org/stamp/stamp.jsp?arnumber=8635806.

Tompkins, Dave, *How to Wreck a Nice Beach*, Stopsmiling Books, Chicago, Illinois, 2010.

Triantafyllopoulos, Christos, "Intercepted Conversations - Bell Labs A-3 Speech Scrambler and German Codebreakers," *Christos military and intelligence corner*, February 2, 2012, https://chris-intel-corner.blogspot.com/2012/02/intercepted-conversations-bell-labs-3.html.

Turing, Alan M. and Donald Bayley, "Report on Speech Secrecy System DELILAH, A Technical Description Compiled by A. M. Turing and Lieutenant D. Bayley REME, 1945–1946," *Cryptologia*, Vol. 36, No. 4, October 2012, pp. 295–340.

Weadon, Patrick D., *Sigsaly Story*, National Security Agency, Fort George G. Meade, Maryland, 2009, available online at https://www.nsa.gov/about/cryptologic-heritage/historical-figures-publications/publications/wwii/sigsaly-story/.

MODERN CRYPTOLOGY

There are two kinds of cryptography in this world: cryptography that will stop your kid sister from reading your files, and cryptography that will stop major governments from reading your files. This book is about the latter.

– **Bruce Schneier**[1]

---

[1] Schneier, Bruce, *Applied Cryptography*, second edition, John Wiley & Sons, New York, 1996, p. xix.

Chapter 11

Claude Shannon

In Section 7.7, we took a brief look at the life of Alan Turing, who is considered by many to be the father of computer science. If anyone could be considered the American version of Turing, it would be Claude Shannon, who is known as "the father of information theory."

11.1 About Claude Shannon

In addition to their work in the budding field of computer science, Alan Turing and Claude Shannon (Figure 11.1) also had atheism in common and an eccentric nature (Shannon would sometimes juggle while riding a unicycle in the halls of Bell Labs, although he is not reported to have ever had a gas mask on while biking[1]). Shannon spent the 1940–1941 academic year at Princeton's Institute for Advanced Study; Turing had earned his doctorate in mathematics from Princeton University in 1938. They met during World War II, when Turing came to Washington, DC to share cryptanalytic techniques with the Americans. Shannon worked at Bell Labs at the time, and Turing would meet with him in the cafeteria. Both men worked on SIGSALY, the voice encryption system ultimately used for wartime communications between President Roosevelt and Prime Minister Churchill (see Chapter 10).

We've already made use of some of Shannon's results in previous chapters. Recall that it was Shannon who first proved the one-time pad is unbreakable when properly used. Also, he came up with a way of calculating unicity distances, the length of ciphertext at which we can expect a unique solution. To understand this concept more fully, we must first examine how Shannon found a way to measure the information content of a message.

11.2 Measuring Information

How much information does a given message contain? If the message verifies something we expect, it can be argued that the amount of information is less than if it indicates the unexpected.

---

[1] Golomb, Solomon W., Elwyn Berlekamp, Thomas M. Cover, Robert G. Gallager, James L. Massey, and Andrew J. Viterbi, "Claude Elwood Shannon (1916-2001)," *Notices of the American Mathematical Society*, Vol. 49, No. 1, January 2002, pp. 8–16, p. 10 cited here.

Figure 11.1   Claude Shannon (1916–2001). (Attribution 2.0 Generic (CC BY 2.0) by Tekniska museet, https://www.flickr.com/photos/tekniskamuseet/6832884236/sizes/o/.)

For example, suppose the message is a weather report and the various possibilities, along with their probabilities are as follows:

M_1 = Sunny	0.05	
M_2 = Cloudy	0.15	
M_3 = Partly Cloudy	0.70	
M_4 = Rain	0.10	

A report of "Sunny" is more surprising than a report of "Partly Cloudy" and can therefore be said to convey more information.

Let the function that measures the amount of information conveyed by a message be denoted by $I(M)$, where M is the message. In general, if the probability of M_i is greater than the probability of M_j, we should have $I(M_i) < I(M_j)$. Also, if we receive two weather reports (for different days), the total amount of information received, however we measure it, should be the sum of the information provided by each report; that is, $I(M_iM_j) = I(M_i) + I(M_j)$.

The input of the function I is shown to be the message, but I should really just be a function of the probability of the message; that is, if two messages are equally probable, they should be evaluated as containing the same amount of information. We also want the function I to change *continuously* as a function of this probability. Informally, a very small change in the probability of a message should not cause a "jump" (discontinuity) in the graph of I.

Can you think of any functions that fit all of the conditions given above?

This is the manner in which Claude Shannon approached the problem. Rather than guess at a formula, he formulated rules, like those above, and then sought functions that fit the rules.[2] As an extra clue as to what the function could be, consider the fact that if a message M has probability 1, no real information is conveyed; that is, $I(M) = 0$ in this case.

Shannon found that there was essentially just one function that satisfied his conditions, namely the negation of the logarithm.[3] The amount of information contained in a message M is thus given by

$$-K\sum_i \log_2(M_i)$$

where the sum is taken over the individual components of the message, using the probability of each component as the value for M_i. If the message consists of seven weather reports, we would sum seven terms to get the total amount of information. There is some flexibility in that K may be any positive constant, but since this only amounts to choosing units, Shannon simplified matters by taking it to be one.

Shannon presented his result as a weighted average.[4] Using his formula, given below, one can calculate the average amount of information conveyed by a message selected from a set of possibilities with probabilities given by p_i.

$$-K\sum_i p_i \log_2(p_i)$$

For example, a single weather report, selected from the four possibilities given above, in accordance with its probability, will have an average information content of

$$-[(0.05)\log_2(0.05) + (0.15)\log_2(0.15) + (0.70)\log_2(0.70) + (0.10)\log_2(0.10)] \approx 1.319$$

However, a single report may contain more or less information than this average, depending on how likely it is. In probability, such a calculation is known as an *expected value*. The units of this measure will be discussed shortly. The formula is simple, but giving a name to it proved tricky.

Looking back, during a 1961 interview, Shannon remarked,

> My greatest concern was what to call it. I thought of calling it 'information,' but the word was overly used, so I decided to call it 'uncertainty.' When I discussed it with John von Neumann, he had a better idea. Von Neumann told me, 'You should call it entropy, for two reasons. In the first place your uncertainty function has been used in statistical mechanics under that name, so it already has a name. In the second place, and more important, no one knows what entropy really is, so in a debate you will always have the advantage.'[5]

---

[2] Shannon, Claude E., "A Mathematical Theory of Communication," reprinted with corrections from *The Bell System Technical Journal*, Vol. 27, pp. 379–423, 623–656, July, October, 1948, p. 10.

[3] Nyquist and Hartley both made use of logarithms in their work on information theory, prior to Shannon. It seems that it was an idea whose time had arrived.

[4] Shannon, Claude, "A Mathematical Theory of Communication," reprinted with corrections from *The Bell System Technical Journal*, Vol. 27, pp. 379–423, 623–656, July, October, 1948. The result appears on p. 11 of the revised paper, but the proof is given in Appendix II, which is in the second part of the split paper.

[5] Tribus, Myron and Edward C. McIrvine, "Energy and Information," *Scientific American*, Vol. 225, No. 3, September 1971, pp. 179–184, 186, 188, p. 180 cited here. Thanks to Harvey S. Leff for providing this reference!

In physics, this mysterious quantity is denoted by S. Although the formula is the same, Shannon used H for his entropy. The name entropy stuck, but it is sometimes referred to as "information entropy" or "Shannon entropy" to distinguish it from the concept in physics. Another reason that Shannon's original names were problematic is that the results we get from his formula don't always correspond to how we're used to thinking about information or uncertainty.

If you are trying to evaluate Shannon entropy on a calculator or with a computer program, you'll quickly bump into the problem of evaluating a base 2 logarithm. Let's take a quick look at how to get around this difficulty.

$$y = \log_2 x \Leftrightarrow 2^y = x \text{ (by definition)}$$

Now take $\log_b$ of both sides. You may use $b = 10$ (common logarithm) or $b = e$ (natural logarithm) or any other base $b > 0$. Bases 10 and e are the ones that commonly have keys devoted to them (log and ln, respectively) on calculators. I'll use base $b = e$, with the notation ln.

$$\ln\left(2^y\right) = \ln\left(x\right)$$

Making use of one of the properties of logarithms, we may bring the exponent of the argument down in front.

$$y\ln(2) = \ln(x)$$

Dividing through by ln(2) now gives

$$y = \frac{\ln(x)}{\ln(2)}$$

Thus, we have rewritten our base 2 logarithm in terms of logarithms with base e. We have, by the way, just derived the change of base formula for logarithms.

Also, because we can rewrite this last equation to express

$$y = \log_2 x \text{ as } y = \frac{1}{\ln(2)}\ln(x)$$

we see logarithmic functions with different bases only differ by a constant multiplier. Recalling that Shannon's formulation for entropy was only unique up to a constant multiple, we see that this flexibility is equivalent to being able to use any base for the logarithm. Base 2 is especially convenient when considering digital data; thus, 2 was the base chosen. Regardless of what form the data takes, using the base 2 logarithm results in the units of entropy being "bits per message" or "bits per character," if we want an average rate of information transmission.

The average entropy, H, of English may be calculated by using the probability of each of the 26 letters in the following formula:

$$H = -\sum p_i \log_2 (p_i)$$

but this is really only an estimate, as the effects over groups of letters have not been accounted for yet. Rules such as Q being (almost always) followed by U and "I before E, except after C" show there is order in English on the scale of two- and three-character groupings. Entropy approximations based on single letter frequencies are often denoted as H_1. Better estimates are given by H_2 and H_3 where the probabilities used in these are for digraphs and trigraphs. As N grows, H_N/N converges monotonically to a limit (see Table 11.1).

Table 11.1   First-, Second-, and Third-Order Entropy for Various Languages

Language	H_1	H_2	H_3
English			
Contemporary	4.03	3.32	3.1
Poe	4.100	3.337	2.62
Shakespeare	4.106	3.308	2.55
Chaucer	4.00	3.07	2.12
German	4.08	3.18	–
French	4.00	3.14	–
Italian	3.98	3.03	–
Spanish	3.98	3.01	–
Portuguese	3.91	3.11	–
Latin	4.05	3.27	2.38
Greek	4.00	3.05	2.19
Japanese	4.809	3.633	–
Hawaiian	3.20	2.454	1.98
Voynich ms.	3.66	2.22	1.86

Source: Bennett, Jr., William Ralph, *Scientific and Engineering Problem-solving with the Computer,* Prentice Hall, Englewood Cliffs, New Jersey, 1976, p. 140.

As von Neumann indicated, the idea of entropy existed in physics before Claude Shannon applied it to text. To make sense of entropy in physics, one must first understand that imbalances in systems offer usable energy. As an example, consider a hot cup of coffee in a room. It has energy that can be used—one could warm one's hands over it or use it to melt an ice cube. If left alone, its energy will slowly dissipate with no effect other than slightly warming the room. When the coffee and the room settle to an equilibrium temperature, the room will contain the same amount of total energy, but there will no longer be any useable energy. The amount of entropy or "unusable energy" can be said to have increased.

One of the few theories that scientists have enough confidence in to label a "law" is the second law of thermodynamics. This states that entropy must increase in a closed system; that is, if no energy is being added to a system (i.e., the system is closed), then the amount of energy in the system that cannot be used must increase over time. In other words, the amount of usable energy in the system must decrease. To simplify this further, everything winds down. If the cup of coffee described in the previous paragraph had a heating coil in it, which was plugged into an outlet that connected it to an outside power source, then the coffee would not cool and the entropy of the room would not increase. This is not a violation of the second law, however, because the room could no longer be considered a closed system. It receives energy from outside.

Shannon's "text entropy" follows the second law of thermodynamics, as Table 11.1 shows. Over the centuries, the entropy of a particular language increases, as it does when one language springs from another. This is an empirical result; in other words, experiment (measuring the entropy of various texts) indicates it is true, but we do not have a proof. It makes sense though, because, as a language evolves, more exceptions to its rules appear, and more words come in from foreign languages. Hence, the frequency distribution of character groups tends to grow more uniform, increasing the entropy. This phenomenon can be used to roughly date writings; however, not all authors in a particular generation, or even century, will exhibit the same entropy in their texts. Edgar Allan Poe's writings, for example, exhibit higher entropy than those of his peers.[6] This is due to the unusually large vocabulary he commanded.

The maximum possible value for H_1 occurs when all probabilities are equal (1/26). This gives us

$$H_1 = -\sum (1/26) \log_2 (1/26) = -\sum (1/26) \ln(1/26)/\ln(2)$$

$$= -\ln(1/26)/\ln(2) \quad \text{(because we are summing over 26 letters)}$$

$$\approx 4.7$$

The idea of entropy also reveals approximately how many meaningful strings of characters we can expect of length N. The answer is given by 2^{HN}. This can be used, for example, to estimate the keyspace for a running key cipher of a given length in any particular language.

The idea of entropy has also been influential in the arts (literature, in particular) to various degrees over the decades. The best example I've come upon is Isaac Asimov's short story "The Last Question."[7] A more recent example of entropy in pop culture is provided by nerdcore hip hop artist MC Hawking's song "Entropy."[8] This song educates while it entertains!

11.3 One More Time…

Closely related to entropy is Shannon's idea of redundancy, denoted here as D_N. Basically, high entropy (disorder) corresponds to low redundancy and low entropy corresponds to high redundancy. Mathematically, the two are related as follows

$$D_N = \log_2 (26^N) - H_N \text{ is the redundancy in a message of length } N$$

---

[6] Bennett, Jr., William Ralph, *Scientific and Engineering Problem-solving with the Computer*, Prentice Hall, Englewood Cliffs, New Jersey, 1976, p. 140.

[7] The story first appeared in *Science Fiction Quarterly*, November 1956. See the References and Further Reading section at the end of this chapter for more on this tale.

[8] http://www.mchawking.com/ is the main page for the musician. The lyrics to Entropy can be found at http://www.mchawking.com/includes/lyrics/entropy_lyrics.php.

(using a 26-letter alphabet).[9] Just as H_N/N converges down to a limiting value as N increases, D_N/N increases to a limiting value with N.

Shannon found the redundancy of English to be $D \approx 0.7$ decimal (base 10) digits per letter. Dividing this value by log(26), we get the relative redundancy of English, which is about 50%. The value 26 was used in the log, as Shannon chose to omit word spacing and simply use a 26-letter alphabet for his calculation.[10] He explained what his value for the redundancy of English means and how it may be obtained.

> The redundancy of ordinary English, not considering statistical structure over greater distances than about eight letters, is roughly 50%. This means that when we write English half of what we write is determined by the structure of the language and half is chosen freely. The figure 50% was found by several independent methods which all gave results in this neighborhood. One is by calculation of the entropy of the approximations to English. A second method is to delete a certain fraction of the letters from a sample of English text and then let someone attempt to restore them. If they can be restored when 50% are deleted the redundancy must be greater than 50%. A third method depends on certain known results in cryptography.[11]

Cipher Deavours used 1.11 for his approximation of D, which converts to 78%.[12] Although he didn't explain how he came up with this larger value, one possibility is that he included a blank space as a character of the alphabet. As spacing rarely changes the meaning of a sentence, its presence increases the redundancy. In many early examples of writing, ranging from ancient Greek through medieval times, word spacing isn't present. And word spacing isn't the only omission that has been common historically. The original Hebrew version of the Old Testament was written without vowels. The redundancy of this language allows it to be read anyway.

I decided to do an experiment of my own. I asked a student, Josh Gross, to send me random messages with various percentages of letters removed and word spacing preserved. The hardest of his challenges (the ones with the most letters removed) were the following:

65% Removed

```
_n _h_s _i___e _i_e _t _s ___a___t ___t _h _i___e _f _it_
__s _e__ ___n_ r____ f___ t_e _o__ r _f _h_ _d___ _ t_
t_e _k_


_h _i___e _a _h__ _b_e _d_____d _r__ _h _g_e_t ___d_t_o__ _f
_h _s___h. _t _s _p__e _h_____r _o _ _d___i___l __a_a_i_t__
___t_m _f _x___i___e
```

---

9 Shannon, Claude E., "Communication Theory of Secrecy Systems," *The Bell System Technical Journal*, Vol. 28, No. 4, October 1949, pp. 656–715, p. 689 cited here.

10 Shannon, Claude E., "Communication Theory of Secrecy Systems," *The Bell System Technical Journal*, Vol. 28, No. 4, October 1949, pp. 656–715, p. 700 cited here.

11 Shannon, Claude, "A Mathematical Theory of Communication," reprinted with corrections from *The Bell System Technical Journal*, Vol. 27, pp. 379–423, 623–656, July, October, 1948, pp. 14–15.

12 Deavours, Cipher, "Unicity Points in Cryptanalysis, *Cryptologia*," Vol. 1, No. 1, January 1977, pp. 46–68, p.46 cited here. Also see p. 660 of Shannon, Claude E., "Communication Theory of Secrecy Systems," *The Bell System Technical Journal*, Vol. 28, No. 4, October 1949, pp. 656–715.

I sent him my solutions and then asked for messages with more letters removed. Josh sent these:

70% Removed

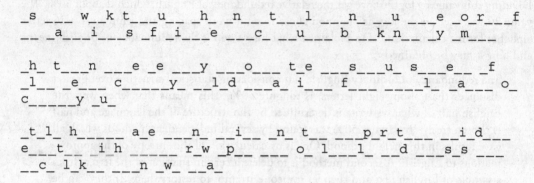

The answers are given at the end of this chapter, after the References and Further Reading list, but I encourage you to try to solve them yourself first. You may use pattern word programs to make your work easier. It doesn't matter how long it takes you to determine the missing letters or what resources you make use of. If the letters are recoverable, they're redundant!

Not solely focused on solving "important problems," Shannon applied the idea of redundancy to a popular recreation for the cryptologically inclined:[13]

> The redundancy of a language is related to the existence of crossword puzzles. If the redundancy is zero any sequence of letters is a reasonable text in the language and any two-dimensional array of letters forms a crossword puzzle. If the redundancy is too high the language imposes too many constraints for large crossword puzzles to be possible. A more detailed analysis shows that if we assume the constraints imposed by the language are of a rather chaotic and random nature, large crossword puzzles are just possible when the redundancy is 50%. If the redundancy is 33%, three-dimensional crossword puzzles should be possible, etc.

The idea of entropy also has important applications to data compression:[14]

> The ratio of the entropy of a source to the maximum value it could have while still restricted to the same symbols will be called its *relative entropy*. This is the maximum compression possible when we encode into the same alphabet.

Speaking of compression, it should be pointed out that Shannon's work has been greatly compressed in this chapter. The reader is encouraged to pursue the references for a fuller treatment.

---

[13] Shannon, Claude, "A Mathematical Theory of Communication," reprinted with corrections from *The Bell System Technical Journal*, Vol. 27, pp. 379–423, 623–656, July, October, 1948. The result appears on p. 15 of the revised paper.

[14] Shannon, Claude, "A Mathematical Theory of Communication," reprinted with corrections from *The Bell System Technical Journal*, Vol. 27, pp. 379–423, 623–656, July, October, 1948. The result appears on p. 14 of the revised paper.

Shannon also provided the theoretical background for error-correcting codes.[15] These codes are the opposite of the sort we are interested in. They seek to make the message easier to read by introducing extra redundancy. In this manner, mutilated messages may still be recovered. Cryptographers aim to minimize redundancy. Patterns are a cryptanalyst's best friend, so a good cipher should mask their presence!

11.4 Unicity Points

In Sections 3.4, 4.4, and 4.5, unicity points (the length for which a ciphertext can be expected to have a unique solution) were given for various ciphers. Shannon presented a general technique for calculating these values. We have the unicity point, U, given by $U = \log_2(K)/D$, where K is the number of possible keys and D is the redundancy per letter of the message. If the message is compressed prior to encipherment, the value of D is decreased, thus raising the value of U. Some messages that would have unique solutions can therefore be made ambiguous by taking this extra step. As was seen above, the value for D is determined empirically. Like entropy, it even varies from author to author.

11.5 Dazed and Confused

Another pair of concepts that Shannon developed that play a major role in modern cryptography is *diffusion* and *confusion*. Diffusion means distributing the influence of each plaintext letter over several ciphertext letters. For example, if matrix encryption is used with a 5×5 matrix, then each plaintext letter affects five ciphertext letters. Cryptographers today want this effect to be present for both plaintext letters and individual bits in a cipher's key. They've taken the idea to its natural extreme and desire that a change to a single message letter, or bit in the key, will change the entire ciphertext, or about half of the bits, if the message takes that form. This is termed an *avalanche effect*, in what is an excellent analogy. This makes cryptanalysis significantly more difficult, as the ciphertext cannot be attacked in small pieces, but must be taken as a whole. To continue the 5×5 matrix encryption example, because each plaintext letter affects five ciphertext letters, we cannot simply pick out the letter E, as we did with monoalphabetic substitution ciphers. Instead, we must work on groups of five letters at a time. With greater diffusion, cryptanalysis may become an all-or-nothing proposition.

Confusion means making the relationship between the plaintext and ciphertext complex. The idea is best illustrated by modern ciphers, for which knowing a large amount of plaintext and the corresponding ciphertext, along with the enciphering algorithm, fails to allow the cryptanalyst to determine the key. In modern systems, confusion is obtained through substitution and diffusion is obtained through transposition. Although older systems, such as the World War I German ciphers

---

[15] We need to be careful in this area, though, not to credit Shannon with too much, as he does make use of previous work by Richard Hamming. This is done on p. 28 (cited on p. 27) in Shannon's revised paper (Shannon, Claude, "A Mathematical Theory of Communication," reprinted with corrections from *The Bell System Technical Journal*, Vol. 27, pp. 379–423, 623–656, July, October, 1948). A very simple means of adding redundancy to create an error-correcting code is to insert extra bits at regular intervals to serve as parity checks (making certain sets of bits have an even sum).

ADFGX and ADFGVX, employed both substitution and transposition, this did not become a standard approach to encryption until the computer era.

Shannon noted a disadvantage in ciphers with high confusion and diffusion.[16]

> Although systems constructed on this principle would be extremely safe they possess one grave disadvantage. If the mix is good then the propagation of errors is bad. A transmission error of one letter will affect several letters on deciphering.

We'll see examples of modern systems that combine substitution and transposition to satisfy Shannon's conditions in Sections 13.1 and 20.3.

In addition to the important work described above, Shannon found time to pursue other projects of a more recreational nature. Examples include rocket-powered Frisbees, a motorized pogo stick, machines to play chess and solve Rubik's cube, a flame-throwing trumpet, and a mysterious box with a switch on it. When someone saw this box sitting on his desk and flipped the switch, the box would open and a mechanical hand would reach out and flip the switch back again. After this, the hand would pull back into the box and the lid would close, returning the system to its original state.[17]

The National Security Agency, not bothered by mild eccentricities, invited Shannon to join their Scientific Advisory Board.

> Perhaps, because of that information theory, it was suggested that he [Shannon] should be on our Advisory Board [NSASAB] – and he was appointed to it. He came down there and was tremendously interested in what he found there. He sort of repudiated his book on secret communications after that. He said that he would never have written it if he knew then what he learned later.[18]

So, what did Shannon see when he joined NSASAB? While this question will remain unanswered, NSA is the focus of the next chapter. The present chapter closes with brief sections on entropy in religion and literature.

11.6 Entropy in Religion

> The first and second laws of thermodynamics have been used, affirmed, rejected, manipulated, exploited, and criticized in order both to further and to censure religion.
>
> – **Erwin N. Hiebert**[19]

Sir Arthur Eddington remarked, "The law that entropy always increases—the second law of thermodynamics—holds, I think, the supreme position among the laws of Nature."[20] As such, it is not

---

[16] Shannon, Claude E., "Communication Theory of Secrecy Systems," *The Bell System Technical Journal*, Vol. 28, No. 4, October 1949, pp. 656–715, p. 713 cited here.

[17] http://en.wikipedia.org/wiki/Claude_Shannon.

[18] Campaigne, Howard H, AFIPS Oral History Interview, 1974, p. 10.

[19] Hiebert, Erwin N., "The Uses and Abuses of Thermodynamics in Religion," *Daedalus*, Vol. 95, No. 4, Fall 1966, pp. 1046–1080, p. 1049 quoted from here.

[20] Eddington, Sir Arthur, *The Nature of the Physical World*, Cambridge University Press, Cambridge, UK, 1928, p. 74.

surprising that it has been used to rationalize previously held beliefs of various individuals, despite these beliefs sometimes being mutually exclusive!

One example is its application in 1951 by Pope Pius XII, who claimed that the second law of thermodynamics confirmed traditional proofs of the existence of God.[21] By his reasoning, if everything must wind down, the Universe cannot be infinitely old or it would have already wound down completely. Therefore, the Universe must have had a beginning, a creation, and therefore a creator. The big bang theory, which many scientists back, also gives a date for the beginning of the Universe, but these scientists tend to see it as a process of creation that does not require a God.

In another direction, some scientists have argued that the idea of an eternal afterlife is a violation of the second law. That is, if there is a heaven or hell, it cannot last forever. The punk rock band Bad Religion recorded a song titled "Cease" that makes this point in a somewhat subtle way. The second law is not explicitly mentioned in the song, but it can be inferred.

Creationists sometimes argue that life on Earth has become more and more complex over time and that the only way this apparent violation of the second law could be possible is if God caused it. The scientific rebuttal to this is that the second law only applies to closed systems. The Earth is not a closed system, because a massive amount of energy is constantly transferred to it by the sun. It is this energy, ultimately, that makes evolution possible.

There's more that could be said on this topic, but we now turn to a less controversial use of entropy.

11.7 Entropy in Literature

There are different ways in which entropy can appear in literature. One is by having a sort of decay or decline take place for a character or a larger group over the course of the tale. Zbigniew Lewicki, chair of the American literature department at the University of Warsaw, Poland, pointed out that "not all literature of despair and catastrophe is necessarily entropic."[22] Some of the authors who produced works that would seem to fit nicely into the category of entropic literature never actually heard of the concept of entropy, much less the second law. A nice example is provided by Edgar Allan Poe's short story "The Fall of the House of Usher." It was published in 1839, prior to the second law being first proposed by Rudolf Clausius in 1850.

Of course, some other authors had a deep acquaintance with such topics. Thomas Pynchon, for example, studied physics, information theory, and literature in college.[23] In 1960, he authored a short story titled "Entropy." His novel *The Crying of Lot 49* makes use of the concept in a much less explicit way. Lewicki explained:

> Entropy in *The Crying of Lot 49* can thus be seen both as its main organizing principle and as Pynchon's basic philosophical assumption… Even if the world described there is entropic, it would seem that by the very act of writing and reading about it, a certain

---

[21] Freese, Peter, *From Apocalypse to Entropy and Beyond: The Second Law of Thermodynamics in Post-War American Fiction*, Die Blaue Eule, Essen, Germany, 1997, p. 126. William Ralph Inge made the same argument in the 1930s. See p. 1069 of Hiebert, Erwin N., "The Uses and Abuses of Thermodynamics in Religion," *Daedalus*, Vol. 95, No. 4, Fall 1966, pp. 1046–1080.

[22] Lewicki, Zbigniew, *The Bang and the Whimper: Apocalypse and Entropy in American Literature*, Greenwood Press, Westport, Connecticut, 1984, p. 115.

[23] Shaw, Deborah and Davis, Charles H., "The Concept of Entropy in the Arts and Humanities," *Journal of Library and Information Science*, No. 9.2, 1983, pp. 135–148, p. 137 cited here.

amount of information is passed on, which would cause a decrease of entropy, at least locally. But – and this seems to be Pynchon's ultimate coup - *The Crying of Lot 49* conveys practically no such information. We do not learn anything about the characters that is not ambiguous.[24]

Many novels make use of mysteries to intrigue readers and keep them turning the pages, but by the end a nice resolution has been presented and everything makes sense. By contrast, at the end of a novel in the postmodern genre, things make less sense than at the start. *The Crying of Lot 49* accomplishes this, providing another style of entropic literature. There is less order at the end. Postmodernism isn't limited to text, however. An example made for television is the series *Lost* (2004–2010). *Mad* magazine suggested that the title referred to the script.

Some authors have attempted to reverse entropy. There was even a journal devoted to this purpose, namely *Extropy: Transhuman Technology, Ideas, and Culture*, which first appeared in 1989. It featured articles such as "The Heat Death of Timothy Leary," in which the deceased was criticized for allowing himself to be cremated instead of fighting entropy by being cryogenically preserved (1996).[25] This was a nonfiction piece. Can fiction fight entropy? Lewicki related the manner in which some authors attempted this:

> It would follow from what has been said that the most improbable messages, namely those composed of words haphazardly put together, could most effectively counter entropy and provide the greatest amount of information. Such works of literature have in fact been created, but common sense tells us that they have neither decreased the level of entropy nor offered much information.[26]

Recall that Shannon considered messages of low probability to contain the most information. Anthony Purdy, a Professor of Romance Languages at the University of Alberta, also addressed such attempts.

> Hence the simplistically reductive (and self-contradictory) belief that, since entropy is a measure of probability, the less predictable and the more 'experimental' a literary work, the more effective it will be in the 'struggle against entropy'.[27]

He noted, however, that "there is no necessary correlation between information and meaning."[28] While the information content may have been high, according to Shannon's calculation, it didn't

---

[24] Lewicki, Zbigniew, *The Bang and the Whimper: Apocalypse and Entropy in American Literature*, Greenwood Press, Westport, Connecticut, 1984, pp. 92–93.

[25] More, Max, "The Heat Death of Timothy Leary," *Extropy: Transhuman Technology, Ideas, and Culture*, #17, August 1996. The full table of contents for this issue can be seen at http://extropians.weidai.com/extropians.96/0133.html. I first saw this particular article referenced in Freese, Peter, *From Apocalypse to Entropy and Beyond: The Second Law of Thermodynamics in Post-War American Fiction*, Die Blaue Eule, Essen, Germany, 1997, pp. 98–99.

[26] Lewicki, Zbigniew, *The Bang and the Whimper: Apocalypse and Entropy in American Literature*, Greenwood Press, Westport, Connecticut, 1984, p. 73.

[27] Bruce, Donald and Anthony Purdy, editors, *Literature and Science*, Rodopi, Amsterdam, Netherlands, 1994, p. 9.

[28] Bruce, Donald and Anthony Purdy, editors, *Literature and Science*, Rodopi, Amsterdam, Netherlands, 1994, p. 11.

correspond to what we are used to thinking of as information. Purdy gave examples of how some attempts failed:

> ...the high information content of such literary works as Marc Saporta's *Composition #1* or Raymond Queneau's *Cent mille milliards de poèmes*, which depend on a randomizing principle akin to the shuffling of a deck of cards, does not generate a corresponding increase in meaning. In fact, the high entropy of the source tends, if anything, to *reduce* the amount of information transmitted in any *single* reading...[29]

The journal devoted to the cause of extropy gave in to entropy in 1996 and folded.

As was explained in Section 11.2, languages follow the second law of thermodynamics. That is, the (information) entropy of a language increases over time. Some novels set in the future do not reflect this and the characters speak in the same manner as the author's contemporaries. One notable exception is Anthony Burgess's *A Clockwork Orange*. This novel begins with

> Our pockets were full of deng, so there was no real need from the point of crasting any more pretty polly to tolchock some old veck in an alley and viddy him swim in his blood while we counted the takings and divided by four, nor to do the ultraviolent on some shivering starry grey haired ptitsa in a shop and go off with the till's guts. But, as they say, money isn't everything.

At least one edition has a small dictionary in the back to help readers comprehend the future slang Burgess introduced. Much of it was Russian in origin. In America, many Spanish words are part of everyday speech, whether it is the speaker's native language or not. We understand what is meant when someone is described as "macho" and when a hate-monger refers to a group of people as "bad hombres."

While there's a large quantity of fiction in which entropy plays an important role, my favorite is a short story by Isaac Asimov, a man who straddled the worlds of literature and science (he had a PhD in chemistry). The tale is titled "The Last Question" and it connects entropy and religion in a very entertaining way. Instead of summarizing it here, I encourage you to seek out and enjoy the original. It has been reprinted often and details are provided in the On Entropy in Literature section of the References and Further Reading list below.

References and Further Reading

On Information Theory

Arndt, Christoph, *Information Measures: Information and its Description in Science and Engineering*, Springer, Berlin, Germany, 2001.

Bennett, Jr., William Ralph, *Scientific and Engineering Problem-solving with the Computer*, Prentice Hall, Englewood Cliffs, New Jersey, 1976.

Deavours, Cipher, "Unicity Points in Cryptanalysis," *Cryptologia*, Vol. 1, No. 1, January 1977, pp. 46–68.

---

[29] Bruce, Donald and Anthony Purdy, editors, *Literature and Science*, Rodopi, Amsterdam, Netherlands, 1994, p. 11.

Elias, Peter, "Two Famous Papers," *IRE Transactions on Information Theory*, Vol. 4, No. 3, September 1958, p. 99. In this humorous piece Elias, the editor of the journal it appeared in, complains about two types of papers that he wishes people would stop writing. One type details premature attempts at revolutionizing various fields using the ideas of information theory. Elias used "Information Theory, Photosynthesis and Religion" (title courtesy of D. A. Huffman) to represent this class of papers.

Gleick, James, *The Information: A History, A Theory, A Flood*, Vintage Books, New York, 2011.

Golomb, Solomon W., Elwyn Berlekamp, Thomas M. Cover, Robert G. Gallager, James L. Massey, and Andrew J. Viterbi, "Claude Elwood Shannon (1916-2001)," *Notices of the American Mathematical Society*, Vol. 49, No. 1, January 2002, pp. 8–16.

Hellman, Martin, "An Extension of the Shannon Theory Approach to Cryptography," *IEEE Transactions on Information Theory*, Vol. 23, No. 3, May 1977, pp. 289–294.

Leff, Harvey S., *Maxwell's Demon, Entropy, Information, Computing*, Princeton University Press, Princeton, New Jersey, 2014.

Pierce, John Robinson, *Symbols, Signals, and Noise: The Nature and Process of Communication*, Harper & Row Publishers, New York, 1961. This is a longer, less mathematical, presentation of Shannon's ideas, intended for a broader audience.

Reeds, James, "Entropy Calculations and Particular Methods of Cryptanalysis," *Cryptologia*, Vol. 1, No. 3, July 1977, pp. 235–254. The author notes the difficulty encountered in solving ciphers that are just slightly over the unicity point in length. He shows how to approximate the length L that will allow solution in practice, as opposed to merely in theory. His calculations rely on using a value for D that represents the amount of redundancy in the language that is actually exploited by the cryptanalytic methods applied.

Roch, Axel, "Biopolitics and Intuitive Algebra in the Mathematization of Cryptology? A Review of Shannon's "A Mathematical Theory of Communication" from 1945," *Cryptologia*, Vol. 23, No. 3, July 1999, pp. 261–266.

Shannon, Claude E., "A Mathematical Theory of Communication," reprinted with corrections from *The Bell System Technical Journal*: Vol. 27, pp. 379–423, 623–656, July, October, 1948.

Shannon, Claude E., "Communication Theory of Secrecy Systems," *The Bell System Technical Journal*, Vol. 28, No. 4, October, 1949, pp. 656–715. Shannon noted, "The material in this paper appeared in a confidential report 'A Mathematical Theory of Cryptography' dated Sept. 1, 1946, which has now been declassified." Following Shannon's death, Jiejun Kong wrote, "Recently I am shocked to find that this paper does not have a typesetted version on the colossal Internet, the only thing people can get is a set of barely-legible scanned JPEG images from photocopies (see http://www3.edgenet.net/dcowley/docs.html). So here is my memorial service to the great man. I spent a lot of time to input and inspect the entire contents of this 60-page paper. During my typesetting I am convinced that his genius is worth the time and effort I spent!" His work may be found at http://netlab.cs.ucla.edu/wiki/files/shannon1949.pdf. Thank you, Jiejun!

Shannon, Claude E., "Prediction and Entropy of Printed English," *The Bell System Technical Journal*, Vol. 30, No. 1, January 1951, pp. 50–64.

Shannon, Claude E., "The Bandwagon," *IRE Transactions on Information Theory*, Vol. 2, No. 1, March 1956, p. 3. In this piece, Shannon wrote,

I personally believe that many of the concepts of information theory will prove useful in other fields… but the establishing of such applications is not a trivial matter of translating words into a new domain, but rather the slow tedious process of hypothesis and experimental verification.

Shannon, Claude E. and Warren Weaver, *The Mathematical Theory of Communication*, University of Illinois, Urbana, 1949. This book reprints Shannon's 1948 paper and Weaver's popularization of it.

Sloane, Neil J. A. and Aaron D. Wyner, editors, *Claude Elwood Shannon: Collected Papers*, IEEE Press, New York, 1993.

Soni, Jimmy and Rob Goodman, *A Mind at Play: How Claude Shannon Invented the Information Age*, Simon & Schuster, New York, 2017. This is a biography of Claude Shannon.

Tribus, Myron and Edward C. McIrvine, "Energy and Information," *Scientific American*, Vol. 225, No. 3, September 1971, pp. 179–184, 186, 188.

"Variations of the 2nd Law of Thermodynamics," *Institute of Human Thermodynamics*, http://www. humanthermodynamics.com/2nd-Law-Variations.html. This page, which is part of a much larger website, gives 118 variations of the famous law.

Weaver, Warren, "The Mathematics of Communication," *Scientific American*, Vol. 181, No. 1, January 1949, pp. 11–15. This is an early popularization of Shannon's work in information theory.

On Entropy and Religion

Hiebert, Erwin N., "The Uses and Abuses of Thermodynamics in Religion," *Daedalus*, Vol. 95, No. 4, Fall 1966, pp. 1046–1080.

On Entropy in Literature

Asimov, Isaac, "The Last Question," *Science Fiction Quarterly*, November 1956. This short story has been reprinted numerous times. Collections of Asimov stories that include it are *Nine Tomorrows* (1959), *Opus 100* (1969), *The Best of Isaac Asimov* (1973), *Robot Dreams* (1986), *Isaac Asimov: The Complete Stories*, Vol. 1 (1990). There's even an audio version narrated by Leonard Nimoy. See http:// bestsciencefictionbooks.com/forums/threads/the-last-question-by-asimov.398/.

Bruce, Donald and Anthony Purdy, editors, *Literature and Science, Rodopi*, Amsterdam, Netherlands, 1994.

Burgess, Anthony, *A Clockwork Orange*, Ballantine Books, New York, 1965. This later paperback edition includes a seven-page glossary, not present in the first edition, to help readers translate the slang used in the novel.

Di Filippo, Paul, *Ciphers: A Post-Shannon Rock-n-Roll Mystery*, Cambrian Publications, Campbell, California, 1997. The first sentence of Chapter 00000001 of this novel gives the reader an idea of what he or she is getting into:

> That this sophic, stochastic, Shannonesque era (which, like most historically identifiable periods, resembled a nervous tyro actor insofar as it had definitely Missed Its Cue, arriving when it did precisely in July 1948, ignoring conventional calendars and expectations, which of course dictated that the *Zeitgeist* should change only concurrently with the decade)—that this era should today boast as one of its most salient visual images the widely propagated photo of a barely post-pubescent actress dry-humping a ten-foot long, steel-grey and olive-mottled python thick as a wrestler's biceps (and what a cruel study for any wrestler, whether to fuck or pinion this opulent opponent)—this fact did not bother Cyril Prothero (who was, after all, a product of this selfsame era) half so much as that it (the era) seemed—the more he learned, the more wickedly perverse information that came flooding into his possession—to be exquisitely poised, trembling, just awaiting A Little Push, on the verge of ending.

I expect that readers will either love or hate this novel, and that all those who fall into the first (smaller) group will be aware of the significance of July 1948.

Freese, Peter, *From Apocalypse to Entropy and Beyond: The Second Law of Thermodynamics in Post-War American Fiction*, Die Blaue Eule, Essen, Germany, 1997.

Lewicki, Zbigniew, *The Bang and the Whimper: Apocalypse and Entropy in American Literature*, Greenwood Press, Westport, Connecticut, 1984.

Poe, Edgar Allan, "The Fall of the House of Usher," *Burton's Gentleman's Magazine*, September 1839. You won't have trouble finding a reprint of this tale.

Pynchon, Thomas, "Entropy," *Kenyon Review*, Vol. 22, No. 2, Spring 1960, pp. 27–92. Despite his studies, Pynchon admitted:

> Since I wrote this story I have kept trying to understand entropy, but my grasp becomes less sure the more I read. I've been able to follow the OED definitions, and the way Isaac Asimov

explains it, and even some of the math. But the qualities and quantities will not come together to form a united notion in my head.[30]

Pynchon, *The Crying of Lot 49*, J.B. Lippincott, Philadelphia, Pennsylvania, 1966.

Shaw, Deborah and Charles H. Davis, "The Concept of Entropy in the Arts and Humanities," *Journal of Library and Information Science*, Vol. 9, No. 2, 1983, pp. 135–148, available online at https://jlis.glis. ntnu.edu.tw/ojs/index.php/jlis/article/viewFile/141/141.

Wells, H. G., *The Time Machine*, Henry Holt and Company, New York, 1895. Prior to appearing in this form, it was serialized in the January through May issues of *The New Review*. Near the end of the novel, the protagonist travels farther into the future and witnesses the consequences of entropy.

Discography

A pair of songs dealing with entropy are:

"Cease," from Bad Religion's 1996 album *The Gray Race*.

"Entropy," from MC Hawking's 2004 album *A Brief History of Rhyme: MC Hawking's Greatest Hits*.

Videography

Claude Shannon: Father of the Information Age, http://www.ucsd.tv/search-details.aspx?showID=6090. This first aired on January 30, 2002, on UCTV, San Diego.

Answers to Missing Letters Challenges

For the challenges with 65% of the letters missing, I came up with:

> In this simple life it is apparent that the image of birth has been transferred from the mother of the individual to the sky.

> The issue may have been debated from the great traditions of the surah. It is opposed however to a distinctly shamanistic system of experience.

Josh had actually started with:

> In this simple rite it is apparent that the image of birth has been transferred from the mother of the individual to the sky.

> The image can have been derived from the great traditions of the south. It is applied however to a distinctly shamanistic system of experience.

For the challenges with 70% of the letters missing, I was only able to come up with a solution that made sense for the first one:

> As you walk through the industrial towns you lose yourself in labyrinths of little tract houses blackened by smoke.

---

[30] Pynchon, Thomas, Introduction to "Entropy," *Slow Learner: Early Stories*, Little, Brown, Boston, Massachusetts, 1984, p. 14.

I noted that "little" could also be "simple" and "tract" could be "ranch" or "shack." I considered other alternatives, but these were my best guesses. Josh had actually started with:

> As you walk through the industrial towns you lose yourself in labyrinths of little brick houses blackened by smoke.

So, I was only off by one word. I had considered "brick" as a possibility, but I'm used to thinking of brick homes as being more expensive and that didn't seem to fit the context.

In each example, I got at least one word wrong. How did you do?

As for the other two challenges with 70% of the letters missing, I managed to string *some* words together that made sense, but I couldn't keep the whole sentence on a single topic! They ended up looking like diverse sentences spliced together. My nonsensical "solutions" were:

> The Texan made me bury Timothy the newsgroup sargent of blog metal cutaways filed metal into flat deadly havoc of cavalry gun.

> Italy harbor are giant ships and the cooperating children exactly fish where raw open bed of leveled mud exists, and such lakes join two seas.

Josh had actually started with:

> The train bore me away, through the monstrous scenery of slag heaps, chimneys, piled scrap iron, foul canals, paths of cindery mud…

> Still houses are being built, and the Corporation building estates, with their row upon row of little red houses, all much liker than two peas.

Probably, you did better than I did! I should note that I resisted the temptation to Google short strings of words I suspected were present in the hope of finding the quotes. To make it a fair test, just work with programs accessing lists of words, not texts.

Chapter 12

National Security Agency

There is a large gap in the literature on cryptology. Following Claude Shannon's papers "A Mathematical Theory of Communication" (1948) and "Communication Theory of Secrecy Systems" (1949), there was virtually nothing new until 1967, when David Kahn's *The Codebreakers*, a historical work, was published. There are some exceptions, such as the papers by Jack Levine, which although original, were not in the direction that Shannon's work pointed. A great deal of new research was being done, but the public was unaware of it because of the National Security Agency (NSA).

Figure 12.1   Providing and protecting vital information through security. (http://www.nsa.gov.)

As Figure 12.1 indicates, NSA does not stand for No Such Agency. Yet, for much of the agency's history this name was appropriate, as the organization was deeply shrouded in secrecy.

> What few histories exist are so highly classified with multiple codewords that almost no one has access to them.

> **—James Bamford[1]**

---

[1] Bamford, James, *Body of Secrets: Anatomy of the Ultra-Secret National Security Agency from the Cold War through the Dawn of a New Century*, Doubleday, New York, 2001.

The situation has changed somewhat, since Bamford made this comment. NSA has released a four-volume history of the agency, but a great deal of material has been redacted from it. Still, this represents a tremendous break from tradition; the agency was literally born in secrecy.

12.1 Origins of NSA

We now know that NSA was established by President Truman, without informing Congress, in 1952, when he issued the top secret directive "Communications Intelligence Activities." This directive was still classified 30 years later.[2] However, NSA didn't arise from nothing; rather it had its roots in the armed forces code and cipher groups of World War II, most prominently, Friedman's Army group, Signal Intelligence Service, later to become the Army Security Agency,[3] and The Navy's OP-20-G. Following World War II, when the Air Force became a separate service, having previously been part of the Army, there was also an Air Force Security Agency. Before NSA existed, a move towards centralization was marked by the creation of the Armed Forces Security Agency (AFSA). Thus, there's continuity such that the birth of NSA is more like a renaming. NSA existed for several years before moving to its current location, Fort Meade, Maryland (Figure 12.2). Over the years, though the trend was toward centralization, SCAs (Service Cryptologic Agencies) continued.

Figure 12.2 The National Security Agency's headquarters at Fort George G. Meade. (Courtesy of the National Security Agency, https://web.archive.org/web/20160325220758/http://www. nsa.gov/about/_images/pg_hi_res/nsa_aerial.jpg.)

---

[2] Bamford, James, *The Puzzle Palace*, Houghton, Mifflin, and Company, New York, 1982, p. 16.

[3] Both fell under the Signal Corps, and many other names were used for the group over the years. Because Friedman obtained Yardley's materials, we may start with the Cipher Bureau, and continue the chain with Signal Intelligence Service, Signal Security Division, Signal Security Branch, Signal Security Service, Signal Security Agency, and finally Army Security Agency. And this was all in a 30 year period (1917–1947)!

Unlike the much smaller Central Intelligence Agency (CIA), NSA is under the Secretary of Defense and is, per Truman's original order, tasked with serving the entire government. Today, NSA's main directorates are the Information Assurance Directorate (IAD) and the Signals Intelligence Directorate (SID). While the SID side (cryptanalysis) is sexier, IAD could be more important. If you could only spend money on one side, which would it be, offense or defense? You can ponder this while reading the history that follows.

12.2 TEMPEST

NSA's precursor, AFSA, failed to predict the outbreak of war in Korea, but America's cryptologists did go on to have some cryptanalytic success later in that conflict.[4] There were also early successes against the Soviets in the Cold War.[5] On the other side of the code battle, there were serious problems, right from the start, in protecting America's own communications, as a once classified history relates.

> At this point, the newly established NSA decided to test all of its equipment. The result—everything radiated. Whether it was mixers, keying devices, crypto equipment, EAM machinery, or typewriters, it sent out a signal…[half a line of text redacted]… Plain text was being broadcast through…[half a line of text redacted]… the electromagnetic environment was full of it.[6]

Various countermeasures were taken to minimize the distance at which emanations could be measured to reveal information. These countermeasures were dubbed TEMPEST (Transient Electromagnetic Pulse Emanation Standard).[7] The term is used for both an equipment specification and the process of preventing usable emanations. If you pursue this topic, you'll see references to TEMPEST attacks, but this is not technically correct. Although it is clear what the writers mean, TEMPEST technology is purely defensive.

Van Eck phreaking is a term that may properly be used for attacking a system by measuring electromagnetic radiation, but only in the special case of when the intent is to reproduce the monitor. This can be done at impressive distances, or from as nearby as the hotel room next door. This type of attack is named after the Dutchman Wim van Eck, who authored the 1985 paper "Electromagnetic Radiation from Video Display Units: An Eavesdropping Risk?" in which he demonstrated the attack for CRTs.[8] In 2004, another researcher revealed that LCD systems are

---

[4] Johnson, Thomas, R., *American Cryptology During the Cold War, 1945–1989, Book I: The Struggle for Centralization, 1945–1960*, Center for Cryptologic History, National Security Agency, Fort George G. Meade, Maryland, 1995, p. 33.

[5] See Section 2.8 for information on Venona.

[6] Johnson, Thomas, R., *American Cryptology During the Cold War, 1945–1989, Book I: The Struggle for Centralization, 1945–1960*, Center for Cryptologic History, National Security Agency, Fort George G. Meade, Maryland, 1995, p. 221.

[7] Some sources say that TEMPEST was simply a codeword and not an acronym. If true, what I provide here must have been made up later. A list of variants, including Tiny Electromagnetic Particles Emitting Secret Things, is given at https://acronyms.thefreedictionary.com/tempest.

[8] Van Eck, Wim, "Electromagnetic Radiation from Video Display Units: An Eavesdropping Risk?," *Computers & Security*, Vol. 4, No. 4, December 1985, pp. 269–286. John Young wrote, "Wim van Eck's article is actually the source of most of the incorrect TEMPEST information out there." See http://cryptome.org/tempest-time.htm, for a TEMPEST timeline with Young's corrections.

also vulnerable to this sort of attack and constructed the necessary equipment to carry out the attack for less than $2,000.[9]

Attacks made possible by electromagnetic emissions aren't limited to monitors. In 1956, a bugged telephone allowed the British to hear a Hagelin machine used by Egyptians in London. The mere sound of the machine allowed the British cryptanalysts to determine its settings and recover messages.[10] In general, electromechanical cipher machines are vulnerable to such acoustical attacks. This one example is used here to represent many incidents.

In October 1960, following a briefing by NSA, the United States Communications Security Board (USCSB) established the Subcommittee on Compromising Emanations (SCOCE) to study the problem. The committee learned that the Flexowriter was the worst emanator, allowing a properly equipped observer to read plaintext from a distance of 3,200 feet.[11]

Non-electronic/electric encryption doesn't offer a safe alternative. Microphones have been hidden in typewriters to allow recordings to be made of sensitive information being typed. The sounds of the keys hitting the paper may be distinguished to reveal the individual letters. This can also be done by placing a tiny microphone between keys on a computer keyboard.[12] It doesn't matter how good a nation's ciphers are if the enemy can get the messages by other means! A great many (possibly) secure methods of encryption are implemented in insecure ways.

12.3 Size and Budget

NSA's budget and the number of employees are both kept secret, creating a cottage industry of finding clever ways to estimate these figures. Such sport has attracted the attention of investigative journalists, as well as historians, and curious mathematicians and computer scientists. One simple way to approach the problem is by counting the number of cars in the parking lot, as revealed by satellite photographs. Other estimates of the number of employees have been based on the number of square feet of workspace and the number of employees belonging to the credit union. Of course, not all agency employees work at Fort Meade. Actually, most work elsewhere. Some early figures have now been declassified (Table 12.1).

Like any large organization, cryptologic agencies need many support people, people who aren't directly involved in codemaking or codebreaking. Of the cryptanalysts, the largest group was focused on Soviet ciphers, but other nations, including many friendly to the United States, were not ignored. By 1961, the NSA had broken the cipher systems of over 40 countries.[13] Third-world countries made easier pickings, and as they became the battlegrounds of the Cold war, the value of the intelligence grew.

> By the time Kennedy arrived at the White House, cryptology had become the elephant in the intelligence closet. McGeorge Bundy discovered that of the 101,900

---

[9] Kuhn, Markus G., "Electromagnetic Eavesdropping Risks of Flat-Panel Displays," in Martin D. and A. Serjantov, editors, *Privacy Enhancing Technologies* (*PET 2004*), Lecture Notes in Computer Science, Vol. 3424. Springer, Berlin, Germany, 2004, pp. 23–25.

[10] http://cryptome.org/tempest-time.htm.

[11] Johnson, Thomas, R., *American Cryptology During the Cold War, 1945–1989, Book II: Centralization Wins, 1960–1972*, Center for Cryptologic History, National Security Agency, Fort George G. Meade, Maryland, 1995, p. 381.

[12] Keefe, Patrick Radden, *CHATTER: Dispatches from the secret world of Global Eavesdropping*, Random House, New York, 2005, p. 71.

[13] Bamford, James *Body of Secrets: Anatomy of the Ultra-Secret National Security Agency from the Cold War through the Dawn of a New Century*, Doubleday, New York, 2001, p. 147.

Table 12.1   Growth of Government Cryptology in America (1949–1960)

Date	Cryptologic Population		
	Armed Forces Security Agency (AFSA)	*National Security Agency (NSA)*	*Total, Including Service Cryptologic Agencies (SCAs)*
Dec. 1949	4,139	–	10,745
Dec. 1952	–	8,760	33,010
Nov. 1956	–	10,380	50,550
Nov. 1960	–	12,120	72,560

Source: Johnson, Thomas, R., *American Cryptology During the Cold War, 1945–1989, Book I: The Struggle for Centralization, 1945–1960,* Center for Cryptologic History, National Security Agency, Fort George G. Meade, Maryland, 1995, p. 64.

Americans engaged in intelligence work, 59,000 were cryptologists of one stripe or another (58 percent). Of those, about half worked in the Continental United States, while the other half plied their trade overseas at collection and processing sites. NSA had 10,200 assigned (17 percent of the total) but only 300 overseas billets.[14]

Two milestones for the American intelligence community were passed by 1967: a SIGINT budget in excess of $1 billion that included over 100,000 employees.[15] In the late 1960s, the computers at NSA occupied 5.5 acres.[16] Lieutenant General William E. Odom, Director of NSA (DIRNSA) from 1985–1988 noted:[17]

> For example, development of modern digital computational means—computers— occurred almost entirely as a result of the National Security Agency's research and development efforts in the late 1950s. IBM and CDC essentially got their start in modern computers from National Security Agency funding, and without it, we might be two decades behind where we are in computers today.

12.4 The *Liberty* and the *Pueblo*

In the late 1960s, budgetary constraints in the United States and increasing nationalism in the third world forced NSA to close some intercept stations. As an alternative method of collection, SIGINT ships saw increased use. They would be positioned close enough to a nation of interest

---

[14] Johnson, Thomas, R., *American Cryptology During the Cold War, 1945–1989, Book II: Centralization Wins, 1960–1972,* Center for Cryptologic History, National Security Agency, Fort George G. Meade, Maryland, 1995, p. 293.

[15] Johnson, Thomas, R., *American Cryptology During the Cold War, 1945–1989, Book II: Centralization Wins, 1960–1972,* Center for Cryptologic History, National Security Agency, Fort George G. Meade, Maryland, 1995, p. 479.

[16] Bamford, James, *Body of Secrets: Anatomy of the Ultra-Secret National Security Agency from the Cold War through the Dawn of a New Century,* Doubleday, New York, 2001, p. 578. Johnson, Thomas, R., *American Cryptology During the Cold War, 1945–1989, Book II: Centralization Wins, 1960–1972,* Center for Cryptologic History, National Security Agency, Fort George G. Meade, Maryland, 1995, p. 368 reveals that the 5-acre mark had almost been hit by 1968.

[17] Odom, William, *Fixing Intelligence for a More Secure America,* second edition, Yale University Press, New Haven, Connecticut, 2004, p. 6.

to capture signals, yet avoided danger by remaining in international waters. The expectation was that the ships would be ignored by those being spied upon, just as the United States ignored Soviet SIGINT ships.[18] That expectation was not met.

During the Six-Day War in June 1967, Israel attacked the spy ship USS *Liberty*. Thirty-four men died as a result and many more were injured. Although the *Liberty* was flying an American flag, the Israeli government claims to this day that they weren't aware that it belonged to the United States until after the attack. There is intense disagreement among researchers as to whether this is true or not. The section on the *Liberty* in the reference list at the end of this chapter offers volumes representing both perspectives. Like the question of whether or not Yardley sold secrets to the Japanese, this debate is likely to continue for many years.

In January 1968, close on the heels of the attack on the *Liberty*, another SIGINT ship, the USS *Pueblo* was captured, along with her crew, by North Korea. This happened so quickly that only a small fraction of the classified material on board could be destroyed. A declassified history described the event: "It was everyone's worst nightmare, surpassing in damage anything that had ever happened to the cryptologic community."[19]

By following Kerckhoffs's rules, however, NSA averted an even greater disaster. All of the NSA's cipher equipment was designed to remain secure, even if the enemy somehow gained the full details of the designs, as happened here. Security truly resided in the keys, and those that were captured would not be used again; hence, only messages from late 1967 and early 1968 were compromised.[20]

Although there was less loss of life with the capture of the *Pueblo* (one death) than in the attack on the *Liberty*, the Koreans held the surviving 82 crewmembers as prisoners, and they proved to be brutal captors. Beatings were carried out in private while the North Koreans provided pictures for public consumption intended to show how well the Americans were being treated. The crew of the *Pueblo* minimized the propaganda value of these pictures by positioning their hands to convey a coded signal the Koreans didn't recognize. Upon being asked, the Americas explained that it was the "Hawaiian good luck sign." See Figure 12.3.

How the North Koreans learned that the true meaning of this gesture was not "good luck" will be related shortly. But understand now that they only learned this after the pictures had been distributed throughout the world! Upon getting the message, the North Koreans retaliated with greater physical abuse. In all, the crew spent 11 months in captivity, before America signed an insincere apology that gained their release.

A reunion picture similar to that in Figure 12.4 served as the cover of the Fall 2008 *CRYPTOLOG* (Vol. 29, No. 4), the journal of the U.S. Naval Cryptologic Veterans Association (NCVA). After seeing this cover, I emailed Jay Browne to tell him that it gave me a chuckle. He responded, in part, with the following:

> The cover photo was somewhat "controversial." I told Bob Payne (our Editor) to standby for heavy seas. In fact, of the 4000 or so copies we print, we received a grand total of 2 negative comments! After we received the first one I drafted an editorial for

---

[18] Spy *planes* were another matter. The Soviets tried, and too often succeeded, in shooting these down, the most famous incident being the U-2 piloted by Francis Gary Powers.

[19] Johnson, Thomas, R., *American Cryptology During the Cold War, 1945–1989, Book II: Centralization Wins, 1960–1972*, Center for Cryptologic History, National Security Agency, Fort George G. Meade, Maryland, 1995, p. 439.

[20] Johnson, Thomas, R., *American Cryptology During the Cold War, 1945–1989, Book II: Centralization Wins, 1960–1972*, Center for Cryptologic History, National Security Agency, Fort George G. Meade, Maryland, 1995, p. 452.

Figure 12.3 *Pueblo* **crewmembers sending a coded message.**

Figure 12.4 **The 40th reunion of *Pueblo* crewmembers. (Courtesy of Karen Pike Photography.)**

the winter issue - attached. Bob chose to ignore the issue all together, and in hindsight he was probably right.

The unpublished editorial saw print for the first time in the first edition of this book and is included again in this new edition.[21]

EDITORIAL

CRYPTOLOG has received several "comments" regarding the cover photograph of the Fall issue. Some readers may have been offended by the display of the assembled PUEBLO crew members and the so-called "Hawaiian Good Luck" sign, but *CRYPTOLOG* believes there is a larger story involved.

---

[21] Thanks to Jay Browne for sharing this.

To appreciate the historical context, the reader must go back to the events surrounding the capture of the United States Ship PUEBLO—the first such capture since the 1880s—and the treatment of her surviving crew. The late Commanding Officer of the ship, Commander Lloyd M. Bucher, wrote in his book, "My officers and I were hauled before [The] Glorious General who suddenly reappeared on the scene and delivered one of his towering histrionic rages, which were both comic and frightening to behold. He confronted us with and let us examine a copy of a page from *Time* magazine [18 October 1968 issue] containing a group picture of our infamous Room 13 gang. The caption fully explained the meaning of the "Hawaiian Good Luck Sign." ... I also knew we were about to pay for their severe loss of face. I had not been beaten yet, but Glorious General kept me after he had dismissed the other officers and during a denunciation lasting several hours, threatened me with speedy execution after a trial which he indicated was now absolutely inevitable. He was pretty convincing about it and I was returned to my cell feeling that my chances for survival had sunk to zero."

"On the following day, men continued to be selected for beatings. Shingleton and Scarborough received brutal ones. Radioman Hayes had his jaw broken. The officers began catching it as well."[22]

So the cost of an expression, a gesture, the men—our men—suffered mightily at the hands of their captors. Other newspapers and magazines printed the "offending" photo but it was *Time* that explained the meaning.

While the cover photo may have offended some readers, *CRYPTOLOG* is offended, even today some 40 years later, by the treatment of the North Koreans and the callous disregard of our people by *Time* magazine.

The cover photo speaks volumes to both—here's to you North Korea and to *Time*!

The losses described above brought an end to the use of slow moving ships for gathering intelligence.

12.5 The Church Committee Investigations

Some figures for the budget and size of America's cryptologic community in the 1950s and 1960s were provided above, but they were not known at the time, or even in the 1970s. Thus, the guessing game continued. In 1975, Tad Szulc, writing for *Penthouse*, estimated the numbers at $10 billion per year and 120,000, respectively.[23] His budget calculation included the budgets of the National Reconnaissance Office (NRO),[24] Army, Navy, and Air Force Intelligence. The intimate connections between these agencies were his justification for treating them as a single unit budget-wise.[25] The number of employees is worldwide. Szulc estimated that 10,000 of these would

---

[22] Bucher, Lloyd M., with Mark Rascovich, *Bucher: My Story*, Doubleday, Garden City, New York, 1970.

[23] Szulc, Tad, "The NSA - America's $10 Billion Frankenstein," *Penthouse*, November 1975, pp. 54–56, 70, 72, 184, 186, 188, 191–192, 194–195, p. 194 cited here.

[24] The National Reconnaissance Office (NRO) once lost track of over $2 billion! See Weiner, Tim, "A Secret Agency's Secret Budgets Yield Lost Billions, Officials Say," *The New York Times*, January 30, 1996, p. 5A.

[25] Some budgets were, in fact, lumped together, but not as Szulc figured it. Beginning with fiscal year 1959, a Consolidated Cryptologic Program (CCP) centralized all cryptologic budgeting (including the three services, NSA, and, to a lesser extent CIA) under the Director of NSA (DIRNSA). See Johnson, Thomas, R., *American Cryptology During the Cold War, 1945–1989, Book I: The Struggle for Centralization, 1945–1960*, Center for Cryptologic History, National Security Agency, Fort George G. Meade, Maryland, 1995, p. 260.

be working at the Fort Meade headquarters with a budget of around one billion dollars per year. Not to be outdone, *Playboy* published an article by David Kahn the following month in which he estimated that NSA employed 100,000 people with a budget of several billion dollars per year.[26] A graph from a declassified history of the agency reveals how many employees were actually under NSA's control (Figure 12.5).

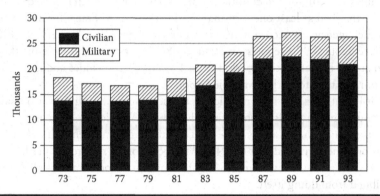

Figure 12.5 Employment figures for NSA from 1973 to 1993. (From Johnson, Thomas R., *American Cryptology During the Cold War, 1945–1989.* Book III. *Retrenchment and Reform, 1972–1980,* Center for Cryptologic History, National Security Agency, Fort Meade, Maryland, 1995, p. 23.)

But why would *Playboy* and *Penthouse* both attempt to expose NSA's size and budget within a month of each other? In addition to their normal distaste for cover-ups, there was a bandwagon to jump on. The U.S. intelligence agencies were being investigated for alleged crimes by a congressional committee and almost every magazine and newspaper had something to say about it. The congressional committee was led by Senator Frank Church, and is therefore often referred to as the Church Committee. It examined ways in which the agencies illegally spied on and disrupted the activities of American citizens. It seems that any group that did not fall in line with the status quo was targeted. War protestors, civil rights activists, feminists, and Native American activists were all harassed under the government sponsored COINTELPRO (counterintelligence program). Many of the victims, such as Martin Luther King, are now generally considered to have helped move the country in the right direction.[27] Other well-known individuals who had their privacy violated included actress Jane Fonda, pediatrician and best-selling author Dr. Benjamin Spock, and folk singer Joan Baez.

The programs uncovered by the Church Committee investigation aroused much public indignation. The CIA and FBI received the greatest scrutiny. NSA might have had a tougher time, but it appears that the committee didn't even want to investigate that particular agency! The paragraph from Johnson's history of NSA that indicates this follows (without any redactions this time).

> To begin with NSA wasn't even on the target list. But in the course of preliminary investigation, two Senate staffers discovered in the National Archives files some Defense paperwork relating to domestic wiretaps which referred to NSA as the source of the request. The committee was not inclined to make use of this material, but the two staffers leaked the documents to Representative Bella Abzug of New York, who

---

[26] Kahn, David, "The Code Battle," *Playboy,* December 1975, pp. 132–136, 224–228.

[27] See http://www.icdc.com/~paulwolf/cointelpro/churchfinalreportIIIb.htm for details on the harassment of King and http://www.icdc.com/~paulwolf/cointelpro/cointel.htm for more general information.

was starting her own investigation. Church terminated the two staffers, but the damage had been done, and the committee somewhat reluctantly broadened its investigation to include the National Security Agency.[28]

NSA programs included SHAMROCK, which involved the interception of private cables from the United States to certain foreign countries,[29] and MINARET, which involved checking all electronic messages that had at least one terminal outside the United States for names on watch lists provided by other agencies.[30] According to NSA Director Lew Allen, Jr., between 1967 and 1973, the Agency gave about 3,900 reports on about 1,680 Americans who were on the watch list.[31] Another estimate for the larger time period from 1962 to 1973 includes about 75,000 Americans and organizations in the group that was spied upon.[32]

Notice that in both cases at least one side of the communication link was outside the United States, even though, in many cases, both individuals were Americans. There is no evidence of the NSA spying on pairs of Americans within the United States. It's been pointed out that Canada can legally spy on Americans and that the NSA has a cozy sort of reciprocal intelligence agreement with Canada and other countries,[33] but NSA maintains that it doesn't ask its allies to do anything that it's prohibited from doing itself.

Some web pages change frequently. At one time, the following was part of NSA's official web presence.

> The agency, the investigations showed, had monitored the domestic conversations of Americans without the proper court warrants. It was chastised and forbidden to overhear such communications, and Congress established a special court to grant national-security wiretaps.

This is typically what happens when some part of the Federal Government is caught breaking the law. An investigation is held, nobody is punished, and legislation is passed to re-outlaw the crime. In another example of crime without punishment, former CIA director Richard Helms committed perjury and got off with just a $2,000 fine.[34]

The special court that was established to grant national-security wiretaps came into being with the Foreign Intelligence Surveillance Act of 1978. Permission slips for eavesdropping thus became known as FISA warrants and were granted by the Foreign Intelligence Surveillance Court. This court approved so many requests that one critic refused to characterize it as a rubber stamp, pointing out that even a rubber stamp runs out of ink sometimes! On the other hand, supporters of the program argue that warrants were almost never refused because the applications were well justified

---

[28] Johnson, Thomas, R., *American Cryptology During the Cold War, 1945–1989, Book III: Retrenchment and Reform, 1972–1980*, Center for Cryptologic History, National Security Agency, Fort George G. Meade, Maryland, 1995, pp. 92–93.

[29] This program predated NSA, having begun with ASA, following World War II.

[30] Halperin, Morton H., Jerry J. Berman, Robert L. Borosage, and Christine M. Marwick, *The Lawless State, The Crimes of the U.S. Intelligence Agencies*, Penguin Books, New York, 1976, p. 173.

[31] Foerstal, Herbert N., *Secret Science: Federal Control of American Science and Technology*, Prager, Westport, Connecticut, 1993, p. 111.

[32] Fitsanakis, Joseph, "National Security Agency: The historiography of concealment," in de Leeuw, Karl, and Jan Bergstra, editors, *The History of Information Security, A Comprehensive Handbook*, Elsevier, Amsterdam, Netherlands, 2007, pp. 523–563, pp. 545–546 cited here.

[33] Constance, Paul, "How Jim Bamford Probed the NSA," *Cryptologia*, Vol. 21, No. 1, January 1997, pp. 71–74.

[34] James Bamford, *Body of Secrets: Anatomy of the Ultra-Secret National Security Agency from the Cold War through the Dawn of a New Century*, Doubleday, New York, 2001, p. 62.

in nearly every case. When FISA was first imposed, NSA decided that there would be no close calls. Applications were only to be made on solid evidence. Also, the statistics are skewed by the fact that in the early days weaker applications were sometimes returned without being officially denied. They could then be strengthened with more information and submitted again or simply forgotten. In this way, some possible rejections never became part of the statistics we see today.

In his four-volume history of NSA, Thomas R. Johnson maintains that the Agency did not act improperly. For example, he states repeatedly that the 1968 "Omnibus Crime Control and Safe Streets Act" overruled Section 605 of the Federal Communications Act of 1934, which forbid eavesdropping.[35] Even if this is true under the letter of the law, it clearly violates the spirit of the constitution. Certainly the founding fathers would not have been amused by this justification.

The second half of the 1970s also marked the beginning of a debate between NSA and American professors, mainly mathematicians, computer scientists, and engineers, who had begun to make important cryptographic discoveries. Previously the NSA had a monopoly on research In these areas, and they did not want to see it end. They feared the loss of control that public pursuits in this field would entail. The various attempts of NSA to stop the academics are discussed in the following chapters of this book, along with the relevant mathematics. Although much time was spent battling congressional committees and academics in the 1970s, the NSA did manage to sign a treaty in 1977 with a long-time enemy, the CIA.[36]

Details are lacking, but a breakthrough was reportedly made in 1979 in deciphering Russia's encrypted voice transmissions.[37] With Venona successes having been disclosed in 1996, we may be able to look forward to revelations on how voice decrypts affected Cold War politics in the 1980s in the not-too-distant future.

12.6 Post Cold War Downsizing

During the 1990s, the budget and the number of employees were reportedly cut by one-third.[38] This would make sense, as the cold war was now over. NSA's overall budget for 1995–1999 was still over $17.5 billion. At the turn of the millennium, there were about 38,000 employees, not counting the 25,000 people who staffed listening posts throughout the world.[39] David Kahn did an excellent job of putting budgetary matters in perspective:[40]

> The real question, however, is whether it is worth the billions spent on it. The answer depends on what the money would otherwise be used for. If the Government were to

---

[35] Johnson, Thomas, R., *American Cryptology During the Cold War, 1945–1989, Book I: The Struggle for Centralization, 1945–1960*, Center for Cryptologic History, National Security Agency, Fort George G. Meade, Maryland, 1995, p. 274 and Johnson, Thomas, R., *American Cryptology During the Cold War, 1945–1989, Book II: Centralization Wins, 1960–1972*, Center for Cryptologic History, National Security Agency, Fort George G. Meade, Maryland, 1995, p. 474.

[36] Johnson, Thomas, R., *American Cryptology During the Cold War, 1945–1989, Book III: Retrenchment and Reform, 1972–1980*, Center for Cryptologic History, National Security Agency, Fort George G. Meade, Maryland, 1995, p. 197.

[37] Bamford, James, *Body of Secrets: Anatomy of the Ultra-Secret National Security Agency from the Cold War through the Dawn of a New Century*, Doubleday, New York, 2001, p. 370.

[38] Bamford, James, *A Pretext for War: 9/11, Iraq, and the Abuse of America's Intelligence Agencies*, Doubleday, New York, 2004, p. 112 and 356.

[39] Bamford, James, *Body of Secrets: Anatomy of the Ultra-Secret National Security Agency from the Cold War through the Dawn of a New Century*, Doubleday, New York, 2001, pp. 481–482.

[40] Kahn, David, "The Code Battle," *Playboy*, December 1975, pp. 132–136, 224–228.

spend it on some more jet fighters or ICBMs, probably the NSA investment is better. Intelligence is cheap and cost-effective. It can often save more than it costs. But if the Government were actually to spend the money on schools and hospitals and transportation, that investment is probably better. For a nation's strength depends far less upon its secret intelligence than upon its human and material resources. No doubt a balance is best. The problem is to strike that balance, and this depends largely on the wisdom and determination of a country's leaders, and of its people.

This did not represent a brand new perspective. Decades earlier, an American President, and World War II general, commented on the opportunity cost of military spending:

> Every gun that is made, every warship launched, every rocket fired, signifies, in the final sense, a theft from those who hunger and are not fed, those who are cold and are not clothed. The world in arms is not spending money alone. It is spending the sweat of its laborers, the genius of its scientists, the hopes of its children.

—**Dwight D. Eisenhower**[41]

Unfortunately, it is very difficult for anyone on the outside to determine whether the people's money is best spent on NSA or elsewhere. Although the agency's successes are typically kept secret (have you heard, for example, that NSA helped prevent a nuclear war between India and Pakistan?), its failures usually receive a great deal of publicity. This leads to a warped perspective, making it difficult to write a balanced account.

NSA's gigantic parking lot, which fills completely early in the morning every weekday, seems to indicate a healthy budget, which, in turn indicates that Congress must be convinced it's getting its money's worth from the Agency. Whatever the budget is, tough choices still need to be made to stay within it. Although more parking spaces are badly needed, huge parking decks are not being constructed. There are other projects that have a stronger need for the funds the decks would require. So, the budget is not unlimited! Nor is it squandered on ridiculous salaries. Many agency employees, with technical skills in high demand, could make much more money on the outside. One employee that I met gave up such a high-paying job to go to work for NSA following 9/11. The rewards of his new career are of a different nature.

12.7 The Crypto AG Connection

The story that is now referred to as "the intelligence coup of the century" begins with the Swiss cipher machine company Crypto AG (see Section 7.1). For decades, it was alleged that NSA was able to convince this trusted manufacturer to place a backdoor, for the use of NSA and Britain's Government Communications Headquarters (GCHQ), into their machines. This project was said to have begun in 1957, when William Friedman traveled to Switzerland to meet with Boris Hagelin.[42] The trip was first reported on in a biography of William Friedman written by

---

[41] Eisenhower, Dwight D., "The Chance for Peace," speech to the American Society of Newspaper Editors, Washington, DC, April 16, 1953. Quoted here from Zinn, Howard, *Terrorism and War*, Seven Stories Press, New York, 2002, p. 96.

[42] Bamford, James, *The Puzzle Palace*, Houghton, Mifflin, and Company, New York, 1982, pp. 321–324.

Ronald W. Clark.[43] NSA got wind of this and exhibited great interest in the manuscript, which seemed to confirm that there was *something* to the story. Clark was not intimidated by the Agency, but he didn't really know much about the trip. His biography merely opened the door on this topic.

As the story went, the deal wasn't made in a single trip. Friedman had to return, and in 1958, Hagelin agreed. Crypto AG machines were eventually adopted for use by 120 nations, but it seemed unlikely that they were all rigged. According to some accounts the security levels provided depended on the country in which the machine was to be used. In any case, there doesn't appear to have been any suspicion of rigging until 1983.[44] Twenty-five years is a very long time to keep such a large-scale project secret. This particular quarter-century includes the switch from electro-mechanical machines to computerized digital encryption, and the belief was that the backdoors remained in place through this transition.

There are various accounts of how knowledge of the allegedly rigged machines leaked out. One version is as follows. The spy Jonathan Pollard betrayed a tremendous amount of material to Israel, including, apparently, details of the Crypto AG machines' backdoors. This information was then given to the Soviets, in 1983, in exchange for allowing a larger number of Jews to leave the Soviet Union for Israel.[45]

However the news got out, it is claimed to have later spread to Iran. The next episode in the story is very well-documented. It is the arrest of the Crypto AG salesman Hans Buehler in March 1992. The Iranians charged him with spying for the Federal Republic of Germany and the United States and imprisoned him for nine months. He was only released when Crypto AG paid a $1 million bail. A few weeks later, the company dismissed Buehler and insisted that he reimburse them for their expense! Buehler initially had no idea why he was arrested, as he relates in his book *Verschlüsselt*.[46] He tells his story in the same style as Franz Kafka's *Der Prozess* (*The Trial*), in which the protagonist, who is arrested early in the story, never learns what the charges against him are. Prior to his arrest, Buehler was not aware of any rigging of the machines he was selling, but he was to eventually conclude, after speaking with several former employees of Crypto AG, that the machines were rigged and that he was paying the price for the duplicity of others.

As you have surely noticed, I am not conveying any details of how exactly the backdoors worked. Full details are not publicly available. Statements such as "The KGB and GRU found out about the 'seed key' used by NSA as a 'master key' to unlock encoded communications transmitted by Crypto AG machines," made by Wayne Madsen, only hint at how it might have worked.[47]

---

[43] Clark, Ronald, *The Man Who Broke Purple*, Little, Brown and Company, Boston, Massachusetts, 1977. As we saw in Section 8.5, Clark's title isn't very accurate. Previous biographies by this author include the very popular *Einstein: the Life and Times* (1971) and another on Bertrand Russell.

[44] Madsen, Wayne, "Crypto AG: The NSA's Trojan Whore?" *Covert Action Quarterly*, Issue 63, Winter 1998, available online at http://mediafilter.org/caq/cryptogate/ and https://web.archive.org/web/20000815214548/http://caq.com:80/CAQ/caq63/caq63madsen.html.

[45] At least, according to Madsen, Wayne, "The Demise of Global Communications Security, The Neocons' Unfettered Access to America's Secrets," *Online Journal™*, September 21, 2005, http://67.225.133.110/~gbppr org/obama/nytimes_ww2/09-21-05_Madsen.pdf. In this piece, Madsen also provides an alternate explanation, that doesn't involve Pollard, for how the Soviets learned the secret: "Ex-CIA agents report that the Russian intelligence successors to the former KGB were actually tipped off about the Crypto AG project by CIA spy Aldrich Ames," Having two different accounts shows how speculative this whole story really is.

[46] Buehler, Hans, *Verschlüsselt*, Werd, Zürich, Switzerland, 1994. This book is in German and no translation is currently available.

[47] Madsen, Wayne, "The Demise of Global Communications Security, The Neocons' Unfettered Access to America's Secrets," *Online Journal™*, September 21, 2005, http://67.225.133.110/~gbpprorg/obama/nytimes_ww2/09-21-05_Madsen.pdf.

Other sources describe the keys as somehow being sent with the messages by the rigged machines. Or, in the early days of mechanical machines, it could have been a simple matter of omitting certain levers in the devices sold to particular customers. Could a mathematical analysis of the machines reveal the secret? This would make an excellent research project, but a reluctance to approach it is understandable. It may merely result in months of wasted effort with no new theorems or results to show for the work.

In 1995, a Swiss engineer spoke to Scott Shane, then of *The Baltimore Sun*,[48] under the condition that his anonymity be maintained. Shane revealed what he learned:

> Sometimes the mathematical formulas that determined the strength of the encryption contained certain flaws making the codes rapidly breakable by a cryptanalyst who knew the technical details.

> In other cases, the designs included a "trapdoor"—allowing an insider to derive the numerical "key" to an encrypted text from certain clues hidden in the text itself.[49]

Again, this is intriguing, but not nearly as detailed as we would desire! Shane provided another piece of evidence in support of the conspiracy, for which Crypto AG failed to provide any alternative explanation. It's a 1975 document that shows NSA cryptographer Nora L. Mackebee attended a meeting with Crypto AG employees to discuss the design of new cipher machines.[50] Motorola engineer Bob Newman recalls Mackebee at several meetings, as she was one of several consultants who was present when Motorola was helping Crypto AG with designs as the company made the switch from mechanical machines to electronic.[51] Shane contacted Mackabee, who had since retired, but she said she couldn't talk about Crypto AG.[52] Crypto AG executives consistently denied the existence of backdoors in any of their machines, as one would expect whether they were present or not.

In 2014 and 2015, NSA released over 52,000 pages of material connected with William F. Friedman.[53] Among these were papers indicating that the NSA-Crypto AG connection was real. They showed that negotiations between Friedman and Hagelin dated back to 1951 and included $700,000 in compensation for Hagelin. This was well before Friedman's 1957 trip, uncovered by his biographer Ronald W. Clark. It took a great deal of time to work out all of the details of the agreement between Friedman and Hagelin, although Hagelin was cooperating from the start.[54] Following the release of these documents, the Swiss company's response to the old allegations

---

[48] Shane now works for *The New York Times*.

[49] Shane, Scott and Tom Bowman, *No Such Agency, America's Fortress of Spies*, Reprint of a six-part series that appeared in *The Baltimore Sun*, December 3–15, 1995, p. 10.

[50] Shane, Scott and Tom Bowman, *No Such Agency, America's Fortress of Spies*, Reprint of a six-part series that appeared in *The Baltimore Sun*, December 3–15, 1995, pp. 9–10.

[51] Shane, Scott and Tom Bowman, *No Such Agency, America's Fortress of Spies*, Reprint of a six-part series that appeared in *The Baltimore Sun*, December 3–15, 1995, p. 10.

[52] Mackebee died on March 3, 2015.

[53] National Security Agency, William F. Friedman Collection of Official Papers, https://www.nsa.gov/News-Features/Declassified-Documents/Friedman-Documents/.

[54] What can be determined from NSA's Friedman release is thoroughly detailed at Simons, Marc and Paul Reuvers, "The gentleman's agreement, Secret deal between the NSA and Hagelin, 1939–1969," Crypto Museum, https://www.cryptomuseum.com/manuf/crypto/friedman.htm, created: July 30, 2015, last changed: May 10, 2020.

shifted from strong denial to "whatever happened in the past, this is certainly not happening today" and "mechanisms have been put in place, to prevent this from happening in the future."[55]

On February 11, 2020, a much more complete story was revealed, to the dismay of the US intelligence community. This time it wasn't a planned release. The investigative team consisted of Greg Miller of *The Washington Post*, and men and women from German and Swiss television.[56] The compromise of Crypto AG was more complete than had been suspected. While NSA was deeply involved, it was the CIA that turned out to be the Victor Kiam of the crypto equipment market. Older readers will remember Kiam from commercials he did for Remington in which he enthusiastically said, "I liked the shaver so much, I bought the company!" This is exactly what the CIA did. On June 12, 1970, they secretly bought Crypto AG, in a joint purchase with the West German Federal Intelligence Service (Bundesnachrichtendienst, or BND for short). In addition to the United States and West Germany, four other countries, Israel, Sweden, Switzerland, and the U.K., knew of the operation, or were given intelligence gathered from it.[57] This group managed to keep the identities of the new owners of Crypto AG secret from the public for 50 years. It makes one wonder what other long-term secrets have been kept.[58]

The researchers based much of their reporting on histories prepared by the CIA and BND, although they did not indicate how they obtained these histories.[59] They also conducted interviews with current and former members of the intelligence community and Crypto AG employees. The CIA history noted that the Crypto AG material "represented over 40 percent of NSA's total machine decryptions, and was regarded as an irreplaceable resource."[60]

Prior to becoming a secret owner, the CIA had made payments to Hagelin. There was one in 1960 for $855,000 to renew the "licensing agreement" that he had made with Friedman. There were also annual payments of $70,000 and cash infusions of $10,000 for marketing Crypto AG products. The latter helped to insure that the company would continue to dominate the world market.[61] Prior to their ownership, the CIA needed the company to stay successful! When CIA and BND became co-owners, the profits were a nice bonus that were then poured into other operations. And what a way to make money! The CIA history noted:

> Foreign governments were paying good money to the U.S. and West Germany for the privilege of having their most secret communications read by at least two (and possibly as many as five or six) foreign countries.[62]

55 "The Crypto Agreement," *BBC Radio 4*, July 28, 2015, available online at https://www.bbc.co.uk/programmes/b0639w3v.

56 Simons, Marc and Paul Reuvers, "Operation RUBICON/THESAURUS, The secret purchase of Crypto AG by BND and CIA," Crypto Museum, https://www.cryptomuseum.com/intel/cia/rubicon.htm, created: December 12, 2019, last changed: May 10, 2020.

57 Miller, Greg, "The intelligence Coup of the Century," *The Washington Post*, February 11, 2020, available online at https://tinyurl.com/yck5xur2.

58 Hint: https://tinyurl.com/y5oy4v78.

59 Miller noted, "The first [history] is a 96-page account of the operation completed in 2004 by the CIA's Center for the Study of Intelligence, an internal historical branch. The second is an oral history compiled by German intelligence officials in 2008."

60 Miller, Greg, "The intelligence Coup of the Century," *The Washington Post*, February 11, 2020, available online at https://tinyurl.com/yck5xur2.

61 Miller, Greg, "The intelligence Coup of the Century," *The Washington Post*, February 11, 2020, available online at https://tinyurl.com/yck5xur2.

62 Miller, Greg, "The intelligence Coup of the Century," *The Washington Post*, February 11, 2020, available online at https://tinyurl.com/yck5xur2.

This operation went under the code name "Thesaurus," later changed to "Rubicon." The technical details will be uncovered in the years to come, but it appears that the idea of backdoors, wasn't quite right. Instead, two versions would be made of a machine: a good one for friendly nations and one that appeared to be good, but was really much less secure, for other nations. Creating the illusion of a secure system was tricky and, sometimes, foreign crypto experts would get suspicious. Still, it worked in the electromechanical machine era, as well as with more advanced devices in the decades that followed. Peter Jenks, of NSA, recognized that a circuit-based system could be designed so that it appeared to generate random streams of characters, while it really had a short enough period to be broken by NSA cryptanalysts with powerful computers at their disposal.[63]

The list of countries that would receive weaker systems kept growing and the West Germans became nervous about how broadly the American's were spying. Nations receiving weaker machines even included members of NATO. Fearing the fall-out, if this were to be exposed, the Germans allowed the CIA to buy them out in 1994 and become the sole owner. But, by 2018, what had been the intelligence coup of the (20th) century, may have been surpassed, for in that year the CIA sold off Crypto AG's assets.[64]

The Crypto-AG connection is tremendously important, but it wasn't the only news-worthy item connected with NSA. The next section looks at some other developments at the agency over the last 20 years.

12.8 2000 and Beyond

While NSA definitely had successes, as the last section detailed, the year 2000 began roughly, with a computer crash occurring in January that put the whole information processing system out of operation for three days. It seems that no one is immune from computer problems. James Bamford estimated that, on 9/11, 16,000 employees worked at NSA and its surrounding facilities.[65] He quotes Lieutenant General Michael Hayden, Director of NSA at the time, as referring to "all of the agency's 38,000 employees."[66] There are also about 25,000 employees at overseas listening posts and one would assume the numbers have increased since 9/11. According to Hayden, the number of new recruits jumped to 1,500 per year by 2004.[67] Relatively few of these were mathematicians. The most common job classifications were (1) Security Guard, (2) Polygraph Examiner, and (3) Linguist. At position 4, we finally get to Analysts.[68]

[63] Miller, Greg, "The intelligence Coup of the Century," *The Washington Post*, February 11, 2020, available online at https://tinyurl.com/yck5xur2. This will make more sense after you read Chapter 19. Alternatively, you can read Simons, Marc and Paul Reuvers, "Operation RUBICON/THESAURUS, The secret purchase of Crypto AG by BND and CIA," Crypto Museum, https://www.cryptomuseum.com/intel/cia/rubicon.htm, created: December 12, 2019, last changed: May 10, 2020.

[64] Miller, Greg, "The intelligence Coup of the Century," *The Washington Post*, February 11, 2020, available online at https://tinyurl.com/yck5xur2.

[65] Bamford, James, *A Pretext for War: 9/11, Iraq, and the Abuse of America's Intelligence Agencies*, Doubleday, New York, 2004, p. 53.

[66] Bamford, James, *A Pretext for War: 9/11, Iraq, and the Abuse of America's Intelligence Agencies*, Doubleday, New York, 2004, p. 113.

[67] Bamford, James, *A Pretext for War: 9/11, Iraq, and the Abuse of America's Intelligence Agencies*, Doubleday, New York, 2004, p. 356.

[68] Keefe, Patrick Radden, *CHATTER: Dispatches from the secret world of Global Eavesdropping*, Random House, New York, 2005, p. 236

There were massive increases in both budget and the number of employees after 9/11, but this was not immediate. First came some reorganizing cuts, which were made by encouraging retirements. In particular, NSA no longer needed so many Soviet linguists or high-frequency specialists.[69] Bamford provides a physical description of NSA (as of 2004):

> Nicknamed Crypto City, it consists of more than fifty buildings containing more than seven million square feet of space. The parking lot alone covers more than 325 acres and have [sic] room for 17,000 cars.[70]

While these stats are impressive, there's a lot more to the agency, former NSA Deputy Director Chris Inglis put matters in perspective near the end of his service in 2014:

> But if you want to really know what the core of NSA is, it's its brain trust. It's its people. All right? We employ some, you know, number of people which includes 1,000 Ph.D.s, which includes a diverse array of disciplines that we bring to bear.[71]

It's likely that much of the information gathered (by all modern nations) is collected by means other than direct mathematical attacks on the mature cipher systems of the 21st century. Backdoors, exploitation of electromagnetic emissions, and hacking attacks are probably the source of much intelligence. A new Cyber Command Center (Cybercom) that carries out such work, in addition to safeguarding American systems from such attacks, has been established and is located at NSA. The director of NSA is now dual-hatted and also directs Cybercom. It is simply too inconvenient to not have important systems online, and too dangerous to do so without making intense efforts to protect these systems.

In recent years there's been a massive amount of media attention on alleged domestic spying by NSA. The majority of the journalists making such claims are likely well-intentioned and simply trying to report accurately; however, it seems that they are (in some cases) confusing mere interception with actually reading messages or listening to conversations. NSA is allowed to *accidentally* intercept domestic conversations, and one must understand that the technological environment is such that it is, in many cases, impossible to intercept the desired targets, in isolation, without also gathering untargeted items. Email messages don't travel like postcards. Instead they're broken into packets, which may follow various paths before being reassembled at the intended destination. Phone conversations are combined in large groups, to which data compression algorithms are applied. Thus, to gather the intercepts NSA legitimately needs to do its job, it must also unintentionally acquire other data. The unintended intercepts are then filtered out.

Of course, the potential for abuse is present. Americans tend to fear big government and secrecy. And NSA is a very big government agency that must also be very secretive! NSA employees swear an oath to protect the constitution and I believe they take this oath much more seriously than recent Presidents have. Certainly, the Agency takes the oath seriously. An NSA employee told me that the people who work there can make mistakes and keep their jobs,

---

[69] Bamford, James, *A Pretext for War: 9/11, Iraq, and the Abuse of America's Intelligence Agencies*, Doubleday, New York, 2004, p. 356.

[70] Bamford, James, *A Pretext for War: 9/11, Iraq, and the Abuse of America's Intelligence Agencies*, Doubleday, New York, 2004, p. 52.

[71] "Transcript: NSA Deputy Director John Inglis," *npr*, https://www.npr.org/2014/01/10/261282601/transcript-nsa-deputy-director-john-inglis, January 10, 2014. Note: Inglis goes by his middle name, Chris, despite how he is named in this piece.

in many cases, but that if they spy on Americans they can be fired the same day. He said that he'd seen it happen.

Everything I've seen and heard at NSA has convinced me that respect for the constitution is a key component of the culture there. A casual conversation I had with an NSA employee helps to illustrate this. I complained about Mitt Romney having said to a protestor, "Corporations are people, my friend"[72] and she responded with something like, "I know and that's a huge pain for us, because if they're American corporations, we can't spy on them, even if they are 45% owned by a foreign country that's really controlling them and they're up to no good." So, American corporations have the same constitutional rights as American citizens and those rights are respected by NSA.

Another NSA employee described how an American President wanted NSA to do things that the director thought were prohibited. The director stood his ground and refused to cooperate. Much of the legislation that presidents and congress have pushed for has not been asked for, or even desired, by the Agency.

12.9 Interviewing with NSA

The NSA is the largest employer of mathematicians in the world and the interview process is guaranteed to be interesting.[73] If you make it past the initial screening, you will be invited to Fort Meade for a much closer look. Your favorite part of the process is not likely to be the polygraph test.[74] The purpose of the polygraph tests at NSA is basically to verify two things:

1. The applicant does not have anything in his or her background that might make him or her subject to blackmail.
2. The applicant is not a spy.

Item number 1 justifies the asking of many potentially embarrassing questions. An acquaintance of mine was asked if he had ever cheated on his wife. If he had, someone who knew about it could blackmail him into revealing classified information in return for his silence on the matter. I assume he was, indeed, faithful, as he was hired; however, the polygraph test was still rough on him. Initially he was so nervous that everything showed up as a lie. Finally, they asked him if he had ever killed anyone. He said, "No," which also showed up as a lie! After a break, during which he managed to calm down, things went more smoothly.

Oddly enough, there are applicants who confessed to crimes such as murder, rape, and whole-sale selling of illegal drugs during the polygraph test.[75] In fact, of the 20,511 applicants between 1974 and 1979, 695 (3.4%) admitted to the commission of a felony, nearly all of which had

---

[72] I know that he is legally correct. Corporate personhood is law. We live in a strange world. The United States once had people as property (slaves) and now has property as people (corporations).

[73] See *Interviewing with an Intelligence Agency (or, A Funny Thing Happened on the Way to Fort Meade)*, by Ralph J. Perro (a pseudonym) available online at http://www.fas.org/irp/eprint/nsa-interview.pdf. The pseudonym was a take-off on Ralph J. Canine, the first director of NSA.

[74] There are two types of polygraph tests. The first is the "lifestyle polygraph," done prior to employment. The second is the "counterintelligence polygraph," which is given 5 years later.

[75] Bamford, James, *Body of Secrets: Anatomy of the Ultra-Secret National Security Agency from the Cold War through the Dawn of a New Century*, Doubleday, New York, 2001, p. 540.

previously gone undetected.[76] Bamford described how these tests received a black eye during the 1950s and early 1960s because of the heavy use of EPQs (embarrassing personal questions).

> These questions are almost inevitably directed toward intimate aspects of a person's sex life and bear little relationship to the person's honesty or patriotism. Following a congressional investigation and an internal crackdown, the personal questions are now somewhat tamer. "Have you ever had an adult homosexual experience?" for example, is one of the standard questions today.[77]

This quote is from 1982. The NSA is now more tolerant. Although EPQs are still used, homosexuality is not necessarily considered a problem. There is even a social club, GLOBE, for Gay, Lesbian, or Bisexual Employees.[78] Of course, the applicant need not worry in any case. The answers to these questions are kept confidential.

It wouldn't be revealed, for example, that Bernon F. Mitchell told his interrogator about certain "sexual experimentations" with dogs and chickens he had carried out when he was between the ages of 13 and 19.[79] Okay, maybe this wasn't the best example. Despite Mitchell's strange history, he was hired along with William H. Martin. Johnson summarized the interview process for these two men with "Certain questions about their psychological health came up on the polygraph and background investigation but were not regarded as serious impediments to employment."[80]

In 1960, these men betrayed the agency to the Russians, sparking a purge of homosexuals. In all, 26 NSA employees were fired because of their sexual conduct.[81] The discrimination was even extended to other government positions. President Eisenhower ordered a secret blacklisting of gays from employment within the federal government. He was aided in this by J. Edgar Hoover, who maintained a list of homosexuals.[82]

---

[76] Bamford, James, *Body of Secrets: Anatomy of the Ultra-Secret National Security Agency from the Cold War through the Dawn of a New Century*, Doubleday, New York, 2001, p. 540. A pair of NSA historians expressed skepticism that such statistics were ever compiled.

[77] Bamford, James, *The Puzzle Palace*, Houghton, Mifflin, and Company, New York, 1982, p. 162.

[78] Bamford, James, *Body of Secrets: Anatomy of the Ultra-Secret National Security Agency from the Cold War through the Dawn of a New Century*, Doubleday, New York, 2001, p. 485.

[79] Bamford, James, *The Puzzle Palace*, Houghton, Mifflin, and Company, New York, 1982, p. 180.

[80] Johnson, Thomas, R., *American Cryptology During the Cold War, 1945–1989, Book I: The Struggle for Centralization, 1945–1960*, Center for Cryptologic History, National Security Agency, Fort George G. Meade, Maryland, 1995, p. 182. Part of my purpose in relating this story is to ease concerns of potential NSA applicants. You don't have to be 100% squeaky clean to get hired. You'll probably do better on the interview than Mitchell and they hired him!

[81] Johnson, Thomas, R., *American Cryptology During the Cold War, 1945–1989, Book I: The Struggle for Centralization, 1945–1960*, Center for Cryptologic History, National Security Agency, Fort George G. Meade, Maryland, 1995, p. 284.

[82] Bamford, James, *Body of Secrets: Anatomy of the Ultra-Secret National Security Agency from the Cold War through the Dawn of a New Century*, Doubleday, New York, 2001, pp. 543–544.

Others who betrayed the NSA and affiliated groups include:[83]

1. *William Weisband*—His betrayal took place before NSA was created and he is thought by some to be responsible for "Black Friday," the day on which all Warsaw Pact encryption systems were changed, shutting out the cryptanalysts, October 29, 1948.

2. *Joseph Sydney Petersen, Jr.*—Caught in 1953, his indictment for betraying secrets to the Dutch got the NSA some unwanted publicity.

3. *Roy A. Rhodes*—He provided the Soviets with cryptographic information and was caught based on information from a NKVD Lieutenant Colonel, who defected in 1957.

4. *Robert Lee Johnson*—Active in the 1950s and early 1960s, he provided the Soviets with key lists for cipher machine and other valuable material.

5. *Jack Dunlap*—As a spy for the Soviets from 1959 to 1963, he stole documents by tucking them under his shirt. He committed suicide while being investigated.

6. *Victor Norris Hamilton*—This former NSA cryptanalyst defected to the Soviet Union in July 1963.

7. *Robert S. Lipka*—He got away with betraying NSA to the Soviets, while he worked there from 1964 to 1967, but in 1993 his ex-wife turned him in.

8. *Christopher Boyce* and *Daulton Lee*—The story of these traitors was told in Robert Lindsey's *The Falcon and the Snowman* (Simon and Schuster, New York, 1979). In 1984, the book was made into an excellent movie of the same title that starred Timothy Hutton and Sean Penn.

9. *William Kampiles*—This CIA employee sold the Soviets the Big Bird Satellite manual in 1978. The system delivered both signals intelligence and photo surveillance.

10. *John Anthony Walker, Jr.*—He began spying for the Soviets in 1967, and before his arrest in 1984, had managed to recruit three other Navy men into what became known as the Walker Spy Ring.

11. *Jonathan Jay Pollard*—Speculation concerning his revelations to Israel was provided earlier in this chapter.

12. *Ronald William Pelton*—He was convicted of espionage in 1986 for providing the Soviets with detailed information on the United States' electronic espionage abilities.

13. *David Sheldon Boone*—This NSA cryptanalyst sold secrets to the Russians and in 1998 received a sentence of 24 years.

12.10 Another Betrayal

Another person joined the list of infamous individuals in Section 12.9 after the first edition of this book went to the printer, namely Edward Snowden. In May 2013, he left his job as a contractor at NSA Hawaii for Hong Kong, and the following month newspaper articles relating classified material he had stolen began appearing. He soon made his way to Russia, where he remains, as of

---

[83] Examples taken from Fitsanakis, Joseph, "National Security Agency: The Historiography of Concealment," in de Leeuw, Karl and Jan Bergstra, editors, *The History of Information Security, A Comprehensive Handbook*, Elsevier, Amsterdam, 2007, pp. 523–563, pp. 535, 538, 543–544 cited here; Johnson, Thomas, R., *American Cryptology During the Cold War, 1945–1989, Book I: The Struggle for Centralization, 1945–1960*, Center for Cryptologic History, National Security Agency, Fort George G. Meade, Maryland, 1995, pp. 277–279; Johnson, Thomas, R., *American Cryptology During the Cold War, 1945–1989, Book II: Centralization Wins, 1960–1972*, Center for Cryptologic History, National Security Agency, Fort George G. Meade, Maryland, 1995, pp. 470–471; Polmar, Norman and Thomas B. Allen, *Spy Book*, Random House, New York, 1997.

this writing. Snowden is considered by some to be a patriotic whistle-blower. Chris Inglis, former Deputy Director, NSA, explained why this appellation is inappropriate.

> I do find it curious that Snowden, who is now kind of in the protective embrace of Russia and who once enjoyed the protective embrace of China, has said nothing about those legal regimes which most independent observers would say runs roughshod over civil liberties, human rights. I find it curious he would not say a word about that. But it's consistent actually with what he said while he was in the United States. Nothing. He made no complaint to anyone about what he now observes, what's, in his view, a violation of US person privacy, said nary a word the whole time he was at the National Security Agency, nary a word the whole time he was with the CIA. When asked about that by Jim Bamford, I believe two years ago, the spring of 2014, he said he had at one time raised a question to an NSA lawyer. When we went back and took a look at that it turns out the question that he asked was "Are the priorities that were kind of articulated to me in a class that I took on the protection of US person privacies, the US constitution, an equal priority between law and executive order, and then policies, regulations, and the like?" Lawyer came back the same day answering that question for Mr. Snowden, saying "No, that's not exactly right. Turns out that a statute, a law, trumps an executive order. They're only on the same line in the priorities table because, in the absence of a law, an executive order stands in." If that's a complaint about the protection of US person privacy in the United States of America. I'm hard pressed to see it. I'm hard pressed to understand it. Having raised not one question about that issue while he was here in the United States, my assumption is that Snowden doesn't have the courage of his convictions when he thinks he might be held personally accountable for standing up and defending those convictions. It's not an official position, but that's how I feel.[84]

> He [Snowden] said he was worried about the violation of US person privacy. Most of the information he released has nothing to do with that. He said that he could prove that the United States violated US person privacy beyond the reach of law, beyond the constitutional norms that are established. There's been no proof of that. Now we as a matter of policy might decide that we're uncomfortable with collecting telephone metadata. That doesn't make it illegal. Bad policy or a different choice about policy doesn't make it illegal or unconstitutional.[85]

Inglis also explained that much of what Snowden claimed is inaccurate. An example will help illustrate this.

> We have to distinguish between what Snowden said and what was true. His allegations are not one in the same as revelations. Much of what he extrapolated from his information was frankly untrue. He said, early on, that any NSA analyst could, sitting at his or her desk, target the communications, the content of the communications, of the President of the United States of America. Quote unquote. Patently untrue. It's not

---

[84] Irari Report, "Edward Snowden: NSA Perspective from former Deputy Director," https://www.youtube.com/watch?v=G5evenZOFU0, March 30, 2016.

[85] Irari Report, "Edward Snowden: NSA Perspective from former Deputy Director," https://www.youtube.com/watch?v=G5evenZOFU0, March 30, 2016.

only illegal to do such a thing, but there are procedural controls and technical controls in place that make that impossible. He said that the National Security Agency targeted the content of the communications of US persons. That's not true. It is absolutely true that in targeting the content of legitimate foreign intelligence targets that sometimes the other end of that conversation is a US person and it's almost impossible to determine that with great precision upfront, but because of that there are procedures in place of what exactly do you do when you encounter that situation. I would describe that as a feature, not as a burden, not as a sin. There had been no *no* evidence, since Snowden has come out, that what he alleged is in fact true, that there have been any violations of law.[86]

The House Permanent Select Committee on Intelligence produced a study on the Snowden betrayal. While it remains classified, a three page executive summary was released. It is reproduced below.[87] You will see that it mirrors Inglis's views and contains the lines "Snowden was not a whistleblower" and "Snowden was, and remains, a serial exaggerator and fabricator."

Executive Summary of Review of the Unauthorized Disclosures of Former National Security Agency Contractor Edward Snowden
September 15, 2016

In June 2013, former National Security Agency (NSA) contractor Edward Snowden perpetrated the largest and most damaging public release of classified information in U.S. intelligence history. In August 2014, the Chairman and Ranking Member of the House Permanent Select Committee on Intelligence (HPSCI) directed Committee staff to carry out a comprehensive review of the unauthorized disclosures. The aim of the review was to allow the Committee to explain to other Members of Congress—and, where possible, the American people—how this breach occurred, what the U.S. Government knows about the man who committed it, and whether the security shortfalls it highlighted had been remedied.

Over the next two years, Committee staff requested hundreds of documents from the Intelligence Community (IC), participated in dozens of briefings and meetings with IC personnel, conducted several interviews with key individuals with knowledge of Snowden's background and actions, and traveled to NSA Hawaii to visit Snowden's last two work locations. The review focused on Snowden's background, how he was able to remove more than 1.5 million classified documents from secure NSA networks, what the 1.5 million documents contained, and the damage their removal caused to national security.

The Committee's review was careful not to disturb any criminal investigation or future prosecution of Snowden, who has remained in Russia

---

[86] Irari Report, "Edward Snowden: NSA Perspective from former Deputy Director," https://www.youtube.com/watch?v=G5evenZOFU0, March 30, 2016.

[87] U.S. House of Representatives, *Executive Summary of Review of the Unauthorized Disclosures of Former National Security Agency Contractor Edward Snowden*, September 15, 2016, available online at https://fas.org/irp/congress/2016_rpt/hpsci-snowden-summ.pdf.

since he fled there on June 23, 2013. Accordingly, the Committee did not interview individuals whom the Department of Justice identified as possible witnesses at Snowden's trial, including Snowden himself, nor did the Committee request any matters that may have occurred before a grand jury. Instead, the IC provided the Committee with access to other individuals who possessed substantively similar knowledge as the possible witnesses. Similarly, rather than interview Snowden's NSA co-workers and supervisors directly, Committee staff interviewed IC personnel who had reviewed reports of interviews with Snowden's co-workers and supervisors. The Committee remains hopeful that Snowden will return to the United States to face justice.

The bulk of the Committee's 36-page review, which includes 230 footnotes, must remain classified to avoid causing further harm to national security; however, the Committee has made a number of unclassified findings. These findings demonstrate that the public narrative popularized by Snowden and his allies is rife with falsehoods, exaggerations, and crucial omissions, a pattern that began before he stole 1.5 million sensitive documents.

First, Snowden caused tremendous damage to national security, and the vast majority of the documents he stole have nothing to do with programs impacting individual privacy interests–they instead pertain to military, defense, and intelligence programs of great interest to America's adversaries. A review of the materials Snowden compromised makes clear that he handed over secrets that protect American troops overseas and secrets that provide vital defenses against terrorists and nation-states. Some of Snowden's disclosures exacerbated and accelerated existing trends that diminished the IC's capabilities to collect against legitimate foreign intelligence targets, while others resulted in the loss of intelligence streams that had saved American lives. Snowden insists he has not shared the full cache of 1.5 million classified documents with anyone; however, in June 2016, the deputy chairman of the Russian parliament's defense and security committee publicly conceded that "Snowden did share intelligence" with his government. Additionally, although Snowden's professed objective may have been to inform the general public, the information he released is also available to Russian, Chinese, Iranian, and North Korean government intelligence services; any terrorist with Internet access; and many others who wish to do harm to the United States.

The full scope of the damage inflicted by Snowden remains unknown. Over the past three years, the IC and the Department of Defense (DOD) have carried out separate reviews–with differing methodologies–of the damage Snowden caused. Out of an abundance of caution, DOD reviewed all 1.5 million documents Snowden removed. The IC, by contrast, has carried out a damage assessment for only a small subset of the documents. The Committee is concerned that the IC does not plan to assess the damage of the vast majority of documents Snowden removed. Nevertheless, even by a conservative estimate, the U.S. Government has spent hundreds of millions of dollars, and will eventually spend billions, to attempt to mitigate the damage Snowden caused. These dollars would have been better spent on combating America's adversaries in an increasingly dangerous world.

Second, Snowden was not a whistleblower. Under the law, publicly revealing classified information does not qualify someone as a whistleblower. However, disclosing classified information that Shows fraud, waste, abuse, or other illegal activity to the appropriate law enforcement or oversight personnel–including to Congress–does make someone a whistleblower and affords them with critical protections. Contrary to his public claims that he notified numerous NSA officials about what he believed to be illegal intelligence collection, the Committee found no evidence that Snowden took any official effort to express concerns about U.S. intelligence activities–legal, moral, or otherwise–to any oversight officials within the U.S. Government, despite numerous avenues for him to do so. Snowden was aware of these avenues. His only attempt to contact an NSA attorney revolved around a question about the legal precedence of executive orders, and his only contact to the Central Intelligence Agency (CIA) Inspector General (IG) revolved around his disagreements with his managers about training and retention of information technology specialists.

Despite Snowden's later public claim that he would have faced retribution for voicing concerns about intelligence activities, the Committee found that laws and regulations in effect at the time of Snowden's actions afforded him protection. The Committee routinely receives disclosures from IC contractors pursuant to the Intelligence Community Whistleblower Protection Act of 1998 (IC WPA). If Snowden had been worried about possible retaliation for voicing concerns about NSA activities, he could have made a disclosure to the Committee. He did not. Nor did Snowden remain in the United States to face the legal consequences of his actions, contrary to the tradition of civil disobedience he professes to embrace. Instead, he fled to China and Russia, two countries whose governments place scant value on their citizens' privacy or civil liberties–and whose intelligence services aggressively collect information on both the United States and their own citizens.

To gather the files he took with him when he left the country for Hong Kong, Snowden infringed on the privacy of thousands of government employees and contractors. He obtained his colleagues' security credentials through misleading means, abused his access as a systems administrator to search his co-workers' personal drives, and removed the personally identifiable information of thousands of IC employees and contractors. From Hong Kong he went to Russia, where he remains a guest of the Kremlin to this day.

It is also not clear Snowden understood the numerous privacy protections that govern the activities of the IC. He failed basic annual training for NSA employees on Section 702 of the Foreign Intelligence Surveillance Act (FISA) and complained the training was rigged to be overly difficult. This training included explanations of the privacy protections related to the PRISM program that Snowden would later disclose.

Third, two weeks before Snowden began mass downloads of classified documents, he was reprimanded after engaging in a workplace spat

with NSA managers. Snowden was repeatedly counseled by his managers regarding his behavior at work. For example, in June 2012, Snowden became involved in a fiery e-mail argument with a Supervisor about how computer updates should be managed. Snowden added an NSA senior executive several levels above the supervisor to the e-mail thread, an action that earned him a swift reprimand from his contracting officer for failing to follow the proper protocol for raising grievances through the chain of command. Two weeks later, Snowden began his mass downloads of classified information from NSA networks. Despite Snowden's later claim that the March 2013 congressional testimony of Director of National Intelligence James Clapper was a "breaking point" for him, these mass downloads *predated* Director Clapper's testimony by eight months.

Fourth, Snowden was, and remains, a serial exaggerator and fabricator. A close review of Snowden's official employment records and submissions reveals a pattern of intentional lying. He claimed to have left Army basic training because of broken legs when in fact he washed out because of shin splints. He claimed to have obtained a high school degree equivalent when in fact he never did. He claimed to have worked for the CIA as a "senior advisor," which was a gross exaggeration of his entry-level duties as a computer technician. He also doctored his performance evaluations and obtained new positions at NSA by exaggerating his résumé and stealing the answers to an employment test. In May 2013, Snowden informed his supervisor that he would be out of the office to receive treatment for worsening epilepsy. In reality, he was on his way to Hong Kong with stolen secrets.

Finally, the Committee remains concerned that more than three years after the start of the unauthorized disclosures, NSA, and the IC as a whole, have not done enough to minimize the risk of another massive unauthorized disclosure. Although it is impossible to reduce the chance of another Snowden to zero, more work can and should be done to improve the security of the people and computer networks that keep America's most closely held secrets. For instance, a recent DOD Inspector General report directed by the Committee found that NSA has yet to effectively implement its post-Snowden security improvements. The Committee has taken actions to improve IC information security in the Intelligence Authorization Acts for Fiscal Years 2014, 2015, 2016, and 2017, and looks forward to working with the IC to continue to improve security.

The House Committee that produced this report consisted of 22 members, a mix of Republicans and Democrats, who were *unanimous* in signing their names. This should help convince you that, once people are made privy to the classified details, the Snowden betrayal is not a partisan issue.

General Michael Hayden, who served as Director of the National Security Agency from 1999 to 2005 and as Director of the Central Intelligence Agency from 2006 to 2009, offered a list of questions he would have liked to have asked Snowden:[88]

---

[88] Hayden, Michael V., *Playing to the Edge: American Intelligence in the Age of Terror*, Penguin Press, New York, 2016, pp. 419–420.

You've cited Jim Clapper's response to Ron Wyden on NSA surveillance as motivating your actions. That was March 2013 but you began offering documents to Greenwald in December 2012 and to Laura Poitras in January 2013. Weren't you already committed?

While you were in Hong Kong fighting extradition, you told the press that NSA was hacking into Chinese computers. On the surface that looks like you were trying to buy safe passage. Were you?

The week before you fled Hong Kong, the London *Guardian* (based on your documents) claimed that the United States had intercepted Russian president Medvedev's satellite phone while he was at a G20 summit in England. What's the civil liberties issue there or is this just trading secrets for passage again?

You said that you raised your concerns within the system and that you were told not to rock the boat. NSA can't find any evidence. You took hundreds of thousands of documents. Do *any* of them show your raising concerns? A single e-mail, perhaps?

You sound pretty authoritative, but the first PRISM stories were wrong, claiming NSA had free access to the server farms of Google, Hotmail, Yahoo!, and the like. *The Washington Post* later walked that back. Did you misread the slides too?

Le Monde and *El País*, based on your documents, claimed that NSA was collecting tens of millions of metadata events on French and Spanish citizens each month. It turns out those events were collected *by* the French and Spanish in war zones and provided to NSA to help military force protection. Did you get that wrong too?

Hayden described Snowden's betrayal as "the greatest hemorrhaging of legitimate American secrets in the history of the republic." He also noted that "the Snowden revelations kept on coming, often timed for maximum embarrassment and crafted for maximum impact."[89] As for the man himself, Hayden remarked, "I think Snowden is an incredibly naive, hopelessly narcissistic, and insufferably self-important defector."[90]

12.11 NSA and the Media

As an example of how misleading Snowden's "revelations" can be, a particular newspaper piece is examined in some detail. On August 15, 2013, *The Washington Post* ran an article by Barton Gellman titled "NSA broke privacy rules thousands of times per year, audit finds."[91] It was based on an internal (classified) NSA study Snowden leaked and it sounds alarming. We should care deeply about our right to privacy, but we should also look closely at the details before drawing conclusions. Fortunately, in January 2014, NSA Deputy Director Chris Inglis, shortly before his retirement, took the time to explain what happened in an interview he did with Steve Inskeep of NPR. The relevant portions are reproduced below.[92]

INSKEEP: I want to ask about mistakes, errors, violations of privacy. You gave a fascinating talk late last year at the University of Pennsylvania in which you referred to

---

[89] Hayden, Michael V., *Playing to the Edge: American Intelligence in the Age of Terror*, Penguin Press, New York, 2016, p. 411.

[90] Hayden, Michael V., *Playing to the Edge: American Intelligence in the Age of Terror*, Penguin Press, New York, 2016, pp. 419–421.

[91] Available online at https://www.washingtonpost.com/world/national-security/nsa-broke-privacy-rules-thousands-of-times-per-year-audit-finds/2013/08/15/3310e554-05ca-11e3-a07f-49ddc7417125_story.html.

[92] "Transcript: NSA Deputy Director John Inglis," *npr*, https://www.npr.org/2014/01/10/261282601/transcript-nsa-deputy-director-john-inglis, January 10, 2014.

a document that had been disclosed that referred to something like 2,700 errors by the NSA. You argued that about 2,000 of those were not really relevant, set them aside. And then acknowledged there were 711 actual errors where you violated someone's privacy in a way that was not authorized. What happened on those 711 times in one year?

INGLIS: Yeah, so if I could clarify that. The report, first and foremost, was written in the early part of 2012. We wrote it ourselves. And we generate these reports essentially to take a hard look at how all the various things that we do to collect a communication of interest, store the communication of interest, query the communication of interest, we want to make sure we do that exactly right. And we determined in that report that on an annualized basis, we extrapolated the numbers that we had essentially had about 2,776 situations that didn't go exactly according to plan. That was immediately interpreted by some press outlets when that was released - again, it was another unauthorized release - but when it was released, some number of press outlets immediately equated that to 2,776 privacy violations and went so far as to say that they were either willful or kind of attributable to the gross lack of conscientious actions on the side of NSA.

Which is why I went then to some pains to explain what that really was. It turned out in 2,065 of those cases, so about 75 percent of those cases, the situation was that the individual, the organization that we were authorized to understand something about, whose communications we were trying to collect, had moved, right. Either they had physically moved or their services had moved and they were in a different location. Our authorities essentially asked the question up front of where is the party of interest? You know, where is the communication of interest? And where is the collection taking place? And if any of those change, we're probably using the wrong authority. And so, 2,065 we notified ourselves that that had changed. They don't consult with us before they change their location.

And so the system actually worked exactly as it should, which is that it figured that out, stopped the collection, purged back to the point where we last knew with precision where they were and then went after the right authority to essentially begin that again. In my view, that would be a feature, right, a positive feature. That leaves then 711. They weren't privacy violations, per se. What they were was that an analyst somewhere across NSA entered the wrong telephone number, the wrong email address when they were attempting to target A, but instead they could have potentially targeted A-prime. In most of those cases the number that they entered because they fingered it, they got a 2 in there instead of a 3, or something of that sort. The number didn't exist and so it returned.

But in all those cases it was caught because we essentially had checks inside the system, almost always a second check to make sure that what we have done is exactly what we intended to do. And we caught all of those things. And essentially took the right action. Whether it was how we formed the selector or whether it was how we queried a database, whether it was how we disseminated a piece of information. And those 711 occurrences have to be considered against all the activities we took that year. And it turns out that the average analyst, if you attributed those errors to an analyst, none of which were willful, all of which were simply accidents, the average analyst at NSA would make a mistake about every 10 years. The accuracy rate at NSA is 99.99984 percent, which is a pretty good record. But that said, we worry enough about making any mistakes that the 711 are a peculiar interest to us.

We're going to fix those. And so we have driven those down quarter by quarter, year by year.

Later in the interview, Inglis discussed some incidents, from other years, where NSA employees broke the rules:[93]

> Of note—you didn't ask me but I'll bring this up. You know, there is – a discussion has taken place where there have, in fact, been some willful abuses of the signet capabilities that NSA brings to bear. There have been 12 cases over the last 10 or so years where individuals made misuse of the signet system. They essentially tried to collect a communication that they were not authorized to collect 12 times.
>
> The vast majority of those were, in fact, overseas. Right? They were NSAers operating in foreign locations trying to collect the communication of an acquaintance so that they could better understand what that acquaintance was doing, but those acquaintances were foreigners. And our capabilities must be applied in a way that essentially meets the requirements imposed on me such that we would protect the privacy of foreign persons as much as we would protect the privacy of U.S. persons.

It was not stated in the interview, but all of those people were fired.

The terser General Michael Hayden summed up the matter as follows:[94]

> all the incidents were inadvertent; no one claimed that any rules were intentionally violated. All of the incidents were discovered, reported and corrected by NSA itself.
>
> Fully two-thirds of the incidents were composed of "roamers"–legitimately targeted foreigners who were temporarily in the United States (and thus temporarily protected by the Fourth Amendment).

He also pointed out that the 115 incidents of queries being incorrectly entered (typos or too-broad search criteria) were out of 61 million inquiries. Hayden then went on to suggest that the headline for *The Washington Post* article "NSA broke privacy rules thousands of times per year, audit finds" should instead have been "NSA Damn Near Perfect."[95]

12.12 BRUSA, UKUSA, and Echelon

Having seen in Section 12.7 how the United States collaborated with West Germany on the intelligence coup of the century, it should not surprise you that there is close collaboration between the English speaking nations. This section explains how that came to be.

During World War II, an uneasy relationship was formed between the British and American cryptologic organizations. In particular, the British were concerned that the Americans wouldn't be able to keep the information that they might be given secret, but an exchange of information made sense: The British had greater success against German ciphers (thanks, in large part, to the Poles) and the Americans, despite the Brits' early lead, went on to have greater success against the Japanese systems. Each truly had something of value to offer the other side.

In February 1941, an American team that consisted of an even mix of Army (Abraham Sinkov and Leo Rosen) and Navy (Robert H. Weeks and Prescott Currier) cryptologists traveled to

---

[93] "Transcript: NSA Deputy Director John Inglis," *npr*, https://www.npr.org/2014/01/10/261282601/transcript-nsa-deputy-director-john-inglis, January 10, 2014.

[94] Hayden, Michael V., *Playing to the Edge: American Intelligence in the Age of Terror*, Penguin Press, New York, 2016, p. 411. Hayden quoted these lines from a piece he wrote for *USA Today*.

[95] Hayden, Michael V., *Playing to the Edge: American Intelligence in the Age of Terror*, Penguin Press, New York, 2016, p. 412.

London to present the British with a Purple analog. This balance of Army and Navy was important, as the distrust between these service branches was an even greater barrier to the budding
intelligence sharing relationship than the distrust between nations!

After World War II, the intelligence sharing continued and in March 1946, the BRUSA
Agreement made it formal. This agreement was renamed UKUSA in 1948. Project BOURBON
was the codename for the work of America and England against the new common enemy, the
Soviet Union.[96] The United States was involved with other nations to varying degrees. American
cooperation with Canada had begun back in 1940.[97] Australia and New Zealand were to become
partners, as well. Although the agreements between NSA and GCHQ became public knowledge
decades earlier, the declassified documents only became available in June 2010.[98]

Europeans have expressed concern about how far the "Five Eyes" partners (United Kingdom,
United States, Canada, Australia, and New Zealand)[99] have gone in regard to violating the privacy
of individuals and businesses. Of particular concern is a program codenamed ECHELON that
allows the agencies to search the worldwide surveillance network for desired information by using
keywords. Two examples of spying on nonmilitary targets are provided below:

> In 1990, the German magazine *Der Spiegel* revealed that the NSA had intercepted mes-
> sages about an impending $200 million deal between Indonesia and the Japanese satel-
> lite manufacturer NEC Corp. After President Bush intervened in the negotiations on
> behalf of American manufacturers, the contract was split between NEC and AT&T.[100]

> In September 1993, President Clinton asked the CIA to spy on Japanese auto manu-
> facturers that were designing zero-emission cars and to forward that information to
> the Big Three U.S. car manufacturers: Ford, General Motors and Chrysler.[101]

Yet it would be naïve to believe the rest of the world was "playing fair." President Barack Obama
said, "some of the folks who have been most greatly offended publicly, we know privately engage
in the same activities directed at us."[102] Another example shows how NSA is sometimes able to
cancel out the treachery of others.

> From a commercial communications satellite, NSA lifted all the faxes and phone calls
> between the European consortium Airbus, the Saudi national airline and the Saudi
> government. The agency found that Airbus agents were offering bribes to a Saudi

---

[96] Johnson, Thomas, R., *American Cryptology During the Cold War, 1945–1989, Book I: The Struggle for Centralization, 1945–1960*, Center for Cryptologic History, National Security Agency, Fort George G. Meade, Maryland, 1995, p. 159.

[97] Johnson, Thomas, R., *American Cryptology During the Cold War, 1945–1989, Book I: The Struggle for Centralization, 1945–1960*, Center for Cryptologic History, National Security Agency, Fort George G. Meade, Maryland, 1995, p. 17.

[98] NSA, *Declassified UKUSA Signals Intelligence Agreement Documents* [Press Release], National Security Agency, Fort George G. Meade, Maryland, June 24, 2010, available online at http://www.nsa.gov/public_info/press_room/2010/ukusa.shtml.

[99] There are also secondary or junior partners, with whom some information is shared. Israel is not among them, although in other ways the relationship between Israel and the United States is very close.

[100] Poole, Patrick S., ECHELON, Part Two: The NSA's Global Spying Network, http://www.bibliotecapleyades.net/ciencia/echelon/echelon_2.htm.

[101] Poole, Patrick S., ECHELON, Part Two: The NSA's Global Spying Network, http://www.bibliotecapleyades.net/ciencia/echelon/echelon_2.htm.

[102] Hayden, Michael V., *Playing to the Edge: American Intelligence in the Age of Terror*, Penguin Press, New York, 2016, p. 413.

official. It passed the information to U.S. officials pressing the bid of Boeing Co. and McDonnell Douglas Corp., which triumphed last year [1994] in the $6 billion competition.[103]

In any case, private intelligence agencies are on the rise. If a large corporation cannot get the government's help, it can turn to one of these.

Figure 12.6   A wall listing the names of those who died serving NSA. (Courtesy of National Security Agency, https://web.archive.org/web/20160325230227/http://www.nsa.gov/about/_images/pg_hi_res/memorial_wall.jpg)

We've taken a look at some people who betrayed NSA, but they are, of course, the rare exceptions. There are likely more than are publicly known, but I doubt that they outnumber those at the other extreme, who gave their lives in service to America through the agency. A wall inside NSA commemorates these men and women (Figure 12.6). Sadly, during my time with NSA's Center for Cryptologic History, I saw this list grow. The rightmost column now extends to the bottom and new names have been added in the triangular space above. While NSA employees are often accused of violating the privacy of Americans, the numbers show that they are much more likely to die in the line of duty than to intentionally break privacy laws these days.

References and Further Reading

On NSA

Aid, Matthew, *The Secret Sentry: The Untold History of the National Security Agency*, Bloomsbury Press, New York, 2009.

Bamford, James, *The Puzzle Palace*, Houghton Mifflin Company, Boston, Massachusetts, 1982.

Bamford, James, "How I Got the N.S.A. Files… How Reagan Tried to Get Them Back," *The Nation*, Vol. 235, November 6, 1982, pp. 466–468.

---

[103] Shane, Scott and Tom Bowman, *No Such Agency, America's Fortress of Spies*, Reprint of a six-part series that appeared in *The Baltimore Sun*, December 3–15, 1995, p. 2.

Bamford, James, *Body of Secrets: Anatomy of the Ultra-Secret National Security Agency from the Cold War through the Dawn of a New Century*, Doubleday, New York, 2001.

Bamford, James, *The Shadow Factory: The Ultra-secret NSA from 9/11 to the Eavesdropping on America*, Anchor Books, New York, 2008.

Barker, Wayne G. and Coffman, Rodney E., *The Anatomy of Two Traitors, The Defection of Bernon F. Mitchell and William H. Martin*, Aegean Park Press, Laguna Hills, California, 1981.

Boak, David G, *A History of U.S. Communications Security, The David G. Boak Lectures*, National Security Agency, Fort George G. Meade, Maryland, Revised July 1973, Declassified December 2008, available online at https://www.nsa.gov/Portals/70/documents/news-features/declassified-documents/cryptologic-histories/history_comsec.pdf.

Boak, David G, *A History of U.S. Communications Security, The David G. Boak Lectures*, Vol. II, National Security Agency, Fort George G. Meade, Maryland, July 1981, Declassified December 2008, available online at https://www.archives.gov/files/declassification/iscap/pdf/2009-049-doc2.pdf.

Breedan II, John, "What a Former NSA Deputy Director Thinks of the Snowden Movie," *Nextgov*, https://www.nextgov.com/ideas/2016/09/former-nsa-deputy-director-calls-out-snowden-movie-grossly-inaccurate/131911/, September 28, 2016.

Briscoe, Sage and Aaron Magid, "The NSA Director's Summer Program," *Math Horizons*, Vol. 13, No. 4, April 2006, p. 24.

Brownell, George A., *The Origin and Development of the National Security Agency*, Aegean Park Press, Laguna Hills, California, 1981. This is a 98-page book.

Buehler, Hans, *Verschlüsselt*, Werd, Zürich, 1994. This book is in German and no translation is currently available.

Churchill, Ward, and Jim Vander Wall, *The COINTELPRO Papers*, South End Press, Boston, Massachusetts, 1990. NSA is barely mentioned in this book, which is referenced here solely for the information it contains on COINTELPRO. More information on NSA's role may be found at http://www.icdc.com/~paulwolf/cointelpro/churchfinalreportIIIj.htm.

Central Intelligence Agency, *Family Jewels*, 1973, available online at https://www.cia.gov/library/readingroom/collection/family-jewels, released on June 25, 2007, during Michael V. Hayden's term as director of CIA. This nearly 700-page document, created by CIA employees in response to a request from then Director of Central Intelligence James Schlesinger details illegal activities carried out by the agency.

Constance, Paul, "How Jim Bamford Probed the NSA," *Cryptologia*, Vol. 21, No. 1, January 1997, pp. 71–74.

de Leeuw, Karl, and Jan Bergstra, editors, *The History of Information Security, A Comprehensive Handbook*, Elsevier, Amsterdam, 2007. Chapter 18 (pp. 523–563) of this large $265 book is titled National Security Agency: The Historiography of concealment. It is by Joseph Fitsanakis, who complains about the lack of study of this topic and then provides a list of 291 references.

Halperin. Morton H., Jerry J. Berman, Robert L. Borosage, and Christine M. Marwick, *The Lawless State, The Crimes of the U.S. Intelligence Agencies*, Penguin Books, New York, 1976.

Hayden, Michael V., "Beyond Snowden: An NSA Reality Check," *World Affairs*, Vol. 176, No. 5, January/February 2014, pp. 13–23.

Hayden, Michael V., *Playing to the Edge: American Intelligence in the Age of Terror*, Penguin Press, New York, 2016. General Hayden served as Director of the National Security Agency from 1999 to 2005 and as Director of the Central Intelligence Agency from 2006 to 2009.

Johnson, Thomas, R., *American Cryptology During the Cold War, 1945–1989, Book I: The Struggle for Centralization, 1945–1960*, Center for Cryptologic History, National Security Agency, Fort George G. Meade, Maryland, 1995 available online at https://www.nsa.gov/Portals/70/documents/news-features/declassified-documents/cryptologic-histories/cold_war_i.pdf. This book, and the next three references, were declassified (with many redactions) beginning in 2008.

Johnson, Thomas, R., *American Cryptology During the Cold War, 1945–1989, Book II: Centralization Wins, 1960–1972*, Center for Cryptologic History, National Security Agency, Fort George G. Meade, Maryland, 1995, available online at https://www.nsa.gov/Portals/70/documents/news-features/declassified-documents/cryptologic-histories/cold_war_ii.pdf.

Johnson, Thomas, R., *American Cryptology During the Cold War, 1945–1989, Book III: Retrenchment and Reform, 1972–1980*, Center for Cryptologic History, National Security Agency, Fort George G. Meade, Maryland, 1998, available online at https://www.nsa.gov/Portals/70/documents/news-features/declassified-documents/cryptologic-histories/cold_war_iii.pdf.

Johnson, Thomas, R., *American Cryptology During the Cold War, 1945–1989, Book IV: Cryptologic Rebirth, 1981–1989*, Center for Cryptologic History, National Security Agency, Fort George G. Meade, Maryland, 1999, available online at https://www.nsa.gov/Portals/70/documents/news-features/declassified-documents/cryptologic-histories/cold_war_iv.pdf.

Kahn, David, *The Codebreakers*, Macmillan, New York, 1967. (A second edition, with a few pages of updates, appeared in 1996.) The NSA wasn't pleased that Kahn devoted a chapter to them in his book, and they considered various means of suppressing it. See Section 5.9.

Keefe, Patrick Radden, *CHATTER: Dispatches from the Secret World of Global Eavesdropping*, Random House, 2005. On page 97, Keefe placed NSAs budget at $6 billion a year with 60,000 employees.

Langmeyer, Navah and Amy M. Grimes, "Mathematical Life at the National Security Agency," *Math Horizons*, Vol. 8, No. 3, February 2001, pp. 30–31.

Madsen, Wayne, "Crypto AG: The NSA's Trojan Whore?" *Covert Action Quarterly*, Issue 63, Winter 1998, available online at http://mediafilter.org/caq/cryptogate/ and https://web.archive.org/web/20000815214548/http://caq.com:80/CAQ/caq63/caq63madsen.html.

Miller, Greg, "The intelligence Coup of the Century," *The Washington Post*, February 11, 2020, available online at https://tinyurl.com/yck5xur2.

National Security Agency, Website, http://www.nsa.gov/.

National Security Agency, *NSA Employee's Security Manual*. This manual (leaked in 1994) can be found online at http://theory.stanford.edu/~robert/NSA.doc.html.

Odom, General William E., *Fixing Intelligence for a More Secure America*, second edition, Yale University Press, New Haven, Connecticut, 2004. General Odom served as Director of the National Security Agency from 1985 to 1988.

Ransom, Harry Howe, *Central Intelligence and National Security*, Harvard University Press, 1958; third printing, 1965. Pages 116 to 118 discuss NSA.

Shane, Scott and Tom Bowman, *No Such Agency, America's Fortress of Spies*, Reprint of a six-part series that appeared in *The Baltimore Sun*, December 3–15, 1995.

Shane, Scott and Tom Bowman, "U.S. Secret Agency Scored World Coup: NSA Rigged Machines for Eavesdropping," *The Baltimore Sun*, January 3, 1996, p. 1A.

Sherman, David, "The National Security Agency and the William F. Friedman Collection," *Cryptologia*, Vol. 41, No. 3, May 2017, pp. 195–238.

Simons, Marc and Paul Reuvers, "Operation RUBICON/THESAURUS, The secret purchase of Crypto AG by BND and CIA," Crypto Museum, https://www.cryptomuseum.com/intel/cia/rubicon.htm, created: December 12, 2019, last changed: May 10, 2020.

Simons, Marc and Paul Reuvers, "The Gentleman's Agreement, Secret Deal between the NSA and Hagelin, 1939–1969," *Crypto Museum*, https://www.cryptomuseum.com/manuf/crypto/friedman.htm, created: July 30, 2015, last changed: May 10, 2020.

Smoot, Betsy Rohaly, "NSA Release and Transfer of Records Related to William F. Friedman," *Cryptologia*, Vol. 39, No. 1, January 2015, pp. 1–2.

Smoot, Betsy Rohaly, "National Security Agency releases Army Security Agency histories covering 1945–1963," *Cryptologia*, Vol. 41, No. 5, September 2017, pp. 476–478.

Tully, Andrew, *The Super Spies*, William Morrow, New York, September 1969.

Wagner, Michelle, "Organizational Profile: The Inside Scoop on Mathematics at the NSA," *Math Horizons*, Vol. 13, No. 4, April 2006, pp. 20–23.

Weiner, Tim, *Blank Check: The Pentagon's Black Budget*, Warner Books, New York, 1990. This book makes a study of undisclosed budgets.

Willemain, Thomas Reed, *Working on the Dark Side of the Moon: Life Inside the National Security Agency*, Mill City Press, Maitland, Florida, 2017.

On the *Liberty*

Borne John E., *The USS Liberty: Dissenting History vs. Official History*, Reconsideration Press, New York, 1995. This doctoral dissertation was submitted in partial fulfillment of the requirements for the degree of Doctor of Philosophy, Department of History, New York University, September 1993.

Cristol, A. Jay, *The Liberty Incident: The 1967 Israeli Attack on the U.S. Navy Spy Ship*, Brassey's, Inc., Washington DC, 2002. Cristol served for many years in the U.S. Navy, and argues that the Israelis didn't know they were attacking an American ship.

Ennes, Jr., James M., *Assault on the Liberty*, Random House, New York, 1979. Ennes, a lieutenant who was on the *Liberty*, thinks the Israelis knew they were attacking an American ship.

Scott, James, *The Attack on the Liberty*, Simon & Schuster, New York, 2009. Scott, the son of a *Liberty* survivor, thinks the Israeli's knew they were attacking an American ship.

Other works that include material on the *Liberty* were listed in the "On NSA" section of the references above; for example, James Bamford's *Body of Secrets* argues that the attack was known at the time to have been on an American ship, whereas Book II of Thomas R. Johnson's history presents the view that it was not.

On the *Pueblo*

Armbrister, Trevor, *A Matter of Accountability, The True Story of the Pueblo Affair*, Coward-McCann, Inc., New York, 1970.

Brandt, Ed, *The Last Voyage of the USS Pueblo*, W.W. Norton & Co., New York, 1969.

Bucher, Lloyd M. and Mark Rascovich, *Bucher: My Story*, Doubleday, Garden City, New York, 1970.

Crawford, Don, *Pueblo Intrigue*, Tyndale House Publishing, Wheaton, Illinois, 1969.

Gallery, Daniel V., *The Pueblo Incident*, Doubleday, Garden City, New York, 1970.

Harris, Stephen R. and James C. Hefley, *My Anchor Held*, Fleming H. Revell Company, Old Tappan, New Jersey, 1970. Harris was the intelligence officer aboard the *Pueblo* at the time of capture.

Lerner, Mitchell B., *The Pueblo Incident*, University Press of Kansas, Lawrence, Kansas, 2002.

Liston, Robert A., *The Pueblo Surrender*, M. Evans and Company, Inc., New York, 1988. This book actually argues that it was intended that the *Pueblo* be captured!

Also of Interest

Bamford, James, *A Pretext for War: 9/11, Iraq, and the Abuse of America's Intelligence Agencies*, Doubleday, New York, 2004. Bamford examines the following questions: Did Saddam Hussein have weapons of mass destruction, as George W. Bush claimed? Was there a connection between Hussein and Al Qaeda? The results of a poll showed that most Americans believed the answer to both question is yes. Bamford clearly shows that the correct answer was no in both cases and details the abuse of the intelligence agencies that led to the public's misinformed beliefs concerning these issues.

Not of Interest

Brown, Dan, *Digital Fortress*, St. Martin's Press, New York, 1998. Dan Brown's breakthrough novel was *The Da Vinci Code*, but his first novel, *Digital Fortress*, dealt with the NSA. It can be read for entertainment, but doesn't offer any insight into NSA.

Videography

America's Most Secret Agency, The History Channel, January 8, 2001. Although supposedly on NSA, this program features material on World War II, as well. As NSA was born in 1952, this can only be background. It turns out that there was more footage shot on NSA, but the agency got cold feet and asked for it to be cut; hence, the filler—stock footage from World War II.

Inside the NSA: America's Cyber Secrets, National Geographic Video, 45 minutes, 2012.

Pueblo (alternate title, *Pueblo Affair*), ABC Theatre, 102 minutes, originally broadcast March 29, 1973. A reviewer for *The New York Times* commented, "Despite network restrictions of the era, *Pueblo* is refreshingly frank, right down to the first-ever TV display of a familiar obscene gesture (which the American prisoners explain away to their captors as a 'salute!')." (http://movies.nytimes.com/movie/128338/Pueblo/overview).

The Spy Factory, Nova, 53 minutes, originally broadcast February 3, 2009. This serves as a companion to James Bamford's book *The Shadow Factory: The Ultra-secret NSA from 9/11 to the Eavesdropping on America*.

Top Secret: Inside the World's Most Secret Agencies, Discovery Channel, 1999. This series explores the National Security Agency, Scotland Yard, and Israel's Mossad and is narrated by Johnny Depp.

Chapter 13

The Data Encryption Standard

We now turn to a cipher far more advanced than anything previously discussed in this book. The only reason it isn't a good choice for use today is because increased computing power allows brute-force solutions.

13.1 How DES Works

The cipher that would eventually become the Data Encryption Standard, or DES for short,[1] is no longer considered secure, but it is important to us for several reasons.

1. It was the standard for decades.
2. It made use of techniques not previously seen in a non-classified environment and set the bar high for all of the ciphers that followed it.
3. It arose from an unprecedented secret collaboration between IBM and the National Security Agency (NSA) for a publicly available cipher. There had been prior interaction between NSA and industry, but not on anything that would go public.

DES also served to inspire Martin E. Hellman, who would go on to be one of the most important cryptographers of the 20th century. He wrote, "I trace my interest in cryptography to three main sources."[2] One was David Kahn's book, *The Codebreakers*. Another was Claude Shannon's 1949 paper.[3] His description of the third follows.[4]

> From 1968-1969 I worked at IBM's Watson Research Center in Yorktown Heights, NY. One of my colleagues was Horst Feistel, who had been brought in from classified government work to seed IBM's research in cryptography. IBM's work culminated in the Data Encryption Standard (DES) in 1975. While my work at IBM was not in cryptography, I had a number of discussions with Feistel that opened my eyes to previously unforeseen possibilities. The fact that IBM was investing in the development of cryptography for commercial applications also indicated the need and value of such work.

---

[1] You may pronounce DES like a word (it rhymes with Pez) or pronounce each letter individually. There is no standard for this!

[2] Hellman, Martin E., "Work on Cryptography," http://www-ee.stanford.edu/~hellman/crypto.html.

[3] Shannon, Claude E., "Communication Theory of Secrecy Systems," *The Bell System Technical Journal*, Vol. 28, No. 4, October 1949, pp. 656–715.

[4] Hellman, Martin E., "Work on Cryptography," http://www-ee.stanford.edu/~hellman/crypto.html.

Figure 13.1   Horst Feistel (1915–1990). (http://crypto-dox.com/History_of_Cryptography).

Horst Feistel (Figure 13.1), an IBM employee born in Germany, is the man credited as the creator of DES (although others were involved—more on this soon). He wanted to call the system Dataseal, but IBM used the term Demonstration Cipher, which was truncated to Demon. Finally, the name was changed to Lucifer, maintaining what Feistel called "the evil atmosphere" of Demon, as well as "cifer" (cipher).[5] DSD-1 was another name used internally for this cipher.[6] Lucifer was used by Lloyds Bank of London for a cash dispensing system in the early 1970s.[7]

The National Bureau of Standards (NBS)[8] held a competition for a cipher system to meet civilian needs. This system was to be called the Data Encryption Standard or DES. The call for algorithms appeared in the *Federal Register* on May 15, 1973 (Vol. 38, No. 93, p. 12763) and again on August 27, 1974 (Vol. 39, No. 167, p. 30961). Lucifer was the only algorithm deemed acceptable by NBS and their NSA advisors.

The algorithm appeared in the *Federal Register* on March 17, 1975 and again on August 1, 1975 with a request for reader comments.[9] Thus, Lucifer was adopted as the standard on July 15, 1977, and had a final name change to DES. IBM agreed to place the relevant patents in the public domain, so anyone who desired could freely use the algorithm; however, this didn't prevent money being made from the system by other companies that manufactured chips implementing the algorithm.[10]

DES can be intimidating when viewed all at once, but the individual pieces it is made out of are very simple. The basic units (called *blocks*) on which the algorithm works are 64 bits (8

5 Kahn, David, *The Codebreakers*, second edition, Scribner, New York, 1996, p. 980.
6 Levy, Steven, *Crypto: How the Code Rebels Beat the Government, Saving Privacy in the Digital Age*, Viking, New York, 2001.
7 Kinnucan, Paul, "Data Encryption Gurus: Tuchman and Meyer," *Cryptologia*, Vol. 2, No. 4, October 1978, pp. 371–381.
8 NBS was founded in 1901, but renamed Bureau of Standards in 1903. It became NBS again in 1934 and then, finally, National Institute of Standards and Technology (NIST) in 1988 (http://www.100.nist.gov/directors.htm).
9 Roberts, Richard W., National Bureau of Standards, "Encryption Algorithm for Computer Data Protection: Requests for Comments," *Federal Register*, Vol. 40, No. 52, March 17, 1975, pp. 12134–12139 and Hoffman, John D., National Bureau of Standards, "Federal Information Processing Data Encryption Proposed Standard," *Federal Register*, Vol. 40, No. 149, August 1, 1975, pp. 32395–32414.
10 Morris, Robert, Neil J. A. Sloane, and Aaron D. Wyner, "Assessment of the National Bureau of Standards Proposed Federal Data Encryption Standard," *Cryptologia*, Vol. 1, No. 3, July 1977, pp. 281–291, p. 284 cited here. Also see Winkel, Brian J., "There and there a department," *Cryptologia*, Vol. 1, No. 4, 1977, pp. 396–397.

characters) long. One operation used in DES consists of breaking the 64-bit message block in half and switching sides, as depicted in the diagram in Figure 13.2.

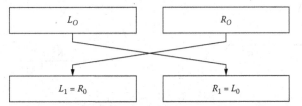

Figure 13.2 L_1, the new left-hand side, is simply R_0, the old right hand side; R_1, the new right-hand side, is L_0, the old left-hand side. (http://csrc.nist.gov/publications/fips/fips46-3/fips46-3.pdf.)

This operation is clearly its own inverse.

Another operation that is its own inverse is adding bits modulo 2. This is also known as XOR (for exclusive OR) and is often denoted by $\oplus$. Figure 13.3 shows a function f, which takes K_1 (derived from the 56-bit key K) and R_0 as inputs and outputs a value that is then XORed with L_0 to obtain R_1.

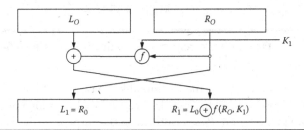

Figure 13.3 Function f. (http://csrc.nist.gov/publications/fips/fips46-3/fips46-3.pdf.)

This combination of two self-inverse operations is referred to as a *round*. DES goes through 16 such rounds. The manner in which the round keys are derived from K will be detailed, but first we examine the function f. In general, we refer to a cipher that uses rounds of the form depicted above (switching sides and applying a function to one half) as a *Feistel system*, or *Feistel cipher*.

The most natural way of combining R_i and K_i would be to XOR them, but R_i is 32 bits long and each of the round keys is 48 bits long. To even things up, R is expanded by repeating some of the bits (their order is changed as well). This is indicated in Figure 13.4, and referred to as E (for expansion). E is given by the following:

32	1	2	3	4	5
4	5	6	7	8	9
8	9	10	11	12	13
12	13	14	15	16	17
16	17	18	19	20	21
20	21	22	23	24	25
24	25	26	27	28	29
28	29	30	31	32	1

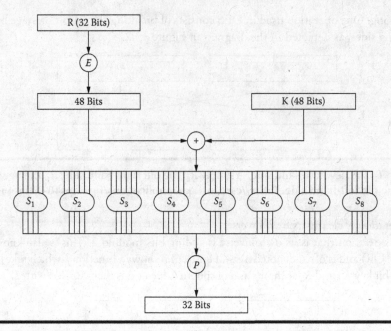

Figure 13.4   The heart of DES. (http://csrc.nist.gov/publications/fips/fips46-3/fips46-3.pdf.)

Once the expanded right hand side and the round key have been XORed, the result is broken up into eight pieces of six bits each, each of which is fed into a substitution box (*S*-box) that returns only four bits. Finally, a permutation, *P*, is performed on the output and the round is complete. See Figure 13.4 for a depiction of these steps. The permutation *P* is given by:

$$
\begin{array}{cccc}
16 & 7 & 20 & 21 \\
29 & 12 & 28 & 17 \\
1 & 15 & 23 & 26 \\
5 & 18 & 31 & 10 \\
2 & 8 & 24 & 14 \\
32 & 27 & 3 & 9 \\
19 & 13 & 30 & 6 \\
22 & 11 & 4 & 25
\end{array}
$$

The nonlinear heart of the algorithm is the *S*-boxes. Each box converts a 6-bit number, $b_1b_2 b_3b_4b_5b_6$, to a 4-bit number by first breaking it up into b_1b_6 and $b_2b_3b_4b_5$. That is, we now have a 2-bit and a 4-bit number. Converting each of these to base 10, the first is between 0 and 3 and the second is between 0 and 15. Thus, a row and column of the *S*-box is referenced. The value at that location is our 4-bit result. All eight *S*-boxes are provided in Table 13.1.

Table 13.1   S-Boxes

S_1							Column Number								
Row No.	0	1	2	3	4	5	6	7	8	9	10	11	12	13	14	15
0	14	4	13	1	2	15	11	8	3	10	6	12	5	9	0	7
1	0	15	7	4	14	2	13	1	10	6	12	11	9	5	3	8
2	4	1	14	8	13	6	2	11	15	12	9	7	3	10	5	0
3	15	12	8	2	4	9	1	7	5	11	3	14	10	0	6	13

S_2							Column Number								
Row No.	0	1	2	3	4	5	6	7	8	9	10	11	12	13	14	15
0	15	1	8	14	6	11	3	4	9	7	2	13	12	0	5	10
1	3	13	4	7	15	2	8	14	12	0	1	10	6	9	11	5
2	0	14	7	11	10	4	13	1	5	8	12	6	9	3	2	15
3	13	8	10	1	3	15	4	2	11	6	7	12	0	5	14	9

S_3							Column Number								
Row No.	0	1	2	3	4	5	6	7	8	9	10	11	12	13	14	15
0	10	0	9	14	6	3	15	5	1	13	12	7	11	4	2	8
1	13	7	0	9	3	4	6	10	2	8	5	14	12	11	15	1
2	13	6	4	9	8	15	3	0	11	1	2	12	5	10	14	7
3	1	10	13	0	6	9	8	7	4	15	14	3	11	5	2	12

S_4							Column Number								
Row No.	0	1	2	3	4	5	6	7	8	9	10	11	12	13	14	15
0	7	13	14	3	0	6	9	10	1	2	8	5	11	12	4	15
1	13	8	11	5	6	15	0	3	4	7	2	12	1	10	14	9
2	10	6	9	0	12	11	7	13	15	1	3	14	5	2	8	4
3	3	15	0	6	10	1	13	8	9	4	5	11	12	7	2	14

(Continued)

Table 13.1 (*Continued*)   S-Boxes

S_5	Column Number															
Row No.	0	1	2	3	4	5	6	7	8	9	10	11	12	13	14	15
0	2	12	4	1	7	10	11	6	8	5	3	15	13	0	14	9
1	14	11	2	12	4	7	13	1	5	0	15	10	3	9	8	6
2	4	2	1	11	10	13	7	8	15	9	12	5	6	3	0	14
3	11	8	12	7	1	14	2	13	6	15	0	9	10	4	5	3

S_6	Column Number															
Row No.	0	1	2	3	4	5	6	7	8	9	10	11	12	13	14	15
0	12	1	10	15	9	2	6	8	0	13	3	4	14	7	5	11
1	10	15	4	2	7	12	9	5	6	1	13	14	0	11	3	8
2	9	14	15	5	2	8	12	3	7	0	4	10	1	13	11	6
3	4	3	2	12	9	5	15	10	11	14	1	7	6	0	8	13

S_7	Column Number															
Row No.	0	1	2	3	4	5	6	7	8	9	10	11	12	13	14	15
0	4	11	2	14	15	0	8	13	3	12	9	7	5	10	6	1
1	13	0	11	7	4	9	1	10	14	3	5	12	2	15	8	6
2	1	4	11	13	12	3	7	14	10	15	6	8	0	5	9	2
3	6	11	13	8	1	4	10	7	9	5	0	15	14	2	3	12

S_8	Column Number															
Row No.	0	1	2	3	4	5	6	7	8	9	10	11	12	13	14	15
0	13	2	8	4	6	15	11	1	10	9	3	14	5	0	12	7
1	1	15	13	8	10	3	7	4	12	5	6	11	0	14	9	2
2	7	11	4	1	9	12	14	2	0	6	10	13	15	3	5	8
3	2	1	14	7	4	10	8	13	15	12	9	0	3	5	6	11

As an example, suppose that just prior to heading into the *S*-boxes, you have the following string of 48 bits:

010100000110101100111101000110110011000011110101.

The first six bits, $b_1b_2b_3b_4b_5b_6 = 010100$, will be substituted for using the first *S*-box, S_1. We have $b_1b_6 = 00$ and $b_2b_3b_4b_5 = 1010$. Converting these to base 10, we get 0 and 10, so we look in row 0 and column 10 of S_1, where we find 6. Converting 6 to base 2 gives 0110. Thus, the first 6 bits of the 48-bit string above are replaced by 0110.

We then move on to the next 6 bits of our original 48-bit string, 000110. If we now label these as $b_1b_2b_3b_4b_5b_6$, we have $b_1b_6 = 00$ and $b_2b_3b_4b_5 = 0011$. Converting these to base 10, we get 0 and 3, so we look in row 0 and column 3 of S_2, where we find 14. Converting 14 to base 2 gives 1110. Thus, the second 6 bits of the 48-bit string above are replaced by 1110. We continue in this manner, 6 bits at a time, until all 48 bits have been replaced by using each of the 8 *S*-boxes, in order. The final result is a 32-bit string.

It might seem that the particular values that fill the substitution boxes aren't important. One substitution is as good as another, right? Wrong! The official description for DES stated, "The choice of the primitive functions *KS*, $S_1, \ldots, S_8$ and *P* is critical to the strength of the encipherment resulting from the algorithm."[11]

Much more will be said about these *S*-boxes in this chapter, but for now, we continue the discussion of how DES works. We are now ready to look at the big picture.

Figure 13.5 illustrates the 16 rounds that were discussed above. Notice that there is no swapping of right and left sides after the last round. This is so that encryption and decryption follow the same steps with the only difference being that the round keys must be used in the opposite order for decryption. It is important to note that the composition of rounds is not a round; that is, in general, two rounds with different keys cannot be realized by a single round using some third key. If this were the case, 16 rounds would be no better than one; they'd only take longer!

The only new elements here are the initial permutation and, at the end, the inverse initial permutation.

The initial permutation is given by:

58	50	42	34	26	18	10	2
60	52	44	36	28	20	12	4
62	54	46	38	30	22	14	6
64	56	48	40	32	24	16	8
57	49	41	33	25	17	9	1
59	51	43	35	27	19	11	3
61	53	45	37	29	21	13	5
63	55	47	39	31	23	15	7

Nobody seems to know why the designers bothered to rearrange the bits of the plaintext—it has no cryptographic effect—but that's how DES is defined.

—**Bruce Schneier**[12]

---

[11] National Institute of Standards and Technology, Data Encryption Standard (DES), Federal Information Processing Standards Publication 46-3, October 25, 1999, p. 17, available online at http://csrc.nist.gov/publications/fips/fips46-3/fips46-3.pdf. Note: *KS* stands for Key Schedule and refers to how the 16 round keys are derived from the 56-bit key *K*. This is detailed later in this chapter.

[12] Schneier, Bruce and Niels Ferguson, *Practical Cryptography*, Wiley, 2003, Indianapolis, Indiana, 2003, p. 52.

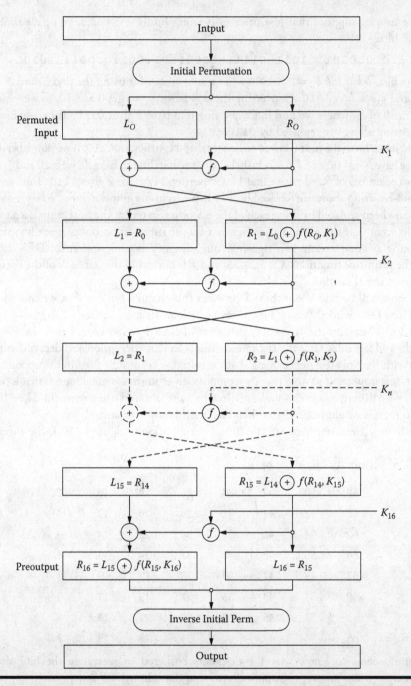

Figure 13.5 The big picture. (http://csrc.nist.gov/publications/fips/fips46-3/fips46-3.pdf.)

Apparently, back in the 1970s, this permutation made it easier to load the chip, when DES encryption is carried out by hardware, rather than software. Hardware implementations of DES work really well, but software implementations aren't very efficient, because software doesn't deal well with permutations of bits. On the other hand, permuting bytes could be done more efficiently. This approach is taken in portions of AES, an algorithm examined in Section 20.3.

The inverse of the Initial Permutation is:

48	8	48	16	56	24	64	32
39	7	47	15	55	23	63	31
38	6	46	14	54	22	62	30
37	5	45	13	53	21	61	29
36	4	44	12	52	20	60	28
35	3	43	11	51	19	59	27
34	2	42	10	50	18	58	26
33	1	41	9	49	17	57	25

These permutations are always as above. They are not part of the key and add nothing to the security of DES. The cipher would be just as strong (or weak) without them, but it wouldn't be DES.

Now, to complete the description of DES, we need to examine how the round keys are obtained from K. It is accurate to refer to the key K as being 56 bits, but an extra 8 bits were used for error detection. These check bits are inserted in positions 8, 16, 24, 32, 40, 48, 56, and 64 in order to make the parity of each byte even. So, when selecting key bits from this 64-bit string, to use as a round key, the positions holding the check digits should be ignored. The relevant 56 bits are selected and permuted as follows.

PC-1 (Permuted Choice 1)

57	49	41	33	25	17	9
1	58	50	42	34	26	18
10	2	59	51	43	35	27
19	11	3	60	42	44	36
63	55	47	39	31	23	15
7	62	54	46	38	30	22
14	6	61	53	45	37	29
21	13	5	28	20	12	4

There is more work to be done before we obtain a round key, though. A blank line was placed in the middle of the 56 bits to indicate that it is split in half, just like the message blocks. To avoid the confusion with L and R, we label these halves C and D.

Each half is individually left shifted cyclically.[13] To illustrate this, if we have

$$c_1 c_2 c_3 c_4 c_5 c_6 c_7 c_8 c_9 c_{10} c_{11} c_{12} c_{13} c_{14} c_{15} c_{16} c_{17} c_{18} c_{19} c_{20} c_{21} c_{22} c_{23} c_{24}$$

and

$$d_1 d_2 d_3 d_4 d_5 d_6 d_7 d_8 d_9 d_{10} d_{11} d_{12} d_{13} d_{14} d_{15} d_{16} d_{17} d_{18} d_{19} d_{20} d_{21} d_{22} d_{23} d_{24}$$

as the two halves and left shift each by two places, we get

$$c_3 c_4 c_5 c_6 c_7 c_8 c_9 c_{10} c_{11} c_{12} c_{13} c_{14} c_{15} c_{16} c_{17} c_{18} c_{19} c_{20} c_{21} c_{22} c_{23} c_{24} c_1 c_2$$

and

$$d_3 d_4 d_5 d_6 d_7 d_8 d_9 d_{10} d_{11} d_{12} d_{13} d_{14} d_{15} d_{16} d_{17} d_{18} d_{19} d_{20} d_{21} d_{22} d_{23} d_{24} d_1 d_2 \,.$$

As Figure 13.6 indicates, the number of shifts the halves undergo depends on which round is being performed.

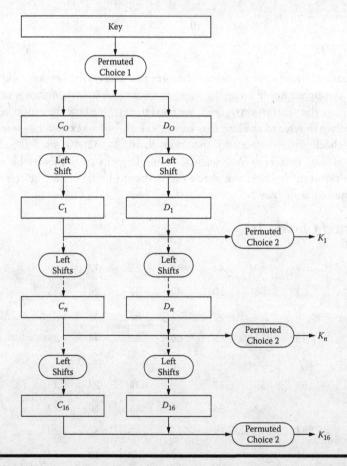

Figure 13.6   Obtaining the 16 round keys from the 56-bit key. (http://csrc.nist.gov/publications/fips/fips46-3/fips46-3.pdf.)

---

[13] Cyclic shifts are often indicated with the notations "≪" or "≫", depending on whether the shift is to the left or to the right.

After the number of left shifts required for a particular round has been performed, the two halves are recombined and 48 bits are selected (and permuted) according to the following table.

PC-2 (Permuted Choice 2)

14	17	11	24	1	5
3	28	15	6	21	10
23	19	12	4	26	8
16	7	27	20	13	2
41	52	31	37	47	55
30	40	51	45	33	48
44	49	39	56	34	53
46	42	50	36	29	32

The last necessary detail is the amount to left shift by in each round. It is usually (but not always!) two units (see Table 13.2).

Table 13.2 Left Shift Schedule

Iteration Number	Number of Left Shifts	Iteration Number	Number of Left Shifts
1	1	9	1
2	1	10	2
3	2	11	2
4	2	12	2
5	2	13	2
6	2	14	2
7	2	15	2
8	2	16	1

Notice that the shifts add up to 28, half the length of the key.

You now have enough information to implement DES in the programming language of your choice. There were many details that required our attention, but each piece is very simple to explain, as well as to code.[14] However, why the *S*-boxes took the form they did was far from clear when Lucifer officially became DES. There'll be more on this soon.

For now, we note that Claude Shannon would have been pleased by the design of DES. The combination of the transpositions and the substitutions made by the *S*-boxes over 16 rounds provides the diffusion and confusion he desired. Suppose we encipher some message *M* with DES using key *K* to

---

[14] It should be kept in mind, though, that DES was designed to be implemented on a chip, not expressed in software.

get ciphertext C. Then, we change a single bit of M or K and encipher again to get a second ciphertext C'. Comparing C and C', we will typically find that they differ in about half of the positions.

See Figure 13.7 for another bit of cryptographic humor.

Figure 13.7   Another cartoon for the cryptologically informed (xkcd.com/153/). If you hover over the cartoon with the mouse you get the alternate text, "If you got a big key space, let me search it."

13.2 Reactions to and Cryptanalysis of DES

There were two main objections to DES right from the start.

13.2.1 Objection 1: Key Size Matters

The published description of DES made it sound like it used a 64-bit key, but, as we saw above, it's really only 56 bits, because 8 bits are used as error checks.

> Diffie, Hellman and others have objected that a 56-bit key may be inadequate to resist a brute force attack using a special purpose computer costing about $20 million. Others have estimated ten times that much. Whichever figure is correct, there is little safety margin in a 56-bit key.[15]

---

[15] Morris, Robert, Neil J. A. Sloane, and Aaron D. Wyner, "Assessment of the National Bureau of Standards Proposed Federal Data Encryption Standard," *Cryptologia*, Vol. 1, No. 3, July 1977, pp. 281–291, p. 281 cited here. References for their objection are: Diffie, Whitfield, *Preliminary Remarks on the National Bureau of Standards Proposed Standard Encryption Algorithm for Computer Data Protection*, unpublished report, Stanford University, May 1975; Diffie, Whitfield and Martin E. Hellman, "A Critique of the Proposed Data Encryption Standard," *Communications of the ACM*, Vol. 19, No. 3, March 1976, pp. 164–165; Diffie, Whitfield and Martin E. Hellman, "Exhaustive Cryptanalysis of the NBS Data Encryption Standard," *Computer*, Vol. 10, No. 6, June 1977, pp. 74–84.

The machine hypothesized by Diffie and Hellman would have 1,000,000 chips, and could break DES, given a single plaintext/ciphertext pair, in about 12 hours.[16] It's not hard to get a plaintext/ciphertext pair. In fact, NBS agreed that this type of attack was reasonable.[17]

Walter Tuchman, part of the IBM design team, argued that the machine would cost $200 million and went on to say that it would be cheaper and easier to get the information via bribery or blackmail than to build the machine. However, he also said, "In our judgment, the 56-bit key length is more than adequate for the foreseeable future, meaning 5 to 10 years."[18] Why not look ahead a bit farther? Whether the machine would cost $20 million or $200 million is a trivial detail. Such costs would be minor for a government with a military intelligence budget in the billions of dollars.

If 56 bits is too small, what size key should be used? Hellman pointed out that the military routinely used key sizes nearly 20 times as large as that of DES.[19] He suggested that NSA was attempting to limit publicly available encryption keys to a size they could break. Because they routinely allowed systems with keys shorter than 64 bits to be exported, but not systems with longer keys, a 56-bit key must be vulnerable to their attacks.[20] Others objecting to the 56-bit key included the following:

Yasaki, E. K., "Encryption Algorithm: Key Size is the Thing," *Datamation*, Vol. 22, No. 3, March 1976, pp. 164–166.

Kahn, David, "Tapping Computers," *The New York Times*, April 3, 1976, p. 27.

Guillen. M., "Automated Cryptography," *Science News*, Vol. 110, No. 12, September 18, 1976, pp. 188–190.

NBS responded to these complaints by holding two workshops:

1. 1976 Workshop on Estimation of Significant Advances in Computer Technology, NBS, August 30-31, 1976. "This was a device-oriented workshop, whose purpose was to examine the feasibility of building the special purpose computer described by Diffie and Hellman." Most of the participants didn't think it was feasible with the current technology, but later an IBM representative not in attendance said it could be done for about $200 million with plaintext recovered in a single day. Diffie and Hellman stuck to their $20 million estimate and responded to skeptics with the paper "Exhaustive Cryptanalysis of the NBS Data Encryption Standard."[21]
2. Workshop on Cryptography in Support of Computer Security, NBS, September 21-22, 1976. This was attended by Morris, Sloane, and Wyner, who had their reactions published in the July 1977 issue of *Cryptologia*.

---

[16] Diffie, Whitfield and Martin E. Hellman, "Exhaustive Cryptanalysis of the NBS Data Encryption Standard," *Computer*, Vol. 10, No. 6, June 1977, pp. 74–84.

[17] Morris, Robert, Neil J. A. Sloane, and Aaron D. Wyner, "Assessment of the National Bureau of Standards Proposed Federal Data Encryption Standard," *Cryptologia*, Vol. 1, No. 3, July 1977, pp. 281–291, p. 286 cited here.

[18] Kinnucan, Paul, "Data Encryption Gurus: Tuchman and Meyer," *Cryptologia*, Vol. 2, No. 4, pp. 371–381, October 1978.

[19] Kolata, Gina Bari, "Computer Encryption and the National Security Agency Connection," *Science*, Vol. 197, No. 4302, July 29, 1977, pp. 438–440.

[20] Hellman, Martin, *Statement to participants at NBS workshop on cryptography in support of computer security*, unpublished memorandum, September 21, 1976.

[21] Diffie, Whitfield and Martin E. Hellman, "Exhaustive Cryptanalysis of the NBS Data Encryption Standard," *Computer*, Vol. 10, No. 6, June 1977, pp. 74–84.

Morris, Sloane, and Wyner concluded that it would "very probably become feasible sometime in the 1980s for those with large resources (e.g., government and very large corporations) to decrypt [the proposed standard] by exhaustive key search."[22] They did admit that they were not aware of any cryptanalytic short-cuts to decrypting; that is, brute-force was the best attack they could find.

It was also observed that if the 56-bit key were obtained from 8 typed characters, as opposed to 56 random bits, then "the cost of surreptitious decryption is lowered by a factor of about 200" and suggested that users be warned of this, as well as advised that security can be increased by enciphering twice with two different keys.[23] This last statement seems premature (although it did turn out to be correct, by a small margin). It wasn't until 1993 that DES keys were proven not to form a group. So, until 1993, one couldn't be sure that double enciphering was any better.[24]

Morris, Sloane, and Wyner suggested the key length should be increased to 64 or even 128 bits.[25] In earlier versions, the algorithm did, in fact, use a 128-bit key.[26] They also wanted at least 32 iterations, instead of just 16.

Addressing concerns of the possibility of a backdoor having been built into DES by NSA collaborators, Tuchman said, "We developed the DES algorithm entirely within IBM using IBMers. The NSA did not dictate a single wire."[27] It is now known that IBM did in fact receive help from NSA. After receiving IBM's work, NBS sent it on to NSA, where changes were made, with the algorithm described here being the final result.[28]

There were even contemporary accounts admitting NSA's involvement:

> Aaron Wyner of Bell Laboratories in Murray Hill, New Jersey, says 'IBM makes no bones about the fact that NSA got into the act before the key size was chosen.' [Alan] Konheim [of IBM at Yorktown Heights, New York] admits that 'IBM was involved with the NSA on an ongoing basis. They [NSA employees] came up every couple of months to find out what IBM was doing.' But, Konheim says, 'The 56-bit key was chosen by IBM because it was a convenient size to implement on a chip.'[29]

More inconvenient would have been an inability to obtain an export license for a version with a larger key!

---

22 Morris, Robert, Neil J. A. Sloane, and Aaron D. Wyner, "Assessment of the National Bureau of Standards Proposed Federal Data Encryption Standard," *Cryptologia*, Vol. 1, No. 3, July 1977, pp. 281–291, p. 281 cited here.

23 Morris, Robert, Neil J. A. Sloane, and Aaron D. Wyner, "Assessment of the National Bureau of Standards Proposed Federal Data Encryption Standard," *Cryptologia*, Vol. 1, No. 3, July 1977, pp. 281–282, 286.

24 Although the claim should not have been stated as a fact, it wasn't based on nothing. For the evidence, see Grossman, Edna, *Group Theoretic Remarks on Cryptographic Systems Based on Two Types of Addition*, Research Report RC-4742, Thomas J. Watson Research Center, IBM, Yorktown Heights, New York, February 26, 1974.

25 Morris, Robert, Neil J. A. Sloane, and Aaron D. Wyner, "Assessment of the National Bureau of Standards Proposed Federal Data Encryption Standard," *Cryptologia*, Vol. 1, No. 3, July 1977, pp. 281–291, p. 282 cited here.

26 Girdansky, M. B., *Data Privacy: Cryptology and the Computer at IBM Research*, IBM Research Reports, Vol. 7, No. 4, 1971, 12 pages; Meyer, Carl H., "Design Considerations for Cryptography," *AFIPS '73 Conference Proceedings*, Vol. 42, AFIPS Press, Montvale, New Jersey, 1973, pp. 603–606 (see Figure 3 on p. 606).

27 Kinnucan, Paul, "Data Encryption Gurus: Tuchman and Meyer," *Cryptologia*, Vol. 2, No. 4, October 1978, pp. 371–381.

28 Trappe, Wade and Lawrence C. Washington, *Introduction to Cryptography with Coding Theory*, Prentice Hall, Upper Saddle River, New Jersey, 2002, p. 97.

29 Kolata, Gina Bari, "Computer Encryption and the National Security Agency Connection," *Science*, Vol. 197, No. 4302, July 29, 1977, pp. 438–440.

From a declassified NSA history, we have the following:[30]

> In 1973 NBS solicited private industry for a data encryption standard (DES). The first offerings were disappointing, so NSA began working on its own algorithm. Then Howard Rosenblum, deputy director for research and engineering discovered that Walter Tuchman of IBM was working on a modification to Lucifer for general use. NSA gave Tuchman a clearance and brought him in to work jointly with the Agency on Lucifer modification.[31]

> NSA worked closely with IBM to strengthen the algorithm against all except brute force attacks and to strengthen substitution tables, called S-boxes. Conversely, NSA tried to convince IBM to reduce the length of the key from 64 to 48 bits. Ultimately, they compromised on a 56-bit key.[32]

> Though its export was restricted, it was known to be widely used outside the United States. According to a March 1994 study, there were some 1,952 products developed and distributed in thirty-three countries.[33]

13.2.2 Objection 2: S-Box Secrecy

Cryptologists also expressed concern over the secret manner in which the S-boxes were designed and suggested that either this should be revealed or new S-boxes should be created in an open manner.[34] At the second NBS workshop, an IBM representative said the S-boxes were constructed to *increase* security. Morris, Sloane, and Wyner commented, "While there is no particular reason to doubt this, some doubt remains."[35] Some regularities in the S-box structure were described in a Lexar Corporation report.[36]

Pointing out that DES had already been proposed for being used as a one-way function to verify computer log-ons and for sending keys over insecure channels, Morris et al. continued, "It is clear that once the DES is approved as a standard, it will find applications in every part of

---

[30] Although not in the original declassified version, these passages were later released thanks to a FOIA request filed by John Young.

[31] Johnson, Thomas, R., *American Cryptology During the Cold War, 1945–1989, Book III: Retrenchment and Reform, 1972–1980*, Center for Cryptologic History, National Security Agency, Fort George G. Meade, Maryland, 1995, p. 232.

[32] Johnson, Thomas, R., *American Cryptology During the Cold War, 1945–1989, Book III: Retrenchment and Reform, 1972–1980*, Center for Cryptologic History, National Security Agency, Fort George G. Meade, Maryland, 1995, p. 232. Another source tells it differently: Foerstel, Herbert N., *Secret Science: Federal Control of American Science and Technology*, Praeger, Westport, Connecticut, 1993, p.129 says the key was chopped from 128 bits (as used in Lucifer) to 56 bits and that NSA really wanted a 32 bit key!

[33] Johnson, Thomas, R., *American Cryptology During the Cold War, 1945–1989, Book III: Retrenchment and Reform, 1972–1980*, Center for Cryptologic History, National Security Agency, fort George G Meade, Maryland, 1995, p. 239.

[34] Morris, Robert, Neil J. A. Sloane, and Aaron D. Wyner, "Assessment of the National Bureau of Standards Proposed Federal Data Encryption Standard," *Cryptologia*, Vol. 1, No. 3, July 1977, pp. 281–291, p. 282 cited here.

[35] For claim and quote about doubt (from the authors), see Morris, Robert, Neil J. A. Sloane, and Aaron D. Wyner, "Assessment of the National Bureau of Standards Proposed Federal Data Encryption Standard," *Cryptologia*, Vol. 1, No. 3, July 1977, pp. 281–291, p. 287 cited here.

[36] Lexar Corporation, *An Evaluation of the NBS Data Encryption Standard*, unpublished report, Lexar Corporation, 11611 San Vicente Boulevard, Los Angeles, California, 1976.

the communications, data processing, and banking industries, by the police and the medical and legal professions, as well as being widely used by the Federal agencies for which it was officially designed." They were right:

> DES quickly took over the encryption market, becoming the code provided by 99 percent of the companies selling equipment.[37]

> Even the Moscow-based Askri company offers a software encryption package called Cryptos for $100. The package is based on the Data Encryption Standard, America's own national standard.[38]

The plea for a longer key wasn't heeded, nor were details of the S-box construction revealed. Although adoption of DES was widespread, these problems prevented it from being universally accepted.

> Robert Morris of Bell Laboratories in Murray Hill, New Jersey, says that officials of the Bell Telephone Company have decided that the DES is too insecure to be used in the Bell System.[39]

In 1985, when DES was used by bankers to encrypt electronic fund transfers, NSA Deputy Director Walter Deeley said he "wouldn't bet a plugged nickel on the Soviet Union not breaking it."[40] Yet, DES remained mandatory until 2002 for all federal agencies, for protection of sensitive unclassified information needing encryption.

A common question over the years since the arrival of DES was whether or not NSA could break it. The quote above seems to indicate an affirmative answer by 1985, but the quote by Matt Blaze reproduced below comes at the question from another perspective.[41]

> An NSA-employed acquaintance, when asked whether the government can crack DES traffic, quipped that real systems are so insecure that they never need to bother.

Blaze went on to give a list of the "Top Ten Threats to Security in Real Systems." Anyone concerned with actual security, and not just the mathematical aspects of the subject should study this list.[42]

13.2.3 S-Boxes Revealed!

In 1990, Eli Biham and Adi Shamir applied a new attack they had developed, called *differential cryptanalysis*, to DES. It turned out that this approach was only better than brute force if there were 15 or fewer rounds and DES had 16. Perhaps this was why! In fact, differential cryptanalysis *was* known to

[37] Foerstel, Herbert N., *Secret Science: Federal Control of American Science and Technology*, Praeger, Westport, Connecticut, 1993, p. 129.

[38] Foerstel, Herbert N., *Secret Science: Federal Control of American Science and Technology*, Praeger, Westport, Connecticut, 1993, p. 138.

[39] Kolata, Gina Bari, "Computer Encryption and the National Security Agency Connection," *Science.* Vol. 197, No. 4302, July 29, 1977, pp. 438–440.

[40] Foerstel, Herbert N., *Secret Science: Federal Control of American Science and Technology*, Praeger, Westport, Connecticut, 1993, p. 129.

[41] Schneier, Bruce, *Applied Cryptography*, second edition, John Wiley & Sons, Inc., New York, 1996, p. 619 from the afterword by Matt Blaze

[42] Schneier, Bruce, *Applied Cryptography*, second edition, John Wiley & Sons, Inc., New York, 1996, pp. 620-621 from the afterword by Matt Blaze.

the DES designers. They called it "T attack." The mysterious *S*-boxes were generated and tested until ones that fit certain criteria were found; they needed to be able to resist T attack and linear cryptanalysis. Neither attack was publicly known at the time. The NSA knew about differential cryptanalysis, but didn't want the information shared; hence, the generation of the *S*-boxes had to remain secret. Susan Landau, among others, believes that NSA had not anticipated linear cryptanalysis.[43] In any case, both of these attacks were rediscovered in the open community and are now available to anyone.

There are some "weak" keys for DES. This term is used to denote keys such that all of the round keys they produce are identical. Obvious examples are a key consisting of 56 zeros and a key consisting of 56 ones, but there are two others. There are also six pairs of semi-weak keys that should be avoided. Rather than generating 16 distinct round keys, these only produce two, each of which is then used for 8 rounds. The result is that a key in each semi-weak pair can decrypt messages enciphered with the other key.[44] Each extra key that can be used reduces the average run-time of a brute-force attack by a factor of two.

13.3 EFF vs. DES

The Electronic Frontier Foundation (EFF), a civil liberties group, eventually buried DES. They did this by designing and building a DES cracker, a specialized piece of hardware that could defeat the system in a reasonable amount of time (see Figures 13.8–13.10). It did so by testing over 90 billion keys per second. It should be noted that DES crackers can be run in parallel; hence, someone with ten of these machines can break DES ten times faster than someone with just one.

Figure 13.8 The EFF DES Cracker sits behind Paul Kocher, the principal designer, who is holding one of the 29 boards, each of which contains 64 custom microchips. (From http://www.cryptography.com/technology/applied-research/research-efforts/des-key-search/des-key-search-photos.html and used with the permission of Paul Kocher.)

[43] In any case, it first appeared publicly in Matsui, Mitsuru, "Linear Cryptanalysis Method for DES Cipher," in Helleseth, Tor, editor, *Advances in Cryptology – EUROCRYPT '93 Proceedings*, Lecture Notes in Computer Science, Vol. 765, Springer, Berlin, Germany, 1994, pp. 386–397.

[44] For more on this topic see Moore, Judy H. and Gustavus Simmons, "Cycle Structure of the DES with Weak and Semiweak Keys," in Odlyzko, Andrew M., editor, *Advances in Cryptology – CRYPTO '86 Proceedings*, Lecture Notes in Computer Science, Vol. 263, Springer, Berlin, Germany, 1987, pp. 9–32.

Figure 13.9 **A close-up view of one of the DES Cracker circuit boards. (From http://www.cryptography.com/technology/applied-research/research-efforts/des-key-search/des-key-search-photos.html and used with the permission of Paul Kocher.)**

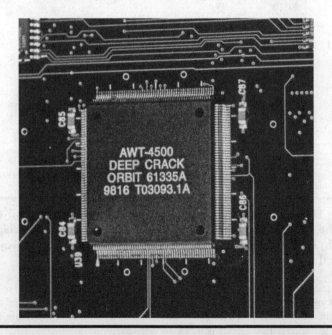

Figure 13.10 **A close-up view of one of the "Deep Crack" custom microchips. (From http://www.cryptography.com/technology/applied-research/research-efforts/des-key-search/des-key-search-photos.html and used with permission of Paul Kocher.)**

To prove the insecurity of DES, EFF built the first unclassified hardware for cracking messages encoded with it. On Wednesday, July 17, 1998 the EFF DES Cracker, which was built for less than $250,000, easily won RSA Laboratory's DES Challenge II contest and a $10,000 cash prize. It took the machine less than 3 days to complete the challenge, shattering the previous record of 39 days set by a massive network of tens of thousands of computers.[45]

Six months later, on Tuesday, January 19, 1999, Distributed.Net, a worldwide coalition of computer enthusiasts, worked with EFF's DES Cracker and a worldwide network of nearly 100,000 PCs on the Internet, to win RSA Data Security's DES Challenge III in a record-breaking 22 hours and 15 minutes. The worldwide computing team deciphered a secret message encrypted with the United States government's Data Encryption Standard (DES) algorithm using commonly available technology. From the floor of the RSA Data Security Conference & Expo, a major data security and cryptography conference being held in San Jose, Calif., EFF's DES Cracker and the Distributed.Net computers were testing 245 billion keys per second when the key was found.[46]

Nevertheless, the broken system was reaffirmed as the standard in 1999! This statement should be qualified—it was reaffirmed in the Triple DES implementation, which neither the EFF machine nor Distributed.Net could break. In 2002, a new standard was finally named: the Advanced Encryption Standard. It is described in Section 20.3.

For now, the record for the least expensive DES cracking machine is held jointly by team members from the German universities of Bochum and Kiel. Dubbed COPACOBANA (Cost-Optimized Parallel Code Breaker), their $10,000 device cracked a DES message in 9 days in 2006. Modifications made since then have improved COPACOBANA's efficiency and it now produces plaintext in less than a day.

13.4 A Second Chance

One way to possibly improve the security of DES is to compose it with itself using different keys. But enciphering twice with two different DES keys does not help much! Merkle and Hellman found that a meet-in-the middle attack reduces the keyspace for a brute-force attack from the expected $(2^{56})(2^{56}) = 2^{112}$ to 2^{57}, only twice that of single DES![47] The meet-in-the-middle attack isn't much more difficult conceptually than a brute-force attack, but it does require a plaintext/ciphertext pair. If the double encryption is represented by

$$C = E_{key2}(E_{key1}(M))$$

we simply form two columns of partial decipherments.

Column $1 = E_k(M)$ for all possible values of the key k.

Column $2 = D_k(C)$ for all possible values of the key k.

---

[45] ""EFF DES Cracker" Machine Brings Honesty to Crypto Debate, Electonic Frontier Foundation Proves that DES is not Secure," *Electronic Frontier Foundation*, https://tinyurl.com/y8fzjymd, July 17, 1998.

[46] "Cracking DES," *Electronic Frontier Foundation*, https://tinyurl.com/y9eqmwdc.

[47] Merkle, Ralph, and Martin Hellman, "On the Security of Multiple Encryption," *Communications of the ACM*, Vol. 24, No. 7, July 1981, pp. 465–467.

These two columns will have an entry in common. When key 1 is used in Column 1 and key 2 is used in Column 2, the entries will both match the result following the first encipherment of the original message. Thus, after calculating 2 columns of 2^{56} values each, the keys must be revealed. This is how we get 2^{57} as the size of the space that must be brute-forced.

When looking for a match between Columns 1 and 2, we might find several. If we have more than a single block of text to work with, we can apply the potential keys to each of these to see which is actually the correct key pair.

With Triple DES, we gain more of an advantage. We may carry out the triple encryption with just two keys by applying the following steps.[48]

1. Encipher with key 1.
2. Decipher with key 2.
3. Encipher with key 1 again.

An obvious question is "Why is the second step a decipherment instead of an encipherment?" Triple DES would be equally strong with the second step being an encipherment, but then the system wouldn't be backward compatible with single DES, a feat that can be achieved, using the scheme provided above, by simply letting key 1 = key 2. Thus, the decision to have step 2 be a decipherment was made for convenience rather than security.

Another good question is "Why not use three different keys instead of just two?" There is actually an attack for the two-key version, but it is not practical.[49] It requires an unrealistic amount of memory. One can, if concerned, use three different keys, and this attack will be even less of a threat. Because three encryptions are done, it doesn't take any longer to use three different keys than to repeat one of them. The only advantage of using just two keys is that you don't need as many keys. The disadvantage is a smaller keyspace.

In the discussion above, we are assuming that repeated encryption is stronger, but this is not the case for all systems. If we encipher a text three times, using a different monoalphabetic substitution system each time, the net effect is the same as enciphering once with some monoalphabetic substitution system different from the three actually used. The same holds true for matrix encryption and many other systems. This is because the set of keys forms a group for these ciphers. Any composition of encipherments is just some other element of the keyspace. A very important question to ask when evaluating the security of Triple DES is whether or not DES is a group. Not only would Triple DES be equivalent to single DES, but single DES would be considerably weakened, if this were true. If DES is a group, a meet in the middle attack could reveal the key in about 2^{28} operations.[50] Fortunately DES is not a group. This was determined in 1993.[51]

Surprisingly, for such an important and longstanding problem, the proof is very easy to follow. Let E_0 and E_1 denote DES encryption using the keys consisting solely of 0s and 1s, respectively.

---

[48] This procedure was suggested in Tuchman, Walter, "Hellman Presents No Shortcut Solutions to DES," *IEEE Spectrum*, Vol. 16, No. 7, July 1979, pp. 40–41.

[49] van Oorschot, Paul C. and Michael J. Wiener, "A Known-plaintext Attack on Two-key Triple Encryption," in Damgård, Ivan B., editor, *Advances in Cryptology – EUROCRYPT '90 Proceedings*, Lecture Notes in Computer Science, Vol. 473, Springer, Berlin, Germany, 1991, pp. 318–325.

[50] Kaliski, Jr., Burton S., Ronald L. Rivest, and Alan T. Sherman, "Is the Data Encryption Standard a Group? (Results of Cycling Experiments on DES)," *Journal of Cryptology*, Vol. 1, No. 1, 1988, pp. 3–36.

[51] Campbell, Keith W. and Michael J. Wiener, "DES is Not a Group," in Brickell, Ernest F., *Advances in Cryptology – Crypto '92*, Springer, Berlin, Germany, 1993, pp. 512–520. This paper credits Don Coppersmith for the proof. It's available online at http://dsns.csie.nctu.edu.tw/research/crypto/HTML/PDF/C92/512.PDF.

We may double encipher a given message M by using both keys: $E_1E_0(M)$. This double encipherment may then be applied to the ciphertext, $E_1E_0(E_1E_0(M))$. For convenience, we denote this as $(E_1E_0)^2(M)$. What is of interest to us is that there are choices for M such that $(E_1E_0)^n(M) = M$, where n is about 2^{32}. This power is small enough that a cryptanalyst can investigate and determine the exact value of n without having a ridiculously long wait. The lowest value of n yielding the original message is referred to as the *cycle length* of M. The lowest value for n that will work for all messages is the order of E_1E_0. The cycle length of any particular message M must divide the order of E_1E_0, which in turn divides the order of the group formed by the DES keys (if it is indeed a group). So, to show that DES is not a group, all that is necessary is to calculate the cycle lengths for various choices of n and look at their least common multiple, which provides a lower bound on the order of E_1E_0 (and hence, the DES keys themselves).

Don Coppersmith was the first to do this.[52] The cycle lengths of 33 messages he examined implied the order of E_1E_0 to be at least 10^{777}. Keith W. Campbell and Michael J. Wiener followed up on this with 295 more messages showing that the subgroup generated by the DES permutation was bounded below by 1.94×10^{2499}. Campbell and Wiener noted that back in 1986 cycle lengths were published by Moore and Simmons[53] that, when taken with the argument above, were sufficient to show DES was not a group; however, this point was missed and the question remained open for years!

DES was not to be the last cipher that NSA had a hand in designing for outside use. A less successful attempt was made with the Clipper chip in 1993.[54] It differed from DES in that the algorithm was classified; all the user would get would be a tamperproof chip. Even worse, the proposal came with the idea of "key escrow," meaning that the built-in key for each chip would be kept on file, so that it could be accessed by law enforcement agents who had obtained warrants. The government's past history of warrantless eavesdropping didn't inspire much confidence in this proposal. Following some heated debate, Clipper vanished. Key escrow attempts proved to be even less welcome in the European Union, where there are, in general, more laws protecting privacy.

13.5 An Interesting Feature

One semester I worked with a student, Austen Duffy, on an attack on DES that had no justification for working. Not surprisingly, it didn't. However, it did reveal something interesting. The idea was that if there existed a message whose ciphertext had a bitsum (aka Hamming weight) that correlated with the bitsum of the key, a brute force attack could be launched on the greatly reduced keyspace. Basically, this special message would be enciphered and its bitsum would, for example, indicate that the bitsum of the key was, say, between 25 and 26 with probability 90%.

[52] Don Coppersmith, In Defense of DES, personal communication to author(s) of DES is not a Group, July 1992. The work was also described briefly in a posting to sci.crypt on *Usernet News*, May 18, 1992.

[53] Moore, Judy H. and Gustavus Simmons, "Cycle Structure of the DES with Weak and Semiweak Keys," in Odlyzko, Andrew M., editor, *Advances in Cryptology – CRYPTO '86 Proceedings*, Lecture Notes in Computer Science, Vol. 263, Springer, Berlin, Germany, 1987, pp. 9–32.

[54] For a non-technical history of the Clipper chip, see pages 226–268 of Levy, Steven, Crypto: *How the Code Rebels Beat the Government, Saving Privacy in the Digital Age*, Viking, New York, 2001. An analysis of the algorithm used by Clipper (it was eventually released) is given in Kim, Jongsung, and Raphaël C.–W. Phan "Advanced Differential-Style Cryptanalysis of the NSA's Skipjack Block Cipher," *Cryptologia*, Vol. 33, No. 3, July 2009, pp. 246–270.

As I said, we were hoping for *some* correlation, not necessarily a perfect correlation. We investigated in a much unsophisticated way. We wrote a program that generated a random message and enciphered it with 100 random keys and then calculated the correlation between the bitsums of the ciphertexts and the keys. Actually, this was all done inside a large loop, so many different random messages were tested in this manner. Every time a message yielded a higher correlation than any previous messages (or tied the current high), it was displayed on the screen. Thus, when run, we saw a list of ever better (but never actually *good*) correlation values displayed beside various messages. Take a look at Figure 13.11, which shows some results, and see if you notice anything unusual, before reading the explanation that follows.

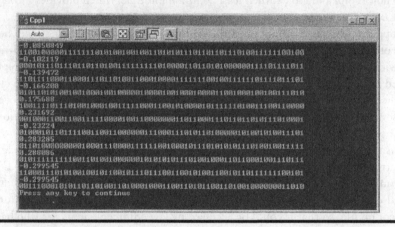

Figure 13.11   Display of correlation values and messages.

We thought it very strange that the last two messages displayed were complementary! Because the messages were randomly generated it seemed unlikely that the complement of one would appear so quickly, and why would they have the same correlation value? I went back and read a bit more about DES and learned that replacing every bit of a message with its complement and doing the same for the key yields a ciphertext that is the complement of the original ciphertext.[55] Algebraically,

$$E(\overline{K},\overline{P}) = \overline{E(K,P)}$$

This result was known (to others) well before we began our attack, but it was new to us. Taking the message yielding the "best" correlation, we tried it repeatedly with random sets of keys and computed the correlation for each set. As the output in Figure 13.12 indicates, the correlation didn't remain at the value found above.

The situation is analogous to taking a large number of coins (as opposed to messages) and flipping them 100 times each. Perhaps one of them will land head side up 60% of the time. The distribution of the number of heads the coins shows should follow a bell curve, so it's not surprising that some coins yield many more heads than others. However, if we take the coin yielding the most heads and flip it another 100 times, repeatedly, we can expect, on average, that there will be 50 heads each time.

---

[55] For a proof see Schneier, Bruce and Niels Ferguson, *Practical Cryptography*, Wiley, Indianapolis, Indiana, 2003, p. 54.

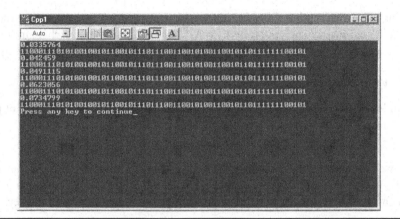

Figure 13.12   Result of retrying the "best" message with more random key sets.

So, a purely investigatory line of thought, not backed by any theory, pointed us to an interesting feature of DES that we had not anticipated. Though the result was not new, I think it serves as a good illustration of how research may go. Even failures may prove interesting and useful.

Several DES variants have been offered to address attacks. The most notable (and simplest) is Triple DES, but there is also DES with Independent Subkeys, DESX, CRYPT(3), Generalized DES (GDES), DES with Alternate *S*-Boxes, RDES, *s*^n DES (a number of variants, with *n* taking the values 2, 3, 4, and 5), DES with Key-Dependent *S*-Boxes, and NEWDES.[56]

13.5.1 Cryptologic Humor

In the Vol. 0, No. 0, issue of *Journal of Craptology*, the paper "Terms Used in the disciplines of Cryptography, IT Security and Risk Analysis" offered the following definition:

> **DES.** *n. Abbr.* Cryptographic algorithm with key size chosen so that it can be **D**ecrypted **E**asily by **S**ecurity agencies.

The Electronic Frontier Foundation included the following claim in their book on the EFF's DES cracker.

> If a civil liberties group can build a DES Cracker for $200,000, it's pretty likely that governments can do the same thing for under a million dollars.[57]

13.6 Modes of Encryption

DES can be implemented in many different modes. This isn't a new idea. Many of the ciphers examined in previous chapters may be implemented in a variety of ways; for example, in Section 2.5, we saw the autokey of Blaise de Vigenère (1523–1596). After a short "priming key" encryption was continued in the manner of a running key cipher, but using previous portions the message itself (or the ciphertext that was being generated) as the key. This was an important contribution, but little was done with it for hundreds of years. In 1958, Jack Levine presented a pair of new methods for

---

[56] For some details on these variants (and further references for those desiring yet more detail) see Schneier, Bruce, *Applied Cryptography*, second edition, John Wiley & Sons, Inc., New York, 1996, pp. 294–300, and 306–308.

[57] Electronic Frontier Foundation, *Cracking DES: Secrets of Encryption Research, Wiretap Politics & Chip Design*, O'Reilly, Sebastopol, California, 1998, pp. 1–16.

implementing matrix encryption.[58] Before showing Levine's approach, we set our notation. We will represent traditional matrix encryption by $C_i = \mathbf{A}P_i$, where $\mathbf{A}$ is the (never changing) enciphering matrix and P_i and C_i are the ith groups of plaintext and ciphertext characters, respectively.

13.6.1 Levine's Methods

Levine introduced a second invertible matrix $\mathbf{B}$, and made the ciphertext a function of not just the current plaintext block, but also the one that went before:

$$C_i = \mathbf{A}P_i + \mathbf{B}P_{i-1}$$

This works well for enciphering the second, third, and so on, groups of plaintext, but for the first group of plaintext characters, P_1, there is no previous group to use for P_0. Thus, in addition to the matrices $\mathbf{A}$ and $\mathbf{B}$, Levine needed to define a priming plaintext (P_0) as part of the key.

Example 1

Our key will be

$$\mathbf{A} = \begin{pmatrix} 5 & 5 \\ 6 & 11 \end{pmatrix} \quad \mathbf{B} = \begin{pmatrix} 13 & 10 \\ 3 & 3 \end{pmatrix} \quad P_0 = \begin{pmatrix} 17 \\ 6 \end{pmatrix}$$

Let the message be the following quote from Nobel Prize-winning physicist Richard Feynman:[59]

```
I TOLD HIM, OF COURSE, THAT I DIDN'T KNOW – WHICH IS
MY ANSWER TO ALMOST EVERY QUESTION.
```

Converting the message to numbers, using the scheme A = 0, B = 1,..., Z = 25, and ignoring all punctuation, we get

8	19	14	11	3	7	8	12	14	5	2
14	20	17	18	4	19	7	0	19	8	3
8	3	13	19	10	13	14	22	22	7	8
2	7	8	18	12	24	0	13	18	22	4
17	19	14	0	11	12	14	18	19	4	21
4	17	24	16	20	4	18	19	8	14	13

Enciphering the first plaintext pair and reducing modulo 26 gives us

$$C_1 = \begin{pmatrix} 5 & 5 \\ 6 & 11 \end{pmatrix} \begin{pmatrix} 8 \\ 19 \end{pmatrix} + \begin{pmatrix} 13 & 10 \\ 3 & 3 \end{pmatrix} \begin{pmatrix} 17 \\ 6 \end{pmatrix} = \begin{pmatrix} 0 \\ 14 \end{pmatrix}$$

The second and third ciphertext pairs are found below.

$$C_2 = \begin{pmatrix} 5 & 5 \\ 6 & 11 \end{pmatrix} \begin{pmatrix} 14 \\ 11 \end{pmatrix} + \begin{pmatrix} 13 & 10 \\ 3 & 3 \end{pmatrix} \begin{pmatrix} 8 \\ 19 \end{pmatrix} = \begin{pmatrix} 3 \\ 0 \end{pmatrix}$$

$$C_3 = \begin{pmatrix} 5 & 5 \\ 6 & 11 \end{pmatrix} \begin{pmatrix} 3 \\ 7 \end{pmatrix} + \begin{pmatrix} 13 & 10 \\ 3 & 3 \end{pmatrix} \begin{pmatrix} 14 \\ 11 \end{pmatrix} = \begin{pmatrix} 4 \\ 14 \end{pmatrix}$$

---

[58] Levine, Jack, "Variable Matrix Substitution in Algebraic Cryptography," *American Mathematical Monthly*, Vol. 65, No. 3, March 1958, pp. 170–179.

[59] Feynman, Richard, *"What Do You Care What Other People Think?"* W. W. Norton & Company, New York, 1988, p. 61.

So, the ciphertext begins 0 14 3 0 4 14, or AODAEO, in terms of letters. This should be enough to make Levine's matrix encryption mode clear.

Levine also presented a version that used previous ciphertext to vary the encryption:

$$C_i = \mathbf{A}P_i + \mathbf{B}C_{i-1}$$

Another method used by Levine to vary the substitutions will only be mentioned in passing, as it didn't lead in the direction of the modern modes of encryption. It took the form $C_i = \mathbf{A}_i P_i$. That is, the matrix $\mathbf{A}$ changed with every encipherment. One only need be careful that the $\mathbf{A}_i$ are generated in a manner that ensures each will be invertible; otherwise, the ciphertext will not have a unique decipherment.

13.6.2 Modern Modes

These ideas surfaced again in the 1970s with modern block ciphers, such as DES. In the following pages, a few modes of encryption currently in use are detailed.

13.6.2.1 Electronic Code Book Mode

If DES (or any other block cipher) is used to encipher one block at a time and no other input is considered for the enciphering process, then the cipher is said to be implemented in Electronic Code Book (ECB) mode. In this mode, repeated blocks will be enciphered identically. Thus, for lengthy messages, or even short ones with stereotyped beginnings, cribs might be easily obtained, in which case the cipher would have to be able to resist a known-plaintext attack (This is the sort of attack Diffie had in mind for a \$20 million DES cracker machine.) This mode has its uses (see Chapter 18), but it is not recommended for enciphering messages longer than a single block. For those, one of the following modes is preferred. The modes that follow each have advantages and disadvantages. For the modes where it is not immediately obvious, the manner in which error propagates will be discussed. For ECB, an error will only alter the block in which it occurs.

13.6.2.2 Cipher Block Chaining Mode

In Cipher Block Chaining (CBC) mode, we start by selecting an initialization vector (IV). This serves as the first ciphertext block. The following ciphertext blocks are formed by XORing each block of the message with the previous ciphertext block prior to enciphering. In symbols, $C_0 = \text{IV}$ and $C_i = E(C_{i-1} \oplus M_i)$ for $i \geq 1$. To decipher we take $E^{-1}(C_i) \oplus C_{i-1}$. Because $E^{-1}(C_i) \oplus C_{i-1} = (C_{i-1} \oplus M_i) \oplus C_{i-1}$ and $\oplus$ is commutative, the C_{i-1} terms drop out to give us our original message block. Encryption is dependent on a previous block of ciphertext, so repeated blocks in the message are likely to be enciphered differently. Unlike ECB, an interceptor cannot change the order of the enciphered blocks without detection.

Error propagation is only slightly worse for CBC than for ECB mode. If an error is present in a given ciphertext block (due to transmission rather than an error made by the encipherer), that block will decipher incorrectly, as will the one that follows. However, the next ciphertext block will be correctly recovered, as its plaintext is only dependent on itself and the previous ciphertext block. The fact that the previous plaintext block is garbled is irrelevant. The incorrect bits in the two damaged blocks will be of the same number and in the same positions. This mode was invented in 1976 by researchers at IBM. They were granted a patent two years later.[60]

13.6.2.3 Cipher Feedback Mode

Cipher Feedback (CFB) mode allows any block cipher to be used as a stream cipher. Stream ciphers are typically used when the data must be encrypted on-the-fly, or in real-time, as it is often put. An example is provided by secure voice communication or streaming data of any kind. Chapter 19 examines stream ciphers in more detail. Stream ciphers usually act on the plaintext in smaller pieces than block ciphers. CFB allows us to generate ciphertext in groups shorter than the block size the cipher normally works with. As an example, we'll encipher 8 bits at a time.

We start by selecting an initialization vector (IV) of the same size as that used by the block cipher. We then compute E(IV) and XOR the leftmost 8 bits of this with the first 8 bits of our message. This provides us with the first 8 bits of the ciphertext, which may then be sent. Now, the IV is changed by appending the 8 bits of ciphertext to the right and discarding the 8 leftmost bits. We then repeat the process to encrypt the next 8 bits of the message. Figure 13.13 should help make this process clearer.

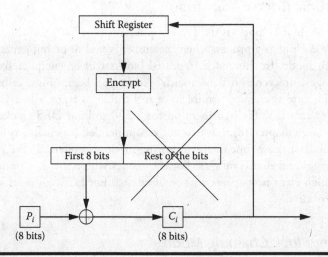

Figure 13.13   Cipher Feedback (CFB) mode.

The two C_i paths heading out indicate the ciphertext goes to the intended recipient, as well as back up to the right-hand side of the shift register. When the 8 bits of ciphertext hit the shift register, they push out the 8 bits on the left-hand side of the register; that is, all the bits in the register shift eight positions to the left and the leftmost 8 bits fall off the edge (get discarded).

If an error is present in a plaintext block, it will change all the ciphertext blocks that follow, but this isn't as bad as it sounds. The error will undo itself upon deciphering, until the original flawed plaintext block is reached. On the other hand, if an error creeps into a ciphertext block, there will

[60] Ehrsam, William F., Carl H. W. Meyer, John L. Smith, and Walter L. Tuchman, *Message Verification and Transmission Error Detection by Block Chaining*, U.S. Patent 4,074,066, February 14, 1978, https://patents.google.com/patent/US4074066A/en.

be an error in the corresponding plaintext block, and it will then creep into the shift register, where it will cause further errors until it is shifted all the way out of the register.

13.6.2.4 Output Feedback Mode

The operation of Output Feedback (OFB) mode has changed slightly from how it was initially defined. The early version is presented first, and then the update is provided. OFB, like CFB above, allows a block cipher to be used as a stream cipher. In fact, OFB is almost identical to CFB. We can use it to encipher in groups smaller than the block size of the enciphering algorithm. Again, 8 bits will be used for our example. The directions are reproduced below, exactly as for CFB, but with the single change emphasized by being placed in bold text.

We start by selecting an initialization vector (IV) of the same size as used by the block cipher. We then compute $E(IV)$ and XOR the leftmost 8 bits of this with the first 8 bits of our message. This provides us with the first 8 bits of the ciphertext, which may then be sent. Now the IV is changed by appending the **8 bits of the enciphered IV** to the right and discarding the 8 leftmost bits. We then repeat the process to encrypt the next 8 bits of the message. This small change makes it possible to generate all of the bytes that will be XORed with the message in advance.

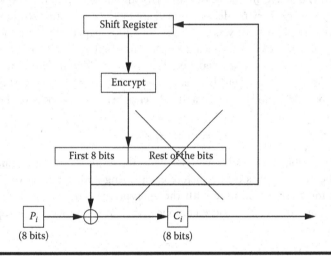

Figure 13.14   Output Feedback (OFB) mode.

The only difference between Figure 13.13 and Figure 13.14 is that in the present mode the bits advancing the shift register arise directly from the encryption step, rather than after the XOR. In this mode, an error that creeps into a ciphertext block will affect only the corresponding plaintext block.

At the Crypto '82 conference held at the University of California, Santa Barbara, a pair of talks addressed weaknesses in using OFB to generate keys in groups of less than 64 bits at a time, as illustrated above. An abstract published in the proceedings of this conference put it bluntly.

> The broad conclusion is reached that OFB with $m = 64$ is reasonably secure but OFB with any other value of m is of greatly inferior security.[61]

---

[61] Davies, Donald W. and Graeme I. P. Parkin, "The Average Cycle Size of the Key Stream in Output Feedback Encipherment (Abstract)," in Chaum, David, Ronald L. Rivest, and Alan T. Sherman (editors), *Advances in Cryptology, Proceedings of Crypto 82*, Plenum Press, New York, 1983, pp. 97–98.

The problem was that smaller values of m, such as the 8 used above, could give rise to cycles of key to be XORed with the text that are simply too short for heavy traffic. So, it shouldn't be a surprise that the update to this mode definition, as given in SP 800-38A, is to have it operate on the entire block, not just a portion of it.[62]

13.6.2.5 Counter Mode

Counter (CTR) mode requires a set of distinct counters (blocks of bits), each having the same size as the plaintext blocks. We label these counters as CNT_i. We then have $C_i = M_i \oplus E(CNT_i)$ for $i \geq 1$. There is no chaining involved in this mode. Repeated plaintext blocks will be enciphered differently, due to the counters all being distinct. The counters are normally defined by choosing a random value for CNT_1 and then incrementing successive counters by one each time. Counter mode was first proposed by Whitfield Diffie and Martin Hellman in 1979.[63]

13.6.2.6 Offset Codebook Mode

There are many other modes that could be described and more will certainly be proposed in the future. The five detailed above are the most important and most popular. However, there is a benefit to detailing one 21st-century mode, as you will see by the end of this section.

Offset Codebook mode (OCB) was put forth in a 2001 paper[64] by four computer scientists, Phillip Rogaway (University of California at Davis), Mihir Bellare (University of California at San Diego), John Black (University of Nevada), and Ted Krovetz (Digital Fountain[65]).

This mode breaks the message into blocks, M_1, M_2, M_3, ..., M_m, usually of size 128 bits each, although the last block, M_m, can be shorter, and then enciphers each block with the following simple formula.

$$C_i = E(M_i \oplus Z_i) \oplus Z_i$$

Although each message block is XORed with Z_i twice, the result is not the original message block. This is because one of the XORs occurs before enciphering and the other occurs after. As long as the XOR operator doesn't commute with the encryption algorithm, the Z_is won't drop out. Determining the values of the Z_i is more complicated than the process of enciphering. The five steps are detailed below.

$$\text{Step 1:} \quad L = E(0^n)$$

0^n designates an n-bit string of all 0s. This is enciphered and the result is called L. Because there will be a sequence of Ls, of which this is the first, it can also be denoted by L_0.

$$\text{Step 2:} \quad R = E(N \oplus L)$$

---

[62] See SP 800-38A, http://csrc.nist.gov/groups/ST/toolkit/BCM/current_modes.html.

[63] Diffie, Whitfield and Martin Hellman, "Privacy and Authentication: An Introduction to Cryptography," *Proceedings of the IEEE*, Vol. 67, No. 3, March 1979, pp. 397–427, p. 417 cited here.

[64] Rogaway, Phillip, Mihir Bellare, John Black, and Ted Krovetz, "OCB: A Block-Cipher Mode of Operation for Efficient Authenticated Encryption," *ACM Conference on Computer and Communications Security (CCS '01)*, ACM Press, pp. 196–205, 2001, available online at http://krovetz.net/csus/papers/ocb.pdf. The journal version is in *ACM Transactions on Information and System Security (TISSEC)*, Vol. 6, No. 3, August 2003, pp. 365–403, available online at https://dl.acm.org/doi/pdf/10.1145/937527.937529 and https://www.cs.ucdavis.edu/~rogaway/papers/ocb-full.pdf.

[65] Digital Fountain was founded to commercialize encoding technology for efficient reliable asynchronous multicast network communication.

N stands for nonce, which is another name for an initialization vector. It is a random string of bits.

Step 3 : $L_i = 2 \cdot L_{i-1}$ for $1 \le i \le m$

This step, performed repeatedly, generates the sequence $L_1, L_2, L_3, \ldots$ The symbol $\cdot$ is not used to represent the simple kind of multiplication that everyone is familiar with. Instead, it designates multiplication over the finite field GF(2^n) with $n = 128$. This requires some explanation. A much smaller example than applies to OCB will be used to convey the idea.

Suppose we want to multiply two values over the finite field GF(2^n) for the case $n = 3$. Then the values being multiplied will be 3-bits each. But before the multiplication is carried out, the bits will become the coefficients of polynomials. For example, to multiply 111 and 011, we multiply $1x^2 + 1x + 1$ and $0x^2 + 1x + 1$. Of course, we don't need to write a coefficient that is 1 (unless it's the constant term) and the terms with 0 as a coefficient drop out. We are really multiplying $x^2 + x + 1$ and $x + 1$. This product is $x^3 + 2x^2 + 2x + 1$. But in our binary world we reduce the coefficients modulo 2, so that they can only be 0 or 1. Our product becomes $x^3 + 1$. Now there is one last step to make. We need to divide by an irreducible polynomial (i.e. one that cannot be factored in our system where the coefficients must be 0 or 1) and take the remainder as our answer. Because we're looking at the case $n = 3$, the irreducible polynomial must be of degree 3. There are only two such polynomials. The one that will be used here is $x^3 + x + 1$. When we divide our product by this number (again adjusting coefficients modulo 2), the remainder is x. Converting this back to a binary representation, we get 010.

For OCB, $n = 128$, so the bits that make up L_{i-1} become the coefficients of a polynomial with as many as 128 terms (i.e. a maximum degree of 127, if the leading coefficient is 1). In Step 4, this value is multiplied by 2, which corresponds to the binary number 10 (ignoring the 126 leading 0s) and the polynomial x. The irreducible polynomial that the product is then divided by in OCB is $m(x) = x^{128} + x^7 + x^2 + 1$. The remainder (which is what we are interested in) is converted back to bits (by taking the coefficients) and is our final answer. This weird way of multiplying strings of bits will be seen again when we get to the Advanced Encryption Standard in Chapter 20.

Step 4 : $Z_1 = L \oplus R$

This step generates the first block of key to be XORed with the first message block.

Step 5 : $Z_i = Z_{i-1} \oplus L_{ntz(i)}$ for $1 \le i \le m$

This last step introduces a new function, $ntz(i)$, which gives the number of trailing 0s in the binary representation of i. If i is read from right to left, this is the number of 0s that are encountered before coming to a 1. For even values of i, we have $ntz(i) = 0$. The result of $ntz(i)$ gives us the subscript for L. That is, it tells us which term in the L sequence is to be XORed with Z_{i-1} to give us Z_i. This step must be performed repeatedly, until enough key is generated to encipher the entire message.

Notice that none of the five steps detailed above require any knowledge of the message that is to be enciphered. These calculations can be carried out in advance. Once the sequence Z_i is known, the message blocks can be enciphered. It should also be noted that if identical messages are enciphered with the same encryption key, but different nonces are used to generate the sequence Z_i, then the final results will be completely different. The nonce needs to be communicated to the recipient, who is only assumed to know the key used by the encryption algorithm E.

There are a few more aspects of OCB that need to be detailed. The first concerns enciphering the last message block, which may be shorter than the rest. A new value must be calculated like so:

$$X = \text{len}(M_m) \oplus (L \cdot 2^{-1}) \oplus Z_m$$

Where $\text{len}(M_m)$ is the length of the last message block expressed as an n-bit number.

The weird polynomial multiplication plays a role in this step. This time, instead of multiplying by $2 = 10 = x$, we are multiplying by 2^{-1}, which is x^{-1}, the multiplicative inverse of x modulo $m(x) = x^{128} + x^7 + x^2 + 1$. That is, we must multiply by the polynomial that, when multiplied by x and divided by $m(x)$ yields a remainder of 1. But how do we find this polynomial? More generally, how can we find the multiplicative inverse of a polynomial modulo another polynomial? The steps for the simpler case of finding the multiplicative inverse of a number mod another number (using the extended Euclidean algorithm) are shown in Section 14.3. To solve this more intimidating sounding problem, there is no difference other than using polynomials instead of integers. While it might seem very tedious to carry out the multiplication by whatever x^{-1} turns out to be, there are great shortcuts that can be taken and it is not hard at all to code this up on computer (or to put it on a chip).[66]

After X is determined, another new value is calculated:

$$Y = E(X)$$

Y is likely longer than M_m, but the appropriate number of bits from the right side of Y are deleted so that the result is the same length as M_m. That is, some of the least significant bits of Y are removed. The truncated Y, $\text{trunc}(Y)$ is then XORed with M_m to get the last ciphertext block. That is:

$$C_m = \text{trunc}(Y) \oplus M_m$$

Thus, the final ciphertext block is the same length as the final message block.

Although the entire message has now been enciphered, a few more calculations are performed to give this mode a special property. The first is called a checksum and is calculated as

$$\text{checksum} = M_1 \oplus M_2 \oplus \cdots \oplus M_{m-1} \oplus Y \oplus C_m 0^*$$

$C_m 0^*$ is simply C_m with 0s appended to the end (as least significant bits) to make this block the same length as the others. The process of adding 0s is called *padding*. In some other systems, padded bits follow a pattern, such as alternating 1s and 0s.

Next, an authentication tag must be calculated, like so

$$\text{tag} = \text{first } \tau \text{ bits of } E(\text{checksum} \oplus Z_m)$$

The number, τ, of most significant bits retained for the tag varies from application to application.

The checksum is an intermediate calculation that is not included in what is sent to the intended recipient. All that is transmitted is the ciphertext and the tag.

Now, what is the advantage of doing the extra calculations needed to determine the tag? The tag gives the recipient a way to confirm the authenticity of the message. After the message is completely deciphered, the recipient can calculate the checksum, and then the tag. If the tag thus

---

[66] The shortcut is detailed in the appendix of Stallings, William, "The Offset Codebook (OCB) Block Cipher Mode of Operation for Authenticated Encryption," *Cryptologia*, Vol. 42, No. 2, March 2018, pp. 135–145.

calculated matches the one he received, then the recipient can conclude that the message is authentic. That is, it came from the person claiming to have sent it, without alteration.

Ciphers don't typically have this property. For example, if DES is implemented in Electronic Code Book (ECB) mode, someone who is able to control the communication channel might be able rearrange the enciphered blocks so that the deciphered message has a different meaning. Perhaps the message was an order for the purchase of 1,000 shares of stock X and 50,000 shares of stock Y. The rearranged blocks might decipher to 1,000 shares of stock Y and 50,000 shares of stock X. If the same key is used for more than one message, an attacker could insert blocks from one message into another as additions or replacements. Both sorts of mischief are blocked by OCB. Changes that could go undetected in ECB will be revealed in OCB, when the tag the recipient calculates doesn't match the tag sent with the ciphertext. If the tag does match, then the message had to come from the person claiming to have sent it. This is what is meant by authenticity. The fact the message could not have been altered without being detected is called message integrity. Usually, integrity follows as a consequence of authenticity. Of course, what authenticity really means in this instance is that the message came from the person who has the key. If the key is stolen, even OCB will be unable to distinguish between authentic and unauthentic messages.

In addition to the great feature of authenticity, OCB is very fast and requires little energy to implement. Hence, it is popular in devices for which these issues matter. To be precise, what was detailed here is OCB1. There is also OCB3, which is slightly more efficient and allows the incorporation of data that must be authenticated, but doesn't require encryption.[67] To answer an obvious question, there was an OCB2, but it was shown to be insecure in 2018.[68]

References and Further Reading

On DES

Babinkostova, Liljana, Alyssa M. Bowden, Andrew M. Kimball, and Kameryn J. Williams, "A Simplified and Generalized Treatment of DES-Related Ciphers," *Cryptologia*, Vol. 39, No. 1, January 2015, pp. 3–24.

Biham, Eli and Adi Shamir, "Differential Cryptanalysis of DES-like Cryptosystems (invited talk)," in Menezes, Alfred J., and Scott A. Vanstone, editors, *Advances in Cryptology – CRYPTO '90 Proceedings*, Lecture Notes in Computer Science, Vol. 537, Springer, Berlin, Germany, 1991, pp. 2–21. This piece, soon to grow, was labeled as an "extended abstract" at this stage.

Biham, Eli and Adi Shamir, "Differential Cryptanalysis of DES-like Cryptosystems," *Journal of Cryptology*, Vol. 4, No. 1, January 1991, pp. 3–72. This marks the appearance of the full paper.

Biham, Eli and Adi Shamir, *Differential Cryptanalysis of the Data Encryption Standard*, Springer, New York 1993. This is an entire book devoted to what began as an "extended abstract."

Campbell, Keith W. and Michael J. Wiener, "Des is Not a Group," in Brickell, Ernest F., editor, *Advances in Cryptology – CRYPTO '92 Proceedings*, Lecture Notes in Computer Science, Vol. 740, Springer, Berlin, Germany, 1993, pp. 512–520.

---

[67] Stallings, William, "The Offset Codebook (OCB) Block Cipher Mode of Operation for Authenticated Encryption," *Cryptologia*, Vol. 42, No. 2, March 2018, pp. 135–145, details both OCB1 and OCB3.

[68] See Inoue, Akiko and Kazuhiko Minematsu, "Cryptanalysis of OCB2,", https://eprint.iacr.org/2018/1040; Poettering, Bertram, "Breaking the confidentiality of OCB2," https://eprint.iacr.org/2018/1087; Iwata, Tetsu, "Plaintext Recovery Attack of OCB2," https://eprint.iacr.org/2018/1090; and Inoue, Akiko, Tetsu Iwata, Kazuhiko Minematsu, and Bertram Poettering, "Cryptanalysis of OCB2: Attacks on authenticity and confidentiality," https://eprint.iacr.org/2019/311.

De Meyer, Lauren and Serge Vaudenay, "DES S-box generator," *Cryptologia*, Vol. 41, No. 2, March 2017, pp. 153–171.

Diffie, Whitfield, and Martin E. Hellman, "A Critique of the Proposed data Encryption Standard," *Communications of the ACM*, Vol. 19, No. 3, March 1976, pp. 164–165.

Electronic Frontier Foundation, *Cracking DES: Secrets of Encryption Research, Wiretap Politics & Chip Design,* O'Reilly, Sebastopol, California, 1998. This book is in the public domain. The bulk of it is source code, intended to be scanned by OCR. The purpose was to get around export laws on software that did not apply to code in print form. The Electronic Frontier Foundation's website is http://www. eff.org/.

Güneysu, Tim, Timo Kasper, Martin Novotný, Christof Paar, and Andy Rupp, "Cryptanalysis with COPACOBANA," *IEEE Transactions on Computers*, Vol. 57, No. 11, November 2008, pp. 1498–1513.

Hellman, Martin, Ralph Merkle, Richard Schroeppel, Lawrence Washington, Whitfield Diffie, Stephen Pohlig, and Peter Schweitzer, *Results of an Initial Attempt to Cryptanalyze the NBS Data Encryption Standard*, Information Systems Laboratory SEL 76-042, Stanford University, September 1976. This is the paper that found a symmetry that cut the keyspace in half under a chosen plaintext attack.

Hellman, Martin, http://cryptome.org/hellman/hellman-ch1.doc. The beginning of an autobiography can be found here.

Juels, Ari, moderator, RSA Conference 2011 Keynote – The Cryptographers' Panel, video available online at http://www.youtube.com/watch?v=0NlZpyk3PKI. The panel consisted of Whitfield Diffie, Martin Hellman, Ron Rivest, Adi Shamir, and Dickie George. George was involved with DES from the NSA side, as a technical director. The conversation is wide-ranging, but much of it concerns DES.

Kahn, David, *The Codebreakers*, second edition, Scribner, New York, 1996, p. 980. Do not consult the first edition for information on DES. This cipher didn't exist when the first edition was published.

Katzman, Jr., Harry, *The Standard Data Encryption Algorithm*, Petrocelli Books, New York, 1977.

Kinnucan, Paul, "Data Encryption Gurus: Tuchman and Meyer," *Cryptologia*, Vol. 2, No. 4, pp. 371–381, October 1978, reprinted from *Mini-Micro Systems*, Vol. 2 No. 9, October 1978, pp. 54, 56–58, 60. This paper quotes Walter Tuchman as saying, "The DES algorithm is for all practical purposes unbreakable." Feistel is mentioned, but this paper gives nearly all of the credit for DES to Walter Tuchman and Carl Meyer. Tuchman was also quoted as saying, "The NSA told us we had inadvertently reinvented some of the deep secrets it uses to make its own algorithms."

Landau, Susan, "Standing the Test of Time: The Data Encryption Standard," *Notices of the AMS*, Vol. 47, No. 3, March 2000, pp. 341–349, available online at http://www.ams.org/notices/200003/fea-landau.pdf.

Levy, Steven, *Crypto: How the Code Rebels Beat the Government, Saving Privacy in the Digital Age*, Viking, New York, 2001.

Matsui, Mitsuru, "Linear Cryptanalysis Method for DES Cipher," in Helleseth, Tor, editor, *Advances in Cryptology – EUROCRYPT '93 Proceedings*, Lecture Notes in Computer Science, Vol. 765, Springer, Berlin, Germany, 1994, pp. 386–397.

Merkle, Ralph and Martin Hellman, "On the Security of Multiple Encryption," *Communications of the ACM*, Vol. 24, No. 7, July 1981, pp. 465–467.

Morris, Robert, Neil J. A. Sloane, and Aaron D. Wyner, "Assessment of the National Bureau of Standards Proposed Federal Data Encryption Standard," *Cryptologia*, Vol. 1, No. 3, July 1977, pp. 281–291. The authors attended the second of two workshops held by NBS on September 21–22, 1976 to evaluate the proposed Data Encryption Standard. This paper presents their conclusions. It also provides many references on the reports, papers, and patents leading up to DES.

Simovits, Mikael J., *The DES: an Extensive Documentation and Evaluation*, Aegean Park Press, Laguna Hills, California, 1996.

Solomon, Richard J., "The Encryption Controversy," *Mini-Micro Systems*, Vol. 2, No. 2, February 1978, pp. 22–26.

U.S. Department of Commerce, *Data Encryption Standard*, FIPS Pub. 46-3, National Institute of Standards and Technology, Washington, DC, 1999, available online at http://csrc.nist.gov/publications/fips/fips46-3/fips46-3.pdf. This is the final version of the government document detailing the Data Encryption Standard.

van Oorschot, Paul C. and Michael J. Wiener, "A Known-plaintext Attack on Two-key Triple Encryption," in Damgård, Ivan B., editor, *Advances in Cryptology – EUROCRYPT '90 Proceedings*, Lecture Notes in Computer Science, Vol. 473, Springer, Berlin, Germany, 1991, pp. 318–325.

On Modes of Encryption

Davies, Donald W. and Graeme I. P. Parkin, "The Average Cycle Size of the Key Stream in Output Feedback Encipherment," [abstract] in Chaum, David, Ronald L. Rivest, and Alan T. Sherman, editors, *Advances in Cryptology, Proceedings of Crypto 82*, Plenum Press, New York, 1983, pp. 97–98.

de Vigenere, Blaise, *Traicté des Chiffres, ou, Secretes Manieres D'escrire*, Abel l'Angelier, Paris, 1586. Although Cardano had previously hacked at the problem, this is where the first working autokeys were presented.

Diffie, Whitfield and Martin Hellman, "Privacy and Authentication: An Introduction to Cryptography," *Proceedings of the IEEE*, Vol. 67, No. 3, March 1979, pp. 397–427.

Ehrsam, William F., Carl H. W. Meyer, John L. Smith, and Walter L. Tuchman, *Message Verification and Transmission Error Detection by Block Chaining*, U.S. Patent 4,074,066, February 14, 1978, https://patents.google.com/patent/US4074066A/en.

Hwang, Tzonelih and Prosanta Gope, "RT-OCFB: Real-Time Based Optimized Cipher Feedback Mode," *Cryptologia*, Vol. 40, No. 1, January 2016, pp. 1–14.

Hwang, Tzonelih and Prosanta Gope, "PFC-CTR, PFC-OCB: Efficient Stream Cipher Modes of Authencryption," *Cryptologia*, Vol. 40, No. 3, 2016, pp. 285–302. The authors argue that "both of the proposed stream cipher modes of Authencryption [in the title of the paper] are quite robust against several active attacks (e.g., message stream modification attacks, known-plain-text attacks, and chosen-plain-text attacks) [… and] can efficiently deal with other issues like "limited error propagation," and so on, existing in several conventional stream cipher modes of operation like CFB, OFB, and CTR."

Levine, Jack, "Variable Matrix Substitution in Algebraic Cryptography," *American Mathematical Monthly*, Vol. 65, No. 3, March 1958, pp. 170–179. Levine, being familiar with classical cryptology, applied autokeys to matrix encryption in this paper.

Rogaway, Phillip, *OCB: Documentation*, https://www.cs.ucdavis.edu/~rogaway/ocb/ocb-doc.htm. This website provides a list of papers by Phillip Rogaway on Offset Codebook mode.

Stallings, William, "NIST Block Cipher Modes of Operation for Confidentiality," *Cryptologia*, Vol. 34, No 2, April 2010, pp. 163–175.

Stallings, William, "NIST Block Cipher Modes of Operation for Authentication and Combined Confidentiality and Authentication," *Cryptologia*, Vol. 34, No. 3, July 2010, pp. 225–235.

Stallings, William, "The offset codebook (OCB) block cipher mode of operation for authenticated encryption," *Cryptologia*, Vol. 42, No. 2, March 2018, pp. 135–145.

Chapter 14

The Birth of Public Key Cryptography

A major problem with all of the methods of encipherment examined thus far is that the sender and receiver must agree on a key prior to the creation and delivery of the message. This is often inconvenient or impossible. There were some failed attempts to overcome this problem before an elegant solution was found. One is detailed below, before examining current solutions.

14.1 A Revolutionary Cryptologist

During the American Revolutionary War, James Lovell attempted to deal with the problem within the context of a cipher system of his own creation.[1] His system was similar to the Vigenère cipher and is best explained through an example. Suppose our message is I HAVE NOT YET BEGUN TO FIGHT and the key is WIN. We form three alphabets by continuing from each of the three key letters like so:

1	W	I	N		15	J	W	A
2	X	J	O		16	K	X	B
3	Y	K	P		17	L	Y	C
4	Z	L	Q		18	M	Z	D
5	&	M	R		19	N	&	E
6	A	N	S		20	O	A	F
7	B	O	T		21	P	B	G
8	C	P	U		22	Q	C	H
9	D	Q	V		23	R	D	I
10	E	R	W		24	S	E	J
11	F	S	X		25	T	F	K
12	G	T	Y		26	U	G	L
13	H	U	Z		27	V	H	M
14	I	V	&					

---

[1] Weber, Ralph E., "James Lovell and Secret Ciphers During the American Revolution," *Cryptologia*, Vol. 2, No. 1, January 1978, pp. 75–88.

Notice that Lovell included & as one of the characters in his alphabet.

The first letter of our message is I, so we look for I in the column headed by W (our first alphabet). It is found in position 14, so our ciphertext begins with 14. Our next plaintext letter is H. We look for H in the column headed by I (our second alphabet) and find it in position 27. Thus, 27 is our next ciphertext number. Then we come to plaintext letter A. Looking at our third alphabet, we find A in position 15. So far, our ciphertext is 14 27 15. We've now used all three of our alphabets, so for the fourth plaintext letter we start over with the first alphabet, in the same manner as the alphabets repeat in the Vigenère cipher. The complete ciphertext is 14 27 15 27 24 1 20 12 12 10 12 16 10 26 8 19 12 2 11 1 21 13 12.

Lovell explained this system to John Adams and Ben Franklin and attempted to communicate with them using it. He avoided having to agree on keys ahead of time by prefacing his ciphertexts with clues to the key, such as "You begin your Alphabets by the first 3 letters of the name of that family in Charleston, whose Nephew rode in Company with you from this City to Boston."[2] Thus, for every message Lovell sent using this scheme, if a key hadn't been agreed on ahead of time, he had to think of some bit of knowledge that he and the recipient shared that could be hinted at without allowing an interceptor to determine the answer. This may have typically taken longer than enciphering! Also, Lovell's cipher seems to have been too complicated for Adams and Franklin even without the problem of key recovery, as both failed to read messages Lovell sent. Abigail Adams was even moved to write Lovell in 1780, "I hate a cipher of any kind."[3]

14.2 Diffie–Hellman Key Exchange

In order to find a satisfactory solution to the problem of key exchange, we must jump ahead from the Revolutionary War to America's bicentennial.[4] The early (failed) attempt to address the problem was presented first to give a greater appreciation for the elegance of the solutions that were eventually discovered.

In 1976, Whitfield Diffie (Figure 14.1) and Martin Hellman (Figure 14.2) presented their solution in a wonderfully clear paper titled "New Directions in Cryptography."[5] Neal Koblitz, whom we will meet later in this book, described Diffie as "a brilliant, offbeat and unpredictable libertarian" and the paper whose results are detailed below as "the most famous paper in the history of cryptography."[6]

Hellman's home page[7] informs the reader that "he enjoys people, soaring, speed skating and hiking." Following the links also reveals a passionate (and well-reasoned!) effort, sustained for over

---

[2] Weber, Ralph E., "James Lovell and Secret Ciphers During the American Revolution," *Cryptologia*, Vol. 2, No. 1, January, 1978, pp. 75–88, p. 83 cited here.

[3] Weber, Ralph E., "James Lovell and Secret Ciphers During the American Revolution," *Cryptologia*, Vol. 2, No. 1, January, 1978, pp. 75–88.

[4] The *idea* of public-key cryptography was presented in an earlier paper by Diffie and Hellman titled "Multiuser Cryptographic Techniques." An actual method wasn't yet ready at the time of that publication. The authors wrote "At present, we have neither a proof that public key systems exist, nor a demonstration system." Ralph Merkle, an undergrad at Berkeley had found a system in 1974, but it lacked elegance. It is of high historical value, yet it never found use. It's described along with a second method (knapsack) developed by Merkle in Chapter 16.

[5] Diffie, Whitfield and Hellman, Martin, "New Directions in Cryptography," *IEEE Transactions on Information Theory*, Vol. IT-22, No. 6, November 1976, pp. 644–654.

[6] Koblitz, Neal, *Random Curves: Journeys of a Mathematician*, Springer, New York, 2007, pp. 301–302.

[7] http://www-ee.stanford.edu/~hellman/.

Figure 14.1   Whitfield Diffie (1944–). (Courtesy of Whitfield Diffie.)

Figure 14.2   Martin Hellman (1945–). (Photograph supplied by Martin Hellman, who dates it as 1976 plus or minus a couple of years.)

a quarter century by Hellman, against war and the threat of nuclear annihilation. He traced his interest in cryptography to three main sources, one of which was the appearance of David Kahn's *The Codebreakers*.[8] As this book was also an influence on Diffie (and many others!), its importance in the field is hard to overestimate. The scheme developed by Diffie and Hellman works as follows.

Alice and Bob must first agree on a prime number (p) and a generator (g) of the multiplicative group of units modulo p. If you haven't yet had an abstract algebra course, the concept of a

---

[8]   Hellman, Martin E., *Work on Cryptography*, http://www-ee.stanford.edu/~hellman/crypto.html.

generator is probably new to you. It is simply a number that, when raised to consecutive powers (mod p), results in the entire group being generated. That's why it's called a *generator* and often denoted by the letter g. If someone is eavesdropping on the line and obtains the values p and g, that's okay! After Alice and Bob agree on these numbers, each person (except the eavesdropper!) selects another number. Let's say Alice chooses x and Bob chooses y. They keep *these* values secret. Alice calculates g^x (mod p) and Bob calculates g^y (mod p). They then exchange these new values. Alice would have great difficulty calculating y from g^y. Similarly, Bob is unlikely to be able to determine Alice's x. However, both can form g^{xy}. Alice simply raises the number Bob sent her to the power of x and Bob raises the number Alice sent him to the power of y. The eavesdropper cannot do this as she knows neither x nor y. She may have intercepted g, p, g^x, and g^y, but this won't help her to find g^{xy}. Thus, Alice and Bob now share a secret number. This will serve as their key for future communication, using whatever system they choose.

When g, p, and g^x (mod p) are known, but not x, determining x is known as the *discrete log problem*. The security of Diffie–Hellman key exchange sounds like it should be equivalent to this problem, but this has yet to be proven. To recap, the two problems are:

1. Discrete Log: Given g^x, find g.
2. Diffie–Hellman: Given g^x and g^y, find g^{xy}.

It is possible that someone may find a way to defeat the key exchange (problem 2) without solving the discrete log problem (problem 1). This is a very big open problem. More will be said about the difficulty of the discrete log problem in Section 16.8, which follows a discussion of complexity theory. In any case, we have a lovely system built on nothing more complicated than the laws of exponents and the idea of remainder arithmetic!

Malcolm Williamson discovered Diffie–Hellman key exchange before the individuals it is named after, but he was employed by GCHQ at the time, and his work was classified. It was only released in 1997.

Implemented as described above, the key exchange requires several messages be sent. If we are willing to do this, even some of the cryptosystems discussed earlier in this text may be used to securely send a message to someone with whom a key exchange has not taken place. An analogy will make the process clear, after which it will be described more mathematically.

Alice can send Bob a message she places in a box and secures with a padlock. Bob doesn't try to open it, because he doesn't have the key. Instead, he simply adds his own padlock and mails it back to Alice. Anyone wishing to see the message at this stage must be able to remove both padlocks. Alice can only remove her own, which she does. She then mails it back to Bob, who removes his padlock and reads the message.

Now for the mathematical version of the physical process described above:

- Let E_A, E_B, D_A, and D_B represent the enciphering and deciphering algorithms (along with the keys A and B) used by Alice and Bob.
- Alice sends Bob $E_A(M)$.
- Bob sends Alice $E_B(E_A(M))$.
- Alice sends Bob $D_A(E_B(E_A(M)))$

Note: If D_A and E_B commute, D_A and E_A will then cancel out and Alice's final message will amount to $E_B(M)$

- Bob then applies D_B to Alice's last message and reads the plaintext.

Commutativity is important here! This process won't work without it.

Diffie and Hellman obviously realized the great importance of their paper (hence, the title[9]), but they also realized that theirs was far from the last word on the topic; the paper included the line "We propose some techniques for developing public key cryptosystems, but the problem is still largely open."[10]

All of the systems that we have looked at up to this point are examples of *symmetric* algorithms; that is, the decryption key is the same or can be easily calculated from the encryption key (as in matrix encryption). More often than not, in classical cryptography the two keys are identical. The Diffie–Hellman scheme detailed above results in such a key being generated, even though it is done openly. That this could be done was revolutionary and inspired others to search for more such techniques. Although there are several "public-key" algorithms, only one will be described in this chapter; others will be seen later.[11] What they all have in common is *asymmetric keys*. That is the enciphering and deciphering keys differ and it is not feasible to calculate one from the other.

14.3 RSA: A Solution from MIT

Figure 14.3 Adi Shamir (1952–), Ron Rivest (1947–), and Len Adleman (1945–). Between the necks of Rivest and Adleman, the chalkboard shows ∴P = NP. (Courtesy of Len Adleman, https://web.archive.org/web/20160305203545/http://www.usc.edu/dept/molecular-science/RSApics.htm.)

Three MIT professors, Ron Rivest, Adi Shamir, and Len Adleman (Figure 14.3), developed the (asymmetric) public key cryptosystem now known as RSA (after their last initials). It doesn't require any messages to be sent prior to the intended ciphertext. With RSA the enciphering key

---

[9] Levy, Steven, *Crypto: How the Code Rebels Beat the Government, Saving Privacy in the Digital Age*, Viking, New York, 2001, p. 88. Levy points out that the title of the paper brings to mind the "mind-blowing paperbacks of the New Directions publishing house," which issued *Waiting for Godot*, *Siddhartha*, and *In the American Grain*.

[10] Diffie, Whitfield and Hellman, Martin, *New Directions in Cryptography*, IEEE Transactions on Information Theory, Vol. IT-22, No. 6, November 1976, pp. 644–654, p. 644 cited here.

[11] Knapsack systems are discussed in Chapter 16.

for any recipient can be made public and his deciphering key cannot be obtained from it without extremely time consuming calculations.

The creators of RSA did not have immediate success in their attack on the problem. In fact, they were able to break their first 32 attempts themselves.[12] Finally, three years after beginning, a workable scheme was found. There may have been an earlier breakthrough, as the following anecdote from a ski trip the researchers went on suggests.

> On only the second day that the Israeli [Shamir] had ever been on skis, he felt he'd cracked the problem. "I was going downhill and all of a sudden I had the most remarkable new scheme," he later recalled. "I was so excited that I left my skis behind as I went downhill. Then I left my pole. And suddenly... *I couldn't remember what the scheme was.*" To this day he does not know if a brilliant, still-undiscovered cryptosystem was abandoned at Killington.[13]

The details of RSA will soon be laid out, but first, we must review some (old!) results from number theory.

14.3.1 Fermat's Little Theorem (1640)

If p is a prime and a is not a multiple of p, then $a^{p-1} = 1 \pmod{p}$.

Fermat stated the theorem in a different form and did not offer a proof. It is presented here in a manner that will prove useful, but it first needs to be generalized a bit. This was done by Leonhard Euler (with proof!) in 1760.[14] Euler's generalization may be stated tersely as

$$(m,n) = 1 \Rightarrow m^{\varphi(n)} = 1 \pmod{n}$$

The notation $(m, n) = 1$ means that m and n are relatively prime; that is, their greatest common divisor is 1. It is sometimes written using the more descriptive notation $\gcd(m, n) = 1$. The exponent $\varphi(n)$ is called the "Euler φ function" with φ pronounced as *fee* rather than *figh*. It's also sometimes referred to as Euler's totient function.[15] However you read it, $\varphi(n)$ is defined to be the number of positive integers less than n and relatively prime to n. For example, $\varphi(8) = 4$, because 1, 3, 5, and 7 are the only positive integers less than 8 that have no positive factors in common with 8 other than 1. It is easy to see that $\varphi(p) = p - 1$ if p is a prime. Hence, Fermat's little theorem is just a special case of Euler's theorem.

---

[12] Levy, Steven, *Crypto: How the Code Rebels Beat the Government, Saving Privacy in the Digital Age*, Viking, New York, 2001, p. 97.

[13] Levy, Steven, *Crypto: How the Code Rebels Beat the Government, Saving Privacy in the Digital Age*, Viking, New York, 2001, p. 96. Was this what prompted Hughes-Hallett et al. to put a skier on the cover of their calculus book? An attempt to jog Shamir's memory? They even used the first person perspective as if attempting to recreate Shamir's field of vision at the moment that the great idea was passing through his mind!

[14] Sandifer, C. Edward, *The Early Mathematics of Leonhard Euler*, Mathematical Association of America, Washington, DC, 2007, p. 203. Euler's phi function was used at this time, but not given a name until three years later (see the next footnote). Carl Friedrich Gauss introduced the "mod n" notation, the congruence symbol, and φ in 1801.

[15] *Totient* is from Latin and means "to count." Euler named this function in Euler, Leonhard, "Theoremata Arithmetica Nova Methodo Demonstrata," *Novi Commentarii academiae scientiarum Petropolitanae*, Vol. 8, 1763, pp. 74–104, reprinted in *Opera Omnia*, Series 1, Vol. 2, B. G. Teübner, Leipzig, 1915, pp. 531–555.

Proof

Observing that the multiplicative group modulo n has order $\varphi(n)$, and recalling that the order of a group element must divide the order of the group, we see that $m^{\varphi(n)} = 1$ (mod n). The requirement that $(m, n) = 1$ is necessary to guarantee that m is invertible modulo n or, in other words, to ensure that m is an element of the group of units modulo n.

Although Fermat and Euler had not realized it, their work would find application in cryptography, making RSA possible. The eventual applications of once pure mathematics are typically impossible to predict. To see how RSA works, we first multiply both sides of Euler's equation by m to get:

$$m^{\varphi(n)+1} = m \text{ (mod } n).$$

Any message can easily be converted to blocks of numbers. We could, for example, replace each letter with its numerical value A = 01, B = 02, ..., Z = 26, where the leading zeros eliminate ambiguity when the values are run together. Or we may replace each character with its ASCII representation in bits for a base 2 number (which could be converted to base 10). So, let m denote a block of text in some numerical form. Choose a positive integer e such that $(e, \varphi(n)) = 1$. We can then compute d, the multiplicative inverse of e (mod $\varphi(n)$). That is, the product ed will be one more than a multiple of $\varphi(n)$. The result is $m^{ed} = m$ (mod n).

If I want to be able to receive an enciphered message from someone, we don't have to secretly meet to exchange keys. I simply publicly reveal my values for e and n. Anyone who wants to send me a message m can compute and send m^e (mod n). All I have to do is raise this value to the d power and mod out again: $(m^e)^d = m^{ed} = m$ (mod n). I get the original message back. The idea is that I can tell everyone e and n, allowing them to send messages, but it will be very hard for them to compute d; hence, only I can read the messages.

We now take a look at a toy example. The modulus is not large enough to offer any security, but it does help illustrate the key steps.

The first thing Alice must do, if she is to receive RSA enciphered messages, is generate the keys. She starts by selecting two primes, $p = 937$ and $q = 1,069$. Thus, her modulus is $n = pq = 1,001,653$. Next, Alice needs to come up with an enciphering exponent e and a deciphering exponent d. Not just any value will do for e. If e fails to have an inverse modulo $\varphi(n)$, she will not be able to decipher the messages she receives uniquely! For e to be invertible modulo $\varphi(n)$, it must be relatively prime to $\varphi(n)$. Because, $n = pq$, where p and q are distinct primes, we have $\varphi(n) = (p - 1)(q - 1) = (936)$ $(1,068) = 999,648$.[16] Alice tries $e = 125$. To be sure that 125 and 999,648 have no common divisors greater than 1, we may apply the Euclidean algorithm. This is a simple procedure used to calculate gcd values. It does, in fact, go all the way back to Euclid (c. 371–285 BCE).[17]

14.3.2 The Euclidean Algorithm

We begin by dividing the smaller number into the larger, noting the quotient and remainder:

$$999,648 = 125(7,997) + 23$$

---

[16] Proving $\varphi(n) = (p - 1)(q - 1)$, when n is the product of two distinct primes, p and q, is left as an exercise (with some hints). See the online exercises. All exercises are available online only.

[17] Dates are according to Wikipedia; however, http://www.gap-system.org/~history/Biographies/Euclid.html gives (c. 325–265 BCE), but then, both are indicated to be estimates!

We then repeat the procedure using 125 and 23 in place of 999,648 and 125:

$$125 = 23(5) + 10$$

Repeating the process gives:

$$23 = 10(2) + 3$$

And again:

$$10 = 3(3) + 1$$

One more iteration yields a remainder of 0:

$$3 = 1(3) + 0$$

This algorithm may require more or fewer iterations for another pair of numbers, but no matter what values are investigated the algorithm always ends with a remainder of 0 and the last nonzero remainder is the gcd. That we obtained 1 shows the two numbers are relatively prime, so 125 was a good choice for the enciphering key. It is invertible modulo 999,648.

Now that Alice has her enciphering exponent (part of her *public* key), how can she compute her deciphering exponent (her *private* key)?

Rewriting the equations we obtained from the Euclidean algorithm allows us to express the gcd as a linear combination of the two numbers that yielded it. We rewrite the equations as:

$$1 = 10 - (3)(3) \tag{14.1}$$

$$3 = 23 - (10)(2) \tag{14.2}$$

$$10 = 125 - (23)(5) \tag{14.3}$$

$$23 = 999,648 - (125)(7,997) \tag{14.4}$$

Substituting for 3 in Equation 14.1 with the value given by Equation 14.2 gives:

$$1 = 10 - (23 - (10)(2))(3) = 10 - (23)(3) + (10)(6)$$

We now use Equation 14.3 above to substitute for the 10 (twice) to get:

$$1 = 10 - (23)(3) + (10)(6) = 125 - (23)(5) - (23)(3) + [(125) - (23)(5)](6)$$

Collecting all multiples of 125 and 23 gives:

$$1 = 125(7) - 23(38)$$

Finally we substitute for 23 using Equation 14.4 and collect multiples of 125 and 999,648:

$$1 = 125(7) - 23(38) = 125(7) - [999,648 - (125)(7,997)](38)$$

$$1 = 125(303,893) - 999,648(38)$$

Thus, $999,648(-38) + 125(303,893) = 1$.

The procedure detailed above is referred to as the *extended Euclidean algorithm*. This may seem like another digression, but it isn't! If we now mod out both sides by 999,648, we get

0 + 125(303,893) = 1; that is, 125(303,893) = 1 (modulo 999,648), so 303,893 is the inverse we were looking for.

Alice can now post the pair $e = 125$, $n = 1,001,653$ on her website and wait for Bob to send a message. She keeps her deciphering exponent $d = 303,893$ secret.

Once Bob sees Alice's public key, he can begin enciphering his message. For the plaintext, he uses a quote from Martin Hellman:[18]

```
Nuclear war is not war. It is suicide and genocide rolled into one.
```

Ignoring case, he converts the characters to numbers using the following scheme.

A	B	C	D	E	F	G	H	I	J	K	L	M	N	O	P	Q	R	S	T	U	V	W	X	Y	Z
01	02	03	04	05	06	07	08	09	10	11	12	13	14	15	16	17	18	19	20	21	22	23	24	25	26

He gets

```
142103120501182301180919141520230118092009191921090309040501 14
040705141503090405181512120504091420151514 05
```

Because the modulus is only a little over a million, the text must be split into pieces less than a million; otherwise the encryption could fail to be one-to-one. That is, a given ciphertext block could have more than one possible decipherment. Groups of six numbers will do; thus, we have

```
142103  120501  182301  180919  141520
230118  092009  191921  090309  040501
140407  051415  030904  051815  121205
040914  201515  1405
```

The fact that the last number is only four digits does not matter, nor do leading zeros in the seventh and other number groups.

To get the final ciphertext, Bob simply raises each number to the enciphering exponent, 125, and reduces the answer modulo 1,001,653. He gets

```
753502  318690  768006  788338  100328
180627  202068  566203  925899  520764
026563  950434  546436  025305  256706
218585  831923  414714
```

It should be noted that it is not guaranteed that all ciphertext blocks will be nice six-digit numbers (after adding leading zeros when necessary). Moding out by 1,001,653 could yield a remainder larger than 999,999, which would then require 7 digits. If the ciphertext were run together with the occasional block of length 7, decipherment would be ambiguous. Alice wouldn't know where to insert the breaks (i.e., when to take six digits and when to take seven). The problem could be resolved by writing all ciphertext numbers using seven digits. For the example above (because we never got a number in excess of 999,999), all of the ciphertext blocks would begin with a leading zero.

---

[18] The plaintext was taken from *The Nazi Within*, an essay by Martin E. Hellman, which can be found online at http://www-ee.stanford.edu/~hellman/opinion/bitburg.html

If one can factor the modulus, RSA encryption is easy to break, but the security of RSA is not known to be equivalent to factoring. There may be a way to break RSA by some other means that would not reveal the factorization of n. This is an important open problem. So, as with Diffie–Hellman key exchange, security is based on, if not equivalent to, the inability of an attacker to solve a mathematical problem (factoring for RSA and the discrete log problem for Diffie–Hellman). Other public key systems are built on other supposedly hard problems. There will be more on this in later chapters. It's possible that someone could come up with a quick solution for one (or all) of these problems tomorrow or that someone already has and we just don't know it!

The RSA algorithm has an important practical drawback—it is slow. We'll take a quick look at a good way to compute powers, but it is still tedious compared to extremely fast operations like XORing used in other encryption algorithms. Thus, RSA is typically used only to encipher a key for some symmetric system (like DES). The (enciphered) key is then sent along with a message enciphered with that key in the symmetric system. This will be discussed in greater detail in Chapter 18. Because RSA is just applied to the relatively short (compared to most messages) key, the delay is tolerable!

Even if we are only enciphering a tiny amount of text with RSA, we'd still like to do it as efficiently as possible. The technique of repeated squaring, demonstrated below, is a nice way to carry out the exponentiation. As an example, consider 385^{1563} (mod 391). Obviously this is too small of a modulus, but the same technique applies for other values and smaller numbers make for clearer explanations. Naturally, we do not wish to repeatedly multiply by 385, moding out as we go, 1,563 times. Instead, we calculate the following squares:

$$385^2 = 148,225 = 36 (\text{mod } 391)$$

Squaring both sides of the equation above and reducing gives:

$$(385^2)^2 = 36^2 = 1,296 = 123 (\text{mod } 391)$$

Thus, $385^4 = 123$ (mod 391).
We square both sides again to get:

$$385^8 = 123^2 = 15,129 = 271 (\text{mod } 391)$$

Continuing as before, we obtain:

$$385^{16} = 271^2 = 73,441 = 324 \ (\text{mod } 391)$$

$$385^{32} = 324^2 = 104,976 = 188 \ (\text{mod } 391)$$

$$385^{64} = 188^2 = 35,344 = 154 \ (\text{mod } 391)$$

$$385^{128} = 154^2 = 23,716 = 256 \ (\text{mod } 391)$$

$$385^{256} = 256^2 = 65,536 = 239 \ (\text{mod } 391)$$

$$385^{512} = 239^2 = 57,121 = 35 \ (\text{mod } 391)$$

$$385^{1024} = 35^2 = 1,225 = 52 \ (\text{mod } 391)$$

We may stop here. We only need to go up to the power of 2 closest to the desired power (1,563 in this example) without exceeding it.

Now, to compute 385^{1563}, we observe that $1,563 = 1024 + 512 + 16 + 8 + 2 + 1$, so we may write:

$$385^{1563} = (385^{1024})(385^{512})(385^{16})(385^{8})(385^{2})(385) \pmod{391}$$

Substituting the values obtained above:

$$385^{1563} = (52)(35)(324)(271)(36)(385) = 2,214,873,460,800 = 63 \pmod{391}$$

Thus, our answer is 63.

14.4 Government Control of Cryptologic Research

Martin Gardner, a prolific popularizer of mathematics, described the RSA scheme in his "Mathematical Games" column in the August 1977 issue of *Scientific American*. The formal paper by Rivest, Shamir, and Adleman was not slated to appear in *Communications of the ACM* until the late fall of 1977; however, readers of Gardner's column didn't expect to have to wait so long for the full details. Gardner wrote "Their work, supported by grants from the NSF and the Office of Naval Research, appears in "On Digital Signatures and Public-Key Cryptosystems" (Technical Memo 82, April, 1977), issued by the Laboratory for Computer Science, Massachusetts Institute of Technology, 545 Technology Square, Cambridge, Mass. 02139. The memorandum is free to anyone who writes Rivest at the above address enclosing a self-addressed, 9-by-12-inch clasp envelope with 35 cents in postage."[19]

Rivest quickly received 4,000 requests for the paper,[20] and eventually 7,000 requests were piled up in Shamir's office.[21] The papers were finally mailed in December 1977. Meanwhile, Gus Simmons and Mike Norris had a paper published in the October 1977 issue of *Cryptologia* in which they improved upon the system.[22] Why it took so long for the RSA team to mail out their paper (and get it published in a journal) is not simply due to the volume of requests.

August 1977 is also marked in the history of the politics of cryptology by a letter received by the Institute of Electrical and Electronics Engineers (IEEE) Information Theory Group. The letter claimed that publications dealing with cryptology could be in violation of the 1954 Munitions Control Act, the Arms Export Control Act, and the International Traffic in Arms Regulations (ITAR). In other words, cryptologic research was restricted in the same manner as work dealing

---

[19] Gardner, Martin, "Mathematical Games, A New Kind of Cipher that Would Take Millions of Years to Break," *Scientific American*, Vol. 237, No. 2, August 1977, pp. 120–124.

[20] Deavours, Cipher, "A Special Status Report The Ithaca Connection: Computer Cryptography in the Making," *Cryptologia*, Vol 1, No. 4, October 1977, pp. 312–317, p. 312 cited here.

[21] Levy, Steven, *Crypto: How the Code Rebels Beat the Government, Saving Privacy in the Digital Age*, Viking, New York, 2001, p. 114.

[22] Simmons, Gustavus J. and Michael J. Norris, "Preliminary Comments on the M.I.T. Public-Key Cryptosystem," *Cryptologia*, Vol. 1, No. 4, October 1977, pp. 406–414. The improvement consisted of showing how poor choices for the key can allow message recovery without factoring. Over the years a number of special cases that need to be avoided have been recognized. These are detailed in Section 15.1.

with nuclear weapons! Indeed, this was made explicit in the letter: "[A]tomic weapons and cryptology are also covered by special secrecy laws."[23]

The author's provocation appears to have been a symposium on cryptology that the IEEE had slated for October 10, 1977, at a conference at Cornell University in Ithaca, New York. Hellman, Rivest, and others were scheduled as speakers. The letter was reported on in *Science*, *The New York Times*, and elsewhere, generating a great deal of publicity for the conference.[24]

Basically the claim was that such publications and lectures should first be cleared by the National Security Agency (NSA). If this sort of voluntary censorship could be achieved, the constitutionality of the matter need not arise. The full letter is provided in Figure 14.4.[25]

It was eventually revealed by an NSA spokesman that the letter was written (unofficially) by Joseph A. Meyer (Figure 14.5), an NSA employee.[26]

This wasn't the first time Meyer had angered supporters of civil liberties. Paul Dickson's *The Electronic Battlefield* (1976) summarizes a January 1971 article "Crime Deterrent Transponder System" by Meyer in *IEEE Transactions on Aerospace and Electronics Systems*:[27]

> The article by Joseph Meyer, an engineer in the employ of the National Security Agency, recommends a system in which tiny electronic tracking devices (transponders) are attached to those 20 million Americans who have been in trouble with the law—in fact, wearing one of the devices would be a condition of parole. The transponders would be linked by radio to a computer that would monitor the wearer's (or "subscriber's" in the words of Meyer) location and beep out a warning when the person in question was about to violate his territorial or curfew restrictions. In addition, these little boxes would be attached to people in such a manner that they could not be removed without the computer taking note of the act. Taking off or tinkering with your transponder would, in Meyer's world, be a felony. Good engineer that he is, Meyer has also thought out some of the other applications of these portable units, which include monitoring aliens and political minorities. Robert Barkan, the writer who first brought the Meyer proposal and other such ideas to a broader audience through his articles, had this to say about the transponder system in *The Guardian*, "'1984' is still fiction, but no longer *science* fiction. The technology of the police state is ready. All that remains is for the government to implement it."

Many of the requests generated by Martin Gardner's column for the RSA paper came from outside the United States. In light of Meyer's letter, Rivest remarked, "If I were more of a skeptic, I'd think I was being set up."[28]

---

[23] Levy, Steven, *Crypto: How the Code Rebels Beat the Government, Saving Privacy in the Digital Age*, Viking, New York, 2001, p. 109.

[24] Shapley, Deborah and Gina Bari Kolata, "Cryptology: Scientists Puzzle Over Threat to Open Research, Publication," *Science*, Vol. 197, No. 4311, September 30, 1977, pp. 1345–1349; Browne, Malcolm W., "Harassment Alleged over Code Research," *The New York Times*, October 19, 1977, available online at https://www.nytimes.com/1977/10/19/archives/harassment-alleged-over-code-research-computer-scientists-say-us.html.

[25] My attempt to find the full letter led to John Young, who runs www.cryptome.org. He was able to obtain it from Martin Hellman and posted it on his website in January 2010.

[26] Foerstel, Herbert N., *Secret Science: Federal Control of American Science and Technology*, Praeger, Westport, Connecticut, 1993, p. 114.

[27] Dickson, Paul, *The Electronic Battlefield*, Indiana University Press, Bloomington, Indiana, 1976, p. 141.

[28] Shapley, Deborah and Gina Bari Kolata, "Cryptology: Scientists Puzzle Over Threat to Open Research, Publication," *Science*, Vol. 197, No. 4311, September 30, 1977, pp. 1345–1349.

JUL 11 '977

IEEE · · · · · · ·
NE\.

J. A. Meyer
5600 Namakagan Rd.
Bethesda, Md. 20016

7 July 77

Mr. E. K. Gannet
Staff Secretary, IEEE Publications Board
IEEE Hq.
345 East 47th Street
New York, N.Y. 10017

Dear Mr. Gannet,

I have noticed in the past months that various IEEE Groups have been publishing and exporting technical articles on encryption and cryptology --- a technical field which is covered by Federal Regulations, viz: ITAR (International Traffic in Arms Regulations, 22 CFR 121-128). I assume that the IEEE Groups are unfamiliar with the ITAR, which apply to publication and export of unclassified as well as classified technical data, and I thought I would draw your attention to them. I have enclosed a few pages of the ITAR which are pertinent.

The key points of ITAR are that unclassified technical data are covered (22 CFR 125.01). All forms of export, including publications and symposia, are covered (22 CFR 125.03). Licences are required unless the material is exempted (22 CFR 125.04). Prior approval by a cognizant government agency is required before publication within the U.S. (22 CFR 125.11, footnote 3). Encryption and cryptologic and related systems are covered by ITAR (Categories XI(c), XIII(b)). The regulations are issued under law and hence have force of law (Mutual Security Act of 1954, Section 414 = 22 USC 1934).

Although ITAR covers a very wide range of weapons technologies, atomic weapons and cryptology are also covered by special secrecy laws (42 USC 2274-77 and 18 USC 798), an indication of the importance of these technologies.

The June 1977 Information Theory Group Newsletter contains minutes of a meeting 19 Oct 76 at which it was proposed that the IT Group become an advisor to NBS on cryptologic secrecy and security, and this led to a call for papers on encryption for the 1977 International Symposium at Ithaca. The reason given was that NBS had only one Federal agency to refer to for cryptologic advice. However, Executive Order 11905 defined that consolidation as government policy. The IT Group seems active in cryptology, and have published several papers in the Nov 76 and May 77 Transactions-IT. The June 1977 issue of Computer also had an article in the same technologic area. One of the papers was presented at an IEEE symposium at Ronneby, Sweden. Several papers on encryption were given at ICC-77. A paper on speech scramblers was given at a VTG meeting at Orlando, this Spring. Another paper on speech scramblers is scheduled for the Cybernetics meeting in September in Washington, D.C.. The International Symposium on Information Theory at Ithaca in October 1977 will have more papers on encryption. The Facilitator for the IEEE-USSR IT exchange program, Prof. Ephremides, declared on page 7 of the June 77 ITC newsletter (enclosed) that he would forward preprints of new work directly to the USSR in accord with an IEEE-USSR agreement. If any technical papers on encryption or cryptology are sent to USSR before they have been published, a difficulty could arise because, according to ITAR, an export licence is required (22 CFR 125.04). Apparently at Ronneby, Sweden this formality was skipped.

Superficially it appears that a small number of authors are providing most of the papers, and most of the motivation. They may not be aware of the full burden of government controls. Some of the topics addressed, e.g. the DES algorithm, are intended for U.S. government activities and hence may be covered by 18 USC 798 as well as by ITAR. Unless clearances or export licences are obtained from the State Department, or there is some special exemption, the IEEE could find itself in possible technical violation of the ITAR (22 USC 1934(c)). As an IEEE member, I suggest that IEEE might wish to review this situation, for these modern weapons technologies, uncontrollably disseminated, could have more than academic effect.

Yours faithfully,

J. A. Meyer

1684794 M

Figure 14.4   Letter received by the IEEE Information Theory Group.

Figure 14.5   Joseph A. Meyer. (Adapted from Meyer, Joseph A., "Crime deterrent transponder system," *IEEE Transactions on Aerospace and Electronics Systems,* **Vol. AES-7, No. 1, January 1971, pp. 2-22, p. 22 cited here. With permission from IEEE.)**

Hellman explained how he dealt with the threat:

> On the advice of Stanford's general counsel, I even presented two papers at a 1977 symposium at Cornell University, instead of my usual practice of having the student co-authors do the presentations. The attorney told me that if the ITAR were interpreted broadly enough to include our papers, he believed they were unconstitutional. But a court case could drag on for years, severely hindering a new Ph.D.'s career (especially if the attorney's belief was not shared by the jury), whereas I was already a tenured professor.
>
> I presented these thoughts to Ralph Merkle and Steve Pohlig, the students in question, but left the final decision to them. Initially they wanted to take the risk and give the papers, but eventually concern from their parents won out. Fortunately, the presentations went off without incident, though it was dramatic having Ralph and Steve stand mute by the podium, so they would get the recognition they deserved, as I gave the papers.[29]

Although not scheduled to present, Diffie made a point of delivering a talk at an informal session, showing that he wasn't easily intimidated.[30]

---

[29] Hellman, Martin E., *Work on Cryptography,* http://www-ee.stanford.edu/~hellman/crypto.html.
[30] Levy, Steven, *Crypto: How the Code Rebels Beat the Government, Saving Privacy in the Digital Age,* Viking, New York, 2001, p. 113.

DR. DWIVEDI

THE **I**NSTITUTE OF
ELECTRICAL AND
ELECTRONICS
ENGINEERS, INC.

345 EAST 47th STREET, NEW YORK, NEW YORK 10017

DIRECT NUMBER
(212) 644- 7548

July 20, 1977

Mr. J. A. Meyer
5600 Namakagan Road
Bethesda, MD 20016

Dear Mr. Meyer:

I appreciate your calling my attention to the current International Traffic in Arms
Regulations (ITAR) as they relate to the exporting of unclassified technical data
outside the U. S. We had looked into this question about 10 years ago, and I was
glad to have the opportunity you gave us to confirm that our basic exemption from
these regulations has not changed since then.

All IEEE conference publications and journals are exempted from export license
requirements under Section 125.11(a)(1). In addition, footnote 3 to that Section
places the burden of obtaining any required Government approval for publication of
technical data on the person or company seeking publication.

It appears to me that the forwarding of preprints to the USSR under the IEEE-USSR IT
exchange program is the one activity that needs to be examined further in the light
of the ITAR regulations. I am therefore bringing your letter and attachments to the
attention of potentially interested parties within the IEEE to review this further.

Again, I am extremely grateful to you for bringing this potentially important question
to our attention.

Sincerely yours,

E. K. Gannett
Staff Director
Publishing Services

EKG/mk
bcc: Dr. N. P. Dwivedi
Dr. R. M. Emberson

Figure 14.6 *(Continued)* **Responses to Meyer's letter.**

Did the reaction of the IEEE, as an organization, mirror that of the individuals discussed above? To see the official reaction, look at Figure 14.6 on this and the next two pages.[31] The first is a reply to the Meyer letter that started it all.

The combination of the cryptographers' brave stance and the fact that Meyer's claims were not actually backed up by the ITAR, which contained an exemption for published material, resulted in no real changes in the manner in which cryptologic research was now being carried out—publicly. Indeed, the publicity the letter generated served to *promote* public interest! The threats did, however, delay the RSA team's distribution of their paper. As mentioned earlier, Simmons and Norris described it in the pages of *Cryptologia* before the technical report was mailed to the *Scientific American* readers or appeared in a journal. Ironically, prior to all of this, Simmons, who managed the Applied Mathematics Department at Sandia Laboratories, attempted to hire Rivest,

---

[31] These letters, like the original Meyer letter, were obtained from Martin Hellman via John Young.

THE **I**NSTITUTE OF
ELECTRICAL AND
ELECTRONICS
ENGINEERS, INC.

345 EAST 47th STREET, NEW YORK, NEW YORK 10017

DR. NARENDRA P. DWIVEDI
Director of Technical Activities

August 8, 1977

(212) 644-7890

Dr. F. Jeninek
IT Group President
IBM Watson Research Center
Room 14-147, P.O.Box 218
Yorktown Heights, N.Y. 10598

Prof. Anthony Ephemerides
Elec. Engr. Department
University of Maryland
College Park, Md. 20742

Dr. Martin E. Hellman
Bd. of Governors, IT Group
730 Alvarado Court
Stanford, Ca. 94305

Prof. J.L. Massey
Publications, IT Group
Dept. of Elec. Engr.
Massachusetts Institute of Tech.
Cambridge, Mass. 02139

Dr. A.D. Wyner
Nominations Com. IT Group
Bell Telephone Labs.
Room 2-357
600 Mountain Ave.
Murray Hill, N.J. 07974

Dr. M.G. Smith,C-16 President
T.J. Watson Research Lab.
Box 218
Yorktown Heights,N.Y. 10598

SUBJECT: Export Control of Technical Information and Data
           — Encryption, Cryptology, etc.

REF: 1.   International Traffic in Arms (ITAR)
                Regulations, 22 CFR 121-128, excerpts attached

           2.   IT Group Newsletter of June 1977

Dear Friends:

A concerned and good meaning member has drawn our attention to a possible violation
by authors of ITAR regulations in some subjects which can be linked to be of possible
military use. It appears that IEEE and its Groups/Societies/Councils are exempt but
the individuals (and/or their employers) have to watch out. I am enclosing the
correspondence and excerpts of ITAR.

Based on my experience of working with NASA for a decade, I have the following practical suggestions:

1.   If anyone of the authors of a paper is working for a Defense/NASA Con-
           tractor Company, then that author should get the paper cleared as
           "unclassified suitable for foreign publication and presentation".

2.   If the sole author of a paper has no clearing facility with the employing
           institution or is self-employed then the author should refer the paper
           to the Office of Munitions Control, Dept. of State, Washington, D.C. for
           their ruling.

           (I have no personal experience of dealing with them.)

                                                                        (Cont'd)

Figure 14.6 *(Continued)*   **Responses to Meyer's letter.**   *(Continued)*

but Rivest declined because he thought he would have more freedom at MIT! Meyer's letter was
far from the last attempt to control public cryptologic research.[32]

Whitfield Diffie relates:[33]

> A more serious attempt occurred in 1980, when the NSA funded the American
> Council on Education to examine the issue with a view to persuading Congress to give
> it legal control of publications in the field of cryptography. The results fell far short

---

[32] Nor was it the first! Back in 1975, the NSA attempted to intimidate other organizations, such as the National
Science Foundation (NSF), out of providing funding for cryptologic research (for details, see Levy, Steven,
Crypto: How the Code Rebels Beat the Government, Saving Privacy in the Digital Age, Viking, New York, 2001, p.
107). If grant money for important cryptographic work is only available through NSA, censorship is achieved
in a more subtle (but still apparent) manner.

[33] From p. xvii of the foreword by Whitfield Diffie in Schneier, Bruce, *Applied Cryptography*, second edition,
Wiley, New York, 1996.

Messrs. Jeninek, Ephemerides, Hellman
Massey, Wyner and Smith   -2-   August 8, 1977

If you are beginning to feel that it is not always easy to carry out good-intentioned projects, I welcome you to the club and wish you the best.

Yours sincerely,

Narendra P. Dwivedi

Narendra P. Dwivedi

Encl. (as above)
NPD:mgc

cc: Dr. F.H. Blecher, V.P., Technical Activities
    Dr. R.M. Emberson, Acting General Manager
    Mr. E.J. Gannott, Staff Dir., Publishing Services
    Dr. J. Sevick, Chairman, Transnational Relations Com.
    Transnational Relations Com. File
    Dr. R.B. Saunders, President IEEE
    Dr. H. Sherman, Principal Investigator, IEEE/PRC Exchange Program
    Dr. S.S. Yau, Director, Div. V
    Dr. M. Sloan, Governing Bd., C-16
    Dr. T. Kaileth
    Publications File

Figure 14.6 *(Continued)*   Responses to Meyer's letter.

of NSA's ambitions and resulted in a program of voluntary review of cryptographic papers; researchers were requested to ask the NSA's opinion on whether disclosure of results would adversely affect the national interest before publication.

In addition to this sort of intimidation, another attempt at government control of cryptographic research is illustrated by an episode from 1980. On August 14 of that year, Leonard Adleman received a phone call from Bruce Barnes of the National Science Foundation (NSF). Barnes explained that the NSF couldn't fund parts of his grant proposal because of an "interagency matter." A day later, Adleman heard from Vice Admiral Bobby Ray Inman, director of NSA. Inman offered NSA funding for the proposal, but Adleman was disturbed by this turn of events and declined. He remarked, "It's a very frightening collusion between agencies."[34]

Many researchers resented the government's attempts to control their work. The rhetoric became strong on both sides with Inman saying at a meeting of the American Association for the Advancement of Science, "The tides are moving, and moving fast, toward legislated solutions that in fact are likely to be much more restrictive, not less restrictive, than the voluntary [censorship system]."[35]

Although NSA made all of these attempts to control cryptographic research in the open community, none worked, and, that may have been to their advantage! David Kahn, distinguished historian of cryptology, points out that publicly conducted cryptologic research improves the nation's overall skills in the field; leads to advances in communications, mathematics, and computer science; and provides security for databases throughout the country. "For all of these reasons, then," he concluded, "I feel that no limitations should be placed on the study of cryptology. Besides all

---

[34] Kolata, Gina Bari, "Cryptography: A New Clash between Freedom and National Security," *Science*, Vol. 209, No. 4460 August 29, 1980, pp. 995–996.

[35] Bamford, James, *The Puzzle Palace*, Houghton Mifflin Company, Boston, 1982, p. 363.

of these reasons, it seems to me that something more fundamental will in the end prevent any restrictions anyway. It is called the First Amendment."[36]

14.5 RSA Patented; Alice and Bob Born Free

As was mentioned earlier, Diffie and Hellman seemed to have realized the importance of their landmark paper, but this is not always the case for authors.

> I thought this would be the least important paper my name would ever appear on.

> —Len Adleman[37]

Adleman actually argued that he shouldn't be listed as an author, claiming he had done little—mainly breaking previous attempts. The others insisted and he agreed as long as his name would be put last. It seems logical that the authors' names should appear on papers in the order of their contributions, but this is very rarely the case in mathematics. The convention is to list the authors alphabetically. So, RSA could easily have become known as ARS instead, if Adleman hadn't asked to be put last.

Unlike many of the systems we've examined, RSA was patented[38] in 1983 and was therefore not royalty-free; however, it was placed in the public domain in 2000, before the patent was set to expire in 2003. It's likely that you've used RSA, whether you were aware of the fact or not. Internet transactions in which you provide a credit card number are often secured in this manner. Although it happens behind the scenes, your computer uses the vendor's public key to secure the message you send.

The creators of RSA all made other contributions to cryptology that will be discussed in later chapters, as will the way in which RSA may be used to sign messages. It is worth mentioning now that Rivest invented Alice and Bob,[39] the two characters who are typically used to illustrate cryptographic protocols.[40] Previously, messages were sent from person A to person B. The creation of Alice and Bob helped to humanize the protocols. Over the years, other characters have been added to the cast. These characters are so convenient, I chose to use them in explaining earlier systems, even though the original descriptions did not include them.

---

[36] Quoted in Foerstel, Herbert N., *Secret Science: Federal Control of American Science and Technology*, Praeger, Westport, Connecticut, 1993, p. 125, which cites *The Government's Classification of Private Ideas*, hearings before a subcommittee of the Committee on Government Operations, U.S. House of Representatives, 96th Congress, Second Session, 28 February, 20 March, 21 August 1980, U.S. Government Printing Office, Washington, DC, 1981, p. 410.

[37] Levy, Steven, *Crypto: How the Code Rebels Beat the Government, Saving Privacy in the Digital Age*, Viking, New York, 2001, p. 101.

[38] Rivest, Ronald L., Adi Shamir, and Leonard M. Adleman, *Cryptographic Communications System and Method*, U.S. Patent 4,405,829, September 20, 1983, available online at https://patentimages.storage.googleapis.com/49/43/9c/b155bf231090f6/US4405829.pdf.

[39] This was in the paper describing RSA: Rivest, Ronald L., Adi Shamir, and Leonard Adleman, *On Digital Signatures and Public-Key Cryptosystems* (There was soon a title change to *A Method for Obtaining Digital Signatures and Public-key Cryptosystems*. The date is the same for both.), MIT Laboratory for Computer Science Report MIT/LCS/TM-82, April 1977, Cambridge, Massachusetts.

[40] Levy, Steven, *Crypto: How the Code Rebels Beat the Government, Saving Privacy in the Digital Age*, Viking, New York, 2001, p. 102.

Schneier presents the characters as actors in a play:[41]

Dramatis Personae	
Alice	First participant in all the protocols
Bob	Second participant in all the protocols
Carol	Participant in the three- and four-party protocols
Dave	Participant in the four-party protocols
Eve	Eavesdropper
Mallory[42]	Malicious active attacker
Trent	Trusted arbitrator
Walter	Warden; he'll be guarding Alice and Bob in some protocols
Peggy	Prover
Victor	Verifier

Alice and Bob are also handy characters for those working in coding theory. This does not refer to the study of the sort of codes discussed in this text. Rather than provide a definition, I encourage you to read the transcript (which does provide one) of The Alice and Bob After Dinner Speech, given at the Zurich Seminar, April 1984, by John Gordon, by invitation of Professor James Massey. To intrigue you, a few paragraphs are reproduced below.[43]

> Coding theorists are concerned with two things. Firstly and most importantly they are concerned with the private lives of two people called Alice and Bob. In theory papers, whenever a coding theorist wants to describe a transaction between two parties he doesn't call them A and B. No. For some longstanding traditional reason he calls them Alice and Bob.
>
> Now there are hundreds of papers written about Alice and Bob. Over the years Alice and Bob have tried to defraud insurance companies, they've played poker for high stakes by mail, and they've exchanged secret messages over tapped telephones.
>
> If we put together all the little details from here and there, snippets from lots of papers, we get a fascinating picture of their lives. This may be the first time a definitive biography of Alice and Bob has been given.

---

[41] Schneier, Bruce, Applied *Cryptography, second edition,* John Wiley & Sons, New York, p. 23.

[42] In the first edition of Schneier's *Applied Cryptography,* this character was named Mallet. Although he has since been replaced by Mallory, he served as the inspiration for the pen-name of an American Cryptogram Association (ACA) member.

[43] The transcript is available at http://downlode.org/Etext/alicebob.html.

Whatever the details of their personal lives may be, Alice and Bob have certainly helped cryptologists. In Malgorzata Kupiecka's paper, "Cryptanalysis of Caesar Cipher,"[44] she formally acknowledged their help by writing at the end, "I wish to thank Alice and Bob."

However, it turns out that Alice and Bob weren't always Alice and Bob. Recently, Leonard Adleman shared their origin story with me. The famous pair did not appear in early draft versions of the RSA paper. Rivest had sent one of these drafts to Richard Schroeppel, asking for comments and information on the current state of factoring. Schroeppel's August 1, 1977 reply included the following:

> On p. 4, I would suggest a notation change. Using p and q for primes is quite standard, but you shouldn't continue assigning letters in the same sequence to nonprimes. Perhaps m would be a good symbol for pq, and e for the public exponent you have called s. Another literary suggestion: name your protagonists, perhaps Adolf and Bertholt or somesuch. This would reserve isolated letters for mathematical quantities.

In sharing this with me, Adleman commented, "While the names Richard suggests are problematic, I never liked Ron's choice either."[45]

14.6 History Rewritten

On December 17, 1997, Government Communications Headquarters (GCHQ), the British equivalent of the National Security Agency, revealed that public key cryptography was developed within their top-secret environment prior to its discovery by the American academics.[46] The Brits referred to it as "non-secret encryption." James H. Ellis was the first to show that it was possible in 1970[47] and, sometime before 1976, Malcolm Williamson discovered the approach credited to Diffie and Hellman.[48] It was also revealed at this time that Clifford Cocks had a version of the "RSA" scheme in 1973.[49]

---

[44] Kupiecka, Malgorzata, "Cryptanalysis of Caesar Cipher," *Journal of Craptology*, Vol. 3, November 2006. Available online at http://www.anagram.com/~jcrap/Volume_3/caesar.pdf/. You may recall from Section 13.2 that when differential cryptanalysis was discovered, it was found that DES, an older system by that time, was optimized against it. Hence, it was concluded that NSA was aware of the attack years earlier. In a similar vein, Malgorzata demonstrates that differential cryptanalysis is not a good attack against the Caesar shift cipher and concludes that the Romans must have been aware of this approach and optimized their encryption against it. *Journal of Craptology* specializes in humorous papers of this sort. See http://www.anagram.com/~jcrap/ for complete electronic contents.

[45] Adleman, Leonard, email to the author, January 5, 2020.

[46] The Story of Non-Secret Encryption, http://web.archive.org/web/19980415070754/http://www.cesg.gov.uk/ellisint.htm, December 17, 1997.

[47] Ellis, James H., *The possibility of secure non-secret digital encryption*, CESG Report 3006, January 1970, available online at http://fgrieu.free.fr/histnse/CESG_Research_Report_No_3006_0.pdf; Ellis, James H., *The possibility of secure non-secret analogue encryption*, CESG Report 3007, May 1970, available online at http://fgrieu.free.fr/histnse/CESG_Research_Report_No_3007_1.pdf.

[48] Williamson, Malcolm J., *Non-Secret Encryption Using a Finite Field*, CESG Report, 21 January 1974; Williamson, Malcolm J., *Thoughts on Cheaper Non-Secret Encryption*, CESG Report, 10 August 1976. Williamson had the idea put forth in this second paper long before it was published.

[49] Cocks, Clifford C., *Note on "Non-Secret Encryption,"* CESG Report, 20 November 1973, available online at http://fgrieu.free.fr/histnse/Cliff%20Cocks%20paper%2019731120.pdf.

However, Bobby Ray Inman, a director of NSA, claimed in congressional testimony, that NSA had discovered public key cryptography 10 years before the academics.[50] If correct, this would mean that NSA had it before GCHQ. There was no evidence publicly available in support of Inman's claim until 2019, when a Freedom of Information Act request led to the release of a 100+ page, previously top secret, history titled *Fifty Years of Mathematical Cryptanalysis* (1937–1987). Although the released version was heavily redacted, the lines reproduced below survived.

> Rick Proto, [188], had suggested the exponentiation scheme as an "irreversible" transformation, prior to Ellis' paper on nonsecret encryption. No public key cryptosystem has yet been used operationally,[51]

Whatever followed the comma at the end of the quoted lines was redacted and remains a mystery, as do the details of the reference [188]. In the bibliography at the end of the history, this entire reference was redacted. Recognizing a transformation as apparently irreversible and building a public key system out of it are two different things, but if all Proto did was the former, why would the title of his paper still be secret in 2019, when the history was released?

Back in 2009, NSA recognized Proto's importance by naming a facility at Fort Meade the "Richard C. Proto Symposium Center." Prior to this, the only named facility was Friedman Auditorium.[52] A pair of references on Proto are given in the list below, but the specifics of his contributions to NSA are sorely lacking.

References and Further Reading

Anon., "Richard Proto, 2013 Hall of Honor Inductee," *Cryptologic Hall of Honor*, National Security Agency, https://www.nsa.gov/about/cryptologic-heritage/historical-figures-publications/hall-of-honor/Article/1621556/richard-proto/.

Cocks, C. C., *Note on "Non-Secret Encryption"*, CESG Report, November 20, 1973, available online at http://fgrieu.free.fr/histnse/Cliff%20Cocks%20paper%2019731120.pdf.

Diffie, Whitfield, "The First Ten Years of Public-Key Cryptography," *Proceedings of the IEEE*, Vol. 76, No. 5, May 1988, pp. 560–577, available online at http://www.rose-hulman.edu/class/csse/csse442/201020/Homework/00004442.pdf.

Diffie, Whitfield and Martin Hellman, "Multiuser Cryptographic Techniques," in Hammer, Carl, editor, *AFIPS '76: Proceedings of the June 7-10, 1976, National Computer Conference and Exposition*, ACM Press, New York, 1976, pp. 109–112. This is, by the way, the meeting where *Cryptologia's* founding editors first met.

Diffie, Whitfield, and Hellman, Martin, "New Directions in Cryptography," *IEEE Transactions on Information Theory*, Vol. IT-22, No. 6, November 1976, p. 644–654.

Ellis, James H., *The possibility of secure non-secret digital encryption*, CESG Report 3006, January 1970, available online at http://fgrieu.free.fr/histnse/CESG_Research_Report_No_3006_0.pdf.

---

[50] Diffie, Whitfield and Susan Landau, *Privacy on the Line: The Politics of Wiretapping and Encryption*, The MIT Press, Cambridge, Massachusetts, 1998, p. 253, note 15 for Chapter 3.

[51] Stahly, Glenn F., *Fifty Years of Mathematical Cryptanalysis (1937-1987)*, National Security Agency, August 1988, p. 79, available online at https://cryptome.org/2019/10/nsa-50-years-crypto-et-al.pdf.

[52] "Tribute to Richard Proto," Congressional Record, Vol. 155, No. 81, June 2, 2009, Extensions of Remarks, pp. E1269–E1270, from the Congressional Record Online through the Government Publishing Office, available online at https://www.govinfo.gov/content/pkg/CREC-2009-06-02/html/CREC-2009-06-02-pt1-PgE1269-4.htm.

Ellis, James H., *The possibility of secure non-secret analogue encryption*, CESG Report 3007, May 1970, available online at http://fgrieu.free.fr/histnse/CESG_Research_Report_No_3007_1.pdf.

Ellis, James H., *The Story of Non-Secret Encryption*, 1987, available online at http://web.archive.org/web/20030610193721/http://jya.com/ellisdoc.htm. This paper details the discovery of non-secret encryption at the U.K. Government Communications Headquarters (GCHQ). References are provided to the now declassified technical papers. It was reprinted as Ellis, James H., "The History of Non-Secret Encryption," *Cryptologia*, Vol. 23, No. 3, July 1999, pp. 267–273.

Foerstel, Herbert N., *Secret Science: Federal Control of American Science and Technology*, Praeger, Westport, Connecticut, 1993. This book takes a broad look at the topic, but Chapter 4 focuses on cryptography.

Gardner, Martin, "Mathematical Games, A New Kind of Cipher That Would Take Millions of Years to Break," *Scientific American*, Vol. 237, No. 2, August 1977, pp. 120–124.

Hellman, Martin, Homepage, http://www-ee.stanford.edu/~hellman/.

Kahn, David, "Cryptology Goes Public," *Foreign Affairs*, Vol. 58, No. 1, Fall 1979, pp. 141–159.

Levy, Steven, *Crypto: How the Code Rebels Beat the Government, Saving Privacy in the Digital Age*, Viking, New York, 2001. Aimed at a general audience, this book is the best single source on the history of modern cryptology up to 2001. Of course, more technical surveys exist for those wishing to see the mathematics in greater detail.

Reilly, Larry, "Top-Secret Famous," *Fairfield Now*, Fall 2009, pp. 22–26, available online at http://saccovanzettiexperience.com/site/wp-content/uploads/2015/05/FairfieldNow_ALifeinSecrets.pdf. This article is on Richard C. Proto.

Rivest, Ron, Homepage, http://theory.lcs.mit.edu/~rivest/.

Rivest, Ronald L., Adi Shamir, and Leonard Adleman, *On Digital Signatures and Public-Key Cryptosystems* (There was soon a title change to *A Method for Obtaining Digital Signatures and Public-key Cryptosystems*. The date is the same for both.), MIT Laboratory for Computer Science Report MIT/LCS/TM-82, April 1977, Cambridge, Massachusetts. This report later appeared as cited in the reference below.

Rivest, Ronald L., Adi Shamir, and Leonard Adleman, "A Method for Obtaining Digital Signatures and Public-key Cryptosystems," *Communications of the ACM*, Vol. 21, No. 2, February 1978, pp. 120–126.

Simmons, Gustavus J. and Michael J. Norris, "Preliminary Comments on the MIT Public-Key Cryptosystem," *Cryptologia*, Vol. 1, No. 4, October 1977, pp. 406–414.

Williamson, Malcolm J., *Non-Secret Encryption Using a Finite Field*, CESG Report, 21 January 1974.

Williamson, Malcolm J., *Thoughts on Cheaper Non-Secret Encryption*, CESG Report, 10 August 1976.

Breaking News!

On December 15, 2020, when the present book was in the proof stage, Whit Diffie was inducted into NSA's Hall of Honor in what he described as "the most unlikely honor of my life." David Kahn was also inducted into the Hall on this great day. Thus, work done in the academic community (not subject to NSA's control) was recognized as being of great value.

Chapter 15

Attacking RSA

The most obvious, most direct, and most difficult way to attack RSA is by factoring the modulus. The paragraph below indicates how this approach works. This chapter then details 12 non-factoring attacks, before returning to examine various factoring algorithms.

One way to compute d (given e and n) is to first factor n. We will have $n = pq$. Because p and q are distinct primes, they are relatively prime; hence, $\varphi(n) = (p-1)(q-1)$. Having calculated the value of $\varphi(n)$, we may easily find the multiplicative inverse of e modulo $\varphi(n)$. Because this inverse is d, the deciphering exponent, we have now broken the system; however, the "factor n" step is highly nontrivial!

15.1 A Dozen Non-Factoring Attacks

As you examine the 12 attacks below, please note that they are almost all easy to "patch" against. One of the best and most recent attacks is presented last. It's probably the most difficult to patch.

15.1.1 Attack 1. Common Modulus Attack

This attack was first demonstrated by Gus Simmons (Figure 15.1) in 1983.[1]

Imagine that a common message is sent to two individuals who share the same value for n, but use distinct values for e. Suppose Eve intercepts both enciphered messages:

$$C_1 = M^{e_1} (\bmod\, n) \quad \text{and} \quad C_2 = M^{e_2} (\bmod\, n).$$

If e_1 and e_2 are relatively prime, she may then use the Euclidean algorithm to find integers x and y such that $xe_1 + ye_2 = 1$. Exactly one of x and y must be negative. Assume it is x. Eve then calculates

$$\left(C_1^{-1}\right)^{-x} C_2^y = C_1^x C_2^y = \left(M^{e_1}\right)^x \left(M^{e_2}\right)^y = M^{xe_1 + ye_2} = M^1 = M (\bmod\, n).$$

Thus, Eve, who hasn't recovered d, can obtain M.

PATCH: Don't have the same modulus for any two users!

---

[1] Simmons, Gustavus J., "A "Weak" Privacy Protocol Using the RSA Crypto Algorithm," *Cryptologia*, Vol. 7, No. 2, April 1983, pp. 180–182.

Figure 15.1   Gustavus Simmons (1930–). (Courtesy of Gustavus Simmons.)

Imagine the malicious hacker Mallory controls Alice and Bob's communication channel. When Alice requests Bob's public key, Mallory changes the e that Bob tries to send her by a single bit. Instead of (e, n), Alice receives (e', n). When Alice enciphers her message, Mallory lets it pass unchanged to Bob, who is unable to read it. After some confusion, Bob sends his public key to Alice again, since she clearly didn't use the right values. Alice then sends the message again using (e, n). Mallory may then use the attack described above to read M.[2]

PATCH: Never resend the same message enciphered two different ways. If you must resend, alter the message first.

15.1.2 Attack 2. Man-in-the-Middle

In the attack described above, where a hacker controls the communications, you may well ask why he doesn't simply keep Bob's public key and send Alice his own. When Alice encrypts a message, thinking Bob will get it, Mallory can read it using her own key and then re-encipher it with Bob's key before passing it on. She can even make changes first, if she desires. Similarly, if Bob requests Alice's key, Mallory can keep it and send Bob another key she has made for herself. In this manner, Mallory has complete control over the exchanges. For obvious reasons, this is known as a *man-in-the-middle* attack. Studying ways to prevent attacks like these falls under the "protocols" heading of cryptography. We do not pursue this line here, but the reader will find the subject treated nicely by Schneier.[3]

---

[2] Joye, Marc, and Jean-Jaque Quisquater, "Faulty RSA Encryption," *UCL Crypto Group Technical Report CG-1997/8*, Université catholique de Louvain, Louvain-la-Neuve, Belgium, 1987. This attack was first presented in the paper cited here, although Mallory was not present. It was assumed that an accidental error necessitated the resending of the message.

[3] Schneier, Bruce, *Applied Cryptography*, second edition, John Wiley & Sons, New York, 1996.

15.1.3 Attack 3. Low Decryption Exponent

In 1990, Michael J. Wiener presented an attack for when the decryption exponent, d, is small.[4] To be more precise, the attack applies when

$$q < p < 2q \quad \text{and} \quad d < \frac{\sqrt[4]{n}}{3}$$

In this case, d may be computed efficiently. To see how this is done,[5] we begin with $ed = 1$ (mod $\varphi(n)$) and rewrite it as $ed - k\varphi(n) = 1$ for some k in the set of integers. We then divide both sides by $d\varphi(n)$ to get

$$\frac{e}{\varphi(n)} - \frac{k}{d} = \frac{1}{d\varphi(n)}$$

$\varphi(n) \approx n$, so we have

$$\frac{e}{\varphi(n)} - \frac{k}{d} \approx \left|\frac{e}{n} - \frac{k}{d}\right| = \left|\frac{ed - kn}{nd}\right| = \left|\frac{ed - k\varphi(n) - kn + k\varphi(n)}{nd}\right|$$

$\varphi(n)$ is actually a bit smaller than n, so we need to introduce absolute value signs following the $\approx$ to be sure the quantity remains positive.

In the last step above, we added 0 to the numerator in the form $-k\varphi(n) + k\varphi(n)$.

Because $ed - k\varphi(n) = 1$, the numerator simplifies, and we get the above

$$= \left|\frac{1 - kn + k\varphi(n)}{nd}\right| = \left|\frac{1 - k(n - \varphi(n))}{nd}\right|$$

Previously it was noted that $\varphi(n) \approx n$, but how good is this approximation? The difference $n - \varphi(n)$ now appears in our numerator, so we'd like to find a bound for it. We have

$$n - \varphi(n) = n - (n - p - q + 1) = p + q - 1$$

But for this attack, we required $q < p < 2q$. The second half of this double inequality gives us

$$p + q - 1 < 3q - 1$$

and the first half of the double inequality tells us that q is the smaller factor and therefore

$$q < \sqrt{n}$$

Hence,

$$3q - 1 < 3\sqrt{n} - 1 < 3\sqrt{n}$$

The last few steps thus establish that

$$n - \varphi(n) < 3\sqrt{n}$$

---

[4] Wiener, Michael J., "Cryptanalysis of Short RSA Secret Exponents," *IEEE Transactions on Information Theory*, Vol. 36, No. 3, 1990, pp. 553–558.

[5] Adapted from p. 206 of Boneh, Dan, "Twenty Years of Attacks on the RSA Cryptosystem," *Notices of the American Mathematical Society*, Vol. 46, No. 2, February 1999, pp. 203–213.

and we plug this in to our previous equation to get

$$\left| \frac{1 - k(n - \varphi(n))}{nd} \right| \le \left| \frac{3k\sqrt{n}}{nd} \right| = \frac{3k}{d\sqrt{n}}$$

Now recall that we have $ed - k\varphi(n) = 1$. This may be rewritten as $k\varphi(n) = ed - 1$, which is less than ed; that is, $k\varphi(n) < ed$. Because $e < \varphi(n)$, this last inequality requires $k < d$. We assumed that

$$d < \frac{\sqrt[4]{n}}{3}$$

Stringing these inequalities together, we get

$$k < \frac{\sqrt[4]{n}}{3}$$

From this, and the work above, we have

$$\left| \frac{e}{n} - \frac{k}{d} \right| \le \frac{3k}{d\sqrt{n}} < \frac{3\left(\frac{\sqrt[4]{n}}{3} \right)}{d\sqrt{n}} = \frac{1}{d\sqrt[4]{n}}$$

Again using

$$d < \frac{\sqrt[4]{n}}{3}$$

we see

$$\frac{1}{d} > \frac{3}{\sqrt[4]{n}}$$

which gives

$$\frac{1}{3d} > \frac{1}{\sqrt[4]{n}}$$

Making this substitution in the inequality above yields

$$\left| \frac{e}{n} - \frac{k}{d} \right| < \frac{1}{3d^2}$$

which in turn is

$$< \frac{1}{2d^2}.$$

The weaker bound is indicated because it allows us to apply a theorem concerning the number of solutions k/d satisfying the inequality. Namely,

$$\left| \frac{e}{n} - \frac{k}{d} \right| < \frac{1}{2d^2}$$

implies that there are fewer than $\log_2(n)$ fractions k/d that approximate e/n this closely. A technique is available to efficiently check the possibilities until the correct d is found.[6]

PATCH 1: Using a small value for d allows for quicker decryption, but don't do it!

PATCH 2: Use a small value for d, but increase e to the point that the attack above won't work. We can increase e by multiples of $\varphi(n)$, without making any difference other than increasing the time needed for encryption. Making $e > n^{1.5}$ will block the attack above.

15.1.4 Attack 4. Partial Knowledge of p or q[7]

If n has m digits, and we can somehow determine the first or last $m/4$ digits of either p or q, then we can factor n efficiently. How on Earth would we know so many digits of one of the factors? Well, if the method used to generate the primes needed for RSA is such that a large portion of p or q is predictable or guessable, this can be done.

PATCH: Generate p and q wisely!

15.1.5 Attack 5. Partial Knowledge of d[8]

Suppose the modulus has m digits. If we are able to determine the last $m/4$ digits of d, then we may be able to efficiently find d. The "may" depends on the value of e. If e is small, this attack works wonderfully, but if e is close to n in size, then the attack isn't any better than a trial-and-error hunt for d. Of course, e is public, so the cryptanalyst knows whether this attack is worth the trouble before beginning.

PATCH: Why would you let anyone see any portion of d in the first place? If you're going to be this careless, you need to use a large value for e.

15.1.6 Attack 6. Low Encryption Exponent Attack

If e, the value the message is raised to, is very small, and the message is also small, then we might have $M^e < n$, thee modulus. If this happens, the attacker may simply take the eth root of the ciphertext (over the integers) to obtain M. Thus, the discrete log problem never arises.

PATCH: It is tempting to have either e or d be small, so that encryption or decryption may be carried out more rapidly, but there is no point in sacrificing security for speed.

15.1.7 Attack 7. Common Enciphering Exponent Attack

This attack makes use of a theorem seen in every introductory number theory text. Its first known appearance was in a work by Sun Zi (c. 400–460). We'll take a look at it before examining its application to cryptanalysis.

---

[6] For a little more detail, see p. 206 of Boneh, Dan, "Twenty Years of Attacks on the RSA Cryptosystem," *Notices of the American Mathematical Society*, Vol. 46, No. 2, February 1999, pp. 203–213. For a lot more detail, see Wiener, Michael J., "Cryptanalysis of Short RSA Secret Exponents," *IEEE Transactions on Information Theory*, Vol. 36, No. 3, 1990, pp. 553–558.

[7] Coppersmith, Don, "Small Solutions to Polynomial Equations, and Low exponent RSA Vulnerabilities," *Journal of Cryptology*, Vol. 10, No. 4, September 1997, pp. 233–260.

[8] Boneh, Dan, Glenn Durfee, and Yair Frankel, "An Attack on RSA Given a Fraction of the Private Key Bits," in Ohta, Kazuo and Dingyi Pei, editors, *Advances in Cryptology – ASIACRYPT '98*, Lecture Notes in Computer Science, Vol. 1514, Springer, Berlin, Germany, 1998, pp. 25–34.

15.1.7.1 The Chinese Remainder Theorem

Given k congruences

$$x = a_1 \;(\text{mod } n_1)$$
$$x = a_2 \;(\text{mod } n_2)$$
$$x = a_3 \;(\text{mod } n_3)$$
$$\vdots$$
$$x = a_k \;(\text{mod } n_k)$$

such that the n_i are positive integers and pairwise relatively prime, and the a_i are integers, there exists values for x satisfying all of the equations. If one solution is given by x', then the set of all solutions is given by $x = x' + yn$, where y takes every integer value and $n = n_1 n_2 n_3 \ldots n_k$.

Gauss expressed the solution as

$$x = \sum_{i=1}^{k} a_i N_i M_i \left(\text{mod } n\right)$$

where $N_i = n/n_i$ and $M_i = N_i^{-1} \;(\text{mod } n_i)$. The easiest way to see how a solution may be obtained is through an example.

Example 1

Suppose we have

$$x = 5 \left(\text{mod } 9\right)$$
$$x = 8 \left(\text{mod } 11\right)$$
$$x = 6 \left(\text{mod } 14\right).$$

For convenience, we label the moduli as $n_1 = 9$, $n_2 = 11$, and $n_3 = 14$. We compute the product of the moduli $n = n_1 n_2 n_3 = (9)(11)(14) = 1{,}386$. Using Gauss's formula, we have

$$x = \sum_{i=1}^{k} a_i N_i M_i \left(\text{mod } n\right) = a_1 N_1 M_1 + a_2 N_2 M_2 + a_3 N_3 M_3 \left(\text{mod } n\right)$$

The values for the a_i were given in the problem, and the N_i are very easy to calculate. To get the M_i, we could use a technique from Section 14.3 (the Euclidean algorithm), but for such small values, it is easier to do the calculations in one's head. For large numbers we'd resort to the algorithm. We have

$$M_1 = N_1^{-1} = (154)^{-1} = (1)^{-1} = 1 \;(\text{mod } 9)$$
$$M_2 = N_2^{-1} = (126)^{-1} = (5)^{-1} = 9 \;(\text{mod } 11)$$
$$M_3 = N_3^{-1} = (99)^{-1} = (1)^{-1} = 1 \;(\text{mod } 14)$$

So, plugging in these values along with the n_i and N_i, we have

$$x = (5)(154)(1) + (8)(126)(9) + (6)(99)(1) \;(\text{mod } 1{,}386)$$

$$= 10{,}436 \;(\text{mod } 1{,}386)$$

$$= 734 \;(\text{mod } 1{,}386)$$

Checking against our original three equations, we see that the answer, $x = 734$, works.

Now that the Chinese Remainder Theorem is clear, we're ready to examine how it can be used to attack RSA encryption. We already saw the danger in users destined to receive the same message sharing a modulus. It is also a disaster if the same message goes to users who have different values for n, but share the same e. The Chinese remainder theorem may allow the message to be recovered in this case. It only depends on how many copies of the message are sent out relative to the value of e. If there are m copies sent, all using the same e and distinct moduli, and $m \geq e$, then the message can be recovered. As a small example, we use $e = m = 3$. Our three intercepted ciphertexts follow below:

$$C_1 = M^3 \pmod{n_1}$$
$$C_2 = M^3 \pmod{n_2}$$
$$C_3 = M^3 \pmod{n_3}$$

This looks a bit different than the example of the Chinese remainder theorem above, but we may rewrite these equations as

$$M^3 = C_1 \pmod{n_1}$$
$$M^3 = C_2 \pmod{n_2}$$
$$M^3 = C_3 \pmod{n_3}$$

Now they match the example, with M^3 taking the place of x.

If the moduli are not all pairwise relatively prime, then we may compute the gcd of two of them to arrive at a prime factor of one of the moduli. Because this allows us to then factor a modulus, and go on to recover d very easily, we assume the moduli are relatively prime. This being the case, we can apply the Chinese remainder theorem to the equations above to get an integer C' such that

$$M^3 = C' \pmod{n_1 n_2 n_3}$$

We have $M < n_1$, $M < n_2$, and $M < n_3$, so it follows that $M^3 < n_1 n_2 n_3$. Therefore, we may solve for M in the equation above by taking the normal cube root of C' over the integers, without worrying about modular arithmetic.

A generalization of this attack was found by Johan Håstad, who realized that the messages needn't be identical, as long as they are related linearly.[9] Don Coppersmith made further improvements.[10]

PATCH: Using a value for e that greatly exceeds the largest number of copies of a message that would ever be sent will block this attack. An alternate patch is to pad messages with random bits, so that no two are alike.

---

[9] Håstad, Johan, "On using RSA with Low Exponent in a Public Key Network," in Williams, Hugh C., editor, *Advances in Cryptology – CRYPTO '85 Proceedings*, Lecture Notes in Computer Science, Vol. 218, Springer, Berlin, Germany, 1986, pp. 403–408.

[10] Coppersmith, Don, "Small Solutions to Polynomial Equations, and Low Exponent RSA Vulnerabilities," *Journal of Cryptology*, Vol. 10, No. 4, December 1997, pp. 233–260.

15.1.8 Attack 8. Searching the Message Space

If the sender is in the habit of sending one of a small number of stereotyped messages, a cryptanalyst can simply encrypt each and see which one matches the intercepted ciphertext.

PATCH: Make sure the potential message space is large! Or, as in the alternative patch for Attack 7, pad the messages with random bits. This way, if only a small set of messages will typically be sent, repeats can be padded differently to create too many possibilities to check in the manner of the attack described above. For example, if the attacker guesses the message might be "Nothing to report," but 50 bits of padding have been used, he'd have to encipher all 2^{50} different versions of "Nothing to report," before being able to decide if that was or was not the actual message. One standard for padding put out by RSA is PKCS #1. It stands for Public Key Cryptography Standard #1. This standard is now at version 2.2, as modifications were demanded by attacks (such as the adaptive chosen ciphertext attack) found against earlier versions.

15.1.9 Attack 9. Adaptive Chosen Ciphertext Attacks

Figure 15.2 Daniel Bleichenbacher (1964–). (Courtesy of Daniel Bleichenbacher and Kit August.)

A chosen ciphertext attack is one in which the attacker gets to create any ciphertext he likes and then get the corresponding plaintext. Adding the adjective "adaptive" means that he can do this repeatedly, using knowledge he gained in previous iterations to adapt his next ciphertext in such a way as to optimize the information it yields. This is the sort of attack Daniel Bleichenbacher (Figure 15.2) launched against RSA padded in accordance with PKCS #1. Actually, his attack was a bit less demanding in that he didn't need the corresponding plaintexts, but rather just the knowledge of whether or not these plaintexts corresponded to some block of data encrypted using PKCS #1. Describing his attack, he wrote, "we expect that the attack needs roughly 2^{20} chosen ciphertexts to succeed."[11] This sounds daunting, but Bleichenbacher points out that the attack is practical, because there are servers that will accept ciphertexts and return error messages when

---

[11] Bleichenbacher, Daniel, "Chosen Ciphertext Attacks against Protocols Based on the RSA Encryption Standard PKCS #1," in Krawczyk, Hugo, editor, Advances in Cryptology – *CRYPTO '98 Proceedings*, Lecture Notes in Computer Science, Vol. 1462, Springer, Berlin, Germany, 1998, pp. 1–12.

they are not PKCS conforming. So, it's not like he's expecting Bob to respond to his million-plus ciphertexts!

Reporting on experiments carried out for his new attack, Bleichenbacher wrote, "We tested the algorithm with different 512-bit and 1024-bit keys. The algorithm needed between 300 thousand and 2 million chosen ciphertexts to find the message." But improved systems give new attacks a tough race, and Bleichenbacher admitted that version 2 of PKCS #1, which made use of results published in 1995, was not vulnerable to his attack.[12]

15.1.10 Attack 10. Timing Attack[13]

Figure 15.3 Paul Kocher (1973–). (http://www.cryptography.com/company/profiles/paul-kocher.html).

Why is this man smiling? Paul Kocher (Figure 15.3) made his mark on cryptology in 1995, while still an undergraduate at Stanford, by discovering timing attacks. These are not limited to RSA but may be applied to a variety of systems. Kocher was also responsible for leading the design of the Electronic Frontier Foundation's DES cracker (see Section 13.3). I look forward to seeing what he'll do next. The manner in which his timing attack applies to RSA follows.

For this attack to work, the attacker needs some access to the recipient's machine. If the attacker can obtain the ciphertexts, as well as the amount of time taken by the recipient's machine to decipher the messages, he can work backward to find what the decryption exponent must be. That is, if decryption is done by, for instance, the repeated squaring method detailed in Section 14.3, then

---

[12] The 1995 paper is Bellare, Mihir and Phillip Rogaway, "Optimal asymmetric encryption," in De Santis, Alfredo, editor, *Advances in Cryptology – EUROCRYPT '94 Proceedings*, Lecture Notes in Computer Science, Vol. 950, Springer, Berlin, Germany, 1995, pp. 92–111.

[13] Kocher, Paul, "Timing attacks on implementations of Diffie-Hellman, RSA, DSS, and other systems," in Koblitz, Neal, editor, *Advances in Cryptology – CRYPTO '96 Proceedings*, Lecture Notes in Computer Science, Vol. 1109, Springer, Berlin, Germany, 1996, pp. 104–113.

each multiplication will take a certain amount of time, dependent on the machine doing it. These times then correspond to particular keys.

PATCH: In the paper that introduced his attack, Kocher also put himself in the defensive position and suggested patches. He explained how random delays may be inserted into the decryption algorithm to throw the timing off, but moving back to offense indicated that an attacker could still break the system by collecting more data. Another patch he proposed is called *blinding*. Blinding had previously been used in another context. Here it meant multiplying the received ciphertext by a random, but invertible number r raised to the value of the enciphering exponent (modulo n), prior to any deciphering. So, one is then left to decipher r^eC. This is done in the normal manner, by exponentiation to the power d. We get $(r^eC)^d = r^{ed}C^d = rM$. The recipient must then multiply by the inverse of r to recover the message. Using a different r each time will prevent an attacker from being able to find a correlation between the original ciphertext block and the time needed to decipher.

15.1.11 Attack 11. Textbook RSA Attack[14]

If one is unable to factor the modulus n, there's still a way that factoring can pay off. Simply shift the focus from n to the message M. Because, $C = M^e$ (mod n), if M can be factored as $M = M_1M_2$, then $C = (M_1M_2)^e = M_1^eM_2^e$ (mod n). It follows that

$$C/M_2^e = M_1^e \,(\text{mod}\, n)$$

For the moment, let's assume that $M_1 \leq 2^{m_1}$ and $M_2 \leq 2^{m_2}$ for some positive integers m_1 and m_2.

Now, if all we have is the intercepted ciphertext C, we don't know M and cannot attempt to factor it directly, but the equality above allows us to determine the factorization. We begin by building a table of M_1^e (mod n) values for all $M_1 = 2, 3, \ldots, 2^{m_1}$. Next, we start calculating C/M_2^e (mod n) for $M_2 = 2, 3, \ldots, 2^{m_2}$. As each of these latter values is calculated, it is checked for in the M_1^e (mod n) table. When a match is found, we have the values of M_1 and M_2 satisfying $C/M_2^e = M_1^e$ (mod n). Multiplying the values M_1 and M_2 together gives us the message M.

This attack should remind you of the attack on double DES presented in Section 13.4. Both are meet-in-the-middle attacks.

A similar attack works against Elgamal, a system detailed in Sections 16.8 and 17.2.4.

PATCH: Randomly pad the message prior to encryption, so that M is not small compared to n. This is also a way to patch against attacks 7 and 8. It is very important to pad! The term "textbook RSA" basically means RSA *without* padding, because that is how RSA is often presented in textbooks.

15.1.12 Attack 12. Ron Was Wrong, Whit Is Right Attack

In February 2012, a group of six researchers released a paper titled "Ron was wrong, Whit is Right."[15] The paper ended with

> Factoring one 1024-bit RSA modulus would be historic. Factoring 12720 such moduli is a statistic. The former is still out of reach for the academic community (but

[14] Boneh, Dan, Antoine Joux and Phong Q. Nguyen, "Why textbook ElGamal and RSA encryption are insecure," in Okamoto, Tatsuaki, editor, *Advances in Cryptology – ASIACRYPT 2000 Proceedings*, Lecture Notes in Computer Science, Vol. 1976, Springer, Berlin, Germany, 2000, pp. 30–43.

[15] Lenstra, Arjen K., James P. Hughes, Maxime Augier, Joppe W. Bos, Thorsten Kleinjung and Christophe Wachter, "Ron was wrong, Whit is right," February 2012, https://anonymous-proxy-servers.net/paper/064.pdf.

anticipated). The latter comes as an unwelcome warning that underscores the difficulty of key generation in the real world.

The quote is a reworking of a remark commonly attributed to Joseph Stalin, "A single death is a tragedy; a million deaths is a statistic." Just as Stalin was responsible for millions of deaths, so the team that reworked the quote was responsible for destroying thousands of RSA moduli, by factoring them. Their attack worked on real-world public keys, yet if handed some randomly chosen public key, they would be unlikely to be able to break it. This seeming paradox will soon become clear. Their attack is extremely simple!

The researchers simply gathered millions of public keys, then took the moduli two at a time and used the Euclidean algorithm to find the greatest common divisors. Nothing was achieved when the gcd turned out to be 1, but in thousands of cases it was one of the prime factors of the moduli. This happened because, in these cases, the two moduli were of the form $n_1 = pq$ and $n_2 = pr$, for some primes p, q, and r. Once the common factor p is found, both moduli could then be very easily factored. Some of the moduli thus factored were 2048 bits long. Greater size would tend to improve security for most attacks, but the vulnerability resulting from different users selecting a common prime still exists.

On average, 2 out of every 1,000 moduli were factored. So, RSA, as implemented is 98.8% secure against this attack. The authors concluded that RSA is "significantly riskier" than systems based on Diffie-Hellman.

In a classic case of understatement, the authors of the "Ron was wrong, Whit is right" paper pointed out that their results "may indicate that proper seeding of random number generators is still a problematic issue." They went on to note

> The lack of sophistication of our methods and findings make it hard for us to believe that what we have presented is new, in particular to agencies and parties that are known for their curiosity in such matters. It may shed new light on NIST's 1991 decision to adopt DSA as digital signature standard as opposed to RSA, back then a public controversy.

Indeed, the potential problem was noticed earlier by Don Johnson. In a 1999 paper, he considered the effect of a random number generator that had been intentionally "chilled" so that it generated primes from a smaller set than if it was functioning properly.[16] However, the 2012 paper does not assign blame to a malevolent insider, but rather to unintentionally poor algorithms being used to generate the primes.

PATCH: Somehow make sure you are choosing the primes p and q as randomly as possible. This is much harder than it sounds.

There are many other non-factoring attacks on RSA. Some of these may be found in the references for this chapter and others are discussed in Section 17.2. Factoring attacks are presented below; however, none of these (other than attack 12, above) represent any real threat to RSA, provided that various special cases are avoided when selecting primes and exponents and when implementing the system. Of course, the minimum size of the modulus required for decent security increases constantly with computing speed and the occasional improved factoring algorithm.

---

[16] Johnson, Don, *ECC, Future Resiliency and High Security Systems*, Certicom Whitepaper, March 30, 1999, revised July 6, 1999, available online at http://web.archive.org/web/20040215121823/www.comms.engg.susx.ac.uk/fft/crypto/ECCFut.pdf, pp. 12–14 are relevant here.

15.2 A Factoring Challenge

The first part of this chapter was concerned with nonfactoring attacks on RSA. We now take the direct approach. Various algorithms for factoring are considered, although none are quick enough to break RSA with the size primes presently used. Factoring is especially hard when the number is a product of two large primes. To help convince yourself that this is a difficult problem, try to factor

25195908475657893494027183240048398571429282126204
03202777713783604366202070759555626401852588078440
69182906412495150821892985591491761845028084891200
72844992687392807287776735971418347270261896375014
97182469116507761337985909570009733045974880842840
17974291006424586918171951187461215151726546322822
16869987549182422433637259085141865462043576798423
38718477444792073993423658482382428119816381501067
48104516660377306056201619676256133844143603833904
14952634432190114657544454178424020924616515723350
77870774981712577246796292638635637328991215483143
81678998850404453640235273819513786365643912120103
97122822120720357

This number has 617 digits in base 10, but it's called RSA-2048, because it has that many bits in base 2. There was once a $200,000 prize offered by RSA Security for anyone who could find the two prime factors and explain how they did it. The RSA factoring challenge has expired, but they did give away prizes for the factorization of other, smaller products. See Table 15.1 for sizes and prizes.

Table 15.1   RSA Factoring Challenge Prizes

Challenge Number	Prize ($US)	Status	Submission Date	Submitters
RSA-576	$10,000	Factored	December 3, 2003	J. Franke et al.
RSA-640	$20,000	Factored	November 2, 2005	F. Bahr et al.
RSA-704	$30,000	Not factored		
RSA-768	$50,000	Not factored		
RSA-896	$75,000	Not factored		
RSA-1024	$100,000	Not factored		
RSA-1536	$150,000	Not factored		
RSA-2048	$200,000	Not factored		

Note: The website that was the source for this table has been archived at http://web.archive.org/web/20071126084720/http://www.rsa.com:80/rsalabs/node.asp?id=2093.

RSA Security answers an obvious question in the FAQ section of their website:[17]

Why is the RSA Factoring Challenge no longer active?

Various cryptographic challenges—including the RSA Factoring Challenge—served in the early days of commercial cryptography to measure the state of progress in practical cryptanalysis and reward researchers for the new knowledge they have brought to the community. Now that the industry has a considerably more advanced understanding of the cryptanalytic strength of common symmetric-key and public-key algorithms, these challenges are no longer active. The records, however, are presented here for reference by interested cryptographers.

I think the challenge also served as a show of strength for the company. Not many businesses will put their products to the test in such a straight-forward way. Attempts to factor the larger numbers continued even after the financial incentive went away. RSA-768 was factored in 2009 by an international team.[18] The smaller RSA-704 held out until 2012, when it was factored by Shi Bai, Emmanuel Thomé, and Paul Zimmermann.[19] The rest remain unsolved as of this writing.

15.2.1 An Old Problem

The problem of distinguishing prime numbers from composite numbers and of resolving the latter into their prime factors is known to be one of the most important and useful in arithmetic... The dignity of the science itself seems to require that every possible means be explored for the solution of a problem so elegant and celebrated.

—**Carl Friedrich Gauss (1801)[20]**

There was no practical reason for wanting to find an efficient factoring algorithm in Gauss's day, but despite this fact, the problem already had a long history. We begin our brief survey of factoring algorithms with a method from the ancient world that lives on, in a modified form, as a piece in the best non-quantum methods of today.

15.3 Trial Division and the Sieve of Eratosthenes (c. 284–204 BCE)

To factor a number, one may simply try dividing by each number less than that number. For example, to factor 391, one would first try 2 as a factor. It doesn't work. One would then go on to try 3, 4, 5, 6, 7, ..., 390, until a factor is found. We quickly find ways to improve this approach. If 2 is not a factor, then 4, 6, 8, etc., won't be factors. If 3 is not a factor, then 6, 9, 12, 15, etc., won't be factors. Hence, we only need to check prime numbers when seeking factors; that is, we need

[17] http://www.rsa.com/rsalabs/node.asp?id=2094#WhyIs.
[18] Kleinjung, Thorsten, Kazumaro Aoki, Jens Franke, Arjen K. Lenstra, Emmanuel Thomé, Joppe W. Bos, Pierrick Gaudry, Alexander Kruppa, Peter L. Montgomery, Dag Arne Osvik, Herman te Riele, Andrey Timofeev, and Paul Zimmermann, "Factorization of a 768-Bit RSA Modulus," in Rabin, Tal, editor, *Advances in Cryptology – CRYPTO 2010 Proceedings*, Lecture Notes in Computer Science, Vol. 6223, Springer, Berlin, Germany, 2010, pp. 333–350, available online at https://link.springer.com/content/pdf/10.1007%2F978-3-642-14623-7_18.pdf.
[19] Bai, Shi, Emmanuel Thomé, and Paul Zimmermann "Factorisation of RSA-704 with CADO-NFS," 2012, hal-00760322f, 4 pages, available online at https://hal.inria.fr/file/index/docid/760322/filename/369.pdf.
[20] Gauss, Carl Friedrich, *Disquisitiones Arithmeticae*, Gerhard Fleischer, Leipzig, 1801, Article 329.

only test the numbers 2, 3, 5, 7, 11, 13, 17, and 19. You'll notice that we stopped far short of 391 this time. We really only need to check up to the square root of the number we're trying to factor. If a number, *n*, is the product of two smaller numbers, both numbers cannot exceed the square root of *n* or the product would exceed *n*.

Figure 15.4 Eratosthenes (c. 275–192 BCE). (http://www.livius.org/gi-gr/greeks/scientists. html).

This sort of approach is the idea behind the sieve of Eratosthenes (Figure 15.4), the purpose of which is to eliminate the composite numbers from a list of consecutive integers. To begin we write out the first few natural numbers:

```
 1   2   3   4   5   6   7   8   9  10
11  12  13  14  15  16  17  18  19  20
21  22  23  24  25  26  27  28  29  30
31  32  33  34  35  36  37  38  39  40
41  42  43  44  45  46  47  48  49  50
51  52  53  54  55  56  57  58  59  60
61  62  63  64  65  66  67  68  69  70
71  72  73  74  75  76  77  78  79  80
81  82  83  84  85  86  87  88  89  90
91  92  93  94  95  96  97  98  99 100
```

In order to be prime, a number must have exactly two distinct positive divisors. The first natural number, 1, has only one positive divisor, so it is not prime and we cross it out:

```
 ■   2   3   4   5   6   7   8   9  10
11  12  13  14  15  16  17  18  19  20
21  22  23  24  25  26  27  28  29  30
31  32  33  34  35  36  37  38  39  40
41  42  43  44  45  46  47  48  49  50
51  52  53  54  55  56  57  58  59  60
61  62  63  64  65  66  67  68  69  70
71  72  73  74  75  76  77  78  79  80
81  82  83  84  85  86  87  88  89  90
91  92  93  94  95  96  97  98  99 100
```

Our next entry, 2, is prime. We now cross out all multiples of 2 (except 2 itself). The number 2 is boldfaced to stress its primality:

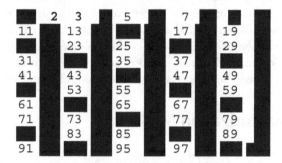

After 2, the next remaining entry, 3, is prime. We now cross out all multiples of 3:

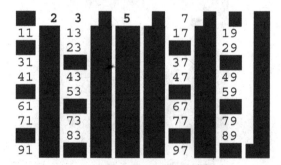

The next prime is 5. We now cross out all multiples of 5:

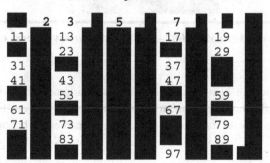

The next prime is 7. We now cross out all multiples of 7:

The next prime is 11. We could now cross out all multiples of 11, but this has already been done, as these numbers, less than 100, all have smaller prime divisors. In fact, all of the numbers that remain are prime. If one weren't, it would have to factor into a pair of numbers, both of which exceed 10. This is an impossibility, as the product must be less than 100. As explained earlier, we only need to search up to the square root of the number.

> One of the students of the great poet Callimachus was *Eratosthenes of Cyrene* (c. 275–192 BCE), who became librarian in the Museum, the scientific institute of Alexandria. He invented a new method to calculate prime numbers, drew a famous world map, catalogued several hundreds of stars, but became especially famous for his calculation of the circumference of the earth, based on the angle of the shadow that the sun made over a vertical pole at Alexandria at noon and the fact that at the same time, the sun light fell straight into a well as Syene in southern Egypt. He concluded that the circumference was 45,460 kilometers, which is pretty close to the real figure. He also wrote a treatise on chronology and a book on musical theory, composed poems and comedies, and was responsible for two dictionaries and a book on grammar. As an ethnologist, he suggested that the common division between civilized people and barbarians was invalid. Eratosthenes was nicknamed *bêta* or 'number two', because in no branch of science was he ever the best, although he excelled in nearly every one of them.[21]

15.4 Fermat's Factorization Method

Figure 15.5 Pierre de Fermat (1601–1665). (http://en.wikipedia.org/wiki/File:Pierre_de_Fermat.png.)

We now jump ahead to the 17th century and a factorization method due to Pierre de Fermat (Figure 15.5). Let n be the number we are attempting to factor. If we can find a representation of n as the difference of two squares (i.e., $n = x^2 - y^2$), then we can easily factor n as $n = (x - y)(x + y)$. How practical is this? The first important question to ask is whether or not every composite

---

[21] Biography by Jona Lendering taken from http://www.livius.org/gi-gr/greeks/scientists.html.

number n can be represented in this manner. Consider $n = ab$, where a and b are both greater than one and neither need be prime. Letting $x = (a + b)/2$ and $y = (a - b)/2$, we see

$$x^2 - y^2 = \left(\frac{a+b}{2}\right)^2 - \left(\frac{a-b}{2}\right)^2 = \left(\frac{a^2 + 2ab + b^2}{4}\right) - \left(\frac{a^2 - 2ab + b^2}{4}\right) = \frac{4ab}{4} = ab$$

Thus, such a representation always exists.

An example will make it clear how this observation may be applied. Consider $n = 391$. We start with

$$x = \lceil \sqrt{n} \rceil = 2$$

The $\lceil \; \rceil$ notation is referred to as the *ceiling function*. It indicates that the quantity within should be rounded up to the closest integer. In Section 14.7, a related function called the *floor function* will also be needed. It is notated with $\lfloor \; \rfloor$ and means that the quantity within should be rounded down to the closest integer. Most programming languages have commands built in to perform these operations.

Using $x = 20$, we continue like so:

$$x^2 - n = 400 - 391 = 9 = \text{a perfect square, namely } 3^2$$

Thus, $y = 3$ and we have $n = 20^2 - 3^2 = (20 - 3)(20 + 3) = (17)(23)$. The factorization is revealed! If the first value for x yielded a value for $x^2 - n$ that wasn't a perfect square, we would increment x by one and try again until a solution is found. In general, quicker factorization algorithms now exist, but Fermat's method has an advantage when the prime factors are close together. Thus, when picking primes p and q for RSA, we need to avoid values such that $p - q < 2n^{1/4}$. There's nothing magical about this particular bound. It just roughly indicates for which primes Fermat's method will be practical.

15.5 Euler's Factorization Method

Figure 15.6 Leonhard Euler (1707–1783). (http://en.wikipedia.org/wiki/File:Leonhard_Euler.jpg).

I tell math majors that Leonhard Euler (Figure 15.6) should be mentioned at least once in every mathematics course, and if he isn't, they should go to the registrar's office and ask for a tuition refund, for something important was left out! We've already seen how Euler's generalization of Fermat's little theorem paved the way for RSA encryption. We now examine Euler's method for factoring.

Instead of looking at *n* as the difference of two squares, we may often view it as the sum of two squares (in two different ways); for example $130 = 11^2 + 3^2 = 7^2 + 9^2$. Let us assume that we can do this for a number that we are attempting to factor. That is,

$$n = a^2 + b^2 = c^2 + d^2.$$
$$\Rightarrow a^2 - c^2 = d^2 - b^2$$
$$\Rightarrow (a-c)(a+c) = (d-b)(d+b)$$

Letting *g* denote the greatest common divisor of $a - c$ and $d - b$, we have $a - c = gx$ and $d - b = gy$ for some relatively prime integers *x* and *y*. So, our previous line now gives us (by substitution)

$$(gx)(a+c) = (gy)(d+b)$$
$$\Rightarrow (x)(a+c) = (y)(d+b) \tag{15.1}$$

The right-hand side is divisible by *y*, so the left-hand side must also be divisible by *y*. However, *y* doesn't divide *x*, as *x* and *y* are relatively prime. Therefore *y* must divide $a + c$. That is, there exists *m* such that $ym = a + c$. By substituting for $(a + c)$ in equation 15.1, we get $(y)(d + b) = (x)(ym)$. We can then divide through by *y* to get $d + b = xm$.

Now consider the product of

$$\left[\left(\frac{g}{2} \right)^2 + \left(\frac{m}{2} \right)^2 \right] \quad \text{and} \quad (x^2 + y^2).$$

$$\left[\left(\frac{g}{2} \right)^2 + \left(\frac{m}{2} \right)^2 \right](x^2 + y^2) = \left[\frac{g^2}{4} + \frac{m^2}{4} \right](x^2 + y^2)$$

$$= \frac{1}{4}(g^2 + m^2)(x^2 + y^2)$$

$$= \frac{1}{4}\left[(gx)^2 + (gy)^2 + (mx)^2 + (my)^2 \right]$$

$$= \frac{1}{4}\left[(a-c)^2 + (d-b)^2 + (d+b)^2 + (a+c)^2 \right]$$

$$= \frac{1}{4}\left[a^2 - 2ac + c^2 + d^2 - 2bd + b^2 + d^2 + 2bd + b^2 + a^2 + 2ac + c^2 \right]$$

$$= \frac{1}{4}\left[2a^2 + 2b^2 + 2c^2 + 2d^2 \right]$$

$$= \frac{1}{4}\left[2(a^2 + b^2) + 2(c^2 + d^2) \right]$$

$$= \frac{1}{4}(2n + 2n)$$

$$= \frac{1}{4}(4n)$$

$$= n$$

So, the product we were considering is a factorization of n, as desired. Thus, once the representation of n as a sum of two squares, in two ways, is obtained, one may calculate g, m, x, and y to obtain a factorization. The problem is efficiently finding the necessary sums!

We now make another big leap in time to a living mathematician, John Pollard, who has several factorization algorithms named after him.

15.6 Pollard's $p - 1$ Algorithm

Recall Fermat's Little Theorem:

> If p is a prime and a is not a multiple of p, then $a^{p-1} = 1 \pmod{p}$.

So, if m is a multiple of $p - 1$, then $a^m = 1 \pmod{p}$. In other words, p divides $a^m - 1$. So, $\gcd(a^m - 1, n)$, where $a^m - 1$ is first reduced mod n, might reveal a factorization of n, because p divides both $a^m - 1$ and n.

But how can we find a multiple $m > 1$ of $p - 1$? Well, we *hope* that $p - 1$ is B-smooth (i.e., all prime factors of $p - 1$ are less than B)[22] for some small B. We let m be defined as the product of all primes q less than B. Just as a capital sigma denotes a summation, a capital pi denotes a product. We have

$$m = \prod_{q \leq B} q \quad q \text{ prime.}$$

Example 2

Using $n = 713$ as a small example to illustrate this method, we may take $a = 2$ and $B = 5$. Then $m = (2)(3)(5) = 30$. It follows that $\gcd(a^m - 1, n) = \gcd(2^{30} - 1, 713) = 31$. Division then reveals that $713 = (31)(23)$.

This approach won't work with $n = 85$, no matter how large we make B. The reason for this is that the prime factors of 85 are $p = 5$ and $q = 17$, so when we look at $p - 1$ and $q - 1$, we get 4 and 16. No matter how many primes we multiply together, we'll never get a multiple of either of these numbers, unless we repeat the prime factor 2. In case $p - 1$ and $q - 1$ both have a repeated prime factor, we must modify the way in which we defined m. Possible patches include

$$m = B! \quad \text{and} \quad m = \prod_{q \leq B} q^n$$

where the product runs over all prime factors q less than B, and n is some positive integer greater than 1.

There are other twists we can put on this algorithm, but the above gives the general idea. The time needed for this method is roughly proportional to the largest prime factor of $p - 1$. Hence, it is efficient for p such that $p - 1$ is smooth. To resist this attack, choose primes of the form $2q + 1$, where q is prime. That way, $p - 1$ will have at least one large factor. There is also Pollard's ρ (*rho*)

---

[22] There are a few terms in mathematics that sound like the names of gangsta rappers: B-smooth, 2-pi, cube root, etc.

Algorithm (1975) for factoring, but it is strongest when the number being attacked has small factors, which is certainly not the case for RSA, so it won't be detailed here.[23]

15.7 Dixon's Algorithm[24]

Figure 15.7   John D. Dixon. (Courtesy of John D. Dixon.)

John D. Dixon's (Figure 15.7) algorithm has its roots in Fermat's method of factorization. Rather than insist on finding x and y such that $x^2 - y^2 = n$, in order to factor n, we could content ourselves with an x and y such that $x^2 - y^2 = kn$. This may also be expressed as $x^2 = y^2 \pmod{n}$. So, if we can find two squares, x^2 and y^2, that are equal modulo n, we may have a factorization given by $(x - y)(x + y)$. It's only "may," because this broadening of Fermat's method allows the possibility that $(x - y) = k$ and $(x + y) = n$. This idea goes back to Maurice Kraitchik in the 1920s.[25]

We can find potential x and y values quicker by not insisting that y be a perfect square.

$$20^2 = 3^2 \pmod{391} \Rightarrow 20^2 - 3^2 = 391 \ \left(\text{from the Fermat example}\right)$$

But, if this didn't work we could investigate further:

$$21^2 = 50 = (2)(5)^2 \pmod{391}$$
$$22^2 = 93 = (3)(31) \pmod{391}$$
$$23^2 = 138 = (2)(3)(23) \pmod{391}$$
$$24^2 = 185 = (5)(537) \pmod{391}$$
$$25^2 = 234 = (2)(117) \pmod{391}$$
$$26^2 = 285 = (5)(57) \pmod{391}$$
$$27^2 = 338 = (2)(13)^2 \pmod{391}$$
$$28^2 = 2 = (2) \pmod{391}$$

[23] See Pollard, John M., "A Monte Carlo Method for Factorization," *BIT Numerical Mathematics*, Vol. 15, No. 3, September 1975, pp. 331–334.

[24] Dixon, John D., "Asymptotically Fast Factorization of Integers," *Mathematics of Computation*, Vol. 36, No. 153, January 1981, pp. 255–260.

[25] Pomerance, Carl, "A Tale of Two Sieves," *Notices of the American Mathematical Society*, Vol. 43, No. 12, December 1996, pp. 1473–1485, p. 1474 cited here.

Multiplying these last two together, we have

$$(27)^2(28)^2 = (2)(13)^2(2) = (2)^2(13)^2 \pmod{391}$$
$$(27 \cdot 28)^2 = (2 \cdot 13)^2 \pmod{391}$$
$$(756)^2 = (26)^2 \pmod{391}$$
$$(365)^2 = (26)^2 \pmod{391}$$

So we get

$$(365 - 26)(365 + 26) = 132{,}549 = kn$$
$$132{,}549 / 391 = 339$$

We get one factor equaling k and the other equaling n in this "factorization"—not what we wanted! We can get even more possible combinations by using negatives, as well.

$$21^2 = 50 = (2)(5)^2 \pmod{391} \qquad 21^2 = -341 = (-1)(11)(31) \pmod{391}$$
$$22^2 = 93 = (3)(31) \pmod{391} \qquad 22^2 = -298 = (-1)(2)(149) \pmod{391}$$
$$23^2 = 138 = (2)(3)(23) \pmod{391} \qquad 23^2 = -253 = (-1)(11)(23) \pmod{391}$$
$$24^2 = 185 = (5)(537) \pmod{391} \qquad 24^2 = -206 = (-1)(2)(103) \pmod{391}$$
$$25^2 = 234 = (2)(117) \pmod{391} \qquad 25^2 = -157 = (-1)(157) \pmod{391}$$
$$26^2 = 285 = (5)(57) \pmod{391} \qquad 26^2 = -106 = (-1)(2)(53) \pmod{391}$$
$$27^2 = 338 = (2)(13)^2 \pmod{391} \qquad 27^2 = -53 = (-1)(53) \pmod{391}$$
$$28^2 = 2 = (2) \pmod{391} \qquad 28^2 = -389 = (-1)(389) \pmod{391}$$
$$29^2 = 59 = (59) \pmod{391} \qquad 29^2 = -332 = (-1)(2)^2(83) \pmod{391}$$
$$30^2 = 118 = (2)(59) \pmod{391} \qquad 30^2 = -273 = (-1)(3)(7)(13) \pmod{391}$$
$$31^2 = 179 = (179) \pmod{391} \qquad 31^2 = -212 = (-1)(2)^2(53) \pmod{391}$$

Piecing together 28^2 from the first column and 26^2 and 31^2 from the second column, we get

$$(28)^2(26)^2(31)^2 = (2)(-1)(2)(53)(-1)(2)^2 (53) \pmod{391}$$
$$= (-1)^2(2)^4(53)^2 = \left[(2)^2(53)\right]^2 \pmod{391}$$

That is,

$$(28 \cdot 26 \cdot 31)^2 = \left(2^2 \cdot 53\right)^2 \pmod{391}.$$

$$(22{,}568)^2 = (212)^2 \pmod{391}$$

$$(281)^2 = (212)^2 \pmod{391}$$

$$(281 - 212) = 69$$

$$\gcd(391, 69) = 23$$

We've found a factor! But we'd like a better way to pick values to test. Ideally, we'd pick "most likely" candidates first. These are

$$\left\lfloor \sqrt{kn} \right\rfloor \text{ and } \left\lceil \sqrt{kn} \right\rceil, \text{ for } k = 1, 2, 3, \ldots$$

Notice that to generate this list we make use of the floor and ceiling functions, introduced in Section 15.4. Once we have the list, we'd like a quicker way to pull out potential solutions than just visual inspection. For a serious factoring problem, the list might contain thousands or even millions[26] of entries—visual inspection just won't do!

We accomplish this with linear algebra. We first establish a factor base that consists of −1 and the first few primes. For the example above, our factor base could be taken as {−1, 2, 3, 5, 7, 11, 13, 17, 19, 23, 29, 31, 37, 41, 43, 47, 51, 53}. Each number in our table may then be represented as a vector. If the power of a factor is odd, we place a 1 in that position of the vector. If a given factor is not present, or has an even power, we place a 0 in that position. A few examples follow:

```
                    {-1, 2, 3, 5, 7,11,13,17,19,23,29,31,37,41,43,47,51,53}
21² = (2)(5)²       [ 0, 1, 0, 0, 0, 0, 0, 0, 0, 0, 0, 0, 0, 0, 0, 0, 0, 0]
21² = (-1)(11)(31)  [ 1, 0, 0, 0, 0, 1, 0, 0, 0, 0, 0, 1, 0, 0, 0, 0, 0, 0]
22² = (3)(31)       [ 0, 0, 1, 0, 0, 0, 0, 0, 0, 0, 0, 1, 0, 0, 0, 0, 0, 0]
23² = (2)(3)(23)    [ 0, 1, 1, 0, 0, 0, 0, 0, 0, 1, 0, 0, 0, 0, 0, 0, 0, 0]
23² = (-1)(11)(23)  [ 1, 0, 0, 0, 0, 1, 0, 0, 0, 1, 0, 0, 0, 0, 0, 0, 0, 0]
```

If we stop our factor base at 53, we cannot include values such as our second factorization for 22^2, because it contained the factor 149. If we made our factor base larger, it could be accommodated.

Finding a potential solution is equivalent to selecting vectors whose sum modulo 2 is the zero vector. To find such solutions efficiently we may construct a matrix **M** whose columns are the vectors above and then look for a solution to the matrix equation

$$\mathbf{M}X = 0 \ (\text{modulo } 2),$$

where X is the column vector

$$\begin{pmatrix} x_1 \\ x_2 \\ \vdots \\ x_k \end{pmatrix}$$

for a factor base with k elements.

---

[26] For factoring a record-breakingly large prime, the factor base will now contain about a million values, according to Pomerance, Carl, "A Tale of Two Sieves," *Notices of the American Mathematical Society*, Vol. 43, No. 12, December 1996, pp. 1473–1485, p. 1483 cited here.

This approach was discovered by Michael Morrison and John Brillhart and published in 1975.[27] The pair used it, with some help from continued fractions, to factor the seventh Fermat number

$$F_7 = 2^{2^7} + 1$$

Having a large factor base increases the chances of finding a solution after a given number of values has been investigated, but it slows down the linear algebra step.

We look at this refinement with a second example. Suppose we wish to factor 5,141. We select the base {−1, 2, 3, 5, 7, 11, 13} and make a table of values. Note that one representation was chosen for each value. We could have listed two, but going with just the smaller of the two in absolute value gives us a factorization more likely to be useful.

k	$\lfloor \sqrt{kn} \rfloor$	$\lceil \sqrt{kn} \rceil$	*Previous Value Squared*
1	71		$5{,}041 = -100 = (-1)(2)^2(5)^2$
1		72	$5{,}184 = 40 = (2)^3(5)$
2	101		$10{,}201 = -81 = (-1)(3)^4$
2		102	$10{,}404 = 122 = (2)(61)$
3	124		$15{,}376 = -47 = (-1)(47)$
3		125	$15{,}625 = 202 = (2)(101)$
4	143		$20{,}449 = -115 = (-1)(5)(23)$
4		144	$20{,}736 = 172 = (2)^2(43)$
5	160		$25{,}600 = -105 = (-1)(3)(5)(7)$
5		161	$25{,}921 = 216 = (2)^3(3)^3$
6	175		$30{,}625 = -221 = (-1)(13)(17)$
6		176	$30{,}976 = 130 = (2)(5)(13)$
7	189		$35{,}721 = -266 = (-1)(2)(7)(19)$
7		190	$36{,}100 = 113 = (113)$
8	202		$40{,}804 = -324 = (-1)(2)^2(3)^4$

---

[27] Morrison, Michael A. and John Brillhart, "A Method of Factoring and the Factorization of F_7," *Mathematics of Computation*, Vol. 29, No. 129, January 1975, pp. 183–205.

We now delete all factorizations in the right-most column that aren't 13-smooth. We're left with:

k	$\lfloor \sqrt{kn} \rfloor$	$\lceil \sqrt{kn} \rceil$	Previous Value Squared
1	71		$5{,}041 = -100 = (-1)(2)^2(5)^2$
1		72	$5{,}184 = 40 = (2)^3(5)$
2	101		$10{,}201 = -81 = (-1)(3)^4$
5	160		$25{,}600 = -105 = (-1)(3)(5)(7)$
5		161	$25{,}921 = 216 = (2)^3(3)^3$
6		176	$30{,}976 = 130 = (2)(5)(13)$
8	202		$40{,}804 = -324 = (-1)(2)^2(3)^4$

Now we form the appropriate matrix equation.

$$\begin{pmatrix} 1 & 0 & 1 & 1 & 0 & 0 & 1 \\ 0 & 1 & 0 & 0 & 1 & 1 & 0 \\ 0 & 0 & 0 & 1 & 1 & 0 & 0 \\ 0 & 1 & 0 & 1 & 0 & 1 & 0 \\ 0 & 0 & 0 & 1 & 0 & 0 & 0 \\ 0 & 0 & 0 & 0 & 0 & 0 & 0 \\ 0 & 0 & 0 & 0 & 0 & 1 & 0 \end{pmatrix} \begin{pmatrix} x_1 \\ x_2 \\ x_3 \\ x_4 \\ x_5 \\ x_6 \\ x_7 \end{pmatrix} = \begin{pmatrix} 0 \\ 0 \\ 0 \\ 0 \\ 0 \\ 0 \\ 0 \end{pmatrix}$$

A little bit of linear algebra leads us to the following solutions for the column vector of xs.

$$\begin{pmatrix} 1 \\ 0 \\ 1 \\ 0 \\ 0 \\ 0 \\ 0 \end{pmatrix}, \begin{pmatrix} 1 \\ 0 \\ 0 \\ 0 \\ 0 \\ 0 \\ 1 \end{pmatrix}, \begin{pmatrix} 1 \\ 0 \\ 1 \\ 0 \\ 0 \\ 0 \\ 1 \end{pmatrix}.$$

The first solution uses the relations

$$(71)^2 = 5{,}041 = -100 = (-1)(2)^2(5)^2 \pmod{5{,}141}$$

and

$$(101)^2 = 10{,}201 = -81 = (-1)(3)^4 \pmod{5{,}141}$$

Combining them, we have

$$(71 \cdot 101)^2 = (-1)^2(2 \cdot 5 \cdot 9)^2 \pmod{5{,}141}$$

$$(71 \cdot 101)^2 = (2 \cdot 5 \cdot 9)^2 (\text{mod } 5{,}141)$$

$$(7{,}171)^2 = (90)^2 (\text{mod } 5{,}141)$$

$$(2{,}030)^2 = (90)^2 (\text{mod } 5{,}141)$$

$$2{,}030 - 90 = 1{,}940$$

$$\gcd(1940, 5141) = 97$$

We may then divide 5,141/97 = 53 to get the complete factorization: 5,141 = (97)(53). If the gcd had been 1, we would have gone on to try the second, and possibly the third solution.

Although the method detailed here is named after Dixon, Dixon himself humbly pointed out that[28]

> …the method goes back much further than my paper in 1981. References to my paper sometimes do not seem to appreciate this fact. For example, the entry in Wikipedia seems to credit the idea to me. The fact is the idea in one form or another had been around for a much longer time, but no-one had been able to give a rigorous analysis of the time complexity of the versions which were used. What I did in my paper was to show that a randomized version of the method can be analyzed and that (at least qualitatively and asymptotically) it is faster than other known methods (in particular, subexponential in log N). As far as I know, except for improved constants, this is still true…

> I did not suggest that the randomized version which I described would be competitive in practice with algorithms which were currently in use (but most of which still have no rigorous analysis). I do not think anyone has seriously tried to factor a large number using random squares.

Dixon didn't know the whole history when he published his 1981 paper, but he did include it in a later paper.[29] In what seems to be a theme with important work in cryptology in recent decades, Dixons's 1981 paper was rejected by the first journal to which he submitted it.[30]

15.7.1 The Quadratic Sieve

When using Dixon's algorithm, much time may be wasted in factoring numbers that end up not being B-smooth, and are therefore discarded prior to the matrix step. The quadratic sieve eliminates much of this inefficiency by quickly discarding numbers that aren't B-smooth by using the Sieve of Eratosthenes with a twist. When running through the numbers, instead of crossing them out if they are divisible by a prime, we divide them by the highest power of the particular prime that goes into them. After sieving through in this manner with all of the primes in the factor

---

[28] Email from John Dixon to the author, October 27, 2010.
[29] Dixon, John D., "Factorization and Primality Tests," *American Mathematical Monthly*, Vol. 91, No. 6, June-July 1984, pp. 333–352 (see Section 11 for the historical background).
[30] Email from John D. Dixon to the author, October 27, 2010.

base, there is a 1 in every position of our list that contained a *B*-smooth number. The numbers represented by other values are discarded. This is an oversimplification; there are several shortcuts that make the process run much quicker, but it conveys the general idea. The interested reader can pursue the references for further details.

When Martin Gardner provided the first published description of RSA, he gave his readers a chance to cryptanalyse the new system:

> As a challenge to *Scientific American* readers the M.I.T. group has encoded another message, using the same public algorithm. The ciphertext is:
>
> 9686 9613 7546 2206
> 1477 1409 2225 4355
> 8829 0575 9991 1245
> 7431 9874 6951 2093
> 0816 2982 2514 5708
> 3569 3147 6622 8839
> 8962 8013 3919 9055
> 1829 9451 5781 5154
>
> Its plaintext is an English sentence. It was first changed to a number by the standard method explained above, then the entire number was raised to the 9,007th power (modulo *r*) by the shortcut method given in the memorandum. To the first person who decodes this message the M.I.T. group will give $100.[31]

The modulus, which Gardner labeled as *r*, would now be written as *n*. In any case, it was the following number, which became known as RSA-129, as it is 129 digits long.

114381625757888867669235779976146612010218296721242362562561842935
706935245733897830597123563958705058989075147599290026879543541

Gardner didn't expect a solution in his lifetime, but a combination of increased computing power, improved factoring algorithms, and Gardner's own longevity resulted in his seeing a solution on April 26, 1994. The factors were

3490529510847650949147849619903898133417764638493387843990820577

and

32769132993266709549961988190834461413177642967992942539798288533.

They were determined using a quadratic sieve and the plaintext turned out to be[32]

    The magic words are squeamish ossifrage.

For numbers up to about 110 digits (in base 10), the quadratic sieve is the best general method for factoring presently available. For larger values, the number field sieve is superior.[33]

---

[31] Gardner, Martin, "Mathematical Games, A New Kind of Cipher That Would Take Millions of Years to Break," *Scientific American*, Vol. 237, No. 2, August 1977, pp. 120–124, text from p. 123, ciphertext from p. 121.

[32] Hayes, Brian P., "The Magic Words are Squeamish Ossifrage," *American Scientist*, Vol. 82. No. 4, July–August 1994, pp. 312–316.

[33] Stamp, Mark and Richard M. Low, *Applied Cryptanalysis: Breaking Ciphers in the Real World*, John Wiley & Sons, Hoboken, New Jersey, 2007, p. 316.

15.8 Pollard's Number Field Sieve[34]

New methods aren't always quickly embraced, even in mathematics. Some were skeptical about Pollard's latest effort, but when the ninth Fermat number, $F_9 = 2^{2^9} + 1$, was factored using the number field sieve in 1990, its value became apparent. Carl Pomerance summed up the significance of this event:[35]

> This sensational achievement announced to the world that Pollard's number field sieve had arrived.

A description of the algorithm requires a background in modern algebra that is beyond the scope of this text; however, there are some elements in common with simpler methods. For example, sieving remains the most time-consuming step in this improved algorithm.

In 2003, Adi Shamir (the "S" in RSA), along with Eran Tromer, published designs for specialized hardware to perform factorizations based on the number field sieve.[36] They named it TWIRL, which is short for The Weizmann Institute Relation Locator, with the Weizmann Institute being their employer. "Relation Locator" refers to finding factoring relations in a matrix, like the one described in Section 15.7. The pair estimated that $10 million worth of hardware would be sufficient for a machine of this design to complete the sieving step for a 1,024-bit RSA key in less than a year.

RSA remains as secure against factoring attacks as it was when it was first created. Improved factoring techniques and improved hardware have simply forced users to use longer keys. If TWIRL concerns you, simply use a 2,048-bit key. Your biggest concern, as seen in Section 15.1.12, is making sure the primes used were generated in as random a manner as possible!

15.8.1 Other Methods

There are many other algorithms for factoring, including ones that make use of continued fractions (used in some of the above!) and elliptic curves. Perhaps most intriguing is an approach put forth by Peter Shor in 1994 that factors in polynomial time, provided that one has a quantum computer to run it on.[37] The reader is encouraged to pursue the references to learn more.

---

[34] Lenstra, Arjen K., Hendrik W. Lenstra, Jr., Mark S. Manasse, and John M. Pollard, "The Number Field Sieve," in Lenstra, Arjen K. and Hendrik W. Lenstra, Jr., editors, *The Development of the Number Field Sieve*, Lecture Notes in Mathematics, Vol. 1554, Springer, Berlin, Germany, 1993, pp. 11–42.

[35] Pomerance, Carl, "A Tale of Two Sieves," *Notices of the American Mathematical Society*, Vol. 43, No. 12, December 1996, pp. 1473–1485, p. 1480 cited here.

[36] Shamir, Adi and Eran Tromer, "Factoring Large Numbers with the Twirl Device," in Boneh, Dan, editor, *Advances in Cryptology – CRYPTO 2003 Proceedings*, Lecture Notes in Computer Science, Vol. 2729, Springer, Berlin, Germany, 2003, pp. 1–27.

[37] Shor, Peter, "Algorithms for quantum computation: discrete logarithms and factoring," in Goldwasser, Shafi, editor, *35th Annual IEEE Symposium on Foundations of Computer Science (FOCS)*, IEEE Computer Society Press, Los Alamitos, California, 1994, pp. 124–134.

15.8.2 Cryptological Humor

Although 2 is not a large prime, it is the only even prime. I learned this is in a boring book title *Even Prime Numbers*.[38] Of course, being the *only* even prime, makes it the oddest prime of all.

References and Further Reading

On Non-Factoring RSA Attacks

Acıiçmez, Onur, Çetin Kaya Koç, and Jean-Pierre Seifert, "On the Power of Simple Branch Prediction Analysis," in Deng, Robert and Pierangela Samarati, editors, *ASIACCS '07: Proceedings of the 2nd ACM Symposium on Information, Computer and Communications Security*, ACM, New York, 2007, pp. 312–320. This is one of the attacks against RSA that wasn't discussed in this chapter. Like the timing attack, it assumes access to the intended recipient's machine.

Blakley, G. Robert, and Itshak Borosh, "Rivest-Shamir-Adleman Public Key Cryptosystems do not Always Conceal Messages," *Computers & Mathematics with Applications*, Vol. 5, No. 3, 1979, pp. 169–178.

Boneh, Dan, "Twenty Years of Attacks on the RSA Cryptosystem," *Notices of the American Mathematical Society*, Vol. 46, No. 2, 1999, pp. 203–213. This is a nice survey paper.

Boneh, Dan and Glenn Durfee, "Cryptanalysis of RSA with Private Key d Less than $N^{0.292}$," in Stern, Jacques, editor, *Advances in Cryptology – EUROCRYPT '99 Proceedings*, Lecture Notes in Computer Science, Vol. 1592, Springer, Berlin, Germany, 1999, pp. 1–11.

Boneh, Dan, Glenn Durfee, and Yair Frankel, "An Attack on RSA Given a Small Fraction of the Private Key Bits," in Ohta, Kazuo and Dingyi Pei, editors, *Advances in Cryptology – ASIACRYPT '98*, Lecture Notes in Computer Science, Vol. 1514, Springer, Berlin, Germany, 1998, pp. 25–34.

Boneh, Dan, Antoine Joux, and Phong Q. Nguyen, "Why Textbook ElGamal and RSA Encryption are Insecure," in Okamoto, Tatsuaki, editor, *Advances in Cryptology – ASIACRYPT 2000 Proceedings*, Lecture Notes in Computer Science, Vol. 1976, Springer, Berlin, Germany, 2000, pp. 30–43.

Coppersmith, Don, "Small Solutions to Polynomial Equations, and Low Exponent RSA Vulnerabilities," *Journal of Cryptology*, Vol. 10, No. 4, December 1997, pp. 233–260.

DeLaurentis, John M., "A Further Weakness in the Common Modulus Protocol for the RSA Cryptoalgorithm," *Cryptologia*, Vol. 8, No. 3, July 1984, pp. 253–259.

Diffie, Whitfield and Hellman, Martin, "New Directions in Cryptography," *IEEE Transactions on Information Theory*, Vol. IT-22, No. 6, November 1976, pp. 644–654.

Hayes, Brian P., "The Magic Words Are Squeamish Ossifrage," *American Scientist*, Vol. 82, No. 4, July-August 1994, pp. 312–316. This paper discusses factoring techniques and includes some history. The title celebrates the recovery of the message concealed by the RSA challenge cipher in Martin Gardner's *Scientific American* column.

Johnson, Don, *ECC, Future Resiliency and High Security Systems*, Certicom Whitepaper, March 30, 1999, revised July 6, 1999, available online at http://web.archive.org/web/20040215121823/www.comms.engg.susx.ac.uk/fft/crypto/ECCFut.pdf.

Joye, Marc and Jean-Jacques Quisquater, *Faulty RSA Encryption*, UCL Crypto Group Technical Report CG-1997/8, Université Catholique de Louvain, Louvain-La-Neuve, Belgium, 1987.

Kaliski, Burt and Matt Robshaw, "The Secure Use of RSA," *CryptoBytes*, Vol. 1, No. 3, Autumn 1995, pp. 7–13, available online at ftp://ftp.rsa.com/pub/cryptobytes/crypto1n3.pdf. This paper summarizes various attacks on RSA.

Kocher, Paul, "Timing Attacks on Implementations of Diffie-Hellman, RSA, DSS, and Other Systems," in Koblitz, Neal, editor, *Advances in Cryptology – CRYPTO '96 Proceedings*, Lecture Notes in Computer Science, Vol. 1109, Springer, Berlin, Germany, 1996, pp. 104–113.

---

[38] By the author of *Groups of Order One*.

Lenstra, Arjen K., James P. Hughes, Maxime Augier, Joppe W. Bos, Thorsten Kleinjung, and Christophe Wachter, "Ron was wrong, Whit is right," February 2012, https://anonymous-proxy-servers.net/paper/064.pdf.

Levy, Steven, *Crypto: How the Code Rebels Beat the Government, Saving Privacy in the Digital Age*, Viking, New York, 2001.

Rivest, Ronald L., Adi Shamir, and Leonard Adleman, *On Digital Signatures and Public-key Cryptosystems* (There was soon a title change to *A Method for Obtaining Digital Signatures and Public-key Cryptosystems*. The date is the same for both.), MIT Laboratory for Computer Science Report MIT/LCS/TM 82, Cambridge, Massachusetts, April 1977. This report later appeared as cited in the reference below.

Rivest, Ronald L., Adi Shamir, and Leonard Adleman, "A Method for Obtaining Digital Signatures and Public-key Cryptosystems," *Communications of the ACM*, Vol. 21, No. 2, February 1978.

Robinson, Sara, "Still Guarding Secrets after Years of Attacks, RSA Earns Accolades for its Founders," *SIAM News*, Vol. 36, No. 5, June 2003, pp. 1–4.

Simmons, Gustavus J., "A "Weak" Privacy Protocol Using the RSA Crypto Algorithm," *Cryptologia*, Vol. 7, No. 2, April 1983, pp. 180–182.

Wiener, Michael J., "Cryptanalysis of Short RSA Secret Exponents," *IEEE Transactions on Information Theory*, Vol. 36, No. 3, May 1990, pp. 553–558.

On Factoring

Bach, E. and J. Shallit, "Factoring with Cyclotomic Polynomials," *Mathematics of Computation*, Vol. 52, No. 185, January 1989, pp. 201–209.

Dixon, John D., "Asymptotically Fast Factorization of Integers," *Mathematics of Computation*, Vol. 36, No. 153, January 1981, pp. 255–260.

Dixon, John D., "Factorization and Primality Tests," *American Mathematical Monthly*, Vol. 91, No. 6, June–July 1984, pp. 333–352. See Section 11 of this paper for the historical background.

Lenstra, Jr., Hendrik W., "Factoring Integers with Elliptic Curves," *Annals of Mathematics*, Vol. 126, No. 3, November 1987, pp. 649–673.

Lenstra, Arjen K., Hendrik W. Lenstra, Jr., Mark S. Manasse, and John M. Pollard, "The Number Field Sieve," in Lenstra, Arjen K. and Hendrik W. Lenstra, Jr., editors, *The Development of the Number Field Sieve*, Lecture Notes in Mathematics, Vol. 1554, Springer, 1993, pp. 11–42.

Lenstra, Arjen K., "Factoring," in Tel, Gerard and Paul Vitányi, editors, *Distributed Algorithms, WDAG 1994*, Lecture Notes in Computer Science, Vol. 857, Springer, Berlin, Germany, 1994, pp. 28–38.

Lenstra, Arjen K., Eran Tromer, Adi Shamir, Wil Kortsmit, Bruce Dodson, James Hughes, and Paul Leyland, "Factoring Estimates for a 1024-Bit RSA Modulus," in Laih, Chi Sung, editor, *Advances in Cryptology – ASIACRYPT 2003 Proceedings*, Lecture Notes in Computer Science, Vol. 2894, Springer, Berlin, Germany, 2003, pp. 55–74. This was a follow-up paper to the one above.

Montgomery, Peter L. and Robert D. Silverman, "An FFT Extension to the $P - 1$ Factoring Algorithm," *Mathematics of Computation*, Vol. 54, No. 190, April 1990, pp. 839–854.

Morrison, Michael A. and John Brillhart, "A Method of Factoring and the Factorization of F_7," *Mathematics of Computation*, Vol. 29, No. 129, January 1975, pp. 183–205. This paper describes factoring with continued fractions.

Pollard, John M., "Theorems on Factorization and Primality Testing," *Proceedings of the Cambridge Philosophical Society*, Vol. 76, No. 3, November 1974, pp. 521–528.

Pollard, John M., "A Monte-Carlo Method for Factorization," *Bit Numerical Mathematics*, Vol. 15, No. 3, 1975, pp. 331–334.

Pollard, John M., Home Page, https://sites.google.com/site/jmptidcott2/.

Pomerance, Carl, and Samuel S. Wagstaff, Jr., "Implementation of the Continued Fraction Integer Factoring Algorithm," *Congressus Numerantium*, Vol. 37, 1983, pp. 99–118.

Pomerance, Carl, "The Quadratic Sieve Factoring Algorithm," in Beth, Thomas, Norbert Cot, and Ingemar Ingemarsson, editors, *Advances in Cryptology, Proceedings of EUROCRYPT 84*, Lecture Notes in Computer Science, Vol. 209, Springer, Berlin, Germany, 1985, pp. 169–182, available online at www.math.dartmouth.edu/~carlp/PDF/paper52.pdf.

Pomerance, Carl, "A Tale of Two Sieves," *Notices of the American Mathematical Society*, Vol. 43, No. 12, December 1996, pp. 1473–1485.

Shamir, Adi and Eran Tromer, "Factoring Large Numbers with the Twirl Device," in Boneh, Dan, editor, *Advances in Cryptology – CRYPTO 2003 Proceedings*, Lecture Notes in Computer Science, Vol. 2729, Springer, Berlin, Germany, 2003, pp. 1–27.

Shor, Peter, "Algorithms for Quantum Computation: Discrete Logarithms and Factoring," in Goldwasser, Shafi, editor, *35th Annual IEEE Symposium on Foundations of Computer Science (FOCS)*, IEEE Computer Society Press, Los Alamitos, California, 1994, pp. 124–134.

Williams, Hugh C. and Jeffrey O. Shallit, "Factoring Integers before Computers," in Gautschi, Walter, editor, *Mathematics of Computation 1943–1993: A Half-Century of Computational Mathematics, Proceedings of Symposia in Applied Mathematics*, Vol. 48, American Mathematical Society, Providence, Rhode Island, 1994, pp. 481–531.

Chapter 16

Primality Testing and Complexity Theory

Thus, even starting with the most fundamental and ancient ideas concerning prime numbers, one can quickly reach the fringe of modern research. Given the millennia that people have contemplated prime numbers, our continuing ignorance concerning the primes is stultifying.

—**Richard Crandall and Carl Pomerance**[1]

16.1 Some Facts about Primes

The primes p and q used in public key cryptography must be large, as the method is insecure if an attacker is able to factor their product. Fortunately, plenty of large primes exist. It has been known since the time of Euclid that there are infinitely many primes, and therefore infinitely many larger than any given size. A proof follows.

Theorem: There are infinitely many primes.

Proof by contradiction: Suppose there are only finitely many primes. Let S be the set of all primes, p_1 through p_k. Now consider $n = p_1 p_2 \ldots p_k + 1$ (the product of all primes + 1). This number is certainly larger than any in S, so if it is prime we have a contradiction. We also have a contradiction if it is not prime. Because all of the numbers in S leave a remainder of 1 when divided into n, the prime factors of n cannot be in S. Hence, the set S cannot contain every prime, and we see that there must be infinitely many primes. Q.E.D.

There are several other proofs.[2] One of these follows immediately from establishing that

$$\sum_i \frac{1}{p_i}$$

---

[1] Crandall, Richard E. and Carl Pomerance, *Prime Numbers: A Computational Perspective*, Springer, New York, 2001, pp. 6–7.
[2] See Ribenboim, Paulo, *The New Book of Prime Number Records*, Springer, New York, 1996, pp. 3–18.

diverges. If this series converged, the number of primes could be infinite or finite, but divergence leaves only one option. If there were finitely many primes, the series would have to converge. So, there cannot be a largest prime. There is, however, a largest *known* prime. Table 16.1 lists the top 10 largest known primes. The meaning of GIMPS, Seventeen or Bust, and Mersenne in this table are explained Section 16.4.2.

Table 16.1   Top 10 Largest Known Primes (as of October 12, 2020)

Rank	Prime	Digits	Discoverer	Year	Reference
1	$2^{82589933} - 1$	24862048	GIMPS	2018	Mersenne 51?
2	$2^{77232917} - 1$	23249425	GIMPS	2018	Mersenne 50?
3	$2^{74207281} - 1$	22338618	GIMPS	2016	Mersenne 49?
4	$2^{57885161} - 1$	17425170	GIMPS	2013	Mersenne 48?
5	$2^{43112609} - 1$	12978189	GIMPS	2008	Mersenne 47
6	$2^{42643801} - 1$	12837064	GIMPS	2009	Mersenne 46
7	$2^{37156667} - 1$	11185272	GIMPS	2008	Mersenne 45
8	$2^{32582657} - 1$	9808358	GIMPS	2006	Mersenne 44
9	$10223 \times 2^{31172165} + 1$	9383761	Seventeen or Bust	2016	
10	$2^{30402457} - 1$	9152052	GIMPS	2005	Mersenne 43

Source: http://primes.utm.edu/largest.html.

Even though there are infinitely many primes, there are still arbitrarily long sequences of integers (without skipping any) that do not contain any primes. If you want n integers in a row, all of which are composite (nonprime), here you are!

$$(n+1)! + 2, \quad (n+1)! + 3, \quad (n+1)! + 4, \quad \ldots, \quad (n+1)! + n, \quad (n+1)! + n + 1$$

The first is divisible by 2, the second by 3, and so on.

The first gap between primes that contains 1,000 composites occurs right after the prime 1,693,182,318,746,371. This prime is followed by 1,131 composites. This fact was discovered by Bertil Nyman, a Swedish nuclear physicist.[3] Even though, gaps aside, we have plenty of primes to choose from (recall there are infinitely many), they become increasingly rare as a percent of the total, as the length of the number we desire grows.

The function $\pi(n)$ is defined to be the number of primes less than or equal to n. Some sample values are provided in the table below:

n	$\pi(n)$
10	4
100	25
1,000	168

---

[3] Caldwell, Chris K. and G. L. Honaker, Jr., *Prime Curios! The Dictionary of Prime Number Trivia*, CreateSpace, Seattle, Washington, 2009, p. 218.

So, 40% of the first 10 integers are prime, but only 16.8% of the first 1,000 are prime. $\pi(n)$ may be calculated for any value of n by simply testing the primality of each positive integer less than or equal to n. However, if n is large, this is a very time-consuming method. It would be nice if there were some expression of $\pi(n)$ that is easier to evaluate such as $p(n) = \lfloor 1.591 + 0.242n - 0.0000752n^2 \rfloor$. Recall that $\lfloor n \rfloor$ denotes the greatest integer less than or equal to n. By plugging in values, we see that $p(n)$ works great for 10, 100, and 1,000, but it becomes inaccurate for higher values of n, as can be seen by expanding the table above. Also, it doesn't do well for most values under 1,000, other than the ones in the table![4]

n	$\pi(n)$
10	4
100	25
1,000	168
10,000	1,229
100,000	9,592
1,000000	78,498
10,000,000	664,579
100,000,000	5,761,455
1,000,000,000	50,847,534
10,000,000,000	455,052,511

German mathematician Carl Friedrich Gauss (1777–1855) came up with the *prime number theorem*, but was unable to prove it. It says:

$$\pi(n) \sim \frac{n}{\ln(n)}$$

We saw the $\sim$ notation previously in Section 1.6, where Stirling's formula was presented. Recall that it is pronounced "asymptotically approaches" and means that the ratio of the two quantities, $\pi(n)$ and $\dfrac{n}{\ln(n)}$, in this case, approaches 1 as n approaches infinity.

Gauss came up with another estimate, the logarithmic integral of n, which is written $li(n)$. It also asymptotically approaches $\pi(n)$ as $n \rightarrow \infty$, and it gives more accurate estimates when n is small.

$$li(n) = \int_{2}^{n} \frac{1}{\ln(x)}\, dx$$

That these functions converge to $\pi(n)$ was proven in 1896, independently, by Jacques Hadamard and C. J. de la Vallée-Poussin. Their proofs used the Riemann Zeta function.

---

[4] The function $p(n)$ was obtained by using Lagrangian interpolation on the values for which it was seen to work perfectly.

Testing values, it appears that $li(n) > \pi(n)$ for all n; however, this is not true. Although it is ridiculously large, there is a point at which $li(n)$ switches from being an overestimate to being an underestimate. Stanley Skewes, a South African mathematician, investigated where this change-over occurs. He could not establish an exact value, but he did find a bound. The switch takes place somewhere before $e^{e^{e^{79}}}$. This bound assumes that the Riemann hypothesis holds.[5] Without this assumption, Skewes found the larger bound $10^{10^{10^{963}}}$.[6] It was learned, before Skewes made his calculations, that $li(n)$ eventually switches back to being an underestimate again. In fact, it switches between an overestimate and an underestimate infinitely many times![7] For many years, Skewes's numbers held the record for being the largest numbers that ever served a useful purpose (i.e., used in a proof), but they have since been dwarfed by other values. They've also been diminished in another way—much smaller bounds have been found for where $li(n)$ first transitions to an underestimate.

Still the question remains, how can we find or generate large prime numbers? *Primality testing* is concerned with deciding whether or not a given number is prime. For the quickest tests, the revelation that a number is not prime is made without revealing any of the factors. They're like existence theorems for nontrivial, proper factors.

We first look at some probabilistic tests. These tests can sometimes prove a number is composite, but they can never quite prove primality. The best they can do is suggest primality, with arbitrarily high probabilities.

16.2 The Fermat Test

Recall Fermat's Little Theorem (1640):

If p is a prime and a is not a multiple of p, then $a^{p-1} = 1 \pmod{p}$.

This offers a test that may show a number to be composite, for if $(a, n) = 1$ and $a^{n-1} \neq 1 \pmod{n}$, then n cannot be prime. However, if the result is 1, we can't conclude that n is prime. Fermat's Little Theorem is not an "if and only if" theorem.

As usual, a proof was not offered by Fermat. The first to verify this with a proof was Gottfried von Leibniz. The special case for $a = 2$ was known to the Chinese as far back as 500 BCE, but they didn't have the generalization Fermat provided.[8]

To test a number n for primality, we can use values for a from the set $\{2, 3, 4, \ldots, n-1\}$. If one of these yields anything other than 1, we know n is composite. The repeated squaring technique of exponentiation can be used to make these calculations go a bit quicker.

[5] Skewes, Stanley, "On the Difference $\pi(x)$ – Li(x)," *Journal of the London Mathematical Society*, Vol. 8 (Series 1), No. 4, 1933, pp. 277–283.

[6] Skewes, Stanley, "On the Difference $\pi(x)$ – Li(x) (II)," *Proceedings of the London Mathematical Society*, Vol. 5 (Series 3), No. 1, 1955, pp. 48–70.

[7] Littlewood, John Edensor, "Sur la Distribution des Nombres Premiers," *Comptes Rendus*, Vol. 158, 1914, pp. 1869–1872.

[8] McGregor-Dorsey, Zachary Strider, "Methods of primality testing," *MIT Undergraduate Journal of Mathematics*, Vol. 1, 1999, pp. 133–141.

Example 1 ($n = 391$)

To calculate 2^{390} modulo 391, we use the repeated squaring technique introduced in Section 14.3. We first calculate 2 to various powers of 2 modulo 391:

$$2^2 = 4$$
$$2^4 = 16$$
$$2^8 = 256$$
$$2^{16} = 65,536 = 239 \,(\text{mod } 391)$$
$$2^{32} = 57,121 = 35 \,(\text{mod } 391)$$
$$2^{64} = 1,225 = 52 \,(\text{mod } 391)$$
$$2^{128} = 2,704 = 358 \,(\text{mod } 391)$$
$$2^{256} = 128,164 = 307 \,(\text{mod } 391)$$

and then multiply appropriate values to get the desired power:

$$2^{390} = \left(2^2\right)\left(2^4\right)\left(2^{128}\right)\left(2^{256}\right) = (4)(16)(358)(307) = 7,033,984 = 285 \,(\text{mod } 391).$$

Because 2^{390} did not simplify to 1, we can conclude that 391 is not prime.

This test gives us no indication what the factors of 391 are. Also, this test doesn't always work so nicely! The base 2 is nice to use for testing purposes, but it won't unmask all composites.

Example 2 ($n = 341$)

$2^{340} = 1 \,(\text{mod } 341)$, so we cannot draw an immediate conclusion. We then check $3^{340} \,(\text{mod } 341)$ and get 56. We may now conclude that 341 is composite. Because 341 was able to sneak by the base 2 test, we call 341 a *base 2 pseudoprime*. Base 3 revealed the composite nature of 341, but sometimes neither 2 nor 3 will reveal composites.

Example 3 ($n = 1,729$)

For this example, $2^{1,728} = 1 \,(\text{mod } 1,729)$, so we cannot draw an immediate conclusion. We then check $3^{1728} \,(\text{mod } 1,729)$ and get 1 again. We still cannot draw a conclusion. Continuing on with other bases, we *always* get 1, if the base is relatively prime to 1,729. It's tempting to conclude that 1,729 is prime. But, because Fermat's Little Theorem isn't "if and only if," we haven't proven anything. In fact there is strong evidence that 1,729 is not prime, such as the fact that $1,729 = (7)(13)(19)$!

Using a base that's less than 1,729, but not relatively prime to it, will reveal 1,729 to be composite, but such a number would be a factor of 1,729. It would be quicker to use trial division, if we need to find a base that is a factor of a number to prove it is composite.

A composite number n for which every base relatively prime to n yields 1 is called a *Carmichael number* after Robert Carmichael (Figure 16.1).

For years it was an open problem to determine the number of Carmichael numbers, but in 1994 it was shown that there are infinitely many.[9] Carmichael found the first of the numbers that

---

[9] Alford, W. R., A. Granville and Carl Pomerance, "There are Infinitely Many Carmichael Numbers," *Annals of Mathematics*, Vol. 139, No. 3, May 1994, pp. 703–722.

would be named after him in 1910. The first few Carmichael numbers[10] are 561, 1105, 1729, 2465, 2821, 6601, 8911, 10585,…

Figure 16.1 Robert Carmichael (1879–1967). (From https://web.archive.org/web/20121003061628/ http://www.maa.org/aboutmaa/carmichael.gif.[11] Copyright Mathematical Association of America, 2012. All rights reserved.)

There are only 20,138,200 Carmichael numbers less than 10^{21}, which is about 0.000000000002% of the total.[12] Thus, the odds that a randomly chosen number will be a Carmichael number are very small. Nevertheless, we'd like to have a primality test that isn't fooled by these numbers.

16.3 The Miller–Rabin Test[13]

This is the most popular scheme for testing the large numbers needed for RSA encryption for primality. To test whether or not n is prime, we begin by finding the highest power of 2 that divides $n - 1$. We label this power t and then define $d = (n - 1)/(2^t)$. We have $2^t d = n - 1$. (Note that d

---

[10] "A002997, Carmichael numbers: composite numbers n such that a^(n – 1) == 1 (mod n) for every a coprime to n." *The On-Line Encyclopedia of Integer Sequences®*, http://oeis.org/A002997.

[11] This website is the result of following a link from https://web.archive.org/web/20120214210612/http://www. maa.org/aboutmaa/maaapresidents.html. Other links lead to images of each of the Mathematical Association of America's presidents from 1916 to 2010. Carmichael was president in 1923.

[12] Pinch, Richard G. E., *The Carmichael Numbers up to 10^{21}*, May 15, 2007, available online at http://s369624816. websitehome.co.uk/rgep/p82.pdf.

[13] Weisstein, Eric W., "Primality Test," *MathWorld*, A Wolfram Web Resource, https://mathworld.wolfram.com/ PrimalityTest.html refers to this as the Rabin-Miller test, as does Bruce Schneier in the second edition of *Applied Cryptography*, p. 259. Richard A. Mollin adds a name to get Miller–Rabin-Selfridge. He explains that John Selfridge deserves this recognition, as he was using the test in 1974, before it was published by Miller. See Mollin, Richard A., *An Introduction to Cryptography*, Chapman & Hall / CRC, Boca Raton, Florida, 2001, p. 191. If you want to avoid names altogether, it is also referred to as the *strong pseudoprimality test*. Whatever you choose to call it, the primary reference is Rabin, Michael O., "Probabilistic Algorithm for Testing Primality," *Journal of Number Theory*, Vol. 12, No. 1, February 1980, pp. 128–138.

must be odd.) We then pick an integer $a < n$. If n is prime, and a is relatively prime to n, then one of the following must hold:

1. $a^d = 1 \pmod{n}$
2. $a^{2^s d} = -1 \pmod{n}$ for some $0 \leq s < t - 1$

Thus, if neither holds true, we know n cannot be prime.

It is known that for every n that is composite, at least 75% of the choices for a will reveal that fact via the test above. Passing the test for a particular base doesn't prove the number is prime, but the test can be repeated with different values for a. Passing the test for a different value represents an independent event, so the probability of passing after m bases have been investigated is less than $(1/4)^m$, if n is composite. Thus, we can test until the probability is vanishingly small.

Example 4

Because $n = 1729$ caused trouble earlier, we'll investigate this value with our new test. We have $n - 1 = 1728$, which is divisible by 2^6, but not by 2^7, so $t = 6$. $1728/(2^6) = 27$, so we have $d = 27$ and $(2^6)(27) = n - 1$. We're now ready to investigate condition 1, above. Picking $a = 2$ and calculating $a^d \pmod{n}$, we get

$$2^{27} \pmod{1729} = 645.$$

Because we didn't get 1, we need to consider condition 2. To do so, we calculate

$$a^{2^s d} \text{ for } 0 \leq t < t - 1 = 5$$

We get the following:

s	$a^{2^s d} \pmod{n}$
0	645 (this was already calculated in the step above)
2	1,065
3	1
4	1
5	1

None, of the values for $a^{2^s d} \pmod{n}$ yield -1, so condition 2 is not satisfied. Because neither condition 1 nor condition 2 holds, 1,729 fails the Miller–Rabin Test and cannot be prime.

Because there are guaranteed to be bases that reveal n to be composite, if it is composite, we could make this a deterministic test by testing every base less than the number. However, that would be silly, as the time required to do so would well exceed the time needed to test for primality by trial division; hence, we refer to the Miller–Rabin test as a probabilistic test. Rabin and Miller are pictured in Figures 16.2 and 16.3. Why is Rabin smiling? Perhaps he could see the future, as revealed in the caption to Figure 16.2

Naturally, some enjoy the challenge of finding numbers that fool primality tests. François Arnault did so for the Miller–Rabin test, as implemented by the computer algebra system

Figure 16.2   Michael O. Rabin (1931–). Rabin split the $1 million Dan David Prize with two others. (From SEAS, Michael O. Rabin Wins Dan David Prize: Computer Science Pioneer Shares $1 Million Prize for Outstanding Achievements in the Field [press release], Harvard School of Engineering and Applied Science, February 16, 2010.)

Figure 16.3   Gary L. Miller. (http://www.cs.cmu.edu/~glmiller/).

ScratchPad.[14] Implementations such as this apply the test to a small set of bases. The composite number that squeaked by all of these was 1195068768795265792518361315725116351898245581. However, Arnault noted that after he found this exception ScratchPad was improved, and renamed Axiom, with a new primality test that isn't just Miller–Rabin, and which recognizes the number above as composite.

---

[14] Arnault, François, "Rabin-Miller Primality Test: Composite Numbers Which Pass It," *Mathematics of Computation*, Vol. 64, No. 209, January 1995, pp. 355–361.

It is an intriguing possibility that there may be a hybrid system that catches all composites. In other words, we may have two tests, each of which misses certain numbers, but if there is no overlap between these sets of missed numbers then passing both tests guarantees primality.

16.3.1 Generating Primes

The actual generation of large primes for use with RSA is usually done as follows:

1. Generate a random string of bits of the desired size. Make sure the first bit is 1 to guarantee the number has a minimum size, and make sure the last bit is 1, so the number isn't even (and therefore composite).
2. Test the number to be sure it isn't divisible by any small primes (less than a million, for example).
3. Apply the Miller–Rabin Test for a sufficient number of rounds to obtain a probability of primality that is comfortably close to 1.

We now take a look at a few tests that are guaranteed to yield a correct answer. These are known as *deterministic tests*. Unlike probabilistic tests, there is no uncertainty. The downside is that they are not as fast as the Miller–Rabin Test.

16.4 Deterministic Tests for Primality

An old result that leads to a deterministic test for primality is Wilson's Theorem:

Let n be a positive integer. Then n is prime if and only if $(n-1)! = -1 \pmod{n}$.

This result is named after British mathematician John Wilson (1741–1793), although he was not the first to find it, publish it, or prove it. Those accomplishments go to Leibniz, Waring,[15] and Lagrange,[16] respectively.

Wilson's theorem is elegant, but it is not a quick way to test for primality! Probabilistic tests, while less satisfying, are more practical, when there is a huge savings in time.

Leonard Adleman, along with Carl Pomerance and, Robert S. Rumely, developed a deterministic test for primality in 1983 that was the quickest algorithm of its kind for a while. It was not as fast as Miller–Rabin, but it did offer absolute certainty.[17] Another deterministic test was provided by Adleman and Huang using elliptic curves, but this test also took unacceptably long compared to probabilistic tests.

16.4.1 The AKS Primality Test (2002)

Finally in 2002, the first deterministic polynomial time test for primality was found by Professor Manindra Agrawal, and two students, Neeraj Kayal and Nitin Saxena (Figure 16.4), at the Indian

---

[15] Waring, Edward, *Meditationes Algebraicae*, Cambridge University Press, Cambridge, UK, 1770.

[16] In 1773, according to Weisstein, Eric W, "Wilson's Theorem." *MathWorld*, A Wolfram Web Resource, http://mathworld.wolfram.com/WilsonsTheorem.html.

[17] Adleman, Leonard, Carl Pomerance and, Robert S. Rumely, "On Distinguishing Prime Numbers From Composite Numbers," *Annals of Mathematics*, Vol. 117, No. 1, January 1983, pp. 173–206.

Institute of Technology in Kanpur.[18] The students were in graduate school when the proof was completed, but were undergraduates when most of the work was done.[19]

Figure 16.4   Manindra Agrawal. (http://www.cse.iitk.ac.in/users/manindra/), Neeraj Kayal (http://research.microsoft.com/en-us/people/neeraka/), and Nitin Saxena (http://www.math. uni-bonn.de/people/saxena/to_photos.html).

Just as the encryption scheme developed by Rivest, Shamir, and Adleman became known as RSA, so the primality test of Agrawal, Kayal, and Saxena became known as AKS (Figure 16.5).

Input: integer $n > 1$.

[1]   If $n = a^b$ for $a \in N$ and $b > 1$, output COMPOSITE.

[2]   Find the smallest r such that $O_r(n) > \log^2 n$.

[3]   If $1 < (a, n) < n$ for some $a \le r$, output COMPOSITE.

[4]   If $n \le r$, output PRIME.

[5]   For $a = 1$ to $\left\lfloor \sqrt{\phi(r)} \log n \right\rfloor$ do

   if $((X + a)^n \ne X^n + a \pmod{X^r - 1, \ n})$, ouput COMPOSITE.

[6]   Output PRIME.

Figure 16.5   The AKS Algorithm (From Agrawal, Manindra, Neeraj Kayal, and Nitin Saxena, "PRIMES Is in P," *Annals of Mathematics*, Second Series, Vol. 160, No. 2, September 2004, pp. 781–793, p. 784 cited here. With permission.)

The algorithm is very simple, but a few notations need to be explained.

- $O_r(n)$ is the order of n modulo r; that is, the smallest value k such that $n^k = 1 \pmod{r}$.
- ϕ is the Euler phi function, represented here as ϕ. See Section 14.3 of this book.
- log denotes a base 2 logarithm.

---

[18] It was posted online in August 2002, and appeared in print over two years later, in September 2004. See Agrawal, Manindra, Neeraj Kayal, and Nitin Saxena, "PRIMES is in P," *Annals of Mathematics*, Second Series, Vol. 160, No. 2, September 2004, pp. 781–793, available online at http://www.math.princeton.edu/~annals/ issues/2004/Sept2004/Agrawal.pdf.

[19] Aaronson, Scott, *The Prime Facts: From Euclid to AKS*, http://www.scottaaronson.com/writings/prime.pdf, 2003, p. 10.

- (mod $X^r - 1$, n) means divide by the polynomial $X^r - 1$ and take the remainder, and also reduce all coefficients modulo n.

Step 5 in Figure 16.5 makes use of what is sometimes called "freshman exponentiation." Usually a student who expands out $(X + a)^n$ as $X^n + a^n$, will lose points, but the work is correct, if the expansion is carried out modulo n, where n is a prime. In fact, we can simplify further, if a is relatively prime to the modulus n. In that case, Fermat's little theorem tells us $a^n = a$ (mod n). Thus, when n is prime, we have $(X + a)^n = X^n + a$. If equality doesn't hold, n must be composite. A close look at Step 5 reveals that we are not just moding out by n, but also by $X^r - 1$. This helps to speed up what would otherwise be a time-consuming calculation, as we cannot apply the shortcuts described above without knowing n is prime!

Example 5 ($n = 77$)

1. We cannot express n in the form a^b, where a is a natural number and b is an integer, so we cannot draw a conclusion at this point.

2. Find the smallest r such that $O_r(n) > \log^2 n = (\log 77)^2 \approx (6.266787)^2 \approx 39.27$.

   We need the order of 77 modulo r to be 40 or higher. The order of an element divides the order of the group, so we need a group with 40 or more elements. We start with $r = 41$, because the (multiplicative) group of integers modulo 41 has 40 elements, and find that it works. The order of 77 (mod 41) is 41.

3. Now check if $1 < (a, 77) < 77$ for some $a \leq r = 41$.

   This happens for $a = 7$ (and other values), so we stop and declare 77 to be composite. It may seem like a silly way to test, because step 3, by itself, takes longer than trial division would, but remember that this is just a small example to illustrate how the test works. For numbers of the size we're interested in, this test is quicker than trial division.

Example 6 ($n = 29$)

1. We cannot express n in the form a^b, where a is a natural number and b is an integer, so we cannot draw a conclusion at this point.

2. Find the smallest r such that $O_r(n) > \log^2 n = (\log 29)^2 \approx (4.85798)^2 \approx 23.6$.

   We need the order of 29 modulo r to be 24 or higher. The order of an element divides the order of the group, so we need a group with 24 or more elements. We find that $r = 31$ is the smallest possibility. The (multiplicative) group of integers modulo 31 has 30 elements, and the order of 29 (mod 31) is 31.

3. Now check if $1 < (a, 29) < 29$ for some $a \leq r = 31$. There is no such a.

4. Because $n < r$, we conclude n is prime.

   Rather than roll out a third example to show how step 5 may come to be used, we continue with Example 6, pretending that r didn't exceed n. To do so, we'll pretend r was 24.[20]

---

[20] We couldn't use $r = 24$ in Example 6, because the (multiplicative) group of integers modulo 24 (after discarding the values that don't have inverses) consists of only eight elements.

5. Our instructions for this step are shown below:

For $a = 1$ to $\lfloor \sqrt{\phi(r)} \log(n) \rfloor$ do

If $((X + a)^n \neq X^n + a \pmod{X^r - 1, n})$, output COMPOSITE;

We have $\phi(r) = \phi(24) = 8$, because there are 8 numbers (1, 5, 7, 11, 13, 17, 19, 23) less than 24 and relatively prime to 24.

$\lfloor \sqrt{8} \log(29) \rfloor \approx \lfloor \sqrt{8} (4.85798) \rfloor \approx 13$

For the longest step, we verify that $(X + a)^{29} = X^{29} + a \pmod{X^{24} - 1, 29}$, for all a such that $1 \leq a \leq 13$.

6. Because we made it to step 6, we conclude that 29 is prime.

Oddly, for someone who made such an important contribution, Agrawal doesn't admit to a strong passion for cryptology:

I have a peripheral interest in Cryptography, Complex Analysis, and Combinatorics.

—**Manindra Agrawal**[21]

I wish my peripheral vision was as good as his!

16.4.2 GIMPS

Some numbers are easier to test for primality than others. Mersenne numbers, which have the form $2^n - 1$, are currently the easiest. For values of n, such as $n = 2$ or 3, that yield primes, we call the numbers *Mersenne primes*, after the French mathematician Marin Mersenne (1588–1648). The ease of testing such numbers is why the top 8 largest known primes are all Mersenne primes. Another advantage numbers of this form have, when it comes to a chance of making it on the top 10 list, is that anyone with a computer may download a program that allows him or her to join in the testing. The program, Great Internet Mersenne Prime Search (GIMPS, for short), allows people to donate otherwise idle time on their personal computers to testing numbers of the form $2^n - 1$ for primality. In the top 10 list, the number referenced as Mersenne 47 is known to be the 47th Mersenne prime. By contrast, the number referenced as Mersenne 48? is known to be a Mersenne prime, but it might not be the *48th* such number. That is, mathematicians have yet to prove that there is no Mersenne prime between Mersenne 47 and this number.

There is one prime on the current top 10 list that is not a Mersenne prime and wasn't discovered through GIMPS. It was credited to "Seventeen or Bust," which is another program that anyone may download to help search for primes taking a special form, in this case the form $k \cdot 2^n + 1$ for certain values of k.[22]

Primes having special forms should not be considered for cryptographic purposes. Also, bigger isn't always better. If your RSA modulus has 578,028,320,322,400 digits, how long do you think it will take an attacker to figure out what two primes you multiplied together, when there are so few primes known with over 20 million digits? Proving extremely large numbers to be prime is not done for the sake of applications, but rather for love of the game. If you need to get something useful out of the search for record-breakingly large primes, how about $100,000?

---

[21] http://www.cse.iitk.ac.in/users/manindra/.

[22] For details of Seventeen or Bust and why some mathematicians care, see https://en.wikipedia.org/wiki/Seventeen_or_Bust, https://primes.utm.edu/bios/page.php?id=429, and http://www.prothsearch.com/sierp.html.

Perhaps the most valuable prime ever found was $2^{43112609} - 1$. When Edson Smith found this mathematical gem in 2008, it was the first one ever found with more than ten million digits, so he won $100,000. He used software provided by the Great Internet Mersenne Prime Search (GIMPS), so he will share the prize with them. And what of his part? He will give that money to University of California at Los Angeles' (UCLA's) mathematics department. It was their computers he used to find the prime.[23]

16.5 Complexity Classes, P vs. NP, and Probabilistic vs. Deterministic

The runtime of an algorithm is, of course, important across the board—not just for primality testing. The idea of measuring runtime as a function of the size of the input was first presented in 1965 by Juris Hartmanis and Richard E. Stearns.[24] If a program's runtime is always the same, regardless of the size of the input, it is said to run in constant time. Runtimes can also be linear, quadratic, cubic, etc. functions of the input size. Efficient programs solve problems in polynomial time (of one degree or another). This means that a polynomial function could be used to measure runtime as a function of the input size. Such problems are said to be in the complexity class **P** (for polynomial time). Of course, a given problem could often be solved in a variety of ways, not all of which are efficient. For some problems, the best known solutions grow exponentially, as a function of the input size. These are labeled **EXP**. We are interested in the most efficient solutions that can be found. Earlier in this chapter we saw that it was not until 2002 that a deterministic polynomial time test for factoring was found.

A problem falls in the complexity class **NP** (nondeterministic polynomial), if a solution can be checked in polynomial time. This class includes all of **P**, because for those problems, a solution could be checked in polynomial time by simply finding the solution and seeing if it matches what has been proposed. However, there are many problems in **NP** that are not known to belong to **P**. The fancy part of the name — nondeterministic — means that guessing is allowed, like in a probabilistic algorithm. It's just another way of expressing the definition I gave. The machine can guess at solutions and check each of them in polynomial time. If the number of possible solutions grows rapidly as a function of the input, the whole process may not run in polynomial time.

There are many other problems for which polynomial time solutions haven't yet been found, but proposed solutions can be checked in polynomial time. For example, you may not have a polynomial time algorithm to crack a ciphertext, but you can quickly check whether a particular key is correct.

A problem, is said to be **NP**-complete if it is **NP** and finding a polynomial time solution for it would imply there's a polynomial time solution for every **NP** problem; that is, a solution to one could be adapted to solve anything in **NP**. The complexity class **NP**-complete was first identified in 1972.[25]

---

[23] Caldwell, Chris and G. L. Honaker, Jr., *Prime Curios! The Dictionary of Prime Number Trivia*, CreateSpace, Seattle, Washington, 2009, p. 243.

[24] Hartmanis, Juris and Richard E. Stearns. "On the Computational Complexity of Algorithms," *Transactions of the American Mathematical Society*, Vol. 117, No. 5, May 1965, pp. 285–306.

[25] Karp, Richard M., "Reducibility Among Combinatorial Problems," in Miller, Raymond E., James W. Thatcher, and Jean D. Bohlinger, editors, *Complexity of Computer Computations*, Plenum, New York, 1972, pp. 85–103.

Another complexity class is labeled **NP**-hard. These problems are **NP**-complete, but there doesn't need to be a way to verify solutions in polynomial time. Thus, **NP**-hard properly contains **NP**-complete.

Thousands of problems are now known to be **NP**-complete. A few examples follow.

1. *Traveling salesman problem* — Suppose a traveling salesman wants to visit every state capital in the United States by car. What path would be the shortest? We could solve this problem by considering all of the possibilities. If we allow the salesman to start anywhere, there are 50! Possible routes, so although the solution is among them, we're not likely to find it.

2. *Knapsack problem* — The knapsack problem (aka the subset sum problem) consists of finding a selection of numbers from a given set such that the sum matches some desired value. For example, if our set is $S = \{4, 8, 13, 17, 21, 33, 95, 104, 243, 311, 400, 620, 698, 805, 818, 912\}$ and we wish to find values that sum to 666, we may take $620 + 21 + 17 + 8$. There may be no solution. In our example, we cannot obtain a sum of 20. However, a solution may be found, or its nonexistence demonstrated, simply by forming all possible subsets of S and checking their sums. This is clearly not practical for large S, because the number of subsets grows exponentially with the size of S (a set of n elements has 2^n subsets). The name of this problem comes from imagining the desired value as the size of a knapsack that we wish to completely fill, with the numbers representing the sizes of objects.

3. *Hamiltonian graph problem* — This is similar to the first example. Imagine the salesman is restricted from traveling directly from some cities to other cities. Pretend, for example, the highway from Harrisburg, PA to Annapolis, MD is one-way! If the salesman wants to go from Annapolis to Harrisburg, he must first head to some other capital. We'll investigate this problem in greater detail in Section 21.3.

4. *Decoding linear codes* — Let M be a $m \times n$ matrix of 0s and 1s. Let y be a vector with n components (i.e., an n-tuple), each of which is either 0 or 1. Finally, let k be a positive integer. The time-consuming question is this: Is there a vector x with m components, each of which is 0 or 1, but with no more than k 1s such that xM $= y$ (mod 2)?[26] Robert McEliece turned this into a public key cryptosystem in 1978.[27]

5. *Tetris* — Yes, the addictive game is **NP**-complete. A team of three computer scientists proved this in 2002.[28]

Some of the **NP**-complete problems have special cases that are easy to solve. This doesn't matter. Even if almost all instances of a problem can be solved rapidly, a problem could be classified as **NP**-complete or **NP**-hard or **EXP**. Complexity theory considers worst cases and is not concerned with the time required on average or in the best case.

It is possible that polynomial time solutions exist for all **NP** problems, and that mathematicians have just not been clever enough to find them. The fact that deterministic primality testing

---

[26] Talbot, John and Dominic Welsh, *Complexity and Cryptography An Introduction*, Cambridge University Press, Cambridge, UK, 2006, pp. 162–163.

[27] McEliece, Robert J., *A Public-Key Cryptosystem Based on Algebraic Coding Theory*, Deep Space Network Progress Report 42-44, Jet Propulsion Laboratory, California Institute of Technology, January and February 1978, pp. 114–116.

[28] Demaine, Erik D., Susan Hohenberger, and David Liben-Nowell, "Tetris is Hard, Even to Approximate," in Warnow, Tandy and Binhai Zhu, editors, *Computing and Combinatorics*, *9th Annual International Conference* (*COCOON 2003*), Lecture Notes in Computer Science Vol. 2697, 2003, Springer, Berlin, Germany, pp. 351–363.

resisted being place in **P** until the 21st century can be seen as evidence of this; however, it is widely believed that **P ≠ NP**.

A proof showing either **P = NP** or **P ≠ NP** is considered the Holy Grail of computer science. This problem was one of the seven Millennium Prize Problems, so a proof that withstands peer review will net the author a $1,000,000 prize from the Clay Mathematics Institute.[29]

Conferences are a great place to learn a bit of mathematics outside one's own specialty. At the 2008 Joint Mathematics Meetings in San Diego, I greatly enjoyed the AMS Josiah Willard Gibbs lecture delivered by Avi Wigderson of the Institute for Advanced Study. It was titled "Randomness—A Computational Complexity View." The main result was likely well-known to experts in complexity theory, but was new to me. It follows the conjectures below.

Conjecture 1: P ≠ NP. *That is, some* **NP** *problems require exponential time/size.* This seems likely. As there are now thousands of **NP**-complete problems that have been heavily studied and thus far resisted polynomial time solutions, it would be surprising if such solutions exist for them, as well as all other **NP** problems.

Conjecture 2: *There are problems that can be solved in polynomial time with probabilistic algorithms, but not with deterministic algorithms.* Again, this seems very reasonable. At first, the only polynomial time algorithms for primality testing were probabilistic. Eventually a deterministic algorithm was found, but it would be surprising if this could always be done. The weaker probabilistic tests should be quicker in some cases. After all, they don't give as firm an answer, so they should be faster.

And now for the shocker — there's a theorem (it's been proven!) that one of these conjectures must be wrong.[30] Unfortunately, the theorem doesn't tell us which one. Because we can only have one of the above conjectures, I'd bet on **P ≠ NP**. If this is true, the conclusion we can draw from the negation of conjecture 2 is that probabilistic algorithms aren't as powerful as they seem.

There are many other complexity classes. The above is not intended to be a survey of the field, but rather an introduction to some concepts that are especially relevant to cryptology.

16.5.1 Cryptologic Humor

I have a proof for **P = NP** in the special case where **N = 1**.

The paper[31] "Terms Used in the disciplines of Cryptography, IT Security and Risk Analysis" by John Gordon offered the following definition:

> **NP-hard** *a*. Non-Pharmacologically hard, *i.e.* not just due to Viagra.

16.6 Ralph Merkle's Public Key Systems

We'll soon look at an encryption system created by Ralph Merkle (Figure 16.6), based on an **NP**-complete problem, but first we examine Merkle's undergraduate work in cryptography.

---

[29] http://www.claymath.org/ is the general page for the Institute and http://www.claymath.org/millennium/ is the page for the prize problems.

[30] Impagliazzo, Russell and Avi Wigderson, "P=BPP unless E has Subexponential Circuits: Derandomizing the XOR Lemma," in *Proceedings of the 29th annual ACM Symposium on Theory of Computing* (*STOC 1997*), ACM Press, New York, 1997, pp. 220–229, available online at https://dl.acm.org/doi/pdf/10.1145/258533.258590.

[31] Gordon John, "Terms Used in the disciplines of Cryptography, IT Security and Risk Analysis," *Journal of Craptology*, Vol. 0, No. 0, December 1998. See http://www.anagram.com/jcrap/ for more information on this humorous journal, including the complete contents.

Figure 16.6 Ralph Merkle, Martin Hellman, and Whitfield Diffie. (http:engineering.stanford. edu/about/images/memories/pop_timemag.jpg Copyright Chuck Painter/Stanford News Service).

Merkle was enrolled in CS 244 Computer Security at the University of California at Berkeley in the fall of 1974, his last semester as an undergraduate. This course required a project. Each student had to submit two proposals and the professor would then use his broader experience to help steer the students in the right direction. For Project 1, Merkle proposed a scheme for public key cryptography, something that had not yet been done. The work of Diffie and Hellman was discussed earlier in this text, but it came later, historically. Merkle was the first. The first page of his proposal is reproduced in Figure 16.7. Be sure to read the professor's comment at the top.[32]

This proposal continued almost to the end of a sixth page. By contrast, Merkle's second project proposal followed, but weighed in at only 22 words, not counting the final sentence, "At this point, I must confess, that I am not entirely thrilled by the prospect of engaging in this project, and will expand upon it only if prodded."

Following his professor's negative reaction to his proposal, Merkle rewrote it, making it shorter and simpler. He showed the rewrite to the professor, but still failed to convince him of its value. Merkle then dropped the class, but didn't give up on his idea. He showed it to another faculty member who said, "Publish it, win fame and fortune!"[33]

So, in August of 1975, Merkle submitted a paper to *Communications of the ACM*, but a reviewer wrote the following to the editor:[34]

> I am sorry to have to inform you that the paper is not in the main stream of present cryptography thinking and I would not recommend that it be published in the *Communications of the ACM*.

---

[32] The full proposal may be found at Merkle, Ralph C., *Publishing a New Idea*, http://www.merkle.com/1974/. This is the source for the page reproduced here.

[33] Merkle, Ralph C., *Publishing a New Idea*, http://www.merkle.com/1974/.

[34] Merkle, Ralph C., *Publishing a New Idea*, http://www.merkle.com/1974/.

Project 2 looks more reasonable, maybe because your description of Project 1 is huddled terribly. Talk to me about them today.

C.S. 244
FALL 1974

Ralph Merkle

Project Proposal

Topic: Establishing secure communications between seperate secure sites over insecure communication lines.

Assumptions: No prior arrangements have been made between the two sites, and it is assumed that any information known at either site is known to the enemy. The sites, however, are now secure, and any new information will not be divulged.

Method 1: Guessing. Both sites guess at keywords. These guesses are one-way encrypted, and transmitted to the other site. If both sites should chance to guess at the same keyword, this fact will be discovered when the encrypted versions are compared, and this keyword will then be used to establish a communications link.

Discussion: No, I am not joking. If the keyword space is of size N, then the probability that both sites will guess at a common keyword rapidly approaches one after the number of guesses exceeds sqrt(N). Anyone listening in on the line must examine all N possibilities. In more concrete terms, if the two sites can process 1000 guesses per second, and desire to establish a link in roughly 10 seconds, then they can use a keword space of size $N=10,000^2=10^8$. If the enemy is presumed to have a comprable technology, i.e., 1000 guesses/sec, then he can consider all 10^8 possibilities in $10^8/10^3$ seconds, or 10^5 seconds, which is about one day. As the

Figure 16.7 The first page of Merkle's proposal.

The editor, in her rejection letter to Merkle, added that she "was particularly bothered by the fact that there are no references to the literature."[35] There *couldn't* be any references to the literature, because it was a brand-new idea! On a personal note, knowing that reviewers will sometimes look at a paper's reference list before reading it, to see if the author has "done his homework," and then factor that into their decision, I always make sure my papers have many references.

Merkle didn't give up. He revised the paper and eventually (almost three years later!) it was published in *Communications of the ACM*.[36] By this time, Merkle was not the first to publish on the topic of public key cryptography, although he was the first to conceive it, write it up, and submit it. What lessons can we learn from this? I think there are three.

1. Undergraduates can make important contributions. We've seen that repeatedly in the history of cryptology.
2. Be persistent. If Merkle had let the negative reactions of his professor, the reviewer, and the editor discourage him, his discovery would never have been recognized.
3. Communication skills are important, even for math and computer science students. It doesn't do you any good to be the deepest thinker on a topic, if you cannot eventually get your ideas across to duller minds. I think writing skills are important, even if the undergraduates I know don't all agree. Merkle's professor eventually became aware that he had rejected a good idea. He blamed his error on a combination of "Merkle's abstruse writing style and his own failings as a mathematician."[37]

In Merkle's system, two people wishing to communicate over an insecure channel who have not agreed on a key ahead of time, may come to an agreement over the insecure channel. Merkle referred to these people as X and Y (this was before Alice and Bob came on the scene). Person X sends person Y "puzzles." These puzzles can take the form of enciphered messages, using any traditional symmetric system. Merkle's example was Lucifer, a precursor to DES. The key can be weakened by only using 30 bits of it, for example, and fixing the rest. It should be weakened enough that the recipient can break it with some effort. It's assumed that no method quicker than brute-force is available for solving these puzzles.

Upon receiving N of these puzzles, person Y tries to break one. Whichever puzzle Y selects, the plaintext will be a unique puzzle number and a unique key intended for use with a traditional system. Y then sends the puzzle number back to X. Having kept a list of the puzzle numbers and corresponding keys that he sent to Y, X now knows what key Y has revealed by solving that particular puzzle. Now that both X and Y have agreed on a key, they may use it to converse securely over the insecure channel. The only information an eavesdropper will have picked up is what puzzle number led to the key, not the key itself. The eavesdropper could solve all of the puzzles and thereby uncover the key, but if N is large, this may not be practical. On average, an eavesdropper would find the key after cracking half of the messages.

One of the seven references in the published version of Merkle's paper was "Hiding Information and Receipts in Trap Door Knapsacks" by Merkle and Hellman, which had been accepted to appear in *IEEE Transactions on Information Theory*. We'll now examine the idea presented in that paper.

---

[35] Merkle, Ralph C., *Publishing a New Idea*, http://www.merkle.com/1974/.

[36] The final published version had seven references.

[37] Levy, Steven, *Crypto: How the Code Rebels Beat the Government, Saving Privacy in the Digital Age*, Viking, New York, 2001, p. 80.

16.7 Knapsack Encryption

The knapsack problem is known to be **NP**-complete, so it should make a very firm foundation to build a cryptosystem upon. After all, factoring, although believed to be hard, has eluded attempts to pin it down to being **NP**-complete, and look at how successful RSA has been. A system with a firmer foundation might be used with even greater confidence, but how could it be done? Merkle and Hellman worked out the details for their 1978 paper referred to above. Their solution follows. We'll use the knapsack from our brief list of **NP**-complete problems:

$$S = \{4, 8, 13, 17, 21, 33, 95, 104, 243, 311, 400, 620, 698, 805, 818, 912\}$$

Given a selection of elements of S, their sum is easily computed, but the inverse problem appears intractable (not in **P**), so we'll use it for a one-way function.

Our toy example for S has 16 elements. We can represent any combination of them with a 16-bit string. For example 0000000000010100 represents $620 + 805 = 1425$. Defining a function f that takes such strings to the sums they represent, we can write $f(0000000000010100) = 1425$.

Now, if we wish to send the message HELP IS ON THE WAY, we may convert each of the letters to their ASCII bit representations and encipher in 16-bit blocks.[38] Of course, this is just a fancy way of making a digraphic substitution (as each character is 8 bits) and it is not a public key system. Also, we need to be careful! Is it possible that 1425 has more than one solution? If so, decipherment will not be unique and the recipient will have to make choices. If S is chosen carefully, this problem can be avoided. The trick is to make the elements of S such that, when ordered from least to greatest, the value of every element exceeds the sum of all those that came before.[39] However, this ruins our one-way function. Although each number that can be obtained from summing elements of S now has only one such solution, it is easy to find—we simply apply the greedy algorithm of taking the largest element of S that doesn't exceed our given number and, after subtracting it form our desired total, repeat the process as many times as necessary until we get to zero.

Fortunately, there is a way to disguise our knapsack so that an attacker will not be able to take advantage of this simple approach. We multiply every element in our knapsack by some number m and then reduce the result modulo n. The modulus needs to exceed the sum of the elements of S, and m should be relatively prime to n, to guarantee m^{-1} exists modulo n. The disguised knapsack serves as the public key and is used as described above. The private portion of the key is m^{-1} and n. It is also a good idea to keep m secret, but it isn't needed for decryption. The recipient simply multiplies each block by m^{-1} modulo n and uses this value to solve the knapsack problem using the fast algorithm that goes with the original superincreasing knapsack. The answer he gets will also work for the value sent using the disguised knapsack, and thus yield the desired plaintext.

Example 7

We start with the superincreasing knapsack

$$S = \{5, 7, 15, 31, 72, 139, 274, 560, 1659, 3301, 6614, 13248, 26488, 53024, 106152, 225872\}$$

The total of our knapsack elements is 437461. Because n must exceed this sum, we take $n = 462193$. Now we pick an m relatively prime to n, such as $m = 316929$. Multiplying every element in our knapsack by m (mod n) gives the disguised knapsack

---

[38] All you need to know to understand what follows is that ASCII assigns values to each character and those for the capital letters are A = 65, B = 66, C = 67,..., Z = 90. These values may then be converted to binary (base 2).

[39] The technical term for such a set is *superincreasing*.

484 ■ *Secret History*

mS = {198066, 369731, 132005, 118746, 171431, 144796, 408455, 460321, 271770, 239870, 123151, 114180, 3893, 430202, 80931, 10862}

We then randomly scramble the order of our knapsack to further disguise it, using the key

10, 7, 14, 2, 8, 5, 16, 11, 9, 6, 15, 12, 1, 3, 4, 13

to get:

mS = {239870, 408455, 430202, 369731, 460321, 171431, 10862, 123151, 271770, 144796, 80931, 114180, 198066, 132005, 118746, 3893}

We can now reveal the elements of mS and the value of the modulus n to the world and anyone wanting to send a message can, but to decipher, we need m^{-1} (mod n). We may find this by using the Euclidean algorithm, as demonstrated in Section 14.3. We get m^{-1} = 304178.

If someone wants to send the message WELL DONE, he or she would have to encipher the letters in pairs. The first pair WE is represented in ASCII by 87 and 69. In binary this is 01010111 and 01000101. Running all 16 bits together we have 0101011101000101. The positions of the 1s indicate that we should take the reordered mS knapsack values in positions 2, 4, 6, 7, 8, 10, 14, and 16 and add them together. We get

408455 + 369731 + 171431 + 10862 + 123151 + 144796 + 132005 + 3893 = 1364324

Reducing this modulo n = 462193, we get 439938 as the final ciphertext, C. The rest of the pairs of message letters may be enciphered in the same way.

Once the enciphered message is received, the recipient begins the decipherment process by multiplying the first portion of the ciphertext by m^{-1} (mod n), where n = 462193. This is $m^{-1}C$ = (304178)(439938) = 133819460964 = 259481 (mod n).

The recipient then turns to the original knapsack

S = {5, 7, 15, 31, 72, 139, 274, 560, 1659, 3301, 6614, 13248, 26488, 53024, 106152, 225872}

and finds values that sum to 259481 using the greedy algorithm of looking for the largest number in the knapsack that doesn't exceed 259481. Once it is found, that number is subtracted from 259481 and the search is repeated using the reduced number to get

259481 = 225872 + 26488 + 6614 + 274 + 139 + 72 + 15 + 7

Recalling the scrambling of the order of the disguised knapsack elements, the recipient scrambles the secret knapsack in the same way:

S = {3301, 274, 53024, 7, 560, 72, 225872, 6614, 1659, 139, 106152, 13248, 5, 15, 31, 26488}

The knapsack values used to get the sum 259481 are now in positions 7, 16, 8, 2, 10, 6, 14, and 4. Ordering these, the recipient has 2, 4, 6, 7, 8, 10, 14, and 16 and constructs a 16-bit number with 1s in those positions. This is 0101011101000101. Splitting into bytes, converting to base 10, and finally to the letters those numbers represent in ASCII, the recipient gets 01010111 01000101 = 87 69 = WE.

So, we now have a public key cryptosystem based on an **NP**-complete problem, but you should not be overconfident! As was mentioned before, a clever cryptanalyst might find a method of attack that avoids the intractable problem. Indeed, such was the case here. Obviously, if the attacker multiplies the disguised knapsack by m^{-1} and mods out by n, he'll be able to read messages as easily as the intended recipient. That's why m^{-1} and n are kept secret. However, m^{-1} and n aren't the only numbers that will work. It turns out that any multiplier and modulus that yield a superincreasing set will serve to break the system. The attacker doesn't even need any ciphertext to begin his work. He may start the attack as soon as the public key is made public! Once he recovers a superincreasing knapsack (not necessarily the one the user is keeping secret), he's ready to read any message using the original public key.

Even if the attack described above wasn't possible, there would still be a serious flaw. An attacker could simply encipher every 16-bit combination, recording the results, and then crack any ciphertext by doing a reverse look-up in the table thus created. This could be averted by using a larger knapsack, and thus enciphering large blocks that would block the brute-force attack on the message space. However, the attack above still stands.

The moral of this story is that a system built on a secure foundation isn't necessarily secure! Over the years various revisions of the knapsack idea have been put forth and broken. This has happened to other systems, such as matrix encryption. After several rounds of cracks and patches, most cryptographers consider a system too flawed to become workable and lose interest in further iterations.

There's another story connected with knapsack encryption that has a more important moral. At the Crypto '82 conference held at University of California, Santa Barbara, Leonard Adleman gave a dramatic talk. While he was describing an approach that could be used to break a variant of the Merkle-Hellman knapsack cipher, he was running a program on an Apple II personal computer to actually demonstrate the break! I'll let Hellman offer his perspective on this talk:[40]

> He started his lecture by saying, "First the talk, then the public humiliation." I was livid! Cryptography is more an art than a science, and all of the top researchers had come up with at least one system that had been broken. Why did he have to say he was going to humiliate me?
>
> Later, I realized he was talking about himself, not me. He was afraid the computer would crash or some other problem would prevent him from proving that his approach worked. This experience of mine is a great example of why Dorothie [his wife] is right about giving everyone the benefit of the doubt, at least initially.

As it turned out, the computer did not crash. The reactions of some of the participants at the program's success can be seen in Figure 16.8.

Hellman's moral to this story is "Get curious, not furious." When you are upset by someone's words or actions, instead of becoming angry, become curious as to the person's motivations. Perhaps your first assumptions are wrong and the person didn't mean any harm. The book by Hellman from which I excerpted the paragraphs above explains how simple approaches like this can improve your personal relationships, as well as international relationships. If such ideas would only catch on, they'd have a much bigger impact on the world than Hellman's work in cryptology. Spread the word!

---

[40] Hellman, Dorothie and Martin Hellman, *A New Map for Relationships: Creating True Love at Home & Peace on the Planet*, New Map Publishing, 2016, p. 138. The website https://anewmap.com/ includes a link for a free download of the eBook version.

Figure 16.8 Leonard Adleman (left), Adi Shamir (right), and Martin Hellman (by the overhead projector) watch as knapsack encryption is broken. (Courtesy of Len Adleman.)

16.8 Elgamal Encryption

Figure 16.9 Taher Elgamal (1955–) (Creative Commons Attribution 3.0 Unported license, by Wikipedia user Alexander Klink, https://commons.wikimedia.org/wiki/File:Taher_Elgamal_it-sa_2010.jpg.).

The difficulty of solving the discrete log problem (see Section 14.2) was used by Taher Elgamal[41] (Figure 16.9), an Egyptian-born American cryptographer, to create the Elgamal public key cryptosystem in 1985.[42] Alice and Bob will illustrate how Elgamal works.

---

[41] You will see Elgamal spelled El Gamal and ElGamal in the literature. In this text, the name is spelled as Elgamal spells it.

[42] Elgamal, Taher, "A Public Key Cryptosystem and a Signature Scheme Based on Discrete Logarithms," *IEEE Transactions on Information Theory*, Vol. 31, No. 4, July 1985, pp. 469–472.

Alice begins by picking a large prime p and a generator g of the multiplicative group of integers modulo p. Recall from Section 14.2 that taking consecutive powers of a generator (mod p), until we reach 1, will result in the entire group being generated. Alice then selects a private key a and computes $A = g^a$ mod p. She publishes g, p, and A, only keeping the value a secret.

Bob wishes to send a message M, which he must put in the form of a number between 2 and p. If the message is too long for this, it can be broken into pieces. Bob then selects a key k (mod p) and computes $C_1 = g^k$ (mod p) and $C_2 = MA^k$ (mod p). He then sends Alice both C_1 and C_2. Thus, this system has the disadvantage that the ciphertext is twice as long as the message.

To recover the message M, Alice computes $x = C_1^a$ (mod p). From this, she is able to then calculate x^{-1} by using the Euclidean algorithm, as detailed in Section 14.3.

Finally, Alice computes $x^{-1}C_2$ (mod p), which reveals the message, because

$$x^{-1}C_2 = \left(C_1^a\right)^{-1}C_2 = \left(g^{ak}\right)^{-1}MA^k = \left(g^{ak}\right)^{-1}M\left(g^a\right)^k = M\left(g^{ak}\right)^{-1}g^{ak} = M.$$

The security of this system relies on the inability of an attacker to find the value of a when he is provided with g, p, and $A = g^a$ (mod p). This is the discrete log problem.

Example 8

We may use the prime $p = 2687$ and the value $g = 22$ for illustrative purposes. If Alice's private key is $a = 17$, she must calculate $A = g^a$ (mod p). This is $22^{17} = 824$ (mod 2687). She reveals to the world p, g, and A. If Bob wants to send the brief message HI, he must convert it to numbers first. We'll simply replace the letter using the scheme A = 0, B = 1, ..., Z = 25, but ASCII or other methods may be used. We get 0708, or 708. For his key Bob randomly chooses $k = 28$. He computes

$$C_1 = g^k\left(\text{mod } p\right) = 22^{28}\left(\text{mod } 2687\right) = 55$$

and

$$C_2 = MA^k\left(\text{mod } p\right) = \left(708\right)\left(824^{28}\right)\left(\text{mod } 2687\right) = 1601$$

Bob sends Alice 55 and 1601. To read the message, Alice first computes $x = C_1^a$ (mod p) = 55^{17} (mod 2687) = 841. Using the Euclidean algorithm, she finds the multiplicative inverse of 841 (mod 2687) is 2048. She then computes $x^{-1}C_2$ (mod p) = (2048)(1601) (mod 2687) = 708, which under our encoding scheme, translate to HI.

Basing a cryptosystem on a "hard" problem is not an idea new to this chapter. We've already examined a cipher based on the hardness of factoring, namely RSA. Recall that RSA is *based on* the factoring problem, but it is not known to be equivalent to that problem. Nor do we know that factoring is **NP**-complete. It might fall in **P**, like primality testing. The same is true for the discrete log problem. We don't know if it is in **P** or **NP**-complete. The Merkle-Hellman system and the McEliece system (alluded to briefly) are the only systems in this chapter based on problems *known* to be **NP**-complete.

See Figure 16.10 for a photograph of the key players in the world of public key cryptography.

Figure 16.10 The gangs all here! From left to right: Adi Shamir, Ron Rivest, Len Adleman, Ralph Merkle, Martin Hellman, and Whit Diffie. Notice that none of these cryptologic all-stars is wearing a tie. It's what's in your head that matters! (Picture courtesy of Eli Biham, taken at the presentation on August 21 at Crypto 2000, an IACR conference. The 21 at the bottom right is part of a date stamp that was mostly cropped out for reproduction here.)

References and Further Reading

On Primes and Primality Testing

Aaronson, Scott. *The prime facts: From Euclid to AKS*, 2003, http://www.scottaaronson.com/writings/prime.pdf. This is a very nice paper for providing background on how AKS works.

Adleman, Leonard M., Carl Pomerance, and Robert S. Rumely, "On Distinguishing Prime Numbers from Composite Numbers," *Annals of Mathematics*, Vol. 117, No. 1, January 1983, pp. 173–206.

Agrawal, Manindra, Neeraj Kayal, and Nitin Saxena, "PRIMES Is in P," *Annals of Mathematics, Second Series*, Vol. 160, No. 2, September 2004, pp. 781–793.

Bornemann, Folkmar, "PRIMES Is in P: A Breakthrough for "Everyman,"" *Notices of the AMS*, Vol. 50, No. 5, May 2003, pp. 545–552.

Bressoud, David M., *Factorization and primality testing*, Springer, New York, 1989.

Caldwell, Chris and G. L. Honaker, Jr., *Prime Curios! The Dictionary of Prime Number Trivia*, CreateSpace, Seattle, Washington, 2009.

Carmichael, Robert D., "Note on a New Number Theory Function," *Bulletin of the American Mathematical Society*, Vol. 16, No. 5, February 1910, pp. 232–238, available online at http://www.ams.org/journals/bull/1910-16-05/S0002-9904-1910-01892-9/S0002-9904-1910-01892-9.pdf.

Grantham, Jon, "Frobenius Pseudoprimes," *Mathematics of Computation*, Vol. 70, No. 234, 2001, pp. 873–891. This paper offers a method that is slower than Miller-Rabin, but has a lower chance of missing a composite for a random base.

Granville, Andrew, "It is Easy to Determine Whether a Given Integer is Prime," *Bulletin of the American Mathematical Society* (New Series), Vol. 42, No. 1, January 2005, pp. 3–38.

Kranakis, Evangelos, *Primality and Cryptography*, John Wiley & Sons, Chichester (Sussex), UK, 1986. Kranakis details several primality tests not mentioned here. He also discusses pseudorandom generators, RSA, and the Merkle-Hellman knapsack system.

McGregor-Dorsey, Zachary S., "Methods of Primality Testing", *MIT Undergraduate Journal of Mathematics*, Vol. 1, 1999, pp. 133–141. This very nice history is available online at http://www-math.mit.edu/phase2/UJM/vol1/DORSEY-F.PDF.

Miller, Gary L., "Riemann's Hypothesis and Tests for Primality," *Journal of Computer and System Sciences*, Vol. 13, No. 3, December 1976, pp. 300–317.

Rabin, Michael O., "Probabilistic Algorithm for Testing Primality," *Journal of Number Theory*, Vol. 12, No. 1, February 1980, pp. 128–138.

Ramachandran, R., "A Prime Solution," *Frontline*, Vol. 19, No. 17, August 17–30, 2002, available online at http://www.flonnet.com/fl1917/19171290.htm. This is a popular piece on AKS and the men behind it.

Ribenboim, Paulo, *The Book of Prime Number Records*, second edition, Springer, New York, 1989.

Ribenboim, Paulo, *The Little Book of Big Primes*, Springer, New York, 1991. This is a condensed version of *The Book of Prime Number Records*.

Ribenboim, Paulo, *The New Book of Prime Number Records*, Springer, New York, 1996. This is an update of *The Book of Prime Number Records*.

Ribenboim, Paulo, *The Little Book of Bigger Primes*, second edition, Springer, New York, 2004. This is a condensed version of *The New Book of Prime Number Records*.

Robinson, Sara, "Researchers Develop Fast Deterministic Algorithm for Primality Testing," *SIAM News*, Vol. 35, No. 7, September 2002, pp. 1–2.

The following books each list the bracketed number as the first prime. Can anyone continue the sequence?

[1] *The 1986 Information Please Almanac*, 39th edition, Houghton Mifflin Company, Boston, Massachusetts, p. 430.

[2] Ribenboim, Paulo, *The New Book of Prime Number Records*, Springer, New York, 1996, p. 513.

[3] Garrett, Paul, *Making, Breaking Codes: An Introduction to Cryptology*, Prentice Hall, Upper Saddle River, New Jersey, 2001, p. 509.

More Primes? The following books indicate the bracketed number is prime.

[4] Posamentier, Alfred S., and Ingmar Lehmann, *The (Fabulous) Fibonacci Numbers*, Prometheus Books, Amherst, New York, 2007, p. 333. Typos aside, this is a very good book.

[27] King, Stephen, *Dreamcatcher*, Scribner, New York, 2001, p. 211.

On Complexity Theory

Fortnow, Lance and Steve Homer, *A Short History of Computational Complexity*, November 14, 2002, http://people.cs.uchicago.edu/~fortnow/papers/history.pdf.

Garey, Michael R. and David S. Johnson, *Computers and Intractability: A Guide to the Theory of NP-Completeness*, W. H. Freeman and Co., New York, 1979. This book lists over 300 NP-complete problems.

Hartmanis, Juris and Richard E. Stearns. "On the Computational Complexity of Algorithms," *Transactions of the American Mathematical Society*, Vol. 117, No. 5, May 1965, pp. 285–306. Complexity Theory begins here!

Talbot, John and Dominic Welsh, *Complexity and Cryptography an Introduction*, Cambridge University Press, Cambridge, UK, 2006.

On Tetris

Breukelaar, Ron, Erik D. Demaine, Susan Hohenberger, Hendrik Jan Hoogeboom, Walter A. Kosters, and David Liben-Nowell, "Tetris is Hard, Even to Approximate," *International Journal of Computational Geometry and Applications*, Vol. 14, No. 1, April 2004, pp. 41–68. This paper is the merged version of two previous papers: "Tetris is Hard, Even to Approximate" by Erik D. Demaine, Susan Hohenberger, and David Liben-Nowell, and "Tetris is Hard, Made Easy" by Ron Breukelaar, Hendrik Jan Hoogeboom, and Walter A. Kosters.

Breukelaar, Ron, Hendrik Jan Hoogeboom, and Walter A. Kosters, "Tetris is Hard, Made Easy, Technical Report 2003-9, Leiden Institute of Advanced Computer Science, Universiteit Leiden, 2003.

Demaine, Erik D., Susan Hohenberger, and David Liben-Nowell, "Tetris is Hard, Even to Approximate," in Warnow, Tandy and Binhai Zhu, editors, *Computing and Combinatorics, 9th Annual International Conference (COCOON 2003)*, Lecture Notes in Computer Science Vol. 2697, 2003, Springer, Berlin, Germany, 2003, pp. 351–363.

Peterson, Ivars, "Tetris Is Hard," *Ivars Peterson's MathTrek*, Mathematical Association of America, October 28, 2002, http://web.archive.org/web/20120120070205/http://www.maa.org/mathland/mathtrek_10_28_02.html.

On McEliece's System

Chabaud, Florent, "On the security of Some Cryptosystems Based on Error-Correcting Codes," in De Santis, Alfredo, editor, *Advances in Cryptology – EUROCRYPT '94 Proceedings*, Lecture Notes in Computer Science, Vol. 950, Springer, Berlin, Germany, 1995, pp. 131–139.

McEliece, Robert J., *A Public-Key Cryptosystem Based on Algebraic Coding Theory*, Deep Space Network Progress Report 42-44, Jet Propulsion Laboratory, California Institute of Technology, January and February 1978, pp. 114–116.

On Ralph Merkle and Knapsack Encryption (Developed with Martin Hellman)

Chor, Benny, and Ron Rivest, "A Knapsack-type Public-Key Cryptosystem Based on Arithmetic in Finite Fields," in Blakley, G. Robert and David Chaum, editors, *Advances in Cryptology: Proceedings of CRYPTO 84*, Lecture Notes in Computer Science, Vol. 196, Springer, Berlin, Germany, 1985, pp. 54–65.

Chor, Benny, and Ron Rivest, "A Knapsack-type Public-Key Cryptosystem Based on Arithmetic in Finite Fields," *IEEE Transactions on Information Theory*, Vol. 34, No. 5, September 1988, pp. 901–909.

Hellman, Martin E., "The Mathematics of Public-Key Cryptography," *Scientific American*, Vol. 241, No. 2, August 1979, pp. 146–157.

Hellman, Martin E. and Ralph C. Merkle, "Hiding Information and Signatures in Trapdoor Knapsacks," *IEEE Transactions on Information Theory*, Vol. 24, No. 5, pp. 525–530, September 1978. The knapsack system was introduced in this paper.

Lenstra, Jr., Hendrik W., "On the Chor-Rivest Knapsack Cryptosystem," *Journal of Cryptology*, Vol. 3, No. 3, Summer 1991, pp. 149–155.

Merkle, Ralph, "Secure Communication Over Insecure Channels," *Communications of the ACM*, Vol. 21, No. 4, April 1978, pp. 294–299.

Merkle, Ralph, Home Page, www.merkle.com. This page includes a full account of how Merkle discovered public-key encryption and how the world reacted.

Shamir, Adi, *Embedding Cryptographic Trapdoors in Arbitrary Knapsack Systems*, Technical Report 230, MIT, Laboratory for Computer Science, Cambridge, Massachusetts, 1982.

Shamir, Adi, "Embedding cryptographic trapdoors in arbitrary knapsack systems," *Information processing letters*, Vol. 17, No. 2, 1983, pp 77–79,

Shamir, Adi, "A Polynomial Time Algorithm for Breaking the Basic Merkle-Hellman Cryptosystem (Extended Abstract)," in Chaum, David, Ronald L. Rivest, and Alan T. Sherman, editors, *Advances in Cryptology, Proceedings of CRYPTO '82*, Plenum Press, New York, 1983, pp. 279–288.

Shamir, Adi, "A Polynomial Time Algorithm for Breaking the Basic Merkle-Hellman Cryptosystem," in *SFCS '82: Proceedings of the 23rd Annual Symposium on Foundations of Computer Science*, November 1982, pp. 145–152.

Shamir, Adi, "A Polynomial Time Algorithm for Breaking the Basic Merkle-Hellman Cryptosystem," *IEEE Transactions on Information Theory*, Vol. IT-30, No. 5, September 1984, pp. 699–704.

Vaudenay, Serge, "Cryptanalysis of the Chor-Rivest Cryptosystem," *Journal of Cryptology*, Vol. 14, No. 2, Spring 2001, pp. 87–100.

On Elgamal

Elgamal, Taher, "A Public Key Cryptosystem and a Signature Scheme Based on Discrete Logarithms," in Blakley, G. Robert and David Chaum, editors, *Advances in Cryptology: Proceedings of CRYPTO 84*, Lecture Notes in Computer Science, Vol. 196, Springer, Berlin, Germany, 1985, pp. 10–18.

Elgamal, Taher, "A Public Key Cryptosystem and a Signature Scheme Based on Discrete Logarithms," *IEEE Transactions on Information Theory*, Vol. 31, No. 4, July 1985, pp. 469–472.

Niehues, Lucas Boppré, Joachim von zur Gathen, Lucas Pandolfo Perin, and Ana Zumalacárregui, "Sidon Sets and Statistics of the ElGamal Function," *Cryptologia*, Vol. 44, No. 5, September 2020, pp. 438–450.

On Improving Relationships

Hellman, Dorothie and Martin Hellman, *A New Map for Relationships: Creating True Love at Home & Peace on the Planet*, New Map Publishing, 2016. The website https://anewmap.com/ includes a link for a free download of the eBook version.

Chapter 17

Authenticity

There are many situations in which we would like to be sure that a message was actually composed and sent by the person it appears to be from. This is referred to as the authenticity of the message and it is not a new problem.

17.1 A Problem from World War II

During World War II, Special Operations Executive (SOE) agents in Holland captured by the Gestapo were forced to continue sending messages to their handlers indicating that nothing had changed. Some of these messages asked for more supplies, which, if sent, would be used by the Nazis.

Fortunately, every message sent by an agent was to include certain security checks. These consisted of intentional errors of a particular type, or insertions of specific numbers that might vary as a function of the date. Unfortunately, agents who hadn't been captured sometimes forgot to include these. Thus, when they were legitimately omitted, headquarters sometimes thought it was an unintentional oversight. In fact, in the case discussed here, the supplies were sent and used by the Nazis. Nevertheless, such deception couldn't last forever. When he figured the gig was up, the enemy sent a brief note of appreciation:[1]

> We thank you for the large deliveries of arms and ammunitions which you have been kind enough to send us. We also appreciate the many tips you have given us regarding your plans and intentions which we have carefully noted. In case you are concerned about the health of some of the visitors you have sent us you may rest assured they will be treated with the consideration they deserve.

In addition to the security checks mentioned above, there are other ways to verify a user's identity. Just as you can recognize the voice of a friend on the telephone, a telegraph operator has a style that can be recognized by other operators. This is referred to as the *fist* of the sender and can be depicted graphically (Figures 17.1 and 17.2).

---

[1] Marks, Leo, *Between Silk and Cyanide: a Codemaker's War, 1941-1945*. The Free Press, New York, 1998, p. 522.

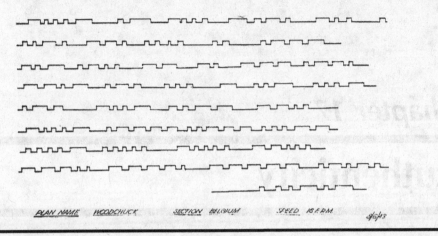

Figure 17.1 Each line represents a distinct "fist" sending the same alphabet and numbers. (From the David Kahn Collection, National Cryptologic Museum, Fort Meade, Maryland.)

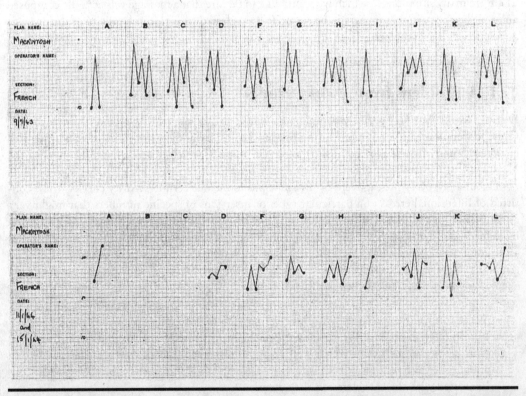

Figure 17.2 The change in this agent's fist was the result of his enlisting another operator's help. (From the David Kahn Collection, National Cryptologic Museum, Fort Meade, Maryland.)

This was known to the Nazis as well as the British. Just as a skilled impersonator might be able to use his voice to fool you into thinking he's someone else, one telegrapher can imitate the fist of another. The Nazis had success in this arena, but fists were used against them to track U-boats by tracking the recognizable style of each of their radio operators. In today's world, the need for certainty of a sender's identity continues to be of great importance, even in times of peace. We'll now look at some modern attempts to address this problem.

17.2 Digital Signatures (and Some Attacks)

In Section 14.3 We saw how RSA encryption can be used to allow secure communication between individuals who have not met to exchange keys. RSA can also be used to verify identities. Suppose Alice wants to be sure that Bob can tell that a message she'll be sending him really came from her. To be very convincing, she could encipher it by raising it to her private key, instead of her public key. She would then mod out by her n. Upon receiving the message $C = M^d$ (mod n), Bob would apply Alice's public key and recover the message, like so: $C^e = (M^d)^e = M^{ed} = M$ (mod n).

The only way a message can be constructed such that applying Alice's public key undoes it is by using Alice's private key. Hence, if Alice kept her private key secret, nobody else could have sent this message. On the downside, anyone who knows Alice's public key can read it; for example, Eve the eavesdropper would have no difficulty recovering the message.

There's a way around this difficulty. Because Bob has his own RSA key set, Alice can encipher one more time using Bob's public key. For clarity, we label the keys of Alice and Bob with the subscripts a and b. Alice would then send the following

$$\left(M^{d_a} \left(\text{mod } n_a \right) \right)^{e_b} \left(\text{mod } n_b \right)$$

Now Bob has twice as much work to do to recover the message, but he is the only one who can do so, and Alice is the only one who could have sent it. Bob first applies his private key, then he applies Alice's public key, and finally he reads the message.

Atari® used RSA "as the basis of its protection scheme on video game cartridges. Only cartridges that have been signed by the company's public key work in the company's video game machines."[2] The signing capability of RSA makes it even more useful, but it does allow attacks not discussed in Chapter 15.

17.2.1 Attack 13. Chosen Ciphertext Attack

If a ciphertext $C = M^e$ (mod n) is intercepted and the interceptor is able to learn the corresponding plaintext for the chosen ciphertext $C' = Cr^e$ (mod n), where r is a random value, then he may solve for M as follows. He starts with the plaintext for C':

$$(C')^d = (Cr^e)^d = (M^e r^e)^d = ((Mr)^e)^d = (Mr)^{ed} = (Mr)^1 = Mr$$

Seeing that it is a multiple of r, the value he randomly selected, he may multiply by the inverse of r to get M.

But how could the attacker get the corresponding plaintext for the ciphertext C'? Simple — he sends it to the person who created C. It will look like any other legitimate enciphered message, so the recipient will raise it to the power d to find the plaintext. It will be gibberish, but the recipient may simply assume it was garbled in transmission. The attacker then need only obtain this "deciphered" message.

The first RSA attack examined in Chapter 15 was the common modulus attack. This can be taken further, now that the signature aspect of RSA had been covered. John M. DeLaurentis did

---

[2] Garfinkel, Simson, *PGP: Pretty Good Privacy*, O'Reilly & Associates, Sebastopol, California, 1995, p. 95.

so in 1984, in a paper in which he showed that an insider posed a much greater threat than Eve. He could, like Eve, read M, but he could also break the system completely, and be able to view all messages and sign with anybody's key.[3]

17.2.2 Attack 14. Insider's Factoring Attack on the Common Modulus

Suppose Mallet is the insider. We represent his enciphering and deciphering exponents by e_M and d_M, respectively. His modulus, which is also held by all other users is represented by n. Mallet can multiply his two exponents together and subtract 1 and then factor out the highest power of 2 that divides this number to get

$$(e_M)(d_M) - 1 = (2^k)(s)$$

where s is an integer that must be odd.

Mallet may then choose a random integer a, such that $1 < a < n - 1$. If a and n have a common divisor greater than 1, then the Euclidean algorithm can be used to find it, and it will be one of the primes, p or q, used to generate n. Knowing the factorization of n would then allow him to determine the private key of anyone in the system. Thus, we assume a and n are relatively prime.

If Mallet can find a value $x \neq \pm 1$ such that $x^2 = 1 \pmod{n}$, he can use x to factor n by writing $x^2 - 1 = 0 \pmod{n}$ and $(x - 1)(x + 1) = 0 \pmod{n}$. A prime factor of n will then be provided by either $\gcd(x - 1, n)$ or $\gcd(x + 1, n)$. We'll return to this idea, but first, we make a few observations.

Because $(e_M)(d_M) = 1 \pmod{\varphi(n)}$, by definition, we have $(e_M)(d_M) - 1 = 0 \pmod{\varphi(n)}$. Mallet can substitute into this last equality using the identity $(e_M)(d_M) - 1 = (2^k)(s)$ to get $(2^k)(s) = 0 \pmod{\varphi(n)}$. In other words $(2^k)(s)$ is a multiple of $\varphi(n)$. Thus,

$$a^{2^k s} = 1 \pmod{n}.$$

It might be possible to replace k with a smaller nonnegative integer such that the equality still holds. Let $k' > 0$ denote the smallest such number. Then,

$$a^{2^{k'} s} = 1 \pmod{n}.$$

So,

$$\left(a^{2^{k'-1} s}\right)^2 = 1 \pmod{n}$$

Thus, Mallet has the desired value x in the form

$$a^{2^{k'-1} s}$$

He could be unlucky and have

$$a^{2^{k'-1} s} = -1 \pmod{n}$$

in which case it is a square root of 1 (mod n), but not one that is useful for factoring n. DeLaurentis showed that this happens at most half of the time. So, if Mallet is unlucky, he can simply randomly

---

[3] DeLaurentis, John M., "A Further Weakness in the Common Modulus Protocol for the RSA Cryptosystem," *Cryptologia*, Vol. 8, No. 3, July 1984, pp. 253–259.

choose another base *a* and try again. Once *n* is factored, Mallet can easily generate all of the private keys.

Like the Miller–Rabin primality test, this is a probabilistic attack, but DeLaurentis went on to show that assuming the Riemann hypothesis makes his attack deterministic. He also presented a second attack that allows Mallet to read and sign messages without factoring *n*.

17.2.3 Attack 15. Insider's Nonfactoring Attack

Mallet knows that his public key e_M must be relatively prime to $\varphi(n)$, so that a private key may be found that satisfies $e_M d_M = 1 \pmod{\varphi(n)}$. This latter equation is equivalent to

$$e_M d_M - 1 = k\varphi(n) \text{ for some positive integer } k$$

Now $e_M d_M - 1$ needn't be relatively prime to Alice's public key e_A.

Let *f* be the greatest common divisor of $e_M d_M - 1$ and e_A. The value of *f* may be found rapidly using the Euclidean algorithm. We define $r = (e_M d_M - 1)/f = (k\varphi(n))/f$ and note that *r* is relatively prime to e_A. Because *f* is a divisor of e_A and e_A is relatively prime to $\varphi(n)$, *f* must also be relatively prime to $\varphi(n)$. We now have two very useful facts:

1. *f* is a factor of $k\varphi(n)$.
2. *f* is relatively prime to $\varphi(n)$.

From this we may conclude that *f* is a factor of *k*. In other words, *r* is a multiple of $\varphi(n)$. That is, $r = (k/f)(\varphi(n))$.

Because *r* is relatively prime to $\varphi(n)$, we may use the extended Euclidean algorithm to find integers *x* and *y* such that $xr + ye_A = 1$. In particular, we can find a pair *x* and *y* such that *y* is positive. Because *r* is a multiple of $\varphi(n)$, when we reduce this last equality modulo $\varphi(n)$, the *xr* term drops out to leave $ye_A = 1 \pmod{\varphi(n)}$.

Thus, *y* fits the definition of a multiplicative inverse of $e_A \pmod{\varphi(n)}$. It might not be the inverse that Alice actually uses, but it will work just as well. Mallet's use of this value will result in encipherments and signatures that are identical to those produced by Alice.

17.2.4 Elgamal Signatures

Like RSA, Elgamal can also be used to sign messages. Unlike RSA, the user needs an extra key component to do so. Recall that for regular Elgamal encryption, Alice only needs a large prime *p*, an element *g* of maximal order in the multiplicative group of integers modulo *p*, and a private key *a*. She computes $A = g^a \pmod{p}$ and then publishes *g*, *p*, and *A*. Only *a* is kept secret.

Let *s* denote Alice's second secret key. She must compute $v = g^s \pmod{p}$. Then to sign a message *M*, she selects a random enciphering key *e* and computes

$$S_1 = g^e \pmod{p} \text{ and } S_2 = (M - sS_1)e^{-1} \pmod{p-1}.$$

Thus, like Bob's Elgamal-enciphered message to Alice (Section 16.8), Alice's signed message consists of two values, S_1 and S_2. These would be sent along with the message.

Alice makes v, g, and p public. From these, her signature may be verified by computing $v^{S_1} S_1^{S_2}$ (mod p) and g^M (mod p). If the signature is valid, both values will be the same. A few simple steps show why these two values must match:

$$v^{S_1} S_1^{S_2} = g^{sS_1} g^{eS_2} = g^{sS_1+eS_2} = g^{sS_1+e(M-sS_1)e^{-1}} = g^{sS_1+(M-sS_1)} = g^M$$

17.3 Hash Functions: Speeding Things Up

Just as RSA is slow for enciphering an entire message, and is therefore typically used only to encipher keys, we also don't want to have to apply RSA to an entire message, in order to sign it. One way of approaching this problem is to somehow condense the message to something much smaller and then sign this representation of the original. The compressed message, or *message digest*, is often referred to as the *hash* of the message. The process of generating this, called *hashing*, should have certain characteristics:

1. The same message must always yield the same hash. That is, we have a hash *function*. This will be important when it comes to verifying signatures based on the hash. We denote the hash function by $h(x)$.
2. Hashing should be fast. After all, speed is the drawback of using the methods previously described.
3. It should be difficult (computationally infeasible) to find two messages that hash to the same value. That is, we should not be able to find distinct x and x' such that $h(x) = h(x')$. If this happens, we say that a *collision* has been found. It would be very bad if a legitimately signed document could have a portion replaced (with the new text having some other meaning) and hash to the same value as the original. This would make the altered document appear to be signed as well. However, any collision is seen as a serious threat, even if it is found using two messages that look more like random text than meaningful messages. This is a very interesting condition, as no hash function can be one to one—the whole idea is to create a shorter message! Thus, collisions must exist, even for the best hash functions. Yet finding them should be very difficult!
4. Computing any preimage should be difficult. That is, given the hash of a message, $h(x)$, we should not be able to recover the message or any other text y such that $h(y) = h(x)$.

Hash functions came about in the 1950s, but they were not then intended to serve cryptographic purposes.[4] The idea of these functions was simply to map values in a dataset to smaller values. In this manner, comparisons could be made more easily and searches could be accelerated. In 1974, researchers who may not have been aware of hash functions, recognized that "hard-to-invert" functions offered a way to check if a message had been changed.[5] They credited the idea to Gus Simmons, who had presented the idea to them "in connection with monitoring the

---

[4] Knuth, Donald, *The Art of Computer Programming – Sorting and Searching*, Vol. 3, Addison-Wesley, Reading, Massachusetts, 1973. See pp. 506–549.
[5] Gilbert, Edgar N., Florence J. MacWilliams and Neil J. A. Sloane, "Codes Which Detect Deception," *The Bell System Technical Journal*, Vol. 53, No. 3, March 1974, pp. 405–424.

production of certain materials in the interest of arms limitation."[6] The cryptographic connection was described formally in 1981 by Mark N. Wegman and Larry Carter, a pair of researchers who did non-cryptographic work with hash functions in the late 1970s.[7]

17.3.1 Rivest's MD5 and NIST's SHA-1, SHA-2, and SHA-3

Message-Digest algorithm 5 (MD5) condenses messages of arbitrary length down to 128 bits.[8] It's not especially encouraging that we're on 5 already.[9] Cipher systems that went through many insecure versions, such as matrix encryption and knapsack encryption, are not viewed with much enthusiasm, when new versions appear, yet MD5 is popular. All of the MD hash functions were designed by Ron Rivest (the R in RSA). Version 5 came out in 1991. Since then, researchers have found quicker and quicker methods of finding collisions. In 2009, Tao Xie and Dengguo Feng reduced the time required to mere seconds, so there's not much point to further reductions.[10] We now look at an alternative.

MD5 was obviously used as the model for SHA-1, since they share many common features.

—Mark Stamp and Richard M. Low[11]

The Secure Hash Algorithm (SHA) offers another series of hash functions. The first, now called SHA-0, was designed by the National Security Agency (NSA) in 1993. It was followed by NSA's SHA-1 in 1995. Both produced a 160-bit hash. In fact, there is only a very small difference between the two and NSA claimed the modification improved security, but they didn't provide an explanation as to how.

Although much cryptanalysis went before, it wasn't until 2004 that a collision was actually found for SHA-0. It took the equivalent of 13 days' work on a supercomputer.[12] The following year, a flaw was found in SHA-1. When the National Security Agency announced their "NSA Suite B Cryptography" in 2005, the recommendations for hashing were SHA-256 and SHA-384, which are two algorithms taken from a set of six that are collectively called SHA-2.[13] Like its predecessors, the SHA-2 algorithms were designed by NSA. Despite the weaknesses of SHA-1 and NSA's recommendation, the first edition of this book, published in 2013, included the following:

SHA-1 forms part of several widely used security applications and protocols, including TLS and SSL, PGP, SSH, S/MIME, and IPsec. Those applications can also use

---

[6] Gilbert, Edgar N., Florence J. MacWilliams, and Neil J. A. Sloane, "Codes Which Detect Deception," *The Bell System Technical Journal*, Vol. 53, No. 3, March 1974, pp. 405–424, p. 406 quoted from here.

[7] Wegman, Mark N., and J. Lawrence Carter, "New Hash Functions and Their Use in Authentication and Set Equality," *Journal of Computer and System Sciences*, Vol. 22, No. 3, June 1981, pp. 265–279.

[8] http://en.wikipedia.org/wiki/MD5 has a clear (but brief) explanation of the algorithm. A more detailed explanation can be found at http://www.freesoft.org/CIE/RFC/1321/3.htm.

[9] A meaningful collision for MD4 was presented in Dobbertin, Hans, "Cryptanalysis of MD4," *Journal of Cryptology*, Vol. 11, No. 4, Fall 1998, pp. 253–271.

[10] Xie, Tao and Dengguo Feng, *How To Find Weak Input Differences For MD5 Collision Attacks*, May 30, 2009. http://eprint.iacr.org/2009/223.pdf.

[11] Stamp, Mark and Richard M. Low, *Applied Cryptanalysis: Breaking Ciphers in the Real World*, Wiley-Interscience, Hoboken, New Jersey, 2007, p. 225.

[12] Biham, Eli, Rafi Chen, Antoine Joux, Patrick Carribault, William Jalby, and Christophe Lemuet, "Collisions of SHA-0 and Reduced SHA-1," in Cramer, Ronald, editor, *Advances in Cryptology – EUROCRYPT 2005 Proceedings*, Lecture Notes in Computer science, Vol. 3494, Springer, Berlin, Germany, 2005, pp. 36–57.

[13] https://web.archive.org/web/20090117004931/http://www.nsa.gov/ia/programs/suiteb_cryptography/index.shtml.

MD5 ... SHA-1 hashing is also used in distributed revision control systems such as Git, Mercurial, and Monotone to identify revisions, and to detect data corruption or tampering. The algorithm has also been used on Nintendo console Wii for signature verification during boot, but a significant implementation flaw allowed an attacker to bypass the security scheme.[14]

Few organizations update their security in a timely manner.

Since 2013, more and more impressive attacks against SHA-1 have been found. While the attacks against SHA-2 are far less impressive, the similarities between this set of algorithms and SHA-1 are enough to concern cryptologists. Therefore, NIST held a competition to find an alternative (not a replacement) to SHA-2. This alternative was to be called SHA-3. Following a lengthy competition with 64 contenders, the winner, announced on October 2, 2012, was Keccak, a hash function designed by Guido Bertoni, Joan Daemen, Michaël Peeters, and Gilles Van Assche.[15] In the abstract of their paper titled "The Making of KECCAK," the designers wrote

Its structure and components are quite different from its predecessors, and at first sight it seems like a complete break with the past. In this article, researchers show that KECCAK is the endpoint of a long learning process involving many intermediate designs, mostly gradual changes, but also some drastic changes of direction.[16]

Daemen remarked, "Personally I consider my greatest result Keccak and not Rijndael. In fact its design contains the best elements of all cryptographic research I've done since I've started in 1989."[17]

While mathematicians are working out security issues, world governments are racing ahead. Bill Clinton was the first U.S. President to use digital signatures. These were applied, most suitably, to an e-commerce treaty with Ireland in September 1998 and, in 2000, a bill that gave digital signatures the same status, legally, as traditional signatures.

You may have noticed that I didn't provide details of any of the algorithms mentioned above. It's not because these details are unimportant. It's simply because this is one of the few areas in cryptology that I could never get too excited about. I think part of the problem for me is that, unlike cipher systems, there's no message to uncover. One usually just looks for collisions. The References and Further Reading list at the end of this chapter provides several books that devote serious space to the topic of hash functions, as well as some important papers, if you're interested. We now move on to how hash functions can be used to protect passwords, a topic I find much more interesting.

17.3.2 Hash Functions and Passwords

Hash functions have applications other than allowing for quicker digital signatures. Passwords should never be stored on a computer, but there needs to be a way to tell if the correct password has been entered. This is typically done by hashing the password after it is entered and comparing

---

[14] https://web.archive.org/web/20110221161831/http://en.wikipedia.org/wiki/SHA-1. The Wikipedia article provided the following reference for the Wii hack: http://debugmo.de/?p=61.

[15] See https://www.nist.gov/news-events/news/2012/10/nist-selects-winner-secure-hash-algorithm-sha-3-competition and http://csrc.nist.gov/groups/ST/hash/sha-3/sha-3_selection_announcement.pdf.

[16] Bertoni, Guido, Joan Daemen, Michaël Peeters, and Gilles Van Assche, "The Making of KECCAK," *Cryptologia*, Vol. 38, No. 1, January 2014, pp. 26–60.

[17] Email from Joan Daemen to the author, March 28, 2012. See Chapter 20 for a description of Rijndael.

the result with a value that is stored. Because computing any preimage should be difficult for a good hash function, someone gaining access to the hashed values shouldn't be able to determine the passwords.

In the first edition of this book, I wrote, "The important idea of storing the password in a disguised form doesn't seem to be strongly associated with anyone. Somebody had to be first to hit on this idea, but I haven't been able to learn who." After quoting what was clearly an incorrect account of this innovation from Wikipedia, I wrote, "Anyone who can provide a definitive first is encouraged to contact me!" I soon heard from Steven Bellovin (see Section 2.12). He wrote:

> While visiting Cambridge (my 1994 visit, I think), I was told by several people that Roger Needham had invented the idea. I asked him why he had never claimed it publicly; he said that it was invented at the Eagle Pub – then the after-work gathering spot for the Computer Laboratory – and both he and the others present had had sufficiently much India Pale Ale that he wasn't sure how much was his and how much was anyone else's...[18]

Ross Anderson noted, "For some time Roger Needham and Mike Guy each credited the other with the invention."[19] In any case, Maurice Wilkes was the first to publish the solution (in 1968), giving Needham full credit.[20] Needham's good night in the pub was back in 1967.[21] A decade later, he continued to pursue his research in the same manner. Sape Mullender, a PhD student at the time, recalled:

> The first time I met Roger was in the late seventies and, at the end of the afternoon, a chunk of the department, including Roger, went to the Eagle, the pub in Bene't Street. Until the Computer Laboratory moved to the outskirts of the city, that was the place where the day's events were discussed almost daily. It's also the pub where Crick and Watson contemplated DNA and came up with the double helix structure.[22]

Mullender checked in again after another decade:

> I spent a sabbatical in Cambridge in 1987/1988 and daily trips to the Eagle were still very much part of the culture. Roger would definitely join a few times a week. During that time, the Eagle was renovated and lost some of its dingy charm, so excursions further afield were also undertaken. The Bath and the Mill became frequent haunts. I think the reason those places became so inspirational was that (1) students were relaxed when there and that's a condition for having brainwaves and (2) mingling took place between different groups and this gave an opportunity for fresh ideas to come to some of the problems.[23]

---

[18] Bellovin, Steven, email to the author, August 5, 2013.
[19] Email from Ross Anderson to the author, May 20, 2020. Anderson credits both in his forthcoming book Anderson, Ross, *Security Engineering*, Third Edition, Wiley, Indianapolis, Indiana, see chapter 3, p. 108, available online at https://www.cl.cam.ac.uk/~rja14/book.html.
[20] Wilkes, Maurice V., *Time-Sharing Computer Systems*, American Elsevier, New York, 1968, pp. 91–92.
[21] Schofield, Jack, "Roger Needham," *The Guardian*, March 10, 2003, available online at https://www.theguard ian.com/news/2003/mar/10/guardianobituaries.microsoft.
[22] Email from Sape Mullender to the author, May 20, 2020.
[23] Email from Sape Mullender to the author, May 20, 2020.

Cipher Deavours, perhaps unaware of Needham's role in this story, wrote, "Non-classified work on such functions dates back to the early 1970s."[24] One would assume that the government had studied the problem before then. Deavours went on to reference a 1977 paper by Peter J. Downey that describes a system in use in 1972.[25] This system's initial step was vaguely described as "some preliminary reductions in [the password's] length." But after that, the next step is crystal clear. The password, p, was enciphered as

$$C = (2^{16})(p) \ (\text{mod } 10^{19} - 1).$$

Downey broke this system, recovering passwords from their enciphered values. Doing so only required computing the multiplicative inverse of 2^{16} (mod $10^{19} - 1$). It's a wonder anyone thought this could be secure!

Hashes used in this manner needn't decrease the size of the data being "hashed," but they must have the property that preimages are difficult to compute. Sometimes such hashes are referred to as one-way functions; however, there is no mathematical proof that one-way functions even exist! Some other references to password protection from the early 1970s are listed below:

> Bartek, Douglas J., "Encryption for Data Security," *Honeywell Computer Journal*, Vol. 8, No. 2, September 1974, pp. 86–89.
>
> Evans, Jr., Arthur, William Kantrowitz and Edwin Weiss, "A User Authentication Scheme not Requiring Secrecy in the Computer," *Communications of the ACM*, Vol. 17, No. 8, August 1974, pp. 437–442.
>
> Purdy, George B., "A High Security Log-in Procedure," *Communications of the ACM*, Vol. 17, No. 8, August 1974, pp. 442–445.
>
> Wilkes, Maurice V., *Time-Sharing Computer Systems*, American Elsevier, New York, 1972. This is a second edition of the work that first presented Needham's idea.

A 1979 paper by Robert Morris and Ken Thompson of Bell Labs is worth looking at in greater detail.[26] It gave the history of password security on the UNIX time-sharing system. At first, the passwords were enciphered by software simulating a cipher machine used by the United States during WW II, namely the M-209 (see the last paragraph of Section 9.1). However, it turned out that the machine wasn't any better in this capacity than it was for keeping secrets long-term during the war.

> It turned out that the M-209 program was usable, but with a given key, the ciphers produced by this program are trivial to invert. It is a much more difficult matter to find out the key given the cleartext input and the enciphered output of the program. Therefore, the password was used not as the text to be encrypted but as the key, and a constant was encrypted using this key. The encrypted result was entered into the password file.[27]

---

[24] Deavours, Cipher, "The Ithica Connection: Computer Cryptography in the Making, A Special Status Report," *Cryptologia*, Vol. 1, No. 4, October, 1977, pp. 312–316, p. 313 cited here.

[25] Downey, Peter J., *Multics Security Evaluation: Password and File Encryption Techniques*, ESD-TR-74-193, Vol. III, Electronic Systems Division, Hanscom Air Force Base, Massachusetts, June 1977.

[26] Morris, Robert and Ken Thompson, "Password Security: A Case History," *Communications of the ACM*, Vol. 22, No. 11, November 1979, pp. 594–597.

[27] Morris, Robert and Ken Thompson, "Password Security: A Case History," *Communications of the ACM*, Vol. 22, No. 11, November 1979, pp. 594–597, quotation taken from online version at https://citeseerx.ist.psu.edu/viewdoc/summary?doi=10.1.1.128.1635.

Morris and Thompson found that the M-209 simulation software had a fatal flaw — it was too fast! This is not a normal complaint to have, but in this case it meant that a brute-force attack could be run rapidly, testing a subset of possible keys that were more likely to be chosen by users. These included short keys and words found in English dictionaries. To patch against this sort of attack, a more up-to-date algorithm was modified (to become even slower) and applied.

> The announcement of the DES encryption algorithm by the National Bureau of Standards was timely and fortunate. The DES is, by design, hard to invert, but equally valuable is the fact that it is extremely slow when implemented in software. The DES was implemented and used in the following way: The first eight characters of the user's password are used as a key for the DES; then the algorithm is used to encrypt a constant. Although this constant is zero at the moment, it is easily accessible and can be made installation-dependent. Then the DES algorithm is iterated 25 times and the resulting 64 bits are repacked to become a string of 11 printable characters.[28]

The brute-force attack was now less of a threat, but an additional precaution was still taken — users were urged to select "more obscure" passwords.[29] Another improvement consisted of making the password a "salted password." This is an important feature that, like carefully choosing the password, is still relevant today. Morris and Thompson explained:

> The key search technique is still likely to turn up a few passwords when it is used on a large collection of passwords, and it seemed wise to make this task as difficult as possible. To this end, when a password is first entered, the password program obtains a 12-bit random number (by reading the real-time clock) and appends this to the password typed in by the user. The concatenated string is encrypted and both the 12-bit random quantity (called the *salt*) and the 64-bit result of the encryption are entered into the password file.[30]
>
> When the user later logs in to the system, the 12-bit quantity is extracted from the password file and appended to the typed password. The encrypted result is required, as before, to be the same as the remaining 64 bits in the password file. This modification does not increase the task of finding any individual password, starting from scratch, but now the work of testing a given character string against a large collection of encrypted passwords has been multiplied by 4096 (2^{12}). The reason for this is that there are 4096 encrypted versions of each password and one of them has been picked more or less at random by the system.[31]

---

[28] Morris, Robert and Ken Thompson, "Password Security: A Case History," *Communications of the ACM*, Vol. 22, No. 11, November 1979, pp. 594–597, quotation taken from online version at https://citeseerx.ist.psu.edu/viewdoc/summary?doi=10.1.1.128.1635.

[29] Morris, Robert and Ken Thompson, "Password Security: A Case History," *Communications of the ACM*, Vol. 22, No. 11, November 1979, pp. 594–597.

[30] Morris, Robert and Ken Thompson, "Password Security: A Case History," *Communications of the ACM*, Vol. 22, No. 11, November 1979, pp. 594–597, quotation taken from online version at https://citeseerx.ist.psu.edu/viewdoc/summary?doi=10.1.1.128.1635.

[31] Morris, Robert and Ken Thompson, "Password Security: A Case History," *Communications of the ACM*, Vol. 22, No. 11, November 1979, pp. 594–597, quotation taken from online version at https://citeseerx.ist.psu.edu/viewdoc/summary?doi=10.1.1.128.1635.

While DES was slow in software, it was fast on a chip. To prevent someone from running the attack through a commercial DES chip, the software used didn't perfectly match the DES algorithm. The expansion table, E, instead of being fixed, was made to depend on the 12-bit random number.[32]

Robert Morris joined the National Security Agency in 1986.[33]

We now turn to a signature scheme that explicitly includes a hash function.

17.4 The Digital Signature Algorithm

The Digital Signature Algorithm, abbreviated DSA, is a modified version of the Elgamal signature scheme. It was proposed by the National Institute of Standards and Technology (NIST) in 1991, and became part of the Digital Signature Standard (DSS) in 1994.[34] It works as follows:

Randomly generate a prime q with at least 160 bits. Then test $nq + 1$ for primality, where n is a positive integer large enough to give the desired level of security. If $nq + 1$ is prime, move on to the next step; otherwise, pick another prime q and try again.[35]

We then need an element g of order q in the multiplicative group modulo p. This element can be found quickly by computing $g = h^{(p-1)/q} \pmod{p}$, where h is an element of maximal order (a primitive root) modulo p. As in Elgamal, another secret value, s, must be chosen, and then we calculate $v = g^s \pmod{p}$; p, q, g, and v are made public, but s is kept secret.

Signing a message consists of calculating two values, S_1 and S_2. The calculation for S_2 requires the hash of the message to be signed and the value of S_1. On the other hand, the calculation of S_1 doesn't depend on the message. Thus, S_1 values can be created in advance, before the messages exist, to save time. Each S_1 does require, though, that we pick a random value k between 1 and $q - 1$. We have

$$S_1 = (g^k \pmod{p}) \pmod{q} \text{ and } S_2 = k^{-1}(\text{hash}(M) + sS_1) \pmod{q}$$

The inverse of k, needed for S_2, is calculated $\pmod{q}$. The two values S_1 and S_2 constitute the signature for message M, and are sent with it.

To verify that a signature is genuine, we compute the following

$$U_1 = S_2^{-1}\text{hash}(M) \pmod{q}$$

$$U_2 = S_1 S_2^{-1} \pmod{q}$$

$$V = (g^{U_1} v^{U_2} \pmod{p}) \pmod{q}$$

---

[32] Morris, Robert and Ken Thompson, "Password Security: A Case History," *Communications of the ACM*, Vol. 22, No. 11, November 1979, pp. 594–597.

[33] Markoff, John, "Robert Morris, Pioneer in Computer Security, Dies at 78," *The New York Times*, June 29, 2011, available online at https://www.nytimes.com/2011/06/30/technology/30morris.html?_r=1&hpw.

[34] National Institute of Standards and Technology (NIST), Digital Signature Standard (DSS), FIPS Publication 186, May 19, 1994, available online at https://web.archive.org/web/20131213131144/http://www.itl.nist.gov/fipspubs/fip186.htm. Some revisions have been made over the years. As of July 2013, we have 186-4, available online at https://nvlpubs.nist.gov/nistpubs/FIPS/NIST.FIPS.186-4.pdf.

[35] There is actually a different scheme for generating p and q that is recommended by NIST. The method given here is an intentional simplification.

U_1 requires calculating the inverse of S_2. This is done modulo q. The signature, $S_1 S_2$, is deemed genuine if $V = S_1$. DSA creates signatures at the same speed as RSA, but requires 10 to 40 times as long to verify signatures.[36] It is quicker than Elgamal, though.

Example 1

As a small example to illustrate DSA, we choose the prime $q = 53$. We then need a much larger prime p, such that $p - 1$ is divisible by q. We try $10q + 1 = 531$, but it is divisible by 3. We are also able to factor $12q + 1 = 637$. Finally, $14q + 1$ gives us the prime 743. So we now have $q = 53$ and $p = 743$ as our two primes. We also need an element g of order q in the multiplicative group modulo p. So, we compute $g = h^{(p-1)/q} \pmod{p}$, where h is an element of maximal order modulo p. We quickly find that $h = 5$ is suitable. We then have $g = 5^{(p-1)/q} \pmod{p} = 5^{14} \pmod{p} = 212$. We randomly set $s = 31$. Then $v = g^s \pmod{p}$, so we get $v = 212^{31} = 128$. p, q, g, and v are made public, but s is kept secret. We're now ready to sign a message.

We'll let the message be the single letter D, perhaps a grade someone will be receiving. We represent it numerically as 3 (using our old scheme A = 0, B = 1,... Z = 25). Our signature requires two numbers be calculated. These numbers require us to pick a random value k between 1 and $q - 1$. We let $k = 7$. Then

$$S_1 = (g^k \pmod{p}) \pmod{q} \text{ and } S_2 = k^{-1}(\text{hash }(M) + sS_1) \pmod{q}$$

becomes (ignoring the hash step and simply using the full message instead)

$$S_1 = (212^7 \pmod{743}) \pmod{53} = 94 \pmod{53} = 41$$

and

$$S_2 = 7^{-1}(3 + (31)(41)) \pmod{53} = 38(1274) \pmod{53} = 23.$$

Recall that all the inverses that need to be calculated for DSA are done modulo q, the smaller prime. These two values constitute the signature for message M, and are sent with it.

To verify that the signature is genuine, we compute the following:

$$U_1 = S_2^{-1} \text{ hash}(M) \pmod{q} = 23^{-1}(3) \pmod{53} = 30(3) \pmod{53} = 37$$

$$U_2 = S_1 S_2^{-1} \pmod{q} = (41)(30) \pmod{53} = 11$$

$$V = (g^{U_1} v^{U_2} \pmod{p}) \pmod{q} = (212^{37} \cdot 128^{11} \pmod{743}) \pmod{53} = 94 \pmod{53} = 41.$$

The signature is deemed genuine, because $V = S_1$.

If the message is being sent to the registrar's office and a student attempts to replace the D with a C, the signature will not be valid. The registrar could then contact the professor and ask that the grade be resent.

Initially, NIST claimed they created the DSA, then they revealed that NSA helped some. Finally, they admitted that NSA designed it.[37] I think the people would place greater trust in

---

[36] Schneier, Bruce, *Applied Cryptography*, second edition, John Wiley & Sons, New York, 1996, p. 485.
[37] Schneier, Bruce, *Applied Cryptography*, second edition, John Wiley & Sons, New York, 1996, p. 486.

government agencies, if they didn't lie to us. The patent for DSA lists David W. Kravitz as the inventor.[38] He holds an undergraduate degree in mathematics and a Ph.D. in Electrical Engineering. He created DSA during an 11-year stint at the NSA.[39]

Like another system NSA had a hand in, experts in the open community felt DSA's key was too small. Originally, the modulus was set at 512 bits, but because of these complaints, it was adjusted so that it could range from 512 to 1024 bits, in 64-bit increments.[40]

References and Further Reading

Anderson, Ross, *Security Engineering*, Third Edition, Wiley, to appear. As of this writing, much of this book is freely available online at https://www.cl.cam.ac.uk/~rja14/book.html. That will change, back and forth. Anderson explained:

> I'm writing a third edition of Security Engineering, and hope to have it finished in time to be in bookstores for Academic Year 2020-1. With both the first edition in 2001 and the second edition in 2008, I put six chapters online for free at once, then added the others four years after publication. For the third edition, I've negotiated an agreement with the publisher to put the chapters online for review as I write them. So the book will come out by instalments, like Dickens' novels. Once the manuscript's finished and goes to press, all except seven sample chapters will disappear for a commercial period of 42 months. I'm afraid the publishers insist on that. But thereafter the whole book will be free online forever.[41]

If you happen to follow the link above when most of the book is off-line, you can scroll down to access earlier editions. Or you could just *buy* the third edition. It's really good. It's clear, comprehensive, entertaining, and contains neat quotes like this one (from Chapter 20):

> Whoever thinks his problem can be solved using cryptography, doesn't understand his problem and doesn't understand cryptography. — Attributed by Roger Needham and Butler Lampson to each other

Biham, Eli, Rafi Chen, Antoine Joux, Patrick Carribault, William Jalby, and Christophe Lemuet. "Collisions of SHA-0 and Reduced SHA-1," in Cramer, Ronald, editor, *Advances in Cryptology – EUROCRYPT 2005 Proceedings*, Lecture Notes in Computer science, Vol. 3494, Springer, Berlin, Germany, 2005, pp. 36–57.

DeLaurentis, John M., "A Further Weakness in the Common Modulus Protocol for the RSA Cryptosystem," *Cryptologia*, Vol. 8, No. 3, July 1984, pp. 253–259.

Dobbertin, Hans, "Cryptanalysis of MD4," *Journal of Cryptology*, Vol. 11, No. 4, Fall 1998, pp. 253–271.

Elgamal, Taher, "A Public Key Cryptosystem and a Signature Scheme Based on Discrete Logarithms," in Blakley, G. Robert and David Chaum, editors, *Advances in Cryptology: Proceedings of CRYPTO 84*, Lecture Notes in Computer Science, Vol. 196, Springer, Berlin, Germany, 1985, pp. 10–18.

Elgamal, Taher, "A Public Key Cryptosystem and a Signature Scheme Based on Discrete Logarithms," *IEEE Transactions on Information Theory*, Vol. 31, No. 4, July 1985, pp. 469–472.

Gilbert, Edgar N., Florence J. MacWilliams, and Neil J. A. Sloane, "Codes Which Detect Deception," *The Bell System Technical Journal*, Vol. 53, No. 3, March 1974, pp. 405–424.

---

[38] Kravitz, David W., Digital signature algorithm, United States Patent 5,231,668, July 27, 1993, available online at https://patents.google.com/patent/US5231668.

[39] "About Dr. David W. Kravitz," *TrustCentral*, https://trustcentral.com/about/about-dr-david-w-kravitz/.

[40] Schneier, Bruce, *Applied Cryptography*, second edition, John Wiley & Sons, New York, 1996, p. 486.

[41] Anderson, Ross, Security Engineering—Third Edition, https://www.cl.cam.ac.uk/~rja14/book.html.

Holden, Joshua, "A Good Hash Function is Hard to Find, and Vice Versa," *Cryptologia*, Vol. 37, No. 2, April 2013, pp. 107–119.

Horng, Gwoboa, "Accelerating DSA Signature Generation," *Cryptologia*, Vol. 39, No. 2, April 2015, pp. 121–125.

Kishore, Neha and Priya Raina, "Parallel Cryptographic Hashing: Developments in the Last 25 Years," *Cryptologia*, Vol. 43, No. 6, November 2019, pp. 504–535.

Menezes, Alfred J., Paul C. van Oorschot, Scott A. Vanstone, *Handbook of Applied Cryptography*, CRC Press, Boca Raton, Florida, 1997. Chapter 9 (pp. 321–383) is focused on hash functions. This book is freely available online in its entirety (780 pages) at http://labit501.upct.es/~fburrull/docencia/SeguridadEnRedes/teoria/bibliography/HandbookOfAppliedCryptography_AMenezes.pdf.

Morris, Robert and Ken Thompson, "Password Security: A Case History," *Communications of the ACM*, Vol. 22, No. 11, November 1979, pp. 594–597, available online at https://citeseerx.ist.psu.edu/viewdoc/summary?doi=10.1.1.128.1635 and https://dl.acm.org/doi/pdf/10.1145/359168.359172.

National Institute of Standards and Technology (NIST), Announcing the Standard for Secure Hash Standard, Federal Information Processing Standards Publication 180-1, April 17, 1995, available online at http://web.archive.org/web/20120320233841/http://www.itl.nist.gov/fipspubs/fip180-1.htm. This describes SHA-1.

National Institute of Standards and Technology (NIST), Announcing the Secure Hah Standard, Federal Information Processing Standards Publication 180-2 (+ Change Notice to include SHA-224), August 1, 2002, available online at http://csrc.nist.gov/publications/fips/fips180-2/fips180-2withchangenotice.pdf. This describes SHA-2.

National Institute of Standards and Technology (NIST), Digital Signature Standard (DSS), FIPS Publication 186, May 19, 1994, available online at https://web.archive.org/web/20131213131144/http://www.itl.nist.gov/fipspubs/fip186.htm. Some revisions have been made over the years. As of July 2013, we have 186-4, available online at https://nvlpubs.nist.gov/nistpubs/FIPS/NIST.FIPS.186-4.pdf.

National Institute of Standards and Technology (NIST), Hash Functions, SHA-3 Project, https://csrc.nist.gov/projects/hash-functions/sha-3-project. This website has many useful links related to SHA-3, including the FIPS documents.

Pfitzmann, Birgit, *Digital Signature Schemes: General Framework and Fail-Stop Signatures*, Lecture Notes in Computer Science, Vol. 1100, Springer, New York, 1991.

Preneel, Bart, "Cryptographic Hash Functions," *European Transactions on Telecommunications*, Vol. 5, No. 4, 1994, pp. 431–448.

Preneel, Bart, *Analysis and Design of Cryptographic Hash Functions*, doctoral dissertation, Katholieke Universiteit Leuven, Belgium, February 2003, available online at http://homes.esat.kuleuven.be/~preneel/phd_preneel_feb1993.pdf.

Preneel, Bart, René Govaerts, and Joos Vandewalle, "Information Authentication: Hash Functions and Digital Signatures," in Preneel, Bart, René Govaerts, and Joos Vandewalle, editors, *Computer Security and Industrial Cryptography: State of the Art and Evolution*, Lecture Notes in Computer Science, Vol. 741, Springer, 1993, pp. 87–131.

Schofield, Jack, "Roger Needham," *The Guardian*, March 10, 2003, available online at https://www.theguardian.com/news/2003/mar/10/guardianobituaries.microsoft.

Stallings, William, "Digital Signature Algorithms," *Cryptologia*, Vol. 37, No. 4, October 2013, pp. 311–327.

Stallings, William, *Cryptography and Network Security: Principles and Practice*, 8th edition, Pearson, Edinburgh Gate, Harlow, Essex, UK, 2020. This is a comprehensive look at cryptography. No cryptanalysis is present, but there is much material on hash functions.

Stamp, Mark, and Richard M. Low, *Applied Cryptanalysis: Breaking Ciphers in the Real World*, Wiley-Interscience, Hoboken, New Jersey, 2007. Chapter 5 of this book (pp. 193–264) discusses cryptanalysis of hash functions. The authors state in the conclusion for this chapter (p. 256), "For many years, it seems that hash functions had been largely ignored by cryptographers. But with the successful attack on MD5, and similar results for SHA-1 pending, hash functions have moved from a sleepy cryptographic backwater to the forefront of research."

Stevens, Marc, Elie Bursztein, Pierre Karpman, Ange Albertini, and Yarik Markov, *Shattered*, https://shattered.io/. This website is devoted to the breaking of SHA-1. It includes a link to the technical paper whose authors are listed here. Others contributed to the work detailed on this page, as well.

Stevens, Marc, Pierre Karpman, and Thomas Peyrin, *The SHAppening: freestart collisions for SHA-1*, https://sites.google.com/site/itstheshappening/.

Wegman, Mark N. and J. Lawrence Carter, "New Hash Functions and Their Use in Authentication and Set Equality," *Journal of Computer and System Sciences*, Vol. 22, No. 3, June 1981, pp. 265–279.

Wikipedia contributors, "Password," *Wikipedia, The Free Encyclopedia*, https://en.wikipedia.org/wiki/Password.

Wilkes, M. V., *Time-Sharing Computer Systems*, American Elsevier, New York, 1968.

Winternitz, Robert S., "Producing a One-way Hash Function from DES," in Chaum, David, editor, *Advances in Cryptology, Proceedings of Crypto '83*, Plenum Press, New York, 1984, pp. 203–207.

Xie, Tao and Dengguo Feng, How To Find Weak Input Differences For MD5 Collision Attacks, May 30, 2009, http://eprint.iacr.org/2009/223.pdf.

Chapter 18

Pretty Good Privacy and Bad Politics

Entire messages can be enciphered with RSA, but it's a slow algorithm. The competition, in the late 1970s, was the Data Encryption Standard (DES), which was a thousand times faster. Yet for DES, a key had to be agreed on ahead of time. So, what's the solution? Which of these two system should be used? The answer is both! Loren M. Kohnfelder suggested a hybrid system in his 1978 undergraduate thesis, written while he was studying electrical engineering at MIT.[1] Whitfield Diffie recalled ten years later how this was "hailed as a discovery in its own right."[2]

18.1 The Best of Both Worlds

Here's how RSA and DES can be combined.

1. Generate a random session key, K, which will only be used once.
2. Use K to encipher the message with DES (or any other traditional symmetric cipher).
3. Encipher the session key using RSA and send it along with the ciphertext.

The recipient can use RSA to recover the session key K. He then applies it to the ciphertext and reads the original message. Because RSA is only ever used on the relatively short (compared to the message) key, its slowness is not much of a disadvantage. The whole process is fast and doesn't pose any serious key management problems.

Sending a session key along with the ciphertext wasn't a new idea. Recall that the Nazis used Enigma in this manner, although they needed a daily key to encipher every session key. Public key encryption did away with that. The same RSA key may be used for years.

---

[1] Kohnfelder, Loren M., *Toward a Practical Public Key Encryption Algorithm*, undergraduate thesis, Department of Electrical Engineering, Massachusetts Institute of Technology, Cambridge, Massachusetts, May 1978, available online at https://dspace.mit.edu/bitstream/handle/1721.1/15993/07113748.pdf?sequence=1. Kohnfelder's thesis advisor was Len Adleman.

[2] Diffie, Whitfield, "The First Ten Years of Public-Key Cryptography," *Proceedings of the IEEE*, Vol. 76, No. 5, May 1988, pp. 560–577, p. 566 cited here.

It should be noted that in having the best of both worlds by combining RSA and DES, we run into the problem of a chain being only as strong as its weakest link. The RSA portion may be ignored if DES can be broken and DES needn't be attacked if the primes used for the RSA portion are too small, or if another weakness is present in the RSA implementation. For example, when using RSA on a short message, like a DES key, some padding (aka salt) needs to be added to the message prior to encryption (see Section 15.1.11). Secure implementation of an otherwise sound system is a highly non-trivial step. In addition to the normal security problems, programmers in the 1980s also had to contend with machines much slower than what we take for granted today.

18.2 The Birth of PGP

It was in the 1980s that Charlie Merritt, along with two friends, made a program to (slowly) run public key encryption on Z-80 computers,[3] but there didn't seem to be any demand. Merritt's friends quit, and Merritt continued on with his wife. Most of the interest was from businesses who wanted to protect their secrets from foreign competitors, but National Security Agency (NSA) representatives kept stopping by to warn Merritt against exporting any crypto software. Trying to find domestic buyers led him to Philip Zimmermann, who owned a small company.[4]

Zimmermann had come up with the idea of a hybrid system on his own in 1984 (he hadn't seen Kohnfelder's Bachelor's Thesis), but he was confronted with many technical problems in implementing it. He was very happy to hear from Merritt, who helped him with these problems. Zimmermann had a degree in computer science, but he had struggled with calculus. He planned to call his program PGP, which stood for Pretty Good Privacy. This was a tribute to Ralph's Pretty Good Groceries, a fictional sponsor of Garrison Keillor's *Prairie Home Companion* radio show. Meanwhile, another hybrid system, constructed by Rivest and Adleman, was about to be marketed by RSA Data Security as "Mailsafe."

With Merritt's help Zimmermann, conquered the necessary mathematics to get RSA implemented as efficiently as possible on the slow machines of the time period. Zimmermann himself moved slowly, working on the project part-time. Finally, in 1990, he pushed other matters aside, risking financial ruin, and spent six months of 12-hour days to get it done.[5] This was not an attempt to get rich. He planned on making it shareware. Interested people could obtain it freely, but they were expected to mail Zimmermann a check, if they decided they wanted to use it. Rather than being financial, Zimmermann's motivations were political. Strong encryption would be a way to keep a prying government out of the affairs of its citizens. Zimmermann's motivations became intensified when he learned in April 1991 that then Senator Joseph Biden had cosponsored the Senate "Anti-Crime" Bill 266, which, if passed, would require "that providers of electronic communications services and manufacturers of electronic communications service equipment

---

[3] It took 10 minutes to create a 256-bit key and 20-30 seconds to encipher a small file. See Garfinkel, Simson, *PGP: Pretty Good Privacy*, O'Reilly & Associates, Sebastopol, California, 1995, p. 88.

[4] Levy, Steven, *Crypto: How the Code Rebels Beat the Government, Saving Privacy in the Digital Age*, Viking, New York, 2001, p. 190

[5] Levy, Steven, *Crypto: How the Code Rebels Beat the Government, Saving Privacy in the Digital Age*, Viking, New York, 2001, p. 195

shall ensure that communications systems permit the government to obtain the plaintext contents of voice, data, and other communications when appropriately authorized by law."[6]

There is no agreement on who initiated the particular provision of the bill quoted above. Some say the Federal Bureau of Investigation was responsible, but according to the Electronic Frontier Foundation (EFF), it was Joe Biden himself.[7]

Fearing that strong crypto was about to be outlawed, Zimmermann quickly finished his program, relabeled it as freeware, and with the help of friends, began to distribute it on American Internet sites in June 1991. He needn't have rushed, because Biden reacted to anger over his proposed anti-privacy legislation, by withdrawing that section from the Bill.[8] Zimmermann still faced legal troubles, though. Although he wasn't directly responsible, PGP left the country within a day of being posted online, in violation of encryption export laws.

Zimmermann's cipher for his first hybrid system was based on work Merritt had done for the Navy.[9] After making his changes, he renamed it Bass-O-Matic, after a blender used in a *Saturday Night Live* skit, in which Dan Aykroyd portrayed a salesman who used the machine to chop up a fish.[10] Bass-O-Matic proved to be the weak link in PGP, but Zimmermann had other problems. RSA was patented and he did not have permission to make use of it.

Back in November 1986, Zimmermann had met with James Bidzos, the president of RSA Data Security. It was mainly a meeting between Bidzos and Merritt, who did contract work for RSA Data Security, but Zimmermann was there. Bidzos and Zimmerman didn't get along at all — they were complete political opposites. Zimmermann refused contract work from Bidzos because his company had a military connection. As for Bidzos's view of the military, he had joined the U.S Marines, even though he was a citizen of Greece.[11]

Despite these differences, Bidzos gave Zimmermann a copy of Mailsafe. What he didn't give him was permission to use RSA in his own encryption program. Bidzos remembered this when PGP appeared. Simson Garfinkel observed that, "What followed could only be described as a low-intensity war by Bidzos against PGP and Zimmermann."[12]

Zimmermann wasn't the only one to infringe upon the RSA patent. The following excerpt from a question-and-answer session at a Computers, Freedom, & Privacy conference in 1992,

---

[6] Quote reproduced here from Levy, Steven, *Crypto: How the Code Rebels Beat the Government, Saving Privacy in the Digital Age*, Viking, New York, 2001, p. 195. Also, Garfinkel, Simson, *PGP: Pretty Good Privacy*, O'Reilly & Associates, Sebastopol, California, 1995, p. 97.

[7] Garfinkel, Simson, *PGP: Pretty Good Privacy*, O'Reilly & Associates, Sebastopol, California, 1995, p. 97. Garfinkel cites the EFF's online newsletter *EFFector*.

[8] The anger came from two groups—civil libertarians and industry. It makes for an unlikely alliance, but big businesses depend on strong encryption to protect themselves from spying competitors, so they found themselves on the same side as the privacy advocates.

[9] Merritt had protested the war in Vietnam, but he didn't see any conflict between that viewpoint and his doing work for the Navy. See Garfinkel, Simson, *PGP: Pretty Good Privacy*, O'Reilly & Associates, Sebastopol, California, 1995, p. 91.

[10] Levy, Steven, *Crypto: How the Code Rebels Beat the Government, Saving Privacy in the Digital Age*, Viking, New York, 2001, p. 194

[11] Garfinkel, Simson, *PGP: Pretty Good Privacy*, O'Reilly & Associates, Sebastopol, California, 1995, p. 93.

[12] Garfinkel, Simson, *PGP: Pretty Good Privacy*, O'Reilly & Associates, Sebastopol, California, 1995, p. 100.

shows that Bidzos could have a sense of humor about it.[13] The question, left out here for brevity, was initially answered by John P. Barlow (cofounder of the Electronic Frontier Foundation):

> **Barlow:** The problem with trying to regulate the flow of information, which is a little like trying to regulate the flow of wind, is that it's quite possible to keep it out of the hands of individuals and small institutions. It's very difficult to keep it out of the hands of large institutions, so you have, in effect, the situation where the Soviets are using RSA in their launch codes and have for a long time, and yet we can't use it as individuals in the United States, you know, and that's just dumb.

> **Bidzos:** My revenue forecasts are being revised downward.

> **Barlow:** You weren't getting any royalties on that anyway, were you Jim?

> **Bidzos:** Maybe.

Apparently, PGP didn't initially bother NSA. Zimmermann soon learned (from Eli Biham) that his system was vulnerable to differential cryptanlaysis.[14] Other flaws were present, including one that prevented the last bit of each byte from being properly encrypted.[15] After the flaws were pointed out, Zimmermann began working on version 2.0, but this time he got help from much stronger cryptographers from around the world,[16] who appreciated what he was trying to do. Bass-O-Matic was replaced by a Swiss cipher, the International Data Encryption Algorithm (IDEA), which offered a 128-bit key. Many other improvements were made and new features introduced. PGP 2.0 was released from Amsterdam and Auckland (the hometowns of two of Zimmermann's new collaborators) in September 1992.[17]

In November 1993, following a deal Zimmermann made in August of that year, ViaCrypt PGP Version 2.4 came out. ViaCrypt had a license for RSA, so their product was legal, but it was indeed a product. Users had to pay $100 for it.[18]

In the meanwhile RSA Data Security had released a free version, RSAREF, for noncommercial use. The first version had restrictions that prevented Zimmermann from making use of it within PGP,[19] but RSAREF 2.0 didn't, and Zimmermann included it in PGP Version 2.5. Thus, this version, while free, was only legal for noncommercial use. Soon thereafter, Version 2.6 appeared, an update made to appease Bidzos who was furious. Version 2.6 was intentionally made incompatible

---

[13] Cryptography and Control: National Security vs. Personal Privacy [VHS], CFP Video Library #210, Topanga, California, March 1992, This 77 minute long tape shows a panel session, with questions and answers, from The Second Conference on Computers, Freedom, & Privacy (CFP). See http://www.cfp.org/ for what this group has done over the years and http://www.forests.com/cfpvideo/ for their full video catalog.

[14] Levy, Steven, *Crypto: How the Code Rebels Beat the Government, Saving Privacy in the Digital Age*, Viking, New York, 2001, p. 200.

[15] Garfinkel, Simson, *PGP: Pretty Good Privacy*, O'Reilly & Associates, Sebastopol, California, 1995, p. 102.

[16] Branko Lankester (Netherlands), Peter Gutmann (New Zealand), Jean-Loup Gailly (France), Miguel Gallardo (Spain), Peter Simons (Germany). See Garfinkel, Simson, *PGP: Pretty Good Privacy*, O'Reilly & Associates, Sebastopol, California, 1995, p. 103; Levy, Steven, *Crypto: How the Code Rebels Beat the Government, Saving Privacy in the Digital Age*, Viking, New York, 2001, p. 200.

[17] Levy, Steven, *Crypto: How the Code Rebels Beat the Government, Saving Privacy in the Digital Age*, Viking, New York, 2001, p. 203

[18] Garfinkel, Simson, *PGP: Pretty Good Privacy*, O'Reilly & Associates, Sebastopol, California, 1995, pp. 105–106.

[19] For example, a license for the free version couldn't be obtained by anyone who had previously infringed upon the RSA patent. See Garfinkel, Simson, *PGP: Pretty Good Privacy*, O'Reilly & Associates, Sebastopol, California, 1995, p. 105.

with previous versions of PGP, beginning on September 1, 1994, so as not to allow the patent infringers to continue without the upgrade.[20] The program quickly appeared in Europe, in violation of the export laws of the time. This was followed by the appearance of PGP 2.6ui, which was an "unofficial international" version that updated an older version in a way that made it compatible with 2.6, for private or commercial use.

So in 1993 and 1994, PGP was legitimized in both commercial and freeware versions, but in the meantime, Zimmermann began to have legal troubles with the U.S. Government. Beginning in February 1993, Zimmermann faced investigators from the U.S. Customs Department. They were concerned about how PGP was exported. It seems that they were looking for someone small to make an example of. It would be easier to gain a prosecution of Zimmermann based on the export laws, than to take on RSA Data Security (whose RSAREF had been exported) or Netcom (whose FTP servers could be used to download PGP from abroad).[21]

This wasn't Zimmermann's first brush with the law. He had been arrested twice before at nuclear freeze rallies in the 1980s. No charges were ever filed, though.[22] Ultimately, the export violation investigation ended the same way. In 1996, the Government closed the case without charges being pressed. Still, it must have been a nerve-wracking experience, as he could have faced serious jail time if convicted.

PGP 3 was not a minor upgrade. It used a new algorithm for encryption CAST-128 and replaced the RSA component with a choice of DSA or Elgamal. Also, for the first time, the program had a nice interface. All previous versions were run from the command line. ViaCrypt was still producing commercial versions. They used even numbers for their versions, while Zimmermann used odd numbers for the free versions, but the commercial version 4 was ready before the free version 3, so Zimmermann renamed the free version PGP 5, for its May 1997 release.[23]

The commercial production of PGP software has changed hands several times since 1997. Most recently, in 2010, Symantec Corp. bought PGP for $300 million.[24] The company would certainly be worth far less if the export laws hadn't been changed in 2000. Thus, all versions of PGP became legal for export. Was this a victory for privacy advocates? Zimmermann wrote:[25]

> The law changed because the entire U.S. computer industry (which is the largest, most powerful industry in the U.S.) was united in favor of lifting the export controls. Money means political influence. After years of fighting it, the government finally had to surrender. If the White House had not lifted the export controls, the Congress and the Judicial system were preparing to intervene to do it for them.

It was described by some as a victory for civil libertarians, but they only won because of their powerful allies in industry.

There is much more that can be said about PGP. Back in Section 1.17 we briefly digressed into data compression. This is relevant to PGP, which compresses files prior to encryption. Because

---

[20] Garfinkel, Simson, *PGP: Pretty Good Privacy*, O'Reilly & Associates, Sebastopol, California, 1995, p. 108.

[21] Garfinkel, Simson, *PGP: Pretty Good Privacy*, O'Reilly & Associates, Sebastopol, California, 1995, p. 112. The opinion concerning why Zmmermann was targeted is my own.

[22] Levy, Steven, *Crypto: How the Code Rebels Beat the Government, Saving Privacy in the Digital Age*, Viking, New York, 2001, p. 190.

[23] http://en.wikipedia.org/wiki/Pretty_Good_Privacy.

[24] Kirk, Jeremy, "Symantec Buys Encryption Specialist PGP for $300M," *Computerworld*, April 29, 2010, available online at http://www.computerworld.com/s/article/9176121/Symantec_buys_encryption_specialist_PGP_for_300M.

[25] http://www.philzimmermann.com/EN/faq/index.html.

compression reduces redundancy, and redundancy is of great value to the cryptanalyst, this step is well worth the extra time it takes. It should also be pointed out that PGP isn't just for email. It includes a feature allowing the user to apply conventional cryptography (no RSA involved here) to compress and encipher files for storage.[26] The user only needs to come up with a random passphrase to use as a key.

18.3 In Zimmermann's Own Words

Although there is some overlap with this chapter (and previous chapters), a short essay by Philip Zimmermann (Figure 18.1) deserves to be reprinted here. The following is part of the Original 1991 *PGP User's Guide* (updated in 1999).[27]

Figure 18.1　Philip Zimmermann (1954–).

Why I Wrote PGP

"Whatever you do will be insignificant, but it is very important that you do it."

—Mahatma Gandhi

It's personal. It's private. And it's no one's business but yours. You may be planning a political campaign, discussing your taxes, or having a secret romance. Or you may be communicating with a political dissident in a repressive country. Whatever it is, you don't want your private electronic mail (email) or confidential documents read by anyone else. There's nothing wrong with asserting your privacy. Privacy is as apple-pie as the Constitution.

The right to privacy is spread implicitly throughout the Bill of Rights. But when the United States Constitution was framed, the Founding Fathers saw no need to explicitly spell out the right to a private conversation. That would have been silly. Two hundred years ago, all conversations were private. If someone else was within earshot, you could just go out behind the barn and have your conversation there. No one could listen in without your knowledge. The right to a private conversation was a natural right, not just in a philosophical sense, but in a law-of-physics sense, given the technology of the time.

---

[26] Zimmermann, Philip R., *The Official PGP User's Guide*, The MIT Press, Cambridge, Massachusetts, 1995, p. 18

[27] Reproduced from http://web.mit.edu/prz/, Zimmermann's homepage.

But with the coming of the information age, starting with the invention of the telephone, all that has changed. Now most of our conversations are conducted electronically. This allows our most intimate conversations to be exposed without our knowledge. Cellular phone calls may be monitored by anyone with a radio. Electronic mail, sent across the Internet, is no more secure than cellular phone calls. Email is rapidly replacing postal mail, becoming the norm for everyone, not the novelty it was in the past.

Until recently, if the government wanted to violate the privacy of ordinary citizens, they had to expend a certain amount of expense and labor to intercept and steam open and read paper mail. Or they had to listen to and possibly transcribe spoken telephone conversation, at least before automatic voice recognition technology became available. This kind of labor-intensive monitoring was not practical on a large scale. It was only done in important cases when it seemed worthwhile. This is like catching one fish at a time, with a hook and line. Today, email can be routinely and automatically scanned for interesting keywords, on a vast scale, without detection. This is like driftnet fishing. And exponential growth in computer power is making the same thing possible with voice traffic.

Perhaps you think your email is legitimate enough that encryption is unwarranted. If you really are a law-abiding citizen with nothing to hide, then why don't you always send your paper mail on postcards? Why not submit to drug testing on demand? Why require a warrant for police searches of your house? Are you trying to hide something? If you hide your mail inside envelopes, does that mean you must be a subversive or a drug dealer, or maybe a paranoid nut? Do law-abiding citizens have any need to encrypt their email?

What if everyone believed that law-abiding citizens should use postcards for their mail? If a nonconformist tried to assert his privacy by using an envelope for his mail, it would draw suspicion. Perhaps the authorities would open his mail to see what he's hiding. Fortunately, we don't live in that kind of world, because everyone protects most of their mail with envelopes. So no one draws suspicion by asserting their privacy with an envelope. There's safety in numbers. Analogously, it would be nice if everyone routinely used encryption for all their email, innocent or not, so that no one drew suspicion by asserting their email privacy with encryption. Think of it as a form of solidarity.

Senate Bill 266, a 1991 omnibus anticrime bill, had an unsettling measure buried in it. If this non-binding resolution had become real law, it would have forced manufacturers of secure communications equipment to insert special "trap doors" in their products, so that the government could read anyone's encrypted messages. It reads, "It is the sense of Congress that providers of electronic communications services and manufacturers of electronic communications service equipment shall ensure that communications systems permit the government to obtain the plain text contents of voice, data, and other communications when appropriately authorized by law." It was this bill that led me to publish PGP electronically for free that year, shortly before the measure was defeated after vigorous protest by civil libertarians and industry groups.

The 1994 Communications Assistance for Law Enforcement Act (CALEA) mandated that phone companies install remote wiretapping ports into their central office digital switches, creating a new technology infrastructure for "point-and-click" wiretapping, so that federal agents no longer have to go out and attach alligator clips to phone lines. Now they will be able to sit in their headquarters in Washington and listen in on your phone calls. Of course, the law still requires a court order for a wiretap.

But while technology infrastructures can persist for generations, laws and policies can change overnight. Once a communications infrastructure optimized for surveillance becomes entrenched, a shift in political conditions may lead to abuse of this new-found power. Political conditions may shift with the election of a new government, or perhaps more abruptly from the bombing of a federal building.

A year after the CALEA passed, the FBI disclosed plans to require the phone companies to build into their infrastructure the capacity to simultaneously wiretap 1 percent of all phone calls in all major U.S. cities. This would represent more than a thousandfold increase over previous levels in the number of phones that could be wiretapped. In previous years, there were only about a thousand court-ordered wire-taps in the United States per year, at the federal, state, and local levels combined. It's hard to see how the government could even employ enough judges to sign enough wiretap orders to wiretap 1 percent of all our phone calls, much less hire enough federal agents to sit and listen to all that traffic in real time. The only plausible way of processing that amount of traffic is a massive Orwellian application of automated voice recognition technology to sift through it all, searching for interesting keywords or searching for a particular speaker's voice. If the government doesn't find the target in the first 1 percent sample, the wiretaps can be shifted over to a different 1 per-cent until the target is found, or until everyone's phone line has been checked for subversive traffic. The FBI said they need this capacity to plan for the future. This plan sparked such outrage that it was defeated in Congress. But the mere fact that the FBI even asked for these broad powers is revealing of their agenda. Advances in technology will not permit the maintenance of the status quo, as far as privacy is concerned. The status quo is unstable. If we do nothing, new technologies will give the government new automatic surveillance capabilities that Stalin could never have dreamed of. The only way to hold the line on privacy in the information age is strong cryptography.

You don't have to distrust the government to want to use cryptography. Your busi-ness can be wiretapped by business rivals, organized crime, or foreign governments. Several foreign governments, for example, admit to using their signals intelligence against companies from other countries to give their own corporations a competitive edge. Ironically, the United States government's restrictions on cryptography in the 1990s have weakened U.S. corporate defenses against foreign intelligence and orga-nized crime.

The government knows what a pivotal role cryptography is destined to play in the power relationship with its people. In April 1993, the Clinton administration unveiled a bold new encryption policy initiative, which had been under develop-ment at the National Security Agency (NSA) since the start of the Bush administra-tion. The centerpiece of this initiative was a government-built encryption device, called the Clipper chip, containing a new classified NSA encryption algorithm. The government tried to encourage private industry to design it into all their secure communication products, such as secure phones, secure faxes, and so on. AT&T put Clipper into its secure voice products. The catch: At the time of manufacture, each Clipper chip is loaded with its own unique key, and the government gets to keep a copy, placed in escrow. Not to worry, though—the government promises that they will use these keys to read your traffic only "when duly authorized by law." Of course, to make Clipper completely effective, the next logical step would be to outlaw other forms of cryptography.

The government initially claimed that using Clipper would be voluntary, that no one would be forced to use it instead of other types of cryptography. But the public reaction against the Clipper chip was strong, stronger than the government anticipated. The computer industry monolithically proclaimed its opposition to using Clipper. FBI director Louis Freeh responded to a question in a press conference in 1994 by saying that if Clipper failed to gain public support, and FBI wiretaps were shut out by non-government-controlled cryptography, his office would have no choice but to seek legislative relief. Later, in the aftermath of the Oklahoma City tragedy, Mr. Freeh testified before the Senate Judiciary Committee that public availability of strong cryptography must be curtailed by the government (although no one had suggested that cryptography was used by the bombers).

The government has a track record that does not inspire confidence that they will never abuse our civil liberties. The FBI's COINTELPRO program targeted groups that opposed government policies. They spied on the antiwar movement and the civil rights movement. They wiretapped the phone of Martin Luther King Jr. Nixon had his enemies list. Then there was the Watergate mess. More recently, Congress has either attempted to or succeeded in passing laws curtailing our civil liberties on the Internet. Some elements of the Clinton White House collected confidential FBI files on Republican civil servants, conceivably for political exploitation. And some over-zealous prosecutors have shown a willingness to go to the ends of the Earth in pursuit of exposing sexual indiscretions of political enemies. At no time in the past century has public distrust of the government been so broadly distributed across the political spectrum, as it is today.

Throughout the 1990s, I figured that if we want to resist this unsettling trend in the government to outlaw cryptography, one measure we can apply is to use cryptography as much as we can now while it's still legal. When use of strong cryptography becomes popular, it's harder for the government to criminalize it. Therefore, using PGP is good for preserving democracy. If privacy is outlawed, only outlaws will have privacy.

It appears that the deployment of PGP must have worked, along with years of steady public outcry and industry pressure to relax the export controls. In the closing months of 1999, the Clinton administration announced a radical shift in export policy for crypto technology. They essentially threw out the whole export control regime. Now, we are finally able to export strong cryptography, with no upper limits on strength. It has been a long struggle, but we have finally won, at least on the export control front in the US. Now we must continue our efforts to deploy strong crypto, to blunt the effects increasing surveillance efforts on the Internet by various governments. And we still need to entrench our right to use it domestically over the objections of the FBI.

PGP empowers people to take their privacy into their own hands. There has been a growing social need for it. That's why I wrote it.

Philip R. Zimmermann
Boulder, Colorado
June 1991 (updated 1999)

18.4 The Impact of PGP

So what difference did exporting strong encryption make? Did it merely support terrorists, drug dealers, and pedophiles, as those favoring a government monopoly on cryptology like to suggest? Messages sent to Zimmermann, now posted on his website, show PGP to have benefited human rights activists in Guatemala and the Balkans and freedom fighters in Kosovo and Romania.[28] It's not documented on his website, but Zimmerman frequently talked about Burmese rebels making use of PGP.[29] Thus, some sympathetic groups benefited from the program. It certainly helped less sympathetic individuals as well, but there is no obvious way to make technology only accessible to the "good guys." It's left to the reader to decide if the trade-off is worth it. Zimmermann believes it is.

Wikipedia references episodes from 2003 through 2009 that indicate that neither the FBI nor British police were able to break PGP during those years.[30] However, there are also instances of the FBI reading PGP messages. They did so in these cases by means other than cryptanalysis. With a warrant, the FBI can break into a suspect's home or business and install keyloggers on his or her computers. This approach revealed the password protecting the PGP enciphered messages of accused Mafia loan shark Nicodemo S. Scarfo Jr. in 1999.[31] The same method was used against alleged Ecstasy manufacturers, and known PGP users, Mark Forrester and Dennis Alba in 2007.[32] The FBI has other, more sophisticated, software at its disposal, as well.

In 2018, a team of researchers announced a flaw that allowed recovery of some PGP enciphered messages sent from 2003 to 2018.[33] This problem was not actually in PGP, but rather how it was implemented in various email programs. As I remarked in Section 18.1, secure implementation of an otherwise sound system is a highly non-trivial step.

18.5 Password Issues

The recipient's private key must be available for deciphering, but ought not be stored on the computer, where it might become accessible to someone else. The similar problem of password storage was discussed in Section 17.3.2. PGP addresses the issue by enciphering the user's private key before storing it on the machine. A password must then be entered to "unlock" the key. A function combines the password and the enciphered key to reveal, temporarily, the private key for the program's use. Now, of course, that password must be carefully chosen.

Poor password selection can be illustrated with a few anecdotes. As an undergraduate, I worked in a computer lab on campus. One day a girl came in and asked how to change her password.

---

[28] http://www.philzimmermann.com/EN/letters/index.html.

[29] Levy, Steven, *Crypto: How the Code Rebels Beat the Government, Saving Privacy in the Digital Age*, Viking, New York, 2001, p. 289.

[30] Wikipedia contributors, "Pretty Good Privacy," *Wikipedia, The Free Encyclopedia*, https://en.wikipedia.org/wiki/Pretty_Good_Privacy#Security_quality.

[31] Chidi, George A., "Federal judge allows keyboard-stroke capture," *cnn.com*, January 7, 2002, http://www.cnn.com/2002/TECH/internet/01/07/fbi.surveillance.idg/index.html.

[32] McCullagh, Declan, "Feds use keylogger to thwart PGP, Hushmail," *c|net*, July 20, 2007, https://www.cnet.com/news/feds-use-keylogger-to-thwart-pgp-hushmail/.

[33] Poddebniak, Damian, Christian Dresen, Jens Müller, Fabian Ising, Sebastian Schinzel, Simon Friedberger, Juraj Somorovsky, and Jörg Schwenk, "Efail: Breaking S/MIME and OpenPGP Email Encryption using Exfiltration Channels," *27th USENIX Security Symposium*, Baltimore, Maryland, August 2018. See https://efail.de/ for a link to this and much more.

Before I could answer, a coworker responded, "Why? Did you break up with him?" She immediately blushed. Her password was not well chosen.

The psychological approach was also applied by Richard Feynman at Los Alamos during the Manhattan Project, when the secrets of the atomic bomb warranted the highest level of protection. He turned to one of the filing cabinets that stored those secrets and guessed that the combination lock that secured it might be set to an important mathematical constant. He tried π first, forwards, backward, every way he could imagine. It didn't work, but e did. The combination was 27-18-28 and Feynman soon found he could open all five filing cabinets with that combination.[34]

A study of 43,713 hacked MySpace account passwords revealed the following top 10 list, with their frequencies in parentheses.[35]

1. password1 (99)
2. iloveyou1 (52)
3. abc123 (47)
4. myspace1 (39)
5. fuckyou1 (31)
6. summer07 (29)
7. iloveyou2 (28)
8. qwerty1 (26)
9. football1 (25)
10. 123abc (22)

MySpace forced users to include at least one non-alphabetic character in their passwords. Many users obviously just tack a 1 onto their original choice. A study of 32 million passwords hacked from RockYou.com, gives a different top 10 list.[36]

1. 123456
2. 12345
3. 123456789
4. Password
5. iloveyou
6. princess
7. rockyou
8. 1234567
9. 12345678
10. abc123

Passwords ought not be vulnerable to dictionary attacks, ought not consist solely of letters, and ought to be long. Ideally, they look random. We remember many random looking numbers (phone numbers, Social Security numbers, etc.), so it ought to be easy enough to remember one more with

[34] Feynman, Richard, *"Surely You're Joking Mr. Feynman!"* W. W. Norton & Company, New York, 1985, pp. 147–151.

[35] "A brief analysis of 40,000 leaked MySpace passwords," November 1, 2007, http://www.the-interweb.com/serendipity/index.php?/archives/94-A-brief-analysis-of-40,000-leaked-MySpace-passwords.html.

[36] Coursey, David, "Study: Hacking Passwords Easy As 123456," *PCWorld.* January 21, 2010, available online at http://www.pcworld.com/businesscenter/article/187354/study_hacking_passwords_easy_as_123456.html.

a mix of numbers and letters. But it really isn't just one more! We need passwords for bank cards, and every website we wish to use for online purchases, and they should all be different! Often this is not the case. Recall, the atomic secrets alluded to earlier were distributed over five filing cabinets, *all* of which had *e* as the combination. Inevitably, in many cases, the passwords are written down and kept near the computer.

Another potential place for implementation flaws is in generating the primes to be used for the RSA portion, as we saw in Section 15.1.12. Back in the 1990s, PGP users were able to select the size of these primes from the low-commercial, high-commercial or "military" grade — up to over 1,000 bits.[37] Increased speed is the only reason to select the smaller sizes. Once a size is selected, the program prompts the user to type some arbitrary text. The text itself is ignored, but the time interval between keystrokes is used to generate random numbers that are then used to generate the primes.[38]

There are many other technical details that need to be addressed if one wants to implement a hybrid system securely. Bearing this in mind, we can better understand why it took Zimmermann so long to code up the first version of PGP.

18.6 History Repeats Itself

In Section 18.2, Senator Joseph Biden's co-sponsorship of Senate "Anti-Crime" Bill 266 was mentioned, along with how its anti-cryptology provision was ultimately withdrawn. Sadly, it was not forgotten. In 2010, Biden was serving as Vice President under Barack Obama, and this administration hoped to bring back the deleted provision. A paragraph from a September 27, 2010 *New York Times* piece summed up the new proposal.

> Essentially, officials want Congress to require all services that enable communications—including encrypted e-mail transmitters like BlackBerry, social networking Web sites like Facebook and software that allows direct "peer to peer" messaging like Skype—to be technically capable of complying if served with a wiretap order. The mandate would include being able to intercept and unscramble encrypted messages.[39]

To put the proposed limitation on communications providers more tersely,

> They can promise strong encryption. They just need to figure out how they can provide us plain text.

> **—FBI General Counsel Valerie Caproni, September 27, 2010.**[40]

According to the *New York Times* article, the administration expected that the bill would be considered in 2011. However, their plan fizzled out. At a congressional hearing on February 17,

---

[37] Zimmermann, Philip R., *The Official PGP User's Guide*, The MIT Press, Cambridge, Massachusetts, 1995, p. 21.

[38] Zimmermann, Philip R., *The Official PGP User's Guide*, The MIT Press, Cambridge, Massachusetts, 1995, p. 22.

[39] Savage, Charlie, "U.S. Tries to Make It Easier to Wiretap the Internet," *New York Times*, September 27, 2010, p. A1, available online at http://www.nytimes.com/2010/09/27/us/27wiretap.html and http://archive.nytimes.com/www.nytimes.com/2010/09/27/us/27wiretap.html.

[40] Savage, Charlie, "U.S. Tries to Make It Easier to Wiretap the Internet," *New York Times*, September 27, 2010, p. A1, available online at http://www.nytimes.com/2010/09/27/us/27wiretap.html and http://archive.nytimes.com/www.nytimes.com/2010/09/27/us/27wiretap.html.

2011, the following exchange took place between Henry C. "Hank" Johnson, Jr., a Republican Congressman from Georgia, and Caproni:

> **Mr. JOHNSON.** What is it exactly that you would want Congress to do, or are you asking Congress for anything?
>
> **Ms. CAPRONI.** Not yet.
>
> **Mr. JOHNSON.** Or did we just simply invite you here to tell us about this?
>
> **Ms. CAPRONI.** You invited me, and we came. But we don't have a specific request yet. We are still—the Administration is considering—I am really here today to talk about the problem. And I think if everyone understands that we have a problem, that is the first step, and then figuring out how we fix it is the second step. The Administration does not yet have a proposal. It is something that is actively being discussed within the Administration, and I am optimistic that we will have a proposal in the near future.[41]

The FBI did not propose a bill in the near future. Apparently, they couldn't figure out exactly what it would be.

Earlier in the hearing, Susan Landau, of the Radcliffe Institute for Advanced Study, Harvard University, gave testimony explaining NSA's point of view.

> I want to step back for a moment and talk about cryptography, a fight we had in the 1990's in which the NSA and the FBI opposed the deployment of cryptography through the communications infrastructure. In 1999, the U.S. Government changed its policy.
>
> The NSA has been firmly behind the change of policy, and endorsed a full set of unclassified algorithms to be used for securing the communications network. The NSA obviously believes that in the conflict between communications surveillance and communications security, we need to have communications security.[42]

18.7 A Terrorist and an iPhone

A terrorist act committed on December 2, 2015 led to the issue being raised again in a big way. On that day, Syed Rizwan Farook and Tashfeen Malik used semi-automatic rifles (variants of the AR-15) and semi-automatic pistols to kill 14 people and seriously wound 22 others. The killers were gunned down by law enforcement later that day. The deceased shooters couldn't be inter-rogated, but the FBI had another potentially useful source of information — an Apple iPhone 5C that Farook used. Unlocking the phone required a four-digit code, which sounds easy to

---

[41] Going Dark: Lawful Electronic Surveillance in the Face of New Technologies, Hearing before the Subcommittee on Crime, Terrorism, and Homeland Security of the Committee on the Judiciary House of Representatives, One Hundred Twelfth Congress, First Session, February 17, 2011, pp. 49–50, available online at https://www.govinfo.gov/content/pkg/CHRG-112hhrg64581/pdf/CHRG-112hhrg64581.pdf.

[42] Going Dark: Lawful Electronic Surveillance in the Face of New Technologies, Hearing before the Subcommittee on Crime, Terrorism, and Homeland Security of the Committee on the Judiciary House of Representatives, One Hundred Twelfth Congress, First Session, February 17, 2011, p. 24, available online at https://www.govinfo.gov/content/pkg/CHRG-112hhrg64581/pdf/CHRG-112hhrg64581.pdf.

brute-force, but the phone was set to make its contents permanently inaccessible (by erasing the stored form of the AES encryption key), if the correct code wasn't entered by the tenth attempt.[43]

On February 9, 2016, FBI Director James Comey claimed that the Bureau couldn't unlock the iPhone.[44] The FBI appealed to Apple to create software that would allow the iPhone's contents to be accessed, but Apple refused and the legal battle began. FBI leadership apparently thought that the emotionally charged issue of terrorism was one that the Bureau could use to rally the American public around their old (lost) cause, to force telecommunications providers to develop techniques to provide the government with access to encrypted data on demand.

Apple refused to comply with the FBI's request. On February 16, in "A Message to our customers," the tech giant's CEO, Tim Cook, explained,[45]

> The United States government has demanded that Apple take an unprecedented step which threatens the security of our customers. We oppose this order, which has implications far beyond the legal case at hand.

Cook also explained why encryption is necessary:[46]

> Smartphones, led by iPhone, have become an essential part of our lives. People use them to store an incredible amount of personal information, from our private conversations to our photos, our music, our notes, our calendars and contacts, our financial information and health data, even where we have been and where we are going.
>
> All that information needs to be protected from hackers and criminals who want to access it, steal it, and use it without our knowledge or permission. Customers expect Apple and other technology companies to do everything in our power to protect their personal information, and at Apple we are deeply committed to safeguarding their data.
>
> Compromising the security of our personal information can ultimately put our personal safety at risk. That is why encryption has become so important to all of us.
>
> For many years, we have used encryption to protect our customers' personal data because we believe it's the only way to keep their information safe. We have even put that data out of our own reach, because we believe the contents of your iPhone are none of our business.

Cook then noted that Apple provided the FBI with all of the information that they could actually access and legally supply to the FBI in connection with the San Bernadino case, and explained why creating the backdoor the FBI requested would be a dangerous move.[47]

> But now the U.S. government has asked us for something we simply do not have, and something we consider too dangerous to create. They have asked us to build a backdoor to the iPhone.
>
> The government suggests this tool could only be used once, on one phone. But that's simply not true. Once created, the technique could be used over and over again,

---

[43] AES is detailed in Section 20.3. https://en.wikipedia.org/wiki/FBI–Apple_encryption_dispute.

[44] Volz, Dustin and Mark Hosenball, "FBI director says investigators unable to unlock San Bernardino shooter's phone content," Reuters, February 9, 2016, https://www.reuters.com/article/us-california-shooting-encryption-idUSKCN0VI22A.

[45] Cook Tim, "A Message to our Customers," Apple, February 16, 2016, https://www.apple.com/customer-letter/.

[46] Cook Tim, "A Message to our Customers," Apple, February 16, 2016, https://www.apple.com/customer-letter/.

[47] Cook Tim, "A Message to our Customers," Apple, February 16, 2016, https://www.apple.com/customer-letter/.

on any number of devices. In the physical world, it would be the equivalent of a master key, capable of opening hundreds of millions of locks — from restaurants and banks to stores and homes. No reasonable person would find that acceptable.

The government is asking Apple to hack our own users and undermine decades of security advancements that protect our customers — including tens of millions of American citizens — from sophisticated hackers and cybercriminals.

Near the end of the letter, Cook wrote, "We are challenging the FBI's demands with the deepest respect for American democracy and a love of our country." As I see it, opposing the government, when it is wrong, is a patriotic act. Cook apparently felt the same way, for he closed with "And ultimately, we fear that this demand [from the FBI] would undermine the very freedoms and liberty our government is meant to protect."

The (online) letter included a link to "Answers to your questions about Apple and security." This Q and A offered a bit more detail to help people understand the implications of a backdoor, in terms of legal precedent, and the possibility of abuse, noting, "The only way to guarantee that such a powerful tool isn't abused and doesn't fall into the wrong hands is to never create it."[48]

On February 18, John McAfee, who has been imprisoned in 11 countries and described as a "cybersecurity legend and psychedelic drug enthusiast,"[49] weighed in on the matter in an op-ed piece.

The fundamental question is this: Why can't the FBI crack the encryption on its own? It has the full resources of the best the US government can provide.

With all due respect to Tim Cook and Apple, I work with a team of the best hackers on the planet. These hackers attend Defcon in Las Vegas, and they are legends in their local hacking groups, such as HackMiami. They are all prodigies, with talents that defy normal human comprehension. About 75% are social engineers. The remainder are hardcore coders. I would eat my shoe on the Neil Cavuto show if we could not break the encryption on the San Bernardino phone. This is a pure and simple fact.

And why do the best hackers on the planet not work for the FBI? Because the FBI will not hire anyone with a 24-inch purple mohawk, 10-gauge ear piercings, and a tattooed face who demands to smoke weed while working and won't work for less than a half-million dollars a year. But you bet your ass that the Chinese and Russians are hiring similar people with similar demands and have been for many years. It's why we are decades behind in the cyber race.[50]

McAfee closed with an offer of assistance.

So here is my offer to the FBI. I will, free of charge, decrypt the information on the San Bernardino phone, with my team. We will primarily use social engineering, and it will take us three weeks. If you accept my offer, then you will not need to ask Apple to place a backdoor in its product, which will be the beginning of the end of America.

---

[48] "Answers to your questions about Apple and security," Apple, https://www.apple.com/customer-letter/answers/.

[49] Hathaway, Jay, "Antivirus Wild Man John McAfee Offers to Solve FBI's iPhone Problem So Apple Doesn't Have To," February 19, 2016, *Intelligencer*, available online at https://nymag.com/intelligencer/2016/02/john-mcafee-says-he-can-crack-that-iphone.html.

[50] McAfee, John, "JOHN MCAFEE: I'll decrypt the San Bernardino phone free of charge so Apple doesn't need to place a back door on its product," *Business Insider*, February 18, 2016, available online at https://www.businessinsider.com/john-mcafee-ill-decrypt-san-bernardino-phone-for-free-2016-2.

If you doubt my credentials, Google "cybersecurity legend" and see whose name is the only name that appears in the first 10 results out of more than a quarter of a million.[51]

While a social engineering attack (manipulating people to get them to provide access or information) is often the easiest approach, it wouldn't be useful in the case of the iPhone. Indeed, McAfee later admitted what he wrote here (and said elsewhere) wasn't exactly right.

I speak through the press, to the press, and to the general public. For example, last night I was on RT, and I gave a *vastly* oversimplified explanation of how you would hack into the iPhone.[52] I can't possibly go in and talk about the secure spaces on the A7 chip. I mean, who's going to understand that crap? Nobody. But you gotta believe me: I understand it. And I do know what I'm doing, else I would not be where I am. This is a fact. Someone who does not understand software cannot start a multibillion dollar company. This is just a fact of life. So, if I look like an idiot, it is because I am speaking *to* idiots.[53]

He also explained his purpose in making intentionally inaccurate statements.

By doing so, I knew that I would get a shitload of public attention, which I did. That video, on my YouTube account, it has 700,000 views. My point is to bring to the American public the problem that the FBI is trying to [fool] the American public. How am I going to do that, by just going off and saying it? No one is going to listen to that crap. So I come up with something sensational. Now, what I did not lie about was my ability to crack the iPhone. I can do it. It's a piece of friggin' cake. You could probably do it.[54]

Many others in the tech community found some of the FBI's statements disingenuous. To them it seemed like the FBI was more interested in setting a legal precedent than in gaining access to the iPhone in question. There were certainly people the FBI could appeal to if all they truly wanted was the contents of that particular phone. When the FBI revealed, on March 28, 2016, that they found a third party who was able to access the iPhone without Apple's help[55] the reaction from many was, "Of course you did." Because the FBI backed down, the case ended without a ruling against them. If the Bureau was confident they would win, would the mysterious third party have even been appealed to for help?

[51] McAfee, John, "JOHN MCAFEE: I'll decrypt the San Bernardino phone free of charge so Apple doesn't need to place a back door on its product," *Business Insider*, February 18, 2016, available online at https://www.businessinsider.com/john-mcafee-ill-decrypt-san-bernardino-phone-for-free-2016-2.

[52] McAfee, John, "John McAfee Reveals To FBI, On National TV, How To Crack The iPhone (RT Interview)," *John McAfee* YouTube Channel, February 29, 2016, https://www.youtube.com/watch?v=MG0bAaK7p9s. This time, the explanation didn't involve social engineering.

[53] Carmichael, Joe, "John McAfee Challenges Reddit," *Inverse*, March 2, 2016, https://www.inverse.com/article/12277-john-mcafee-challenges-reddit.

[54] Turton, William, *daily dot*, "John McAfee lied about San Bernardino shooter's iPhone hack to 'get a s**tload of public attention'," February 29, 2020, https://www.dailydot.com/debug/john-mcafee-lied-iphone-apple-fbi/.

[55] Segall, Laurie, Jose Pagliery, and Jackie Wattles, *cnn.com*, "FBI says it has cracked terrorist's iPhone without Apple's help," March 29, 2016, https://money.cnn.com/2016/03/28/news/companies/fbi-apple-iphone-case-cracked/index.html.

There's an important detail to consider when it comes to assessing how honest the FBI has been on this issue. FBI Director James Comey indicated that the Bureau paid over $1.3 million to the third party for the hack.[56] If so, the FBI was drastically overcharged, as you will soon see. According to another source, the cost was only $15,000.[57] If the latter is true, why did Comey lie?

On September 14, 2016, Sergei Skorobogatov, a computer scientist at the University of Cambridge, published a paper online giving full details of how to hack the iPhone 5C, Skorobogatov noted, "The process does not require any expensive and sophisticated equipment. All needed parts are low cost and were obtained from local electronics distributors."[58] He also posted videos demonstrating the process on YouTube. The timing of this research project was not a coincidence. On March 28, Comey had said that "NAND mirroring" wouldn't be used to hack into the iPhone, and that "It doesn't work."[59] This was the approach that Skorobogatov successfully applied. John Gruber noted, "When the FBI lies it's a "fib". When you lie to the FBI it's a "felony"."[60]

As of this writing (May 23, 2020), McAfee's interview on RT[61] (inaccurately explaining an iPhone 5C hack) has 1,127,246 views, while Skorobogatov's video[62] (actually demonstrating an iPhone5C hack) has only 241,601 views. It looks like McAfee is pretty media savvy after all. The conclusion that the FBI was playing the media was reached by some politicians, in addition to technical people. For example, Democratic Senator Ron Wyden said, "There are real questions about whether [the FBI] has been straight with the public on [the Apple case]."[63]

The San Bernadino shooter case wasn't the first time the FBI initiated such a legal challenge. An earlier instance involved an iPhone 5S, but the crime involved drugs, not terrorism. The FBI lost this case on February 29, 2016, when a federal judge rejected its request to order Apple to open the iPhone.[64] There were several other attempts involving iPhones and iPads, but the terrorist's iPhone got the most attention.

---

[56] Lichtblau, Eric and Katie Benner, *The New York Times*, "F.B.I. Director Suggests Bill for iPhone Hacking Topped $1.3 Million," April 21, 2016, https://www.nytimes.com/2016/04/22/us/politics/fbi-director-suggests-bill-for-iphone-hacking-was-1-3-million.html.

[57] Fontana, John, *ZDNet*, "FBI's strategy in Apple case caught in distortion field," April 6, 2016, https://www.zdnet.com/article/fbis-strategy-in-apple-case-caught-in-distortion-field/.

[58] Skorobogatov, Sergei P., "The bumpy road towards iPhone 5c NAND mirroring," https://arxiv.org/abs/1609.04327. Also, see project page at https://www.cl.cam.ac.uk/~sps32/5c_proj.html. This website includes links to YouTube videos so you can watch the attack being carried out.

[59] Chaffin, Bryan, "FBI Director Comey Denies 'NAND Mirroring' Will Be Used to Unlock Terrorist's iPhone," *The Mac Observer*, March 24, 2016, https://www.macobserver.com/tmo/article/fbi-director-comeydenies-nand-mirroring-will-be-used-to-unlock-terrorists; Keizer, Gregg, "FBI chief shoots down theory that NAND mirroring will be used to crack terrorist's iPhone," *Computer World*, March 24, 2016, http://www.computerworld.com/article/3048243/apple-ios/fbichief-shoots-down-theory-that-nand-mirroring-will-be-used-tocrack-terrorists-iphone.html.

[60] Gruber, John, "Buzzfeed: 'Questions hang over FBI after Apple showdown fizzles'," *Daring Fireball*, March 23, 2016, https://daringfireball.net/linked/2016/03/23/buzzfeed-apple-fbi.

[61] McAfee, John, "John McAfee Reveals To FBI, On National TV, How To Crack The iPhone (RT Interview)," *John McAfee* YouTube Channel, February 29, 2016, https://www.youtube.com/watch?v=MG0bAaK7p9s,

[62] Demonstration of iPhone 5c NAND mirroring, *Presentation* YouTube Channel, https://www.youtube.com/watch?v=tM66GWrwbsY.

[63] Fontana, John, *ZDNet*, "FBI's strategy in Apple case caught in distortion field," April 6, 2016, https://www.zdnet.com/article/fbis-strategy-in-apple-case-caught-in-distortion-field/.

[64] Ackerman, Spencer, Sam Thielman, and Danny Yadron, "Apple case: judge rejects FBI request for access to drug dealer's iPhone," *The Guardian*, February 29, 2016, available online at https://www.theguardian.com/technology/2016/feb/29/apple-fbi-case-drug-dealer-iphone-jun-feng-san-bernardino.

The FBI likely pushed harder in the terrorism case, believing that they would have great support from the public. They miscalculated badly. The FBI was opposed, not only by Apple, but also by Access Now, Amazon.com, the American Civil Liberties Union, Box, the Center for Democracy and Technology, Cisco Systems, Dropbox, the Electronic Frontier Foundation, Evernote, Facebook, Google, Lavabit, LinkedIn, Microsoft, Mozilla, Nest Labs, Pinterest, Slack Technologies, Snapchat, Twitter, WhatsApp, and Yahoo![65]

And it wasn't just industry and liberal and libertarian organizations that opposed the FBI. The Bureau even took knocks from an intelligence community giant. General Michael Hayden, who had served as a director of the National Security Agency, as well as the Central Intelligence Agency, said,

> You can argue this on constitutional grounds. Does the government have the right to do this? Frankly, I think the government does have a right to do it. You can do balancing privacy and security… dead men don't have a right to privacy. I don't use those lenses. My lens is the security lens, and frankly, I think it's a close but clear call that Apple's right on just raw security grounds.[66]
>
> Jim Clapper has said, he's the Director of National Intelligence, that the greatest threat to the United States is the cyber threat and I think Apple is technologically correct when they say doing what the FBI wants them to do in this case will make their technology, their encryption, overall weaker than it would otherwise be. So I get why the FBI wants to get into the phone, but we make tradeoffs like this all the time and this may be a case where we've got to give up some things in law enforcement and even counter terrorism in order to preserve this aspect, our cybersecurity.[67]
>
> Any effort to legislate or to use a court to stop this broad technological trend just isn't going to work. We are going to a world of very high-end encryption that will be used routinely by people around the planet. Now, from my own line of work, signals intelligence, intercepting communications, that represents a challenge, but there are also tools available that you can still get meaningful intelligence out of communications even though you might never read the content.[68]

As for the American public, a CBS poll showed they were split between siding with Apple and the FBI.[69] This is not surprising, given that it's an issue that takes a bit of time to understand. It's likely a tiny percentage that actually have a good understanding of it.

---

[65] https://en.wikipedia.org/wiki/FBI-Apple_encryption_dispute, which cites Brandom, Russell, "Google, Microsoft, and other tech giants file legal briefs in support of Apple," *The Verge*, March 3, 2016, https://www.theverge.com/2016/3/3/11156704/apple-fbi-amicus-briefs-iphone-encryption-fight; Maddigan, Michael M. and Neil Kumar Katyal, "Brief of Amici Curiae Amazon, et. al," Hogan Lovells US LLP, March 2, 2016; and "Amicus Briefs in Support of Apple," *Apple Press Info.* March 3, 2016, https://www.apple.com/newsroom/2016/03/03Amicus-Briefs-in-Support-of-Apple/.

[66] Limitone, Julia, *FOXBusiness*, "Fmr. NSA, CIA Chief Hayden Sides with Apple Over Feds," March 7, 2016, https://www.foxbusiness.com/features/fmr-nsa-cia-chief-hayden-sides-with-apple-over-feds.

[67] Limitone, Julia, *FOXBusiness*, "Fmr. NSA, CIA Chief Hayden Sides with Apple Over Feds," March 7, 2016, https://www.foxbusiness.com/features/fmr-nsa-cia-chief-hayden-sides-with-apple-over-feds.

[68] Limitone, Julia, *FOXBusiness*, "Fmr. NSA, CIA Chief Hayden Sides with Apple Over Feds," March 7, 2016, https://www.foxbusiness.com/features/fmr-nsa-cia-chief-hayden-sides-with-apple-over-feds.

[69] Anon., "CBS News poll: Americans split on unlocking San Bernardino shooter's iPhone," *CBS News*, March 18, 2016, https://www.cbsnews.com/news/cbs-news-poll-americans-split-on-unlocking-san-bernardino-shooters-iphone/.

While the attention didn't go the way the FBI wanted, they did succeed in generating a lot of it. John Oliver devoted an installment of his HBO program *Last Week Tonight with John Oliver* to the topic on March 14, 2016, before the FBI got into the iPhone.[70] As of this writing (May 23, 2020), the episode, which is critical of the FBI's position, has racked up 12,127,673 views on YouTube. It included some quotes from Republican Senator Lindsey Graham. The first, given below in greater context than by Oliver, is from a Republican presidential primary debate held on December 15, 2015.

> The bottom line is, we're at war. They're trying to come here to kill us all and it's up to the government to protect you within constitutional means. Any system that would allow a terrorist to communicate with somebody in our country and we can't find out what they're saying is stupid. If I'm president of the United States, and you join ISIL, you are going to get killed or captured. And the last thing you are going to hear if I'm president is, you've got a right to remain silent.[71]

Then presidential candidate Donald Trump suggested a boycott of Apple until they comply with the FBI.

Three months after Graham's quote above, the Senator had the following exchange with Attorney General Loretta E. Lynch.

> **Attorney General Loretta E. Lynch:** I think for us the issue is about a criminal investigation into a terrorist act and the need to obtain evidence.

> **Graham:** But it's just not so simple and I'll end with this. I thought it was that simple. I was all with you until I actually started getting briefed by people in the intel community and I will say I'm a person who's been moved by the arguments of the precedent we set and the damage we may be doing to our own national security.

This quote was also played by Oliver, who said it was a miracle that Graham had "met the concept of nuance," but did he really? Read on.

Despite the FBI having backed down, some US Senators still had their sights set on forcing backdoors into communications devices. On April 7, 2016, a familiar bit of draft legislation was leaked, followed by an official release on the 13th. As with the bill Biden had cosponsored in 1991, this new proposal, called the Compliance with Court Orders Act of 2016 (CCOA), would require "any person who provides a product or method to facilitate a communication or the processing or storage of data" to "be capable of complying" with court orders to turn over "data in an intelligible format" even if the data was enciphered by the user. Senator Ron Wyden said, "This flawed bill would leave Americans more vulnerable to stalkers, identity thieves, foreign hackers

---

[70] Oliver, John, Encryption: Last Week Tonight with John Oliver (HBO), *LastWeekTonight* YouTube Channel, March 14, 2016, https://www.youtube.com/watch?v=zsjZ2r9Ygzw.

[71] Wofford, Taylor, *Newsweek*, "Full Transcript: CNN Republican Undercard Debate," December 15, 2015, https://www.newsweek.com/cnn-republican-undercard-debate-transcript-405767.

and criminals."[72] This time, the authors of the bill were Republican Richard Burr and Democrat Dianne Feinstein.[73] The Senators were quoted as saying,

> The underlying goal is simple: when there's a court order to render technical assistance to law enforcement or provide decrypted information, that court order is carried out. No individual or company is above the law.[74]

Ultimately, this legislation went nowhere, just like the 2011 attempt.

18.8 Another Terrorist and Another iPhone

Neither the US Senate nor the FBI seems to have learned much from previous battles they lost. Sadly, evidence of these groups containing slow-learners comes from another terrorist attack. This one occurred on December 6, 2019 in Pensacola, Florida, where Mohammed Saeed Alshamrani killed three men and wounded eight more with a 9mm Glock handgun. Alshamrani's iPhones became an issue.

On December 10, the US government attacked representatives in the tech community. In imitation of the failed attempts in 1991, 2011, and 2016, they demanded that backdoors be put in place to allow plaintext to be turned over when a court so orders. Senator Graham said, "You're going to find a way to do this or we're going to do this for you."[75] Republican Senator Marsha Blackburn accused the tech companies of creating a "safe harbor" for criminals and added, "That is why on a bipartisan basis you're hearing us say we've slapped your hand enough and you all have got to get your act together or we will gladly get your act together for you."[76] Apparently these senators aren't aware of the fact that the tech companies are, collectively, more powerful than they are.

On May 18, 2020, Attorney General William Barr made the following comments:

> Thanks to the great work of the FBI — and no thanks to Apple — we were able to unlock Alshamrani's phones.[77]
>
> Apple's decision has dangerous consequences for the public safety and the national security and is, in my judgement, unacceptable. Apple's desire to provide privacy for its customers is understandable, but not at all costs. ... There is no reason why companies like Apple cannot design their consumer products and apps to allow for

---

[72] Hosenball, Mark and Dustin Volz, *Reuters*, "U.S. Senate panel releases draft of controversial encryption bill," April 13, 2016, https://finance.yahoo.com/news/u-senate-panel-releases-draft-192224282.html and Pfefferkorn, Riana, *Just Security*, "Here's What the Burr-Feinstein Anti-Crypto Bill Gets Wrong," April 15, 2016, https://www.justsecurity.org/30606/burr-feinstein-crypto-bill-terrible/.

[73] Volz, Dustin and Mark Hosenball, *Reuters*, "Leak of Senate encryption bill prompts swift backlash," April 8, 2016, https://www.reuters.com/article/us-apple-encryption-legislation-idUSKCN0X52CG.

[74] Volz, Dustin and Mark Hosenball, *Reuters*, "Leak of Senate encryption bill prompts swift backlash," April 8, 2016, https://www.reuters.com/article/us-apple-encryption-legislation-idUSKCN0X52CG.

[75] Feiner, Lauren, "Senators threaten to regulate encryption if tech companies won't do it themselves," *CNBC*, December 10, 2019, https://www.cnbc.com/2019/12/10/senators-threaten-encryption-regulation-for-tech-companies.html.

[76] Feiner, Lauren, "Senators threaten to regulate encryption if tech companies won't do it themselves," *CNBC*, December 10, 2019, https://www.cnbc.com/2019/12/10/senators-threaten-encryption-regulation-for-tech-companies.html.

[77] Welch, Chris, *The Verge*, "The FBI successfully broke into a gunman's iPhone, but it's still very angry at Apple," May 18, 2020, https://www.theverge.com/2020/5/18/21262347/attorney-general-barr-fbi-director-wray-apple-encryption-pensacola.

court-authorized access by law enforcement, while maintaining very high standards of data security. Striking this balance should not be left to corporate board rooms.[78]

FBI Director Christopher A. Wray also pushed hard against Apple, saying,

> Public servants, already swamped with important things to do to protect the American people — and toiling through a pandemic, with all the risk and hardship that entails — had to spend all that time just to access evidence we got court-authorized search warrants for months ago.[79]

Still, they got the data they wanted. And I continue to believe that they are making it sound like a greater challenge than it actually was. One of the iPhones in this case had been shot and the FBI still got Into It![80]

That same day, Apple responded:

> The terrorist attack on members of the US armed services at the Naval Air Station in Pensacola, Florida was a devastating and heinous act. Apple responded to the FBI's first requests for information just hours after the attack on December 6, 2019 and continued to support law enforcement during their investigation. We provided every piece of information available to us, including iCloud backups, account information and transactional data for multiple accounts, and we lent continuous and ongoing technical and investigative support to FBI offices in Jacksonville, Pensacola, and New York over the months since.
>
> On this and many thousands of other cases, we continue to work around-the-clock with the FBI and other investigators who keep Americans safe and bring criminals to justice. As a proud American company, we consider supporting law enforcement's important work our responsibility. The false claims made about our company are an excuse to weaken encryption and other security measures that protect millions of users and our national security.
>
> It is because we take our responsibility to national security so seriously that we do not believe in the creation of a backdoor — one which will make every device vulnerable to bad actors who threaten our national security and the data security of our customers. There is no such thing as a backdoor just for the good guys, and the

---

[78] Welch, Chris, *The Verge*, "The FBI successfully broke into a gunman's iPhone, but it's still very angry at Apple," May 18, 2020, https://www.theverge.com/2020/5/18/21262347/attorney-general-barr-fbi-director-wray-apple-encryption-pensacola and *KTVN*, "AG Barr: Apple's decision has dangerous consequences for public safety," https://www.ktvn.com/clip/15067612/ag-barr-apples-decision-has-dangerous-consequences-for-public-safety for a video clip of this quote.

[79] Welch, Chris, *The Verge*, "The FBI successfully broke into a gunman's iPhone, but it's still very angry at Apple," May 18, 2020, https://www.theverge.com/2020/5/18/21262347/attorney-general-barr-fbi-director-wray-apple-encryption-pensacola.

[80] Pfefferkorn, Riana, *TechCrunch* "The FBI is mad because it keeps getting into locked iPhones without Apple's help," May 22, 2020, https://techcrunch.com/2020/05/22/the-fbi-is-mad-because-it-keeps-getting-into-locked-iphones-without-apples-help/.

American people do not have to choose between weakening encryption and effective investigations.

Customers count on Apple to keep their information secure and one of the ways in which we do so is by using strong encryption across our devices and servers. We sell the same iPhone everywhere, we don't store customers' passcodes and we don't have the capacity to unlock passcode-protected devices. In data centers, we deploy strong hardware and software security protections to keep information safe and to ensure there are no backdoors into our systems. All of these practices apply equally to our operations in every country in the world.[81]

I'm convinced that Apple will come out on top again, but there is yet another update to this tale of history repeating itself that needs to be addressed before closing out this chapter.

18.9 Yet Another Attempt at Anti-Crypto Legislation

On March 5, 2020, Senator Lindsey Graham introduced yet another bill attempting to do away with strong encryption, although with a new, more subtle, approach that does not involve the government actually banning it. This time, instead of appealing to Americans' fear of terrorists (which hasn't worked), the boogeymen are pedophiles. If you won't give up your privacy and security to help catch terrorists, will you give it up to help stop pedophiles? Hint: they won't magically disappear, if you say yes.

The new bill is titled the "Eliminating Abusive and Rampant Neglect of Interactive Technologies Act of 2020" or "EARN IT" for short.[82] Senator Graham's cosponsor, Democrat Senator Richard Blumenthal is quick to point out that the bill doesn't include the word "encryption,"[83] but Riana Pfefferkorn, writing for The Brookings Institution, has characterized it as "a sneak ban on encryption."[84] In a nutshell, the Act would hold tech companies liable for illegal content transferred by their users. Of course, there's no way for tech companies to know much about the content of their users communications, if the data is securely encrypted. Hence, to avoid the risk of being sued into bankruptcy, the companies would have to have a way to access all communications and actually scan all of it looking for objectionable material. Hackers from around the world could also search the communications, looking for other material, exponentially increasing cybercrime.

The title of this book is *Secret History*, but this section is more current events than history. It would be nice if the battle between the tech giants and portions of the government would end in

---

[81] Welch, Chris, *The Verge*, "The FBI successfully broke into a gunman's iPhone, but it's still very angry at Apple," May 18, 2020, https://www.theverge.com/2020/5/18/21262347/attorney-general-barr-fbi-director-wray-apple-encryption-pensacola.

[82] S.3398 — EARN IT Act of 2020, 116th Congress (2019–2020), *congress.gov*, https://www.congress.gov/bill/116th-congress/senate-bill/3398/text.

[83] Mullin, Joe, Electronic Frontier Foundation (EFF), "The EARN IT Bill Is the Government's Plan to Scan Every Message Online," March 12, 2020, https://www.eff.org/deeplinks/2020/03/earn-it-bill-governments-not-so-secret-plan-scan-every-message-online.

[84] Pfefferkorn, Riana, *Brookings*, "The EARN IT Act is a disaster amid the COVID-19 crisis," May 4, 2020, https://www.brookings.edu/techstream/the-earn-it-act-is-a-disaster-amid-the-covid-19-crisis/.

favor of the former, before the next edition, and that this section would truly become history, but I'm not optimistic. It might take several more editions.

References and Further Reading

Anon, "Hardware Hack Defeats iPhone Passcode Security," *BBC News*, https://www.bbc.com/news/technology-37407047, September 19, 2016.

Baker, Jim, "Rethinking Encryption," *Lawfare*, https://www.lawfareblog.com/rethinking-encryption, October 22, 2019.

Encryption Working Group, Carnegie Endowment for International Peace, "Moving the Encryption Policy Conversation Forward," https://carnegieendowment.org/2019/09/10/moving-encryption-policy-conversation-forward-pub-79573, September 10, 2019.

Farivar, Cyrus, "Bill Aims to Thwart Strong Crypto, Demands Smartphone Makers be Able to decrypt," *Ars Technica*, https://arstechnica.com/tech-policy/2016/01/bill-aims-to-thwart-strong-crypto-demands-smartphone-makers-be-able-to-decrypt/, January 14, 2016.

Farivar, Cyrus, "Yet Another Bill Seeks to Weaken Encryption-by-Default on Smartphones," *Ars Technica*, https://arstechnica.com/tech-policy/2016/01/yet-another-bill-seeks-to-weaken-encryption-by-default-on-smartphones/, January 21, 2016.

Feiner, Lauren, "Senators Threaten to Regulate Encryption if Tech Companies won't do it Themselves," *CNBC*, https://www.cnbc.com/2019/12/10/senators-threaten-encryption-regulation-for-tech-companies.html, December 10, 2019.

Garfinkel, Simson, *PGP: Pretty Good Privacy*, O'Reilly & Associates, Sebastopol, California, 1995. Zimmermann notes, "Good technical info on PGP of that era, with mostly correct history of PGP."[85]

Goodin, Dan, "Critical PGP and S/MIME Bugs can Reveal Encrypted Emails—Uninstall Now [Updated]," *Ars Technica*, May 14, 2018.

Howell, Jen Patja, "The Lawfare Podcast: Jim Baker and Susan Landau on 'Moving the Encryption Policy Conversation Forward'," *Lawfare*, https://www.lawfareblog.com/lawfare-podcast-jim-baker-and-susan-landau-moving-encryption-policy-conversation-forward, October 8, 2019. You can listen to the podcast at this website.

International PGP Home Page, The, https://web.archive.org/web/20120303005616/http://www.pgpi.org/. This is an archived page; it is no longer active, but has historical value. Freeware versions of PGP were available here, along with source code and manuals. A later archived version, https://web.archive.org/web/20170328170323/http://pgpi.org:80/, from March 28, 2017, is stripped of useful content and merely says, "The owner of pgpi.org is offering it for sale for an asking price of 8000 USD!"

Levy, Steven, *Crypto: How the Code Rebels Beat the Government, Saving Privacy in the Digital Age*, Viking, New York, 2001.

McAfee, John, "John McAfee and the FBI Finally Face Off On CNN (CNN Interview)," *John McAfee* YouTube Channel, https://www.youtube.com/watch?v=HqI0jbKGaT8, March 1, 2016.

Mullin, Joe, *Electronic Frontier Foundation (EFF)*, "The EARN IT Bill Is the Government's Plan to Scan Every Message Online," https://www.eff.org/deeplinks/2020/03/earn-it-bill-governments-not-so-secret-plan-scan-every-message-online, March 12, 2020.

Newton, Casey, *The Verge*, "A sneaky attempt to end encryption is worming its way through Congress," https://www.theverge.com/interface/2020/3/12/21174815/earn-it-act-encryption-killer-lindsay-graham-match-group, March 12, 2020.

OpenPGP, https://www.openpgp.org/.

Pfefferkorn, Riana, "The EARN IT Act: How to Ban End-to-End Encryption Without Actually Banning It," *The Center for Internet and Society*, Stanford Law School, https://cyberlaw.stanford.edu/blog/2020/01/earn-it-act-how-ban-end-end-encryption-without-actually-banning-it, January 30, 2020.

---

[85] Zimmermann, Philip, Crypto Bibliography, http://www.philzimmermann.com/EN/bibliography/index.html.

Pfefferkorn, Riana, *TechCrunch* "The FBI is mad because it keeps getting into locked iPhones without Apple's help," https://techcrunch.com/2020/05/22/the-fbi-is-mad-because-it-keeps-getting-into-locked-iphones-without-apples-help/, May 22, 2020.

Poddebniak, Damian, Christian Dresen, Jens Müller, Fabian Ising, Sebastian Schinzel, Simon Friedberger, Juraj Somorovsky, and Jörg Schwenk, "Efail: Breaking S/MIME and OpenPGP Email Encryption using Exfiltration Channels," presentation at *27th USENIX Security Symposium*, Baltimore, Maryland, August 2018. See https://efail.de/ for a link to this presentation and much more.

Savage, Charlie, "U.S. Tries to Make It Easier to Wiretap the Internet," *New York Times*, September 27, 2010, p. A1, available online at http://www.nytimes.com/2010/09/27/us/27wiretap.html and http://archive.nytimes.com/www.nytimes.com/2010/09/27/us/27wiretap.html.

Scahill, Jeremy and Josh Begley, *The Intercept*, "The CIA Campaign to Steal Apple's Secrets," https://theintercept.com/2015/03/10/ispy-cia-campaign-steal-apples-secrets/, March 10, 2015.

Schneier, Bruce, *E-Mail Security: How to Keep Your Electronic Messages Private*, John Wiley & Sons, New York, 1995. Schneier covered both PGP and PEM (Privacy Enhanced Mail), which used DES or triple DES, with two keys, along with RSA.

Shwayder, Maya, *Digital Trends*, "The FBI broke Apple's iPhone encryption. Here's why you shouldn't panic," https://www.digitaltrends.com/news/fbi-iphone-hack-encryption-pensacola-shooter-analysis/?itm_source=35&itm_content=1x7&itm_term=2498265, May 18, 2020.

Smith, Ms., "NAND mirroring proof-of-concept Show that FBI could Use it to Crack iPhone," *CSO Online*, http://www.networkworld.com/article/3048488/security/nandmirroring-proof-of-concept-show-that-fbi-could-use-it-to-crackiphone.html, March 28, 2016.

Stallings, William, *Protect Your Privacy: the PGP user's guide*, Prentice Hall PTR, Englewood Cliffs, New Jersey, 1995.

U.S. Senate Committee on the Judiciary, "Encryption and Lawful Access: Evaluating Benefits and Risks to Public Safety and Privacy," https://www.judiciary.senate.gov/meetings/encryption-and-lawful-access-evaluating-benefits-and-risks-to-public-safety-and-privacy, December 10, 2019. This website has a video of the hearing and transcripts.

Wikipedia contributors, "Crypto Wars," *Wikipedia, The Free Encyclopedia*, https://en.wikipedia.org/wiki/Crypto_Wars

Wikipedia contributors, "EFAIL," *Wikipedia, The Free Encyclopedia*, https://en.wikipedia.org/wiki/EFAIL.

Wikipedia contributors, "FBI–Apple encryption dispute," *Wikipedia, The Free Encyclopedia*, https://en.wikipedia.org/wiki/FBI–Apple_encryption_dispute.

Wikipedia contributors, "List of the most common passwords," *Wikipedia, The Free Encyclopedia*, https://en.wikipedia.org/wiki/List_of_the_most_common_passwords.

Zimmermann, Philip R., *The Official PGP User's Guide*, MIT Press, Cambridge, Massachusetts, 1995. Even this book is available for free (in an ASCII version online). Zimmermann notes that it's "out of date with current PGP software, but still politically interesting."[86]

Zimmermann, Philip R., *PGP Source Code and Internals*, MIT Press, Cambridge, Massachusetts, 1995. The export laws were strange. Although PGP couldn't legally be exported as software, the source code could. And, of course, source code can be scanned and converted for use by a text editor; hence, this book. On another note, the experts know that trying to keep an algorithm secret is a bad sign! Revealing the details, if the software is any good, will ease concerns, not increase them.

Zimmermann, Philip, R., Home Page, http://www.philzimmermann.com/EN/background/index.html.

---

[86] Zimmermann, Philip, Crypto Bibliography, http://www.philzimmermann.com/EN/bibliography/index.html.

Chapter 19

Stream Ciphers

Anyone who considers arithmetical methods of producing random digits is, of course, in a state of sin.

—John von Neumann (1951)[1]

The tape machine depicted in Section 2.9 can be seen as the beginning of a still ongoing area of cryptographic research — stream ciphers. Such systems attempt to generate random numbers that can be combined with the message in an approximation of the unbreakable one-time pad. The problem is, machines cannot generate random numbers. Thus, we usually refer to such numerical sequences as *pseudorandom* and the devices that create them as *pseudorandom number generators* or PRNGs for short.

It should be noted that a much earlier origin for stream ciphers is offered by the autokey ciphers of the 16th century, as discussed in Section 2.5. In any case, stream ciphers are especially important when we want to encipher and decipher data in real time. Applications such as secure cell phone conversations and encrypted streaming video provide examples. In these cases, the pseudorandom sequences consist of 0s and 1s that are usually generated bit by bit or byte by byte. We begin with an early attempt using modular arithmetic.

19.1 Congruential Generators

One way to generate a pseudorandom sequence is with a *linear congruential generator* (LCG):[2]

$$X_n = (aX_{n-1} + b) \ (\text{mod } m)$$

---

[1] John von Neumann was on the National Security Agency Scientific Advisory Board (NSASAB). He's quoted here from Salomon, David, *Data Privacy and Security*, Springer, New York, 2003, p. 97.

[2] Lehmer, Derrick Henry, "Mathematical Methods in Large-Scale Computing Units," in Aiken, Howard H., *Proceedings of a Second Symposium on Large-Scale Digital Calculating Machinery*, Annals of the Computation Laboratory of Harvard University, Vol. 26, Harvard University Press, Cambridge, Massachusetts, 1951 pp. 141–146. The conference was held on September 1, 1949. This may have been the first attempt to generate pseudorandom numbers with a linear congruential generator.

533

For example, if we take $a = 3$, $b = 5$, $m = 26$, and seed the generator with $X_0 = 2$, we get:

$$X_0 = 2$$

$$X_1 = 3(2) + 5 = 11$$

$$X_2 = 3(11) + 5 = 12 \pmod{26}$$

$$X_3 = 3(12) + 5 = 15 \pmod{26}$$

$$X_4 = 3(15) + 5 = 24 \pmod{26}$$

$$X_5 = 3(24) + 5 = 25 \pmod{26}$$

$$X_6 = 3(25) + 5 = 2 \pmod{26}$$

At this point, we're back at our starting value. The output will now continue, as before, 11, 12, 15, 24, 25, ... Clearly, this is not random! We get stuck in a cycle of period 6. Still, if this could be modified to generate a cycle with a much longer period, longer than any message we might want to encipher, it seems like it could be a reasonable way to generate a key that could then be paired with the message, one letter at a time, modulo 26.

Of course, the modern approach uses bits instead. This isn't a problem, as we could convert these values to bits and XOR them with our message (also expressed in bits). However, the effect is the same as that of a binary Vigenère cipher, because the values to be XORed repeat.

If we select the values of a, b, and m more carefully, we can cycle through all values from 0 to $m - 1$, but we must then repeat the cycle, as in the example above. If m is large enough, this might seem safe; for example, m may be larger than the length of the message. Nevertheless, this technique is not secure. Jim Reeds was the first to publicly break such ciphers in 1977.[3]

An obvious next step for the cryptographer is to try higher power congruential generators, such as quadratics. Notice that each term still only depends on the one that came before:

$$X_n = (aX_{n-1}^2 + bX_{n-1} + c) \pmod{m}$$

These were broken by Joan B. Plumstead, along with cubic generators.[4] In fact, as others showed, no matter what degree we try, such systems can be broken![5]

Daniel Guinier stayed linear, but suggested up to 1,024 LCGs be used, and combined additively, to get an "astronomically" long period![6] He was not the first to consider using more than one, but he did take it to an extreme. Alas, combinations of LCGs, and other variations such as multiplying previous terms, have failed to stand the test of time. At this point we move on to another method for generating pseudorandom sequences.

---

[3] Reeds, James, "Cracking a Random Number Generator," *Cryptologia*, Vol. 1. No. 1, January 1977, pp. 20–26. A later paper on this topic is Plumstead, Joan B., "Inferring a Sequence Generated by a Linear Congruence," in *Proceedings of the 23rd Annual Symposium on Foundations of Computer Science*, IEEE Computer Society Press, Los Alamitos, California, 1982, pp. 153–159.

[4] Boyar, Joan, "Inferring Sequences Produced by Pseudo-random Number Generators," *Journal of the ACM (JACM)*, Vol. 36, No. 1, January 1989, pp. 129-141. Joan B. Plumstead's later papers were published under the name Joan Boyar.

[5] Lagarias, Jeffrey C. and James Reeds, "Unique Extrapolation of Polynomial Recurrences," *SIAM Journal on Computing*, Vol. 17, No. 2, April 1988, pp. 342–362.

[6] Guinier, Daniel, "A Fast Uniform "Astronomical" Random Number Generator," *SIGSAC Review* (ACM Special Interest Group on Security Audit & Control), Vol. 7, No. 1, Spring 1989, pp. 1–13.

19.2 Linear Feedback Shift Registers

When we moved to "degree 2," for congruential generators, we might have written the equation as

$$X_n = (aX_{n-1} + bX_{n-2} + c) \pmod{m}$$

This way each value depends on the two previous values (hence, degree 2) and we can attain longer periods. Nothing is squared. We would, of course, need two seed values X_0 and X_1. The first number we generate would be X_2. This is the basic idea behind *linear feedback shift registers* (LFSRs). They are very fast (in hardware) when working with bits modulo 2. We could indicate mod 2 by setting $m = 2$, but as we've seen before, the convention is to replace $+$ with $\oplus$ to represent XOR, which is the same as addition modulo 2. LFSRs are usually represented diagrammatically rather than algebraically (Figure 19.1).

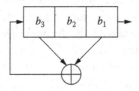

Figure 19.1   A Small LFSR.

The figure is best explained with an example. We may seed the register (the values of the bs) with the bits 101; that is $b_3 = 1$, $b_2 = 0$, and $b_1 = 1$. The diagonal arrows indicate that we get our new bit by taking the XOR of b_3 and b_1, which is $1 \oplus 1 = 0$. Notice that b_2 is not used in this calculation. The bits that are used, b_3 and b_1, are referred to as the taps. The new bit that is calculated, based on the taps, follows the longest arrow and takes the place of b_3, but b_3 doesn't just vanish. Instead, it advances to the right to take the place of b_2, which in turn advances to the right to replace b_1. With nowhere left to go, b_1 "falls of the edge" (indicated by the shortest arrow) and is gone. These steps are then all repeated with the new values. Starting with the seed, our register holds the following values, as we iterate:

$$101$$
$$010$$
$$001$$
$$100$$
$$110$$
$$111$$
$$011$$
$$101$$

which brings us back to the start. Notice that this register cycles through seven different sets of values. We say that it has period 7. The rule depicted diagrammatically may also be represented algebraically as

$$b_{n+3} = b_{n+2} \oplus b_n, \text{ for } n = 1, 2,\ldots$$

To use a LFSR as a stream cipher, we simply take the bits that shift through the register. The LFSR above gives the stream 1010011 (which then repeats). This is the third column in the list of register values given above. XORing our plaintext with a string of only 7 bits, used repeatedly, is not a very secure method of encryption! We'll improve upon this, but what is to come will be clearer if we start out slowly.

Observe that there's a "bad seed" that's useless for cryptographic purposes. If we start off with $b_3 = 0$, $b_2 = 0$, and $b_1 = 0$, we'll never get any nonzero values. XORing the stream of bits that is generated by this seed with the plaintext message will leave it unchanged.

In general, the longest period possible for a LFSR with n elements is $2^n - 1$, so our original example above was maximized in this respect. A different (nonzero) seed will simply cycle through the same values beginning at a different point. We cannot get a longer period by XORing different bits, although this can cause the states to be cycled through in a different order. So, if we want a LFSR that generates a longer key, we must look at those with more elements, but more elements don't guarantee that longer keys will result.

We can investigate whether or not a LFSR produces a maximal period by examining a polynomial associated with the register. For example, for the LFSR pictured in Figure 19.1, the polynomial is $p(x) = x^3 + x^1 + 1$. The powers of x are taken to be the positions of the bits that are made use of in the XOR, and $+ 1$ is always tacked on to the end. This is called a *tap polynomial* or *connection polynomial*.

We first see whether or not the tap polynomial can be factored (mod 2). If it cannot, we say it is irreducible. In this case, if the polynomial is of degree n, the period must divide $2^n - 1$. We can check the polynomial above for reducibility modulo 2, by plugging in 0 and 1 to get

$$p(0) = 0^3 + 0^1 + 1 = 1 \ (\mathrm{mod}\ 2)$$

$$p(1) = 1^3 + 1^1 + 1 = 3 = 1 \ (\mathrm{mod}\ 2)$$

Because neither 0 nor 1 is a root, neither x nor $x + 1$ can be a factor.[7] A third-degree polynomial must have a linear factor, if it is reducible, so the polynomial above must be irreducible. Higher degree polynomials require other methods for checking for reducibility; for example, the fourth degree polynomial $f(x) = x^4 + x^2 + 1$ has no roots modulo 2, but it is not irreducible. We have $x^4 + x^2 + 1 = (x^2 + x + 1)(x^2 + x + 1) \ (\mathrm{mod}\ 2)$.

So, our third-degree example, being irreducible, must have a period that divides $2^3 - 1 = 7$; hence, the period is either 1 or 7. We attain period 1 with the seed consisting of all zeros, and period 7 with any other seed. We'll see another important cryptographic use of irreducible polynomials in Section 20.3.

But what if $2^n - 1$ is not prime? For example, for a LFSR with a register containing 4 bits, and a tap polynomial that is irreducible, all we can conclude is that the period must divide $2^4 - 1$. This doesn't tell us the period is 15, because 3 and 5 are also factors of $2^4 - 1$. Thus, the result above is less useful when $2^n - 1$ is composite. Fortunately, we have another test. In the definition that follows, we let l denote the length of the register.

An irreducible polynomial $p(x)$ is said to be *primitive* if:

1. $p(x)$ is a factor of $x^{2^l - 1} - 1$, and
2. $p(x)$ is not a factor of $x^k - 1$ for any positive divisor k of $2^l - 1$.

We then have the following result: *A LFSR with a tap polynomial that is primitive will have a maximal period.* As an example of an LFSR with a long period, we have $b_{n + 31} = b_{n + 3} \oplus b_n$ for

---

[7] This is not a typo. Modulo 2, $x - 1$ and $x + 1$ are the same, because $-1 = 1 \ (\mathrm{mod}\ 2)$.

$n = 1, 2, \ldots$ This LFSR requires a 31-bit seed, which will then generate a sequence of bits with a period of $2^{31} - 1 = 2,147,483,647$.

19.3 LFSR Attack

With such long periods so easily obtained, a LFSR might seem like a secure system. We only broke the Vigenère cipher by taking advantage of patterns established by the repeating key, and it would take extremely long messages to have that possibility here. However, there are other mathematical options open to us for attacking this system. We will assume that for a portion of the ciphertext the corresponding plaintext is known (i.e., we have a crib). From this, we easily obtain a portion of the key. Suppose this cribbed key is 10101100. We can see that the period is greater than or equal to 8, because there is no repetition in the portion we recovered. Therefore, the LFSR must have at least 4 elements. Assuming it has exactly 4 elements, the LFSR must be of the form

$$b_{n+4} = a_3 b_{n+3} \oplus a_2 b_{n+2} \oplus a_1 b_{n+1} \oplus a_0 b_n$$

where each of the a_i is either 0 or 1. The string of known key bits, 10101100, labeled $b_1 b_2 b_3 b_4 b_5 b_6 b_7 b_8$ for convenience, although they needn't be from the start of the message, tells us

$$1 = a_3 0 \oplus a_2 1 \oplus a_1 0 \oplus a_0 1$$

$$1 = a_3 1 \oplus a_2 0 \oplus a_1 1 \oplus a_0 0$$

$$0 = a_3 1 \oplus a_2 1 \oplus a_1 0 \oplus a_0 1$$

$$0 = a_3 0 \oplus a_2 1 \oplus a_1 1 \oplus a_0 0$$

From this system of equations, we may solve for the a_i. This may be done without the techniques of linear algebra, but for larger examples we'd really want to use matrices, so we'll use one here. We have

$$\begin{pmatrix} 1 & 0 & 1 & 0 \\ 0 & 1 & 0 & 1 \\ 1 & 0 & 1 & 1 \\ 0 & 1 & 1 & 0 \end{pmatrix} \begin{pmatrix} a_0 \\ a_1 \\ a_2 \\ a_3 \end{pmatrix} = \begin{pmatrix} 1 \\ 1 \\ 0 \\ 0 \end{pmatrix}$$

The four by four matrix has the inverse

$$\begin{pmatrix} 0 & 1 & 1 & 1 \\ 1 & 1 & 1 & 0 \\ 1 & 1 & 1 & 1 \\ 1 & 0 & 1 & 0 \end{pmatrix}.$$

So our solution is

$$\begin{pmatrix} a_0 \\ a_1 \\ a_2 \\ a_3 \end{pmatrix} = \begin{pmatrix} 0 & 1 & 1 & 1 \\ 1 & 1 & 1 & 0 \\ 1 & 1 & 1 & 1 \\ 1 & 0 & 1 & 0 \end{pmatrix} \begin{pmatrix} 1 \\ 1 \\ 0 \\ 0 \end{pmatrix} = \begin{pmatrix} 1 \\ 0 \\ 0 \\ 1 \end{pmatrix}$$

Thus, the equation for the LFSR appears to be $b_{n+4} = b_{n+3} \oplus b_n$. This equation may then be used to generate all future key bits, as well as previous key bits, if the crib occurred in a position other than the start of the message

If the equation fails to yield meaningful text beyond the crib, we'd have to consider a five-element LFSR, and if that doesn't work out, then a six-element LFSR, etc. However, we'd need more key bits to uniquely determine the coefficients for anything beyond 4 elements. In general, we need $2n$ bits of key for an n-element LFSR. As n grows, $2n$ quickly becomes tiny, as a percentage, compared to the maximal period of $2^n - 1$ for the n-element LFSR. So, although we needed a little over half of the repeating key to uniquely solve the 4 element LFSR, the period $2^{31} - 1 = 2,147,483,647$ LFSR defined by $b_{n+31} = b_n \oplus b_{n+3}$ could be recovered from only 62 bits of key, a tiny percentage of the whole (and less than 8 characters, as each keyboard character translates to 8 bits). This is not an unreasonable crib! As mentioned before, modern ciphers are expected to hold strong against known-plaintext attacks, so the LFSRs described above are not useful for cryptographic purposes. However, they are incorporated as components of stronger systems.

19.4 Cell Phone Stream Cipher A5/1

There are several ways to strengthen LFSRs. One is to remove the linearity constraint and use other methods of combining elements in the register, such as multiplication. Another approach is to combine LFSRs, as in the following cell phone cipher designed in 1987, which is shown in Figure 19.2.

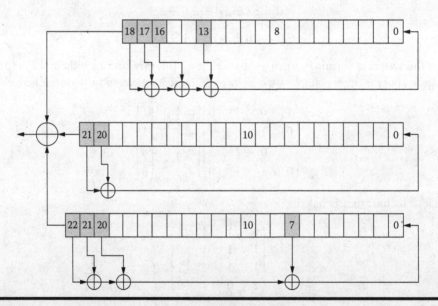

Figure 19.2   A5/1 Stream Cipher. (http://en.wikipedia.org/wiki/Stream_cipher.)

Figure 19.2 indicates that the A5/1 stream cipher consists of three linear feedback shift registers. The first XORs the bits in positions 13, 16, 17, and 18 to get a new bit, which is then placed at the end, forcing all of the bits to shift one position to the left. The last bit, formerly in position 18, shifts off the register and is XORed with bits from the other two LFSRs to finally provide the bit that is XORed with the message to yield a bit of ciphertext.

Because all three LFSRs must be seeded, the key is $19 + 22 + 23 = 64$ bits long. Notice that we count the bit in position 0 for each LFSR, along with the rest. Each of the three LFSRs has a length that is relatively prime to the lengths of the others. This would generate a period that's the product of all three. However, there's another feature that lengthens the period. Notice that the diagram for A5/1 has bits labeled in positions 8, 10, and 10. These are called *clocking bits*. In each cycle, the bits in the clocking positions are examined. Because there is an odd number of clocking bits, there must be either more 1s than 0s or more 0s than 1s in these positions. The registers that have the more popular bit in their clocking positions advance. If all three bits match, all of the registers advance.

In defiance of Kerckhoffs's rules, the algorithm provided above was kept secret, while it was being placed in over 100 million cell phones. In compliance with Kerckhoffs's rules the public learned it anyway! It was part of the Global System for Mobile Communications (GSM) cellphone standard. Various attacks have made it clear that the system is insecure. Details may be found in the papers in the References and Further Reading list at the end of this chapter.

A5/2 made use of four LFSRs that advance in an irregular manner, like those of A5/1. Although this might make A5/2 sound stronger than A5/1 (4 is bigger than 3, right?), it isn't. It was purposely made weaker, intended for use in certain countries, while Americans and Europeans used the stronger A5/1. A5/2 was made public in August 1999, and before the month ended, Ian Goldberg, David A. Wagner, and Lucky Green broke it.[8] For details, see the References and Further Reading section at the end of this chapter.

19.5 RC4

RC4 (Rivest Cipher 4), designed by Ron Rivest in 1987, was a very popular stream cipher. Again, in denial of Kerckhoffs's rules, the details of this cipher were kept secret and could only be obtained by signing a nondisclosure agreement with RSA Data Security Inc. In September 1994, however, the source code was anonymously posted to the Cypherpunks mailing list.[9]

The cipher starts off with a list of all 8-bit numbers, in order. These bytes are

$$S_0 = 00000000$$

$$S_1 = 00000001$$

$$S_2 = 00000010$$

$$S_3 = 00000011$$

$$S_4 = 00000100$$

$$S_5 = 00000101$$

$$\vdots$$

$$S_{255} = 11111111.$$

Each S_i is just the binary expression for the base-10 number i.

---

[8] Goldberg, Ian, David Wagner, and Lucky Green, "The (Real-Time) Cryptanalysis of A5/2," paper presented at the Rump Session of the CRYPTO '99 conference, Santa Barbara, California, August 15–19, 1999.

[9] Schneier, Bruce, *Applied Cryptography*, second edition, John Wiley & Sons, New York, 1996, p. 397.

These bytes are then shuffled so that their new order appears random. To do this, another set of 256 bytes is initialized using the key. The key may be any length up to 256 bytes. At the low end, there are attacks that can break RC4 if the key is just 40 bits.

Whatever length key is selected, we simply split it into bytes and label them K_0, K_1, K_2, K_3,... K_{255}. If we reach the end of our key before we fill 256 bytes, we continue filling bytes using our key over again, from the start. For example, if our key was only 64 bytes long, we'd have to lay it end to end four times in order to have enough bytes to fill K_0 through K_{255}. The shuffling of the S_i is then carried out by the following loop:

$$j = 0$$

$$\text{for } i = 0 \text{ to } 255$$

$$j = (j + S_i + K_i) \,(\text{mod } 256)$$

$$\text{Swap } S_i \text{ and } S_j$$

$$\text{next } i$$

After resetting the index variables i and j to 0, we're ready to generate our key stream for enciphering. The key, K, actually used for encryption is then generated byte by byte using the following equations, applied repeatedly:

$$i = (i + 1) \,(\text{mod } 256)$$

$$j = (j + S_i) \,(\text{mod } 256)$$

$$\text{Swap } S_i \text{ and } S_j$$

$$t = (S_i + S_j) \,(\text{mod } 256)$$

$$K_i = S_t$$

$$C_i = M_i \oplus K_i$$

We apply the steps above to each byte, M_i, of the message, until they are all enciphered.

RC4 is a simple cipher and easy to program. It also marks a departure from the other methods discussed in this section. The exact period of RC4 is not known, but analysis thus far indicates that it is very likely in excess of 10^{100}.[10] This lower bound is a familiar number. Mathematicians were referring to 10^{100} as a googol, long before an Internet search engine appropriated the name in a misspelled form.

RC4 was used in the Secure Socket Layer (SSL) and Wired Equivalent Privacy (WEP), both of which were found to be insecure. Because of its weaknesses, WEP is sometimes said to stand for White Elephant Protection. WEP implemented RC4 in a manner similar to how Enigma was used, as described in Chapter 7. A 24-bit initialization vector (IV) placed at the front of WEP ciphertexts helped to generate a session key.

When it came time to fill the key bytes, the IV bits were used first. They were then followed by a key that can be used many times, because the randomly generated IV results in the scrambled

---

[10] RSA Laboratories, "3.6.3 What is RC4?" https://web.archive.org/web/20130627230107/http://www.rsa.com/rsalabs/node.asp?id=2250.

key differing each time. However, given a sufficient depth of messages (just like the Poles needed to recover Enigma keys), these initialization vectors allow WEP to be broken.[11]

Other software packages that made use of RC4 included Microsoft® Windows® and Lotus Notes®. RC4 was actually the most popular stream cipher in software. Today, it is no longer considered secure. Several papers describing attacks on it are listed in the References and Further Reading section at the end of this chapter.

You may come upon references to RC5 and RC6. Although these sound like newer versions of the system described above, they are not. Both RC5 and RC6 are block ciphers. The numbering simply indicates the order in which Rivest developed these unrelated ciphers. It's analogous to the numbering of Beethoven's symphonies.

Although RC4 is broken, no other stream cipher has yet been recognized as the new champion. One contender that is gaining in popularity and may capture the title is ChaCha20, designed by Daniel J. Bernstein in 2008.[12] It has been adopted by Google as a replacement for RC4 in Transport Layer Security (TLS), the successor to Secure Sockets Layer (SSL).

Caution! Even if a stream cipher is mathematically secure, it can be broken when misused. For example, if the same initial state (seed) is used twice, the security is no better than a binary running key cipher. Keys should never be reused in stream ciphers!

References and Further Reading

On Congruential Generators

Bellare, Mihir, Shafi Goldwasser, and Daniele Micciancio, ""Pseudo-Random" Number Generation Within Cryptographic Algorithms: The DDS Case," in Kaliski, Jr., Burton S., editor, *Advances in Cryptology – Crypto '97 Proceedings*, Lecture Notes in Computer Science, Vol. 1294, Springer, New York, 1997, pp. 277–291.

Boyar, Joan, "Inferring Sequences Produced by Pseudo-random Number Generators," *Journal of the ACM* (JACM), Vol. 36, No. 1, January 1989, pp. 129–141.

Guinier, Daniel, "A Fast Uniform "Astronomical" Random Number Generator," *SIGSAC Review* (ACM Special Interest Group on Security Audit & Control), Vol. 7, No. 1, Spring 1989, pp. 1–13.

Knuth, Donald, "Deciphering a Linear Congruential Encryption," *IEEE Transactions on Information Theory*, Vol. 31, No. 1, January 1985, pp. 49–52. The Jedi Knight of computer algorithms pays cryptanalysis a visit!

Krawczyk, Hugo, "How to Predict Congruential Generators," in Brassard, Gilles, editor, *Advances in Cryptology — Crypto '89 Proceedings*, Lecture Notes in Computer Science, Vol. 435, Springer, Berlin, Germany, 1990, pp. 138–153.

Krawczyk, Hugo, "How to Predict Congruential Generators," *Journal of Algorithms*, Vol. 13, No. 4, December 1992, pp. 527–545.

Lagarias, Jeffrey C. and James Reeds, "Unique Extrapolation of Polynomial Recurrences," *SIAM Journal on Computing*, Vol. 17, No. 2, April 1988, pp. 342–362.

Marsaglia, George and Thomas Bray, "One-Line Random Number Generators and Their Use in Combination," *Communications of the ACM*, Vol. 11, No. 11, November 1968, pp. 757–759.

Park, Stephen and Keith Miller, "Random Number Generators: Good Ones Are Hard to Find," *Communications of the ACM*, Vol. 31, No. 10, October 1988, pp. 1192–1201.

---

[11] For more details of this attack presented in a clear manner see Stamp, Mark and Richard M. Low, *Applied Cryptanalysis: Breaking Ciphers in the Real World*, Wiley-Interscience, Hoboken, New Jersey, 2007, pp. 105–110.

[12] Bernstein, Daniel J., "ChaCha, a variant of Salsa20," January 28, 2008, http://cr.yp.to/chacha/chacha-20080120.pdf.

Plumstead, Joan B., "Inferring a Sequence Generated by a Linear Congruence," in *Proceedings of the 23rd Annual Symposium on Foundations of Computer Science*, IEEE Computer Society Press, Los Alamitos, California, 1982, pp. 153–159.

Reeds, James, ""Cracking" a Random Number Generator," *Cryptologia*, Vol. 1. No. 1, January 1977, pp. 20–26. Reeds shows how a crib can be used to break a linear congruential random number generator.

Reeds, James, "Solution of a Challenge Cipher," *Cryptologia*, Vol. 3, No. 2, April 1979, pp. 83–95.

Reeds, James, "Cracking a Multiplicative Congruential Encryption Algorithm," in Wang, Peter C. C., Arthur L. Schoenstadt, Bert I. Russak, and Craig Comstock, editors, *Information Linkage Between Applied Mathematics and Industry*, Academic Press, New York, 1979, pp. 467–472.

Vahle, Michael O. and Lawrence F. Tolendino, "Breaking a Pseudo Random Number Based Cryptographic Algorithm," *Cryptologia*, Vol. 6, No. 4, October 1982, pp. 319–328.

Wichmann, Brian and David Hill, "Building a Random-Number Generator," *Byte*, Vol. 12, No. 3, March 1987, pp. 127–128.

On LFSRs

Barker, Wayne G., *Cryptanalysis of Shift-Register Generated Stream Cipher Systems*, Aegean Park Press, Laguna Hills, California, 1984.

Golomb, Solomon, *Shift Register Sequences*, second edition, Aegean Park Press, Laguna Hills, California, 1982. This edition is a reprint of one from Holden-Day, San Francisco, California, 1967. Golomb worked for the National Security Agency.

Goresky, Mark and Andrew Klapper, *Algebraic Shift Register Sequences*, Cambridge University Press, Cambridge, UK, 2012.

Selmer, Ernst S., *Linear Recurrence Relations Over Finite Fields*, mimeographed lecture notes, 1966, Department of Mathematics, University of Bergen, Norway. Selmer was the Norwegian government's chief cryptographer.

Zierler, Neal, "Linear Recurring Sequences," *Journal of the Society for Industrial and Applied Mathematics*, Vol. 7, No. 1, March 1959, pp. 31–48.

On A5/1

Barkan, Elad and Eli Biham, "Conditional Estimators: An Effective Attack on A5/1," in Preneel, Bart, and Stafford Tavares, editors, *Selected Areas in Cryptography 2005*, Springer, Berlin, Germany, 2006, pp. 1–19.

Barkan, Elad, Eli Biham, and Nathan Keller, "Instant Ciphertext-Only Cryptanalysis of GSM Encrypted Communication," in Boneh, Dan, editor, *Advances in Cryptology — CRYPTO 2003 Proceedings*, Lecture Notes in Computer Science, Vol. 2729, Springer, Berlin, Germany, 2003, pp. 600–616.

Barkan, Elad, Eli Biham, and Nathan Keller, "Instant Ciphertext-Only Cryptanalysis of GSM Encrypted Communication," *Journal of Cryptology*, Vol. 21, No. 3, July 2008, pp. 392–429.

Biham, Eli and Orr Dunkelman, "Cryptanalysis of the A5/1 GSM Stream Cipher," in Roy, Bimal and Eiji Okamoto, editors, *Progress in Cryptology: INDOCRYPT 2000*, Lecture Notes in Computer Science, Vol. 2247, Springer, Berlin, Germany, 2000, pp. 43–51.

Biryukov, Alex, Adi Shamir, and David Wagner, "Real Time Cryptanalysis of A5/1 on a PC," in Schneier, Bruce, editor, *Fast Software Encryption, 7th International Workshop, FSE 2000*, Lecture Notes in Computer Science, Vol. 1978, Springer, Berlin, Germany, 2001, pp. 1–18.

Ekdahl, Patrik and Thomas Johansson, "Another attack on A5/1," *IEEE Transactions on Information Theory*, Vol. 49, No. 1, January 2003, pp. 284–289, available online at http://www.it.lth.se/patrik/papers/a5full.pdf.

Golic, Jovan Dj, "Cryptanalysis of Alleged A5 Stream Cipher," in Fumy, Walter, editor, *Advances in Cryptology — EUROCRYPT '97 Proceedings*, Lecture Notes in Computer Science, Vol. 1233, Springer, Berlin, Germany, 1997, pp. 239–255, available online at https://link.springer.com/content/pdf/10.1007/3-540-69053-0_17.pdf.

Gueneysu, Tim, Timo Kasper, Martin Novotný, Christof Paar, and Andy Rupp, *"Cryptanalysis with COPACOBANA,"* *IEEE Transactions on Computers*, Vol. 57, No. 11, November 2008, pp. 1498–1513.

Maximov, Alexander, Thomas Johansson, and Steve Babbage, "An Improved Correlation Attack on A5/1," in Handschuh, Helena and M. Anwar Hasan, editors, *Selected Areas in Cryptography 2004*, Lecture Notes in Computer Science, Vol. 3357, Springer, Berlin, Germany, 2004, pp. 1–18.

Stamp, Mark, *Information Security: Principles and Practice*, Wiley-Interscience, Hoboken, New Jersey, 2006. Several GSM security flaws are detailed in this book.

On RC4

AlFardan, Nadhem, Daniel J. Bernstein, Kenneth G. Paterson, Bertram Poettering, and Jacob C. N. Schuldt, "On the Security of RC4 in TLS," 22nd USENIX Security Symposium, August 2013, https://www.usenix.org/conference/usenixsecurity13/technical-sessions/paper/alFardan.

Arbaugh, William A., Narendar Shankar, Y. C. Justin Wan, and Kan Zhang, "Your 802.11 Wireless Network has No Clothes," *IEEE Wireless Communications*, Vol. 9, No. 6, December 2002, pp. 44–51. The publication date was given in this formal reference; the paper itself was dated March 30, 2001.

Borisov, Nikita, Ian Goldberg, and David Wagner, *Security of the WEP Algorithm*, ISAAC, Computer Science Department, University of California, Berkeley, http://www.isaac.cs.berkeley.edu/isaac/wep-faq.html. This page contains a summary of the findings of Brisov, Goldberg, and Wagner, as well as links to their paper and slides from a pair of presentations.

Fluhrer, Scott, Itsik Mantin, and Adi Shamir, "Weaknesses in the Key Scheduling Algorithm of RC4," in Vaudenay, Serge and Amr M. Youssef, editors, *Selected Areas in Cryptography 2001*, Lecture Notes in Computer Science, Vol. 2259, Springer, Berlin, Germany, 2002, pp. 1–24.

Jindal, Poonam and Brahmjit Singh, "RC4 Encryption-A Literature Survey," *International Conference on Information and Communication Technologies (ICICT 2014)*, *Procedia Computer Science*, Vol. 46, 2015, pp. 697–705.

Kundarewich, Paul D., Steven J. E. Wilton, and Alan J. Hu, "A CPLD-based RC4 Cracking System," in Meng, Max, editor, *Engineering Solutions for the Next Millennium, 1999 IEEE Canadian Conference on Electrical and Computer Engineering*, Vol. 1, IEEE, Piscataway, New Jersey, 1999, pp. 397–402.

Rivest, Ronald L. and Jacob Schuldt, "Spritz — a spongy RC4-like stream cipher and hash function," October 27, 2014, https://people.csail.mit.edu/rivest/pubs/RS14.pdf. Rivest confirmed the history of RC4 and its code in this paper.

Mantin, Itsik, "Analysis of the Stream Cipher RC4," master's thesis under the supervision of Adi Shamir, Weizmann Institute of Science, Rehovot, Israel, November 27, 2001, available online at https://tinyurl.com/yc9upxmu. This thesis is sometimes referenced under the title "The Security of the Stream Cipher RC4." The website referenced also contains other papers on RC4 and WEP.

Mantin, Itsik and Adi Shamir, "A Practical Attack on Broadcast RC4," in Matsui, Mitsuru, editor, *Fast Software Encryption, 8th International Workshop, FSE 2001*, Lecture Notes in Computer Science, Vol. 2355, Springer, Berlin, Germany, 2002, pp 152–164.

Paul, Goutam and Subhamoy Maitra, "Permutation after RC4 Key Scheduling Reveals the Secret Key," in Adams, Carlisle, Ali Miri, and Michael Wiener, editors, *Selected Areas of Cryptography, 14th International Workshop, SAC 2007*, Lecture Notes in Computer Science, Vol. 4876, Springer, Berlin Germany, 2007, pp 360–337.

Stubblefield, Adam, John Ioannidis, and Aviel D. Rubin, "Using the Fluhrer, Mantin and Shamir Attack to Break WEP," AT&T Labs Technical Report TD-4ZCPZZ, Revision 2, August 21, 2001, available online at https://tinyurl.com/y8eunft9.

Walker, Jesse R., *IEEE P802.11 Wireless LANs, Unsafe at any Key Size; an Analysis of the WEP Encapsulation*, IEEE Document 802.11-00/362, submitted October 27, 2000, available online at https://tinyurl.com/y9b888vl.

General

Bernstein, Daniel J., *ChaCha, a variant of Salsa20*, January 28, 2008, http://cr.yp.to/chacha/chacha-20080120.pdf.

Cusick, Thomas W., Cunsheng Ding, and Ari Renvall, *Stream Ciphers and Number Theory*, revised edition, North-Holland Mathematical Library, Vol. 66, Elsevier, New York, 2004. The original edition, published in 1998, was Vol. 55 of the same series.

Pommerening, Klaus, "Cryptanalysis of nonlinear feedback shift registers," *Cryptologia*, Vol. 40, No. 4, July 2016, pp. 303–315.

Ritter, Terry, "The Efficient Generation of Cryptographic Confusion Sequences," *Cryptologia*, Vol. 15, No. 2, April 1991, pp. 81–139. This survey paper includes a list of 213 references.

Robshaw, Matt J. B., *Stream Ciphers Technical Report TR-701*, Version 2.0, RSA Laboratories, Bedford, Massachusetts, 1995.

Rubin, Frank, 1978, "Computer Methods for Decrypting Random Stream Ciphers," *Cryptologia*, Vol. 2, No. 3, July 1978, pp. 215–231.

Rueppel, Rainer, A., *Analysis and Design of Stream Ciphers*, Springer, New York, 1986.

van der Lubbe, Jan, *Basic Methods of Cryptography*, Cambridge University Press, Cambridge, UK, 1998.

Chapter 20

Suite B All-Stars

In 2005, The National Security Agency (NSA) made public a list of recommended cryptographic algorithms and protocols. Known as "Suite B," these are believed to be the best of the unclassified schemes of that era. Two of them are covered in the present chapter.

20.1 Elliptic Curve Cryptography

Following Diffie-Hellman and RSA, another method of carrying out public key cryptography was independently discovered in 1985 by Neal Koblitz (Figure 20.1) and Victor S. Miller (Figure 20.2). One of the advantages of their elliptic curve cryptography is the appearance (no proof yet!) of a level of security equivalent to RSA with much smaller keys. For example, it's been estimated that a 313-bit elliptic curve key is as secure as a 4096-bit RSA key.[1]

Before we can detail the cryptographic application of these curves, some background must be provided. An *elliptic curve* is the set of solutions to an equation of the form $y^2 = x^3 + ax + b$, as well as a point at infinity, ∞. We refer to these as *Weierstrass equations*, in honor of Karl Weierstrass (Figure 20.3), who studied them in the 1800s, long before they were suspected of having any cryptologic applications.[2] So, as with the work of Fermat and Euler, we once again see that the eventual uses of mathematical results are impossible to predict! Despite the naming honor, work on elliptic curves goes back much farther than Weierstrass:[3]

> Elliptic curves have a long, rich history in several areas of mathematics. The so-called "chord and tangent" method for adding points on an elliptic curve actually goes back to Diophantus in the third century.

The long history shouldn't be surprising, because, as Koblitz pointed out, "Elliptic curves are the simplest type of curve that is more complicated than a conic section."[4] It's just a small step from degree 2 to 3.

---

[1] Blake, Ian, Gadiel Seroussi, and Nigel Smart, *Elliptic Curves in Cryptography*, London Mathematical Society Lecture Note Series, Vol. 265, Cambridge University Press, Cambridge, UK, 1999, p. 9.
[2] Other "big names" who studied elliptic curves include Abel, Jacobi, Gauss, and Legendre.
[3] Koblitz, Neal, *Random Curves: Journeys of a Mathematician*, Springer, Berlin, Germany, 2008, p. 313.
[4] Koblitz, Neal, *Random Curves: Journeys of a Mathematician*, Springer, Berlin, Germany, 2008, p. 303.

Figure 20.1   Neal Koblitz (1948–). (Courtesy of Neal Koblitz.)

Figure 20.2   Victor S. Miller (1947–). (Courtesy of Victor S. Miller.)

The graphs of the solutions take two basic forms, depending on whether the elliptic curve has three real roots, or just one.

We make the following work simpler by avoiding elliptic curves that have roots of multiplicity higher than one.[5] When looking at complex solutions (as opposed to the real solutions graphed in Figure 20.4), the visualization takes the form of a torus (donut).

---

[5] Another simplification involves not using fields of characteristic 2 or 3. The interested reader may consult the references for the reasons for these simplifications. These simplifications are just that. We may proceed, with greater difficulty, without them.

Figure 20.3   Karl Weierstrass (1815–1897). (http://en.wikipedia.org/wiki/File:Karl_Weierstrass. jpg.)

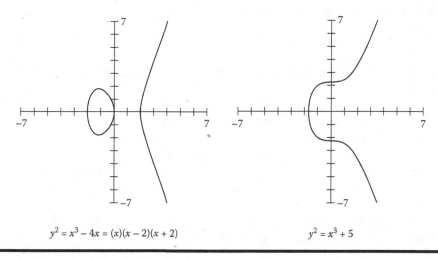

$$y^2 = x^3 - 4x = (x)(x - 2)(x + 2)$$ $$y^2 = x^3 + 5$$

Figure 20.4   Examples of elliptic curves.

Elliptic curves don't resemble ellipses. The name arises from their connection with elliptic integrals, which arose in the 19th century, as mathematicians attempted to find formulas for arc lengths of ellipses. Examples are given by $\int_c^d \dfrac{dx}{\sqrt{x^3 + ax + b}}$ and $\int_c^d \dfrac{x\,dx}{\sqrt{x^3 + ax + b}}$.

Setting the denominator of either integrand above equal to y, then gives us $y^2 = x^3 + ax + b$, an elliptic curve.

We define addition of points on an elliptic curve in a strange way. To add points P_1 and P_2, we draw a line through them and observe that this line passes through a third point on the curve, I. No, I is not the sum! I is merely an intermediate point. We reflect the point I about the x-axis to get a new point P_3. Then we say $P_1 + P_2 = P_3$. This is illustrated in Figure 20.5.

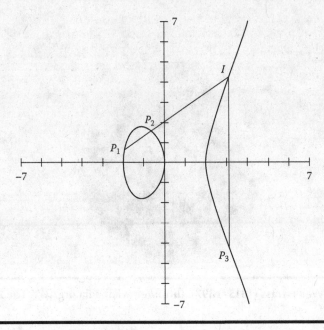

Figure 20.5   Elliptic curve point addition.

If we wish to add a point to itself, we draw a tangent line to the curve at the given point, and then continue as above.

There's still one more case that needs to be considered. Adding two points that lie on a vertical line (or adding a point at which the curve has a vertical tangent line to itself) provides no other point of intersection with our curve. To "patch" this problem, we introduce an extra point, ∞. This is a common trick in algebraic geometry. Reflecting ∞ about the x-axis still gives ∞; that is, we do not want a second −∞. With these definitions, the points on an elliptic curve (together with ∞) form a commutative group. The identity element is ∞, because $P + ∞ = P$ for any point P.

The addition can be carried out algebraically, as follows. $P_1 = (x_1, y_1)$ and $P_2 = (x_2, y_2)$ implies $P_1 + P_2 = (m^2 - x_1 - x_2, m(2x_1 - (m^2 - x_2)) - y_1)$, where

$$m = (y_2 - y_1)/(x_2 - x_1), \text{ if } P_1 \neq P_2$$

and

$$m = (3x_1^2 + a)/2y_1, \text{ if } P_1 = P_2$$

The proof is left as an exercise.

The graphs given above show elliptic curves where the solutions are real numbers, but we may instead examine the same equations modulo n. For example, consider

$$y^2 = x^3 + 3x + 4 \pmod{7}$$

By plugging in the values 0, 1, 2, 3, 4, 5, and 6 for x and seeing which are perfect squares modulo 7, we're able to quickly get the complete set of solutions:

$$(0,4), (0,5), (1,1), (1,6), (2,2), (2,5), (5,2), (5,5), (6,0), \infty$$

There is no point on the curve with an x value of 3, for example, because that would imply $y^2 = 2$ (mod 7) for some y value in the set $\{0, 1, 2, 3, 4, 5, 6\}$ and this is not true.

Notice that with the exception of 0, every perfect square that results from plugging in a value for x leads to two values for y.

It's interesting to note that if we're looking for points (x, y) on an elliptic curve such that x and y are both rational numbers, the number of solutions may be infinite or finite, but if there are only finitely many, there will be no more than 16. On the other hand, we're able to get arbitrarily large, yet still finite, solution sets by working modulo a prime.

So, how many points will satisfy a given elliptic curve modulo p? If we plug in the values, 0, 1, 2, ..., $p - 1$ we can expect to get a value that is a square root modulo p (and hence, a point on the curve) about half the time. This is because of a result from number theory that tells us half of the nonzero integers are perfect squares modulo a prime. But each square root, other than 0, will have two answers, so there should be about p points. Adding in the point ∞, we are now up to $p + 1$ points, but we don't always get this exact value. Letting the correct value be denoted by N, our error will be $|N - (p + 1)|$. German mathematician Helmut Hasse found a bound on this error around 1930. He showed

$$|N - (p + 1)| \leq 2\sqrt{p}.$$

For example, with $p = 101$, our estimate suggests 102 points, and Hasse's theorem guarantees the actual number is between $102 - 2\sqrt{101}$ and $102 + 2\sqrt{101}$. That is, rounding to the appropriate integer, between 82 and 122.

We're now ready to show how elliptic curves can be used to agree on a key over an insecure channel. This is the *elliptic curve cryptography* (ECC) version of Diffie-Hellman key exchange. Normally, Alice and Bob would appear at this point, but Neal Koblitz recalled:[6]

> When I wrote my first book on cryptography I tried to change this anglocentric choice of names to names like Alicia and Beatriz or Aniuta and Busiso. However, the Anglo-American domination of cryptography is firmly entrenched — virtually all books and journals are in English, for example. My valiant attempt to introduce a dollop of multiculturalism into the writing conventions of cryptography went nowhere. Everyone still says "Alice" and Bob."

As a tribute to Koblitz, the key exchange will be carried out by Aïda and Bernardo. They proceed as follows.

1. Aïda and Bernardo agree on an elliptic curve E (mod p), for some prime p.
2. They agree on a point B on their curve E. (this is also done publicly)
3. Aïda selects a random (secret) integer a and computes aB, which she sends to Bernardo.
4. Bernardo selects a random (secret) integer b and computes bB, which he sends to Aïda.
5. Both Aïda and Bernardo are now able to compute abB, the x coordinate of which can be adapted to serve as their secret key for a symmetric system.

---

[6] Koblitz, Neal, *Random Curves: Journeys of a Mathematician*, Springer, Berlin, Germany, 2008, p. 321.

Given points P and B on an elliptic curve, finding an integer x such that $xB = P$ is the elliptic curve analog of the discrete log problem. No efficient method is known for solving this when the values are large. The letter B was chosen, as this point plays the role of the base in this version of the discrete log problem. Although an eavesdropper on an exchange like the one detailed above will know aB and bB, he cannot efficiently find a or b. This is good, because either one of these values would allow recovery of the secret key generated by the exchange between Aïda and Bernardo.

When calculating a large multiple of a point, we'd like to avoid the tedium of adding the number to itself a large number of times. Happily, we can adapt the repeated squaring technique used to raise a number to a large power modulo some integer n. This technique is most easily explained with an example.

To find $100P$, we express it as $2(2(P + 2(2(2(P + 2P)))))$. Thus, in place of 99 additions, we have 2 additions and 6 doublings.[7] But how did we find this representation?

Another example will serve to illustrate the process. Suppose we wish to calculate $86P$. Because 86 is divisible by 2, we start out with

$$2$$

Halving 86 gives 43, which is not divisible by 2, so we continue with a P, instead of another 2:

$$2(P +$$

When we do this, we subtract 1 from the number we are reducing. Now that we are down to 42, we halve it again, by placing a 2 in our representation. We get

$$2(P + 2$$

But half of 42 is 21, which is odd, so we continue with a P:

$$2(P + 2(P +$$

Subtracting 1 from 21, we get 20, which can be halved twice, so we append a pair of 2s:

$$2(P + 2(P + 2(2$$

After halving 20 twice, we're down to 5, which is odd, so we use another P:

$$2(P + 2(P + 2(2(P +$$

Subtracting 1 from 5, leaves 4, which we can halve twice, so we append a pair of 2s:

$$2(P + 2(P + 2(2(P + 2(2$$

We are finally down to 1, so we end with a P, and close off all of the parentheses:

$$2(P + 2(P + 2(2(P + 2(2P))))).$$

This is the most time-consuming portion of the algorithm. In 2008 a much quicker method was developed by V. S. Dimitrov et al, but it only works for elliptic curves with a very special

---

[7] This example was taken from Kobitz, Neal, *A Course in Number Theory and Cryptography*, second edition, Springer, New York, 1994, p. 178. The example also appears on p. 162 of the first edition, but there are a pair of typos present in that edition obscure this simple technique.

equation and does not apply to elliptic curves in general.[8] In this special case, the time required is now sublinear.

To use elliptic curves for enciphering a message, rather than just agreeing on a key, we must be able to represent plaintext characters by points on the curve. Hasse's theorem above shows that for a sufficiently large modulus there will be enough such points, but there is not yet a fast (polynomial time) deterministic way to assign characters to points![9] The problem is, for a given *x* value, there may or may not be a point *y* such that (*x*, *y*) lies on the curve. Koblitz described a method of pairing characters and points as follows.

1. Represent the message as a number, *m*.
2. There will only be a point (*m*, *y*) on the curve if *m* is a perfect square mod *p*.
3. Append a few bits to *m* such that the new number, *x*, is a perfect square mod *p*. This may take a few tries, using different bits, until a solution is found. The more bits you are willing to append, the greater your chance of success, as each attempt has about a 50% chance of yielding a perfect square.

Once the message has been converted to points on the curve, we're ready to begin the enciphering.

20.1.1 Elgamal, ECC Style

Just as Diffie-Hellman has an ECC version, so does Elgamal. An example will show how Bisahalani can send an enciphered message to Aditsan in this system. In order to be able to receive enciphered messages, Aditsan chooses the elliptic curve $y^2 = x^3 - 2x + 3$ (mod 29), and the point $B = (8, 21)$ on it. Aditsan also selects a random (secret) number, say $s = 3$, and computes $sB = 3(8, 21)$. To do this calculation, he first writes $3B = B + 2B$. This indicates that he must first perform a doubling, and then a sum.

1. *Doubling* — Because multiplying a point by 2 is the same as adding it to itself, we use the formula (adapted from what was given here earlier):

$$2B = B + B = (m^2 - 2x, \; m(3x - m^2) - y), \text{ where } m = (3x^2 + a)/2y$$

Referring back to the curve, we have $a = -2$. We calculate $m = [(3)(8)^2 - 2]/(2, 21) = 190/42 = 16/13$ (mod 29). Now division, in this context, means multiplying the numerator by the inverse of the denominator. The inverse of 13 (mod 29) is 9, so the above becomes $(16)(9) = 144 = 28$ (mod 29). But 28 can be conveniently expressed as -1 (mod 29). We then plug this value for *m* into our sum formula to get $2B = B + B = ((-1)^2 - 2(8), (-1)(3(8) - (-1)^2) - 21) = (14, 14)$.

2. *Summing* — We must now add the original *B* to our new doubled value:

$$B + 2B = (8, \; 21) + (14, 14) = (m^2 - x_1 - x_2, \; m(2x_1 - (m^2 - x_2)) - y_1)$$

---

[8] Dimitrov, Vassil S., Kimmo U. Järvinen, Michael J. Jacobson, Jr., Wai Fong Chan, and Zhun Huang, "Provably Sublinear Point Multiplication on Koblitz Curves and its Hardware Implementation," *IEEE Transactions on Computers*, Vol. 57, No. 11, November 2008, pp. 1469–1481.

[9] However, there are probabilistic algorithms that may be applied to make the chance of failure arbitrarily small. And a theorem from Section 16.5 leads me to believe that a deterministic algorithm in polynomial time does exist.

We calculate $m = (y_2 - y_1)/(x_2 - x_1) = (14 - 21)/(14 - 8) = -7/6 = (-7)(5) = -35 = 23$ (mod 29). And then make use of that value in the finding $B + 2B = ((23)^2 - 8 - 14, 23(2(8) - ((23)^2 - 14)) - 21) = (507, -11498) = (14, 15)$.

Aditsan now reveals his public key as follows. The elliptic curve $y^2 = x^3 - 2x + 3$ (mod 29) and the points $B = (8, 21)$ and $sB = (14, 15)$. He keeps s secret. Recall that for larger values, knowing B and sB does not allow us to efficiently find s.

Seeing that Aditsan has posted his public key, Bisahalani prepares his message. He does this by converting it to an x value, adding bits at the end, to guarantee that $x^3 - 2x + 3$ will be a perfect square modulo 29. To make things simpler, for illustrative purposes, we'll just assume his message is represented as 12. He ends up with the point on the curve $M = (12, 5)$. He then selects a random number $k = 5$ and computes

$$kB = 5(8, 21) = (14, 15)$$

and

$$M + k(sB) = (12, 5) + 5(14, 15) = (12, 5) + (8, 5) = (15, 19)$$

Because we've already shown how such calculations are carried out, the work is omitted this time.[10] Upon receiving Bisahalani's two-part message, Aditsan computes

$$s(kB) = 3(14, 15) = (8, 8)$$

followed by

$$[M + k(sB)] - a(kB) = (15, 19) - (8, 8)$$

Now, this is something new. How is subtraction carried out? Very simply! The point $-(8, 8)$ is just a reflection. We have $-(8, 8) = (8, -8)$. Picturing the curve may help you to see this. So we have $(15, 19) - (8, 8) = (15, 19) + (8, -8) = (15, 19) + (8, 21)$ (mod 29). Performing this last addition gives $(12, 5) = M$, and so Aditsan has now recovered the original message.

There is also an ECC analog for Elgamal signatures.

20.2 Personalities behind ECC

Although they could have filed patents and spent the rest of their lives working on the business end of elliptic curves, the creators of ECC don't seem to be motivated by profit. In fact, Neal Koblitz didn't immediately recognize the commercial potential of elliptic curve cryptography. He thought of it as "just a nice theoretical construction to study."[11] ECC co-discover Victor Miller did recognize its practical value, but he was working for IBM and the bureaucracy wasn't interested in promoting any cryptographic systems other than DES at the time.[12] Thus, neither discoverer of this new technique sought a patent.

---

[10] Also, an online elliptic curve calculator available at http://www.csulb.edu/~wmurray/java/WillEllipticApplet. html.

[11] Koblitz, Neal, *Random Curves: Journeys of a Mathematician*, Springer, Berlin, Germany, 2008, p. 299.

[12] Koblitz, Neal, *Random Curves: Journeys of a Mathematician*, Springer, Berlin, Germany, 2008, p. 300.

Scott Vanstone (of the University of Waterloo) was the first to commercialize elliptic curve cryptography, through a company now called the Certicom Corporation.[13] In March 1997, he offered Koblitz $1,000 a month to serve as a consultant. Koblitz accepted and donated the money, first to the University of Washington, but upon discovery of its misuse, redirected it to the Kovalevskaia Fund.[14] Certicom, a Canadian company, is a competitor of RSA and has NSA as its largest customer: "In 2003 NSA paid Certicom a $25 million licensing fee for 26 patents related to ECC."[15] As was mentioned at the start of this chapter, NSA also encouraged others to use the system by including a key agreement and a signature scheme based on ECC in its "Suite B" list of recommendations.[16] To answer the obvious question, a quote from NSA is provided below:[17]

> Another suite of NSA cryptography, Suite A, contains classified algorithms that will not be released. Suite A will be used for the protection of some categories of especially sensitive information.

The Suite B block cipher, AES, is detailed later in this chapter. ECC earned NSA's endorsement by standing the test of time, and massive peer review:[18]

> [E]xcept for a relatively small set of elliptic curves that are easy to avoid, even at present — more than twenty years after the invention of ECC — no algorithm is known that finds discrete logs in fewer than $10^{n/2}$ operations, where n is the number of decimal digits in the size of the elliptic curve group.

Neal Koblitz's political convictions have also stood the test of time. Whereas many activist burn out, Koblitz's autobiography shows him sustaining his radicalism for decades on end. He's been arrested several times, including during his first year teaching at Harvard, but he never worried about it affecting his employment.[19]

> I had read about the history of mathematics and was aware of the long tradition of tolerance of eccentricity and political dissidence among mathematicians.

In June 1997, Koblitz learned that the official RSA website put up a page filled with skeptical remarks about ECC. This was part of the American company's aggressive approach and it included a comment from RSA co-creator Ron Rivest.[20]

> But the security of a cryptosystem based on elliptic curves is not well understood, due in large part to the abstruse nature of elliptic curves. Few cryptographers understand elliptic curves, so… trying to get an evaluation of the security of an elliptic curve cryptosystem is a bit like trying to get an evaluation of some recently discovered Chaldean poetry.

[13] Koblitz, Neal, *Random Curves: Journeys of a Mathematician*, Springer, Berlin, Germany, 2008, p. 302–303.
[14] Koblitz, Neal, *Random Curves: Journeys of a Mathematician*, Springer, Berlin, Germany, 2008, p. 314.
[15] Koblitz, Neal, *Random Curves: Journeys of a Mathematician*, Springer, Berlin, Germany, 2008, p. 319.
[16] NSA Suite B Cryptography, National Security Agency, January 15, 2009, https://web.archive.org/web/20090117004931/http://www.nsa.gov/ia/programs/suiteb_cryptography/index.shtml.
[17] NSA Suite B Cryptography, National Security Agency, January 15, 2009, https://web.archive.org/web/20090117004931/http://www.nsa.gov/ia/programs/suiteb_cryptography/index.shtml.
[18] Koblitz, Neal, *Random Curves: Journeys of a Mathematician*, Springer, Berlin, Germany, 2008, p. 311.
[19] Koblitz, Neal, *Random Curves: Journeys of a Mathematician*, Springer, Berlin, Germany, 2008, p. 23.
[20] Taken here from Koblitz, Neal, *Random Curves: Journeys of a Mathematician*, Springer, Berlin, Germany, 2008, p. 313.

Koblitz's reaction, after asking his wife who the Chaldean's were,[21] was not to post a webpage bashing RSA, but rather to have shirts made featuring an elliptic curve and the text "I Love Chaldean Poetry." He reports that they were a hit with the students, with the exception of those hoping to intern at RSA.[22]

> It is true that a mathematician who is not also something of a poet will never be a perfect mathematician.
>
> — **Karl Weierstrass**[23]

When Weierstrass made the above claim, it is doubtful that he had Chaldean poetry in mind! It's more likely that he was referring to the poet's creativity and sense of beauty. Victor Miller has engaged his artistic side as an actor/singer in 17 community theater productions.[24] He's also sung in many choirs and was one of the winners of a vocal competition at Westminster Conservatory in 2003.

For many years Miller bred and exhibited pedigreed cats. He was a former president of a national breed club, and bred the U.S. national best of breed Colorpoint Shorthair in 1997. Miller himself is a rare breed, being one of the few mathematicians to have a Bacon number. Not strictly limited to community theater, he appeared as an extra in *A Beautiful Mind*, which featured Ed Harris, who starred in *Apollo Thirteen* with Kevin Bacon. Hence, two links connect Miller to Bacon. This connection game began long before Bacon, as mathematicians tried to find their shortest path to Paul Erdös, a number theorist with over 1,500 papers and over 500 coauthors. Miller's Erdös number is also two. Successful at all of these outside pursuits, Miller once faced an unfair rejection in his professional life.

> Similar to Merkle (and Rene Schoof) my paper on the efficient calculation of the "Weil Pairing" was rejected from the 1986 FOCS conference (Schoof's paper on counting points on elliptic curves over finite fields was rejected the previous year). This led Hendrik Lenstra to remark that perhaps that this was a badge of honor.[25]

20.3 The Advanced Encryption Standard (AES)

We're about to examine a block cipher that would eventually become part of Suite B, but the story begins with the weaknesses of DES. As we've seen, DES was criticized from the start for having too short of a key. As computing speeds continued to increase rapidly, the problem only got worse. Finally, in January 1997, another competition was sponsored by the National Institute of Standards and Technology (NIST). The winner would be declared the Advanced Encryption Standard, or AES for short. Unlike the first time around, in 1997 there was a large worldwide community of mathematicians and computer scientists carrying out cryptographic research in the

[21] His wife knows some mathematics, as well, and has authored a biography of Sofia Kovalevskaia: Koblitz, Ann Hibner, *A Convergence of Lives: Sofia Kovalevskaia: Scientist, Writer, Revolutionary*, Rutgers University Press, New Brunswick, New Jersey, 1983.

[22] Koblitz, Neal, *Random Curves: Journeys of a Mathematician*, Springer, Berlin, Germany, 2008, p. 314.

[23] Quoted here from Bell, Eric Temple, *Men of Mathematics*, Dover Publications, New York, 1937, p. 432.

[24] Theater is a passion Miller shares with his daughter, who is, as of this writing, a professional stage manager working on Broadway.

[25] Email from Victor S. Miller to the author, November 10, 2010.

open. It was this community that would be responsible for analyzing the security of the submitted systems. Cryptanalysts could submit their findings to NIST's AES website or present them at AES conferences.

Acceptance of submissions ended on May 15, 1998. The 15 accepted submissions were presented at The First Advanced Encryption Standard Candidate Conference in Ventura, California, August 20–22, 1998.[26] The second conference was held in Rome, Italy, March 22–23, 1999. Five candidates were eliminated at (or prior to) this conference due to various kinds of security problems that were identified.

From the remaining ten candidates, the finalists, announced by NIST a year later, in August 1999, were

1. RC6 (Rivest Cipher 6) from RSA
2. MARS from IBM[27]
3. Twofish from Counterpane (Bruce Schneier, John Kelsey, Doug Whiting, David Wagner, Chris Hall, and Niels Ferguson)
4. Serpent from Ross Anderson, Eli Biham, and Lars Knudsen (an English/Israeli/Danish team)
5. Rijndael from Vincent Rijmen and Joan Daemen (a Belgian team)

NIST justified their selections in a publication[28], but no controversy was expected, as their choices matched the top 5, as voted on at the end of the second AES conference. Cost was a factor in eliminating two candidates and slow runtime was responsible for the failure of another.

New York City hosted the third AES conference on April 13–14, 2000. The attacks made on the finalists proved to be only slightly faster than brute force. Again, attendees voted for their favorites. NIST announced on October 2, 2000 that Rijndael was the winner. Again, controversy was avoided as this matched the results of the attendees' vote. NIST's full justification was provided online once more.[29] The name Rijndael is a combination of the last names of the Belgians who collaborated on its design, Vincent Rijmen and Joan Daemen (Figure 20.6). Various pronunciations have been offered. Many Americans pronounce it Rhine-Doll. At one point the creators had a link on their website for those wishing to hear the authoritative pronunciation; the recording said, "The correct pronunciation is… AES."

The algorithm is royalty-free. This was required of all submissions to NIST, should the algorithm be declared the winner. This fact, combined with endorsement by NIST and the worldwide

---

[26] All 15 are listed in Daemen, Joan and Vincent Rijmen, *The Design of Rijndael: AES—The Advanced Encryption Standard*, Springer, New York, 2002, p. 3.

[27] John Kelsey and Bruce Schneier examined this system in a paper that bore the wonderful title "MARS Attacks! Preliminary Cryptanalysis of Reduced-Round Mars Variants." It's available online at http://www.schneier.com/paper-mars-attacks.html

[28] Nechvatal, James, Elaine Barker, Donna Dodson, Morris Dworkin, James Foti, and Edward Roback, "Status Report on the First Round of the Development of the Advanced Encryption Standard," *Journal of Research of the National Institute of Standards and Technology*, Vol. 104, No. 5, September–October 1999, pp. 435–459, available online at http://nvlpubs.nist.gov/nistpubs/jres/104/5/j45nec.pdf.

[29] Nechvatal, James, Elaine Barker, Lawrence Bassham, William Burr, Morris Dworkin, James Foti, and Edward Roback, "Report on the Development of the Advanced Encryption Standard (AES)," *Journal of Research of the National Institute of Standards and Technology*, Vol. 106, No. 3, May–June 2001, pp. 511–577, available online at https://www.ncbi.nlm.nih.gov/pmc/articles/PMC4863838/.

Figure 20.6 Joan Daemen (1965–) and Vincent Rijmen (1970–). (Courtesy of Vincent Rijmen.)

community of cryptographers, prompted Ron Rivest to remark, somewhat obviously, "It is likely that Rijndael will soon become the most widely-used cryptosystem in the world."[30]

AES offers a choice of key sizes (128, 192, or 256 bits), but always acts on blocks of 128 bits (16 characters). The number of rounds depends on the key size. The different key sizes are often distinguished by referring to AES-128 (10 rounds), AES-192 (12 rounds), or AES-256 (14 rounds). Like DES, AES is derived from earlier systems. Unlike, DES a large safety margin was built into AES. The only attacks the creators could find that were better than brute force apply only to six or fewer rounds (for the 128 bit version).[31]

To help drive home the size of 2^{128} we look at it written out:

$$340,282,366,920,938,463,463,374,607,431,768,211,456.$$

AES is basically composed of four simple and fast operations that act on a 4×4 array of bytes, referred to as "the state." Each of the operations is detailed below.

20.3.1 SubBytes

As the name suggests, this operation makes substitutions for the bytes using the table below. It can be illustrated like the *S*-boxes of DES, and is even known as the Rijndael *S*-box, but there is only one (along with its inverse) and there are other ways to represent it. Below is Rijndael's *S*-box:[32]

---

[30] Daemen, Joan, and Vincent Rijmen, *The Design of Rijndael*, Springer, Berlin, Germany, 2002, p. vi. Although it was behind the scenes, NSA also examined AES on NIST's behalf. They didn't want NIST to be surprised if there was some flaw the open community couldn't find.

[31] Daemen, Joan, and Vincent Rijmen, *The Design of Rijndael*, Springer, Berlin, Germany, 2002, p. 41.

[32] This substitution box is usually presented in hexadecimal. For this book, I thought base-10would be clearer. Finding such a table in Trappe, Wade and Lawrence C. Washington, *Introduction to Cryptography with Coding Theory*, Prentice Hall, Upper Saddle River, New Jersey, 2002, saved me the trouble of conversion. This is one of the most clearly written books covering modern cryptography.

99	124	119	123	242	107	111	197	48	1	103	43	254	215	171	118
202	130	201	125	250	89	71	240	173	212	162	175	156	164	114	192
183	253	147	38	54	63	247	204	52	165	229	241	113	216	49	21
4	199	35	195	24	150	5	154	7	18	128	226	235	39	178	117
9	131	44	26	27	110	90	160	82	59	214	179	41	227	47	132
83	209	0	237	32	252	177	91	106	203	190	57	74	76	88	207
208	239	170	251	67	77	51	133	69	249	2	127	80	60	159	168
81	163	64	143	146	157	56	245	188	182	218	33	16	255	243	210
205	12	19	236	95	151	68	23	196	167	126	61	100	93	25	115
96	129	79	220	34	42	144	136	70	238	184	29	222	94	11	219
224	50	58	10	73	6	36	92	194	211	172	98	145	149	228	121
231	200	55	109	141	213	78	169	108	86	244	234	101	122	174	8
186	120	37	46	28	166	180	198	232	221	116	31	75	189	139	138
112	62	181	102	72	3	246	14	97	53	87	185	134	193	29	158
225	248	152	17	105	217	142	148	155	30	135	233	206	85	40	223
140	161	137	13	191	230	66	104	65	153	45	15	176	84	187	22

The table above is meant to be read from left to right and from top to bottom, like normal English text. Thus, in base 10, 0 goes to 99, 1 goes to 124, ..., 255 goes to 22. These numbers are only expressed in base-10 for convenience (and familiarity). In base-2, each is a byte. These 8 bits may be viewed as the coefficients of a polynomial of degree at most 7. Viewed in this manner, the table above has a much terser representation. It simply sends each polynomial to its inverse modulo $x^8 + x^4 + x^3 + x + 1$, followed by the affine transformation

$$\begin{pmatrix} 1 & 1 & 1 & 1 & 1 & 0 & 0 & 0 \\ 0 & 1 & 1 & 1 & 1 & 1 & 0 & 0 \\ 0 & 0 & 1 & 1 & 1 & 1 & 1 & 0 \\ 0 & 0 & 0 & 1 & 1 & 1 & 1 & 1 \\ 1 & 0 & 0 & 0 & 1 & 1 & 1 & 1 \\ 1 & 1 & 0 & 0 & 0 & 1 & 1 & 1 \\ 1 & 1 & 1 & 0 & 0 & 0 & 1 & 1 \\ 1 & 1 & 1 & 1 & 0 & 0 & 0 & 1 \end{pmatrix} \times \begin{pmatrix} d_7 \\ d_6 \\ d_5 \\ d_4 \\ d_3 \\ d_2 \\ d_1 \\ d_0 \end{pmatrix} \oplus \begin{pmatrix} 0 \\ 1 \\ 1 \\ 0 \\ 0 \\ 0 \\ 1 \\ 1 \end{pmatrix}$$

Example 1

The Rijndael *S*-box sends 53 to 150. We verify that the alternative method does the same. Converting 53 to binary we get 00110101, which has the polynomial representation

$$x^5 + x^4 + x^2 + 1.$$

This gets sent to its inverse modulo $x^8 + x^4 + x^3 + x + 1$, which is $x^5 + x^4 + x^3 + 1$, or 00111001, in binary. This inverse may be calculated using the extended Euclidean algorithm, as shown in Section 14.3. There is no difference other than using polynomials instead of integers. Long division with polynomials reveals

$$(x^8 + x^4 + x^3 + x + 1) = (x^3 + x^2 + x)(x^5 + x^4 + x^2 + 1) + (x^3 + x^2 + 1)$$

and

$$(x^5 + x^4 + x^2 + 1) = (x^2)(x^3 + x^2 + 1) + 1$$

The final remainder of 1 shows that the two polynomials are relatively prime, a necessary condition for the inverse of one to exist modulo the other.

We now solve for the remainder in each of the two equalities above:

$$(x^3 + x^2 + 1) = (x^8 + x^4 + x^3 + x + 1) - (x^3 + x^2 + x)(x^5 + x^4 + x^2 + 1)$$

$$1 = (x^5 + x^4 + x^2 + 1) - (x^2)(x^3 + x^2 + 1)$$

Using the first equality to substitute for $(x^3 + x^2 + 1)$ in the second gives

$$1 = (x^5 + x^4 + x^2 + 1) - (x^2)[(x^8 + x^4 + x^3 + x + 1) - (x^3 + x^2 + x)(x^5 + x^4 + x^2 + 1)]$$

Distributing the x^2 gives

$$1 = (x^5 + x^4 + x^2 + 1) - (x^2)(x^8 + x^4 + x^3 + x + 1) + (x^5 + x^4 + x^3)(x^5 + x^4 + x^2 + 1).$$

Combining the $(x^5 + x^4 + x^2 + 1)$ terms gives

$$1 = -(x^2)(x^8 + x^4 + x^3 + x + 1) + (x^5 + x^4 + x^3 + 1)(x^5 + x^4 + x^2 + 1).$$

Reducing the equality above modulo $(x^8 + x^4 + x^3 + x + 1)$ gives

$$1 = (x^5 + x^4 + x^3 + 1)(x^5 + x^4 + x^2 + 1)$$

So, we see that the inverse of $x^5 + x^4 + x^2 + 1$ (mod $x^8 + x^4 + x^3 + x + 1$) is $x^5 + x^4 + x^3 + 1$, as claimed above. We write this as the binary vector 00111001 and plug it into the matrix equation:

$$
\begin{pmatrix}
1 & 1 & 1 & 1 & 1 & 0 & 0 & 0 \\
0 & 1 & 1 & 1 & 1 & 1 & 0 & 0 \\
0 & 0 & 1 & 1 & 1 & 1 & 1 & 0 \\
0 & 0 & 0 & 1 & 1 & 1 & 1 & 1 \\
1 & 0 & 0 & 0 & 1 & 1 & 1 & 1 \\
1 & 1 & 0 & 0 & 0 & 1 & 1 & 1 \\
1 & 1 & 1 & 0 & 0 & 0 & 1 & 1 \\
1 & 1 & 1 & 1 & 0 & 0 & 0 & 1
\end{pmatrix}
\times
\begin{pmatrix}
a_7 \\ a_6 \\ a_5 \\ a_4 \\ a_3 \\ a_2 \\ a_1 \\ a_0
\end{pmatrix}
\oplus
\begin{pmatrix}
0 \\ 1 \\ 1 \\ 0 \\ 0 \\ 0 \\ 1 \\ 1
\end{pmatrix}
$$

to get

$$
\begin{pmatrix}
1 & 1 & 1 & 1 & 1 & 0 & 0 & 0 \\
0 & 1 & 1 & 1 & 1 & 1 & 0 & 0 \\
0 & 0 & 1 & 1 & 1 & 1 & 1 & 0 \\
0 & 0 & 0 & 1 & 1 & 1 & 1 & 1 \\
1 & 0 & 0 & 0 & 1 & 1 & 1 & 1 \\
1 & 1 & 0 & 0 & 0 & 1 & 1 & 1 \\
1 & 1 & 1 & 0 & 0 & 0 & 1 & 1 \\
1 & 1 & 1 & 1 & 0 & 0 & 0 & 1
\end{pmatrix}
\times
\begin{pmatrix}
0 \\ 0 \\ 1 \\ 1 \\ 1 \\ 0 \\ 0 \\ 1
\end{pmatrix}
\oplus
\begin{pmatrix}
0 \\ 1 \\ 1 \\ 0 \\ 0 \\ 0 \\ 1 \\ 1
\end{pmatrix}
=
\begin{pmatrix}
1 \\ 1 \\ 1 \\ 1 \\ 0 \\ 1 \\ 0 \\ 1
\end{pmatrix}
\oplus
\begin{pmatrix}
0 \\ 1 \\ 1 \\ 0 \\ 0 \\ 0 \\ 1 \\ 1
\end{pmatrix}
=
\begin{pmatrix}
1 \\ 0 \\ 0 \\ 1 \\ 0 \\ 1 \\ 1 \\ 0
\end{pmatrix}
$$

Converting the output, 10010110, to base-10 gives 150, which matches what our table provides.

To invert the affine transformation, we must perform the XOR first and follow it by multiplication with the inverse of the 8×8 matrix above. We then have

$$
\begin{pmatrix}
0 & 1 & 0 & 1 & 0 & 0 & 1 & 0 \\
0 & 0 & 1 & 0 & 1 & 0 & 0 & 1 \\
1 & 0 & 0 & 1 & 0 & 1 & 0 & 0 \\
0 & 1 & 0 & 0 & 1 & 0 & 1 & 0 \\
0 & 0 & 1 & 0 & 0 & 1 & 0 & 1 \\
1 & 0 & 0 & 1 & 0 & 0 & 1 & 0 \\
0 & 1 & 0 & 0 & 1 & 0 & 0 & 1 \\
1 & 0 & 1 & 0 & 0 & 1 & 0 & 0
\end{pmatrix}
\times
\left(
\begin{pmatrix}
a_7 \\ a_6 \\ a_5 \\ a_4 \\ a_3 \\ a_2 \\ a_1 \\ a_0
\end{pmatrix}
\oplus
\begin{pmatrix}
0 \\ 1 \\ 1 \\ 0 \\ 0 \\ 0 \\ 1 \\ 1
\end{pmatrix}
\right)
$$

Distributing the multiplication gives

$$
\begin{pmatrix}
0 & 1 & 0 & 1 & 0 & 0 & 1 & 0 \\
0 & 0 & 1 & 0 & 1 & 0 & 0 & 1 \\
1 & 0 & 0 & 1 & 0 & 1 & 0 & 0 \\
0 & 1 & 0 & 0 & 1 & 0 & 1 & 0 \\
0 & 0 & 1 & 0 & 0 & 1 & 0 & 1 \\
1 & 0 & 0 & 1 & 0 & 0 & 1 & 0 \\
0 & 1 & 0 & 0 & 1 & 0 & 0 & 1 \\
1 & 0 & 1 & 0 & 0 & 1 & 0 & 0
\end{pmatrix}
\times
\begin{pmatrix}
a_7 \\ a_6 \\ a_5 \\ a_4 \\ a_3 \\ a_2 \\ a_1 \\ a_0
\end{pmatrix}
\oplus
\begin{pmatrix}
0 \\ 0 \\ 0 \\ 0 \\ 0 \\ 1 \\ 0 \\ 1
\end{pmatrix}
$$

Knowing how suspicious cryptologists can be, Rijmen and Daemen explained why they chose the polynomial $x^8 + x^4 + x^3 + x + 1$ for this operation.

> The polynomial $m(x)$ ('11B') for the multiplication in $GF(2^8)$ is the first one of the list of irreducible polynomials of degree 8, given in [LiNi86, p. 378].[33]

The reference they provided is [LiNi86] R. Lidl and H. Niederreiter, *Introduction to finite fields and their applications*, Cambridge University Press, 1986.

Any irreducible polynomial of degree 8 could have been used, but by selecting the first from a list provided in a popular (at least among algebraists) book, Rijmen and Daemen allayed suspicion that there was something special about this particular polynomial, that might provide a backdoor. Again, the design process was made transparent.

20.3.2 ShiftRows

In this step, the first row is left unchanged, but the second, third, and fourth rows have their bytes shifted left by one, two, and three bytes, respectively. The shifts are all cyclical. Denoting each

---

[33] Daemen, Joan, and Vincent Rijmen, *AES Proposal: Rijndael*, Document version 2, Date: 03/09/99, p. 25, available online at https://www.cs.miami.edu/home/burt/learning/Csc688.012/rijndael/rijndael_doc_V2.pdf. Thanks to Bill Stallings for providing this reference!

byte as $a_{i,j}$ for some $0 \leq i, j \leq 3$, we get the results shown in Table 20.1 to represent the ShiftRows operation.

Table 20.1 ShiftRows

Before		After	Change
$a_{0,0}\ a_{0,1}\ a_{0,2}\ a_{0,3}$	$\rightarrow$	$a_{0,0}\ a_{0,1}\ a_{0,2}\ a_{0,3}$	(no change)
$a_{1,0}\ a_{1,1}\ a_{1,2}\ a_{1,3}$	$\rightarrow$	$a_{1,1}\ a_{1,2}\ a_{1,3}\ a_{1,0}$	(one byte left shift)
$a_{2,0}\ a_{2,1}\ a_{2,2}\ a_{2,3}$	$\rightarrow$	$a_{2,2}\ a_{2,3}\ a_{2,0}\ a_{2,1}$	(two byte left shift)
$a_{3,0}\ a_{3,1}\ a_{3,2}\ a_{3,3}$	$\rightarrow$	$a_{3,3}\ a_{3,0}\ a_{3,1}\ a_{3,2}$	(three byte left shift)

The inverse of this step is a cyclic right shift of the rows by the same amounts.

20.3.3 MixColumns

In this step, each column of the state is viewed as a polynomial of degree 3 or less. For example, the following column

$$\begin{pmatrix} a_0 \\ a_1 \\ a_2 \\ a_3 \end{pmatrix}$$

is viewed as $a(x) = a_3x^3 + a_2x^2 + a_1x + a_0$. However, the coefficients, a_3, a_2, a_1, and a_0, are all bytes. That is, the coefficients themselves form polynomials that may be added or multiplied modulo the irreducible polynomial $x^8 + x^4 + x^3 + x + 1$ from the SubBytes step.

In the MixColumns step, each column, expressed as a polynomial, is multiplied by the polynomial $c(x) = 3x^3 + x^2 + x + 2$. It is then reduced modulo $x^4 + 1$, so that it may still be expressed as a column (i.e., a polynomial of degree 3 or smaller).

Working modulo $x^4 + 1$ is a bit different than modulo $x^8 + x^4 + x^3 + x + 1$. First of all, $x^4 + 1$ is reducible! So a randomly chosen $c(x)$ needn't be invertible. For this reason, $c(x)$ had to be chosen carefully, but how was $x^4 + 1$ chosen? It was picked so that products could be easily reduced. Moding out by $x^4 + 1$ is the same as defining $x^4 = -1$, but $-1 = 1 \pmod 2$, so we have $x^4 = 1$. This allows us to very easily reduce powers of x. We have $x^5 = x$, $x^6 = x^2$, $x^7 = x^3$, and $x^8 = x^0 = 1$. In general, $x^n = x^{n \pmod 4}$. Thus,

$$\begin{aligned} c(x)a(x) &= (3x^3 + x^2 + x + 2)(a_3x^3 + a_2x^2 + a_1x + a_0) \\ &= 3a_3x^6 + 3a_2x^5 + 3a_1x^4 + 3a_0x^3 \\ &\quad + a_3x^5 + a_2x^4 + a_1x^3 + a_0x^2 \\ &\quad + a_3x^4 + a_2x^3 + a_1x^2 + a_0x \\ &\quad + 2a_3x^3 + 2a_2x^2 + 2a_1x + 2a_0 \end{aligned}$$

reduces to

$$c(x)a(x) = 3a_3x^2 + 3a_2x + 3a_1 + 3a_0x^3$$
$$+ a_3x + a_2 + a_1x^3 + a_0x^2$$
$$+ a_3 + a_2x^3 + a_1x^2 + a_0x$$
$$+ 2a_3x^3 + 2a_2x^2 + 2a_1x + 2a_0.$$

Rearranging the terms, we have

$$c(x)a(x) = 2a_0 + 3a_1 + a_2 + a_3$$
$$+ a_0x + 2a_1x + 3a_2x + a_3x$$
$$+ a_0x^2 + a_1x^2 + 2a_2x^2 + 3a_3x^2$$
$$+ 3a_0x^3 + a_1x^3 + a_2x^3 + 2a_3x^3.$$

This operation may then be represented as

$$\begin{pmatrix} 2 & 3 & 1 & 1 \\ 1 & 2 & 3 & 1 \\ 1 & 1 & 2 & 3 \\ 3 & 1 & 1 & 2 \end{pmatrix} \times \begin{pmatrix} a_0 \\ a_1 \\ a_2 \\ a_3 \end{pmatrix}$$

This needs to be applied to every column. The multiplication is done for each pair of bytes modulo $x^8 + x^4 + x^3 + x + 1$.

However, because the matrix consists only of 1, 2, and 3, which correspond to the polynomials, 1, x, and $x + 1$, respectively, the byte multiplication is especially simple. Multiplying a byte by 1 modulo $x^8 + x^4 + x^3 + x + 1$ changes nothing. Multiplying by x amounts to a left shift of all of the bits. Multiplying by $x + 1$ is just the shift described for x, followed by an XOR with the original value. We need to be careful with the left shifts though! If the leftmost bit was already a 1, it will be shifted out and we must then XOR our result with 00011011 to compensate. This is because the x^7 bit was shifted to x^8, which cannot be represented by bits 0 through 7, but we have $x^8 = x^4 + x^3 + x + 1 \pmod{x^8 + x^4 + x^3 + x + 1}$, so the representation for $x^4 + x^3 + x + 1$ will serve.

For deciphering, one must use the inverse of $c(x)$ modulo $x^4 + 1$. This is given by

$$d(x) = 11x^3 + 13x^2 + 9x + 14.$$

20.3.4 AddRoundKey

Finally, we involve the key! This is simply an XOR (self inverse) of each byte of the state with a byte of the key for the relevant round. Each round uses a distinct key derived from the original key. This is done as follows.

First, the original key is taken 32 bits at a time and placed at the beginning of what will become the "expanded key." This expanded key will eventually be divided into equal size pieces to provide the round keys, in order. For AES-128, the original key will serve to initialize the expanded key blocks k_0, k_1, k_2, k_3. For AES-196, k_4 and k_5 will also be filled at this point; for AES-256, k_6 and k_7 will be filled. Then, more 32 bit blocks are defined recursively. The formulas for each of the three key sizes follow. They all involve a function, f, which will be detailed shortly.

For 128-bit keys:

$$k_i = k_{i-4} \oplus k_{i-1}, \text{ if } i \neq 0 \pmod 4$$
$$k_i = k_{i-4} \oplus f(k_{i-1}), \text{ if } i = 0 \pmod 4$$

For 196-bit keys:

$$k_i = k_{i-6} \oplus k_{i-1}, \text{ if } i \neq 0 \pmod 6$$
$$k_i = k_{i-6} \oplus f(k_{i-1}), \text{ if } i = 0 \pmod 6$$

where f consists of a circular left shift of 1 byte for the input, followed by a substitution using Rijndael's S-box, for each byte, and finally an XOR of this result with the appropriate round constant, RC (to be discussed).

The 256-bit case uses f, but also requires us to introduce a second function f_2.

For 256-bit keys:

$$k_i = k_{i-8} \oplus k_{i-1}, \text{ if } i \neq 0 \pmod 8 \text{ and } i \neq 4 \pmod 8$$
$$k_i = k_{i-8} \oplus f(k_{i-1}), \text{ if } i = 0 \pmod 8$$
$$k_i = k_{i-8} \oplus f_2(k_{i-1}), \text{ if } i = 4 \pmod 8$$

The function f_2 is simpler than f, as it only makes use of the S-box. The shift and XOR are omitted.

The round constants are defined as follows (for any size key).

$$RC_1 = x^0$$
$$RC_2 = x^1$$
$$RC_i = xRC_{i-1} = x^{i-1}, \text{ for } i > 2.$$

As i grows, it will become necessary to reduce the RC_i modulo $x^8 + x^4 + x^3 + x + 1$. Representing the round constants as bytes, the first few are

$$RC_1 = 00000001$$
$$RC_2 = 00000010$$
$$RC_3 = 00000100$$
$$RC_4 = 00001000$$
$$RC_5 = 00010000$$

$$RC_6 = 00100000$$

$$RC_7 = 01000000$$

$$RC_8 = 10000000$$

$$RC_9 = 00011011$$

RC_9 is the first to require reduction modulo $x^8 + x^4 + x^3 + x + 1$.

20.3.5 Putting It All Together: How AES-128 Works

We start by XORing our message with the 0th round key (just the first 128 bits of our expanded key). This is followed by nine rounds, which are identical, except for a different round key being applied each time. Each round consists of the four steps above being applied in the same order as presented. Finally, the 10th round is a bit shorter than the previous 9. The MixColumns step is skipped. That's all there is to it!

20.4 AES Attacks

In October 2000, Bruce Schneier made the incredible comment "I do not believe that anyone will ever discover an attack that will allow someone to read Rijndael traffic."[34] But by 2003, he and coauthor Niels Ferguson were putting doubts into print, "We have one criticism of AES: we don't quite trust the security."[35] They went on to explain the reason for their distrust:[36]

> What concerns us the most about AES is its simple algebraic structure. It is possible to write AES encryption as a relatively simple closed algebraic formula over the finite field with 256 elements. This is not an attack, just a representation, but if anyone can ever solve those formulas, then AES will be broken. This opens up an entirely new avenue of attack. No other block cipher we know of has such a simple algebraic representation. We have no idea whether this leads to an attack or not, but not knowing is reason enough to be skeptical about the use of AES.

They referenced a paper by Ferguson and others, detailing the simple representation of AES.[37] Nevertheless, in 2005, AES was publicly endorsed by the National Security Agency, which made it part of their "Suite B" list of recommendations.

There is, as of this writing, no practical attack against properly implemented AES. There are some theoretical attacks, but this is a completely different matter. If someone finds a way to break a cipher in 2% of the time that a brute-force attack would require, it will be of tremendous interest to cryptologists, but if that 2% still requires millions of years on the world's fastest computer,

---

[34] Schneier, Bruce, "AES Announced," *Crypto-Gram Newsletter*, October 15, 2000, http://www.schneier.com/crypto-gram-0010.html.

[35] Ferguson, Niels and Bruce Schneier, *Practical Cryptography*, Wiley, Indianapolis, Indiana, 2003, p. 56.

[36] Ferguson, Niels and Bruce Schneier, *Practical Cryptography*, Wiley, Indianapolis, Indiana, 2003, p. 57.

[37] Ferguson, Niels, Richard Schroeppel, and Doug Whiting, "A Simple Algebraic Representation of Rijndael," in Vaudenay, Serge and Amr M. Youssef, editors, *Selected Areas in Cryptography: 8th Annual International Workshop*, SAC 2001, Lecture Notes in Computer Science, Vol. 2259, Springer, New York, 2001, pp. 103–111.

it has no importance to someone simply wanting to keep his messages private. A few theoretical attacks and attacks against reduced rounds versions of AES can be found in the papers referenced at the end of this chapter.

20.5 Security Guru Bruce Schneier

Figure 20.7   Bruce Schneier (1963–), Master of Metaphors. (Photograph by Per Ervland; http://www.schneier.com/photo/.)

One of the finalists for the AES competition mentioned above was Twofish, designed by a team that included Bruce Schneier (Figure 20.7). Schneier was also the creator of Blowfish. In addition to his technical skills, Bruce has one of the most entertaining writing styles of anyone working in the field of computer security. You will see his works cited in footnotes throughout part two of this book. In recent years, Bruce has focused on practicalities of security (e.g., implementation issues, non-cryptanalytic attacks, etc.) and broad issues. His style is informal and contains numerous metaphors and examples. His monthly *Crypto-Gram* email newsletter now also exists in blog form.[38] It's packed with links to articles covering all aspects of security. Schneier was a strong critic of policies implemented under George W. Bush following 9/11. A conclusion he and coauthor Niels Ferguson drew from their years of experience is a bit unsettling:[39]

> In all our years working in this field, we have yet to see an entire system that is secure. That's right. Every system we have analyzed has been broken in one way or another.

---

[38] Schneier, Bruce, *Schneier on Security*, A blog covering security and security technology, http://www.schneier.com/. There are links here to many of his published essays, as well as links for purchasing signed copies of his books.

[39] Schneier, Bruce and Niels Ferguson, *Practical Cryptography*, Wiley, Indianapolis, Indiana, 2003, p. 1.

References and Further Reading

On Elliptic Curve Cryptography

Adleman, Leonard and Ming-Deh. Huang, "Recognizing Primes in Random Polynomial Time," in, Aho, Alfred V., editor, *STOC '87: Proceedings of the Nineteenth Annual ACM Symposium on Theory of Computing*, ACM, New York, 1987, pp. 462–469. This paper presents a primality-proving algorithm based on hyperelliptic curves.

Atkin, Arthur Oliver Lonsdale and François Morain, "Elliptic Curves and Primality Proving," *Mathematics of Computation*, Vol. 61, No. 203 (Special Issue Dedicated to Derrick Henry Lehmer), July 1993, pp. 29–68. This survey paper has 97 references. Elliptic curve primality proving is the fastest method for numbers that don't have some special form. Mersenne numbers, for example, can be tested more rapidly by a specialized algorithm. This is why we have the seemingly paradoxical situation where the quickest general method is not responsible for identifying the largest known primes (Mersenne primes).

Blake, Ian, Gadiel Seroussi, and Nigel Smart, *Elliptic Curves in Cryptography*, London Mathematical Society Lecture Note Series 265, Cambridge University Press, Cambridge, UK, 1999.

Cohen, Henri, Cryptographie, "Factorisation et Primalité: L'utilisation des Courbes Elliptiques", *Comptes Rendus de la Journée annuelle de la Société Mathématique de France*, Paris, France, January 1987.

Dimitrov Vassil S., Kimmo U. Järvinen, Michael J. Jacobson, Jr., Wai Fong Chan, and Zhun Huang, "Provably Sublinear Point Multiplication on Koblitz Curves and its Hardware Implementation," *IEEE Transactions on Computers*, Vol. 57, No. 11, November 2008, pp. 1469–1481.

Hankerson, Darrel, Alfred J. Menezes, and Scott Vanstone, *Guide to Elliptic Curve Cryptography*, Springer, New York, 2004.

Koblitz, Neal, *Introduction to Elliptic Curves and Modular Forms*, Graduate Texts in Mathematics No. 97, Springer, New York, 1984; second edition, 1993. The first edition preceded Koblitz's discovery of elliptic curve cryptography.

Koblitz, Neal, "Elliptic Curve Cryptosystems," *Mathematics of Computation*, Vol. 48, No. 177, January 1987, pp. 203–209.

Kobitz, Neal, *A Course in Number Theory and Cryptography*, Springer, New York, 1987; second edition, 1994. This was the first book on cryptography with an introduction to elliptic curves.

Koblitz, Neal, "Hyperelliptic Cryptosystems," *Journal of Cryptology*, Vol. 1, No. 3, Summer 1989, pp. 139–150. "These are curves having even higher powers of x; for example $y^2 = x^5 - x$ is such a curve. In recent years a lot of research, especially in Germany, has been devoted to hyperelliptic curve cryptosystems."[40]

Koblitz, Neal, *Algebraic Aspects of Cryptography*, Springer, New York, 1998.

Koblitz, Neal, "The Uneasy Relationship between Mathematics and Cryptography," *Notices of the AMS*, Vol. 54, No. 8, September 2007, pp. 972–979.

Koblitz, Neal, *Random Curves: Journeys of a Mathematician*, Springer, Berlin, Germany, 2008. This is Koblitz's autobiography. It is primarily concerned with radical politics. Chapter 14 is devoted to cryptography, but for those interested in politics, I recommend the entire work. On p. 312, we have, "In any case, the popularity of randomly generated elliptic curves is my justification for the title of this book."

Koblitz, Neal and Alfred J. Menezes, "Another Look at "Provable Security", *Journal of Cryptology*, Vol. 20, No. 1, January 2007, pp. 3–37. Provable security is an area of mathematics that, as the name indicates, claims to be able to prove the security of certain systems. Koblitz and Menezes reacted with this paper, which essentially observes that the Emperor has no clothes. The assault on provable security claims continued in a pair of papers in the *Notices of the AMS*, referenced in the present list.

Koblitz, Neal and Alfred J. Menezes, "The Brave New World of Bodacious Assumptions in Cryptography," *Notices of the AMS*, Vol. 57, No. 3, March 2010, pp. 357–365.

---

[40] Koblitz, Neal, *Random Curves: Journeys of a Mathematician*, Springer, Berlin, Germany, 2008, p. 313.

Koblitz, Neal, Alfred J. Menezes, and Scott Vanstone, "The State of Elliptic Curve Cryptography," *Designs, Codes and Cryptography*, Vol. 19, Nos. 2–3, March 2000, pp. 173–193.

Lenstra, Jr., Hendrik W., "Factoring Integers with Elliptic Curves," *Annals of Mathematics,* second series, Vol. 126, No. 3, November 1987, pp. 649–673. Lenstra had his result in 1984, before elliptic curves served any other cryptologic purpose. See Koblitz, Neal, *Random Curves: Journeys of a Mathematician*, Springer, Berlin, Germany, 2008, p. 299.

Menezes, Alfred J. *Elliptic Curve Public Key Cryptosystems*, Kluwer Academic Publishers, Boston, Massachusetts, 1993. This was the first book devoted completely to elliptic curve cryptography.

Miller, Victor S., "Use of Elliptic Curves in Cryptography," in Williams, Hugh C., editor, *Advances in Cryptology — CRYPTO '85 Proceedings*, Lecture Notes in Computer Science, Vol. 218, Springer, Berlin, Germany, 1986, pp. 417–426.

NSA/CSS, *The Case for Elliptic Curve Cryptography*, National Security Agency/Central Security Service, Fort George G. Meade, Maryland, January 15, 2009, https://web.archive.org/web/20090117023500/ http://www.nsa.gov/business/programs/elliptic_curve.shtml. The reasons behind the National Security Agency's endorsement of elliptic curve cryptography are detailed in this essay. The concluding paragraph states,

> Elliptic Curve Cryptography provides greater security and more efficient performance than the first generation public key techniques (RSA and Diffie-Hellman) now in use. As vendors look to upgrade their systems they should seriously consider the elliptic curve alternative for the computational and bandwidth advantages they offer at comparable security.

Rosing, Michael, *Implementing Elliptic Curve Cryptography*, Manning Publications Co., Greenwich, Connecticut, 1999.

Smart, Nigel P., "The Discrete Logarithm Problem on Elliptic Curves of Trace One," *Journal of Cryptology*, Vol. 12, No. 3, Summer 1999, pp. 193–196. This paper presents an attack against a very rare sort of elliptic curve that can easily be avoided.

Solinas, Jerome A., "An Improved Algorithm for Arithmetic on a Family of Elliptic Curves," in Kaliski, Jr., Burton S., editor, *Advances in Cryptology — CRYPTO '97 Proceedings*, Lecture Notes in Computer Science, Vol. 1294, Springer, Berlin, Germany, 1997, pp. 357–371. At the Crypto '97 conference, Solinas gave an analysis of an improved algorithm for computations on the curves Koblitz proposed. This was the first paper presented publicly at a cryptography meeting by an NSA employee.[41]

Washington, Lawrence C., *Elliptic Curves: Number Theory and Cryptography*, CRC Press, Boca Rotan, Florida, 2003.

On AES

Biham, Eli and Nathan Keller, "Cryptanalysis of reduced variants of Rijndael," *Proceedings of the Third Advanced Encryption Standard Conference*, National Institute of Standards and Technology (NIST), Washington, DC, 2000, pp. 230–241.

Biryukov, Alex, Dmitry Khovratovich, and Ivica Nikolić, "Distinguisher and related-key attack on the full AES-256," in Halevi, Shai, editor, *Advances in Cryptology —CRYPTO 2009 Proceedings*, Lecture Notes in Computer Science, Vol. 5677, Springer, Berlin, Germany, 2009, pp. 231–249. The abstract includes the following:

> Finally we extend our results to find the first publicly known attack on the full 14-round AES-256: a related-key distinguisher which works for one out of every 2^{35} keys with 2^{120} data and time complexity and negligible memory. This distinguisher is translated into a key-recovery attack with total complexity of 2^{131} time and 2^{65} memory.

---

[41] Koblitz, Neal, *Random Curves: Journeys of a Mathematician*, Springer, Berlin, Germany, 2008, p. 312.

Biryukov, Alex, Orr Dunkelman, Nathan Keller, Dmitry Khovratovich, and Adi Shamir, "Key Recovery Attacks of Practical Complexity on AES-256 Variants with up to 10 Rounds," in Gilbert, Henri, editor, *Advances in Cryptology, EUROCRYPT 2010 Proceedings*, Lecture Notes in Computer Science, Vol. 6110, Springer, Berlin, Germany, 2010, pp. 299–319. This paper presents a practical attack on 10-round AES-256. Because AES-256 has 14 rounds, this attack is not practical against real-world AES.

Biryukov, Alex and Dmitry Khovratovich, "Related-key Cryptanalysis of the Full AES-192 and AES-256," in Matsui, Mitsuru, editor, *Advances in Cryptology — ASIACRYPT 2009 Proceedings*, Lecture Notes in Computer Science, Vol. 5912, Springer, Berlin, Germany, 2009, pp. 1–18.

Bogdanov, Andrey, Dmitry Khovratovich, and Christian Rechberger, "Biclique Cryptanalysis of the Full AES," in Lee, Dong Hoon and Xiaoyun Wang, editors, *Advances in Cryptology — ASIACRYPT 2011 Proceedings*, Lecture Notes in Computer Science, Vol. 7073, Springer, Heidelberg, Germany, 2011, pp. 344–371.

Courtois, Nicolas T. and Josef Pieprzyk, "Cryptanalysis of Block Ciphers with Overdefined Systems of Equations," in Zheng, Yuliang, editor, *Advances in Cryptology — ASIACRYPT 2002 Proceedings*, Lecture Notes in Computer Science, Vol. 2501, Springer, Berlin, Germany, 2002, pp. 267–287.

Daemen, Joan and Vincent Rijmen, *"The Design of Rijndael: AES—The Advanced Encryption Standard"*, Springer, New York, 2002. This is full disclosure, a 238-page book by the creators of AES explaining exactly how the cipher was constructed and tested. Something like this should have accompanied DES.

Ferguson, Niels, John Kelsey, Stefan Lucks, Bruce Schneier, Mike Stay, David Wagner, and Doug Whiting, "Improved Cryptanalysis of Rijndael," in Schneier, Bruce, editor, *Fast Software Encryption, 7th International Workshop, FSE 2000*, Lecture Notes in Computer Science, Vol. 1978, Springer, New York, 2000, pp. 213–230.

Ferguson, Niels, Richard Schroeppel, and Doug Whiting, "A Simple Algebraic Representation of Rijndael," in Vaudenay, Serge and Amr M. Youssef, editors, *Selected Areas in Cryptography, 8th Annual International Workshop, SAC 2001*, Lecture Notes in Computer Science, Vol. 2259, Springer, New York, 2001, pp. 103–111.

Gilbert, Henri and Thomas Peyrin, *"Super-Sbox Cryptanalysis: Improved Attacks for AES-like permutations,"* in Hong, Seokhie and Tetsu Iwata, editors, *Fast Software Encryption, 17th International Workshop, FSE2010*, Springer, Berlin, Germany, 2010, pp. 365–383.

Moser, Jeff, "A Stick Figure Guide to the Advanced Encryption Standard (AES)," *Moserware*, http://www.moserware.com/2009/09/stick-figure-guide-to-advanced.html, September 22, 2009.

Musa, Mohammad A, Edward F. Schaefer, and Stephen Wedig, "A Simplified AES Algorithm and its Linear and Differential Cryptanalyses," *Cryptologia*, Vol. 27, No. 2, April 2003, pp. 148–177. This is useful for pedagogical purposes, but it must be stressed that attacks on the simplified version don't necessarily "scale up" to attacks on AES.

National Institute of Standards and Technology (NIST), Announcing the Advanced Encryption Standard (AES), Federal Information Processing Standards Publication 197, November 26, 2001, available online at http://csrc.nist.gov/publications/fips/fips197/fips-197.pdf.

Phan, Raphaël C.-W., "Impossible Differential Cryptanalysis of 7-Round Advanced Encryption Standard (AES)," *Information Processing Letters*, Vol. 91, No. 1, July 16, 2004, pp. 33–38.

Rijmen, Vincent, "Practical-Titled Attack on AES-128 Using Chosen-Text Relations," January 2010, https://eprint.iacr.org/2010/337.pdf.

Tao, Biaoshuai and Hongjun Wu, "Improving the Biclique Cryptanalysis of AES," in Foo, Ernest and Douglas Stebila, editors, *Information Security and Privacy, ACISP 2015 Proceedings*, Lecture Notes in Computer Science, Vol. 9144, Springer, Cham Switzerland, 2015, pp. 39–56.

Chapter 21

Toward Tomorrow

The history of cryptology can be broken down into eras, with the first being the paper and pencil era. Eventually, the process of encryption was mechanized and we entered the electromechanical machine era. To break these more difficult ciphers, computers were created. But the computers weren't limited to cryptanalysis; they were also used to encipher. Thus, we have the computer era. In recent years, an increasing ability to manipulate quantum particles and DNA has led to new cryptologic techniques and new kinds of computers. As a consequence, we have entered the era of post-quantum cryptology. This chapter tells the story of the birth of this era (at least as far as is publicly known) and speculates on what might come next.

21.1 Quantum Cryptography: How It Works

Quantum cryptography offers an ability to detect someone eavesdropping on a line and, if it is free of such intrusions, securely exchange a key for use with a symmetric cipher, like AES. The foundation of this approach is quantum mechanics. Thus, if there is a flaw in the approach detailed below, then there is also a flaw in the physicists' understanding of quantum theory. The process is best explained through an example.

But before we get to the example, we must understand a little about light. Light comes with a polarization, which may be in any direction. Polarized lenses, often used in sunglasses, serve as filters. Suppose the lens is designed to allow light polarized horizontally to pass through. Then all such photons pass through unchanged. At the other extreme, photons polarized vertically will always be absorbed. Photons polarized at any other angle, θ, relative to the horizontal, will pass through with probability $\cos^2(\theta)$; however, those that make it through will not pass unchanged. On the far side of the filter, the photons will emerge with their polarizations changed to horizontal. Filters are needed to measure polarizations, but this act of measuring may alter the polarization, as described above. It is this feature that makes quantum cryptography possible.

We denote horizontal and vertical polarizations by — and |, respectively. In a similar manner, two of the possible diagonal polarizations are represented by / and \. Now, on to our example.

If Alice wishes to establish a key with Bob for future messages, without meeting him in person, she begins by generating a random string of 0s and 1s and a random string of +s and ×s.[1] For example, such a string might begin

```
0  1  1  0  1  0  1  1  1  0  0  1  0  0  1  0  1  1  0  1
+  ×  +  +  ×  ×  ×  +  ×  +  ×  +  +  ×  +  +  ×  +  ×  ×
```

She will now polarize photons to represent each 0 and 1. The + signs indicate she polarizes those particular photons in either the | direction or the − direction. She will use | to represent 1 and − to represent 0. If a particular bit has an × under it, Alice will use \ to represent 1 and / to represent 0. Adding a third line to our previous list of bits and polarization schemes, we show what Alice actually sends (photons with the indicated polarizations):

```
0  1  1  0  1  0  1  1  1  0  0  1  0  0  1  0  1  1  0  1
+  ×  +  +  ×  ×  ×  +  ×  +  ×  +  +  ×  +  +  ×  +  ×  ×
−  \  |  −  \  /  \  |  \  −  /  |  −  /  |  −  \  |  /  \
```

On the receiving end, Bob sets up a filter for each photon in an attempt to determine its polarization. If he sets his filter up like so |, he'll correctly interpret any photons sent using the + scheme. The | photons will come through and the — photons will reveal themselves by not coming through! Similarly, setting his filter up as — will correctly identify all photons sent according to the + scheme. However, photons sent using the × scheme will come through with probability ½. Once through, they will appear to have the orientation of the filter; thus, Bob only has a fifty percent chance of guessing these correctly.

Similarly, if he uses one of the filters from the × scheme, he'll correctly identify the polarizations of all photons sent according to the × scheme, but err on average on half of the rest. Remember, Bob doesn't know which scheme, + or ×, Alice used for any particular photon; he must guess!

We now add a fourth line to our diagram showing what sort of filter Bob used for each photon.

```
0  1  1  0  1  0  1  1  1  0  0  1  0  0  1  0  1  1  0  1
+  ×  +  +  ×  ×  ×  +  ×  +  ×  +  +  ×  +  +  ×  +  ×  ×
−  \  |  −  \  /  \  |  \  −  /  |  −  /  |  −  \  |  /  \
×  +  +  ×  ×  +  ×  +  ×  ×  +  ×  +  ×  ×  +  +  ×  +  ×
```

Of course, Bob has no way of knowing which recovered bits are correct and which aren't. He calls Alice and tells her what scheme he used for each. In our example, she would tell him that he guessed correctly for positions 3, 5, 7, 8, 9, 13, 14, 16, and 20. It's okay if someone is listening in at this stage. Alice and Bob then discard all positions for which Bob guessed incorrectly. The bits that remain will serve as their key. Thus, their key begins 111110001. These digits should not be revealed; Alice only confirms that Bob guessed the correct filtering scheme for certain positions. This being the case, the pair knows that Bob has correctly recovered the bits in those positions. He may have correctly recovered other bits by chance, but those are ignored. Now suppose someone, say Eve, was eavesdropping on the first communication, when Alice sent the photons. To eavesdrop, Eve would have had to set up filters to measure the photon polarizations. There is no other way to obtain the information she seeks. But Eve's filters will have changed the polarizations on the photons for which she guessed the wrong scheme. These changes would carry through to Bob.

---

[1] Although true randomness is desired, these strings may be pseudorandom in practice.

Thus, to make sure Eve wasn't listening in, Alice and Bob spot check some positions. Alice might ask, "For positions 7, 13, and 16 did you get 0, 1, and 0?" If Bob answers affirmatively, they gain confidence that there was no eavesdropper, discard the bits revealed and use the rest as their key.

To make the example above fit on a single line, only 20 of the photons sent were detailed. In a real implementation, Alice would have to send many more. If she hopes to establish a 128-bit key, sending 256 photons wouldn't be enough. Bob may guess the correct filtering scheme for half of the photons he receives, but no room would be left to check for an eavesdropper. Alice would be wiser to send 300 bits, allowing room for Bob to be an unlucky guesser, as well as allowing enough randomly checked values to determine with a high degree of certainty that no one was listening in.

It may happen that Eve guessed correctly on the filter orientations for all of the bits Alice and Bob used to check for her presence on the line, but the probability of her managing that for n bits is only $(1/2)^n$; however, the probability of her eavesdropping going undetected is a bit higher. She'll guess the correct orientation half the time, but even when she guesses wrong, she'll get the correct value half the time by chance; hence, she draws the correct conclusion 3/4 of the time. Thus, by using n check bits, Alice and Bob reduce the probability of Eve going unnoticed to $(3/4)^n$, which can still be made as small as they desire.

21.2 Quantum Cryptography: Historical Background

The idea outlined above was discovered by Charles H. Bennett (Figure 21.1) and Gilles Brassard (Figure 21.2). These men were inspired by Stephen Wiesner, who had previously used the inability to measure polarizations without risking changes to describe a form of uncounterfeitable currency. Wiesner's idea was so far ahead of its time that it took him many years to find someone willing to publish his paper. The paper finally appeared in 1983,[2] but was written circa 1970. Its publication was triggered by the appearance of another paper, one he wrote with Bennett, Brassard, and Seth Breidbart that introduced the term *quantum cryptography*, namely "Quantum Cryptography, or Unforgeable Subway Tokens."[3]

These papers mark historic milestones, but the idea presented at the start of this chapter hadn't yet appeared. To get to that point in the history, we must recount another rejection first. Brassard recalls:[4]

> At first, we wanted the quantum signal to encode the transmitter's confidential message in such a way that the receiver could decode it if no eavesdropper were present, but any attempt by the eavesdropper to intercept the message would spoil it without revealing any information. Any such futile attempt at eavesdropping would be detected by the legitimate receiver, alerting him to the presence of the eavesdropper. Since this

---

[2] A minor theme in this book, an important paper that was initially rejected: Wiesner, Stephen, "Conjugate Coding," *SIGACT News*, Vol. 15, No. 1, Winter-Spring 1983, pp. 78–88.

[3] Bennett, Charles H., Gilles Brassard, Seth Breidbart, and Stephen Wiesner, "Quantum Cryptography, or Unforgeable Subway Tokens," in Chaum, David, Ronald L. Rivest, and Alan T. Sherman, editors, *Advances in Cryptology, Proceedings of Crypto '82*, Plenum Press, New York, 1983, pp. 267–275. This is the first published paper on Quantum Cryptography. Indeed, the first paper in which those words were even put together.

[4] Brassard, Gilles, "Brief History of Quantum Cryptography: A Personal Perspective," in *Proceedings of IEEE Information Theory Workshop on Theory and Practice in Information Theoretic Security*, Awaji Island, Japan, October 17, 2005, pp. 19–23. A longer (14 pages) version is available online at http://arxiv.org/pdf/quant-ph/0604072v1.pdf. The passage here is taken form page 4 of the online version.

Figure 21.1   Charles H. Bennett (1943–). (Courtesy of Charles H. Bennett.)

Figure 21.2   Gilles Brassard (1955–). (Courtesy of Giles Brassard; http://www.iro.umontreal. ca/~brassard/photos/.)

early scheme was unidirectional, it required the legitimate parties to share a secret key, much as in a one-time pad encryption. The originality of our scheme was that the same one-time pad could be reused safely over and over again if no eavesdropping were detected. Thus, the title of our paper was "Quantum Cryptography II: How to reuse a one-time pad safely even if $\mathbf{P} = \mathbf{NP}$."[5] We submitted this paper to major theoretical computer science conferences, such as STOC (The *ACM Annual Symposium on Theoretical Computer Science*), but we failed to have it accepted. Contrary to Wiesner's "Conjugate Coding", however, our "Quantum Cryptography II" paper has forever remained unpublished (copies are available from the authors).

Undeterred by the rejection, Bennett and Brassard came up with a new way of doing things (the scheme presented at the start of this chapter) and gave a long presentation at the 1983 IEEE

---

[5] Bennett, Charles H., Gilles Brassard, and Seth Breidbart, "Quantum Cryptography II: How to reuse a one-time pad safely even if $\mathbf{P} = \mathbf{NP}$," paper rejected from 15th Annual ACM Symposium on Theory of Computing, Boston, May 1983. Historical document dated "November 1982" available from the first two authors.

Symposium on Information Theory (ISIT), which was held in St-Jovite, Canada (near Brassard's hometown of Montréal). Brassard notes that the one-page abstract that was published[6] provides the official birth certificate for Quantum Key Distribution.[7]

Seeing print and making an impact can be two very different things. Although Wiesner, Bennett, and Brassard were now getting their ideas published, hardly anyone was taking notice. Well, even the best researchers can benefit from social networking. Brassard got an assist when his good friend Vijay Bhargava invited him to give a talk on whatever he wanted (!) at an IEEE conference to be held in Bangalore, India, in December 1984. Brassard accepted the invitation and gave a talk on quantum cryptography. The associated five-page paper, "Quantum Cryptography: Public key Distribution and Coin Tossing," authored by Bennett and Brassard,[8] has, as of this writing (October 13, 2020) earned 8,417 citations according to Google Scholar.

As a side-note, when I ask students to write papers in my cryptology class, someone always asks how long it has to be. Well, a five-page paper could be acceptable… The 1953 paper by Crick and Watson[9] that described the double helical structure of DNA for the first time was only two pages long. So, I could be content with a two page paper… Most students are used to writing to a given length, rather than writing until the topic has been completely covered in as clear a manner as possible. They usually keep asking after I make comments like those above.

Back to Bennett and Brassard! The 8,046 citations referred to above are strongly skewed to more recent years. Brassard noted, "Throughout the 1980's, very few people took quantum cryptography seriously and most people simply ignored it."[10] In fact, in 1987, Doug Wiedemann had a paper published in *Sigact News* that presented the exact same scheme as in the 1984 paper by Bennett and Brassard. He even called it quantum cryptography![11] So, there was someone deeply interested in the topic who didn't know about Bennett and Brassard's work — nor, apparently, did the editor who published the reinvention! One wonders who the reviewers were.

The two researchers decided that they needed to physically demonstrate their scheme to gain some attention. They recruited some help and (without any special budget) had their secret

6   Bennett, Charles H. and Gilles Brassard, "Quantum Cryptography and its Application to Provably Secure Key Expansion, Public-key Distribution, and Cointossing," in *Proceedings of IEEE International Symposium on Information Theory (abstracts)*, St-Jovite, Quebec, Canada, IEEE, New York, 1983, p. 91.

7   Brassard, Gilles, "Brief History of Quantum Cryptography: A Personal Perspective," in *Proceedings of IEEE Information Theory Workshop on Theory and Practice in Information Theoretic Security*, Awaji Island, Japan, October 17, 2005, pp. 19–23. A longer, 14-page version is available online at http://arxiv.org/pdf/quant-ph/0604072v1.pdf.

8   Bennett, Charles H. and Gilles Brassard, "Quantum cryptography: Public key Distribution and coin tossing," in *International Conference on Computers, Systems & Signal Processing*, Vol. 1, Bangalore, India, pp. 175–179, December 1984, available online at https://arxiv.org/ftp/arxiv/papers/2003/2003.06557.pdf.

9   Watson James D. and Crick Francis H. C., "A Structure for Deoxyribose Nucleic Acid," *Nature*, Vol. 171, No. 4356, April 25, 1953, pp. 737–738. This paper was really a single page. Acknowledgments and references caused it to spill over a bit onto a second page.

10 Brassard, Gilles, "Brief History of Quantum Cryptography: A Personal Perspective," in *Proceedings of IEEE Information Theory Workshop on Theory and Practice in Information Theoretic Security*, Awaji Island, Japan, October 17, 2005, pp. 19–23. A longer, 14-page version is available online at http://arxiv.org/pdf/quant-ph/0604072v1.pdf. The passage here is taken from page 5 of the online version.

11 Wiedemann, Doug, "Quantum cryptography," *Sigact News*, Vol. 18, No. 2, 1987, pp. 48–51.

quantum transmission system, over a distance of 32.5 centimeters, working in October 1989. Brassard recalls,[12]

> The funny thing is that, while our theory had been serious, our prototype was mostly a joke. Indeed, the largest piece in the prototype was the power supply needed to feed in the order of one thousand volts to Pockels cells, used to turn photon polarization. But power supplies make noise, and not the same noise for the different voltages needed for different polarizations. So, we could literally hear the photons as they flew, and zeroes and ones made different noises. Thus, our prototype was unconditionally secure against any eavesdropper who happened to be deaf! : -)

Despite the noise, this demonstration marked the turning point for Bennett and Brassard. Physicists were now interested. Physicist Artur K. Ekert found a different way to accomplish the quantum key distribution. He used quantum entanglement, rather than polarity and published his result in a physics journal, which helped spread the idea of quantum cryptography more broadly.[13] The results thus far were even featured in the popular journal *Scientific American*.[14]

Today, there is tremendous worldwide interest in the field, and experiments that transmit actual quantum bits are constantly setting new records in terms of distance. As mentioned above, Bennett and Brassard began in 1989 by measuring distances in centimeters, as photons were sent from a machine to another machine right next to it, and progressed from there to distances around 100 kilometers. Many experts thought this was about the limit, without the use of "quantum repeaters" that could serve as relays, but how could a signal be repeated, if the act of reading it changed it? In 2002, Brassard observed that,[15]

> A repeater that doesn't measure was thought to be impossible in the early 1980s, but since then scientists have shown that it is feasible in principle. But we're nowhere near the technology to build one.

In late October 2010, researchers from Georgia Institute of Technology presented a breakthrough at the annual meeting of the Optical Society of America (OSA). The team had built a quantum repeater that could allow quantum bits to be sent distances of 1,000 kilometers or more.[16]

In the meanwhile, the record for open air transmission of photons hit 144 kilometers, as a key was sent from one of the Canary Islands to another. The team initially used bursts of photons but succeeded with single photons in 2007.[17] This is a very important distinction. Many experiments

---

[12] Brassard, Gilles, "Brief History of Quantum Cryptography: A Personal Perspective," in *Proceedings of IEEE Information Theory Workshop on Theory and Practice in Information Theoretic Security*, Awaji Island, Japan, October 17, 2005, pp. 19–23. A longer, 14-page version is available online at http://arxiv.org/pdf/quant-ph/0604072v1.pdf. The passage here is taken from page 6 of the online version.

[13] Ekert, Artur K., "Quantum Cryptography Based on Bell's Theorem," *Physical Review Letters*, Vol. 67, No. 6, August 5, 1991, pp. 661–663.

[14] Bennett, Charles H., Gilles Brassard and Artur K. Ekert, "Quantum Cryptography," *Scientific American*, Vol. 267, No. 4, October 1992, pp. 50–57.

[15] Quoted in Klarreich, Erica, "Can You Keep a Secret?" *Nature*, Vol. 418, No. 6895, July 18, 2002, pp. 270–272.

[16] Anon., "Long Distance, Top Secret Messages: Critical Component of Quantum Communication Device May Enable Cryptography," *ScienceDaily*, October 20, 2010, http://www.sciencedaily.com/releases/2010/10/101019171803.htm.

[17] Brumfiel, Geoff, "Quantum Cryptography Goes Wireless," *Nature*, published online March 8, 2007, http://www.nature.com/news/2007/070305/full/070305-12.html.

used groups of photons, all having the same polarizations, to represent the individual bits. This makes the system more reliable, but defeats the security of the theoretical model.

On October 21, 2009, quantum key distribution was used to transmit votes securely in a Swiss election.[18]

At the risk of sounding like a travel log, I now present a result from Japan. In 2011, a paper titled "Field test of quantum key distribution in the Tokyo QKD Network" with 46 authors appeared in *Optics Express*.[19] It described a QKD network that achieved quantum OTP (One-Time Pad) encryption for distances up to 135 km at a fast enough rate to allow video conferencing and mobile telephony. The demonstration made in October 2010 included (intentionally) an eavesdropper. The system detected the eavesdropper's presence and rerouted to cut that person out, with no noticeable interruption for the participants (a buffer stored enough key to cover the communication until the rerouting was completed). The authors noted, "These demonstrations suggest that practical applications of QKD in a metropolitan network may be just around the corner." They also noted how such networks could be compromised.

> Many QKD protocols, such as the one-way BB84 protocol, have been proven to be unconditionally secure, which means the *protocol*, which is based on mathematical device model assumptions, cannot be "cracked" as long as the laws of physics remain true. On the other hand real world implementations have unavoidable imperfections and will therefore be susceptible to side-channel attacks.[20]

It's very important not to underestimate the threat of side-channel attacks. They have been growing in importance for decades.

In 2016, China set up a 2,000 km long quantum channel linking Beijing and Shanghai. This long run actually has 32 "trusted nodes" along the way to refresh the signal. These are potential weak spots.[21] The Chinese also launched a satellite in 2016 for the purpose of establishing a link for quantum key distribution in orbit. By 2020, it was successfully exchanging key with a portable station (weighing only 80 kg!) on the ground. This is not just a proof of concept. Industrial and

[18] "Geneva is Counting on Quantum Cryptography as it Counts its Votes," October 11, 2007, https://cordis.europa.eu/docs/projects/cnect/3/506813/080/publishing/readmore/SECOQC-pressrelease2.pdf.

[19] Sasaki, M., M. Fujiwara, H. Ishizuka, W. Klaus, K. Wakui, M. Takeoka, S. Miki, T. Yamashita, Z. Wang, A. Tanaka, K. Yoshino, Y. Nambu, S. Takahashi, A. Tajima, A. Tomita, T. Domeki, T. Hasegawa, Y. Sakai, H. Kobayashi, T. Asai, K. Shimizu, T. Tokura, T. Tsurumaru, M. Matsui, T. Honjo, K. Tamaki, H. Takesue, Y. Tokura, J. F. Dynes, A. R. Dixon, A. W. Sharpe, Z. L. Yuan, A. J. Shields, S. Uchikoga, M. Legré, S. Robyr, P. Trinkler, L. Monat, J.-B. Page, G. Ribordy, A. Poppe, A. Allacher, O. Maurhart, T. Länger, M. Peev, and A. Zeilinger, "Field test of quantum key distribution in the Tokyo QKD Network" *Optics Express*, Vol. 19, No. 11, 2011, pp. 10387–10409.

[20] Sasaki, M., M. Fujiwara, H. Ishizuka, W. Klaus, K. Wakui, M. Takeoka, S. Miki, T. Yamashita, Z. Wang, A. Tanaka, K. Yoshino, Y. Nambu, S. Takahashi, A. Tajima, A. Tomita, T. Domeki, T. Hasegawa, Y. Sakai, H. Kobayashi, T. Asai, K. Shimizu, T. Tokura, T. Tsurumaru, M. Matsui, T. Honjo, K. Tamaki, H. Takesue, Y. Tokura, J. F. Dynes, A. R. Dixon, A. W. Sharpe, Z. L. Yuan, A. J. Shields, S. Uchikoga, M. Legré, S. Robyr, P. Trinkler, L. Monat, J.-B. Page, G. Ribordy, A. Poppe, A. Allacher, O. Maurhart, T. Länger, M. Peev, and A. Zeilinger, "Field test of quantum key distribution in the Tokyo QKD Network" *Optics Express*, Vol. 19, No. 11, 2011, pp. 10387–10409.

[21] Courtland, Rachel, "China's 2,000-km Quantum Link Is Almost Complete," *IEEE Spectrum*, October 26, 2016, available online at https://spectrum.ieee.org/telecom/security/chinas-2000km-quantum-link-is-almost-complete.

Commercial Bank of China (ICBC) and the People's Bank of China use the system, although with heavier, but faster, ground stations.[22]

Quantum particles aren't just for defense, they can also be used to attack ciphers when they are doing their thing in a quantum computer.

21.3 Quantum Computers and Quantum Distributed Key Networks

Traditional computers operate using bits that are either 0 or 1. In the old days, these 0s and 1s were represented by vacuum tubes that were either OFF or ON. The space needed to store bits was reduced dramatically over the decades and the tubes are long gone. But a further reduction has recently been made. In quantum computers it is actually quantum particles that are used to represent the bits. However, there is a fundamental difference. It is not just a matter of a smaller size. Quantum bits, or qubits (pronounced "cue bits") for short, can be 0, 1, or *both*. A description of how quantum computers work is well outside the scope of this book. What's relevant is that these machines can solve some problems that traditional computers cannot and can solve other problems far faster. For example, there's no known polynomial time algorithm for factoring, using a traditional computer, but there is one for a quantum computer. It dates back to 1994 and is known as Shor's algorithm, after Peter Shor who was employed by Bell Labs at the time.[23] Shor also found a polynomial time algorithm for solving the discrete log problem on a quantum computer. As a consequence, RSA, Diffie-Hellman, and elliptic curve cryptography are all vulnerable.

It's not just public key systems that are at risk. Grover's algorithm, discovered by Lov Grover, an Indian-American computer scientist, in 1996,[24] can be used to reduce the number of trials needed to brute-force a symmetric block cipher with an n bit key from 2^n to $2^{n/2}$ on a quantum computer.[25] The ability of a qubit to be both 0 and 1 allows many keys to be tested simultaneously.

The October 23, 2019 issue of *Nature* contained a paper by 77 authors (representing Google). The abstract include a dramatic summary of the power of a quantum computer with 53 qubits:

> Our Sycamore processor takes about 200 seconds to sample one instance of a quantum circuit a million times—our benchmarks currently indicate that the equivalent task for a state-of-the-art classical supercomputer would take approximately 10,000 years. This dramatic increase in speed compared to all known classical

---

[22] Lu, Donna, "China has developed the world's first mobile quantum satellite station," *New Scientist*, January 9, 2020, available online at https://www.newscientist.com/article/2229673-china-has-developed-the-worlds-first-mobile-quantum-satellite-station/#.

[23] Shor, P. W., "Algorithms for quantum computation: discrete logarithms and factoring," in Goldwasser, Shafi, editor, *Proceedings 35th Annual Symposium on Foundations of Computer Science*, IEEE Computer Society Press, Los Alamitos, California, 1994, pp. 124–134.

[24] Grover, Lov K., "A fast quantum mechanical algorithm for database search," in Miller, Gary L., editor, *Proceedings of 28th ACM Symposium on Theory of Computing (STOC '96)*, ACM Press, New York, 1996, pp. 212–219.

[25] Bennett, Charles H., Ethan Bernstein, Gilles Brassard, and Umesh Vazirani, "Strengths and Weaknesses of Quantum Computing," *SIAM Journal on Computing*, Vol. 26, No. 5, October 1997, pp. 1510–1523, available online at https://arxiv.org/pdf/quant-ph/9701001.pdf.

algorithms is an experimental realization of quantum supremacy for this specific computational task, heralding a much anticipated computing paradigm.[26]

IBM, a competitor in the quantum computer development race, objected to this claim, saying that the time on a state-of-the-art classical supercomputer is 2.5 days, not 10,000 years.[27] As of May 2020, IBM has 18 quantum computers, Honeywell has 6, and Google has 5.[28]

One way to protect communications against such new machines (as well as improved versions, yet to be, that will make these look like toys) is by setting up a quantum key distribution network, as described earlier in this chapter. Another is to replace current algorithms with ones believed to be able to resist quantum computer attacks. The next two sections detail how NSA and NIST are slowly prodding people in this direction.

21.4 NSA Weighs In

In Chapter 20, some algorithms recommended by the National Security Agency as part of their "Suite B Cryptography" were detailed. In light of the threat of quantum computers, NSA introduced the "Commercial National Security Algorithm Suite (CNSA Suite)" on August 19, 2015. These algorithms were only intended as a stopgap measure. The agency promised, "IAD [Information Assurance Directorate] will initiate a transition to quantum resistant algorithms in the not too distant future."[29] NSA also gave advice for those who were a bit behind and had not upgraded to Suite B:

> Until this new [quantum resistant algorithms] suite is developed and products are available implementing the quantum resistant suite, we will rely on current algorithms. For those partners and vendors that have not yet made the transition to Suite B elliptic curve algorithms, we recommend not making a significant expenditure to do so at this point but instead to prepare for the upcoming quantum resistant algorithm transition.[30]

The CNSA Suite did not contain any new algorithms. The list had the old popular schemes like AES, Elliptic Curve schemes, SHA, Diffie-Hellman, and RSA. That is RSA was placed in higher esteem than in Suite B and DSA was dropped. The main difference in the retained algorithms

---

[26] Arute, Frank, Kunal Arya, […], and John M. Martinis "Quantum Supremacy Using a Programmable Superconducting Processor," *Nature*, Vol. 574, No. 7779, October 24, 2019, pp. 505–510, available online at https://www.nature.com/articles/s41586-019-1666-5. I hope the 74 authors I represented by […] will forgive me.

[27] Pednault, Edwin, John Gunnels, Dmitri Maslov, and Jay Gambetta, "On "Quantum supremacy"," *IBM Research Blog*, October 21, 2019, available online at https://www.ibm.com/blogs/research/2019/10/on-quantum-supremacy/.

[28] Shankland, Stephen, "IBM now has 18 quantum computers in its fleet of weird machines," *c|net*, May 6, 2020, https://www.cnet.com/news/ibm-now-has-18-quantum-computers-in-its-fleet-of-weird-machines/.

[29] "Commercial National Security Algorithm Suite," National Security Agency | Central Security Service, August 19, 2015, https://apps.nsa.gov/iaarchive/programs/iad-initiatives/cnsa-suite.cfm.

[30] "Commercial National Security Algorithm Suite," National Security Agency | Central Security Service, August 19, 2015, https://apps.nsa.gov/iaarchive/programs/iad-initiatives/cnsa-suite.cfm.

was that the key sizes were much larger. For example, for Diffie-Hellman key exchange, it was "Minimum 3072-bit modulus to protect up to TOP SECRET."[31]

The other newsworthy update was expressed as follows:

> Unfortunately, the growth of elliptic curve use has bumped up against the fact of continued progress in the research on quantum computing, which has made it clear that elliptic curve cryptography is not the long term solution many once hoped it would be. Thus, we have been obligated to update our strategy.[32]

These lines led to much speculation, a summary of which was presented in a paper by Neal Koblitz, a co-discoverer of elliptic curve cryptography, and Alfred J. Menezes.[33] In an email to me, Koblitz noted, "It's interesting that one of the leading contenders for "post-quantum cryptography" is based on elliptic curves, but in a totally different way from ECC. This is the "isogeny-based" approach of Jao and others."[34]

21.5 NIST Responds

The National Institute of Standards and Technology (NIST) responded to the threat of quantum computers in its usual way. On December 20, 2016, the organization announced a competition.[35] Dustin Moody, a NIST mathematician, explained

> We're looking to replace three NIST cryptographic standards and guidelines that would be the most vulnerable to quantum computers. They deal with encryption, key establishment and digital signatures, all of which use forms of public key cryptography.[36]

All of the submitters whose entrees met NIST's acceptability requirements would be invited to present their algorithms at a workshop in early 2018. NIST planned for this to be followed by an evaluation phase that would "take an estimated three to five years" to complete.[37] We are still in this multi-round evaluation phase, as of this writing. Of the original 82 submission received by the November 30, 2017 deadline, 26 made it to round 2. These semi-finalists were announced

---

[31] "Commercial National Security Algorithm Suite," National Security Agency | Central Security Service, August 19, 2015, https://apps.nsa.gov/iaarchive/programs/iad-initiatives/cnsa-suite.cfm.

[32] "Commercial National Security Algorithm Suite," National Security Agency | Central Security Service, August 19, 2015, https://apps.nsa.gov/iaarchive/programs/iad-initiatives/cnsa-suite.cfm.

[33] Koblitz, Neal and Alfred J. Menezes, "A Riddle Wrapped in an Enigma," *IEEE Security & Privacy*, Vol. 14, No. 6, November–December 2016, pp. 34–42, available online at https://eprint.iacr.org/2015/1018.pdf.

[34] Koblitz, Neal, email to the author, January 5, 2020.

[35] See NIST Asks Public to Help Future-Proof Electronic Information, NIST News, December 20, 2016, updated January 8, 2018, https://www.nist.gov/news-events/news/2016/12/nist-asks-public-help-future-proof-electronic-information and Kimball, Kevin, National Institute of Standards and Technology, "Announcing Request for Nominations for Public-Key Post-Quantum Cryptographic Algorithms," *Federal Register*, Vol. 81, No. 244, December 20, 2016, pp. 92787–92788, https://www.federalregister.gov/documents/2016/12/20/2016-30615/announcing-request-for-nominations-for-public-key-post-quantum-cryptographic-algorithms.

[36] NIST Asks Public to Help Future-Proof Electronic Information, NIST News, December 20, 2016, updated January 8, 2018, https://www.nist.gov/news-events/news/2016/12/nist-asks-public-help-future-proof-electronic-information.

[37] NIST Asks Public to Help Future-Proof Electronic Information, NIST News, December 20, 2016, updated January 8, 2018, https://www.nist.gov/news-events/news/2016/12/nist-asks-public-help-future-proof-electronic-information.

on January 30, 2019.[38] On July 22, 2020, NIST announced seven third-round finalists and eight alternates.[39] Which will win is far from obvious. Given this, and the quote from Neal Koblitz and Alfred J. Menezes that follows, I think singling one out to detail here would be inappropriate.

> Most quantum-resistant systems that have been proposed are complicated, have criteria for parameter selection that are not completely clear, and in some cases (such as NTRU) have a history of successful attacks on earlier versions.[40]

21.6 Predictions

Over the course of this text we've seen several inaccurate predictions made by very clever and successful individuals (such as Alan Turing, Martin Gardner, Gilles Brassard). Because turnaround is fair play, I also made a prediction in the first edition. It was "By 2040 quantum computers will have become a reality necessitating a complete rethinking of encryption." This prediction included a footnote that added "I actually think it will happen sooner. I picked a year far enough away to give me a wide safety margin, yet still within my expected lifetime, so I can receive criticism in person, if I'm wrong."

Looking back, this prediction doesn't seem very bold. Here's my new (bolder) prediction for the second edition: By 2050 a computer than bends spacetime will be a reality (if it isn't already, somewhere).

In the meanwhile, let's consider another new type of computer.

21.7 DNA Computing

We now consider another new form of computing that could have a devastating impact on many current encryption algorithms. This tale begins with a familiar face.

Len Adleman (Figure 21.3), the "A" in RSA has another claim to fame. He's the one who came up with the name "computer virus" for those pesky programs that are the bane of everyone who owns a machine, from those who only use it to browse the Internet to the mathematical maestros at the National Security Agency. His graduate student, Fred Cohen, released the first such virus (under carefully controlled conditions!) in 1983.[41]

Should we have been nervous when Adleman decided to try mixing computing science and biology once again in the 1990s? He found biology hard to resist, as it was now becoming "mathematized," as he put it.[42]

> When I was an undergraduate in the '60s, I thought biology was stuff that smelled funny in the refrigerator. Now, biology is finite strings over a four-letter alphabet and functions performed by enzymes on these strings.

---

[38] Alagic, Gorjan, Jacob Alperin-Sheriff, Daniel Apon, David Cooper, Quynh Dang, Carl Miller, Dustin Moody, Rene Peralta, Ray Perlner, Angela Robinson, Daniel Smith-Tone, and Yi-Kai Liu, NISTIR 8240, Status Report on the First Round of the NIST Post-Quantum Cryptography Standardization Process, NIST Information Technology Laboratory, Computer Security Resource Center, Publications, January 2019, https://csrc.nist.gov/publications/detail/nistir/8240/final.

[39] PQC Standardization Process: NIST, Third Round Candidate Announcement, July 22, 2020, https://csrc.nist.gov/News/2020/pqc-third-round-candidate-announcement.

[40] Cook, John D., "Between now and quantum," *John D. Cook Consulting Blog*, May 23, 2019, https://www.johndcook.com/blog/2019/05/23/nsa-recommendations/, which took it from Koblitz, Neal and Alfred J. Menezes, "A Riddle Wrapped in an Enigma," *IEEE Security & Privacy*, Vol. 14, No. 6, November–December 2016, pp. 34–42, available online at https://eprint.iacr.org/2015/1018.pdf.

[41] Bass, Thomas A., "Gene Genie," *Wired*, Vol. 3, No. 8, August 1995, pp. 114–117, 164–168.

[42] Bass, Thomas A., "Gene Genie," *Wired*, Vol. 3, No. 8, August 1995, pp. 114–117, 164–168.

Figure 21.3   Len Adleman (1945–). (Courtesy of Len Adleman; http://usc.edu/dept/molecular-science/fm-adleman.htm.)

His new insight for the 1990s was that DNA could take the place of traditional computing means. It would be well suited for calculations that can be made through the use of parallel processing to a massive degree. Adleman demonstrated the idea of DNA computing by solving an instance of the (noncryptographic) directed Hamiltonian path problem. It is illustrated in Figure 21.4 with the graph he used.

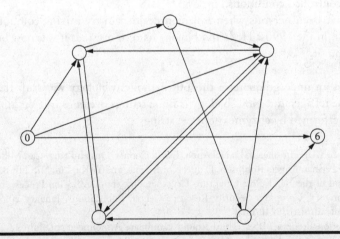

Figure 21.4   Solving an instance of the directed Hamiltonian path problem.

You may imagine the circles as locations on a map. They are called *vertices* (or *nodes* or *points*). The lines connecting them may be thought of as roads connecting the points of interest. They are usually called *edges*. Some of the edges are one-way, in the directions indicated by the arrows. In some cases, there is a separate edge, connecting the same vertices, but in the opposite direction, offering the traveler a simple way back. The challenge is, given a starting vertex v_{in} and an ending vertex v_{out}, to find a path that passes through all of the other vertices exactly once. The path needn't make use of every edge. Such paths do not exist for every directed graph. If one is present, it is called a *Hamiltonian path* after William Rowan Hamilton (1805–1865). Take a few moments to find a Hamiltonian path for the graph in Figure 21.4 with $v_{in} = 0$ and $v_{out} = 6$. Once you have found it, label the intermediate points 1, 2, 3, 4, 5, in the order your solution passes through them. For this particular problem, the solution is unique. Some graphs have several distinct solutions for a given starting point and ending point.

This simple sounding problem is **NP**-complete. We can solve small examples by hand, but there is no known polynomial time solution for such problems in general.

To find the solution you discovered, Adleman began by randomly choosing bases to form strands of DNA, labeled O_0, O_1, O_2, O_3, O_4, O_5, and O_6. Known as *oligonucleotides* (hence, the O used in our notation), these strands are only half of the DNA "ladder" we usually picture when imagining DNA. A few sample values are given below:

$$O_2 = \text{TATCGGATCGGTATATCCGA}$$
$$O_3 = \text{GCTATTCGAGCTTAAAGCTA}$$
$$O_4 = \text{GGCTAGGTACCAGCATGCTT}$$

We may then form the Watson-Crick complement of each of these, using the fact that the complement of A is T and the complement of G is C.

$$\overline{O}_2 = \text{ATAGCCTAGCCATATAGGCT}$$
$$\overline{O}_3 = \text{CGATAAGCTCGAATTTCGAT}$$
$$\overline{O}_4 = \text{CCGATCCATGGTCGTACGAA}$$

There is nothing special about the choices used for the bases or their order. All that matters is that we have as many strands as there are vertices in our graph. Random strings of A, C, G, and T will suffice. However, there is no flexibility in forming the complements (above) or in selecting the bases used to represent the edges (below).

$$O_{2 \to 3} = \text{GTATATCCGAGCTATTCGAG}$$
$$O_{3 \to 4} = \text{CTTAAAGCTAGGCTAGGTAC}$$

Edge $O_{i \to j}$ is created by taking the second half of O_i and appending to it the first half of O_j. These edges are defined so that they may join (by bonding) the oligonucleotides representing the vertices, as shown below:

(edge $O_{2 \to 3}$ joined end to end with edge $O_{3 \to 4}$)
GTATATCCGAGCTATTCGAGCTTAAAGCTAGGCTAGGTAC
CGATAAGCTCGAATTTCGAT
(held together by the vertex $\overline{O}_3$)

Thus, the vertices of our original graph are represented by $\overline{O}_1$, $\overline{O}_2$, $\overline{O}_3$, $\overline{O}_4$, $\overline{O}_5$, $\overline{O}_6$, and $\overline{O}_7$. Notice that the edge $O_{3 \to 2}$ will not be the same as the edge $O_{2 \to 3}$. Starting off with oligonucleotides of length 20 ensures that we'll have more than enough different possibilities to encode all 7 vertices, and from these get the representations of all 14 edges. When the strands representing the vertices and edges are placed close enough to bond, the bonds will represent paths in the graph.

Because the only possible bondings are between C and G or A and T, we can only have vertices linked by edges, in the manner shown above, if the appropriate path is in fact present. A much weaker sort of bonding, like the following, is possible:

(edge $O_{2\to3}$)

GTATATCCGAGCTATTCGAG

　　　　　　　　　CGATAAGCTCGAATTTCGAT

(vertex $\overline{O}_3$)

This doesn't represent anything in terms of the problem we are investigating. Fortunately, such weak bondings typically break apart and do not present any interference with the strong bonds formed to represent partial paths.

When a DNA "soup" is prepared containing many copies of the oligonucleotides representing the vertices and edges of a graph, bonding very quickly forms potential solutions to the Hamiltonian path problem. The investigator may then filter out a valid solution (if one exists for the particular problem) by selecting DNA segments of the appropriate length.

Writing a program to look for a solution to such a problem, for a large graph, using a traditional computer would not take very long, but execution of the program would. DNA computing is quite different in that the setup and final interpretation are the time-consuming parts. The actual runtime of the DNA program is extremely short. Improved lab techniques should eventually cut down on the presently time-consuming portions of this approach, but it is too early to predict how practical this method may become.

Adleman's small example was intended merely as an illustration of his new approach to computing. It is obviously easier to do this particular problem by hand. Even for this small problem, the lab work required a full week. The value of the approach is, as Adleman points out, "that the methods described here could be scaled-up to accommodate much larger graphs."[43]

A key advantage for this approach to larger Hamiltonian path problems is that the number of distinct oligonucleotides needed to represent the graph grows linearly with the size of the graph, although many copies of each are necessary. Adleman used approximately 3×10^{13} copies of the oligonucleotides representing each edge in his small example. This was far more than was necessary and likely led to many copies of the solution being present. Adleman expects that the necessary number of copies of each oligonucleotide grows exponentially with the number of vertices.

Adleman sums up the advantage of this manner of computation, with some plausible improvements: "At this scale, the number of operations per second during the ligation step would exceed that of current supercomputers by more than a thousand fold."[44] Other advantages include dramatically increased efficiency and decreased storage space.

It is important to note that DNA computing is not limited to a special class of problems. In the reference section below you'll see a paper by Boneh, Lipton, and Dunworth that shows how DES can be broken by a DNA computer. This is done by using DNA to code every possible key and then trying them all at once! Such are the possibilities of parallel processing to the degree DNA computing allows.

Back in 1995, Lipton estimated that a DNA computer with trillions of parallel processors could be made for $100,000.[45] This price tag brings to mind the high cost that Diffie and Hellman

---

[43] Adleman, Leonard M., "Molecular Computation of Solutions to Combinatorial Problems," *Science*, Vol. 266, No. 5187, Nov. 11, 1994, pp. 1021–1024, p. 1022 cited here.

[44] Adleman, Leonard M., "Molecular Computation of Solutions to Combinatorial Problems," *Science*, Vol. 266, No. 5187, Nov. 11, 1994, pp. 1021–1024, p. 1023 cited here.

[45] Bass, Thomas A., "Gene Genie," *Wired*, Vol. 3, No. 8, August 1995, pp. 114–117, 164–168.

originally placed on their hypothetical DES cracker ($20 million). When that machine finally appeared, thanks to the EFF, it cost less than $250,000. How much cheaper will DNA computers be in the future? Will you end up owning one?

Just as traditional computers began as specialized machines for solving particular problems, and only later became "universal" programmable machines, so the history of DNA computers goes. A programmable DNA computer was put forth by Israeli researchers from the Weizmann Institute of Science in 2002.[46] It offered a tremendous advantage in terms of speed, efficiency, and storage capacity over traditional machines, but couldn't do everything they can. Who says you can't have it all? Not Melina Kramer *et al*! In 2008, they created a hybrid device, combining biological components and traditional silicon-based chips.[47] With such hybrids, the biological components can take over where they offer an advantage, while we still have old-school silicon technology to handle tasks for which it's better suited.

It's hard to predict where this technology will lead. Should we be surprised that this brand new field was created by someone outside of biology, thinking about it in his spare time? Not at all, according to Adleman. As more and more fields of inquiry are reduced to mathematics, the ability of a single person to comprehend large portions of it becomes more plausible. Adleman predicted that,[48]

> The next generation could produce a scientist in the old sense, a real generalist, who could learn the physics, chemistry, and biology, and be able to contribute to all three disciplines at once.

For a hundred years, it has seemed that science has been growing ever more complicated, but perhaps mathematics will serve as a simplifying force. Andrew Wiles, the man who finally proved Fermat's Last Theorem, has made a similar comment about mathematics itself:[49]

> Mathematics does sometimes give the impression of being spread over such a large area that even one mathematician can't understand another, but if you think back to 18th century mathematics, most modern mathematicians would understand it all and in a much more unified way than the 18th century mathematicians. I think this dispersion that one senses is really just because we don't understand it well enough yet and over the next 200 years all our current methods and proofs will be simplified and people will see it as a whole and it will be much easier. I mean nowadays most high school students will study calculus. That would have been unthinkable in the 17th century, but now it's routine and that will happen to current mathematics in 300 years' time.

---

[46] Benenson, Yaakov, Rivka Adar, Tamar Paz Elizur, Zvi Livneh, and Ehud Shapiro, "DNA Molecule Provides a Computing Machine with Both Data and Fuel," *Proceedings of the National Academy of Sciences*, Vol. 100, No. 5, March 4, 2003 (submitted in 2002), pp. 2191–2196. For a popular account see Lovgren, Stefan, "Computer Made from DNA and Enzymes," *National Geographic News*, February 24, 2003, http://news.nationalgeographic.com/news/2003/02/0224_030224_DNAcomputer.html.

[47] Kramer, Melina, Marcos Pita, Jian Zhou, Maryna Ornatska, Arshak Poghossian, Michael Schoning, and Evgeny Katz, "Coupling of Biocomputing Systems with Electronic Chips: Electronic Interface for Transduction of Biochemical Information," *Journal of Physical Chemistry*, Vol. 113, No. 6, February 12, 2009, pp. 2573–2579. The work was done in 2008 but published in 2009.

[48] Bass, Thomas A., "Gene Genie," *Wired*, Vol. 3, No. 8, August 1995, pp. 114–117, 164–168.

[49] *Fermat's Last Tango*, Clay Mathematics Institute, Cambridge, Massachusetts, 2001, bonus feature May 24, 2000 interview with Andrew Wiles. This is a video. See https://www.imdb.com/title/tt0278443/?ref_=fn_al_tt_1 for details.

References and Further Reading

On Quantum Cryptography

Aaronson, Scott, *Quantum Computing for High School Students*, 2002, http://www.cs.berkeley.edu/~aaronson/highschool.html.

Bennett, Charles H., "Quantum Cryptography: Uncertainty in the Service of Privacy," *Science*, Vol. 257, No. 5071, August 7, 1992, pp. 752–753.

Bennett, Charles H. and Gilles Brassard, "Quantum cryptography: Public key Distribution and coin tossing," *in International Conference on Computers, Systems & Signal Processing*, Vol. 1, Bangalore, India, pp. 175–179, December 1984, available online at https://arxiv.org/ftp/arxiv/papers/2003/2003.06557.pdf. This paper is often cited as the origin of quantum cryptography, although it is not Bennett and Brassard's first published account of their ideas. See Bennet et al., 1983.

Bennett, Charles H. and Gilles Brassard, "Quantum public key distribution reinvented," *Sigact News*, Vol. 18, No. 4, 1987, pp. 51–53. Although Doug Wiedemann didn't know about Bennett and Brassard's 1984 paper when he reinvented their scheme, Bennett and Brassard saw his paper. The paper cited here is the result.

Bennett, Charles H., Gilles Brassard, Seth Breidbart and Stephen Wiesner, "Quantum Cryptography, or Unforgeable Subway Tokens," in Chaum, David, Ronald L. Rivest, and Alan T. Sherman, editors, *Advances in Cryptology, Proceedings of Crypto '82*, Plenum Press, New York, 1983, pp. 267–275. This is the first published paper on Quantum Cryptography.

Bennett, Charles H., François Bessette, Gilles Brassard, Lois Salvail, and John Smolin, "Experimental Quantum Cryptography," *Journal of Cryptology*, Vol. 5, No.1, Winter, 1992, pp. 3–28.

Bennett, Charles H., Gilles Brassard, and Artur K. Ekert, "Quantum Cryptography," *Scientific American*, Vol. 267, No. 4, October 1992, pp. 50–57.

Bennett, Charles H., Gilles Brassard, and N. David Mermin, "Quantum Cryptography Without Bell's Theorem," *Physical Review Letters*, Vol. 68, No. 5, February 3, 1992, pp. 557–559.

Brassard, Gilles, "Brief History of Quantum Cryptography: A Personal Perspective," *Proceedings of IEEE Information Theory Workshop on Theory and Practice in Information Theoretic Security*, Awaji Island, Japan, October 17, 2005, pp. 19–23. A longer, 14-page version is available online at http://arxiv.org/pdf/quant-ph/0604072v1.pdf.

Brassard, Gilles, *A Bibliography of Quantum Cryptography*, http://www.cs.mcgill.ca/~crepeau/CRYPTO/Biblio-QC.html. This gives a much longer bibliography than is provided in the present book.

Brown, Julian, *Minds, Machines, and the Multiverse: The Quest for the Quantum Computer*, Simon & Schuster, New York, 2000. This is an older book (for this subject) aimed at a general audience.

Ekert, Artur K., "Quantum Cryptography Based on Bell's Theorem," *Physical Review Letters*, Vol. 67, No. 6, August 5, 1991, pp. 661–663. This paper presents another (later) scheme to accomplish what Bennett and Brassard did. This time, quantum entanglement was used.

Ekert, Artur K., John G. Rarity, Paul R. Tapster, and G. Massimo Palma, "Practical Quantum Cryptography Based on Two-photon Interferometry," *Physical Review Letters*, Vol. 69, No. 9, August 31, 1992, pp. 1293–1295. This paper presents another (later) scheme to accomplish what Bennett and Brassard did.

Hughes, Richard J., Jane E. Nordholt, Derek Derkacs, and Charles G. Peterson, "Practical Free-space Quantum Key Distribution Over 10 km in Daylight and at Night," *New Journal of Physics*, Vol. 4, No. 43, July 12, 2002, pp. 43.1-43.14. This paper describes how polarized photons were sent through open air and detected 10 kilometers away.

Johnson, George, *A Shortcut Through Time: The Path to the Quantum Computer*, Alfred A. Knopf, New York, 2003. This is a nontechnical introduction to the topic of quantum computing.

Lydersen, Lars, Carlos Wiechers, Christoffer Wittmann, Dominique Elser, Johannes Skaar, and Vadim Makarov, "Hacking Commercial Quantum Cryptography Systems by Tailored Bright Illumination," *Nature Photonics*, Vol. 4, 2010, pp. 686–689. Here is the abstract:

> The peculiar properties of quantum mechanics allow two remote parties to communicate a private, secret key, which is protected from eavesdropping by the laws of physics. So-called

quantum key distribution (QKD) implementations always rely on detectors to measure the relevant quantum property of single photons. Here we demonstrate experimentally that the detectors in two commercially available QKD systems can be fully remote-controlled using specially tailored bright illumination. This makes it possible to tracelessly acquire the full secret key; we propose an eavesdropping apparatus built from off-the-shelf components. The loophole is likely to be present in most QKD systems using avalanche photodiodes to detect single photons. We believe that our findings are crucial for strengthening the security of practical QKD, by identifying and patching technological deficiencies.

The authors noted, "It's patchable of course… just a question of time."[50]

Shor, Peter W., "Algorithms for Quantum Computation: Discrete Logarithms and Factoring," in Goldwasser, Shafi, editor, *Proceedings of the 35th Annual Symposium on Foundations of Computer Science*, IEEE Computer Society Press, Los Alamitos, California, 1994, pp. 124–134. This is a preliminary version of the following reference.

Shor, Peter W., "Polynomial-time Algorithms for Prime Factorization and Discrete Logarithms on a Quantum Computer," *SIAM Journal on Computing*, Vol. 26, No. 5, 1997, pp. 1484–1509.

Wiesner, Stephen, "Conjugate Coding," *SIGACT News*, Vol. 15, No. 1, Winter 1983, pp. 78–88. This paper was written circa 1970. Wiesner was too far ahead of his time and couldn't get it published until 1983!

On Late, but Greatly Deserved Recognition

Charles H Bennett and Gilles Brassard received the highest honors for their invention of quantum cryptography (among other things) after the first edition of this book appeared.

Wolf Prize in Physics for "founding and advancing the fields of Quantum Cryptography and Quantum Teleportation" 2018. https://wolffund.org.il/2018/12/12/gilles-brassard/.

Micius Quantum Prize "For their inventions of quantum key distribution, quantum teleportation, and entanglement purification" 2019. http://miciusprize.org/index/lists/003002.

BBVA Foundation Frontiers of Knowledge Award (with Peter Shor) "for their fundamental role in the development of quantum computation and cryptography" 2019. https://www.frontiersofknowledgeawards-fbbva.es/noticias/the-bbva-foundation-recognizes-charles-h-bennett-gilles-brassard-and-peter-shor-for-their-fundamental-role-in-the-development-of-quantum-computation-and-cryptography/.

On Post-Quantum Cryptography

Researchers didn't wait for quantum computers to come on the market before attempting to develop systems that can resist such machines. It's not true that quantum computers can break any cipher. McEliece's system, alluded to briefly in Section 16.5, and some lattice-based systems, such as recent versions of NTRU,[51] are believed to be secure against such machines… so far. A few references follow.

Bernstein, Daniel J., Johannes Buchmann, and Erik Dahmen, editors, *Post-Quantum Cryptography*. Springer, Berlin, Germany, 2009.

Buchmann, Johannes and Jintai Ding, editors, *Post-Quantum Cryptography: Second International Workshop* (*PQCrypto 2008*), Lecture Notes in Computer Science, Vol. 5299, Springer, Berlin, Germany, 2008.

Ding, Jintai and Rainer Steinwandt, editors, *Post-Quantum Cryptography: 10th International Workshop*, (*PQCrypto 2019*), Lecture Notes in Computer Science, Vol. 11505, Springer, Cham, Switzerland, 2019.

---

[50] "Quantum Hacking," *NTNU* [Norwegian University of Science and Technology] *Department of Electronics and Telecommunications*, http://web.archive.org/web/20120113024714/http://www.iet.ntnu.no/groups/optics/qcr/.

[51] Hoffstein, Jeffrey, Jill Pipher, and Joseph H. Silverman, *An Introduction to Mathematical Cryptography*, Springer, New York, 2008. Chapter 6 of this introductory text is focused on lattice-based cryptosystems, including NTRU, which was created by the authors. Recall, though, that there are attacks on early versions of this cipher. See https://en.wikipedia.org/wiki/NTRUEncrypt for some updates.

Ding, Jintai and Jean-Pierre Tillich, editors, *Post-Quantum Cryptography: 11th International Workshop*, (*PQCrypto 2020*), Lecture Notes in Computer Science, Vol. 12100, Springer, Cham Switzerland, 2020.

Gaborit, Philippe, editor, *Post-Quantum Cryptography: 5th International Workshop*, (*PQCrypto 2013*), Lecture Notes in Computer Science, Vol. 7932, Springer, Berlin, Germany, 2013.

Koblitz, Neal and Alfred J. Menezes, "A Riddle Wrapped in an Enigma," *IEEE Security & Privacy*, Vol. 14, No. 6, November-December 2016, pp. 34–42, available online at https://eprint.iacr.org/2015/1018.pdf.

Lange, Tanja and Rainer Steinwandt, editors, *Post-Quantum Cryptography: 9th International Workshop*, (*PQCrypto 2018*), Lecture Notes in Computer Science, Vol. 10786, Springer, Cham, Switzerland, 2018.

Lange, Tanja and Tsuyoshi Takagi, editors, *Post-Quantum Cryptography: 8th International Workshop*, (*PQCrypto 2017*), Lecture Notes in Computer Science, Vol. 10346, Springer, Cham Switzerland, 2017.

Mosca, Michele, editor, *Post-Quantum Cryptography: 6th International Workshop*, (*PQCrypto 2014*), Lecture Notes in Computer Science, Vol. 8772, Springer, Cham, Switzerland, 2014.

Post-quantum Cryptography: International Workshop (*PQCrypto* 2006). The proceedings for this first *PQCrypto* conference were not published, but the papers are, with only three exceptions, available online by following a link from the conference's webpage, https://postquantum.cr.yp.to/.

Post-Quantum cryptography, http://pqcrypto.org/. This website provides a "one-minute" introduction and useful links.

Sendrier, Nicolas, editor, *Post-Quantum Cryptography: Third International Workshop* (*PQCrypto 2010*), Lecture Notes in Computer Science, Vol. 6061, Springer, Berlin, Germany, 2010.

Takagi, Tsuyoshi, editor, *Post-Quantum Cryptography: 7th International Workshop*, (*PQCrypto 2016*), Lecture Notes in Computer Science, Vol. 9606, Springer, Cham, Switzerland, 2016.

Takagi, Tsuyoshi, Masato Wakayama, Keisuke Tanaka, Noboru Kunihiro, Kazufumi Kimoto, and Dung Hoang Duong, editors, *Mathematical Modelling for Next-Generation Cryptography*, CREST Crypto-Math Project, Springer, Singapore, 2018.

Yang, Bo-Yin, editor, *Post-Quantum Cryptography: 4th International Workshop*, (*PQCrypto 2011*), Lecture Notes in Computer Science, Vol. 7071, Springer, Berlin, Germany, 2011.

On DNA Computing

Adleman, Leonard M., "Molecular Computation of Solutions to Combinatorial Problems," *Science*, Vol. 266, No. 5187, November 11, 1994, pp. 1021–1024, available online at https://www2.cs.duke.edu/courses/cps296.5/spring06/papers/Adleman94.pdf. This is where it all began.

Adleman, Leonard M., "On Constructing a Molecular Computer," in Lipton, Richard J. and Eric B. Baum, editors, *DNA Based Computers: DIMACS workshop, April 4, 1995, DIMACS Series in Discrete Mathematics and Computer Science*, Vol. 27, American Mathematical Society, Providence, Rhode Island, 1996, pp. 1–22. Other papers in this important volume are referenced below.

Adleman, Leonard M., "Computing with DNA," *Scientific American*, Vol. 279, No. 2, August 1998, pp. 54–61. Another article on the same topic, by the same author, but aimed at a wider audience.

Adleman, Leonard M., Paul W. K. Rothemund, Sam Roweis, and Erik Winfree, "On applying molecular computation to the Data Encryption Standard," in Landweber, Laura F. and Eric B. Baum, editors, *DNA Based Computers II: DIMACS workshop, June 10-12, 1996*, DIMACS Series in Discrete Mathematics and Theoretical Computer Science, Vol. 44, American Mathematical Society, Providence, Rhode Island, 1998, pp. 31–44.

Amos, Martyn, *Theoretical and Experimental DNA Computation*, Springer, Berlin, Germany, 2005.

Bass, Thomas A., "Gene Genie," *Wired*, Vol. 3, No. 8, August 1995, pp. 114–117, 164–168. This is a lively account of DNA computing that also gives the reader a glimpse into Adleman's personality.

Benenson, Yaakov, Rivka Adar, Tamar Paz-Elizur, Zvi Livneh, and Ehud Shapiro, "DNA Molecule Provides a Computing Machine with both Data and Fuel," *Proceedings of the National Academy of Sciences*, Vol. 100, No. 5, March 4, 2003, pp. 2191–2196.

Boneh, Dan, Christopher Dunworth, Richard J. Lipton, and Jiri Sgall, "On the Computational Power of DNA," *Discrete Applied Mathematics*, Vol. 71, No. 1–3, December 5, 1996, pp. 79–94, available online at http://www.dna.caltech.edu/courses/cs191/paperscs191/bonehetal.pdf.

Boneh, Dan, Richard J. Lipton, and Christopher Dunworth, "Breaking DES Using a Molecular Computer," in Lipton, Richard J. and Eric B. Baum, editors, *DNA Based Computers: DIMACS workshop, April 4, 1995, DIMACS Series in Discrete Mathematics and Computer Science*, Vol. 27, American Mathematical Society, Providence, Rhode Island, 1996, pp. 37–66. Here is the abstract:

> Recently Adleman has shown that a small traveling salesman problem can be solved by molecular operations. In this paper we show how the same principles can be applied to breaking the Data Encryption Standard (DES). Our method is based on an encoding technique presented by Lipton. We describe in detail a library of operations which are useful when working with a molecular computer. We estimate that given one arbitrary (plain-text, cipher-text) pair, one can recover the DES key in about 4 months of work. Furthermore, if one is given ciphertext, but the plaintext is only known to be one of several candidates then it is still possible to recover the key in about 4 months of work. Finally, under chosen cipher-text attack it is possible to recover the DES key in one day using some preprocessing.

Devlin, Keith, "Test Tube Computing with DNA," *Math Horizons*, Vol. 2, No. 4, April 1995, pp. 14–21. This Mathematical Association of America (MAA) journal consists of articles easily accessible to undergraduates. The article cited here is especially nice and provides a more detailed description of the biochemistry than is given in this book.

Ignatova, Zoja, Israel Martinez-Perez, and Karl-Heinz Zimmermann, *DNA Computing Models*, Springer, Berlin, Germany, 2008.

Kari, Lila, Greg Gloor, and Sheng Yu, "Using DNA to Solve the Bounded Post Correspondence Problem," *Theoretical Computer Science*, Vol. 231, No. 2, January 200, pp. 192–203.

Lipton, Richard J., "DNA Solution of Hard Computational Problems," *Science*, Vol. 268, No, 5210, April 28, 1995, pp. 542–545.

Lipton, Richard J., "Speeding Up Computation via Molecular Biology," in Lipton, Richard J. and Eric B. Baum, editors, *DNA Based Computers: DIMACS workshop, April 4, 1995, DIMACS Series in Discrete Mathematics and Computer Science*, Vol. 27, American Mathematical Society, Providence, Rhode Island, 1996, pp. 67–74.

Lovgren, Stefan, "Computer Made from DNA and Enzymes," *National Geographic News*, February 24, 2003, http://news.nationalgeographic.com/news/2003/02/0224_030224_DNAcomputer.html. This is a popular account of Benenson, Yaakov, Rivka Adar, Tamar Paz-Elizur, Zvi Livneh, and Ehud Shapiro, "DNA Molecule Provides a Computing Machine with both Data and Fuel," *Proceedings of the National Academy of Sciences*, Vol. 100, No. 5, March 4, 2003, pp. 2191–2196.

Păun, Gheorghe, Grzegorz Rozenberg, and Arto Salomaa, *DNA Computing: New Computing Paradigms*, Springer, New York, 1998.

Pelletier, Olivier and André Weimerskirch "Algorithmic Self-Assembly of DNA Tiles and its Application to Cryptanalysis," 2001, available online at https://arxiv.org/abs/cs/0110009. In this paper, the authors took steps towards a DNA computer attack on NTRU. They noted,

> Assuming that a brute force attack can be mounted to break a key security of 2^{40} the described meet-in-the-middle attack in DNA might break systems with a key security of 2^{80}. However, many assumptions are very optimistic for the near future. Furthermore we understand that using a higher security level, e.g., a key security of 2^{285} as proposed in [5] [the 1998 paper proposing NTRU] puts public-key systems like NTRU far out of range for a successful cryptanalysis in DNA.

Index

A

Abel, Rudolf, 94
Abstract Algebra, 8, 103, 223, 230n. 16, 415–416
Abzug, Bella, 353–354
Academia, 355, 423–421
Access Now, 526
Ace pilots, 274
Acoustical attack, 348, 574; *see also* Side channel attacks
Adair, Gilbert, 18–19
Adams, Abigail, 414
Adams, John, 414
Adams, Mike, 208
ADFGVX, 163, 166–169, 173, 182, 335–336
    references, 195, 196
ADFGX, 163, 166–168, 335–336
    cryptanalysis, of 168–182
Aditsan and Bisahalani, 551–552
Adleman, Len
    DNA computing, 579–583, 586, 587
    in group photo, 488,
    and Loren M. Kohnfelder, 509n. 1
    Mailsafe, 510
    Merkle-Hellman knapsack cipher, 485–486
    NSA, 429
    primality testing, 473
    RSA, 417, 423, 430, 432
    universal scientist, 583
Adolf and Bertholt, 432
ADONIS, 305
Advanced Encryption Standard, *see* AES
"The Adventure of the Dancing Men," 14–16, 53–54
Advertisement, 147, 188, 261–263
Aegean Park Press, 187
AES, 397, 407, 553, 554–564, 569, 577
    AddRoundKey, 561–563
    attacks on, 563–564, 566–567
    conferences/competition, 554–555, 564
    history of, 554–556, 559, 563
    in iPhone, 522
    irreducible polynomial, 559–560
    key size choices, 556

MixColumns, 560–561
and NSA, 553, 556n. 30, 563, 577
    references, 566–567
    ShiftRows, 559–560
    SubBytes (Rijndael S-box), 556–559, 560
    workings of, 556–563
AF, 274
Affine cipher, 33–37
Africa, 251, 468
Agony columns, 147
Agrawal, Manindra, 473–474, 476
Agrippa, Cornelius, 62
Aïda and Bernardo, 549–550
Airbus, 373–374
Air Force Security Agency, 346
ALA, *see* American Library Association
Alba, Dennis, 518
Alberti, Leon Battista, 10, 61, 64
Albion College, 60
Aleutian Islands, 274
Alexander's Weekly Messenger, 11, 59
Alexandria, 450
Algebraic cryptography, 21
Algiers, 320
Alice and Bob, 430–432, 482, 549
Alicia and Beatriz, 549
All-America Cable Company, 184
Allen, Jr., Lew, 354
Alphabet of the Magi, 42–43
Alshamrani, Mohammed Saeed, 528
Altland, Nicholas, 190
Amazon.com, 526
American Association for the Advancement of Science
    (AAAS), 429
American Black Chamber, The (aka the Cipher Bureau),
    183–186, 193, 263, 346n. 3
American Black Chamber, The, 186–189, 197
American Civil Liberties Union, 526
American Council on Education, 428–429
American Cryptogram Association, xxv, 4–5, 21, 55n.
    58, 78, 431n. 42
American Library Association, 193

American Mathematical Monthly (appears 10 times per year), 201

American Revolution, 45–47, 57, 124, 413–414

Ames, Aldrich, 357n. 45

Amsterdam, 512

Anagrams, 117–119

Anarchists, 120

Anderson, Jay, xxv

Anderson, Ross, 501, 506, 555

Anglo-Saxon, 7

Angooki Taipu A, *see* Red

Angooki Taipu B, *see* Purple

Aniuta and Busiso, 549

Annapolis, Maryland, 478

Anti-draft pamphlets, 193

Anti-war, 249, 415, 421, 517

Anti-war Council, 249

Apollo 13, 554

Apple, poisoned, 256

Apple computers, 485, 521–530

Arabic cryptology, 19, 54–55

Arc lengths, 547

Argentis, 9n. 13

Aristagoras of Miletus, 7

Arizona, 163, 283

Arkansas, 274

Arlington National Cemetery, 192, 276

Armed Forces Security Agency (AFSA), 346, 347, 349

Arms limitation, 499

Army Navy Journal, 121–122

Army Security Agency, 192, 346, 376

Army Signal Corps, 105, 121–122, 186, 263, 346n. 3

Arnault, François, 471–472

Arnold, Benedict, 48

Ars Magna, 123

Artificial Intelligence, 255–256

Artificial language, 119

ASCII, 419, 483–484, 487, 532

Asimov, Isaac, 332, 339, 341–342

Askri, 394

Assarpour, Ali, 122

Assassination, 45, 194, 274

Aston, Philip, 116–117

Aston, Sir George, 152

Asymmetric keys, 417

Asymptotically approaches, 8–9, 459, 467

Atari, 495

Atbash, 19

Atheists, 258, 327

AT&T, 93, 102, 309, 310, 373, 516

A-3 Scrambler, 310, 311, 312

Atlantis, 130

Atomic bomb, *see* Nuclear weapons/annihilation

Auckland, 512

Australia, 306, 373

Australian coastwatcher, 151

Authenticity, 408–409, 493

Autokey, 64, 79–80, 401, 411, 533

Avalanche effect, 335

Axiom, 472

Axioms, 147, 249

Aykroyd, Dan, 511

B

Babbage, Charles, xix, 66, 213

Babbington, Anthony, 44–45

Babel, 19

Back door

AES, 559

alleged in Crypto AG machines, 356–358, 360

DES, 392

government attempt to require, 520, 527

iPhones, 522–523, 527, 528, 529–530

via side channel, 361

Bacon, Kevin, 554

Bacon, Roger, 42

Bacon, Sir Francis, 129–131, 133–134, 136, 158–159

Bacon number, 554

Bacon's cipher, 130, 133, 158–159

Bad Religion, 337, 342

Baez, Joan, 353

Bai, Shi, 447

Balkans, 518

Baltic Sea, 166

Baltimore Sun, The, 358

Balzac, Honoré de, 14

Bamford, James, 345–346, 360–361, 363, 365, 377

discovery of Yardley manuscript by, 190

government censorship, 192, 194

Bangalore, India, 573

Barkan, Robert, 424

Barlow, John P., 512

Barnes, Bruce, 429

Barr, William, 528–529

Bartek, Douglas J., 502

Baseball, 285

Base 2 pseudoprime, *see* Pseudoprime

Bass-O-Matic, 511, 512

Bataan Death March, 284

Battle of Wits, 224, 271, 272

Battlestar Galactica, 313

Baudot, J. M. E., 100

Baudot code, 98, 100

Bauer, Craig P., 31, 79, 101n. 48, 103, 207–208, 579

Bauer, Friedrich L., 107

Bazeries, Major Etienne, 136

BBVA Foundation Frontiers of Knowledge Award, 585

Beauregard, General Pierre, 8

Beautiful Mind, A, 42n. 48, 554

Beer, 501

Beethoven, 38, 541

Beijing, 575

Belgium, 555
Bellare, Mihir, 406
Bellaso, Giovan Battista, 64
Bell labs, 313, 323, 327, 502, 576
Bellovin, Steven M., xxv, 51, 102–103, 501
Bennett, Charles, 571–574, 585
Berg, Moe, 285
Berlin, 78, 94, 274, 275, 320
Bernstein, Daniel J., 541
Bertoni, Guido, 500
Bertrand, Gustav, 238
Bestiality, 363
Bêta, 450
Bhargava, Vijay, 573
Bible, 19, 48, 72–74, 333
Biblical cryptography, 19
Biden, Joseph, 510–511, 520, 527
Bidzos, James, 511–512
Bierce, Ambrose, 193
Big Bird Satellite Manual, 364
Biham, Eli, 394, 488, 512, 555
Biliteral cipher, 130, 133, 158–159
Billison, Sam, 284
Binns, Jack, 39
Biology, 133, 579, 583; *see also* DNA computing
Bipartisanship, rare example of, 369
Bisexual, 363
Black, John, 406
Black-bag job, 192
BlackBerry, 520
Blackburn, Marsha, 528
Black Friday, 364
Blair, William, 13
Blaze, Matt, 394
Bleichenbacher, Daniel, 442–443
Bletchley Park, 224n. 10, 247, 250–253, 257–258
Blinding, 444
Block ciphers, *see* AES; DES
Blonde Countess, The, 188, 189, 190
Bloodhound Gang, 149
Bloor, Colonel A. W., 276
Blowfish, 564
Blue, 268
Blumenthal, Richard, 530
Blye, P. W., 323
BND, *see* Bundesnachrichtendienst (BND)
Bochum University, 397
Body of Secrets, 56, 194, 377
Boeing Co., 374
Boer War, 151, 285
Boklan, Kent, 74, 122
Bolivia, 95
Bomba, 246, 251
Bombe, 251, 253, 259–260
Bonavoglia, Paolo, xxv, 195, 196
Bond, Raymond T., 13, 53
Boneh, Dan, 582, 587

Book code, 48–50
Boone, David Sheldon, 364
Boone, Pat, 114
Boston, 414
Boucher, Anthony, 193
BOURBON, 373
Bowdler, Harriet and Thomas, 195
Bowdlerize, 195
Box, 526
Boyce, Christopher, 364
Boyd, Carl, 274
Brass, 271
Brassard, Gilles, xxv, 571–574, 579, 585
Breidbart, Seth, 571
Bribery, 184, 373–374, 391
Bride, *see* Venona
Brillhart, John, 457
Brisbane, 320
Brittany Peninsula, 274
Brookings Institution, The, 530
Brown, Dan, 75, 110, 377
Brown, Gordon, 256
Browne, Jay, xxv, 350–352
Brunner, John, 37
BRUSA, 276, 372–373
BTK, Serial Killer, xix
Bucher, Commander Lloyd M., 352
Buck, Friederich Johann, xixn. 2, 213
Budget, 47, 56, 191, 391, 573
 of NSA and overall intelligence community,
  348–349, 352–353, 355–356, 361, 376
Budiansky, Stephen, xxv, 224, 271–272
Buehler, Hans, 357
Bundesnachrichtendienst (BND), 359
Bundy, McGeorge, 348–349
Burgess, Anthony, 339
Burmese rebels, 518
Burr, Richard, 528
Bury, Jan, xxv
Bush, George H. W., 373, 516
Bush, George W., 377, 564

C

Caesar, Julius, 7–8, 33, 130
Caesar cipher, 7–8, 33, 61, 265
 device to implement, 10
 differential cryptanalysis of, 432
 with reversed alphabet, 19
Calculus, 148, 418n. 13, 510, 583
CALEA, 515–516
Callimachus, 450
Calories Don't Count, 194
Campbell, Keith W., 399
Canada, 165, 190, 285, 354, 373, 573
Canadian coin, 38–39
Canary Islands, 574

Can Such Things Be?, 193
Cantor, Georg, 250
Caproni, Valerie, 520–521
Caracristi, Ann, 306
Cardano, Giambattista, 123
Cardano, Girolamo, 122–123
Cardano grille, 122–126
Carmichael, Robert, 469–470
Carmichael number, 469–470
Carroll, Lewis, 26, 74
Carter, Larry, 499
Casanova, 65
Casement, Sir Roger, 165
CAST-128, 513
Castro, Fidel, 95–96
Catholic, Roman, 44, 62
Cats, pedigreed, 554
Cavemen, 3
CBC, *see* Cipher Block Chaining Mode (CBC)
CBS poll, 526
CDC, 349
Ceiling function, 451, 456
Cell phones, 533, 538–539, 542–543
Censorship, 187–188, 192–195, 424, 428n. 32, 429; *see also* Seized manuscripts
    of Cardano, 123
    of Trithemius, 62
Center for Cryptologic History, iii, xxv, 102, 296, 374
Center for Democracy and Technology, The, 526
Central Intelligence Agency, 31, 94, 105, 352n. 25, 373
    Church Committee investigation of, 353
    compared to NSA, 347
    Crypto AG, and, 357n. 45, 359–360
    Family Jewels, 375
    Hayden, General Michael, and, 369, 526
    Helms, Richard, perjury of, 354
    Kryptos, 75–78
    traitors employed by, 364, 365, 368
    treaty signed with NSA, 355
Central limit theorem, 249
Certicom Corporation, 553
CFB, *see* Cipher Feedback Mode (CFB)
ChaCha20, 541
Chaldean poetry, 553–554
Chan, Wing On, 303–304
Charleston, 414
Chaucer, 42, 43, 331
Chess, 336
Childs, J. Rives, 168
Chilled, 445
China, 186n. 29, 468, 523, 575–576
    employment of Yardley, 183, 190
Chinese Black Chamber, The, 190, 197
Chinese remainder theorem, 440–441
Choctaw, 276, 283, 288
Chorus girls, 51–52

Christ, Jesus, 123
Christensen, Chris, xxv, 237n. 18, 311n. 7
Christie, Agatha, 200
Chrysler, 373
Church, Alonzo, 249
Church Committee Investigations, 352–355
Church, Senator Frank, 353, 354
Churchill, Winston, 101, 259
    enciphered communications of, 192, 303, 310, 312, 319, 327
CIA, *see* Central Intelligence Agency
Cifra del Sig. Giovan, La, 64
Cipher
    definition of, xix
    unsolved, *see* Unsolved Ciphers
    use in fiction, 54, 57, 341
        by Boucher, Anthony, 193
        by Brown, Dan, 75, 110n. 3, 377
        by Brunner, John, 37
        by Doyle, Arthur Conan, 14–16
        by Poe, Edgar Allan, 12–14
        by Stephenson, Neal, 257
        by Tolkien, J.R.R., 26, 27
        by Vinge, Vernor, 106
        by Yardley, Herbert O., 189
Cipher Block Chaining Mode (CBC), 395–396
Cipher Bureau (aka The American Black Chamber), 183–186, 193, 263, 346n. 3
Cipher Bureau (feature film), 216
Cipher Bureau (Poland), 245
Cipher disk, 14, 71, 92
Cipher Feedback Mode (CFB), 403–404
Cipher printing telegraph machine, 99
Ciphertext, definition, xxi
Cipher wheel, *see* Wheel cipher
Ciphony, *see* Voice encryption
Cisco Systems, 526
Civil Disobedience, 194, 368
Civil liberties, 370, 424, 511n. 8, 513, 515, 517; *see also* Electronic Frontier Foundation (EFF)
    American Civil Liberties Union, 526
    Russia, China, and, 365, 368
Civil rights activists, 353, 517
Civil War, England, 26
Civil War, U. S., 8, 26, 39, 74, 120–122, 127, 276
Clapper, James, 369, 370, 526
Clark, Ronald, 286, 356–357, 358
Clausius, Rudolf, 337
Clay Mathematics Institute, 479
Cleartext, definition, xxi
Cleomenes, 7
Clinton, William J., 373, 500, 516–517
Clipper chip 399, 516–517
Clocking bits, 539
Clockwork Orange, A, 339
CNSA Suite, *see* Commercial National Security Algorithm Suite (CNSA Suite)

Coast Guard, 137
Cobbe family, 134
Cocks, Clifford, 432
Code; *see also* Zimmermann telegram
    definition, xix, 4
    diplomatic, 94, 183, 188, 196
    two-part, 135
Code Book
    commercial, xix–xx, 51–52, 102–103, 197, 269
    German, 166
Codebreakers, The
    historic importance, 345, 379, 415
    NSA concern about, 192, 376, 429–430
    Pearl Harbor chapter, 273, 286
Code Girls, 296
Code of Love, The, 116–117
Code talkers
    Choctaw, 276, 283, 288
    Comanche, 277, 282, 283, 288
    in Hollywood, 283–284
    Hopi, 283
    Hungarian, 276
    Navajo 261, 276–285
    Navajo Code dictionary, 279–281
    not allowed to vote, 283
    references, 286–289
    Sioux, 277
    in U.S. Civil War, 276
    videography, 289
    in WW I, 276–277, 282
    in WW II, 261, 276–285
Coding Theory; *see also* Error-correcting code
Cohen, Fred, 579
Cohen, Lona, 97–98
Cohen, Morris, 97–98
Coin, 38–39
COINTELPRO (Counterintelligence Program), 353,
    517
Cold War, 218, 347, 348, 348, 353
    action, if turning hot, 94–95, 306
    breaking of Soviet ciphers during, 255, 355
    censorship during, 194
    one-time pad use during, 97, 98
Collision, 498, 499, 500, 506, 508
Colorpoint Shorthair, 554
Colossus, 252–253
Columnar transposition, 110–117, 119–120, 126, 127,
    166
    Jefferson, Thomas, and, 147
Comanche, 277, 282, 283, 288
Combinatorics, 476
Comey, James, 522, 525
Commercial Codes, xix–xx, 51–52, 102–103, 197, 269
Commercial National Security Algorithm Suite (CNSA
    Suite), 577
Communications Assistance for Law Enforcement Act
    (CALEA), *see* CALEA

Communications of the ACM, 423, 480, 482
"Communication theory of secrecy systems," 80, 158,
    340, 345
Competition, 258, 374, 380, 500, 554, 564, 578
Compliance with Court Orders Act of 2016 (CCOA),
    527–528
Complex analysis, 476
Complexity theory, 477–479, 483–484, 487
Composite numbers, 447–448, 469, 472
Compression, *see* Data compression
Coprime, *see* Relatively prime
Computers, Freedom, & Privacy conference, 511–512
Computer virus, 579
Comstock, Ada, 296
Confederacy, 8, 74, 121–122
Confusion, 335–336, 389
Conic section, 545
Connection polynomial, 536
Conner, Major Howard M., 276, 278
Constitution, U.S., 355, 424, 426, 514, 526, 527
    NSA employee oath to protect, 361–362, 365, 372
Continued fractions, 457, 461
Cook, Tim, 522, 523
COPACOBANA, 397
Copier codes, xix
Copper, 133, 271
Coppersmith, Don, 399, 441
Coral, 276
Cornell University, 192, 424, 426
Counter Mode (CTR), 406, 411
Counterpane, 555
Coventry, 259
Cowan, Michael, 156–157
Crandall, Richard, 465
Crawford, Major, 263
Cribs
    applied to matrix encryption, 204, 207, 214
    applied to RSA, 442
    applied to SIGABA, 303–304
    applied to Zodiac cipher, 31
    examples of use, 108–110, 138–139, 170, 403,
        537–538
    for *Kryptos*, 78–79
    references, 159–160, 258, 542
    use in World War II, 246, 270, 271
Crick and Watson, 501, 573
Crimean War, 39
"Crime Deterrent Transponder System," 424
Crossword puzzle, 334
Crowley, Kieran, xxv, 56
CRT, 347
Crying of Lot 49, The, 337–338
Cryptanalysis; *see also* Factoring algorithms; RSA attacks
    of ADFGX and ADFGVX, 168–182
    of affine cipher, 35–37
    by Cipher Bureau (under Yardley), 183, 185–186
    of columnar transposition, 111–114

definition, xix, xxi
of double transposition, 120
of Enigma, *see* Enigma machine, Cryptanalysis
frequency analysis, 17–21
    by Germany, 187, 192, 294, 310
    by Great Britain, 187
hill climbing, 156
isomorphs, 264–265
of matrix encryption, 204–212
meet-in-the-middle attack, 304n. 26, 307, 397–398,
        444, 587
of monoalphabetic substitution cipher, 29–30,
        55–56
of nomenclator, 44–45, 47–48
of one-time pad, 96–98
of Playfair cipher, 152–157
of rectangular transposition, 108–110
of running key cipher, 80–92
as seduction tool, 65
simulated annealing, 126, 127, 156, 161
timing attack, 443–444
of Vigenère cipher, 65–74
of Wheel Cipher, 138–146
Crypto AG, 218, 356–360
Crypto Aktiengesellschaft, *see* Crypto AG
Crypto '82, 405, 485
Crypto Forum, 209
Crypto-Gram online newsletter, 564
Cryptographie Militaire, La, 157
Cryptography, definition, xix
CRYPTOLOG, 350–352
Cryptologia, xxv, 16, 103, 320, 391, 433
    Enigma articles published in, 257
    "Poe Challenge Cipher Solutions," 60
    RSA improved upon in, 423, 427
    wheel cipher cover, 160
Cryptologic History Symposium, 102
Cryptology
    definition, xxi
    reasons to study, xxi–xxii
Cryptonomicon, 257
Cryptos, 394
CRYPT(3), 401
CSP 488, 137; *see* also Wheel cipher
CSP 493, 137; *see* also Wheel cipher
CSP-888/889, *see* SIGABA
CSP-2900, *see* SIGABA
CTR, *see* Counter Mode (CTR)
Cuban missile crisis, 94
Culper spy ring, 45–47, 57
Currency, 135
Currency, uncounterfeitable, 571
Currier, Prescott, 372–373
Cyanide, 256
Cyber Command Center (Cybercom), 361
Cycles, disjoint, 223
Cyclic permutation, *see* Permutation

Cyclometer, 244–245, 246
Cylons, 313
Cypherpunks, 158, 539
Cyrene, 450
Cyrillic alphabet, 5–6, 74
Czars, 120

D

Daemen, Joan, 500, 555–556, 559
Daily sequence, 267
Damm, Arvid Gerhard, 218
Dan David Prize, 472
Dark ages, 42
Data compression, 40–42, 312, 334, 361, 513–514
Data Encryption Standard, *see* DES
Dataseal 380; *see also* DES
Da Vinci Code, The, 75, 377
Dawson, Jr., John W., xxv
Dayton, Ohio, 253, 259–260
D-Day, 252, 274–275, 282
Deadly Double, The, 261–263
Deavours, Cipher, 36, 333, 502
Decimated alphabets, 69
Decipher, definition, xxi
Declaration of Independence, 146, 147
Decoding linear codes, 478
Decoding the IRA, 114–116
Decrypt, definition, xxi
Deeley, Walter, 394
Defcon, 523
Deep Crack, 396
DeLaurentis, John M., 495–497
Delilah, 320
Demon, 380; *see also* DES
Demonstration cipher, 380; *see also* DES
Denton, Jeremiah, 38
Depp, Johnny, 378
Derangement, 6
DES; *see also* COPACOBANA; *DES variants*; Electronic
        Frontier Foundation; Triple DES
    complementary property, 399–400
    cracker 395–397, 443
    DNA computer attack, 582, 587
    flowchart, 386
    history, 379–380
    initial permutation, 385, 387
    keysize controversy, 390–393, 394, 554
    not a group, 398–399
    NSA involvement, 379, 380, 391–395, 399
    objections to, 390–394, 554
    for passwords, 503–504
    references, 409–411
    round key generation, 387–389
    S-boxes, 382–385, 389, 393–395, 401
    Semi-weak keys, 395n. 44
    specialized machine to attack, 390–391, 395–397

variants, 401, 503–504
weak keys, 395
workings of, 380–390
workshops, 391–392, 393, 410
Desalination plant, 274
Desch, Joe, 259–260
Determinant of a matrix, 203, 209
Deterministic test, 471, 473–476, 477–479, 489, 497, 551
Deutsch, Harold, 253–254
Dewey, Thomas E., 187, 286
Dice, 261–263
Dickson, Paul, 424
Dictionary attack, 55, 120, 503, 519
Differential cryptanalysis, 394–395, 432n. 44, 512
Diffie, Whitfield, 406, 414–415, 417, 430, 432, 433, 509
    NSA Hall of Honor, 434
    objections to DES, 390, 391, 403, 410, 582–583
    pictures of, 415, 480, 488
    reaction to attempted intimidation, 426, 428–429
Diffie–Hellman key exchange, 414–417, 422, 445
    elliptic curve cryptography (ECC) version of, 549, 566
    quantum computer threat to, 576, 577, 578
Diffusion, 335–336, 389
Digital Fountain, 406
Digital Signature Algorithm (DSA), 445, 504–506, 513, 577
Digital signatures, 500, 553, 577
    DSA, 445, 504–507, 513
    Elgamal, 497–498, 513, 552
    RSA, 423, 445, 495–497, 513
Digital Signature Standard (DSS), *see* Digital Signature Algorithm (DSA)
Digraph frequencies, 20, 22
Digraphic cipher, 64, 147–148; *see also* Playfair cipher
Diophantus, 545
Dimitrov, Vassil S., 550–551
Diplomatic ciphers, 96, 311; *see also* Red; Purple
Diplomatic codes, 94, 183, 188, 196
Discrete log problem, 416, 422, 439, 486–487, 550, 553, 576
Disparation, La, 18–19
Disquisitiones Arithmeticae, 447
Distinguishing languages, 74
Distributed.Net, 397
Dixon, John D., xxv, 454, 459
DNA, double helical structure, 501, 573
DNA computers, 569, 579–583
    background, 579–580
    DES attack, 582, 587
    example, 580–582
    history, 579–580, 582–583
    programmable, 583
    references, 586–587
Dodgson, Charles, 26, 74, 104
Donnelly, Ignatius, 130–131
Donut, 546

Dooley, John F., 10n. 18, 54, 57, 191–192
Dorabella cipher, 33; *see also* Elgar, Edward
Double transposition, 116, 119–120, 126, 127, 182
Downey, Peter J., 502
Doyle, Arthur Conan, 10, 14–15, 16, 48, 53–54
Dr. Dennis F. Casey Heritage Center on Joint Base San Antonio, 306
Dropbox, 526
Drug, *see* Venona
DSA, 445, 504–506, 513, 577
DSD-1, 380; *see also* DES
DSS, *see* Digital Signature Algorithm (DSA)
Dublin, 134
Duffy, Austen, 399
Dunin, Elonka, 104
Dunlap, Jack, 364
Dunworth, Christopher, 582, 587
d'Urfé, Madame, 65
Durrett, Deanne, 284

E

e, 9, 110, 468, 519, 520
EARN IT, *see* Eliminating Abusive and Rampant Neglect of Interactive Technologies Act of 2020 (EARN IT), 530
Earth, circumference, 450
Easter Island, 16, 17
ECB, *see* Electronic Code Book Mode (ECB)
Echelon, 373
von Eckhardt, Felix, 163
ECM (Electric Cipher Machine) II, *see* SIGABA
e-commerce treaty, 500
Ecstasy, 518
Eddington, Sir Arthur, 249, 336
Edge (of a graph), 580
Education of a Poker Player, The, 189
EFF, *see* Electronic Frontier Foundation
Egypt, 443
Egyptians, 115, 348, 450, 486
Einstein, 249
Eisenhower, Dwight D., 304–305, 322n. 22, 356, 363
Ekert, Artur K., 574
Election, 186, 516, 575
Electric Cipher Machine, *see* SIGABA
Electromagnetic emanations, *see* TEMPEST
Electronic Battlefield, The, 424
Electronic Code Book Mode (ECB), 403, 409
Electronic Frontier Foundation (EFF), xixn. 1, 410, 511, 512, 526
    DES Cracker 395–397, 401, 443, 583
Elgamal, Taher, 486
Elgamal encryption, 486–487, 504–505, 513, 551–552
    attack on, 444
    references, 491, 506
    signatures, 497–498

Elgar, Edward, 31–32
Eliminating Abusive and Rampant Neglect of Interactive
        Technologies Act of 2020 (EARN IT),
        530
Elizabeth I, Queen, 44–45
Elizabethtown College, 211–212
Elliott, Missy, 390
Ellipses, 547
Elliptic curve cryptography (ECC)
    background, 545–549
    demise and possible resurrection of, 578
    example of ECC style Diffie–Hellman, 549–551
    example of ECC style Elgamal, 551–552
    personalities behind, 552–554
    point addition, 547–548
    references, 565–566
    scalar multiplication in, 550–551
Elliptic curve factoring, 461, 463, 566
Elliptic curves, 545–549
Elliptic integrals, 547
Ellis, James H., 432, 433
Elvis, 60–61
Email, 361, 371, 520, 564
    PGP, and, 514, 515, 518
    Snowden, Edward, and, 369, 370
Embarrassing Personal Questions (EPQs), 363
Encipher, definition, xxi
Enciphered code, 94, 96, 163
Encrypt, definition, xxi
Encryption modes; *see* Modes of encryption
England; *see also* Bletchey Park, and Turing, Alan
    bombe rebuilt in, 253
    BOURBON and, 373
    Civil Wars in, 26
    cryptologic museum in, 254
    formal apology to Turing, 256
    G20 summit in, 370
    persecution of Catholics in, 44
    Polish cryptanalysts, and, 246–247
    Soviet Spies captured in, 97
    Yardley breaks ciphers of, 186n. 29
    Zimmerman Telegram and, 164–165
ENIAC, 252
Enigma, An (poem by Poe), 14, 57
Enigma machine; *see also* Cyclometer; Permutations;
        Plugboard; Polish cryptanalysts; Rotors; Rotor
        wirings; Schmidt, Hans Thilo; Steckerbrett;
        The theorem that won the war
    characteristic structure, 229, 236–237
    cryptanalysis I, recovering the wiring, 226–242
    cryptanalysis II, recovering the keys, 243–246
    cryptanalytic notation, 227
    factoring permutations, 229–234
    four-rotor Naval Enigma, 251
    history, 217–218, 220, 226–227, 238, 242, 246–247
    keyspace of, 225–226
    kits for sale, 260
        period, 223–224
        psychological method, 230, 234, 236
        reflector, 222, 223–224, 226, 227, 246
        schematic, 222
        workings of, 220–224
Entropy, 74, 327–339
Entscheidungsproblem, 249–250
Enzymes, 579
Equatorie of the Planetis, The, 42–43
Eratosthenes, 447–448, 450
Erdös, 554
Erdös number, 554
Error-correcting code, 335, 431–432, 490
Error propagation, 80, 336, 403, 405–406, 411
Espionage Act, 193
Estrogen treatment, 256
Euclid, 419, 465
Euclidean algorithm, 408, 419–421, 484, 487, 557; *see
        also* Extended Euclidean algorithm
    used in attacks on RSA, 435, 440, 496, 497
    used to factor RSA moduli, 445
Euler, Leonhard, 33, 418–419, 451–452, 474, 545
Euler's generalization of Fermat's little theorem,
        418–419, 452
Euler's totient function (aka Euler's φ function), 33, 211,
        212, 418, 474
European black chamber, 44, 136
Eve, 94, 431, 435, 495, 496, 570–571, 575
Even Prime Numbers, 462
Evernote, 526
Exclusive OR, *see* XOR
EXP (exponential time), 477, 478
Expected value, 68, 72, 329
Explosives, 193
Export restrictions, 391, 392, 393, 423, 510, 511, 513
    loophole in, 410
    relaxing of, 517–518
Extended Euclidean algorithm, 408, 420, 497, 557
Extropy: Transhuman Technology, Ideas, and Culture, 338,
        339

F

Fabyan, George, 133
Facebook, 520, 526
Factorial, 8–9
    connection with long sequence of composite
        numbers, 466
    primality testing, and, 473
    traveling salesman problem, and, 478
    used in factoring algorithm, 453
    used to find keyspace, 33, 110, 111, 139, 173,
        225–226, 269, 303, 310
Factoring algorithms
    continued fractions, 457, 461, 463
    difference of squares, 450–451, 454
    Dixon's algorithm, 454–459

elliptic curves, 461
Euler's method, 451–453
Fermat's method, 450–451, 454
Pollard's number field sieve, 460, 461
Pollard's $p-1$ algorithm, 453
Pollard's ρ (rho) algorithm, 453–454
quadratic sieve, 459–460
references, 463–464
Shor's algorithm, 576
sieve of Eratosthenes, 447–450, 459
sum of two squares in two ways, 451–453
trial division, 447, 469, 471, 475
Factoring problem, 422, 435, 444–445, 460; *see also*
    RSA Factoring challenge
Falcon and the Snowman, The, 364
Family Limitation, 194
Family Shakespeare, The, 195
Farago, Ladislas, 190–191, 198
Farook, Syed Rizwan, 521
Faustus, Dr., 62
FBI, *see* Federal Bureau of Investigation
FDA, *see* Food and Drug Administration
Federal Bureau of Investigation (FBI), 31, 94, 116, 263,
    353, 518
    abuse of power by, 517
    anti-crypto actions of, 511, 517, 520–529
    wiretapping expansion, and, 516
    World War II-era censorship, and, 193
Federal Communications Act, 355
Federal Register, 380
Feinstein, Dianne, 528
Feistel, Horst, 379–380, 410
Feistel cipher, 381
Feistel system, 381
Feminists, 353
Feng, Dengguo, 499
Ferguson, Niels, 555, 563, 564
Fermat's Last Tango, 583n. 49
Fermat's last theorem, 583
Fermat's little theorem, 418, 452, 453
Fermat number, 457, 461
de Fermat, Pierre, 450, 545, 583
    factoring method of, 450–451, 452, 454
    Fermat number, 457, 461
    Fermat's little theorem, 418–419, 453, 468, 469,
        475
    Fermat test, 468–469
Feynman, Richard, 110n. 4, 402, 519
Fifty Years of Mathematical Cryptanalysis (1937–1987),
    433
Filing/file cabinet, 263, 519, 520
Fingers, scraped raw and bloody, 245
Finite strings over a four-letter alphabet, 579
Finland, 97n. 39
First Amendment, 429–430
First Day Cover, 226–227, 254
FISA warrants, 354

Fist, 493–494
Five Eyes, The, 373
Five-unit printing telegraph code, 98–100
Flame-throwing trumpet, 336
Flexowriter, 348
Flicke, Wilhelm F., 192
Floor function, 451
Flynn's Weekly, 199–200
FOCS Conference, 554
Food and Drug Administration, 194
Folk Medicine, 194
Fonda, Jane, 353
Ford, 373
Foreign Intelligence Surveillance Act, 354, 368; *see also*
    FISA warrants
Forrestel, Mark, 518
Fortener, Sarah, xxv
Fort George G. Meade, 346, 348, 353, 362, 433; *see*
    also National Security Agency; National
    Cryptologic Museum
Foss, Hugh, 267
Four-letter alphabet, 579
Fractionating cipher, 168
France, 44, 168, 186n. 29
    laws regulating cryptology in, xxi–xxii
    World War II, and, 246, 274–275, 276, 304
Francis, The Dauphin of France, 44
Frankfurt, 320
Franklin & Marshall College, xxv, 221
Franklin, Ben, 217, 255, 414
Freedom fighters, 518
Freeh, Louis, 517
Freemasons, 18, 26, 28
Freeware, 511, 513, 531
French coast, 274–275
French counterintelligence and SIGABA, 304
French cryptanalysts, 42n. 51, 126, 168–169
French effort against Enigma, 238, 246
French language
    books used by American cryptanalysts 263
    ciphertext 30
    Disparation, La, 18–19
    entropy, 331
    frequency of letter E in, 18
    Index of Coincidence, 74
French Ministry of War, 136
French national military academy, 10n. 17
French Wheel Cipher 136
Frequencies
    of digraphs, 20, 22
    of doubled letters, 20–21
    of initial letters, 20
    of letters, 17–18, 81
    of passwords 519
    of terminal letters, 20
    of trigraphs 91, 112, 175, 331
    of two-letter combinations, 20, 22

Friedman, Elizebeth, 133
Friedman, William 22, 48, 50, 138, 346, 433
    anagrams, and, 117, 118–119
    bizarre comment by, 192
    buried at Arlington National Cemetery, 276
    cipher machine designed by, 291–293
    contrasted with Yardley, 182–183
    Crypto AG, and, 356–357, 358, 359
    on double transposition, 120
    index of coincidence, and, 67, 74, 81, 96
    Medal of Merit winner, 276
    nervous breakdown, 271
    on Playfair, 156
    Purple, and, 286, 287
    at Riverbank laboratories, 133
    on running key ciphers, 81, 86, 91
    on Shakespeare vs. Bacon, 133
    team assembled by, 263, 264
    Voynich ms., and, 117, 118–119, 159
    Yardley's files obtained by, 186
Furtivis literarum notis, de, 64

G

Gadsby, 17–18
Galileo, 117, 118
Gallian, Joseph A., 114
Gallic Wars, 7n. 8
Gallup, Elizabeth, 133
Gandhi, Mahatma, 514
Gardner, Martin, 60, 423, 424, 460, 579
Garfinkel, Simson, 511
Gatti, Benjamin, xxv
de Gaulle, General Charles, 304
Gauss, Carl Friedrich, 249, 418n. 14, 440, 447, 467,
        545n. 2
Gaussian distribution, 249
Gay, *see* Homosexual
Gay marriage, 256
GCD, *see* Greatest common divisor
GCHQ, 194, 356, 373, 416, 432–433, 434
GCHQ: The Negative Asset, 194
General Motors, 373
Generator of a group, 415–416, 487
Geneva, Illinois, 133
Genie, 65
Genocide, 277, 421
"Gentlemen do not read each other's mail.," 186
George, Dickie, 410
Georgia Institute of Technology, 574
German Foreign Office, 96, 160
German language, 14, 74, 263, 331, 357, 375
Germany; *see also* ADFGX; ADFGVX; Enigma; World
        War I; World War II
    ciphers solved by, 184, 187, 192, 294, 310
    columnar transposition, 120
    Crypto AG, 357, 359–369, 372

Der Spiegel, 373
Feistel, Horst, 380
Gauss, Carl Friedrich, 467
Hasse, Helmut, 549
One-time pad, 94, 96
Purple, and, 274–276
Simons, Peter, 512n. 16
study of Native American languages, 277, 282
transatlantic cable, 184
universities, 397
vs. SIGSALY, 322
Zimmermann telegram, 163–166
"Get curious, not furious." 485–486, 491
Get Smart, 322
Gibbs, Josiah Willard, 479
Gibson, William, 3
Giessen river, 304
Gifford, Gilbert, 44
GI Joe, 284–285
Gillogly, James J., 78, 114–115
GIMPS, 466, 476–477
Git, 500
Glienick bridge, 94
Global System for Mobile Communications (GSM),
        539
God(s), 3–4, 18, 30, 131–132, 167, 337
Gödel, Kurt, 249
Gödel's incompleteness theorem 249
Goldberg, Ian, 539
"The Gold Bug," 12–14
Google, 343, 370, 524, 526, 541, 576, 577
Google Scholar, 573
Googol, 540
Gordon, John, 431, 479
Gorgo, 7
Götterzahlen, 4n. 1
Gott mit uns, 167
Gott mit uns ADFGVX example 167
Government Code and Cypher School (GC&CS), 94n.
        28, 250
Government Communications Headquarters, *see* GCHQ
Government control of cryptologic research, 423–430
Graham, Lindsey, 527, 528, 530
Graham's Magazine, 11–12, 59n. 1
Grand International Mersenne Prime Search, *see* GIMPS
Graysmith, Robert, 56
*Great Cryptogram: Francis Bacon's Cipher in the So-Called
        Shakespeare Plays, The*, 130
Greatest common divisor, 33, 203, 418, 445, 452, 497
Greedy algorithm, 483–484
Greek cryptography, 4–5, 7, 52, 74n. 10, 122, 333
Greek language, xix, xxin. 3
    entropy 331
    word spacing 333
Greek steganography, *see* Steganography, Greek
Green, 268, 276
Green, Lucky, 539

Green Berets, 94
Green Hornet, The (book), 318, 319n. 16
Green Hornet, The (machine), 311; *see also* SIGSALY
Green Hornet (radio show), 322
Greenwald, Glenn, 370
Greenwood, Lloyd, 146, 159–160
Griffing, Alexander, 91–92, 105
Grille 124; *see also* Cardano grille
Grosjean, Major, 42
Gross, Josh, xxv, 28, 333–334, 342–343
Grotjan, Genevieve, 270–272
Grothouse, Brett, 127
Group Theory, 223
Grover, Lov, 576
GRU, 357
GSM, 539
G20 summit, 370
Gu, Ting, 212
Guadalcanal, 278
Guam, 320
Guardian, The, 370, 424
Guatemala, 518
Guevara, Ché, 95–96, 98
Guinier, Daniel, 534
Guns, 285, 356, 521, 528
Guy, Mike, 501

H

Hacked passwords, *see* Passwords, hacked
HackMiami, 523
Hadamard, Jacques, 467
Hagelin, Boris, 218, 356–359
Hagelin machine, 218, 264n. 5, 348; *see also* M-209
Hale, Nathan, 45
Half-rotors, 267
Hall, Chris, 555
Hallas, Sam, xxv, 100
Hamilton, Victor Norris, 364
Hamilton, William Rowan, 581
Hamiltonian graph/path problem, 478, 580–582
Hamlet, 191
Hamming, Richard, 335n. 15
Hamming weight, 399
Hanyok, Robert J. 103
Harden, Bettye, 31
Harden, Donald, 31
Harding, Walter, 194
Harper, John, 253
Harris, Barbara, 96
Harris, Ed, 554
Harrisburg, Pennsylvania, 478
Hartmanis, Juris, 477
Hartwig, Robert E., xxv
Harvard, 296, 472, 521, 553
Hash functions, 498–504, 504–505
Hasse, Helmut, 549

Hasse's theorem, 549, 551
Håstad, Johan, 441
Hatch, David, 283
Hawaii, 320, 364, 366
Hawaiian (language), 6, 331
"Hawaiian good luck sign," 350–352
Hayden, Lieutenant General Michael, 360, 369–370, 372, 375, 526
HBO, 527
Hebern, Edward Hugh, 217
Hebern Electric Code, 217
Hebrew language, 19, 333
Hedges, S. Blair, 133
Hellman, Dorothie, 485, 491
Hellman, Martin E., xxv, 379, 406, 410, 414–415, 417, 434
    "Get curious, not furious." 485–486, 491
    Knapsack encryption, and, 482, 483, 485–486, 487, 490
    on nuclear war, 421
    objections to DES, 390, 391, 397, 410, 582–583
    pictures of, 415, 480, 486, 488
    reaction to attempted intimidation, 424, 426
Helms, Richard, 354
Hemorrhoids, 123
Hepburn, James, 44
Herodotus, 7
Hiebert, Erwin N., 336
Highlander: The Raven, 283
Hilbert, David, 249
Hill cipher, *see* Matrix Encryption
Hill, Donald, 130–131
Hill, Lester, 116–117
Hill climbing, 156
von Hindenburg, Paul, 226
Hiroshi, Ōshima, 274–276
Histiaeus, 7
Historic Doubts, 130, 159
Hitler, 16, 218, 226, 276
Hitt, Captain Parker, 136, 195–196
Hodges, Andrew, 256, 257, 258, 319n. 16
Holland, 493
Hollywood, 190, 283
Holmes, Sherlock, 10, 14–15, 16, 48, 53, 54
Holocaust, 254
Holstein, Otto, 75
Holtwick, Jr., Lieutenant Jack S., 267
Holy Grail (of computer science), 479
Home Guard, 257
Homophones, 44
Homosexuality
    Alan Turing and, 256, 258
    National Security Agency and, 363
Honeywell, 577
Hong Kong, 116, 364, 368, 369, 370
Hopi, 283
Hoover, Herbert, 186

Hoover, J. Edgar, 263, 363
Horner, Captain E. W., 276
Hot line, between Washington, DC and Moscow, 94–95
Hotmail, 370
House Permanent Select Committee on Intelligence, 366–369
Huang, Ming-Deh A., 473
Huffman, David A., 39–40, 42
Huffman coding, 39–42
Human rights activists, 518
Humor, 256n. 65, 257, 340, 401, 462, 479, 512
    Journal of Craptology, 401, 432n. 44
    xkcd, 285, 390
Hungarian code talkers, 276
Hunnicutt, Tom "Captain T," 274
Hurt, John B., 263, 264
Hutton, Timothy, 364
Huygens, Christian, 118
Hybrid DNA/silicon computer, 583
Hybrid system, 305, 473, 509, 510, 511, 520

I

Iberian Peninsula, 275
IBM, 349, 379–380, 391–393, 404, 552, 555, 577
IDEA, 512
IEEE, 423–427, 572–573
IEEE Symposium on Information Theory, 572–573
IEEE Transactions on Aerospace and Electronics Systems, 424, 426
IEEE Transactions on Information Theory, 482
Illinois, 133
iloveyou, 519
Immortality, 74
Index of Coincidence (IC), 67–72, 74, 81, 96, 172, 211, 264,
Index of prohibited books, 62, 194
India, 356, 473–474, 573, 576
Indian Institute of Technology in Kanpur, 473–474
India Pale Ale, 501
Indigo, 268
Indonesia, 373
Induction, 184
Industrial and Commercial Bank of China (ICBC), 575–576
Indus Valley script, 16–17
Infinity, 467, 545
Information Assurance Directorate (IAD), 347, 577
Information theory, 327, 336, 340, 423, 425, 482, 573
    Nyquist, Harry, and, 317, 329n. 3
    Pynchon, Thomas, and, 337
Inglis, Chris, 365–366, 370–372
Initialization Vector (IV), 403, 404, 405, 407, 540–541
Inman, Bobby Ray, 306, 429, 433
Inouye, Captain Kingo, 191
Inquisition, 123
Institute for Advanced Study, Princeton, 327, 479
Institute for Advanced Study, Radcliffe, 521

Institute of Electrical and Electronics Engineers, *see* IEEE
International Data Encryption Algorithm (IDEA), *see* IDEA
International Organization for Standardization (ISO), xxi
International Traffic in Arms Regulations, *see* ITAR
Internet, 340, 367, 515, 517, 540, 579
    distributed attack using the, 397
    Great Internet Mersenne Prime Search (GIMPS), 476–477
    PGP distributed on the, 511
    RSA, and, 430
    similarity of to print media, 242
Introduction to Finite Fields and their Applications, 559
Invertible matrix
    in AES, 557–559, 561
    in matrix encryption 201, 203, 204, 207, 208, 209, 210
Invisible inks, *see* Secret inks
iPads, 525
iPhones, 521–523, 527, 528, 529–530
IPsec, 499
IRA, *see* Irish Republican Army
Iran, 357, 367
Irreducible polynomial, 407, 536, 559, 560
Ireland, 114–116, 165, 500
Irish Republican Army, 114–116
ISO, *see* International Organization for Standardization
Isograms, 23, 55, 56
Isomorphs, 264–265
Israel, 364, 373n. 99, 378, 418, 555, 583
    Crypto AG, and, 357, 359
    USS *Liberty*, and, 350, 377
Italian language, 74, 331
Italy, 280n. 2, 555
ITAR, 423, 425–427
Ithaca, New York, 424
IV, *see* Initialization Vector (IV)
Iwo Jima, 276, 278, 281, 285

J

Jade, 97, 276; *see also* Venona (Jade was used as a codename for both a Japanese cipher and the decipherment of Soviet OTPs)
Jao, David, 578
Japan; *see also* Red; Purple; World War II
    attack on Hong Kong, 116
    one-time pad use, 94–95
    QKD Network in, 575
Japanese auto manufacturer, 373
Japanese ciphers, *see* Coral; Green; Jade; Orange; Purple; Red
Japanese codes; *see also* JN-25
    broken by Yardley, 185–186, 263
Japanese Diplomatic Secrets, 187–188, 192

Japanese Foreign Office, 264
Japanese language, 263, 267
Jarvis, C. D., 194
Jefferson, Thomas, 134–136, 146–147, 159
Jefferson wheel cipher, *see* Wheel Cipher
Jenks, Peter, 360
Jeremiah, Biblical book of, 19
JN-25, 273–274, 287
Johnson, Don, 445
Johnson, Henry C. "Hank" 521
Johnson, Lyndon, 276
Johnson, Robert Lee, 364
Johnson, Thomas R., 349, 353, 355, 363, 377
Johnston, General Albert S., 8
Johnston, TSgt. Philip, 277–278, 281–283
Joint Mathematics Meetings, 479
Journal of Craptology, 401, 432n. 44, 479n. 31
Joyce, James, 194
JPEGs, 40, 340

K

Kaczynski, Theodore (aka Unabomber), 30, 31, 57, 116
Kafka, Franz, 357
Kahn, David, xxv; *see also Codebreakers, The*
　blurb, 106
　Ché Guevara cipher, 95–96
　donation, 138, 323
　on *Japanese Diplomatic Secrets*, 187–188
　on Levine and Hill, 199
　on NSA, 353, 355–356
　NSA Hall of Honor, 434
　on *The Gold Bug*, 14
　on World War II, 247, 251, 258, 274, 276
　wreck of the *Magdeburg*, 196
　on Yardley, 186–187
Kama Sutra, xxii
Kampiles, William, 364
Kana syllables, 267
Kane, Jock, 194
Kanpur, India, 473–474
Kasiski, Friedrich W., 66
Kasiski test, 66, 69–72, 74, 81, 265, 350
Kayal, Neeraj, 473–474
Keating, John P., 165
Keccak, 500
Keillor, Garrison, 510
Kelley, Stephen J., 269, 286
Kelsey, John, 555
Kennedy, John F., 152, 322, 348
Kennedy, Steve, 256–257
Kerckhoffs, Auguste, 157
Kerckhoffs's rules, 157–158, 350, 539
Kesselring, Field Marshall, 252
Key escrow, 399, 516
Key Tape, 95, 98–99, 291–293, 299
KGB, 357

Kiam, Victor, 359
Kiel University, 397
Kieyoomia, Joe, 284
Killington, 418
King, Jr., Martin Luther, 353, 517
King's College, Cambridge, 249
Kladstrup, Regan, xxv
Klooz, Marie Stuart, 188n. 38
Knapsack encryption, 482, 483–486, 488 (*see the last reference*), 490, 499
Knapsack problem, 478, 483
Knudsen, Lars, 555
Koblitz, Neal, xxv, 414, 545–546, 551, 552–554, 565–566,
　on Alice and Bob, 549
　on future of ECC, 570
　on post-quantum crypto 579
Koch, Hugo Alexander, 217, 218
Kocher, Paul, 395–396, 443–444
Kohnfelder, Loren M., 509, 510
Kong, Jiejun, 340
Konheim, Alan, 392
Koran, 19
Korean war, 212, 305
Kosovo, 518
Kovalevskaia Fund, 553
Kovalevskaia, Sofia, 554n. 21
Kozaczuk, Władysław, 245
Kraitchik, Maurice, 454
Kramer, Melina, 583
Kramp, Christian, 8
Kroger, Helen, 97–98
Kroger, Peter, 97–98
Krovetz, Ted, 406
Kryha, 264n. 5
Kryptos, 75–79, 103–104, 114, 213
Kuhl, Alex, 245–247
Kullback, Solomon, 120, 263, 264, 267
Kupiecka, Malgorzata, 432

L

Lacey, Captain James, 26
Laconia, 165
Lagrange, Joseph-Louis, 473
Lagrangian interpolation, 467n. 4
Landau, Susan, xxv, 395, 521
Langen, Henry E., 3
Lasry, George, xxv, 126, 304
Lasswell, Captain Alva, 274
"The Last Question," 332, 339, 341
Last Week Tonight with John Oliver, 527
Latin, 12, 64, 267, 285, 331, 418n. 15
Latin America, German operatives in, 120
Latin code talkers, 285
Lattice-based cryptosystems, 585
Launch codes, 512

Lavabit, 526
Laws (actual and proposed); *see also* Export restrictions
   "Anti-Crime" Bill 266, 510–511, 515, 520
   Anti-crypto, xxi–xxii, 423–429; *see also* Biden,
      Joseph
   Compliance with Court Orders Act of 2016
      (CCOA), 527–528
   Eliminating Abusive and Rampant Neglect of
      Interactive Technologies Act of 2020 (EARN
      IT), 530
   France, xxi–xxii
   position of NSA on, 521
   pro-crypto, xxii, 429–430, 512, 513, 514–517
LCD, 347–348
Leap, Tom, 211
Leary, Timothy, 338
Lebensraum, 226
Lecter, Hannibal, 48
Lederer, Richard, 9
Lee, Daulton, 364
Lee, Pil Joong, 211
Leeson, James, 26, 28
Leff, Harvey S., xxv
Legion of Merit, 213
Legrand, 13
Lehr, Jessica, 212
von Leibniz, Gottfried, 118, 148, 468, 473
Lenstra, Hendrick, 554, 566
Leonidas, 7
Lesbian, 363
Levine, Jack, 199–201, 204, 212–216, 345
   modes of encryption, and, 401–403, 411
   pattern word lists, and, 21, 23, 55, 56
Lewand, Robert, xxv
Lewicki, Zbigniew, 337, 338
Lewis and Clark expedition, 134
Leyte, 283
Lexar Corporation, 393
Libertarian, 414, 511n. 8, 513, 515, 526
Librarians, xxv, 193, 450; *see also* Casanova; Simpson,
      Robert; Stein, René
Library of Congress, 46, 102–103
Lidl, Rudolf, 559
Ligation, 582
Lightnings, 274
li(n), 467–468
Lincoln, Abraham, 121, 194
Lindsey, Robert, 364
Linear algebra, 35, 204, 456, 457, 458, 537; *see also*
      Matrix encryption
Linear congruential generator, *see* Stream cipher,
      congruential generators
Linear Feedback Shift Registers (LFSRs), *see* Stream
      cipher, linear feedback shift registers (LFSRs)
Link, Greg, 79
LinkedIn, 526
Lipka, Robert S., 364

Lipton, Richard J., 582, 587
Literature of entropy, 337–339, 341; *see also* Cipher, use
      in fiction; Science Fiction cover art
Little Richard, 114
Little Cryptogram, The, 131, 159
Lloyds Bank of London, 380
Logarithmic integral, 467–468
Logarithms, 309, 316, 329, 330, 467, 474
Logarithms and modular arithmetic helped win World
      War II, 309, 316–317
Log-ons, 393, 500–504
London, 116, 147, 348, 370, 372–373, 380
   SIGSALY installation in, 319, 320
Lord, Robert, xxv
Lord of the Rings, 26–27
Lorenz machine, 252–253
Los Alamos, 98, 110, 519
Lotus Notes®, 541
Louisiana, 281
Lovell, James, 413–414
Low, Richard M., 499
Lucifer, 380, 389, 393, 482; *see also* DES
Luftwaffe, 218
Lusitania, 163, 165
Lynch, Loretta E., 527
Lysander, 4
Lyster, Mark, 60

M

MacGarrity, Joseph, 165
Mackebee, Nora L., 358
Madsen, Wayne, 357
Mafia, 518
Magdeburg, 166, 196
Magi, alphabet of the, 42–43
Magic (cryptanalysis), 273, 287
Magic/magicians (supernatural), 42–43, 62, 64n. 4, 65,
      460, 530
Magic words, 460
Mahon, Tom, 114–115
Mailsafe, 510–511
Malik, Tashfeen, 521
Mallet, 431, 436, 496–497
Mallory, 431, 436, 496–497
Man Called Intrepid, A, 242, 259
Manhattan Project, 519
Manila, 320
Manson, Charles, 109
Marine cemetery at Iwo Jima, 281
Marine Corps, 276, 277, 278, 284, 285, 511
   cemetery, 281
   linguist, 274
   references, 287–289
Marks, Leo, 15–16, 101–102, 106, 120, 278
MARS, 555
Marshall, General George C., 311

Marshall Islands, 283
Martin, William H., 363
Mary, Queen of Scots, 44–45, 57
MASC, *see* Monoalphabetic substitution cipher
Masons' cipher, 26, 28
Massey, James, 431
Masterkey for Indian Lore and History, The, 282
Mathematical Games, 423
Mathematics, as a simplifying force, 583
Mathematization, 340, 579, 583
Mathews, David, 45
Matrix encryption, 79, 80, 199–204, 335, 398, 417
   cryptanalysis of, 204–212
   determinant, 203, 209
   devices, 215–216
   example of, 201–204
   history of, 199–201, 212
   keyspace of, 202, 204
   loss of faith in, 485, 499
   modes, 401–403, 411
   references, 213–215
Mauborgne, Joseph O., 93–94, 99, 102, 106, 287
   on the Playfair cipher 151, 152
   unsolved cipher created by, 101
Maximal period LFSR, 536–537, 538
Maximilian I, Emperor, 62
McAfee, John, 523–524, 525
McCarthy, Joseph, 194
McDaniels, Dennis, 78
McDevitt, Tim, 211–212
McDonnell Douglas Corp., 374
McEliece, Robert, 478, 487, 490, 585
MC Hawking, 332, 342
MD5, 499–500, 507, 508
Meadows, William C., 282, 288, 289
Medal of Merit, 276
de Medici, Guiliano, 117
Medvedev, Dmitry, 370
Meet-in-the-middle attack, 304n. 26, 397, 444, 587
Mehl, Donald E., 318, 319n. 16, 320
Mein Kampf, 226
Mellen, Greg, 146, 159
Menezes, Alfred J., 578, 579
Mercurial, 500
Meredith, Dan, 244
Merkle, Ralph, 414n. 4, 426, 479–486, 487, 488, 490, 554
   Attack on DES, 397, 410
Merritt, Charlie, 510–511
Mersenne primes, 466, 476–477, 565
Message digest, 498, 499
Message-Digest Algorithm 5, *see* MD5
Mexico, 163–165, 186n. 29
Meyer, Andrea, xxv
Meyer, Joseph A., 424–429
Michigan, 52, 282
Micius Quantum Prize 585
Microsoft®, 526

Microsoft® Windows®, 541
Middle finger, 350–352
Midway Island, 274
MI-8, *see* Military Intelligence Section 8
Military Intelligence Section 8, 183
Miller, Frank, 93n. 27, 102–103
Miller, Gary L., 472
Miller, Greg, 359
Miller, Ken, 78
Miller, Victor S., xxv, 545–546, 552, 554, 566
Millennium prize problem, 479
$1,000,000 prize, 472, 479
Millward, Katherine, 207
MINARET, 354
Mind, 255
MIT, 39, 270, 417, 428, 460, 509
Mitchell, Bernon F., 363
Mitchell, Major John, 274
M-94, 137, 138, 159–160; *see also* Cipher wheel
Modes of encryption, 79–80, 212, 215, 401–409, 411
Mod 26 multiplication table, 34
Modular Arithmetic, 11, 33–34, 201–202, 309, 316–317, 381, 418–419; *see also* SIGSALY, senary scale, which is mod 6, and XOR, which is mod 2. This isn't a comprehensive listing. Modular arithemetic is used widely in almost all modern systems, and is therefore used throughout Part II of this book.
Modular arithmetic and logarithms helped win World War II, 309, 316–317
Molle, Dante, 79
Mollin, Richard A., 470n. 13
Monde, Le, 370
Mongé, Alf, 152–156
Monitoring aliens and political minorities, 424
Monoalphabetic substitution cipher (MASC), 3, 5–7, 8–10, 26–29, 42–43; *see also* Poe, Edgar Allan; Doyle, Arthur Conan
   affine cipher, 33–37
   Agony Columns, 147
   Biblical use of, 19
   Caesar cipher, 7–8
   cryptanalysis of, 17–18, 20–26, 29–30, 55–56
   keyspace of, 8
   keyword used for, 9
   Mason's use of, 26, 28
   References, 56–57
   World War I, 166
M-134, 291, 292, 299
Monotone, 309, 500
Montréal, 573
Moody, Dustin, 578
Moore, Judy H., 399
Morehouse, Lyman F., 99
Moriarty, 48
Morris, Robert, 391–392, 393, 394, 410, 502–504
Morrison, Michael, 457
Morse, Samuel, 148 Morse code, 15, 37–40, 166

Moscow, 94–95, 394
Motorized pogo stick, 336
Motorola, 358
Mount Suribachi, 281
Mozart, Wolfgang Amadeus, 26, 28
Mozilla, 526
M-209, 218–219, 252, 294, 502–503
Mullender, Sape, 501
Multiplex system, 138, 146, 159; *see also* Wheel cipher
Mundy, Liza, 296
Munitions Control Act, 423
Murray Hill, New Jersey, 392, 394
Music, 18, 113, 268n. 15, 311, 450, 522
    Bad Religion, 337, 342
    Hunnicutt, Tom "Captain T" 274
    MC Hawking, 332, 342
    Mozart, Wolfgang Amadeus, 26, 28
    Osbourne, Ozzy, 26
    vocoder use in, 313–315
Myer, Albert J., 121–122
MySpace, 519

N

NAND mirroring, 525
Narbonne, 275
Nash, John, 42
Nathes, Robert C., 323
National Archives of Australia, The, 306
National Bureau of Standards (NBS), 380, 391, 392,
    393, 503 This group later had a name change
    to National Institute of Standards and
    Technology (NIST).
National Cryptologic Museum, xxv, 138, 190, 198,
    219, 222, 252, 253, 255, 258, 260, 272, 273,
    275–276, 306, 323
National Institute of Standards and Technology (NIST)
    Advanced Encryption Standard, and, 554–555, 556n.
        30
    Data Encryption Standard, and, 410
    DSA, and, 445, 504–506, 507
    DSA vs. RSA, and, 445
    Post-quantum cryptology, and, 577, 578–579
    previously known as NBS, 380n. 8
    SHA, and, 499, 500, 507
National Museum of the U.S. Air Force, 306
National Reconnaissance Office, 352
National Science Foundation, 423, 428n. 32, 429
National Security Agency (NSA), 31, 42, 78, 135, 192,
    345–378; *see also* Center for Cryptologic
    History; Export restrictions; Crypto AG;
    National Cryptologic Museum
    Certicom Corporation, and, 553
    Clipper chip, 399, 516–517
    DES involvement, 379, 380, 391–395, 399
    domestic spying accusations, 361–362
    DSA involvement, 445, 505–506
    computer crash, 360
    Hall of Honor, 192, 259, 433, 434
    interference with academia, 423–430
    interviewing with, 362–363
    memorial, "They Served in Silence," 374
    origins of, 346–347
    physical description, 361
    references, 374–378
    SHA-0, 499
    size and budget, 348–349, 352–353, 355–356, 361, 376
    Suite A, 553
    Suite B, 499, 545–567, 577
    TEMPEST, 347–348
    traitors to, 355; *see also* Snowden, Edward
    treaty signed with CIA, 355
    videography, 377–378
National Security Agency Scientific Advisory Board
    (NSASAB), *see* NSASAB
National Security Letters, 193
Native American activists, 353
Native Americans, *see* Code talkers
NATO, 360
Nature, 576
Nature of the Physical World, The, 249, 336
Navajo, 261, 276–285, 287–289, 323; *see also* Code
    talkers
Naval Cryptologic Veterans Association (NCVA), 350–352
Nazi Within, The, 421n. 18
NBS, *see* National Bureau of Standards (NBS)
Nebel, Fritz, 166, 182, 195, 198
NEC Corp., 373
Needham, Roger, 501–502, 506
Nerdcore hip hop, 332, 342
Nerfherder, scruffy, 182–183
Nero, 123
Nest Labs, 526
Netcom, 513
Netherlands, 217, 275, 512n. 16
New Caledonia, 283
"New Directions in Cryptography," 414, 417
New Guinea, 304
New Jersey, 392, 394
Newman, Bob, 358
New Mexico, 163, 283
Newton, Isaac, 117, 118, 148, 268n. 15
New York City, 26, 28, 45, 94, 185, 555
New Yorker, The, 261–263
New York Post, 16
New York Times, 103, 358n. 48, 378, 424, 520–521
New Zealand, 151, 373, 512n. 16
Niederreiter, Harald, 559
Nietzsche, Friedrich, 119
Nihilist cipher, 120
Nimitz, Admiral, 274
9/11, 356, 360, 361, 377, 564
1984, 424
Nintendo, 500

NIST, *see* National Institute of Standards and Technology (NIST)
Nixon, Richard, 517
NKVD, 364
Nobel Prize, 402
Nodes, 575, 580
Nomenclator, 44–48, 49, 57, 134–135, 196
Non-cryptanalytic attacks, 347–348, 361, 564, 574
Noninvertible matrices in matrix encryption, 214
Non-pattern words, 23, 55, 56
Nonsecret encryption, *see* Public Key cryptography
Normandy, 274, 282
Norris, Mike, 423, 427
North Carolina State University, 21, 91, 200, 215–216
Northern Kentucky University, 245
North Korea, 350–352, 367
Not of interest, 53, 110n. 3, 377
Novak, Kayla 211
Noyes, Rear Admiral Leigh, 296
NP-complete, 477–479, 483–484, 487, 489, 581
NP-hard, 478–479
NP (nondeterministic polynomial time), 417, 477–479, 572
NSA, *see* National Security Agency
NSASAB, 336, 533n. 1
NSF, *see* National Science Foundation
NTRU, 579, 585, 587
Nuclear weapons/annihilation, 98, 166, 311, 356, 423–424, 519, 520
    protests against 415, 421, 513
    RSA, and, 512
Nulls, 44, 104, 116, 121, 126, 147
Nyman, Bertil, 466
Nyquist, Harry, 317, 329n. 3

O

Oakland, 320
Obama, Barack, 373, 520
OCB, *see* Offset Codebook Mode (OCB)
Odensholm, 166
Odom, Lieutenant General William E., 349
OFB, *see* Output Feedback Mode (OFB)
Office of Naval Research, 423
Office of Strategic Services, 15n. 27, 94, 186n. 30
Official Secrets Act, 194
Offset Codebook Mode (OCB), 406–409
Ohaver, M. E., 199, 214
Ohio, 253, 259–260
Oklahoma City, 517
O'Leary, Jeremiah, 165
Oligonucleotides, 581–582
Oliver, John, 527
OL-31, 320
Omnibus Crime Control and Safe Streets Act, 355
"On Computable Numbers," 249–250

"On Digital Signatures and Public-Key Cryptosystems," 423, 434, 463
$100,000 prize, 477
One-time pad, 92–96, 101–103, 116, 120
    breakable if misused, 96–98
    discovery of, 93–94, 98–100, 102–103
    Guevara, Ché, 95–96
    German use of, 94, 96
    Japanese use of, 94–95
    19th century discovery of, 93n. 27, 102–103
    OSS use of, 94
    quantum, 572n. 5, 575
    references, 105–106
    Soviet use of, 94, 96–98
    as unbreakable cipher, 92–94, 327, 533
    for voice, 311, 313, 318
One-way function, 393, 483, 502
Open problems, 23,192, 212–213, 416, 422, 446; *see also* Unsolved ciphers
Optical Society of America (OSA), 574
Optics Express, 575
OP-20-G, 346
Opus 100, 341
Orange, 267–268, 287
Orgies, 183
Orwell, George, 424, 516
Osbourne, Ozzy, 26, 42
Osmussaar, 166
OSS, *see* Office of Strategic Services
Ossifrage, squeamish, 460
OTP, *see* One-time pad
Our Fighting Navy, 249
Output Feedback Mode (OFB), 405–406
Oyobi, 267
Ozeki, Naoshi, 191

P

P (polynomial time), 461, 473, 477–479, 551, 576, 581
P = **NP**, 417, 478–479, 572
P = **NP**, proof for special case of, 479
P ≠ **NP**, 479
Padding, 94, 408, 442, 444, 503, 510
Painvin, Georges, 168–169, 172–173, 182, 196, 198
País, El, 370
Paracelsus, 62
Parallel processing, 236, 245, 580, 582
Paris, 168, 320
Parrish, Thomas, 255
Party line, 310n. 3
Passwords
    hacked, 519
    hash functions and, 500–504
    key encryption, and, 500–504, 518
    selection of, 518–520
Pattern words, 20–21, 23, 30, 55–56, 60; *see also* Non-pattern words

Patterson, Robert, 147
Pearl Harbor, 106, 116, 186, 261–263, 273, 286, 311
Peer to peer, 520
Peeters, Michaël, 500
Pelton, Ronald William, 364
Penmanship, Friedman, 183
Penmanship, Yardley, 183
Penn, Sean, 364
Pennsylvania State University, 133
Pennypacker, Morton, 45, 47, 57
Pensacola, Florida, 528–530
Penthouse opposes cover-up, 352–353
People's Bank of China, 576
Perec, Georges, 18–19
Peripheral vision, 476
Perjury, 354
Permutations
    Data Encryption Standard (DES), 382, 385, 386,
        387, 399
    Enigma, 221, 223
        conjugate, 240–241, 243
        notations, 227, 229
        from plugboard to rotor assembly, 228, 242
        products and factoring, 229–237
        reflector, 223
        rotor advancement, 228
        rotors, 221, 223
    matrix encryption cryptanalysis, 210
    Purple, 268–269
Persia, 4, 7, 196
Peterson, Jr., Sydney, 364
Pfefferkorn, Riana, 530
PGP
    history of, 510–514, 531, 532
    impact of, 518
    and NSA, 510, 512, 516
    references, 531–532
    SHA-1, used by, 499
    "Why I Wrote PGP" by Zimmermann, 514–517
Pharnabazus, 4
Phelippes, Thomas, 44–45, 57
$\varphi(n)$, 33, 211–221, 418
Philip II, King of Spain, 42n. 51, 45
Phillips code, 269
Photons, 569–571, 574–575, 584, 585
Pigpen cipher, 26, 28
π, 8–9, 519
$\pi(n)$, 466–468
Pinterest, 526
PKCS #1, 442–443
Playboy supports transparency (in government), 353
Playfair, Baron Lyon, 147, 148
Playfair cipher
    cryptanalysis, 152–157
    history, 147–149, 151–152
    references, 161
    workings of, 149–150

Plugboard
    Enigma, 220–221, 225, 226, 227, 238, 243–244, 246
    Purple, 268–269
    Red, 267
Plumstead, Joan B., 534
Poe, Edgar Allan, 9n. 14, 10–14, 90, 93
    "An Enigma," poem by, 14, 57
    high entropy in writings of, 331, 332, 337
    references 52–53
    vs. Vigenère cipher, 59–60, 104
Poets/poetry, 450, 553–554
Pohlig, Steve, 426
Poitras, Laura, 370
Poland
    commemoration of codebreaking by, 227, 254
    cryptanalysts of, 226–227, 242, 244, 246, 251
    invaded by Germany, 246
    University of Warsaw, 337
Polarized light, 569–571, 574, 575, 584
Police, 31, 114, 256, 394, 424, 515, 518
Polish cryptanalysts, 226–227, 246–247; *see also*
        Rejewski, Marian; Różycki, Jerzy; Zygalski,
        Henryk
Pollard, John, 455, 461
Pollard, Jonathan, 357, 364
POLLUX, 305
Polyalphabetic substitution cipher, 60, 61
Polybius cipher, 42, 51
    ADFGX and ADFGVX use of, 166–167, 168, 196
    Greek use of, 5–6
    Guevara, Ché, and, 95–96, 98
    Viking use of, 6–7,
Polygraphiae, 62–63
Polygraphic cipher, 227; *see also* Digraphic cipher; Matrix
        encryption
Polynomial time, *see* **P**
Polyphonic cipher, 12, 43
Pomerance, Carl, 461, 465, 473
Porlock, 48
Porta, Giovanni Battista, 64, 75, 147–148
Postage stamp, 226–227, 254
Post-quantum cryptography, 569, 577–579, 585–586
Postulates, 147
Potassium cyanide, 256
Potter, Ralph K., 323
POW, *see* Prisoner of war
Powers, Francis Gary, 94, 350n. 18
Poznan, Poland, 254
Prairie Home Companion, 510
Pratt, Fletcher, 130
Prediction
    by Adleman, Len, 583
    by Bauer, Craig P., 79, 195, 530–531, 579
    by Brassard, Gilles, 574
    by Gardner, Martin, 460
    by Turing, Alan, 256
    by Wiles, Andrew, 583

Preimage computation, 498, 501, 502
Pretty Good Privacy (PGP), *see* PGP
Primality testing
    AKS algorithm, 473–476
    definition, 468
    deterministic, 473–477
    elliptic curve, 473, 565
    Fermat test, 468–470
    Miller–Rabin–Selfridge test, 465–473
    Miller–Rabin test, 470–473
    Rabin–Miller test, 470–473
    References, 488–489, 565
    strong pseudoprimality test, 470–473
Prime numbers, 465–468; *see also* Factoring algorithms;
        Primality testing; RSA Factoring challenge
    arbitrarily large runs without any, 466
    in Diffie–Hellman key exchange, 415
    in DSA, 504, 505
    in ECC, 549
    in Elgamal, 487, 497
    in Fermat's little theorem, 418
    generation of, 473
    humor, 462
    infinitely many (proof), 465–466
    Mersenne primes, 466, 476–477, 565
    in RSA 419, 432
    in RSA attacks, 435, 439, 441, 445, 496, 497, 520
    references, 488–489
    top 10 largest known, 466, 476–477
Prime number theorem, 467
Primitive polynomial, 536–537
Primitive root (aka generator), 487, 504
Princeton's Institute for Advanced Study, 327, 479
Princeton University, 201, 250, 285, 327
Printer Codes, xix
PRISM, 370
Prisoner of war, 38, 116–117
Privacy, *see* Laws (actual and proposed)
Privacy lock, 310
Prize money, 133, 397, 446, 472, 476–477, 479, 585
PRNG, *see* Pseudorandom number generator
Probable Word, *see* Cribs
Proceedings of the Engineers' Club of Philadelphia,
        75
Project X, 311; *see also* SIGSALY
Project X-61753, 311; *see also* SIGSALY
Propaganda, failed, 350–351
Proto, Richard "Rick", 433, 434
Protocols, 430–431, 436, 499, 545, 575
Prozess, Der, 357
Psalm 46, 131–133
Pseudoprime, 469–470
Pseudorandom number generator, 98, 533; *see also*
        Stream cipher
Psychological method, 230, 234, 236, 519
PT Boat, 151–152
P-38 Lightnings, 274

Public Key cryptography; *see also* Diffie–Hellman key
        exchange; Elgamal encryption; Elliptic Curve
        Cryptography; RSA
    classified discovery at GCHQ, 432
    developers, group photo of, 488
    knapsack encryption, 482–486, 490, 499
    knapsack problem, 478
    linear codes, 478, 487, 490, 565, 566, 585
    McEliece system, 478, 487, 490, 585
    Merkle's first scheme, 479–482, 490
    prehistory, 413–414
    puzzle scheme, 479–482, 490
    references, 433–434, 490–491
Public Key Cryptography Standard #1 (PKCS #1),
        442–443
Punitive expedition, 163
Purdy, Anthony, 338–339
Purple, 106, 186, 268–276, 286–287, 311, 372–373
    analog, 271, 273, 372–373
    cryptanalysis, of 270–273
    fragment of, 275–276
    intelligence from, 273, 274–276
    keyspace, 269–270
    period of, 269
    references, 286–287
    schematics, 268
    workings of, 268–270
Puzzle Palace, The, 192
Pyle, Joseph Gilpin, 131
Pynchon, Thomas, 337–338, 341–342
Pyrenees, 275

Q

QL69.C9, 193
QKD Network, *see* Quantum Key Distribution Network
        (QKD Network)
Quadratic sieve, 459–460
Quagmire III, 78
Quaker, 45
Quantum computers, 105, 461, 576–579, 584–585
Quantum Cryptography
    background, 569
    devices, 573–576
    example, 569–571
    history, 571–576
Quantum Key Distribution Network (QKD Network),
        575–576, 577, 584–585
Quantum repeater, 574
Queen Elizabeth, 44–45
Queneau, Raymond, 339

R

Rabin, Michael O., 471–472
Radcliffe, 296
Rader, Dennis, xix, 31

Radicalism, 553
RAF, *see* Royal Air Force
Rail fence transposition, 107–108
Ralph's Pretty Good Groceries, 510
"Randomness—A Computational Complexity View," 479
Random number generator, 98, 105, 445, 533
RCA, 310
RC4 (Rivest cipher 4), 158, 539–541, 543
RC5 (Rivest cipher 5), 541
RC6 (Rivest cipher 6), 541, 555
RC–220–T–1, 311; *see also* SIGSALY
Red, 264–267, 268, 270
Red analog, 267
Red Sun of Nippon, 189
Redundancy, 42, 42, 228, 228, 311, 311, 312, 312,
    332–335, 332–335, 342–343, 342–343, 514,
    514
References, importance of, 482
Refrigerator, 579
Rejected (classic) papers, 459, 479–482, 554, 571–572
Rejewski, Marian, 227, 230, 234, 236, 242, 246, 254
Relatively prime (aka coprime)
    connection with Euler's totient function, 435
    connection with theorems of Fermat and Euler, 418, 419
    definition, 33, 203, 418
    proof of using Euclidean algorithm, 420
    use in AES, 558
    use in attacking matrix encryption, 211
    use in attacking RSA, 435, 440, 441, 496, 497
    use in factoring algorithm, 452
    use in knapsack encryption, 483
    use in LFSRs, 539
    use in primality testing, 469, 471, 475, 476
Religion, 336–337; *see also* Bible; Koran
Remington, 359
Rendezvous, 190
Repeated squaring, 422–423, 443, 468–469, 550
Republicans, 369, 517, 521, 527, 528
Reeds, Jim, 52, 57, 340, 534, 542
Rees, Abraham, 13
Reifsneider, Adam, xxv, 82n. 24
Reflector, Enigma, 222, 223–224, 226, 227, 246
Relationships, 485–486, 491
Reverse Caesar cipher, 19
Rhoades, Dustin, 138
Rhodes, Roy A., 364
Rice–Young, Karen, xxv
Riemann hypothesis, 468, 489, 497
Riemann Zeta function, xxii, 467
Rijmen, Vincent, 555–556, 559, 567
Rijndael, *see* AES
Ringle, Barbara, xxv
Riverbank laboratories, 133, 182
Riverbank publications, 105, 133, 159, 263
Rivest, Ron, 417–418, 423, 427–428, 434, 510, 556
    Alice and Bob, and, 430, 432
    MD5, and, 499

pictures of, 417, 488
RC4, RC5, RC6, and, 539, 541, 543, 555
reaction to attempted intimidation, 423, 424
vs. ECC, 553–554
Rochefort, Captain James, 192, 274, 286
Rocket-powered Frisbee, 336
RockYou.com, 519
Rogaway, Phillip, 406
de Rohan, Chevalier, 30
Romaji, 267
Roman Catholic church, 44, 62
Roman cryptography, 7–8, 432n. 44
Romania, 212, 518
Rome, 555
Rommel, 251–252
Romney, Mitt, 362
Rongorongo script, 16–17, 54
Room 40, 166, 195, 196
Roosevelt, Franklin Delano, 192, 286, 303, 310, 312,
    319, 327
Roosevelt, W. E., 184
Rosenblum, Howard, 393
Rosenberg, Julius, 98
Rosenheim, Shawn, 53
Rosen, Kenneth, xxv
Rosen, Leo, 270–272, 372–373
Rotbrunna stone, 6
Rotors 217–218; *see also* Half-rotors
    Enigma, 217–218, 221–225, 242, 246, 251,
        260
    Red analog, 267
    SIGABA, 292, 293, 294–300, 301, 303, 305
Rotor wirings, Enigma, 221, 223
Route transposition, 110, 120–121
Rowlett, Frank
    National Security Medal awarded to, 276
    pictures of, 272, 291, 306
    SIGABA, and, 291–294, 299, 302
    vs. Japanese ciphers, 264, 267, 270, 271, 287
    work relationship with Friedman, 263, 286, 287,
        291–293
    on Yardley, 187
Royal Air Force, 116
Royal Society, 129–130
Różycki, Jerzy, 226–227, 254
RSA; *see also* Factoring algorithms; RSA attacks
    backstory, 417–418
    controversy, 445
    example, 419–423
    factoring challenge 446–447
    mathematics of, 418–419
    patent, 430
    in PGP, 510–514, 520
    RSAREF, 512–513
    seeing print, 423–430
    signatures, 430, 495
    Soviet use, 512

RSA attacks; *see also* Factoring algorithms
  adaptive chosen ciphertext, 442–443
  chosen ciphertext, 442–443, 495–496
  common enciphering exponent, 439–441
  common modulus, 435–436, 495–497
  insider's factoring attack on common modulus, 495–497
  insider's nonfactoring attack, 497
  low decryption exponent, 437–439
  low encryption exponent, 439
  man-in-the-middle, 436
  partial knowledge of *d*, 439
  partial knowledge of *p* or *q*, 439
  references, 462–463
  Ron was wrong, Whit is right, 444–445; *see also* 520 (related)
  searching the message space, 442
  textbook RSA, 444
  timing, 443–444
RSA Data Security, Inc., 158, 397, 510, 511, 512, 513, 539
RSA Factoring challenge, 446–447
RSA Laboratory, 397
RSA-129, 460
RSA-2048, 446
Rubicon, 360
Rubik's Cube, 127, 336
Rumely, Robert S., 473
Runes, 6–7, 26–27
Running key cipher, 80–94, 96, 104–105, 332, 401, 541
  One-time pad used twice, equivalent to, 96, 541
Runtime, 157, 212, 477, 555, 582
Russia, xxi, 94, 98, 186n. 29, 357, 363, 523; *see also* VENONA
  broken voice encryption of, 355
  World War I, and, 166, 196
  World War II, and, 96
  Snowden, Edward, and, 364–365, 366–368, 370
Russian language, 74, 339, 361

S

Sabotage, 165
Safe/Safecracking, 110, 302, 304, 305, 519, 520
Sahu, Neelu, xxv
Sale, Tony, 252
Salt, *see* Padding
San Bernadino shooter, 521–528
Sanborn, James, 75, 78–79
Sandia Laboratories, 427
Sanger, Margaret, 194
Santa Barbara, 405, 485
Sanyam (pen name), *see* Ohaver, M. E.
Saporta, Marc, 339
Satan, agent of, 78
Saturday Night Live, 511

Saudi government, 373–374
Saudi national airline, 373–374
Sawada, Setsuzo, 191
Saxena, Nitin, 473–474
Sayers, Dorothy, 26, 161
S-boxes, *see* DES, *S*-boxes
Scarfo, Jr., Nicodemo S., 518
Scherbius, Arthur, 217–218
Schmeh, Klaus, xxv
Schmidt, Hans Thilo, 238, 242, 243
Schneier, Bruce, 325, 385, 431, 436, 555, 563, 564
Schoof, Rene, 554
Schroeppel, Richard, 432
Science, 424
Science Fiction cover art, 37
Science Service, 105, 282
Scientific American, 60, 75, 423, 427, 460, 462, 574
Scientist, universal, 583
Scrabble, 117, 270
ScratchPad, 471–472
Scytale, *see* Skytale
Second law of thermodynamics, 74, 331–332, 336–337, 339
Secret archives, 263
Secretary of Defense, 342
Secret inks, 101, 184, 193, 197
  dangerous, 184 (*see* hand in Figure 5.7)
Secret romance, 514
Secure Hash Standard (SHA), 499–500, 506–508; *see also* SHA-0; SHA-1; SHA-2; SHA-3
Secure Socket(s) Layer (SSL), 499, 540, 541
Security check, 493
Security Engineering, 506
Seed/seeding, 357, 445, 534, 535, 536, 537, 539, 541
Seized manuscripts, 187–188, 192–195
Selfridge, John, 470n. 13
Senate Bill 266, 510–511, 515, 520
Senate Judiciary Committee, 517
September 11, 2001, 356, 360, 361, 377, 564
Serpent, 555
Seventeen or Bust, 466, 476
Sexual content, censored, 194, 195
"Sexual experimentations," 363
Sexual indiscretions, 517
Shallit, Jeffrey, 53
Shakespeare, 8, 88, 129–134, 158–159, 194–195, 331
Shamir, Adi, 417, 418, 423, 474
  design of hardware for factoring, 461
  differential cryptanalysis, and, 394, 409
  pictures of, 417, 486, 488
SHAMROCK, 354
Shane, Scott, 358
Shanghai, 575
Shannon, Claude, 80, 119, 158, 327–342, 345, 389
  Hellman inspired by, 379
  unbreakability of OTP proven by, 106
Shannon's maxim, 158

Shareware, 510
SHA-0, 499, 506
SHA-1, 499, 500, 506, 507, 508
SHA-2, 499, 500, 507
SHA-3, 500, 507
Shelly, Mary, xxv
Sherborne school, 247–249
Sheshach, 19
Shidehara, Kijuro, 192
Shivers, Robert L., 263
Shor, Peter, 461, 576, 585
Shor's algorithm, 461, 576
Shorthand systems, 184
Side channel attacks, 347–348, 361, 574
Siermine, Nicolette, 211
SIGABA 291–307
    crib applied to, 303–304
    cryptanalysis of, 303–304
    enciphering rate, 299–300
    factory 301
    French counterintelligence and, 304
    keyspace of, 302–304
    missing, 304–305
    patent, 306
    physical security of, 302
    references, 307
    retired, 305–306
    rotors 292, 293, 294–300, 301, 303, 305
    stepping action of, 297, 298, 307
    women's contribution to, 294–296, 301
    workings of, 297–301
Sigact News, 571n. 2, 573, 585
SIGINT ships, 349–352, 377
"The Sigint Sniper," 274
Signal Corps, 105, 121–122, 186, 263, 346n. 3
Signaling, 5
Signals Intelligence Budget 349
Signals Intelligence Directorate (SID), 347
Signals Intelligence Section, 263
Signal Intelligence Service, 186, 213, 286, 346
Signal Security Agency, 96, 346n. 3
SIGROD, 305
SIGSALY, 311–324
    alternate names of, 311
    cooling system, 320
    cost, 322
    efficiency, 322
    experimental station, 320
    installation locations, 320
    key phonograph, 318
    Mehl, Donald E., 318, 319n. 16, 320
    modular arithmetic and logarithms helped win
        World War II, 309, 316–317
    patents, 323
    phoneme, 318
    Pulse Code Modulation, 320
    references, 324

senary scale, 314, 316–317
Shannon, Claude 327
SIGBUSE, 319–320
SIGGRUV, 318
    simplified schematic, 317
    size and weight, 312, 320, 322
    Turing, Alan 319–320, 327
    vocoder, 312–315, 320
Silence of the Lambs, 48
Silhouette, 47, 285
Silk keys, 101–102
Simmons, Gustavus (Gus) J., 399, 423, 427, 435–436, 498
Simpson, Robert, xxv
Simulated annealing, 126, 127, 156
Sinkov, Abraham, 263, 264, 372–373
Six-Day War, 350
Sixes, 268, 269, 270
Skewes, Stanley, 468
Skewes's numbers, 468
Skiing, 418
Skorobogatov, Sergei, 525
Skype, 520
Skytale, 4–5, 7, 122
Slack Technologies, 526
Slaves, 7, 362n. 72
Sloane, Neil, 391–392, 393
Smartphones, *see* iPhones
S/MIME, 499
Smith, Commander Dudley, 101
Smith, Edson, 477
Smithline, Lawren, 146–147
Smoot, Betsy Rohaly, 103
Snapchat, 526
Snowden, Edward, 364–370
Social Security number, 519
S.O.E., *see* Special Operations Executive
Solo, Han, 182
SOS, 39
South Africa, 468
South America, 190
Soviet diplomatic cipher, 96
Soviet Union
    ciphers broken by Germany, 255
    ciphers broken by United States, 186n. 29, 347, 348
    Crypto AG, and, 357
    joint opposition to, 373
    one-time pad use by, 94, 96–98, 105
    RSA, and, 512
    secrets betrayed to, 364
    SIGABA, and, 305–306
    SIGINT ships, 350
    vs, DES, 394
    vs. spy planes, 350n. 18
Spain, 42n. 51, 45, 186n. 29, 275, 370, 512
Spanish language, 13, 74, 263, 331, 339
Spartans, 4, 7
Special Customer, 311; *see also* SIGSALY

Special Operations Executive (SOE), 15–16, 101–102, 106, 116, 120, 278, 493
Spiegel, Der, 373
Spies
    American, 94, 133, 165, 182, 350, 360, 364, 373, 377–378; *see also* Berg, Moe; Culpers (below)
    Ames, Aldrich, 357n. 45
    Cold War, 97, 98
    Culpers, 45–47, 57
    fictional, 189
    German, 184, 190, 277
    Hale, Nathan, 45
    Pollard, Jonathan, 357, 364
    Russian, 94, 95, 97–98
    Swiss, 304
Spock, Dr. Benjamin, 353
Spook Country, 3
S.S. Florida, 39
SSH, 499
SSL, 499, 540, 541
S.S. Republic, 39
Stalin, Joseph, 445, 516
Stallings, William, xxv, 507, 559n. 33
Stamp, 226–227, 254
Stamp, Mark, 269, 303–304, 499, 507, 541n. 11, 543
Stanford, 426, 443
St.-Cyr slide, 10
Stearns, Richard E., 477
Steckerbrett, 220, 227
Steganographia, 57, 62
Steganography, xix, 7, 57, 62, 124, 261–263; *see also* Secret inks
    biliteral cipher, 130, 133, 158–159
    Greek, 7
    and Pearl Harbor attack, 261–263
Stein, David, 78
Stein, René, xxv, 190, 222, 260, 272, 273
Stephenson, Neal, 257
Stepping action
    Enigma, 223–224
    Purple, 268, 269, 270
    Red, 266, 267
    SIGABA, 297, 298, 307
Stern, Bob, xxv
Stevenson, William, 242, 259
Stewart, Henry, 44
Stilwell, Captain, 278n. 42
Stimson, Henry L., 186, 193, 263, 313, 319n. 16
Stirling's formula, 8–9, 467
Stitzinger, Ernie, xxv
St-Jovite, Canada, 573
"Stories of the Black Chamber," 190
Strachey, Oliver, 267
Strait of Dover, 274
Stream cipher, 98, 105, 404, 405, 411, 533–544; *see also* Pseudorandom number generator
    A5/1, 538–539, 542–543

A5/2, 539
cellphone, *see* Stream cipher, A5/1 and A5/2
congruential generators, 533–534, 541–542
linear feedback shift registers (LFSRs), 535–537, 538–539, 542–543
linear feedback shift register attacks, 523–525, 537–538
RC4 (Rivest Cipher 4), 158, 539–541, 543
references, 541–544
Strip cipher, 136, 159–160
St. Paul's Churchyard, 26, 28
Studly cryptologist, 65, 135
Substitution boxes, *see* DES, *S*-boxes
Suetonius, 7
Suicide, 256, 364, 421
Sukhotin's method, 24–26, 55
Sumer/Sumerians, 3–4
Sunglasses, 569
Superencipherment, 167, 195
Sweden, 218, 359
Switch, Telephone, *see* Telephone switch
Switzerland, 218, 356, 359, 575
Syene, 450
Symantec Corp., 513
Szulc, Tad, 352

T

Taller, Dr. Herman, 194
Tap polynomial, 536
Tartaglia, 123
Tate, Christian N. S., 91, 105
T attack, 395, *see* Differential cryptanalysis
Tattoos, 7, 523
Telephone (stepping) switches, 268, 270
TEMPEST, 347–348, 574
Teşeleanu, George, 212
Tetris, 478, 489–490
"Thank you" from enemy, 493
Theater, community, 554
Thermite, 302
The theorem that won the war, 243
Theoretical attack, 563–564
Thermodynamics, second law, 74, 331–332, 336–337, 339
Thesaurus, 360
Thomé, Emmanuel, 447
Thompson, Ken, 502–504
Thoreau, Henry David, 194
Time magazine, 352
Times, The, 147
Timing attack, 443–444
TLS (Transport Layer Security), 499, 541
Tokyo, 320, 575
Tolkien, J. R. R., 26, 27
Tombstone, 26, 28, 158, 281
Tompkins, Dave, xxv, 313–314
Torus, 546

Torture, 16, 38, 256, 283, 284, 350, 352

Townsend, Robert, 45, 47, 57

Traicté des chiffres, 64–65, 411

Training exercise, 264–267

Traitors, 363–370

Transatlantic cable, 184

Transient Electromagnetic Pulse Emanation Standard,
   see TEMPEST

Transponders, 424, 426

Transport Layer Security (TLS), 499, 541

Transposition cipher, 78, 101, 122, 190, 389; *see also*
      Anagrams; Cardano grille; Skytale
   ADFGX and ADFGVX use of columnar, 166–168,
      182, 335–336
   columnar, 110–117, 147
   double, 116, 119–120, 126–127, 182
   rail fence, 107–108
   rectangular, 108–110
   references, 126–127, 195, 196
   Rubik's cube, 127
   word transposition, 120–121

Traveling salesman problem, 478, 587

Treatise on the Astrolabe, 42

Treaty, 355, 500

Trinity Churchyard, 26, 28

Triple DES, 397, 398, 401, 532

Trithemius, Johannes, 57, 62–64

Tromer, Eran, 461

Truman, Harry S., 276, 346, 347

Trump, Donald, 527

TSEC/KL-7 (ADONIS/POLLUX), 305

Tsosie, Harry, 283

Tuchman, Walter, 391, 392, 393, 410

Tuition refund, 451

Tunny, 252

Turing, Alan, 247–251, 255–259, 319–320, 324, 327, 579

Turing, John F., 256, 259

Turing, Sara, 256, 259

Turing, Sir Dermot, xxv, 256

Turing machine, 249–250

Turing test, 256

Turning grille, *see* Cardano grille

Tutti Frutti, 114

Twenties, 268, 269, 270

TWIRL, 461

Twitter, 526

Twofish, 555, 564

$200,000 prize, 446

U

UCLA, 477

UKUSA, 372–373

Ultra Americans, The, 255

Ultra Secret, 255, 286

Ultra Secret, The, 259

Ulysses, 194

Unabomber (Kaczynski, Theodore), 30, 31, 57, 116

Unconstitutional, 365, 426

Uncounterfeitable currency, 571

Undergraduate contributions
   AKS primality test, 474
   first hybrid system, 509
   first public key system, 414n. 4, 479–482
   matrix encryption attacks, 207, 211–212
   Poe challenge solved, 60
   reconstruction of Polish work on Enigma, 245, 258
   running key cipher attack, 91
   timing attack 443
   Vigenère Cipher on a TI-83, 104

Unforgeable subway tokens, 571, 584

Unicity point, 92, 119, 148, 152, 327, 335, 340

Unicycle, 327

United Nations, xxii

Universal language, 119

Universal machine, 250, 583

Universal Product Code (UPC), xix

Universal scientist, 583

University of Alberta, 338

University of California, Berkeley, 480

University of California, Davis, 406

University of California, Los Angeles, 477

University of California, San Diego, 406

University of California, Santa Barbara, 405, 485

University of Cambridge, 501, 525

University of Nevada, 406

University of Virginia, 146

University of Warsaw, Poland, 337

University of Washington, 553

University of Waterloo, 53, 553

Unix, 502

Unsolved Ciphers, 31, 33, 75–79, 101, 251

UPC, *see* Universal Product Code

U.S. Customs Department, 513

USS *Liberty*, 349–350, 377

USS *Pueblo*, 349–352, 377, 378

U-2 spy plane, 94, 350n. 18

V

Vader, Darth (similar to the German *Der Vater* - a
      codename?), 125–126

de la Vallée-Poussin, C. J., 467

Valley of Fear, The, 48

Van Assche, Gilles 500

Van Eck, Wim, 347

Van Eck phreaking, 347; *see also* TEMPEST

Vanstone, Scott, 507, 553

Vault, 263

Venona, 96–97, 105, 355

Ventura, California, 555

de Vere, Edward, 133–134, 158

Vernam, Gilbert, 93–94, 98–100, 102, 105

Vernam cipher, *see* One-time pad

Verne, Jules, 26
Verschlüsselt, 357, 375
Vertex/vertices, 580–582
V for victory, 38
ViaCrypt PGP Version 2.4, 512, 513
Viagra, 479
Viète, François, 42n. 51
Vietnam, 38, 212, 283n. 53, 511n. 9
de Vigenère, Blaise, 60, 64–65, 79–80, 401
Vigenère cipher, 59–80, 92, 94, 99, 134; *see also*
       Running key cipher
  autokey, 64, 79–80, 104, 401, 411, 503
  comparison with Enigma, 223
  comparison with modern stream ciphers, 534, 537
  comparison with Red, 265, 267
  references, 103–104
  revolutionary war variant of, 413–414
Viking cryptography, 6–7, 52
Violet, 268
Virus, computer, 579
Vocoder, 312–315, 320
Vogel, Major General Clayton B., 277
Voice encryption; *see also* SIGSALY
  advantages of, 323
  analog system, 309
  AT&T, 309, 310
  A-3 Scrambler, 310, 311, 312
  broken by Germans, 192, 310
  cell phone stream cipher A5/1, 538–539, 542–543
  cell phone stream cipher A5/2, 539
  Cipher Feedback (CFB) mode, and, 404
  cost of insecure 311
  Delilah, 320
  inverter, 309–310
  mock-up, 323
  RCA, 310
  references, 324, 542–543
  Russian system, 355
  used by JFK, 322
Void, A, 18–19
Voluntary review, 424, 429
Von Neumann, John, 329, 331, 533
Vowel recognition algorithm, 14, 23–26, 55, 181
Vowels enciphered separately, *see* Training exercise
Vowels not present, 264–267, 333
Voynich manuscript, 117, 118–119, 127, 159, 331

W

WACs (Women Army Corps), 294–296, 301
Wadsworth, Decius, 149
Wagner, David A., 539, 555
Walker, Jr., John Anthony, 364
Walker Spy Ring, 364
Walpole, Horace, 130, 159
Walsingham, Sir Francis, 44–45, 57
Walton Athletic Club, 255

Waring, Edward, 473
War protestors, 353
War Secrets of the Ether, 192
Washington, General George, 45, 46, 47, 57
Washington Disarmament Conference, 185
Washington Post, The, 359, 370, 372
Watch list, 354
Watergate, 517
Watson, Dr. (John H.), 48
Watson-Crick complement, 581
Watson Research Center, 379
WAVES (Women Accepted for Volunteer Emergency),
      294–296, 301
Weadon, Patrick, xxv, 219, 252
Weeks, Robert H., 372–373
Wegman, Mark N., 499
Weight–loss program, 168
Weierstrass, Karl, 545, 547, 554
Weierstrass equations, 545
Weil Pairing, 554
Weisband, William, 364
Weiss, Bob, xxv
Weizmann Institute of Science, The, 461, 583
Weizmann Institute Relations Locator, The, *see* TWIRL
Welchman, Gordon, 224n. 10, 251
WEP, 540–541
Westminster Conservatory, 554
West Point, Benedict Arnold and, 48
West Virginia, 135–136
WhatsApp, 526
Wheatstone, Charles, 147, 148–149
Wheel cipher; *see also* M-94
  cryptanalysis of, 138–146
  history of, 101, 135–138
  references, 159–160
  workings of, 136–138
Whip, 108–109
Whistleblower, definition, 368
White Elephant Protection (WEP), 540
Whiting, Doug, 555
Whorehouse, Chinese, 183
Wiedemann, Doug, 573, 584
Wiener, Michael J., 399, 437
Wiesner, Stephen, 571, 572, 573, 585
Wigderson, Avi, xxv, 479
Wii, 500
Wikipedia, 459, 501, 518
Wilde, Oscar, 201
Wiles, Andrew, 583
Wilkes, Maurice V., 501
Williamson, Malcolm, 416, 432
Wilson, John, 473
Wilson, Woodrow, 163, 165, 183
Wilson's theorem, 473
Window Rock, 277
Windtalkers, 283–284
Winkel, Brian J., xxiii, 60

Wired Equivalent Privacy (WEP), 540–541

Wiretaps, domestic, 353, 354, 515–517, 520

Wisconsin, 282

Wolff, Heinrich, 14

Wolf Prize in Physics, 585

Wollheim, Betsy, xxv

Women Accepted for Volunteer Emergency (WAVES), 294–296, 301

Women Army Corps (WACs), 294–296, 301

Word transposition, 120–122

World War I, 136, 163–185, 198, 217, 226; *see also* ADFGX; ADFGVX; Zimmermann telegram
  censorship, 193
  Friedman, William, and, 133, 192
  Lester Hill's Service in, 201
  Native American codetalkers and, 276–277, 282, 283, 285, 287, 288
  references, 195–198
  use of turning grilles in, 126

World War II, 110, 126, 160, 187, 212, 356, 377; *see also* M-209; Special Operations Executive (SOE). In addition to these, Chapters 7 through 10 are focused on World War II.
  American and British cooperation, 372–373
  censorship, and, 192–193
  M-94, 137
  National Security Agency roots in, 342, 346
  as physicists' war, 166
  Propaganda, 38–39
  training exercise, 264–267
  use of one-time pads in, 94, 96
  use of pattern word lists in, 23
  use of Playfair cipher in, 151–152
  use of transposition in, 116–117

World War III, 166

Woytak, Richard A., 242

Wray, Christopher A., 529

Wright, E. V., 17–18

Wright, Steven, 8

Wright-Patterson Air Force Base, 306

Wutka, Mark, 209, 214

Wyden, Ron, 370, 525, 527–528

Wyner, Aaron D., 391–392, 393, 410

X

Xerxes, King, 7

X-files, 283

Xie, Tao, 499

xkcd, 285, 390

XOR, 381

X-Ray, 311–312; *see also* SIGSALY

Y

Yahoo!, 370, 526

Yamamoto, Admiral, 273, 274

Yardley, Herbert O., 106, 163, 165, 182–192, 197–198; *see also The American Black Chamber*
  alleged treason, 186, 187, 190–192, 350
  legacy, 263, 346n. 3
  and Painvin, 198
  references, 197–198
  successes of, 183–186
  work for Canada, 190
  work for China, 183, 190, 197

Yellow, 268

Yorktown Heights, New York, 379, 392

Yoshida, Isaburo, 192

Young, John, xxv, 347n. 8, 393n. 30, 424n. 25, 427n. 31

YouTube, 410, 524, 525, 527

Yum, Dae Hyun, 211–212

Z

Z-80 computer, 510

Zero-emission cars, 373

Zi, Sun, 439

Zimmermann, Arthur, 163

Zimmermann, Paul, 447

Zimmermann, Philip, xxv, 510–514, 518, 520; *see also* PGP
  in his own words, 514–517
  references, 531–532

Zimmermann telegram, 163–165

Zip Codes, xix

Zip files, 40

Zodiac killer, 31, 32, 44, 56

Zodiac II, 16, 56

Zurich, 431

Zygalski, Henryk, 226–227, 254

Zygalski sheets, 246

Printed in the United States
By Bookmasters